International Series in Operations Research & Management Science

Volume 264

Series Editor
Camille C. Price
Stephen F. Austin State University, TX, USA

Associate Series Editor
Joe Zhu
Worcester Polytechnic Institute, MA, USA

Founding Series Editor
Frederick S. Hillier, Stanford University, CA, USA

More information about this series at http://www.springer.com/series/6161

Bhimasankaram Pochiraju • Sridhar Seshadri
Editors

Essentials of Business Analytics

An Introduction to the Methodology and its Applications

Editors
Bhimasankaram Pochiraju
Applied Statistics and Computing Lab
Indian School of Business
Hyderabad, Telangana, India

Sridhar Seshadri
Gies College of Business
University of Illinois at Urbana Champaign
Champaign, IL, USA

ISSN 0884-8289   ISSN 2214-7934 (electronic)
International Series in Operations Research & Management Science
ISBN 978-3-319-68836-7   ISBN 978-3-319-68837-4 (eBook)
https://doi.org/10.1007/978-3-319-68837-4

© Springer Nature Switzerland AG 2019
This work is subject to copyright. All rights are reserved by the Publisher, whether the whole or part of the material is concerned, specifically the rights of translation, reprinting, reuse of illustrations, recitation, broadcasting, reproduction on microfilms or in any other physical way, and transmission or information storage and retrieval, electronic adaptation, computer software, or by similar or dissimilar methodology now known or hereafter developed.
The use of general descriptive names, registered names, trademarks, service marks, etc. in this publication does not imply, even in the absence of a specific statement, that such names are exempt from the relevant protective laws and regulations and therefore free for general use.
The publisher, the authors, and the editors are safe to assume that the advice and information in this book are believed to be true and accurate at the date of publication. Neither the publisher nor the authors or the editors give a warranty, express or implied, with respect to the material contained herein or for any errors or omissions that may have been made. The publisher remains neutral with regard to jurisdictional claims in published maps and institutional affiliations.

This Springer imprint is published by the registered company Springer Nature Switzerland AG.
The registered company address is: Gewerbestrasse 11, 6330 Cham, Switzerland

Professor Bhimasankaram: With the divine blessings of Bhagawan Sri Sri Sri Satya Sai Baba, I dedicate this book to my parents—Sri Pochiraju Rama Rao and Smt. Venkata Ratnamma.

Sridhar Seshadri: I dedicate this book to the memory of my parents, Smt. Ranganayaki and Sri Desikachari Seshadri, my father-in-law, Sri Kalyana Srinivasan Ayodhyanath, and my dear friend, collaborator and advisor, Professor Bhimasankaram.

Contents

1   **Introduction** 1
     Sridhar Seshadri

Part I   Tools

2   **Data Collection** 19
     Sudhir Voleti

3   **Data Management—Relational Database Systems (RDBMS)** 41
     Hemanth Kumar Dasararaju and Peeyush Taori

4   **Big Data Management** 71
     Peeyush Taori and Hemanth Kumar Dasararaju

5   **Data Visualization** 111
     John F. Tripp

6   **Statistical Methods: Basic Inferences** 137
     Vishnuprasad Nagadevara

7   **Statistical Methods: Regression Analysis** 179
     Bhimasankaram Pochiraju and Hema Sri Sai Kollipara

8   **Advanced Regression Analysis** 247
     Vishnuprasad Nagadevara

9   **Text Analytics** 283
     Sudhir Voleti

Part II   Modeling Methods

10 **Simulation** 305
     Sumit Kunnumkal

11 **Introduction to Optimization** 337
     Milind G. Sohoni

12	**Forecasting Analytics**	381
	Konstantinos I. Nikolopoulos and Dimitrios D. Thomakos	
13	**Count Data Regression**	421
	Thriyambakam Krishnan	
14	**Survival Analysis**	439
	Thriyambakam Krishnan	
15	**Machine Learning (Unsupervised)**	459
	Shailesh Kumar	
16	**Machine Learning (Supervised)**	507
	Shailesh Kumar	
17	**Deep Learning**	569
	Manish Gupta	

Part III Applications

18	**Retail Analytics**	599
	Ramandeep S. Randhawa	
19	**Marketing Analytics**	623
	S. Arunachalam and Amalesh Sharma	
20	**Financial Analytics**	659
	Krishnamurthy Vaidyanathan	
21	**Social Media and Web Analytics**	719
	Vishnuprasad Nagadevara	
22	**Healthcare Analytics**	765
	Maqbool (Mac) Dada and Chester Chambers	
23	**Pricing Analytics**	793
	Kalyan Talluri and Sridhar Seshadri	
24	**Supply Chain Analytics**	823
	Yao Zhao	
25	**Case Study: Ideal Insurance**	847
	Deepak Agrawal and Soumithri Mamidipudi	
26	**Case Study: AAA Airline**	863
	Deepak Agrawal, Hema Sri Sai Kollipara, and Soumithri Mamidipudi	
27	**Case Study: InfoMedia Solutions**	873
	Deepak Agrawal, Soumithri Mamidipudi, and Sriram Padmanabhan	
28	**Introduction to R**	889
	Peeyush Taori and Hemanth Kumar Dasararaju	

29	**Introduction to Python**	917
	Peeyush Taori and Hemanth Kumar Dasararaju	
30	**Probability and Statistics**	945
	Peeyush Taori, Soumithri Mamidipudi, and Deepak Agrawal	

Index 965

Disclaimer

This book contains information obtained from authentic and highly regarded sources. Reasonable efforts have been made to publish reliable data and information, but the author and publisher cannot assume responsibility for the validity of all materials or the consequences of their use. The authors and publishers have attempted to trace the copyright holders of all material reproduced in this publication and apologize to copyright holders if permission to publish in this form has not been obtained. If any copyright material has not been acknowledged please write and let us know so we may rectify in any future reprint.

Acknowledgements

This book is the outcome of a truly collaborative effort amongst many people who have contributed in different ways. We are deeply thankful to all the contributing authors for their ideas and support. The book belongs to them. This book would not have been possible without the help of *Deepak Agrawal*. *Deepak* helped in every way, from editorial work, solution support, programming help, to coordination with authors and researchers, and many more things. *Soumithri Mamidipudi* provided editorial support, helped with writing summaries of every chapter, and proof-edited the probability and statistics appendix and cases. *Padmavati Sridhar* provided editorial support for many chapters. Two associate alumni—*Ramakrishna Vempati* and *Suryanarayana Ambatipudi*—of the *Certificate Programme in Business Analytics* (CBA) at *Indian School of Business* (ISB) helped with locating contemporary examples and references. They suggested examples for the Retail Analytics and Supply Chain Analytics chapters. *Ramakrishna* also contributed to the draft of the Big Data chapter. Several researchers in the *Advanced Statistics and Computing Lab* (ASC Lab) at ISB helped in many ways. *Hema Sri Sai Kollipara* provided support for the cases, exercises, and technical and statistics support for various chapters. *Aditya Taori* helped with examples for the machine learning chapters and exercises. *Saurabh Jugalkishor* contributed examples for the machine learning chapters. The ASC Lab's researchers and *Hemanth Kumar* provided technical support in preparing solutions for various examples referred in the chapters. *Ashish Khandelwal, Fellow Program student at ISB*, helped with the chapter on Linear Regression. *Dr. Kumar Eswaran* and *Joy Mustafi* provided additional thoughts for the Unsupervised Learning chapter. The editorial team comprising *Faith Su*, *Mathew Amboy* and series editor *Camille Price* gave immense support during the book proposal stage, guidance during editing, production, etc. The ASC Lab provided the research support for this project.

We thank our families for the constant support during the 2-year long project. We thank each and every person associated with us during the beautiful journey of writing this book.

Contributors

Deepak Agrawal Indian School of Business, Hyderabad, Telangana, India

S. Arunachalam Indian School of Business, Hyderabad, Telangana, India

Chester Chambers Carey Business School, Johns Hopkins University, Baltimore, MD, USA

Maqbool (Mac) Dada Carey Business School, Johns Hopkins University, Baltimore, MD, USA

Manish Gupta Microsoft Corporation, Hyderabad, India

Hema Sri Sai Kollipara Indian School of Business, Hyderabad, Telangana, India

Thriyambakam Krishnan Chennai Mathematical Institute, Chennai, India

Shailesh Kumar Reliance Jio, Navi Mumbai, Maharashtra, India

Hemanth Kumar Dasararaju Indian School of Business, Hyderabad, Telangana, India

Sumit Kunnumkal Indian School of Business, Hyderabad, Telangana, India

Soumithri Mamidipudi Indian School of Business, Hyderabad, Telangana, India

Vishnuprasad Nagadevara IIM-Bangalore, Bengaluru, Karnataka, India

Konstantinos I. Nikolopoulos Bangor Business School, Bangor, Gwynedd, UK

Sriram Padmanabhan New York, NY, USA

Bhimasankaram Pochiraju Applied Statistics and Computing Lab, Indian School of Business, Hyderabad, Telangana, India

Ramandeep S. Randhawa Marshall School of Business, University of Southern California, Los Angeles, CA, USA

Sridhar Seshadri Gies College of Business, University of Illinois at Urbana Champaign, Champaign, IL, USA

Amalesh Sharma Texas A&M University, College Station, TX, USA

Milind G. Sohoni Indian School of Business, Hyderabad, Telangana, India

Kalyan Talluri Imperial College Business School, South Kensington, London, UK

Peeyush Taori London Business School, London, UK

Dimitrios D. Thomakos University of Peloponnese, Tripoli, Greece

John F. Tripp Clemson University, Clemson, SC, USA

Krishnamurthy Vaidyanathan Indian School of Business, Hyderabad, Telangana, India

Sudhir Voleti Indian School of Business, Hyderabad, Telangana, India

Yao Zhao Rutgers University, Newark, NJ, USA

Chapter 1
Introduction

Sridhar Seshadri

Business analytics is the science of posing and answering data questions related to business. Business analytics has rapidly expanded in the last few years to include tools drawn from statistics, data management, data visualization, and machine learning. There is increasing emphasis on big data handling to assimilate the advances made in data sciences. As is often the case with applied methodologies, business analytics has to be soundly grounded in applications in various disciplines and business verticals to be valuable. The bridge between the tools and the applications are the modeling methods used by managers and researchers in disciplines such as finance, marketing, and operations. This book provides coverage of all three aspects: tools, modeling methods, and applications.

The purpose of the book is threefold: to fill the void in the graduate-level study materials for addressing business problems in order to pose data questions, obtain optimal business solutions via analytics theory, and ground the solution in practice. In order to make the material self-contained, we have endeavored to provide ample use of cases and data sets for practice and testing of tools. Each chapter comes with data, examples, and exercises showing students what questions to ask, how to apply the techniques using open source software, and how to interpret the results. In our approach, simple examples are followed with medium to large applications and solutions. The book can also serve as a self-study guide to professionals who wish to enhance their knowledge about the field.

The distinctive features of the book are as follows:

- The chapters are written by experts from universities and industry.
- The major software used are R, Python, MS Excel, and MYSQL. These are all topical and widely used in the industry.

S. Seshadri (✉)
Gies College of Business, University of Illinois at Urbana Champaign, Champaign, IL, USA
e-mail: sridhar@illinois.edu

- Extreme care has been taken to ensure continuity from one chapter to the next. The editors have attempted to make sure that the content and flow are similar in every chapter.
- In Part A of the book, the tools and modeling methodology are developed in detail. Then this methodology is applied to solve business problems in various verticals in Part B. Part C contains larger case studies.
- The Appendices cover required material on Probability theory, R, and Python, as these serve as prerequisites for the main text.

The structure of each chapter is as follows:

- Each chapter has a business orientation. It starts with business problems, which are transformed into technological problems. Methodology is developed to solve the technological problems. Data analysis is done using suitable software and the output and results are clearly explained at each stage of development. Finally, the technological solution is transformed back to a business solution. The chapters conclude with suggestions for further reading and a list of references.
- Exercises (with real data sets when applicable) are at the end of each chapter and on the Web to test and enhance the understanding of the concepts and application.
- Caselets are used to illustrate the concepts in several chapters.

1 Detailed Description of Chapters

Data Collection: This chapter introduces the concepts of data collection and problem formulation. Firstly, it establishes the foundation upon which the fields of data sciences and analytics are based, and defines core concepts that will be used throughout the rest of the book. The chapter starts by discussing the types of data that can be gathered, and the common pitfalls that can occur when data analytics does not take into account the nature of the data being used. It distinguishes between primary and secondary data sources using examples, and provides a detailed explanation of the advantages and constraints of each type of data. Following this, the chapter details the types of data that can be collected and sorted. It discusses the difference between nominal-, ordinal-, interval-, and ratio-based data and the ways in which they can be used to obtain insights into the subject being studied.

The chapter then discusses problem formulation and its importance. It explains how and why formulating a problem will impact the data that is gathered, and thus affect the conclusions at which a research project may arrive. It describes a framework by which a messy real-world situation can be clarified so that a mathematical toolkit can be used to identify solutions. The chapter explains the idea of decision-problems, which can be used to understand the real world, and research-objectives, which can be used to analyze decision-problems.

The chapter also details the challenges faced when collecting and collating data. It discusses the importance of understanding what data to collect, how to collect it, how to assess its quality, and finally the most appropriate way of collating it so that it does not lose its value.

The chapter ends with an illustrative example of how the retailing industry might use various sources of data in order to better serve their customers and understand their preferences.

Data Management—Relational Database Management Systems: This chapter introduces the idea of data management and storage. The focus of the chapter is on relational database management systems or RDBMS. RDBMS is the most commonly used data organization system in enterprises. The chapter introduces and explains the ideas using MySQL, an open-source structural query language used by many of the largest data management systems in the world.

The chapter describes the basic functions of a MySQL server, such as creating databases, examining data tables, and performing functions and various operations on data sets. The first set of instructions the chapter discusses is about the rules, definition, and creation of relational databases. Then, the chapter describes how to create tables and add data to them using MySQL server commands. It explains how to examine the data present in the tables using the SELECT command.

Data Management—Big Data: This chapter builds on some of the concepts introduced in the previous chapter but focuses on big data tools. It describes what really constitutes big data and focuses on some of the big data tools. In this chapter, the basics of big data tools such as Hadoop, Spark, and surrounding ecosystem are presented.

The chapter begins by describing Hadoop's uses and key features, as well as the programs in its ecosystem that can also be used in conjunction with it. It also briefly visits the concepts of distributed and parallel computing and big data cloud.

The chapter describes the architecture of the Hadoop runtime environment. It starts by describing the cluster, which is the set of host machines, or nodes for facilitating data access. It then moves on to the YARN infrastructure, which is responsible for providing computational resources to the application. It describes two main elements of the YARN infrastructure—the Resource Manager and the Node Manager. It then details the HDFS Federation, which provides storage, and also discusses other storage solutions. Lastly, it discusses the MapReduce framework, which is the software layer.

The chapter then describes the functions of MapReduce in detail. MapReduce divides tasks into subtasks, which it runs in parallel in order to increase efficiency. It discusses the manner in which MapReduce takes lists of input data and transforms them into lists of output data, by implementing a "map" process and a "reduce" process, which it aggregates. It describes in detail the process steps that MapReduce takes in order to produce the output, and describes how Python can be used to create a MapReduce process for a word count program.

The chapter briefly describes Spark and an application using Spark. It concludes with a discussion about cloud storage. The chapter makes use of Cloudera virtual machine (VM) distributable to demonstrate different hands-on exercises.

Data Visualization: This chapter discusses how data is visualized and the way that visualization can be used to aid in analysis. It starts by explaining that humans use visuals to understand information, and that using visualizations incorrectly can lead to mistaken conclusions. It discusses the importance of visualization as a cognitive aid and the importance of working memory in the brain. It emphasizes the role of data visualization in reducing the load on the reader.

The chapter details the six meta-rules of data visualization, which are as follows: use the most appropriate chart, directly represent relationships between data, refrain from asking the viewer to compare differences in area, never use color on top of color, keep within the primal perceptions of the viewer, and chart with integrity.

Each rule is expanded upon in the chapter. The chapter discusses the kinds of graphs and tables available to a visualizer, the advantages and disadvantages of 3D visualization, and the best practices of color schemes.

Statistical Methods—Basic Inferences: This chapter introduces the fundamental concepts of statistical inferences, such as population and sample parameters, hypothesis testing, and analysis of variance. It begins by describing the differences between population and sample means and variance and the methods to calculate them. It explains the central limit theorem and its use in estimating the mean of a population.

Confidence intervals are explained for samples in which variance is both known and unknown. The concept of standard errors and the t- and Chi-squared distributions are introduced. The chapter introduces hypothesis testing and the use of statistical parameters to reject or fail to reject hypotheses. Type I and type II errors are discussed.

Methods to compare two different samples are explained. Analysis of variance between two samples and within samples is also covered. The use of the F-distribution in analyzing variance is explained. The chapter concludes with discussion of when we need to compare means of a number of populations. It explains how to use a technique called "Analysis of Variance (ANOVA)" instead of carrying out pairwise comparisons.

Statistical Methods—Linear Regression Analysis: This chapter explains the idea of linear regression in detail. It begins with some examples, such as predicting newspaper circulation. It uses the examples to discuss the methods by which linear regression obtains results. It describes a linear regression as a functional form that can be used to understand relationships between outcomes and input variables and perform statistical inference. It discusses the importance of linear regression and its popularity, and explains the basic assumptions underlying linear regression.

The modeling section begins by discussing a model in which there is only a single regressor. It explains why a scatter-plot can be useful in understanding single-regressor models, and the importance of visual representation in statistical inference. It explains the ordinary least squares method of estimating a parameter, and the use of the sum of squares of residuals as a measure of the fit of a model. The chapter then discusses the use of confidence intervals and hypothesis testing in a linear regression model. These concepts are used to describe a linear regression model in which there are multiple regressors, and the changes that are necessary to adjust a single linear regression model to a multiple linear regression model.

The chapter then describes the ways in which the basic assumptions of the linear regression model may be violated, and the need for further analysis and diagnostic tools. It uses the famous Anscombe data sets in order to demonstrate the existence of phenomena such as outliers and collinearity that necessitate further analysis. The methods needed to deal with such problems are explained. The chapter considers the ways in which the necessity for the use of such methods may be determined, such as tools to determine whether some data points should be deleted or excluded from the data set. The possible advantages and disadvantages of adding additional regressors to a model are described. Dummy variables and their use are explained. Examples are given for the case where there is only one category of dummy, and then multiple categories.

The chapter then discusses assumptions regarding the error term. The effect of the assumption that the error term is normally distributed is discussed, and the Q-Q plot method of examining the truth of this assumption for the data set is explained. The Box–Cox method of transforming the response variable in order to normalize the error term is discussed. The chapter then discusses the idea that the error terms may not have equal variance, that is, be homoscedastic. It explains possible reasons for heteroscedasticity, and the ways to adapt the analysis to those situations.

The chapter considers the methods in which the regression model can be validated. The root mean square error is introduced. Segmenting the data into training and validation sets is explained. Finally, some frequently asked questions are presented, along with exercises.

Statistical Methods—Advanced Regression: Three topics are covered in this chapter. In the main body of the chapter the tools for estimating the parameters of regression models when the response variable is binary or categorical is presented. The appendices to the chapter cover two other important techniques, namely, maximum likelihood estimate (MLE) and how to deal with missing data.

The chapter begins with a description of logistics regression models. It continues with diagnostics of logistics regression, including likelihood ratio tests, Wald's and the Hosmer–Lemeshow tests. It then discusses different R-squared tests, such as Cox and Snell, Nagelkerke, and McFadden. Then, it discusses how to choose the cutoff probability for classification, including discussion of discordant and concordant pairs, the ROC curve, and Youden's index. It concludes with a similar discussion of Multinomial Logistics Function and regression. The chapter contains a self-contained introduction to the maximum likelihood method and methods for treating missing data. The ideas introduced in this chapter are used in several following chapters in the book.

Text Analytics: This is the first of several chapters that introduce specialized analytics methods depending on the type of data and analysis. This chapter begins by considering various motivating examples for text analysis. It explains the need for a process by which unstructured text data can be analyzed, and the ways that it can be used to improve business outcomes. It describes in detail the manner in which Google used its text analytics software and its database of searches to identify vectors of H1N1 flu. It lists out the most common sources of text data, with social media platforms and blogs producing the vast majority.

The second section of the chapter concerns the ways in which text can be analyzed. It describes two approaches: a "bag-of-words" approach, in which the structure of the language is not considered important, and a "natural-language" approach, in which structure and phrases are also considered.

The example of a retail chain surveying responses to a potential ice-cream product is used to introduce some terminology. It uses this example to describe the problems of analyzing sentences due to the existence of grammatical rules, such as the abundance of articles or the different tense forms of verbs. Various methods of dealing with these problems are introduced. The term-document matrix (TDM) is introduced along with its uses, such as generation of wordclouds.

The third and fourth sections of the chapter describe how to run text analysis and some elementary applications. The text walks through a basic use of the program R to analyze text. It looks at two ways that the TDM can be used to run text analysis—using a text-base to cluster or segment documents, and elementary sentiment analysis.

Clustering documents is a method by which similar customers are sorted into the same group by analyzing their responses. Sentiment analysis is a method by which attempts are made to make value judgments and extract qualitative responses. The chapter describes the models for both processes in detail with regard to an example.

The fifth section of the chapter then describes the more advanced technique of latent topic mining. Latent topic mining aims to identify themes present in a corpus, or a collection of documents. The chapter uses the example of the mission statements of Fortune-1000 firms in order to identify some latent topics.

The sixth section of the chapter concerns natural-language processing (NLP). NLP is a set of techniques that enables computers to understand nuances in human languages. The method by which NLP programs detect data is discussed. The ideas of this chapter are further explored in the chapter on Deep Learning. The chapter ends with exercises for the student.

Simulation: This chapter introduces the uses of simulation as a tool for analytics, focusing on the example of a fashion retailer. It explains the use of Monte Carlo simulation in the presence of uncertainty as an aid to making decisions that have various trade-offs.

First, the chapter explains the purposes of simulation, and the ways it can be used to design an optimal intervention. It differentiates between computer simulation, which is the main aim of the chapter, and physical simulation. It discusses the advantages and disadvantages of simulations, and mentions various applications of simulation in real-world contexts.

The second part of the chapter discusses the steps that are followed in making a simulation model. It explains how to identify dependent and independent variables, and the manner in which the relationships between those variables can be modeled. It describes the method by which input variables can be randomly generated, and the output of the simulation can be interpreted. It illustrates these steps using the example of a fashion retailer that needs to make a decision about production.

The third part of the chapter describes decision-making under uncertainty and the ways that simulation can be used. It describes how to set out a range of possible interventions and how they can be modeled using a simulation. It discusses how to use simulation processes in order to optimize decision-making under constraints, by using the fashion retailer example in various contexts.

The chapter also contains a case study of a painting business deciding how much to bid for a contract to paint a factory, and describes the solution to making this decision. The concepts explained in this chapter are applied in different settings in the following chapters.

Optimization: Optimization techniques are used in almost every application in this book. This chapter presents some of the core concepts of constrained optimization. The basic ideas are illustrated using one broad class of optimization problems called linear optimization. Linear optimization covers the most widely used models in business. In addition, because linear models are easy to visualize in two dimensions, it offers a visual introduction to the basic concepts in optimization. Additionally, the chapter provides a brief introduction to other optimization models and techniques such as integer/discrete optimization, nonlinear optimization, search methods, and the use of optimization software.

The linear optimization part is conventionally developed by describing the decision variables, the objective function, constraints, and the assumptions underlying the linear models. Using geometric arguments, it illustrates the concept of feasibility and optimality. It then provides the basic theorems of linear programming. The chapter then develops the idea of shadow prices, reduced costs, and sensitivity analysis, which is the underpinning of any post-optimality business analysis. The solver function in Excel is used for illustrating these ideas. Then, the chapter explains how these ideas extend to integer programming and provides an outline of the branch and bound method with examples. The ideas are further extended to nonlinear optimization via examples of models for linear regression, maximum likelihood estimation, and logistic regression.

Forecasting Analytics: Forecasting is perhaps the most commonly used method in business analytics. This chapter introduces the idea of using analytics to predict the outcomes in the future, and focuses on applying analytics tools for business and operations. The chapter begins by explaining the difficulty of predicting the future with perfect accuracy, and the importance of accepting the uncertainty inherent in any predictive analysis.

The chapter begins by defining forecasting as estimating in unknown situations. It describes data that can be used to make forecasts, but focuses on time-series forecasting. It introduces the concepts of point-forecasts and prediction intervals, which are used in time-series analysis as part of predictions of future outcomes. It suggests reasons for the intervention of human judgment in the forecasts provided by computers. It describes the core method of time-series forecasting—identifying a model that forecasts the best.

The second part of the chapter describes quantitative approaches to forecasting. It begins by describing the various kinds of data that can be used to make forecasts, such as spoken, written, numbers, and so on. It explains some methods of dealing with outliers in the data set, which can affect the fit of the forecast, such as trimming and winsorizing.

The chapter discusses the effects of seasonal fluctuations on time-series data and how to adjust for them. It introduces the autocorrelation function and its use. It also explains the partial autocorrelation function.

A number of methods used in predictive forecasting are explained, including the naïve method, the average and moving average methods, Holt exponential smoothing, and the ARIMA framework. The chapter also discusses ways to predict stochastic intermittent demand, such as Croston's approach, and the Syntetos and Boylan approximation.

The third section of the chapter describes the process of applied forecasting analytics at the operational, tactical, and strategic levels. It propounds a seven-step forecasting process for operational tasks, and explains each step in detail.

The fourth section of the chapter concerns evaluating the accuracy of forecasts. It explains measures such as mean absolute error, mean squared error, and root mean squared error, and how to calculate them. Both Excel and R software use is explained.

Advanced Statistical Methods: Count Data: The chapter begins by introducing the idea of count variables and gives examples of where they are encountered, such as insurance applications and the amount of time taken off by persons that fall sick.

It first introduces the idea of the Poisson regression model, and explains why ordinary least squares are not suited to some situations for which the Poisson model is more appropriate. It illustrates the differences between the normal and Poisson distributions using conditional distribution graphs.

It defines the Poisson distribution model and its general use, as well as an example regarding insurance claims data. It walks through the interpretation of the regression's results, including the explanation of the regression coefficients, deviance, dispersion, and so on.

It discusses some of the problems with the Poisson regression, and how overdispersion can cause issues for the analysis. It introduces the negative binomial distribution as a method to counteract overdispersion. Zero-inflation models are discussed. The chapter ends with a case study on Canadian insurance data.

Advanced Statistical Methods—Survival Analysis: Like the previous chapter, this one deals with another specialized application. It involves techniques that analyze time-to-event data. It defines time-to-event data and the contexts in which it can be used, and provides a number of business situations in which survival analysis is important.

The chapter explains the idea of censored data, which refers to survival times in which the event in question has not yet occurred. It explains the differences between survival models and other types of analysis, and the fields in which it can be used. It defines the types of censoring: right-censoring, left-censoring, and interval-censoring, and the method to incorporate them into the data set.

1 Introduction

The chapter then defines the survival analysis functions: the survival function and the hazard function. It describes some simple types of hazard functions. It describes some parametric and nonparametric methods of analysis, and defines the cases in which nonparametric methods must be used. It explains the Kaplan–Meier method in detail, along with an example. Semiparametric models are introduced for cases in which several covariate variables are believed to contribute to survival. Cox's proportional hazards model and its interpretation are discussed.

The chapter ends with a comparison between semiparametric and parametric models, and a case study regarding churn data.

Unsupervised Learning: The first of the three machine learning chapters sets out the philosophy of machine learning. This chapter explains why unsupervised learning—an important paradigm in machine learning—is akin to uncovering the proverbial needle in the haystack, discovering the grammar of the process that generated the data, and exaggerating the "signal" while ignoring the "noise" in it. The chapter covers methods of projection, clustering, and density estimation—three core unsupervised learning frameworks that help us perceive the data in different ways. In addition, the chapter describes collaborative filtering and applications of network analysis.

The chapter begins with drawing the distinction between supervised and unsupervised learning. It then presents a common approach to solving unsupervised learning problems by casting them into an optimization framework. In this framework, there are four steps:

- Intuition: to develop an intuition about how to approach the problem as an optimization problem
- Formulation: to write the precise mathematical objective function in terms of data using intuition
- Modification: to modify the objective function into something simpler or "more solvable"
- Optimization: to solve the final objective function using traditional optimization approaches

The chapter discusses principal components analysis (PCA), self-organizing maps (SOM), and multidimensional scaling (MDS) under projection algorithms. In clustering, it describes partitional and hierarchical clustering. Under density estimation, it describes nonparametric and parametric approaches. The chapter concludes with illustrations of collaborative filtering and network analysis.

Supervised Learning: In supervised learning, the aim is to learn from previously identified examples. The chapter covers the philosophical, theoretical, and practical aspects of one of the most common machine learning paradigms—supervised learning—that essentially learns to map from an observation (e.g., symptoms and test results of a patient) to a prediction (e.g., disease or medical condition), which in turn is used to make decisions (e.g., prescription). The chapter then explores the process, science, and art of building supervised learning models.

The first part explains the different paradigms in supervised learning: classification, regression, retrieval, recommendation, and how they differ by the nature of their input and output. It then describes the process of learning, from features description to feature engineering to models to algorithms that help make the learning happen.

Among algorithms, the chapter describes rule-based classifiers, decision trees, k-nearest neighbor, Parzen window, and Bayesian and naïve Bayes classifiers. Among discriminant functions that partition a region using an algorithm, linear (LDA) and quadratic discriminant analysis (QDA) are discussed. A section describes recommendation engines. Neural networks are then introduced followed by a succinct introduction to a key algorithm called support vector machines (SVM). The chapter concludes with a description of ensemble techniques, including bagging, random forest, boosting, mixture of experts, and hierarchical classifiers. The specialized neural networks for Deep Learning are explained in the next chapter.

Deep Learning: This chapter introduces the idea of deep learning as a part of machine learning. It aims to explain the idea of deep learning and various popular deep learning architectures. It has four main parts:

- Understand what is deep learning.
- Understand various popular deep learning architectures, and know when to use which architecture for solving a business problem.
- How to perform image analysis using deep learning.
- How to perform text analysis using deep learning.

The chapter explains the origins of learning, from a single perceptron to mimic the functioning of a neuron to the multilayered perceptron (MLP). It briefly recaps the backpropagation algorithm and introduces the learning rate and error functions. It then discusses the deep learning architectures applied to supervised, unsupervised, and reinforcement learning. An example of using an artificial neural network for recognizing handwritten digits (based on the MNIST data set) is presented.

The next section of the chapter describes Convolutional Neural Networks (CNN), which are aimed at solving vision-related problems. The ImageNet data set is introduced. The use of CNNs in the ImageNet Large Scale Visual Recognition Challenge is explained, along with a brief history of the challenge. The biological inspiration for CNNs is presented. Four layers of a typical CNN are introduced—the convolution layer, the rectified linear units layer, the pooling layers, and the fully connected layer. Each layer is explained, with examples. A unifying example using the same MNIST data set is presented.

The third section of the chapter discusses recurrent neural networks (RNNs). It begins by describing the motivation for sequence learning models, and their use in understanding language. Traditional language models and their functions in predicting words are explained. The chapter describes a basic RNN model with three units, aimed at predicting the next word in a sentence. It explains the detailed example by which an RNN can be built for next word prediction. It presents some uses of RNNs, such as image captioning and machine translation.

The next seven chapters contain descriptions of analytics usage in different domains and different contexts. These are described next.

Retail Analytics: The chapter begins by introducing the background and definition of retail analytics. It focuses on advanced analytics. It explains the use of four main categories of business decisions: consumer, product, human resources, and advertising. Several examples of retail analytics are presented, such as increasing book recommendations during periods of cold weather. Complications in retail analytics are discussed.

The second part of the chapter focuses on data collection in the retail sector. It describes the traditional sources of retail data, such as point-of-sale devices, and how they have been used in decision-making processes. It also discusses advances in technology and the way that new means of data collection have changed the field. These include the use of radio frequency identification technology, the Internet of things, and Bluetooth beacons.

The third section describes methodologies, focusing on inventory, assortment, and pricing decisions. It begins with modeling product-based demand in order to make predictions. The penalized L1 regression LASSO for retail demand forecasting is introduced. The use of regression trees and artificial neural networks is discussed in the same context. The chapter then discusses the use of such forecasts in decision-making. It presents evidence that machine learning approaches benefit revenue and profit in both price-setting and inventory-choice contexts.

Demand models into which consumer choice is incorporated are introduced. The multinomial logit, mixed multinomial logit, and nested logit models are described. Nonparametric choice models are also introduced as an alternative to logit models. Optimal assortment decisions using these models are presented. Attempts at learning customer preferences while optimizing assortment choices are described.

The fourth section of the chapter discusses business challenges and opportunities. The benefits of omnichannel retail are discussed, along with the need for retail analytics to change in order to fit an omnichannel shop. It also discusses some recent start-ups in the retail analytics space and their focuses.

Marketing Analytics: Marketing is one of the most important, historically the earliest, and fascinating areas for applying analytics to solve business problems. Due to the vast array of applications, only the most important ones are surveyed in this chapter. The chapter begins by explaining the importance of using marketing analytics for firms. It defines the various levels that marketing analytics can apply to: the firm, the brand or product, and the customer. It introduces a number of processes and models that can be used in analyzing and making marketing decisions, including statistical analysis, nonparametric tools, and customer analysis. The processes and tools discussed in this chapter will help in various aspects of marketing such as target marketing and segmentation, price and promotion, customer valuation, resource allocation, response analysis, demand assessment, and new product development.

The second section of the chapter explains the use of the interaction effect in regression models. Building on earlier chapters on regression, it explains the utility of a term that captures the effect of one or more interactions between other variables. It explains how to interpret new variables and their significance. The use of curvilinear relationships in order to identify the curvilinear effect is discussed. Mediation analysis is introduced, along with an example.

The third section describes data envelopment analysis (DEA), which is aimed at improving the performance of organizations. It describes the manner in which DEA works to present targets to managers and can be used to answer key operational questions in Marketing: sales force productivity, performance of sales regions, and effectiveness of geomarketing.

The next topic covered is conjoint analysis. It explains how knowing customers' preference provides invaluable information about how customers think and make their decisions before purchasing products. Thus, it helps firms devise their marketing strategies including advertising, promotion, and sales activities.

The fifth section of the chapter discusses customer analytics. Customer lifetime value (CLV), a measure of the value provided to firms by customers, is introduced, along with some other measures. A method to calculate CLV is presented, along with its limitations. The chapter also discusses two more measures of customer value: customer referral value and customer influence value, in detail. Additional topics are covered in the chapters on retail analytics and social media analytics.

Financial Analytics: Financial analytics like Marketing has been a big consumer of data. The topics chosen in this chapter provide one unified way of thinking about analytics in this domain—valuation. This chapter focuses on the two main branches of quantitative finance: the risk-neutral or "Q" world and the risk-averse or "P" world. It describes the constraints and aims of analysts in each world, along with their primary methodologies. It explains Q-quant theories such as the work of Black and Scholes, and Harrison and Pliska. P-quant theories such as net present value, capital asset pricing models, arbitrage pricing theory, and the efficient market hypothesis are presented.

The methodology of financial data analytics is explained via a three-stage process: asset price estimation, risk management, and portfolio analysis.

Asset price estimation is explained as a five-step process. It describes the use of the random walk in identifying the variable to be analyzed. Several methods of transforming the variable into one that is identical and independently distributed are presented. A maximum likelihood estimation method to model variance is explained. Monte Carlo simulations of projecting variables into the future are discussed, along with pricing projected variables.

Risk management is discussed as a three-step process. The first step is risk aggregation. Copula functions and their uses are explained. The second step, portfolio assessment, is explained by using metrics such as Value at Risk. The third step, attribution, is explained. Various types of capital at risk are listed.

Portfolio analysis is described as a two-stage process. Allocating risk for the entire portfolio is discussed. Executing trades in order to move the portfolio to a new risk/return level is explained.

A detailed example explaining each of the ten steps is presented, along with data and code in MATLAB. This example also serves as a stand-alone case study on financial analytics.

Social Media Analytics: Social-media-based analytics has been growing in importance and value to businesses. This chapter discusses the various tools available to gather and analyze data from social media and Internet-based sources, focusing on the use of advertisements. It begins by describing Web-based analytical tools and the information they can provide, such as cookies, sentiment analysis, and mobile analytics.

It introduces real-time advertising on online platforms, and the wealth of data generated by browsers visiting target websites. It lists the various kinds of advertising possible, including video and audio ads, map-based ads, and banner ads. It explains the various avenues in which these ads can be displayed, and details the reach of social media sites such as Facebook and Twitter. The various methods in which ads can be purchased are discussed. Programmatic advertising and its components are introduced. Real-time bidding on online advertising spaces is explained.

A/B experiments are defined and explained. The completely randomized design (CRD) experiment is discussed. The regression model for the CRD and an example are presented. The need for randomized complete block design experiments is introduced, and an example for such an experiment is shown. Analytics of multivariate experiments and their advantages are discussed. Orthogonal designs and their meanings are explained.

The chapter discusses the use of data-driven search engine advertising. The use of data in order to help companies better reach consumers and identify trends is discussed. The power of search engines in this regard is discussed. The problem of attribution, or identifying the influence of various ads across various platforms is introduced, and a number of models that aim to solve this problem are elucidated. Some models discussed are: the first click attribution model, the last click attribution model, the linear attribution model, and algorithmic attribution models.

Healthcare Analytics: Healthcare is once again an area where data, experiments, and research have coexisted within an analytical framework for hundreds of years. This chapter discusses analytical approaches to healthcare. It begins with an overview of the current field of healthcare analytics. It describes the latest innovations in the use of data to refine healthcare, including telemedicine, wearable technologies, and simulations of the human body. It describes some of the challenges that data analysts can face when attempting to use analytics to understand healthcare-related problems.

The main part of the chapter focuses on the use of analytics to improve operations. The context is patient flow in outpatient clinics. It uses Academic Medical Centers as an example to describe the processes that patients go through when visiting clinics that are also teaching centers. It describes the effects of the Affordable Care Act, an aging population, and changes in social healthcare systems on the public health infrastructure in the USA.

A five-step process map of a representative clinic is presented, along with a discrete event simulation of the clinic. The history of using operations research-based methods to improve healthcare processes is discussed. The chapter introduces a six-step process aimed at understanding complex systems, identifying potential improvements, and predicting the effects of changes, and describes each step in detail.

Lastly, the chapter discusses the various results of this process on some goals of the clinic, such as arrivals, processing times, and impact on teaching. Data regarding each goal and its change are presented and analyzed. The chapter contains a hands-on exercise based on the simulation models discussed. The chapter is a fine application of simulation concepts and modeling methodologies used in Operations Management to improve healthcare systems.

Pricing Analytics: This chapter discusses the various mechanisms available to companies in order to price their products. The topics pertain to revenue management, which constitutes perhaps the most successful and visible area of business analytics.

The chapter begins by introducing defining two factors that affect pricing: the nature of the product and its competition, and customers' preferences and values. It introduces the concept of a price optimization model, and the need to control capacity constraints when estimating customer demand.

The first type of model introduced is the independent class model. The underlying assumption behind the model is defined, as well as its implications for modeling customer choice. The EMSR heuristic and its use are explained.

The issue of overbooking in many service-related industries is introduced. The trade-off between an underutilized inventory and the risk of denying service to customers is discussed. A model for deciding an overbooking limit, given the physical capacity at the disposal of the company, is presented. Dynamic pricing is presented as a method to better utilize inventory.

Three main types of dynamic pricing are discussed: surge pricing, repricing, and markup/markdown pricing. Each type is comprehensively explained. Three models of forecasting and estimating customer demand are presented: additive, multiplicative, and choice.

A number of processes for capacity control, such as nested allocations, are presented. Network revenue management systems are introduced. A backward induction method of control is explained. The chapter ends with an example of a hotel that is planning allocation of rooms based on a demand forecast.

Supply Chain Analytics: This chapter discusses the use of data and analytical tools to increase value in the supply chain. It begins by defining the processes that constitute supply chains, and the goals of supply chain management. The uncertainty inherent in supply chains is discussed. Four applications of supply chain analytics are described: demand forecasting, inventory optimization, supply chain disruption, and commodity procurement.

A case study of VASTA, one of the largest wireless services carriers in the USA, is presented. The case study concerns the decision of whether the company should change its current inventory strategy from a "push" strategy to a "pull" strategy. The advantages and disadvantages of each strategy are discussed. A basic model to evaluate both strategies is introduced. An analysis of the results is presented. Following the analysis, a more advanced evaluation model is introduced. Customer satisfaction and implementation costs are added to the model.

The last three chapters of the book contain case studies. Each of the cases comes with a large data set upon which students can practice almost every technique and modeling approach covered in the book. The Info Media case study explains the use of viewership data to design promotional campaigns. The problem presented is to determine a multichannel ad spots allocation in order to maximize "reach" given a budget and campaign guidelines. The approach uses simulation to compute the viewership and then uses the simulated data to link promotional aspects to the total reach of a campaign. Finally, the model can be used to optimize the allocation of budgets across channels.

The AAA airline case study illustrates the use of choice models to design airline offerings. The main task is to develop a demand forecasting model, which predicts the passenger share for every origin–destination pair (O–D pair) given AAA, as well as competitors' offerings. The students are asked to explore different models including the MNL and machine learning algorithms. Once a demand model has been developed it can be used to diagnose the current performance and suggest various remedies, such as adding, dropping, or changing itineraries in specific city pairs. The third case study, Ideal Insurance, is on fraud detection. The problem faced by the firm is the growing cost of servicing and settling claims in their healthcare practice. The students learn about the industry and its intricate relationships with various stakeholders. They also get an introduction to rule-based decision support systems. The students are asked to create a system for detecting fraud, which should be superior to the current "rule-based" system.

2 The Intended Audience

This book is the first of its kind both in breadth and depth of coverage and serves as a textbook for students of first year graduate program in analytics and long duration (1-year part time) certificate programs in business analytics. It also serves as a perfect guide to practitioners.

The content is based on the curriculum of the Certificate Programme in Business Analytics (CBA), now renamed as Advanced Management Programme in Business Analytics (AMPBA) of Indian School of Business (ISB). The original curriculum was created by Galit Shmueli. The curriculum was further developed by the coeditors, Bhimasankaram Pochiraju and Sridhar Seshadri, who were responsible for starting and mentoring the CBA program in ISB. Bhimasankaram Pochiraju has been the Faculty Director of CBA since its inception and was a member of the Academic Board. Sridhar Seshadri managed the launch of the program and since then has chaired the academic development efforts. Based on the industry needs, the curriculum continues to be modified by the Academic Board of the Applied Statistics and Computing Lab (ASC Lab) at ISB.

Part I
Tools

Chapter 2
Data Collection

Sudhir Voleti

1 Introduction

Collecting data is the first step towards analyzing it. In order to understand and solve business problems, data scientists must have a strong grasp of the characteristics of the data in question. How do we collect data? What kinds of data exist? Where is it coming from? Before beginning to analyze data, analysts must know how to answer these questions. In doing so, we build the base upon which the rest of our examination follows. This chapter aims to introduce and explain the nuances of data collection, so that we understand the methods we can use to analyze it.

2 The Value of Data: A Motivating Example

In 2017, video-streaming company Netflix Inc. was worth more than $80 billion, more than 100 times its value when it listed in 2002. The company's current position as the market leader in the online-streaming sector is a far cry from its humble beginning as a DVD rental-by-mail service founded in 1997. So, what had driven Netflix's incredible success? What helped its shares, priced at $15 each on their initial public offering in May 2002, rise to nearly $190 in July 2017? It is well known that a firm's [market] valuation is the sum total in today's money, or the net present value (NPV) of all the profits the firm will earn over its lifetime. So investors reckon that Netflix is worth tens of billions of dollars in profits over its lifetime. Why might this be the case? After all, companies had been creating television and

---

S. Voleti (✉)
Indian School of Business, Hyderabad, Telangana, India
e-mail: sudhir_voleti@isb.edu

© Springer Nature Switzerland AG 2019
B. Pochiraju, S. Seshadri (eds.), *Essentials of Business Analytics*, International Series in Operations Research & Management Science 264,
https://doi.org/10.1007/978-3-319-68837-4_2

cinematic content for decades before Netflix came along, and Netflix did not start its own online business until 2007. Why is Netflix different from traditional cable companies that offer shows on their own channels?

Moreover, the vast majority of Netflix's content is actually owned by its competitors. Though the streaming company invests in original programming, the lion's share of the material available on Netflix is produced by cable companies across the world. Yet Netflix has access to one key asset that helps it to predict where its audience will go and understand their every quirk: data.

Netflix can track every action that a customer makes on its website—what they watch, how long they watch it for, when they tune out, and most importantly, what they might be looking for next. This data is invaluable to its business—it allows the company to target specific niches of the market with unerring accuracy.

On February 1, 2013, Netflix debuted House of Cards—a political thriller starring Kevin Spacey. The show was a hit, propelling Netflix's viewership and proving that its online strategy could work. A few months later, Spacey applauded Netflix's approach and cited its use of data for its ability to take a risk on a project that every other major television studio network had declined. Casey said in *Edinburgh*, at the *Guardian Edinburgh International Television Festival*[1] on August 22: "Netflix was the only company that said, 'We believe in you. We have run our data, and it tells us our audience would watch this series.'"

Netflix's data-oriented approach is key not just to its ability to pick winning television shows, but to its global reach and power. Though competitors are springing up the world over, Netflix remains at the top of the pack, and so long as it is able to exploit its knowledge of how its viewers behave and what they prefer to watch, it will remain there.

Let us take another example. The technology "cab" company Uber has taken the world by storm in the past 5 years. In 2014, Uber's valuation was a mammoth 40 billion USD, which by 2015 jumped another 50% to reach 60 billion USD. This fact begs the question: what makes Uber so special? What competitive advantage, strategic asset, and/or enabling platform accounts for Uber's valuation numbers? The investors reckon that Uber is worth tens of billions of dollars in profits over its lifetime. Why might this be the case? Uber is after all known as a ride-sharing business—and there are other cab companies available in every city.

We know that Uber is "asset-light," in the sense that it does not own the cab fleet or have drivers of the cabs on its direct payroll as employees. It employs a franchise model wherein drivers bring their own vehicles and sign up for Uber. Yet Uber does have one key asset that it actually owns, one that lies at the heart of its profit projections: data. Uber owns all rights to every bit of data from every passenger, every driver, every ride and every route on its network. Curious as to how much data are we talking about? Consider this. Uber took 6 years to reach one billion

---

[1]Guardian Edinburgh International Television Festival, 2017 (https://www.ibtimes.com/kevin-spacey-speech-why-netflix-model-can-save-television-video-full-transcript-1401970) accessed on Sep 13, 2018.

rides (Dec 2015). Six months later, it had reached the two billion mark. That is one billion rides in 180 days, or 5.5 million rides/day. How did having consumer data play a factor in the exponential growth of a company such as Uber? Moreover, how does data connect to analytics and, finally, to market value?

Data is a valuable asset that helps build sustainable competitive advantage. It enables what economists would call "supernormal profits" and thereby plausibly justify some of those wonderful valuation numbers we saw earlier. Uber had help, of course. The nature of demand for its product (contractual personal transportation), the ubiquity of its enabling platform (location-enabled mobile devices), and the profile of its typical customers (the smartphone-owning, convenience-seeking segment) has all contributed to its success. However, that does not take away from the central point being motivated here—the value contained in data, and the need to collect and corral this valuable resource into a strategic asset.

3 Data Collection Preliminaries

A well-known management adage goes, "We can only manage what we can measure." But why is measurement considered so critical? Measurement is important because it precedes *analysis*, which in turn precedes *modeling*. And more often than not, it is *modeling* that enables *prediction*. Without *prediction* (determination of the values an outcome or entity will take under specific conditions), there can be no *optimization*. And without *optimization*, there is no *management*. The quantity that gets measured is reflected in our records as "data." The word data comes from the Latin root *datum* for "given." Thus, data (datum in plural) becomes facts which are given or known to be true. In what follows, we will explore some preliminary conceptions about data, types of data, basic measurement scales, and the implications therein.

3.1 Primary Versus Secondary Dichotomy

Data collection for research and analytics can broadly be divided into two major types: primary data and secondary data. Consider a project or a business task that requires certain data. Primary data would be data that is collected "at source" (hence, primary in form) and specifically for the research at hand. The data source could be individuals, groups, organizations, etc. and data from them would be actively elicited or passively observed and collected. Thus, surveys, interviews, and focus groups all fall under the ambit of primary data. The main advantage of primary data is that it is tailored specifically to the questions posed by the research project. The disadvantages are cost and time.

On the other hand, secondary data is that which has been previously collected for a purpose that is *not* specific to the research at hand. For example, sales records, industry reports, and interview transcripts from past research are data that would continue to exist whether or not the project at hand had come to fruition. A good example of a means to obtain secondary data that is rapidly expanding is the API (Application Programming Interface)—an interface that is used by developers to securely query external systems and obtain a myriad of information.

In this chapter, we concentrate on data available in published sources and websites (often called secondary data sources) as these are the most commonly used data sources in business today.

4 Data Collection Methods

In this section, we describe various methods of data collection based on sources, structure, type, etc. There are basically two methods of data collection: (1) data generation through a designed experiment and (2) collecting data that already exists. A brief description of these methods is given below.

4.1 Designed Experiment

Suppose an agricultural scientist wants to compare the effects of five different fertilizers, A, B, C, D, and E, on the yield of a crop. The yield depends not only on the fertilizer but also on the fertility of the soil. The consultant considers a few relevant types of soil, for example, clay, silt, and sandy soil. In order to compare the fertilizer effect one has to control for the soil effect. For each soil type, the experimenter may choose ten representative plots of equal size and assign the five fertilizers to the ten plots at random in such a way that each fertilizer is assigned to two plots. He then observes the yield in each plot. This is the design of the experiment. Once the experiment is conducted as per this design, the yields in different plots are observed. This is the data collection procedure. As we notice, the data is not readily available to the scientist. He designs an experiment and generates the data. This method of data collection is possible when we can control different factors precisely while studying the effect of an important variable on the outcome. This is quite common in the manufacturing industry (while studying the effect of machines on output or various settings on the yield of a process), psychology, agriculture, etc. For well-designed experiments, determination of the causal effects is easy. However, in social sciences and business where human beings often are the instruments or subjects, experimentation is not easy and in fact may not even be feasible. Despite the limitations, there has been tremendous interest in behavioral experiments in disciplines such as finance, economics, marketing, and operations management. For a recent account on design of experiments, please refer to Montgomery (2017).

4.2 Collection of Data That Already Exists

Household income, expenditure, wealth, and demographic information are examples of data that already exists. Collection of such data is usually done in three possible ways: (1) *complete enumeration*, (2) *sample survey*, and (3) through available sources where the data was collected possibly for a different purpose and is available in different published sources. *Complete enumeration* is collecting data on all items/individuals/firms. Such data, say, on households, may be on consumption of essential commodities, the family income, births and deaths, education of each member of the household, etc. This data is already available with the households but needs to be collected by the investigator. The census is an example of complete enumeration. This method will give information on the whole population. It may appear to be the best way but is expensive both in terms of time and money. Also, it may involve several investigators and investigator bias can creep in (in ways that may not be easy to account for). Such errors are known as *non-sampling errors*. So often, a *sample survey* is employed. In a sample survey, the data is not collected on the entire population, but on a representative sample. Based on the data collected from the sample, inferences are drawn on the population. Since data is not collected on the entire population, there is bound to be an error in the inferences drawn. This error is known as the *sampling error*. The inferences through a sample survey can be made precise with error bounds. It is commonly employed in market research, social sciences, public administration, etc. A good account on sample surveys is available in Blair and Blair (2015).

Secondary data can be collected from two sources: internal or external. *Internal data* is collected by the company or its agents on behalf of the company. The defining characteristic of the internal data is its proprietary nature; the company has control over the data collection process and also has exclusive access to the data and thus the insights drawn on it. Although it is costlier than external data, the exclusivity of access to the data can offer competitive advantage to the company. *The external data*, on the other hand, can be collected by either third-party data providers (such as IRI, AC Nielsen) or government agencies. In addition, recently another source of external secondary data has come into existence in the form of social media/blogs/review websites/search engines where users themselves generate a lot of data through C2B or C2C interactions. Secondary data can also be classified on the nature of the data along the dimension of structure. Broadly, there are three types of data: *structured, semi-structured (hybrid),* and *unstructured* data. Some examples of *structured data* are sales records, financial reports, customer records such as purchase history, etc. A typical example of *unstructured data* is in the form of free-flow text, images, audio, and videos, which are difficult to store in a traditional database. Usually, in reality, data is somewhere in between structured and unstructured and thus is called *semi-structured* or *hybrid* data. For example, a product web page will have product details (structured) and user reviews (unstructured).

The data and its analysis can also be classified on the basis of whether a single unit is observed over multiple time points (*time-series data*), many units observed once (*cross-sectional data*), or many units are observed over multiple time periods (*panel data*). The insights that can be drawn from the data depend on the nature of data, with the richest insights available from panel data. The panel could be *balanced* (all units are observed over all time periods) or *unbalanced* (observations on a few units are missing for a few time points either by design or by accident). If the data is not missing excessively, it can be accounted for using the methods described in Chap. 8.

5 Data Types

In programming, we primarily classify the data into three types—*numerals*, *alphabets*, and *special characters* and the computer converts any data type into binary code for further processing. However, the data collected through various sources can be of types such as numbers, text, image, video, voice, and biometrics.

The data type helps analyst to evaluate which operations can be performed to analyze the data in a meaningful way. The data can limit or enhance the complexity and quality of analysis.

Table 2.1 lists a few examples of data categorized by type, source, and uses. You can read more about them following the links (all accessed on Aug 10, 2017).

5.1 Four Data Types and Primary Scales

Generally, there are four types of data associated with four primary scales, namely, *nominal*, *ordinal*, *interval*, and *ratio*. Nominal scale is used to describe categories in which there is no specific order while the ordinal scale is used to describe categories in which there is an inherent order. For example, green, yellow, and red are three colors that in general are not bound by an inherent order. In such a case, a nominal scale is appropriate. However, if we are using the same colors in connection with the traffic light signals there is clear order. In this case, these categories carry an ordinal scale. Typical examples of the ordinal scale are (1) sick, recovering, healthy; (2) lower income, middle income, higher income; (3) illiterate, primary school pass, higher school pass, graduate or higher, and so on. In the ordinal scale, the differences in the categories are not of the same magnitude (or even of measurable magnitude). Interval scale is used to convey relative magnitude information such as temperature. The term "Interval" comes about because rulers (and rating scales) have intervals of uniform lengths. Example: "I rate A as a 7 and B as a 4 on a scale of 10." In this case, we not only know that A is preferred to B, but we also have some idea of how much more A is preferred to B. Ratio scales convey information on an absolute scale. Example: "I paid $11 for A and $12 for B." The 11 and 12 here are termed "absolute" measures because the corresponding zero point ($0) is understood in the same way by different people (i.e., the measure is independent of subject).

Another set of examples for the four data types, this time from the world of sports, could be as follows. The numbers assigned to runners are of nominal data type, whereas the rank order of winners is of the ordinal data type. Note in the latter case that while we may know who came first and who came second, we would not know by how much based on the rank order alone. A performance rating on a 0–10

Table 2.1 A description of data and their types, sources, and examples

Category	Examples	Type	Sources[a]
Internal data			
Transaction data	Sales (POS/online) transactions, stock market orders and trades, customer IP and geolocation data	Numbers, text	http://times.cs.uiuc.edu/~wang296/Data/ https://www.quandl.com/ https://www.nyse.com/data/transactions-statistics-data-library https://www.sec.gov/answers/shortsalevolume.htm
Customer preference data	Website click stream, cookies, shopping cart, wish list, preorder	Numbers, text	C:\Users\username\AppData\Roaming\Microsoft\Windows\Cookies, Nearbuy.com (advance coupon sold)
Experimental data	Simulation games, clinical trials, live experiments	Text, number, image, audio, video	https://www.clinicaltrialsregister.eu/ https://www.novctrd.com/ http://ctri.nic.in/
Customer relationship data	Demographics, purchase history, loyalty rewards data, phone book	Text, number, image, biometrics	
External data			
Survey data	Census, national sample survey, annual survey of industries, geographical survey, land registry	Text, number, image, audio, video	http://www.census.gov/data.html http://www.mospi.gov.in/ http://www.csoisw.gov.in/ https://www.gsi.gov.in/ http://landrecords.mp.gov.in/
Biometric data (fingerprint, retina, pupil, palm, face)	Immigration data, social security identity, Aadhar card (UID)	Number, text, image, biometric	http://www.migrationpolicy.org/programs/migration-data-hub https://www.dhs.gov/immigration-statistics

(continued)

Table 2.1 (continued)

Category	Examples	Type	Sources[a]
Third party data	RenTrak, A. C. Nielsen, IRI, MIDT (Market Information Data Tapes) in airline industry, people finder, associations, NGOs, database vendors, Google Trends, Google Public Data	All possible data types	http://aws.amazon.com/datasets https://www.worldwildlife.org/pages/conservation-science-data-and-tools http://www.whitepages.com/ https://pipl.com/ https://www.bloomberg.com/ https://in.reuters.com/ http://www.imdb.com/ http://datacatalogs.org/ http://www.google.com/trends/explore https://www.google.com/publicdata/directory
Govt and quasi govt agencies	Federal governments, regulators—Telecom, BFSI, etc., World Bank, IMF, credit reports, climate and weather reports, agriculture production, benchmark indicators—GDP, etc., electoral roll, driver and vehicle licenses, health statistics, judicial records	All possible data types	http://data.gov/ https://data.gov.in/ http://data.gov.uk/ http://open-data.europa.eu/en/data/ http://www.imf.org/en/Data https://www.rbi.org.in/Scripts/Statistics.aspx https://www.healthdata.gov/ https://www.cibil.com/ http://eci.nic.in/ http://data.worldbank.org/
Social sites data, user-generated data	Twitter, Facebook, YouTube, Instagram, Pinterest Wikipedia, YouTube videos, blogs, articles, reviews, comments	All possible data types	https://dev.twitter.com/streaming/overview https://developers.facebook.com/docs/graph-api https://en.wikipedia.org/ https://www.youtube.com/ https://snap.stanford.edu/data/web-Amazon.html http://www.cs.cornell.edu/people/pabo/movie-review-data/

[a] All the sources are last accessed on Aug 10, 2017

scale would be an example of an interval scale. We see this used in certain sports ratings (i.e., gymnastics) wherein judges assign points based on certain metrics. Finally, in track and field events, the time to finish in seconds is an example of ratio data. The reference point of zero seconds is well understood by all observers.

5.2 Common Analysis Types with the Four Primary Scales

The reason why it matters what primary scale was used to collect data is that downstream analysis is constrained by data type. For instance, with nominal data, all we can compute are the mode, some frequencies and percentages. Nothing beyond this is possible due to the nature of the data. With ordinal data, we can compute the median and some rank order statistics in addition to whatever is possible with nominal data. This is because ordinal data retains all the properties of the nominal data type. When we proceed further to interval data and then on to ratio data, we encounter a qualitative leap over what was possible before. Now, suddenly, the arithmetic mean and the variance become meaningful. Hence, most statistical analysis and parametric statistical tests (and associated inference procedures) all become available. With ratio data, in addition to everything that is possible with interval data, ratios of quantities also make sense.

The multiple-choice examples that follow are meant to concretize the understanding of the four primary scales and corresponding data types.

6 Problem Formulation Preliminaries

Even before data collection can begin, the purpose for which the data collection is being conducted must be clarified. Enter, problem formulation. The importance of problem formulation cannot be overstated—it comes first in any research project, ideally speaking. Moreover, even small deviations from the intended path at the very beginning of a project's trajectory can lead to a vastly different destination than was intended. That said, problem formulation can often be a tricky issue to get right. To see why, consider the musings of a decision-maker and country head for XYZ Inc.

> Sales fell short last year. But sales would've approached target except for 6 territories in 2 regions where results were poor. Of course, we implemented a price increase across-the-board last year, so our profit margin goals were just met, even though sales revenue fell short. Yet, 2 of our competitors saw above-trend sales increases last year. Still, another competitor seems to be struggling, and word on the street is that they have been slashing prices to close deals. Of course, the economy was pretty uneven across our geographies last year and the 2 regions in question, weak anyway, were particularly so last year. Then there was that mess with the new salesforce compensation policy coming into effect last year. 1 of the 2 weak regions saw much salesforce turnover last year ...

These are everyday musings in the lives of business executives and are far from unusual. Depending on the identification of the problem, data collection strategies, resources, and approaches will differ. The difficulty in being able to readily pinpoint any one cause or a combination of causes as specific problem highlights the issues that crop up in problem formulation. Four important points jump out from the above example. First, that reality is messy. Unlike textbook examples of problems, wherein irrelevant information is filtered out a priori and only that which is required to solve "the" identified problem exactly is retained, life seldom simplifies issues in such a clear-cut manner. Second, borrowing from a medical analogy, there are symptoms—observable manifestations of an underlying problem or ailment—and then there is the cause or ailment itself. Symptoms could be a fever or a cold and the causes could be bacterial or viral agents. However, curing the symptoms may not cure the ailment. Similarly, in the previous example from XYZ Inc., we see symptoms ("sales are falling") and hypothesize the existence of one or more underlying problems or causes. Third, note the pattern of connections between symptom(s) and potential causes. One symptom (falling sales) is assumed to be coming from one or more potential causes (product line, salesforce compensation, weak economy, competitors, etc.). This brings up the fourth point—How can we diagnose a problem (or cause)? One strategy would be to narrow the field of "ailments" by ruling out low-hanging fruits—ideally, as quickly and cheaply as feasible. It is not hard to see that the data required for this problem depends on what potential ailments we have shortlisted in the first place.

6.1 Towards a Problem Formulation Framework

For illustrative purposes, consider a list of three probable causes from the messy reality of the problem statement given above, namely, (1) product line is obsolete; (2) customer-connect is ineffective; and (3) product pricing is uncompetitive (say). Then, from this messy reality we can formulate decision problems (D.P.s) that correspond to the three identified probable causes:

- D.P. #1: "Should new product(s) be introduced?"
- D.P. #2: "Should advertising campaign be changed?"
- D.P. #3: "Should product prices be changed?"

Note what we are doing in mathematical terms—if messy reality is a large multidimensional object, then these D.P.s are small-dimensional subsets of that reality. This "reduces" a messy large-dimensional object to a relatively more manageable small-dimensional one.

The D.P., even though it is of small dimension, may not contain sufficient detail to map directly onto tools. Hence, another level of refinement called the research objective (R.O.) may be needed. While the D.P. is a small-dimensional object, the R.O. is (ideally) a one-dimensional object. Multiple R.O.s may be needed to completely "cover" or address a single D.P. Furthermore, because each R.O. is one-dimensional, it maps easily and directly onto one or more specific tools in the analytics toolbox. A one-dimensional problem formulation component better be

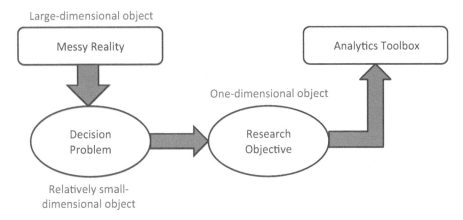

Fig. 2.1 A framework for problem formulation

well defined. The R.O. has three essential parts that together lend necessary clarity to its definition. R.O.s comprise of (a) an action verb and (b) an actionable object, and typically fit within one handwritten line (to enforce brevity). For instance, the active voice statement "Identify the real and perceived gaps in our product line vis-à-vis that of our main competitors" is an R.O. because its components action verb ("identify"), actionable object ("real and perceived gaps"), and brevity are satisfied.

Figure 2.1 depicts the problem formulation framework we just described in pictorial form. It is clear from the figure that as we impose preliminary structure, we effectively reduce problem dimensionality from large (messy reality) to somewhat small (D.P.) to the concise and the precise (R.O.).

6.2 Problem Clarity and Research Type

A quotation attributed to former US defense secretary Donald Rumsfeld in the run-up to the Iraq war goes as follows: "There are known-knowns. These are things we know that we know. There are known-unknowns. That is to say, there are things that we know we don't know. But there are also unknown-unknowns. There are things we don't know we don't know." This statement is useful in that it helps discern the differing degrees of the awareness of our ignorance about the true state of affairs.

To understand why the above statement might be relevant for problem formulation, consider that there are broadly three types of research that correspond to three levels of clarity in problem definition. The first is exploratory research wherein the problem is at best ambiguous. For instance, "Our sales are falling .... Why?" or "Our ad campaign isn't working. Don't know why." When identifying the problem is itself a problem, owing to unknown-unknowns, we take an exploratory approach to trace and list potential problem sources and then define what the problems may be. The second type is descriptive research wherein the problem's identity is somewhat clear. For instance, "What kind of people buy our products?" or "Who is perceived as competition to us?" These are examples of known-unknowns. The third type is causal research wherein the problem is clearly defined. For instance, "Will changing this particular promotional campaign raise sales?" is a clearly identified known-unknown. Causal research (the cause in causal comes from the cause in because) tries to uncover the "why" behind phenomena of interest and its most powerful and practical tool is the experimentation method. It is not hard to see that the level of clarity in problem definition vastly affects the choices available in terms of data collection and downstream analysis.

7 Challenges in Data Collection

Data collection is about data and about collection. We have seen the value inherent in the right data in Sect. 1. In Sect. 3, we have seen the importance of clarity in problem formulation while determining what data to collect. Now it is time to turn to the "collection" piece of data collection. What challenges might a data scientist typically face in collecting data? There are various ways to list the challenges that arise. The approach taken here follows a logical sequence.

The first challenge is in knowing what data to collect. This often requires some familiarity with or knowledge of the problem domain. Second, after the data scientist knows what data to collect, the hunt for data sources can proceed apace. Third, having identified data sources (the next section features a lengthy listing of data sources in one domain as part of an illustrative example), the actual process of mining of raw data can follow. Fourth, once the raw data is mined, data quality assessment follows. This includes various data cleaning/wrangling, imputation, and other data "janitorial" work that consumes a major part of the typical data science project's time. Fifth, after assessing data quality, the data scientist must now judge the relevance of the data to the problem at hand. While considering the above, at each stage one has to take into consideration the cost and time constraints.

Consider a retailing context. What kinds of data would or could a grocery retail store collect? Of course, there would be point-of-sale data on items purchased, promotions availed, payment modes and prices paid in each market basket, captured by UPC scanner machines. Apart from that, retailers would likely be interested in (and can easily collect) data on a varied set of parameters. For example, that may include store traffic and footfalls by time of the day and day of the week, basic segmentation (e.g., demographic) of the store's clientele, past purchase history of customers (provided customers can be uniquely identified, that is, through a loyalty or bonus program), routes taken by the average customer when navigating the store, or time spent on an average by a customer in different aisles and product departments. Clearly, in the retail sector, the wide variety of data sources and capture points to data are typically large in the following three areas:

- Volume
- Variety (ranges from structured metric data on sales, inventory, and geo location to unstructured data types such as text, images, and audiovisual files)
- Velocity—(the speed at which data comes in and gets updated, i.e., sales or inventory data, social media monitoring data, clickstreams, RFIDs—Radio-frequency identification, etc.)

These fulfill the three attribute criteria that are required to being labeled "Big Data" (Diebold 2012). The next subsection dives into the retail sector as an illustrative example of data collection possibilities, opportunities, and challenges.

8 Data Collation, Validation, and Presentation

Collecting data from multiple sources will not result in rich insights unless the data is collated to retain its integrity. Data validity may be compromised if proper care is not taken during collation. One may face various challenges while trying to collate the data. Below, we describe a few challenges along with the approaches to handle them in the light of business problems.

- No common identifier: A challenge while collating data from multiple sources arises due to the absence of common identifiers across different sources. The analyst may seek a third identifier that can serve as a link between two data sources.
- Missing data, data entry error: Missing data can either be ignored, deleted, or imputed with relevant statistics (see Chap. 8).
- Different levels of granularity: The data could be aggregated at different levels. For example, primary data is collected at the individual level, while secondary data is usually available at the aggregate level. One can either aggregate the data in order to bring all the observations to the same level of granularity or can apportion the data using business logic.
- Change in data type over the period or across the samples: In financial and economic data, many a time the base period or multipliers are changed, which needs to be accounted for to achieve data consistency. Similarly, samples collected from different populations such as India and the USA may suffer from inconsistent definitions of time periods—the financial year in India is from April to March and in the USA, it is from January to December. One may require remapping of old versus new data types in order to bring the data to the same level for analysis.
- Validation and reliability: As the secondary data is collected by another user, the researcher may want to validate to check the correctness and reliability of the data to answer a particular research question.

Data presentation is also very important to understand the issues in the data. The basic presentation may include relevant charts such as scatter plots, histograms, and pie charts or summary statistics such as the number of observations, mean, median, variance, minimum, and maximum. You will read more about data visualization in Chap. 5 and about basic inferences in Chap. 6.

9 Data Collection in the Retailing Industry: An Illustrative Example

Bradlow et al. (2017) provide a detailed framework to understand and classify the various data sources becoming popular with retailers in the era of Big Data and analytics. Figure 2.2, taken from Bradlow et al. (2017), "organizes (an admittedly incomplete) set of eight broad retail data sources into three primary groups, namely, (1) traditional enterprise data capture; (2) customer identity, characteristics, social graph and profile data capture; and (3) location-based data capture." The claim is that insight and possibilities lie at the intersection of these groups of diverse, contextual, and relevant data.

Traditional enterprise data capture (marked #1 in Fig. 2.2) from UPC scanners combined with inventory data from ERP or SCM software and syndicated databases (such as those from IRI or Nielsen) enable a host of analyses, including the following:

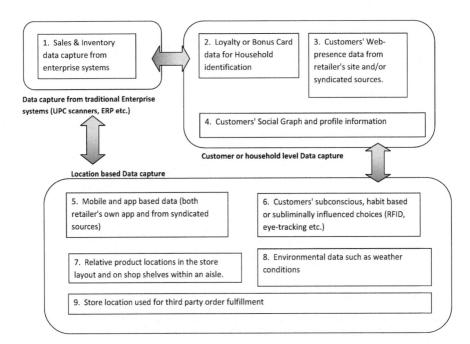

Fig. 2.2 Data sources in the modern retail sector

- Cross-sectional analysis of market baskets—item co-occurrences, complements and substitutes, cross-category dependence, etc. (e.g., Blattberg et al. 2008; Russell and Petersen 2000)
- Analysis of aggregate sales and inventory movement patterns by stock-keeping unit
- Computation of price or shelf-space elasticities at different levels of aggregation such as category, brand, and SKU (see Bijmolt et al. (2005) for a review of this literature)
- Assessment of aggregate effects of prices, promotions, and product attributes on sales

In other words, traditional enterprise data capture in a retailing context enables an overview of the four P's of Marketing (product, price, promotion, and place at the level of store, aisle, shelf, etc.).

Customer identity, characteristics, social graph, and profile data capture identify consumers and thereby make available a slew of consumer- or household-specific information such as demographics, purchase history, preferences and promotional response history, product returns history, and basic contacts such as email for email marketing campaigns and personalized flyers and promotions. Bradlow et al. (2017, p. 12) write:

> Such data capture adds not just a slew of columns (consumer characteristics) to the most detailed datasets retailers would have from previous data sources, but also rows in that household-purchase occasion becomes the new unit of analysis. A common data source for customer identification is loyalty or bonus card data (marked #2 in Fig. 2.2) that customers sign up for in return for discounts and promotional offers from retailers. The advent of household specific "panel" data enabled the estimation of household specific parameters in traditional choice models (e.g., Rossi and Allenby 1993; Rossi et al. 1996) and their use thereafter to better design household specific promotions, catalogs, email campaigns, flyers, etc. The use of household- or customer identity requires that a single customer ID be used as primary key to link together all relevant information about a customer across multiple data sources. Within this data capture type, another data source of interest (marked #3 in Fig. 2.2) is predicated on the retailer's web-presence and is relevant even for purely brick-and-mortar retailers. Any type of customer initiated online contact with the firm—think of an email click-through, online browser behavior and cookies, complaints or feedback via email, inquiries, etc. are captured and recorded, and linked to the customer's primary key. Data about customers' online behavior purchased from syndicated sources are also included here. This data source adds new data columns to retailer data on consumers' online search, products viewed (consideration set) but not necessarily bought, purchase and behavior patterns, which can be used to better infer consumer preferences, purchase contexts, promotional response propensities, etc.

Marked #4 in Fig. 2.2 is another potential data source—consumers' social graph information. This could be obtained either from syndicated means or by customers volunteering their social media identities to use as logins at various websites. Mapping the consumer's social graph opens the door to increased opportunities in psychographic and behavior-based targeting, personalization and hyper-segmentation, preference and latent need identification, selling, word of mouth, social influence, recommendation systems, etc. While the famous AIDA framework in marketing has four conventional stages, namely, awareness, interest, desire, and action, it is clear that the "social" component's importance in data collection, analysis, modeling, and prediction is rising. Finally, the third type of data capture—location-based data capture—leverages customers' locations to infer customer preferences, purchase propensities, and design marketing interventions on that basis. The biggest change in recent years in location-based data capture and use has been enabled by customer's smartphones (e.g., Ghose and Han 2011, 2014). Figure 2.2 marks consumers' mobiles as data source #5. Data capture here involves mining location-based services data such as geo location, navigation, and usage data from those consumers who have installed and use the retailer's mobile shopping apps on their smartphones. Consumers' real-time locations within or around retail stores potentially provide a lot of context that can be exploited to make marketing messaging on deals, promotions, new offerings, etc. more relevant and impactful to consumer attention (see, e.g., Luo et al. 2014) and hence to behavior (including impulse behavior).

Another distinct data source, marked #6 in Fig. 2.2, draws upon habit patterns and subconscious consumer behaviors that consumers are unaware of at a conscious level and are hence unable to explain or articulate. Examples of such phenomena include eye-movement when examining a product or web-page (eye-tracking studies started with Wedel and Pieters 2000), the varied paths different shoppers take inside physical stores which can be tracked using RFID chips inside shopping carts (see, e.g., Larson et al. 2005) or inside virtual stores using clickstream data (e.g., Montgomery et al. 2004), the distribution of first-cut emotional responses to varied product and context stimuli which neuro-marketing researchers are trying to understand using functional magnetic resonance imaging (fMRI) studies (see, e.g., Lee et al. (2007) for a survey of the literature), etc.

Data source #7 in Fig. 2.1 draws on how retailers optimize their physical store spaces for meeting sales, share, or profit objectives. Different product arrangements on store shelves lead to differential visibility and salience. This results in a heightened awareness, recall, and inter-product comparison and therefore differential purchase propensity, sales, and share for any focal product. More generally, an optimization of store layouts and other situational factors both offline (e.g., Park et al. 1989) as well as online (e.g., Vrechopoulos et al. 2004) can be considered given the physical store data sources that are now available. Data source #8 pertains to environmental data that retailers routinely draw upon to make assortment, promotion, and/or inventory stocking decisions. For example, that weather data affects consumer spending propensities (e.g., Murray et al. 2010) and store sales has been known and studied for a long time (see, e.g., Steele 1951). Today, retailers can access a well-oiled data collection, collation, and analysis ecosystem that regularly takes in weather data feeds from weather monitoring system APIs and collates it into a format wherein a rules engine can apply, and thereafter output either recommendations or automatically trigger actions or interventions on the retailer's behalf.

Finally, data source #9 in Fig. 2.2 is pertinent largely to emerging markets and lets small, unorganized sector retailers (mom-and-pop stores) to leverage their physical location and act as fulfillment center franchisees for large retailers (Forbes 2015).

10 Summary and Conclusion

This chapter was an introduction to the important task of data collection, a process that precedes and heavily influences the success or failure of data science and analytics projects in meeting their objectives. We started with why data is such a big deal and used an illustrative example (Uber) to see the value inherent in the right kind of data. We followed up with some preliminaries on the four main types of data, their corresponding four primary scales, and the implications for analysis downstream. We then ventured into problem formulation, discussed why it is of such critical importance in determining what data to collect, and built a simple framework against which data scientists could check and validate their current problem formulation tasks. Finally, we walked through an extensive example of the various kinds of data sources available in just one business domain—retailing—and the implications thereof.

Exercises

Ex. 2.1 Prepare the movie release dataset of all the movies released in the last 5 years using IMDB.

(a) Find all movies that were released in the last 5 years.
(b) Generate a file containing URLs for the top 50 movies every year on IMDB.
(c) Read in the URL's IMDB page and scrape the following information:
     Producer(s), Director(s), Star(s), Taglines, Genres, (Partial) Storyline, Box office budget, and Box office gross.
(d) Make a table out of these variables as columns with movie name being the first variable.
(e) Analyze the movie-count for every Genre. See if you can come up with some interesting hypotheses. For example, you could hypothesize that "Action Genres occur significantly more often than Drama in the top-250 list." or that "Action movies gross higher than Romance movies in the top-250 list."
(f) Write a markdown doc with your code and explanation. See if you can *storify* your hypotheses.

Note: You can web-scrape with the rvest package in R or use any platform that you are comfortable with.

Ex. 2.2 Download the movie reviews from IMDB for the list of movies.

(a) Go to www.imdb.com and examine the page.
(b) Scrape the page and tabulate the output into a data frame with columns "name, url, movie type, votes."
(c) Filter the data frame. Retain only those movies that got over 500 reviews. Let us call this Table 1.
(d) Now for each of the remaining movies, go to the movie's own web page on the IMDB, and extract the following information:
   Duration of the movie, Genre, Release date, Total views, Commercial description from the top of the page.
(e) Add these fields to Table 1 in that movie's row.
(f) Now build a separate table for each movie in Table 1 from that movie's web page on IMDB. Extract the first five pages of reviews of that movie and in each review, scrape the following information:
   Reviewer, Feedback, Likes, Overall, Review (text), Location (of the reviewer), Date of the review.
(g) Store the output in a table. Let us call it Table 2.
(h) Create a list (List 1) with as many elements as there are rows in Table 1. For the ith movie in Table 1, store Table 2 as the ith element of a second list, say, List 2.

Ex. 2.3 Download the Twitter data through APIs.

(a) Read up on how to use the Twitter API (https://dev.twitter.com/overview/api). If required, make a twitter ID (if you do not already have one).
(b) There are three evaluation dimensions for a movie at IMDB, namely, Author, Feedback, and Likes. More than the dictionary meanings of these words, it is interesting how they are used in different contexts.
(c) Download 50 tweets each that contain these terms and 100 tweets for each movie.
(d) Analyze these tweets and classify what movie categories they typically refer to. Insights here could, for instance, be useful in designing promotional campaigns for the movies.

P.S.: R has a dedicated package twitteR (note capital R in the end). For additional functions, refer twitteR package manual.

Ex. 2.4 Prepare the beer dataset of all the beers that got over 500 reviews.

(a) Go to (https://www.ratebeer.com/beer/top-50/) and examine the page.
(b) Scrape the page and tabulate the output into a data frame with columns "name, url, count, style."
(c) Filter the data frame. Retain only those beers that got over 500 reviews. Let us call this Table 1.
(d) Now for each of the remaining beers, go to the beer's own web page on the ratebeer site, and scrape the following information:

"Brewed by, Weighted Avg, Seasonal, Est.Calories, ABV, commercial description" from the top of the page.

Add these fields to Table 1 in that beer's row.

(e) Now build a separate table for each beer in Table 1 from that beer's ratebeer web page. Scrape the first three pages of reviews of that beer and in each review, scrape the following info:

"rating, aroma, appearance, taste, palate, overall, review (text), location (of the reviewer), date of the review."

(f) Store the output in a dataframe, let us call it Table 2.
(g) Create a list (let us call it List 1) with as many elements as there are rows in Table 1. For the ith beer in Table 1, store Table 2 as the ith element List 2.

Ex. 2.5 Download the Twitter data through APIs.

(a) Read up on how to use the twitter API here (https://dev.twitter.com/overview/api). If required, make a twitter ID (if you do not already have one).
(b) Recall three evaluation dimensions for beer at ratebeer.com, viz., aroma, taste, and palate. More than the dictionary meanings of these words, what is interesting is how they are used in context.

So pull 50 tweets each containing these terms.
(c) Read through these tweets and note what product categories they typically refer to. Insights here could, for instance, be useful in designing promotional campaigns for the beers. We will do text analysis, etc. next visit.

P.S.: R has a dedicated package twitteR (note capital R in the end). For additional functions, refer twitteR package manual.

Ex. 2.6 WhatsApp Data collection.

(a) Form a WhatsApp group with few friends/colleagues/relatives.
(b) Whenever you travel or visit different places as part of your everyday work, share your location to the WhatsApp group.

For example, if you are visiting an ATM, your office, a grocery store, the local mall, etc., then send the WhatsApp group a message saying: "ATM, [share of location here]."

Ideally, you should share a handful of locations every day. Do this DC exercise for a week. It is possible you may repeat-share certain locations.

P.S.: We assume you have a smartphone with google maps enabled on it to share locations with.
(c) Once this exercise is completed export the WhatsApp chat history of DC group to a text file. To do this, see below:

Go to WhatsApp > Settings > Chat history > Email Chat > Select the chat you want to export.
(d) Your data file should look like this:

28/02/17, 7:17 pm—fname lname: location: https://maps.google.com/?q=17.463869,78.367403

28/02/17, 7:17 pm—fname lname: ATM

(e) Now compile this data in a tabular format. Your data should have these columns:

- Sender name
- Time
- Latitude
- Longitude
- Type of place

(f) Extract your locations from the chat history table and plot it on google maps. You can use the spatial DC code we used on this list of latitude and longitude co-ordinates or use leaflet() package in R to do the same. Remember to extract and map only your own locations not those of other group members.

(g) Analyze your own movements over a week *AND* record your observations about your travels as a story that connects these locations together.

References

Bijmolt, T. H. A., van Heerde, H. J., & Pieters, R. G. M. (2005). New empirical generalizations on the determinants of price elasticity. *Journal of Marketing Research, 42*(2), 141–156.

Blair, E., & Blair, C. (2015). *Applied survey sampling*. Los Angeles: Sage Publications.

Blattberg, R. C., Kim, B.-D., & Neslin, S. A. (2008). Market basket analysis. *Database Marketing: Analyzing and Managing Customers*, 339–351.

Bradlow, E., Gangwar, M., Kopalle, P., & Voleti, S. (2017). The role of big data and predictive analytics in retailing. *Journal of Retailing, 93*, 79–95.

Diebold, F. X. (2012). On the origin (s) and development of the term 'Big Data'.

Forbes. (2015). From Dabbawallas to Kirana stores, five unique E-commerce delivery innovations in India. Retrieved April 15, 2015, from http://tinyurl.com/j3eqb5f.

Ghose, A., & Han, S. P. (2011). An empirical analysis of user content generation and usage behavior on the mobile Internet. *Management Science, 57*(9), 1671–1691.

Ghose, A., & Han, S. P. (2014). Estimating demand for mobile applications in the new economy. *Management Science, 60*(6), 1470–1488.

Larson, J. S., Bradlow, E. T., & Fader, P. S. (2005). An exploratory look at supermarket shopping paths. *International Journal of Research in Marketing, 22*(4), 395–414.

Lee, N., Broderick, A. J., & Chamberlain, L. (2007). What is 'neuromarketing'? A discussion and agenda for future research. *International Journal of Psychophysiology, 63*(2), 199–204.

Luo, X., Andrews, M., Fang, Z., & Phang, C. W. (2014). Mobile targeting. *Management Science, 60*(7), 1738–1756.

Montgomery, C. (2017). *Design and analysis of experiments* (9th ed.). New York: John Wiley and Sons.

Montgomery, A. L., Li, S., Srinivasan, K., & Liechty, J. C. (2004). Modeling online browsing and path analysis using clickstream data. *Marketing Science, 23*(4), 579–595.

Murray, K. B., Di Muro, F., Finn, A., & Leszczyc, P. P. (2010). The effect of weather on consumer spending. *Journal of Retailing and Consumer Services, 17*(6), 512–520.

Park, C. W., Iyer, E. S., & Smith, D. C. (1989). The effects of situational factors on in-store grocery shopping behavior: The role of store environment and time available for shopping. *Journal of Consumer Research, 15*(4), 422–433.

Rossi, P. E., & Allenby, G. M. (1993). A Bayesian approach to estimating household parameters. *Journal of Marketing Research, 30*, 171–182.

Rossi, P. E., McCulloch, R. E., & Allenby, G. M. (1996). The value of purchase history data in target marketing. *Marketing Science, 15*(4), 321–340.
Russell, G. J., & Petersen, A. (2000). Analysis of cross category dependence in market basket selection. *Journal of Retailing, 76*(3), 367–392.
Steele, A. T. (1951). Weather's effect on the sales of a department store. *Journal of Marketing, 15*(4), 436–443.
Vrechopoulos, A. P., O'Keefe, R. M., Doukidis, G. I., & Siomkos, G. J. (2004). Virtual store layout: An experimental comparison in the context of grocery retail. *Journal of Retailing, 80*(1), 13–22.
Wedel, M., & Pieters, R. (2000). Eye fixations on advertisements and memory for brands: A model and findings. *Marketing Science, 19*(4), 297–312.

Chapter 3
Data Management—Relational Database Systems (RDBMS)

Hemanth Kumar Dasararaju and Peeyush Taori

1 Introduction

Storage and management of data is a key aspect of data science. Data, simply speaking, is nothing but a collection of facts—a snapshot of the world—that can be stored and processed by computers. In order to process and manipulate data efficiently, it is very important that data is stored in an appropriate form. Data comes in many shapes and forms, and some of the most commonly known forms of data are numbers, text, images, and videos. Depending on the type of data, there exist multiple ways of storage and processing. In this chapter, we focus on one of the most commonly known and pervasive means of data storage—relational database management systems. We provide an introduction using which a reader can perform the essential operations. References for a deeper understanding are given at the end of the chapter.

2 Motivating Example

Consider an online store that sells stationery to customers across a country. The owner of this store would like to set up a system that keeps track of inventory, sales, operations, and potential pitfalls. While she is currently able to do so on her own, she knows that as her store scales up and starts to serve more and more people, she

---

H. K. Dasararaju
Indian School of Business, Hyderabad, Telangana, India

P. Taori (✉)
London Business School, London, UK
e-mail: taori.peeyush@gmail.com

© Springer Nature Switzerland AG 2019
B. Pochiraju, S. Seshadri (eds.), *Essentials of Business Analytics*, International Series in Operations Research & Management Science 264,
https://doi.org/10.1007/978-3-319-68837-4_3

will no longer have the capacity to manually record transactions and create records for new occurrences. Therefore, she turns to relational database systems to run her business more efficiently.

A database is a collection of organized data in the form of rows, columns, tables and indexes. In a database, even a small piece of information becomes data. We tend to aggregate related information together and put them under one gathered name called a Table. For example, all student-related data (student ID, student name, date of birth, etc.) would be put in one table called STUDENT table. It decreases the effort necessary to scan for a specific information in an entire database. Since a database is very flexible, data gets updated and extended when new data is added and the database shrinks when data is deleted from the database.

3 Database Systems—What and Why?

As data grows in size, there arises a need for a means of storing it efficiently such that it can be found and processed quickly. In the "olden days" (which was not too far back), this was achieved via systematic filing systems where individual files were catalogued and stored neatly according to a well-developed data cataloging system (similar to the ones you will find in libraries or data storage facilities in organizations). With the advent of computer systems, this role has now been assumed by database systems. Plainly speaking, a database system is a digital record-keeping system or an electronic filing cabinet. Database systems can be used to store large amounts of data, and data can then be queried and manipulated later using a querying mechanism/language. Some of the common operations that can be performed in a database system are adding new files, updating old data files, creating new databases, querying of data, deleting data files/individual records, and adding more data to existing data files. Often pre processing and post-processing of data happen using database languages. For example, one can selectively read data, verify its correctness, and connect it to data structures within applications. Then, after processing, write it back into the database for storage and further processing.

With the advent of computers, the usage of database systems has become ubiquitous in our personal and work lives. Whether we are storing information about personal expenditures using an Excel file or making use of MySQL database to store product catalogues for a retail organization, databases are pervasive and in use everywhere. We also discuss the difference between the techniques discussed in this chapter compared to methods for managing big data in the next chapter.

3.1 Database Management System

A database management system (DBMS) is the system software that enables users to create, organize, and manage databases. As Fig. 3.1 illustrates, The DBMS serves as an interface between the database and the end user, guaranteeing that information is reliably organized and remains accessible.

Fig. 3.1 Relating databases to end users

The main objectives of DBMS are mass storage; removal of duplicity—DBMS makes sure that same data has not been stored earlier; providing multiple user access—two or more users can work concurrently; data integrity—ensuring the privacy of the data and preventing unauthorized access; data backup and recovery; nondependence on a particular platform; and so on. There are dozens of DBMS products available. Popular products include Microsoft Access, MYSQL, Oracle from Oracle Corporation, SQL Server from Microsoft, and DB2 from IBM.

3.2 Relational Database Management System

Relational database management system (RDBMS) is a database management system (DBMS) that is based on the relational model of data. DBMS tells us about the tables but Relational DBMS specifies about relations between different entities in the database. The two main principles of the RDBMS are entity integrity and referential integrity.

- *Entity integrity*: Here, all the data should be organized by having a unique value (primary key), so it cannot accept null values.
- *Referential integrity*: Referential integrity must have constraints specified between two relations and the relationship must always be consistent (e.g., foreign key column must be equal to the primary key column).
  - *Primary key*: Primary key is a column in a table that uniquely identifies the rows in that relation (table).
  - *Foreign key*: Foreign keys are columns that point to primary key columns of another table.

Normalization:

Normalization is the database design technique that is used to efficiently organize the data, optimize the table structures, and remove duplicate data entries. It separates the larger tables into smaller tables and links them using the relationships. Normalization is used to improve the speed, for efficient usage of space, and to increase the data integrity. The important normalizations that are used to organize the database are as follows:

- *First normal form (1NF)*: The table must contain "atomic" values only (should not contain any duplicate values, and cannot hold multiple values).

  Example: Suppose the university wants to store the details of students who are finalists of a competition. Table 3.1 shows the data.

Three students (Jon, Robb, and Ken) have two different parents numbers so the university put two numbers in the same field as you see in Table 3.1. This table is not in 1NF as it does not follow the rule "Only atomic values in the field" as there are multiple values in parents_number field. To make the table into 1NF we should store the information as shown in Table 3.2.

- *Second normal form (2NF)*: Must follow first normal form and no non-key attributes are dependent on the proper subset of any candidate key of the table.

  Example: Assume a university needs to store the information of the instructors and the topics they teach. They make a table that resembles the one given below (Table 3.3) since an instructor can teach more than one topic.

Table 3.1 Students in a university competition

Student_ID	Student_Name	Address	Parents_number
71121	Jon	New York	75430105417540
71122	Janet	Chicago	1915417
71123	Robb	Boston	63648014889636
71124	Zent	Los Angeles	7545413
71125	Ken	Atlanta	40136924016371

Table 3.2 Students in a university competition sorted efficiently

Student_ID	Student_Name	Address	Parents_number
71121	Jon	New York	7543010
71121	Jon	New York	5417540
71122	Janet	Chicago	1915417
71123	Robb	Boston	6364801
71123	Robb	Boston	4889636
71124	Zent	Los Angeles	7545413
71125	Ken	Atlanta	4013692
71125	Ken	Atlanta	4016371

Table 3.3 Instructors in a university

Instructor_ID	Topic	Instructor_Age
56121	Neural Network	37
56121	IoT	37
56132	Statistics	51
56133	Optimization	43
56133	Simulation	43

Table 3.4 Breaking tables into two in order to agree with 2NF

Instructor_ID	Instructor_Age
56121	37
56132	51
56133	43

Instructor_ID	Topic
56121	Neural Network
56121	IoT
56132	Statistics
56133	Optimization
56133	Simulation

Table 3.5 Students in a university competition

Student_ID	Student_Name	Student_ZIP	Student_State	Student_city	Student_Area
71121	Jon	10001	New York	New York	Queens Manhattan
71122	Janet	60201	Illinois	Chicago	Evanston
71123	Robb	02238	Massachusetts	Boston	Cambridge
71124	Zent	90089	California	Los Angeles	Trousdale

Here Instructor_ID and Topic are key attributes and Instructor_Age is a non-key attribute. The table is in 1NF but not in 2NF because the non-key attribute Instructor_Age is dependent on Instructor_ID. To make the table agree to 2NF, we can break the table into two tables like the ones given in Table 3.4.

- *Third normal form (3NF)*: Must follow second normal form and none of the non-key attributes are determined by another non-key attributes.
  Example: Suppose the university wants to store the details of students who are finalists of a competition. The table is shown in Table 3.5.

Here, student_ID is the key attribute and all other attributes are non-key attributes. Student_State, Student_city, and Student_Area depend on Student_ZIP and Student_ZIP is dependent on Student_ID that makes the non-key attribute transitively dependent on the key attribute. This violates the 3NF rules. To make the table agree to 3NF we can break into two tables like the ones given in Table 3.6.

3NF is the form that is practiced and advocated across most organizational environments. It is because tables in 3NF are immune to most of the anomalies associated with insertion, updation, and deletion of data. However, there could be specific instances when organizations might want to opt for alternate forms of table normalization such as 4NF and 5NF. While 2NF and 3NF normalizations focus on functional aspects, 4NF and 5NF are more concerned with addressing multivalued dependencies. A detailed discussion of 4NF and 5NF forms is beyond the scope

Table 3.6 Breaking tables into two in order to agree with 3NF

Student table:

Student_ID	Student_Name	Student_ZIP
71121	Jon	10001
71122	Janet	60201
71123	Robb	02238
71124	Zent	90089

Student_zip table:

Student_ZIP	Student_State	Student_city	Student_Area
10001	New York	New York	Queens Manhattan
60201	Illinois	Chicago	Evanston
02238	Massachusetts	Boston	Cambridge
90089	California	Los Angeles	Trousdale

of discussion for this chapter, but interested reader can learn more online from various sources.[1] It should be noted that in many organizational scenarios, the focus is mainly on achieving 3NF.

3.3 Advantages of RDBMS over EXCEL

Most businesses today need to record and store information. Sometimes this may be only for record keeping and sometimes data is stored for later use. We can store the data in Microsoft Excel. But why is RDBMS the most widely used method to store data?

Using Excel we can perform various functions like adding the data in rows and columns, sorting of data by various metrics, etc. But Excel is a two-dimensional spreadsheet and thus it is extremely hard to make connections between information in various spreadsheets. It is easy to view the data or find the particular data from Excel when the size of the information is small. It becomes very hard to read the information once it crosses a certain size. The data might scroll many pages when endeavoring to locate a specific record.

Unlike Excel, in RDBMS, the information is stored independently from the user interface. This separation of storage and access makes the framework considerably more scalable and versatile. In RDBMS, data can be easily cross-referenced between multiple databases using relationships between them but there are no such options in Excel. RDBMS utilizes centralized data storage systems that makes backup and maintenance much easier. Database frameworks have a tendency to be significantly faster as they are built to store and manipulate large datasets unlike Excel.

---

[1](http://www.bkent.net/Doc/simple5.htm (accessed on Feb 6, 2019))

4 Structured Query Language (SQL)

SQL (structured query language) is a computer language exclusive to a particular application domain in contrast to some other general-purpose language (GPL) such as C, Java, or Python that is broadly applicable across domains. SQL is text oriented, and designed for managing (access and manipulate) data. SQL was authorized as a national standard by the ANSI (American National Standards Institute) in 1992. It is the standard language for relational database management systems. Some common relational database management systems that operate using SQL are Microsoft Access, MySQL, Oracle, SQL Server, and IBM DB2. Even though many database systems make use of SQL, they also have their unique extensions that are specific to their systems.

SQL statements are used to select the particular part of the data, retrieve data from a database, and update data on the database using CREATE, SELECT, INSERT, UPDATE, DELETE, and DROP commands. SQL commands can be sliced into four categories: DDL (data definition language), which is used to define the database structures; DML (data manipulation language), which is used to access and modify database data; DCL (data control language); and TCL (transaction control language).

DDL (Data Definition Language):
DDL deals with the database schemas and structure. The following statements are used to take care of the design and storage of database objects.

1. CREATE: Creates the database, table, index, views, store, procedure, functions, and triggers.
2. ALTER: Alters the attributes, constraints, and structure of the existing database.
3. DROP: Deletes the objects (table, view, functions, etc.) from the database.
4. TRUNCATE: Removes all records from a table, including the space allocated to the records.
5. COMMENT: Associates comments about the table or about any objects to the data dictionary.
6. RENAME: Renames the objects.

DML (Data Manipulation Language):
DML deals with tasks like storing, modifying, retrieving, deleting, and updating the data in/from the database.

1. SELECT: The only data retrieval statement in SQL, used to select the record(s) from the database.
2. INSERT: Inserts a new data/observation into the database.
3. UPDATE: Modifies the existing data within the database.
4. DELETE: Removes one or more records from the table.

Note: There is an important difference between the DROP, TRUNCATE, and DELETE commands. DELETE (Data alone deleted) operations can be recalled back (undo), while DROP (Table structure + Data are deleted) and TRUNCATE operations cannot be recalled back.

DCL (Data Control Language):

Data control languages are used to uphold the database security during multiple user data environment. The database administrator (DBA) is responsible for "grant/revoke" privileges on database objects.

1. GRANT: Provide access or privilege on database objects to the group of users or particular user.
2. REVOKE: Remove user access rights or privilege to the database objects.

TCL (Traction Control Language):

Transaction control language statements enable you to control and handle transactions to keep up the trustworthiness of the information within SQL statements.

1. BEGIN: Opens a transaction.
2. COMMIT: Saves the transaction on the database.
3. ROLLBACK: Rollback (undo the insert, delete, or update) the transaction in the database in case of any errors.
4. SAVEPOINT: Rollback to the particular point (till the savepoint marked) of transaction. The progression done until the savepoint will be unaltered and all transaction after that will be rolled back.

4.1 Introduction to MySQL

In this section, we will walk through the basics of creating a database using MySQL[2] and query the database using the MySQL querying language. As described earlier in the chapter, a MySQL database server is capable of hosting many databases. In databases parlance, a database is often also called a schema. Thus, a MySQL server can contain a number of schemas. Each of those schemas (database) is made up of a number of tables, and every table contains rows and columns. Each row represents an individual record or observation, and each column represents a particular attribute such as age and salary.

When you launch the MySQL command prompt, you see a command line like the one below (Fig. 3.2).

---

[2]MySQL Workbench or Windows version can be downloaded from https://dev.mysql.com/downloads/windows/ (accessed on Feb 15, 2018) for practice purpose.

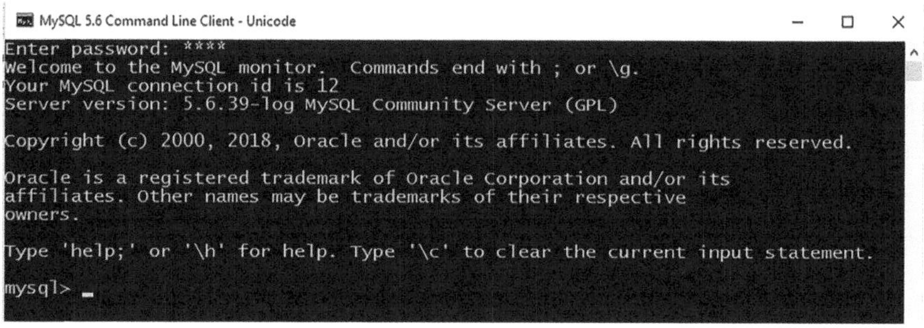

3 Data Management—Relational Database Systems (RDBMS)

Fig. 3.2 MySQL command prompt Interface

The command line starts with "mysql>" and you can run SQL scripts by closing commands with semicolon (;).

4.2 How to Check the List of Databases Available in MySQL?

In order to get started we will first check the databases that are already present in a MySQL server. To do so, type "**show databases**" in the command line. Once you run this command, it will list all the available databases in the MySQL server installation. The above-mentioned command is the first SQL query that we have run. Please note that keywords and commands are case-insensitive in MySQL as compared to R and Python where commands are case-sensitive in nature.

```
mysql> SHOW DATABASES;

Output:
+--------------------+
| Database           |
+--------------------+
| information_schema |
| mysql              |
| performance_schema |
| test               |
+--------------------+
4 rows in set (0.00 sec)
```

You would notice that there are already four schemas listed though we have not yet created any one of them. Out of the four databases, "information_schema", "mysql", and "performance_schema" are created by MySQL server for its internal monitoring and performance optimization purposes and should not be used when we are creating our own database. Another schema "test" is created by MySQL during the installation phase and it is provided for testing purposes. You can remove the "test" schema or can use it to create your own tables.

4.3 Creating and Deleting a Database

Now let us create our own database. The syntax for creating a database in MySQL is:

```
CREATE DATABASE databasename;
```

Let us create a simple inventory database. We shall create a number of tables about products and their sales information such as customers, products, orders, shipments, and employee. We will call the database "product_sales". In order to create a database, type the following SQL query:

```
mysql> CREATE DATABASE product_sales;

Output:
Query OK, 1 row affected (0.00 sec)

mysql> SHOW DATABASES;

Output:
+--------------------+
| Database           |
+--------------------+
| information_schema |
| mysql              |
| performance_schema |
| product_sales      |
| test               |
+--------------------+
5 rows in set (0.00 sec)
```

In the above-mentioned query, we are creating a database called "product_sales." Once the query is executed, if you issue the "show databases" command again, then it will now show five databases (with "product_sales" as the new database). As of now, "product_sales" will be an empty database, meaning there would be no tables inside it. We will start creating tables and populating them with data in a while.

In order to delete a database, you need to follow the following syntax:

```
DROP DATABASE databasename;
```

In our case, if we need to delete "product_sales", we will issue the command:

```
mysql> DROP DATABASE product_sales;

Output:
Query OK, 0 rows affected (0.14 sec)

mysql> SHOW DATABASES;

Output:
+--------------------+
| Database           |
+--------------------+
| information_schema |
| mysql              |
```

```
| performance_schema |
| test               |
+--------------------+
4 rows in set (0.00 sec)
```

Oftentimes, when you have to create a database, you might not be sure if a database of the same name exists already in the system. In such cases, conditions such as "IF EXISTS" and "IF NOT EXISTS" come in handy. When we execute such query, then the database is created if there is no other database of the same name. This helps us in avoiding overwriting of the existing database with the new one.

```
mysql> CREATE DATABASE IF NOT EXISTS product_sales;

Output:
Query OK, 1 row affected (0.00 sec)
```

One important point to keep in mind is the use of SQL DROP commands with extreme care, because once you delete an entity or an entry, then there is no way to recover the data.

4.4 Selecting a Database

There can be multiple databases available in the MySQL server. In order to work on a specific database, we have to select the database first. The basic syntax to select a database is:

```
USE databasename;
```

In our case, if we have to select "product_sales" database, we will issue the command:

```
mysql> USE product_sales;

Output:
Database changed
```

When we run the above query, the default database now is "product_sales". Whatever operations we will now perform will be performed on this database. This implies that if you have to use a specific table in the database, then you can simply do so by calling the table name. If at any point of time you want to check which your selected database is then issue the command:

```
mysql> SELECT DATABASE();

Output:
+---------------+
| DATABASE()    |
+---------------+
| product_sales |
+---------------+
1 row in set (0.00 sec)
```

If you want to check all tables in a database, then issue the following command:

```
mysql> SHOW TABLES;
```

```
Output:
Empty set (0.00 sec)
```

As of now it is empty since we have not yet created any table. Let us now go ahead and create a table in the database.

4.5 Table Creation and Deletion

The syntax for creating a new table is:

```
CREATE TABLE tablename (IF EXISTS);
```

The above command will create the table with table name as specified by the user. You can also specify the optional condition IF EXISTS/IF NOT EXISTS similar to the way you can specify them while creating a database. Since a table is nothing but a collection of rows and columns, in addition to specifying the table name, you would also want to specify the column names in the table and the type of data that each column can contain. For example, let us go ahead and create a table named "products." We will then later inspect it in greater detail.

```
mysql> CREATE TABLE products (productID INT 10 UNSIGNED NOT NULL
    AUTO_INCREMENT, code CHAR(6) NOT NULL DEFAULT ", productname
    VARCHAR(30) NOT NULL DEFAULT ", quantity INT UNSIGNED NOT NULL
    DEFAULT 0, price DECIMAL(5,2) NOT NULL DEFAULT 0.00, PRIMARY
    KEY (productID) );
```

```
Output:
Query OK, 0 rows affected (0.41 sec)
```

In the above-mentioned command, we have created a table named "*products.*" Along with table name, we have also specified the columns and the type of data that each column contains within the parenthesis. For example, "*products*" table contains five columns—productID, code, productname, quantity, and price. Each of those columns can contain certain types of data. Let us look at them one by one:

- *productID* is INT 10 UNSIGNED (INT means integer, it accepts only integer values for productID). And the number 10 after INT represents the size of the integer; here in this case productID accepts an integer of maximum size 10. And the attribute UNSIGNED means nonnegative integers, which means the productID will accept only positive integers. Thus, if you enter any non-integer value, negative value, or number great than that of size 10, it will throw you an error. If you do not specify the attribute UNSIGNED in the command, by default it will take SIGNED attribute, which accepts both positive and negative integers.
- *code* is CHAR(6)—CHAR(6) means a fixed-length alphanumeric string that can contain exactly six characters. It accepts only six characters.

- *productname* is VARCHAR(30). Similar to CHAR, VARCHAR stands for a variable length string that can contain a maximum of 30 characters. The contrast between CHAR and VARCHAR is that whereas CHAR is a fixed length string, VARCHAR can vary in length. In practice, it is always better to use VARCHAR unless you suspect that the string in a column is always going to be of a fixed length.
- *quantity* INT. This means that quantity column can contain integer values.
- *price* DECIMAL(5,2). Price column can contain floating point numbers (decimal numbers) of length 5 and the length of decimal digits can be a maximum of 2. Whenever you are working with floating point numbers, it is advisable to use DECIMAL field.

There are a number of additional points to be noted with regard to the above statement.

For a number of columns such as productID, productname you would notice the presence of NOT NULL. NOT NULL is an attribute that essentially tells MySQL that the column cannot have null values. NULL in MySQL is not a string and is instead a special character to signify absence of values in the field. Each column also contains the attribute DEFAULT. This essentially implies that if no value is provided by the user then use default value for the column. For example, default value for column quantity will be 0 in case no values are provided when inputting data to the table.

The column productID has an additional attribute called AUTO_INCREMENT, and its default value is set to 1. This implies that whenever there is a null value specified for this column, a default value would instead be inserted but this default value will be incremented by 1 with a starting value of 1. Thus, if there are two missing productID entries, then the default values of 1 and 2 would be provided.

Finally, the last line of table creation statement query is PRIMARY KEY (productID). Primary key for a table is a column or set of columns where each observation in that column would have a unique value. Thus, if we have to look up any observation in the table, then we can do so using the primary key for the table. Although it is not mandatory to have primary keys for a table, it is a standard practice to have one for every table. This also helps during indexing the table and makes query execution faster.

If you would now run the command SHOW TABLES, then the table would be reflected in your database.

```
mysql> SHOW TABLES;

Output:
+------------------------+
| Tables_in_product_sales |
+------------------------+
| products               |
+------------------------+
1 row in set (0.00 sec)
```

You can always look up the schema of a table by issuing the "DESCRIBE" command:

```
mysql> DESCRIBE products;
```

```
Output:
+-------------+------------------+------+-----+---------+----------------+
| Field       | Type             | Null | Key | Default | Extra          |
+-------------+------------------+------+-----+---------+----------------+
| productID   | int(10) unsigned | NO   | PRI | NULL    | auto_increment |
| code        | char(6)          | NO   |     |         |                |
| productname | varchar(30)      | NO   |     |         |                |
| quantity    | int(10) unsigned | NO   |     | 0       |                |
| price       | decimal(5,2)     | NO   |     | 0.00    |                |
+-------------+------------------+------+-----+---------+----------------+
5 rows in set (0.01 sec)
```

4.6 Inserting the Data

Once we have created the table, it is now time to insert data into the table. For now we will look at how to insert data manually in the table. Later on we will see how we can import data from an external file (such as CSV or text file) in the database. Let us now imagine that we have to insert data into the products table we just created. To do so, we make use of the following command:

```
mysql> INSERT INTO products VALUES (1, 'IPH', 'Iphone 5S Gold',
    300, 625);
```

```
Output:
Query OK, 1 row affected (0.13 sec)
```

When we issue the above command, it will insert a single row of data into the table "*products.*" The parenthesis after VALUES specified the actual values that are to be inserted. An important point to note is that values should be specified in the same order as that of columns when we created the table "*products.*" All numeric data (integers and decimal values) are specified without quotes, whereas character data must be specified within quotes.

Now let us go ahead and insert some more data into the "*products.*" table:

```
mysql> INSERT INTO products VALUES(NULL, 'IPH',
    'Iphone 5S Black', 8000, 655.25),(NULL, 'IPH',
    'Iphone 5S Blue', 2000, 625.50);
```

```
Output:
Query OK, 2 rows affected (0.13 sec)
Records: 2  Duplicates: 0  Warnings: 0
```

In the above case, we inserted multiple rows of data at the same time. Each row of data was specified within parenthesis and each row was separated by a comma (,). Another point to note is that we kept the productID fields as null when inserting the data. This is to demonstrate that even if we provide null values, MySQL will make use of AUTO_INCREMENT operator to assign values to each row.

Sometimes there might be a need where you want to provide data only for some columns or you want to provide data in a different order as compared to the original one when we created the table. This can be done using the following command:

```
mysql> INSERT INTO products (code, productname, quantity, price)
    VALUES ('SNY', 'Xperia Z1', 10000, 555.48),('SNY', 'Xperia S',
    8000, 400.49);

Output:
Query OK, 2 rows affected (0.13 sec)
Records: 2  Duplicates: 0  Warnings: 0
```

Notice here that we did not specify the productID column for values to be inserted in, but rather explicitly specified the columns and their order in which we want to insert the data. The productID column will be automatically populated using AUTO_INCREMENT operator.

4.7 Querying the Database

Now that we have inserted some values into the products table, let us go ahead and see how we can query the data. If you want to see all observations in a database table, then make use of the SELECT * FROM tablename query:

```
mysql> SELECT * FROM products;

Output:
+-----------+------+-----------------+----------+--------+
| productID | code | productname     | quantity | price  |
+-----------+------+-----------------+----------+--------+
|         1 | IPH  | Iphone 5S Gold  |      300 | 625.00 |
|         2 | IPH  | Iphone 5S Black |     8000 | 655.25 |
|         3 | IPH  | Iphone 5S Blue  |     2000 | 625.50 |
|         4 | SNY  | Xperia Z1       |    10000 | 555.48 |
|         5 | SNY  | Xperia S        |     8000 | 400.49 |
+-----------+------+-----------------+----------+--------+
5 rows in set (0.00 sec)
```

SELECT query is perhaps the most widely known query of SQL. It allows you to query a database and get the observations matching your criteria. SELECT * is the most generic query, which will simply return all observations in a table. The general syntax of SELECT query is as follows:

```
SELECT column1Name, column2Name, ... FROM tableName
```

This will return selected columns from a particular table name. Another variation of SELECT query can be the following:

```
SELECT column1Name, column2Name ....from tableName where
    somecondition;
```

In the above version, only those observations would be returned that match the criteria specified by the user. Let us understand them with the help of a few examples:

```
mysql> SELECT productname, quantity FROM products;

Output:
+-----------------+----------+
| productname     | quantity |
+-----------------+----------+
| Iphone 5S Gold  |      300 |
| Iphone 5S Black |     8000 |
| Iphone 5S Blue  |     2000 |
| Xperia Z1       |    10000 |
| Xperia S        |     8000 |
+-----------------+----------+
5 rows in set (0.00 sec)

mysql> SELECT productname, price FROM products WHERE price < 600;

Output:
+-------------+--------+
| productname | price  |
+-------------+--------+
| Xperia Z1   | 555.48 |
| Xperia S    | 400.49 |
+-------------+--------+
2 rows in set (0.00 sec)
```

The above query will only give name and price columns for those records whose price <600.

```
mysql> SELECT productname, price FROM products
       WHERE price >= 600;

Output:
+-----------------+--------+
| productname     | price  |
+-----------------+--------+
| Iphone 5S Gold  | 625.00 |
| Iphone 5S Black | 655.25 |
| Iphone 5S Blue  | 625.50 |
+-----------------+--------+
3 rows in set (0.00 sec)
```

The above query will only give name and price columns for those records whose price >= 600.

In order to select observations based on string comparisons, enclose the string within quotes. For example:

```
mysql> SELECT productname, price FROM products
    WHERE code = 'IPH';

Output:
+-----------------+--------+
| productname     | price  |
+-----------------+--------+
| Iphone 5S Gold  | 625.00 |
| Iphone 5S Black | 655.25 |
| Iphone 5S Blue  | 625.50 |
+-----------------+--------+
3 rows in set (0.00 sec)
```

The above command gives you the name and price of the products whose code is "IPH."

In addition to this, you can also perform a number of string pattern matching operations, and wildcard characters. For example, you can make use of operators LIKE and NOT LIKE to search if a particular string contains a specific pattern. In order to do wildcard matches, you can make use of underscore character "_" for a single-character match, and percentage sign "%" for multiple-character match. Here are a few examples:

- "phone%" will match strings that start with phone and can contain any characters after.
- "%phone" will match strings that end with phone and can contain any characters before.
- "%phone%" will match strings that contain phone anywhere in the string.
- "c_a" will match strings that start with "c" and end with "a" and contain any single character in-between.

```
mysql> SELECT productname, price FROM products WHERE productname
    LIKE 'Iphone%';

Output:
+-----------------+--------+
| productname     | price  |
+-----------------+--------+
| Iphone 5S Gold  | 625.00 |
| Iphone 5S Black | 655.25 |
| Iphone 5S Blue  | 625.50 |
+-----------------+--------+
3 rows in set (0.00 sec)

mysql> SELECT productname, price FROM products WHERE productname
    LIKE '%Blue%';

Output:
+----------------+--------+
| productname    | price  |
+----------------+--------+
| Iphone 5S Blue | 625.50 |
+----------------+--------+
1 row in set (0.00 sec)
```

Additionally, you can also make use of Boolean operators such as AND, OR in SQL queries to create multiple conditions.

```
mysql> SELECT * FROM products WHERE quantity >= 5000 AND
  productname LIKE 'Iphone%';

Output:
+-----------+------+----------------+----------+--------+
| productID | code | productname    | quantity | price  |
+-----------+------+----------------+----------+--------+
|         2 | IPH  | Iphone 5S Black|     8000 | 655.25 |
+-----------+------+----------------+----------+--------+
1 row in set (0.00 sec)
```

This gives you all the details of products whose quantity is >=5000 and the name like 'Iphone'.

```
mysql> SELECT * FROM products WHERE quantity >= 5000 AND price >
  650 AND productname LIKE 'Iphone%';

Output:
+-----------+------+----------------+----------+--------+
| productID | code | productname    | quantity | price  |
+-----------+------+----------------+----------+--------+
|         2 | IPH  | Iphone 5S Black|     8000 | 655.25 |
+-----------+------+----------------+----------+--------+
1 row in set (0.00 sec)
```

If you want to find whether the condition matches any elements from within a set, then you can make use of IN operator. For example:

```
mysql> SELECT * FROM products WHERE productname IN ('Iphone 5S
  Blue', 'Iphone 5S Black');

Output:
+-----------+------+----------------+----------+--------+
| productID | code | productname    | quantity | price  |
+-----------+------+----------------+----------+--------+
|         2 | IPH  | Iphone 5S Black|     8000 | 655.25 |
|         3 | IPH  | Iphone 5S Blue |     2000 | 625.50 |
+-----------+------+----------------+----------+--------+
2 rows in set (0.00 sec)
```

This gives the product details for the names provided in the list specified in the command (i.e., "Iphone 5S Blue", "Iphone 5S Black").

Similarly, if you want to find out if the condition looks for values within a specific range then you can make use of BETWEEN operator. For example:

3 Data Management—Relational Database Systems (RDBMS)

```
mysql> SELECT * FROM products WHERE (price BETWEEN 400 AND 600)
    AND (quantity BETWEEN 5000 AND 10000);

Output:
+-----------+------+-------------+----------+--------+
| productID | code | productname | quantity | price  |
+-----------+------+-------------+----------+--------+
|         4 | SNY  | Xperia Z1   |    10000 | 555.48 |
|         5 | SNY  | Xperia S    |     8000 | 400.49 |
+-----------+------+-------------+----------+--------+
2 rows in set (0.00 sec)
```

This command gives you the product details whose price is between 400 and 600 and quantity is between 5000 and 10000, both inclusive.

4.8 ORDER BY Clause

Many a times when we retrieve a large number of results, we might want to sort them in a specific order. In order to do so, we make use of ORDER BY in SQL. The general syntax for this is:

```
SELECT ... FROM tableName
WHERE criteria
ORDER BY columnA ASC|DESC, columnB ASC|DESC

mysql> SELECT * FROM
    products WHERE productname LIKE 'Iphone%' ORDER BY price DESC;

Output:
+-----------+------+-----------------+----------+--------+
| productID | code | productname     | quantity | price  |
+-----------+------+-----------------+----------+--------+
|         2 | IPH  | Iphone 5S Black |     8000 | 655.25 |
|         3 | IPH  | Iphone 5S Blue  |     2000 | 625.50 |
|         1 | IPH  | Iphone 5S Gold  |      300 | 625.00 |
+-----------+------+-----------------+----------+--------+
3 rows in set (0.00 sec)
```

If you are getting a large number of results but want the output to be limited only to a specific number of observations, then you can make use of LIMIT clause. LIMIT followed by a number will limit the number of output results that will be displayed.

```
mysql> SELECT * FROM products ORDER BY price LIMIT 2;

Output:
+-----------+------+-------------+----------+--------+
| productID | code | productname | quantity | price  |
+-----------+------+-------------+----------+--------+
|         5 | SNY  | Xperia S    |     8000 | 400.49 |
|         4 | SNY  | Xperia Z1   |    10000 | 555.48 |
+-----------+------+-------------+----------+--------+
2 rows in set (0.00 sec)
```

Oftentimes, we might want to display the columns or tables by an intuitive name that is different from the original name. To be able to do so, we make use of AS alias.

```
mysql> SELECT productID AS ID, code AS productCode , productname
   AS Description, price AS Unit_Price FROM products ORDER
   BY ID;

Output:
+----+-------------+------------------+------------+
| ID | productCode | Description      | Unit_Price |
+----+-------------+------------------+------------+
|  1 | IPH         | Iphone 5S Gold   |     625.00 |
|  2 | IPH         | Iphone 5S Black  |     655.25 |
|  3 | IPH         | Iphone 5S Blue   |     625.50 |
|  4 | SNY         | Xperia Z1        |     555.48 |
|  5 | SNY         | Xperia S         |     400.49 |
+----+-------------+------------------+------------+
5 set (0.00 sec)
```

4.9 Producing Summary Reports

A key part of SQL queries is to be able to provide summary reports from large amounts of data. This summarization process involves data manipulation and grouping activities. In order to enable users to provide such summary reports, SQL has a wide range of operators such as DISTINCT, GROUP BY that allow quick summarization and production of data. Let us look at these operators one by one.

4.9.1 DISTINCT

A column may have duplicate values. We could use the keyword DISTINCT to select only distinct values. We can also apply DISTINCT to several columns to select distinct combinations of these columns. For example:

```
mysql> SELECT DISTINCT code FROM products;

Output:
+------+
| Code |
+------+
| IPH  |
| SNY  |
+------+
2 rows in set (0.00 sec)
```

4.9.2 GROUP BY Clause

The GROUP BY clause allows you to *collapse* multiple records with a common value into groups. For example,

3 Data Management—Relational Database Systems (RDBMS)

```
mysql> SELECT * FROM products ORDER BY code, productID;

Output:
+-----------+------+-----------------+----------+--------+
| productID | code | productname     | quantity | price  |
+-----------+------+-----------------+----------+--------+
|         1 | IPH  | Iphone 5S Gold  |      300 | 625.00 |
|         2 | IPH  | Iphone 5S Black |     8000 | 655.25 |
|         3 | IPH  | Iphone 5S Blue  |     2000 | 625.50 |
|         4 | SNY  | Xperia Z1       |    10000 | 555.48 |
|         5 | SNY  | Xperia S        |     8000 | 400.49 |
+-----------+------+-----------------+----------+--------+
5 rows in set (0.00 sec)mysql> SELECT * FROM products GROUP BY
  code; #-- Only first record in each group is shown

Output:
+-----------+------+-----------------+----------+--------+
| productID | code | productname     | quantity | price  |
+-----------+------+-----------------+----------+--------+
|         1 | IPH  | Iphone 5S Gold  |      300 | 625.00 |
|         4 | SNY  | Xperia Z1       |    10000 | 555.48 |
+-----------+------+-----------------+----------+--------+
2 rows in set (0.00 sec)
```

We can apply GROUP BY clause with aggregate functions to produce group summary report for each group.

The function COUNT(*) returns the rows selected; COUNT(*columnName*) counts only the non-NULL values of the given column. For example,

```
mysql> SELECT COUNT(*) AS `Count` FROM products;

Output:
+-------+
| Count |
+-------+
|     5 |
+-------+
1 row in set (0.00 sec)

mysql> SELECT code, COUNT(*) FROM products
   GROUP BY code;

Output:
+------+----------+
| code | COUNT(*) |
+------+----------+
| IPH  |        3 |
| SNY  |        2 |
+------+----------+
2 rows in set (0.00 sec)
```

We got "IPH" count as 3 because we have three entries in our table with the product code "IPH" and similarly two entries for the product code "SNY." Besides

COUNT(), there are many other aggregate functions such as AVG(), MAX(), MIN(), and SUM(). For example,

```
mysql> SELECT MAX(price), MIN(price), AVG(price), SUM(quantity)
  FROM products;
Output:
+------------+------------+------------+---------------+
| MAX(price) | MIN(price) | AVG(price) | SUM(quantity) |
+------------+------------+------------+---------------+
|     655.25 |     400.49 | 572.344000 |         28300 |
+------------+------------+------------+---------------+
1 row in set (0.00 sec)
```

This gives you MAX price, MIN price, AVG price, and total quantities of all the products available in our products table. Now let us use GROUP BY clause:

```
mysql> SELECT code, MAX(price) AS 'Highest Price', MIN(price) AS
  'Lowest Price' FROM products GROUP BY code;

Output:
+------+---------------+--------------+
| code | Highest Price | Lowest Price |
+------+---------------+--------------+
| IPH  |        655.25 |       625.00 |
| SNY  |        555.48 |       400.49 |
+------+---------------+--------------+
2 rows in set (0.00 sec)
```

This means, the highest price of an IPhone available in our database is 655.25 and the lowest price is 625.00. Similarly, the highest price of a Sony is 555.48 and the lowest price is 400.49.

4.10 Modifying Data

To modify the existing data, use UPDATE, SET command, with the following syntax:

```
UPDATE tableName SET columnName = {value|NULL|DEFAULT}, ... WHERE
  criteria

mysql> UPDATE products SET quantity = quantity + 50,
  price = 600.5 WHERE productname = 'Xperia Z1';

Output:
Query OK, 1 row affected (0.14 sec)
Rows matched: 1  Changed: 1  Warnings: 0
```

Let us check the modification in the products table.

```
mysql> SELECT * FROM products WHERE productname = 'Xperia Z1';
```

3 Data Management—Relational Database Systems (RDBMS)

Output:
```
+-----------+------+--------------+----------+--------+
| productID | code | productname  | quantity | price  |
+-----------+------+--------------+----------+--------+
|         4 | SNY  | Xperia Z1    |    10050 | 600.50 |
+-----------+------+--------------+----------+--------+
1 row in set (0.00 sec)
```

You can see that the quantity of Xperia Z1 is increased by 50.

4.11 Deleting Rows

Use the DELETE FROM command to delete row(s) from a table; the syntax is:
  DELETE FROM tableName # to delete all rows from the table.
  DELETE FROM tableName WHERE criteria # to delete only the row(s) that meets the *criteria*. For example,

mysql> DELETE FROM products WHERE productname LIKE 'Xperia%';

Output:
Query OK, 2 rows affected (0.03 sec)

mysql> SELECT * FROM products;

Output:
```
+-----------+------+------------------+----------+--------+
| productID | code | productname      | quantity | price  |
+-----------+------+------------------+----------+--------+
|         1 | IPH  | Iphone 5S Gold   |      300 | 625.00 |
|         2 | IPH  | Iphone 5S Black  |     8000 | 655.25 |
|         3 | IPH  | Iphone 5S Blue   |     2000 | 625.50 |
+-----------+------+------------------+----------+--------+
3 rows in set (0.00 sec)
```

mysql> DELETE FROM products;

Output:
Query OK, 3 rows affected (0.14 sec)

mysql> SELECT * FROM products;

Output:
Empty set (0.00 sec)

Beware that "DELETE FROM *tableName*" without a WHERE clause deletes ALL records from the table. Even with a WHERE clause, you might have deleted some records unintentionally. It is always advisable to issue a SELECT command with the same WHERE clause to check the result set before issuing the DELETE (and UPDATE).

4.12 Create Relationship: One-To-Many

4.12.1 PRIMARY KEY

Suppose that each product has one supplier, and each supplier supplies one or more products. We could create a table called *"suppliers"* to store suppliers' data (e.g., name, address, and phone number). We create a column with unique value called supplierID to identify every supplier. We set supplierID as the *primary key* for the table suppliers (to ensure uniqueness and facilitate fast search).

In order to relate the suppliers table to the products table, we add a new column into the *"products"* table—the supplierID.

We then set the supplierID column of the products table as a *foreign key* which references the supplierID column of the *"suppliers"* table to ensure the so-called *referential integrity*. We need to first create the *"suppliers"* table, because the *"products"* table references the *"suppliers"* table.

```
mysql> CREATE TABLE suppliers (supplierID INT UNSIGNED NOT NULL
    AUTO_INCREMENT, name VARCHAR(30) NOT NULL DEFAULT ", phone
    CHAR(8) NOT NULL DEFAULT ", PRIMARY KEY (supplierID));

Output:
Query OK, 0 rows affected (0.33 sec)

mysql> DESCRIBE suppliers;

Output:
+------------+------------------+------+-----+---------+----------------+
| Field      | Type             | Null | Key | Default | Extra          |
+------------+------------------+------+-----+---------+----------------+
| supplierID | int(10) unsigned | NO   | PRI | NULL    | auto_increment |
| name       | varchar(30)      | NO   |     |         |                |
| phone      | char(8)          | NO   |     |         |                |
+------------+------------------+------+-----+---------+----------------+
3 rows in set (0.01 sec)
```

Let us insert some data into the suppliers table.

```
mysql> INSERT INTO suppliers VALUE (501, 'ABC Traders',
    '88881111'), (502, 'XYZ Company', '88882222'), (503, 'QQ Corp',
    '88883333');

Output:
Query OK, 3 rows affected (0.13 sec)
Records: 3  Duplicates: 0  Warnings: 0

mysql> SELECT * FROM suppliers;
```

Output:
```
+------------+--------------+----------+
| supplierID | name         | phone    |
+------------+--------------+----------+
|        501 | ABC Traders  | 88881111 |
|        502 | XYZ Company  | 88882222 |
|        503 | QQ Corp      | 88883333 |
+------------+--------------+----------+
3 rows in set (0.00 sec)
```

4.12.2 ALTER TABLE

The syntax for ALTER TABLE is as follows:

```
ALTER TABLE tableName
{ADD [COLUMN] columnName columnDefinition}
{ALTER|MODIFY [COLUMN] columnName columnDefinition
{SET DEFAULT columnDefaultValue} | {DROP DEFAULT}}
{DROP [COLUMN] columnName [RESTRICT|CASCADE]}
{ADD tableConstraint}
{DROP tableConstraint [RESTRICT|CASCADE]}
```

Instead of deleting and re-creating the products table, we shall use the statement "ALTER TABLE" to add a new column supplierID into the products table. As we have deleted all the records from products in recent few queries, let us rerun the three INSERT queries referred in the Sect. 4.6 before running "ALTER TABLE."

```
mysql> ALTER TABLE products ADD COLUMN supplierID INT UNSIGNED
    NOT NULL;

Output:
Query OK, 0 rows affected (0.43 sec)
Records: 0  Duplicates: 0  Warnings: 0

mysql> DESCRIBE products;

Output:
+-------------+------------------+------+-----+---------+----------------+
| Field       | Type             | Null | Key | Default | Extra          |
+-------------+------------------+------+-----+---------+----------------+
| productID   | int(10) unsigned | NO   | PRI | NULL    | auto_increment |
| code        | char(6)          | NO   |     |         |                |
| productname | varchar(30)      | NO   |     |         |                |
| quantity    | int(10) unsigned | NO   |     | 0       |                |
| price       | decimal(5,2)     | NO   |     | 0.00    |                |
| supplierID  | int(10) unsigned | NO   |     | NULL    |                |
+-------------+------------------+------+-----+---------+----------------+
6 rows in set (0.00 sec)
```

4.12.3 FOREIGN KEY

Now, we shall add a *foreign key constraint* on the supplierID columns of the "*products*" child table to the "*suppliers*" parent table, to ensure that every supplierID in the "*products*" table always refers to a *valid* supplierID in the "*suppliers*" table. This is called *referential integrity*.

Before we add the foreign key, we need to set the supplierID of the existing records in the "*products*" table to a valid supplierID in the "*suppliers*" table (say supplierID = 501).

Now let us set the supplierID of the existing records to a valid supplierID of "*supplier*" table. As we have deleted the records from "*products*" table, we can add or update using UPDATE command.

```
mysql> UPDATE products SET supplierID = 501;

Output:
Query OK, 5 rows affected (0.04 sec)
Rows matched: 5  Changed: 5  Warnings: 0
```

Let us add a foreign key constraint.

```
mysql> ALTER TABLE products ADD FOREIGN KEY (supplierID)
    REFERENCES suppliers (supplierID);

Output:
Query OK, 0 rows affected (0.56 sec)
Records: 0  Duplicates: 0  Warnings: 0

mysql> DESCRIBE products;

Output:
+-------------+------------------+------+-----+---------+----------------+
| Field       | Type             | Null | Key | Default | Extra          |
+-------------+------------------+------+-----+---------+----------------+
| productID   | int(10) unsigned | NO   | PRI | NULL    | auto_increment |
| code        | char(6)          | NO   |     |         |                |
| productname | varchar(30)      | NO   |     |         |                |
| quantity    | int(10) unsigned | NO   |     | 0       |                |
| price       | decimal(5,2)     | NO   |     | 0.00    |                |
| supplierID  | int(10) unsigned | NO   | MUL | NULL    |                |
+-------------+------------------+------+-----+---------+----------------+
6 rows in set (0.00 sec)

mysql> SELECT * FROM products;

Output:
+-----------+------+-----------------+----------+--------+------------+
| productID | code | productname     | quantity | price  | supplierID |
+-----------+------+-----------------+----------+--------+------------+
|         1 | IPH  | Iphone 5S Gold  |      300 | 625.00 |        501 |
|         2 | IPH  | Iphone 5S Black |     8000 | 655.25 |        501 |
|         3 | IPH  | Iphone 5S Blue  |     2000 | 625.50 |        501 |
|         4 | SNY  | Xperia Z1       |    10000 | 555.48 |        501 |
|         5 | SNY  | Xperia S        |     8000 | 400.49 |        501 |
+-----------+------+-----------------+----------+--------+------------+
5 rows in set (0.00 sec)
```

```
mysql> UPDATE products SET supplierID = 502 WHERE productID = 1;

Output:
Query OK, 1 row affected (0.13 sec)
Rows matched: 1  Changed: 1  Warnings: 0

mysql> SELECT * FROM products;

Output:
+-----------+------+-----------------+----------+--------+------------+
| productID | code | productname     | quantity | price  | supplierID |
+-----------+------+-----------------+----------+--------+------------+
|         1 | IPH  | Iphone 5S Gold  |      300 | 625.00 |        502 |
|         2 | IPH  | Iphone 5S Black |     8000 | 655.25 |        501 |
|         3 | IPH  | Iphone 5S Blue  |     2000 | 625.50 |        501 |
|         4 | SNY  | Xperia Z1       |    10000 | 555.48 |        501 |
|         5 | SNY  | Xperia S        |     8000 | 400.49 |        501 |
+-----------+------+-----------------+----------+--------+------------+
5 rows in set (0.00 sec)
```

4.13 SELECT with JOIN

SELECT command can be used to query and join data from two related tables. For example, to list the product's name (in products table) and supplier's name (in suppliers table), we could join the two tables using the two common supplierID columns:

```
mysql> SELECT products.productname, price, suppliers.name FROM
   products JOIN suppliers ON products.supplierID
   = suppliers.supplierID WHERE price < 650;

Output:
+-----------------+--------+--------------+
| productname     | price  | name         |
+-----------------+--------+--------------+
| Iphone 5S Gold  | 625.00 | XYZ Company  |
| Iphone 5S Blue  | 625.50 | ABC Traders  |
| Xperia Z1       | 555.48 | ABC Traders  |
| Xperia S        | 400.49 | ABC Traders  |
+-----------------+--------+--------------+
4 rows in set (0.00 sec)
```

Here we need to use products.name and suppliers.name to differentiate the two "names."

Join using WHERE clause (legacy method) is not recommended.

```
mysql> SELECT products.productname, price, suppliers.name FROM
   products, suppliers WHERE products.supplierID =
   suppliers.supplierID AND price < 650;
```

```
Output:
+----------------+---------+--------------+
| productname    | price   | name         |
+----------------+---------+--------------+
| Iphone 5S Gold | 625.00  | XYZ Company  |
| Iphone 5S Blue | 625.50  | ABC Traders  |
| Xperia Z1      | 555.48  | ABC Traders  |
| Xperia S       | 400.49  | ABC Traders  |
+----------------+---------+--------------+
4 rows in set (0.00 sec)
```

In the above query result, two of the columns have the same heading "name." We could create *aliases* for headings. Let us use aliases for column names for display.

```
mysql> SELECT products.productname AS 'Product Name', price,
    suppliers.name AS 'Supplier Name' FROM products JOIN suppliers
    ON products.supplierID = suppliers.supplierID WHERE price < 650;

Output:
+----------------+---------+----------------+
| Product Name   | price   | Supplier Name  |
+----------------+---------+----------------+
| Iphone 5S Gold | 625.00  | XYZ Company    |
| Iphone 5S Blue | 625.50  | ABC Traders    |
| Xperia Z1      | 555.48  | ABC Traders    |
| Xperia S       | 400.49  | ABC Traders    |
+----------------+---------+----------------+
4 rows in set (0.00 sec)
```

5 Summary

The chapter describes the essential commands for creating, modifying, and querying an RDBMS. Detailed descriptions and examples can be found in the list of books and websites listed in the reference section (Elmasri and Navathe 2014; Hoffer et al. 2011; MySQL using R 2018; MySQL using Python 2018). You can also refer various websites such as w3schools.com/sql, sqlzoo.net (both accessed on Jan 15, 2019), which help you learn SQL in gamified console. The practice would help you learn to query large databases, which is quite a nuisance.

Exercises

Ex. 3.1 Print list of all suppliers who do not keep stock for IPhone 5S Black.

Ex. 3.2 Find out the product that has the biggest inventory by value (i.e., the product that has the highest value in terms of total inventory).

Ex. 3.3 Print the supplier name who maintains the largest inventory of products.

Ex. 3.4 Due to the launch of a newer model, prices of IPhones have gone down and the inventory value has to be written down. Create a new column (new_price) where price is marked down by 20% for all black- and gold-colored phones, whereas it has to be marked down by 30% for the rest of the phones.

Ex. 3.5 Due to this recent markdown in prices (refer to Ex. 3.4), which supplier takes the largest hit in terms of inventory value?

References

Elmasri, R., & Navathe, S. B. (2014). *Database systems: Models, languages, design and application*. England: Pearson.

Hoffer, J. A., Venkataraman, R., & Topi, H. (2011). *Modern database management*. England: Pearson.

MySQL using R. Retrieved February, 2018., from https://cran.r-project.org/web/packages/RMySQL/RMySQL.pdf.

MySQL using Python. Retrieved February, 2018., from http://mysql-python.sourceforge.net/MySQLdb.html.

Chapter 4
Big Data Management

Peeyush Taori and Hemanth Kumar Dasararaju

1 Introduction

The twenty-first century is characterized by the digital revolution, and this revolution is disrupting the way business decisions are made in every industry, be it healthcare, life sciences, finance, insurance, education, entertainment, retail, etc. The Digital Revolution, also known as the Third Industrial Revolution, started in the 1980s and sparked the advancement and evolution of technology from analog electronic and mechanical devices to the shape of technology in the form of machine learning and artificial intelligence today. Today, people across the world interact and share information in various forms such as content, images, or videos through various social media platforms such as Facebook, Twitter, LinkedIn, and YouTube. Also, the twenty-first century has witnessed the adoption of handheld devices and wearable devices at a rapid rate. The types of devices we use today, be it controllers or sensors that are used across various industrial applications or in the household or for personal usage, are generating data at an alarming rate. The huge amounts of data generated today are often termed big data. We have ushered in an age of big data-driven analytics where big data does not only drive decision-making for firms

---

Electronic supplementary material The online version of this chapter (https://doi.org/10.1007/978-3-319-68837-4_4) contains supplementary material, which is available to authorized users.

P. Taori (✉)
London Business School, London, UK
e-mail: taori.peeyush@gmail.com

H. K. Dasararaju
Indian School of Business, Hyderabad, Telangana, India

but also impacts the way we use services in our daily lives. A few statistics below help provide a perspective on how much data pervades our lives today:

Prevalence of big data:

- The total amount of data generated by mankind is 2.7 Zeta bytes, and it continues to grow at an exponential rate.
- In terms of digital transactions, according to an estimate by IDC, we shall soon be conducting nearly 450 billion transactions per day.
- Facebook analyzes 30+ peta bytes of user generated data every day.

(Source: https://www.waterfordtechnologies.com/big-data-interesting-facts/, accessed on Aug 10, 2018.)

With so much data around us, it is only natural to envisage that big data holds tremendous value for businesses, firms, and society as a whole. While the potential is huge, the challenges that big data analytics faces are also unique in their own respect. Because of the sheer size and velocity of data involved, we cannot use traditional computing methods to unlock big data value. This unique challenge has led to the emergence of big data systems that can handle data at a massive scale. This chapter builds on the concepts of big data—it tries to answer what really constitutes big data and focuses on some of big data tools. In this chapter, we discuss the basics of big data tools such as Hadoop, Spark, and the surrounding ecosystem.

2 Big Data: What and Why?

2.1 Elements of Big Data

We live in a digital world where data continues to grow at an exponential pace because of ever-increasing usage of Internet, sensors, and other connected devices. The amount of data[1] that organizations generate today is exponentially more than

---

[1]Note: When we say large datasets that means data size ranging from petabytes to exabytes and more. Please note that 1 byte = 8 bits

Metric	Value
Byte (B)	$2^0 = 1$ byte
Kilobyte (KB)	2^{10} bytes
Megabyte (MB)	2^{20} bytes
Gigabyte (GB)	2^{30} bytes
Terabyte (TB)	2^{40} bytes
Petabyte (PB)	2^{50} bytes
Exabyte (EB)	2^{60} bytes
Zettabyte (ZB)	2^{70} bytes
Yottabyte (YB)	2^{80} bytes

what we were generating collectively even a few years ago. Unfortunately, the term big data is used colloquially to describe a vast variety of data that is being generated.

When we describe traditional data, we tend to put it into three categories: *structured, unstructured*, and *semi-structured*. Structured data is highly organized information that can be easily stored in a spreadsheet or table using rows and columns. Any data that we capture in a spreadsheet with clearly defined columns and their corresponding values in rows is an example of structured data. Unstructured data may have its own internal structure. It does not conform to the standards of structured data where you define the field name and its type. Video files, audio files, pictures, and text are best examples of unstructured data. Semi-structured data tends to fall in between the two categories mentioned above. There is generally a loose structure defined for data of this type, but we cannot define stringent rules like we do for storing structured data. Prime examples of semi-structured data are log files and Internet of Things (IoT) data generated from a wide range of sensors and devices, e.g., a clickstream log from an e-commerce website that gives you details about date and time of classes/objects that are being instantiated, IP address of the user where he is doing transaction from, etc. But, in order to analyze the information, we need to process the data to extract useful information into a structured format.

2.2 Characteristics of Big Data

In order to put a structure to big data, we describe big data as having four characteristics: volume, velocity, variety, and veracity. The infographic in Fig. 4.1 provides an overview through example.

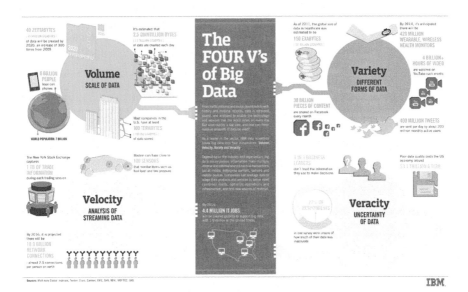

Fig. 4.1 Characteristics of big data. (Source: http://www.ibmbigdatahub.com/sites/default/files/infographic_file/4-Vs-of-big-data.jpg (accessed on Aug 9, 2018))

We discuss each of the four characteristics briefly (also shown in Fig. 4.2):

1. *Volume*: It is the amount of the overall data that is already generated (by either individuals or companies). The Internet alone generates huge amounts of data. It is estimated that the Internet has around 14.3 trillion live web pages, which amounts to 672 exabytes of accessible data.[2]
2. *Variety*: Data is generated from different types of sources that are internal and external to the organization such as social and behavioral and also comes in different formats such as structured, unstructured (analog data, GPS tracking information, and audio/video streams), and semi-structured data—XML, Email, and EDI.
3. *Velocity*: Velocity simply states the rate at which organizations and individuals are generating data in the world today. For example, a study reveals that videos that are 400 hours of duration are uploaded onto YouTube every minute.[3]
4. *Veracity*: It describes the uncertainty inherent in the data, whether the obtained data is correct or consistent. It is very rare that data presents itself in a form that is ready to consume. Considerable effort goes into processing of data especially when it is unstructured or semi-structured.

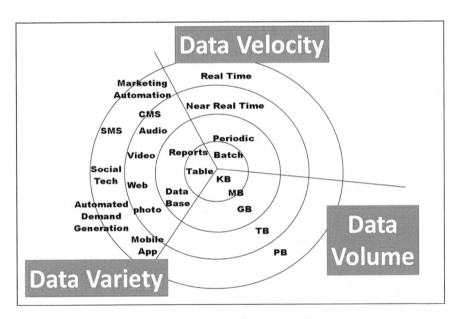

Fig. 4.2 Examples to understand big data characteristics. (Source: https://velvetchainsaw.com/2012/07/20/three-vs-of-big-data-as-applied-conferences/ (accessed on Aug 10, 2018))

---

[2]https://www.iste.org/explore/articleDetail?articleid=204 (accessed on Aug 9, 2018).

[3]https://www.statista.com/statistics/259477/hours-of-video-uploaded-to-youtube-every-minute/ (accessed on Aug 9, 2018).

2.3 Why Process Big Data?

Processing big data for analytical purposes adds tremendous value to organizations because it helps in making decisions that are data driven. In today's world, organizations tend to perceive the value of big data in two different ways:

1. *Analytical usage of data*: Organizations process big data to extract relevant information to a field of study. This relevant information then can be used to make decisions for the future. Organizations use techniques like data mining, predictive analytics, and forecasting to get timely and accurate insights that help to make the best possible decisions. For example, we can provide online shoppers with product recommendations that have been derived by studying the products viewed and bought by them in the past. These recommendations help customers find what they need quickly and also increase the retailer's profitability.
2. *Enable new product development*: The recent successful startups are a great example of leveraging big data analytics for new product enablement. Companies such as Uber or Facebook use big data analytics to provide personalized services to its customers in real time.

    Uber is a taxi booking service that allows users to quickly book cab rides from their smartphones by using a simple app. Business operations of Uber are heavily reliant on big data analytics and leveraging insights in a more effective way. When passengers request for a ride, Uber can instantly match the request with the most suitable drivers either located in nearby area or going toward the area where the taxi service is requested. Fares are calculated automatically, GPS is used to determine the best possible route to avoid traffic and the time taken for the journey using proprietary algorithms that make adjustments based on the time that the journey might take.

2.4 Some Applications of Big Data Analytics

In today's world, every business and industry is affected by, and benefits from, big data analytics in multiple ways. The growth in the excitement about big data is evident everywhere. A number of actively developed technological projects focus on big data solutions and a number of firms have come into business that focus solely on providing big data solutions to organizations. Big data technology has evolved to become one of the most sought-after technological areas by organizations as they try to put together teams of individuals who can unlock the value inherent in big data. We highlight a couple of use cases to understand the applications of big data analytics.

1. *Customer Analytics in the Retail industry*

    Retailers, especially those with large outlets across the country, generate huge amount of data in a variety of formats from various sources such as POS transactions, billing details, loyalty programs, and CRM systems. This data needs to be organized and analyzed in a systematic manner to derive meaningful insights. Customers can be segmented based on their buying patterns and spend at every transaction. Marketers can use this information for creating personalized promotions. Organizations can also combine transaction data with customer preferences and market trends to understand the increase or decrease in demand for different products across regions. This information helps organizations to determine the inventory level and make price adjustments.

2. *Fraudulent claims detection in Insurance industry*

    In industries like banking, insurance, and healthcare, fraudulent transactions are mostly to do with monetary transactions, those that are not caught might cause huge expenses and lead to loss of reputation to a firm. Prior to the advent of big data analytics, many insurance firms identified fraudulent transactions using statistical methods/models. However, these models have many limitations and can prevent fraud up to limited extent because model building can happen only on sample data. Big data analytics enables the analyst to overcome the issue with volumes of data—insurers can combine internal claim data with social data and other publicly available data like bank statements, criminal records, and medical bills of customers to better understand consumer behavior and identify any suspicious behavior.

3 Big Data Technologies

Big data requires different means of processing such voluminous, varied, and scattered data compared to that of traditional data storage and processing systems like RDBMS (relational database management systems), which are good at storing, processing, and analyzing structured data only. Table 4.1 depicts how traditional RDBMS differs from big data systems.

Table 4.1 Big data systems vs traditional RDBMS

RDBMS	Big data systems
These systems are best at processing structured data. Semi-structured and unstructured data like photos, videos, and messages posted on Social Media cannot be processed by RDBMS	These systems have the capability to handle a diverse variety of data (structured, semi-structured, and unstructured data)
These systems are very efficient in handling small amounts of data (up to GBs to TB). Becomes less suitable and inefficient for data in the range of TBs or PBs	These systems are optimized to handle large volumes of data. These systems are used where the amount of data created every day is huge. Example—Facebook, Twitter
Cannot handle the speed with which data arrives on sites such as Amazon and Facebook. The performance of these systems degrades as the velocity of data increases	Since these systems use a distributed computing architecture, they can easily handle high data velocities

There are a number of technologies that are used to handle, process, and analyze big data. Of them the ones that are most effective and popular are distributed computing and parallel computing for big data, Hadoop for big data, and big data cloud. In the remainder of the chapter, we focus on Hadoop but also briefly visit the concepts of distributed and parallel computing and big data cloud.

Distributed Computing and Parallel Computing

Loosely speaking, distributed computing is the idea of dividing a problem into multiple parts, each of which is operated upon by an individual machine or computer. A key challenge in making distributed computing work is to ensure that individual computers can communicate and coordinate their tasks. Similarly, in parallel computing we try to improve the processing capability of a computer system. This can be achieved by adding additional computational resources that run parallel to each other to handle complex computations. If we combine the concepts of both distributed and parallel computing together, the cluster of machines will behave like a single powerful computer. Although the ideas are simple, there are several challenges underlying distributed and parallel computing. We underline them below.

Distributed Computing and Parallel Computing Limitations and Challenges

- *Multiple failure points*: If a single computer fails, and if other machines cannot reconfigure themselves in the event of failure then this can lead to overall system going down.
- *Latency*: It is the aggregated delay in the system because of delays in the completion of individual tasks. This leads to slowdown in system performance.
- *Security*: Unless handled properly, there are higher chances of an unauthorized user access on distributed systems.
- *Software*: The software used for distributed computing is complex, hard to develop, expensive, and requires specialized skill set. This makes it harder for every organization to deploy distributed computing software in their infrastructure.

3.1 Hadoop for Big Data

In order to overcome some of the issues that plagued distributed systems, companies worked on coming up with solutions that would be easier to deploy, develop, and maintain. The result of such an effort was Hadoop—the first open source big data platform that is mature and has widespread usage. Hadoop was created by Doug Cutting at Yahoo!, and derives its roots directly from the Google File System (GFS) and MapReduce Programming for using distributed computing.

Earlier, while using distributed environments for processing huge volumes of data, multiple nodes in a cluster could not always cooperate within a communication system, thus creating a lot of scope for errors. The Hadoop platform provided an improved programming model to overcome this challenge and for making distributed systems run efficiently. Some of the key features of Hadoop are its ability to store and process huge amount of data, quick computing, scalability, fault tolerance, and very low cost (mostly because of its usage of commodity hardware for computing).

Rather than a single piece of software, Hadoop is actually a collection of individual components that attempt to solve core problems of big data analytics—storage, processing, and monitoring. In terms of core components, Hadoop has Hadoop Distributed File System (HDFS) for file storage to store large amounts of data, MapReduce for processing the data stored in HDFS in parallel, and a resource manager known as Yet Another Resource Negotiator (YARN) for ensuring proper allocation of resources. In addition to these components, the ecosystem of Hadoop also boasts of a number of open source projects that have now come under the ambit of Hadoop and make big data analysis simpler. Hadoop supports many other file systems along with HDFS such as Amazon S3, CloudStore, IBM's General Parallel File System, ParaScale FS, and IBRIX Fusion FS.

Below are few important terminologies that one should be familiar with before getting into Hadoop ecosystem architecture and characteristics.

- *Cluster*: A cluster is nothing but a collection of individual computers interconnected via a network. The individual computers work together to give users an impression of one large system.
- *Node*: Individual computers in the network are referred to as nodes. Each node has pieces of Hadoop software installed to perform storage and computation tasks.
- *Master–slave architecture*: Computers in a cluster are connected in a master–slave configuration. There is typically one master machine that is tasked with the responsibility of allocating storage and computing duties to individual slave machines.
- *Master node*: It is typically an individual machine in the cluster that is tasked with the responsibility of allocating storage and computing duties to individual slave machines.
- *DataNode*: DataNodes are individual slave machines that store actual data and perform computational tasks as and when the master node directs them to do so.
- *Distributed computing*: The idea of distributed computing is to execute a program across multiple machines, each one of which will operate on the data that resides on the machine.
- *Distributed File System*: As the name suggests, it is a file system that is responsible for breaking a large data file into small chunks that are then stored on individual machines.

Additionally, Hadoop has in-built salient features such as scaling, fault tolerance, and rebalancing. We describe them briefly below.

- *Scaling*: At a technology front, organizations require a platform to scale up to handle the rapidly increasing data volumes and also need a scalability extension

for existing IT systems in content management, warehousing, and archiving. Hadoop can easily scale as the volume of data grows, thus circumventing the size limitations of traditional computational systems.
- *Fault tolerance*: To ensure business continuity, fault tolerance is needed to ensure that there is no loss of data or computational ability in the event of individual node failures. Hadoop provides excellent fault tolerance by allocating the tasks to other machines in case an individual machine is not available.
- *Rebalancing*: As the name suggests, Hadoop tries to evenly distribute data among the connected systems so that no particular system is overworked or is lying idle.

3.2 Hadoop Ecosystem

As we mentioned earlier, in addition to the core components, Hadoop has a large ecosystem of individual projects that make big data analysis easier and more efficient. While there are a large number of projects built around Hadoop, there are a few that have gained prominence in terms of industry usage. Such key projects in Hadoop ecosystem are outlined in Fig. 4.3.

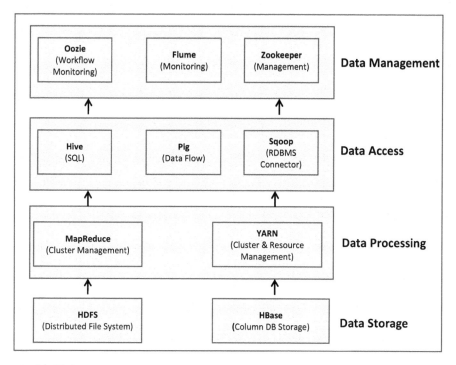

Fig. 4.3 Hadoop ecosystem

We provide a brief description of each of these projects below (with the exception of HDFS, MapReduce, and YARN that we discuss in more detail later).

HBase: HBASE is an open source NoSql database that leverages HDFS. Some examples of NoSql databases are HBASE, Cassandra, and AmazonDB. The main properties of HBase are strongly consistent read and write, Automatic sharding (rows of data are automatically split and stored across multiple machines so that no single machine has the burden of storing entire dataset. It also enables fast searching and retrieval as a search query does not have to be performed over entire dataset, and can rather be done on the machine that contains specific data rows), Automatic Region Server failover (feature that enables high availability of data at all times. If a particular region's server goes down, the data is still made available through replica servers), and Hadoop/HDFS Integration. It supports parallel processing via MapReduce and has an easy to use API.

Hive: While Hadoop is a great platform for big data analytics, a large number of business users have limited knowledge of programming, and this can become a hindrance in widespread adoption of big data platforms such as Hadoop. Hive overcomes this limitation, and is a platform to write SQL-type scripts that can be run on Hadoop. Hive provides an SQL-like interface and data warehouse infrastructure to Hadoop that helps users carry out analytics on big data by writing SQL queries known as Hive queries. Hive Query execution happens via MapReduce—the Hive interpreter converts the query to MapReduce format.

Pig: It is a procedural language platform used to develop Shell-script-type programs for MapReduce operations. Rather than writing MapReduce programs, which can become cumbersome for nontrivial tasks, users can do data processing by writing individual commands (similar to scripts) by using a language known as *Pig Latin*. Pig Latin is a data flow language, Pig translates the Pig Latin script into MapReduce, which can then execute within Hadoop.

Sqoop: The primary purpose of Sqoop is to facilitate data transfer between Hadoop and relational databases such as MySQL. Using Sqoop users can import data from relational databases to Hadoop and also can export data from Hadoop to relational databases. It has a simple command-line interface for transforming data between relational databases and Hadoop, and also supports incremental import.

Oozie: In simple terms, Oozie is a workflow scheduler for managing Hadoop jobs. Its primary job is to combine multiple jobs or tasks in a single unit of workflow. This provides users with ease of access and comfort in scheduling and running multiple jobs.

With a brief overview of multiple components of Hadoop ecosystem, let us now focus our attention on understanding the Hadoop architecture and its core components.

Fig. 4.4 Components of Hadoop architecture

3.3 Hadoop Architecture

At its core, Hadoop is a platform that primarily comprises three components to solve each of the core problems of big data analytics—storage, processing, and monitoring. To solve the data storage problem, Hadoop provides the Hadoop Distributed File System (HDFS). HDFS stores huge amounts of data by dividing it into small chunks and storing across multiple machines. HDFS attains reliability by replicating the data over multiple hosts. For the computing problem, Hadoop provides MapReduce, a parallel computing framework that divides a computing problem across multiple machines where each machine runs the program on the data that resides on the machine. Finally, in order to ensure that different components are working together in a seamless manner, Hadoop makes use of a monitoring mechanism known as Yet Another Resource Negotiator (YARN). YARN is a cluster resource management system to improve scheduling and to link to high-level applications.

The three primary components are shown in Fig. 4.4.

3.4 HDFS (Hadoop Distributed File System)

HDFS provides a fault-tolerant distributed file storage system that can run on commodity hardware and does not require specialized and expensive hardware. At its very core, HDFS is a hierarchical file system where data is stored in directories. It uses a master–slave architecture wherein one of the machines in the cluster is the master and the rest are slaves. The master manages the data and the slaves whereas the slaves service the read/write requests. The HDFS is tuned to efficiently handle large files. It is also a favorable file system for Write-once Read-many (WORM) applications.

HDFS functions on a master–slave architecture. The Master node is also referred to as the NameNode. The slave nodes are referred to as the DataNodes. At any given time, multiple copies of data are stored in order to ensure data availability in the event of node failure. The number of copies to be stored is specified by replication factor. The architecture of HDFS is specified in Fig. 4.5.

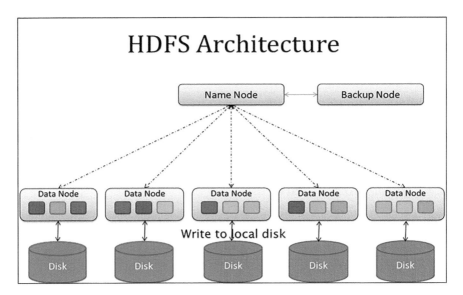

Fig. 4.5 Hadoop architecture (inspired from Hadoop architecture available on https://technocents.files.wordpress.com/2014/04/hdfs-architecture.png (accessed on Aug 10, 2018))

Fig. 4.6 Pictorial representation of storing a 350 MB file into HDFC (**BnRn = Replica n of Block n)

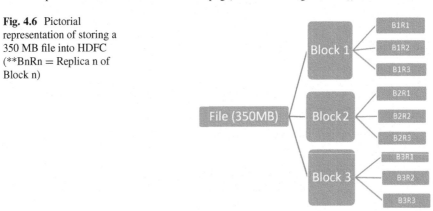

The NameNode maintains HDFS metadata and the DataNodes store the actual data. When a client requests folders/records access, the NameNode validates the request and instructs the DataNodes to provide the information accordingly. Let us understand this better with the help of an example.

Suppose we want to store a 350 MB file into the HDFS. The following steps illustrate how it is actually done (refer Fig. 4.6):

(a) The file is split into blocks of equal size. The block size is decided during the formation of the cluster. The block size is usually 64 MB or 128 MB. Thus, our file will be split into three blocks (Assuming block size = 128 MB).

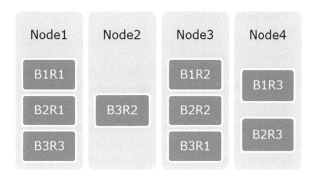

Fig. 4.7 Pictorial representation of storage of blocks in the DataNodes

(b) Each block is replicated depending on the replication factor. Assuming the factor to be 3, the total number of blocks will become 9.
(c) The three copies of the first block will then be distributed among the DataNodes (based on the block placement policy which is explained later) and stored.

Similarly, the other blocks are also stored in the DataNodes.

Figure 4.7 represents the storage of blocks in the DataNodes. Nodes 1 and 2 are part of **Rack1**. Nodes 3 and 4 are part of **Rack2**. A rack is nothing but a collection of data nodes connected to each other. Machines connected in a node have faster access to each other as compared to machines connected across different nodes. The block replication is in accordance with the block placement policy. The decisions pertaining to which block is stored in which DataNode is taken by the NameNode.

The major functionalities of the NameNode and the DataNode are as follows:
NameNode Functions:

- It is the interface to all files read/write requests by clients.
- Manages the file system namespace. Namespace is responsible for maintaining a list of all files and directories in the cluster. It contains all metadata information associated with various data blocks, and is also responsible for maintaining a list of all data blocks and the nodes they are stored on.
- Perform typical operations associated with a file system such as file open/close, renaming directories and so on.
- Determines which blocks of data to be stored on which DataNodes.

Secondary NameNode Functions:

- Keeps snapshots of NameNode and at the time of failure of NameNode the secondary NameNode replaces the primary NameNode.
- It takes snapshots of primary NameNode information after a regular interval of time, and saves the snapshot in directories. These snapshots are known as checkpoints, and can be used in place of primary NameNode to restart in case if it fails.

DataNode Functions:

- DataNodes are the actual machines that store data and take care of read/write requests from clients.
- DataNodes are responsible for the creation, replication, and deletion of data blocks. These operations are performed by the DataNode only upon direction from the NameNode.

Now that we have discussed how Hadoop solves the storage problem associated with big data, it is time to discuss how Hadoop performs parallel operations on the data that is stored across multiple machines. The module in Hadoop that takes care of computing is known as MapReduce.

3.5 MapReduce

MapReduce is a programming framework for analyzing datasets in HDFS. In addition to being responsible for executing the code that users have written, MapReduce provides certain important features such as parallelization and distribution, monitoring, and fault tolerance. MapReduce is highly scalable and can scale to multi-terabyte datasets.

MapReduce performs computation by dividing a computing problem in two separate phases, map and reduce. In the map phase, DataNodes run the code associated with the mapper on the data that is contained in respective machines. Once all mappers have finished running, MapReduce then sorts and shuffle the data and finally the reducer phase carries out a run that combines or aggregates the data via user-given-logic. Computations are expressed as a sequence of distributed tasks on key–value pairs. Users generally have to implement two interfaces. Map (in-key, in-value) → (out-key, intermediate-value) list, Reduce (out-key, intermediate-value) list → out-value list.

The reason we divide a programming problem into two phases (map and reduce) is not immediately apparent. It is in fact a bit counterintuitive to think of a programming problem in terms of map and reduce phases. However, there is a good reason why programming in Hadoop is implemented in this manner. Big data is generally distributed across hundreds/thousands of machines and it is a general requirement to process the data in reasonable time. In order to achieve this, it is better to distribute the program across multiple machines that run independently. This distribution implies parallel computing since the same tasks are performed on each machine, but with a different dataset. This is also known as a shared-nothing architecture. MapReduce is suited for parallel computing because of its shared-nothing architecture, that is, tasks have no dependence on one other. A MapReduce program can be written using JAVA, Python, C++, and several other programming languages.

MapReduce Principles

There are certain principles on which MapReduce programming is based. The salient principles of MapReduce programming are:

- Move code to data—Rather than moving data to code, as is done in traditional programming applications, in MapReduce we move code to data. By moving code to data, Hadoop MapReduce removes the overhead of data transfer.
- Allow programs to scale transparently—MapReduce computations are executed in such a way that there is no data overload—allowing programs to scale.
- Abstract away fault tolerance, synchronization, etc.—Hadoop MapReduce implementation handles everything, allowing the developers to build only the computation logic.

Figure 4.8 illustrates the overall flow of a MapReduce program.

MapReduce Functionality

Below are the MapReduce components and their functionality in brief:

- Master (Job Tracker): Coordinates all MapReduce tasks; manages job queues and scheduling; monitors and controls task trackers; uses checkpoints to combat failures.
- Slaves (Task Trackers): Execute individual map/reduce tasks assigned by Job Tracker; write information to local disk (not HDFS).

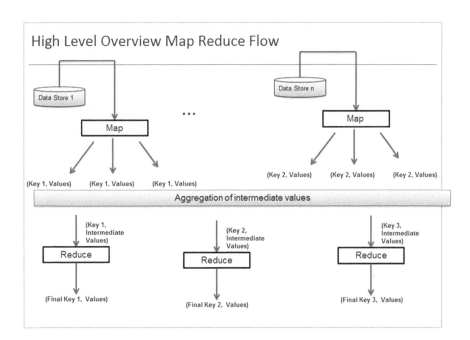

Fig. 4.8 MapReduce program flow. (Source: https://www.slideshare.net/acmvnit/hadoop-map-reduce (accessed on Aug 10, 2018))

- Job: A complete program that executes the Mapper and Reducer on the entire dataset.
- Task: A localized unit that executes code on the data that resides on the local machine. Multiple tasks comprise a job.

Let us understand this in more detail with the help of an example. In the following example, we are interested in doing the word count for a large amount of text using MapReduce. Before actually implementing the code, let us focus on the pseudo logic for the program. It is important for us to think of a programming problem in terms of MapReduce, that is, a mapper and a reducer. Both mapper and reducer take (key,value) as input and provide (key,value) pairs as output. In terms of mapper, a mapper program could simply take each word as input and provide as output a (key,value) pair of the form (word,1). This code would be run on all machines that have the text stored on which we want to run word count. Once all the mappers have finished running, each one of them will produce outputs of the form specified above. After that, an automatic shuffle and sort phase will kick in that will take all (key,value) pairs with same key and pass it to a single machine (or reducer). The reason this is done is because it will ensure that the aggregation happens on the entire dataset with a unique key. Imagine that we want to count word count of all word occurrences where the word is "Hello." The only way this can be ensured is that if all occurrences of ("Hello," 1) are passed to a single reducer. Once the reducer receives input, it will then kick in and for all values with the same key, it will sum up the values, that is, 1,1,1, and so on. Finally, the output would be (key, sum) for each unique key.

Let us now implement this in terms of pseudo logic:

Program: Word Count Occurrences
Pseudo code:

```
input-key: document name
input-value: document content
Map (input-key, input-value)
                 For each word w in input-value
                    produce (w, 1)
output-key: a word
output-values: a list of counts
Reduce (output-key, values-list);
                    int result=0;
                 for each v in values-list;
                       result+=v;
                       produce (output-key, result);
```

Now let us see how the pseudo code works with a detailed program (Fig. 4.9).
Hands-on Exercise:
For the purpose of this example, we make use of Cloudera Virtual Machine (VM) distributable to demonstrate different hands-on exercises. You can download

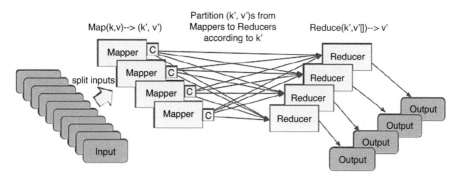

Fig. 4.9 MapReduce functionality example

a free copy of Cloudera VM from Cloudera website[4] and it comes prepackaged with Hadoop. When you launch Cloudera VM the first time, close the Internet browser and you will see the desktop that looks like Fig. 4.10.

This is a simulated Linux computer, which we can use as a controlled environment to experiment with Python, Hadoop, and some other big data tools. The platform is CentOS Linux, which is related to RedHat Linux. Cloudera is one major distributor of Hadoop. Others include Hortonworks, MapR, IBM, and Teradata.

Once you have launched Cloudera, open the command-line terminal from the menu (Fig. 4.11): Accessories → System Tools → Terminal

At the command prompt, you can enter Unix commands and hit ENTER to execute them one at a time. We assume that the reader is familiar with basic Unix commands or is able to read up about them in tutorials on the web. The prompt itself may give you some helpful information.

A version of Python is already installed in this virtual machine. In order to determine version information, type the following:

```
python -V
```

In the current example, it is Python 2.6.6. In order to launch Python, type "`python`" by itself to open the interactive interpreter (Fig. 4.12).

Here you can type one-line Python commands and immediately get the results. Try:

```
print ("hello world")
```

Type `quit()` when you want to exit the Python interpreter.

The other main way to use Python is to write code and save it in a text file with a ".py" suffix.

Here is a little program that will count the words in a text file (Fig. 4.13).

After writing and saving this file as "wordcount.py", do the following to make it an executable program:

---

[4]https://www.cloudera.com/downloads/quickstart_vms/5-13.html (accessed on Aug 10, 2018).

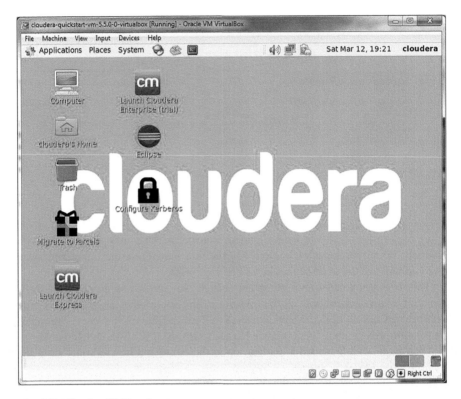

Fig. 4.10 Cloudera VM interface

```
chmod a+x wordcount.py
```

Now, you will need a file of text. You can make one quickly by dumping the "help" text of a program you are interested in:

```
hadoop --help > hadoophelp.txt
```

To pipe your file into the word count program, use the following:

```
cat hadoophelp.txt | ./wordcount.py
```

The "./" is necessary here. It means that wordcount.py is located in your current working directory.

Now make a slightly longer pipeline so that you can sort it and read it all on one screen at a time:

```
cat hadoophelp.txt | ./wordcount.py | sort | less
```

With the above program we got an idea how to use python programming scripts.

Now let us implement the code we have discussed earlier using MapReduce. We implement a Map Program and a Reduce program. In order to do this, we will

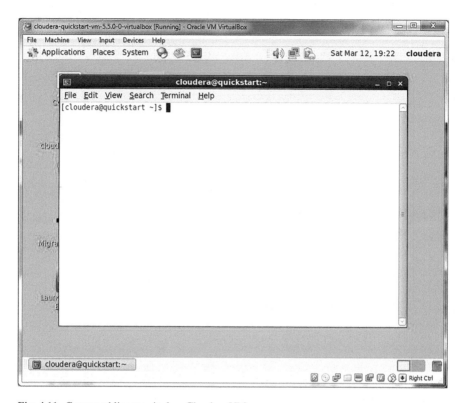

Fig. 4.11 Command line terminal on Cloudera VM

need to make use of a utility that comes with Hadoop—Hadoop Streaming. Hadoop streaming is a utility that comes packaged with the Hadoop distribution and allows MapReduce jobs to be created with any executable as the mapper and/or the reducer. The Hadoop streaming utility enables Python, shell scripts, or any other language to be used as a mapper, reducer, or both. The Mapper and Reducer are both executables that read input, line by line, from the standard input (stdin), and write output to the standard output (stdout). The Hadoop streaming utility creates a MapReduce job, submits the job to the cluster, and monitors its progress until it is complete. When the mapper is initialized, each map task launches the specified executable as a separate process. The mapper reads the input file and presents each line to the executable via stdin. After the executable processes each line of input, the mapper collects the output from stdout and converts each line to a key–value pair. The key consists of the part of the line before the first tab character, and the value consists of the part of the line after the first tab character. If a line contains no tab character, the entire line is considered the key and the value is null. When the reducer is initialized, each reduce task launches the specified executable as a separate process. The reducer converts the input key–value pair to lines that are presented to the executable via stdin. The reducer collects the executables result from stdout and converts each line

Fig. 4.12 Launch Python on Cloudera VM

to a key–value pair. Similar to the mapper, the executable specifies key–value pairs by separating the key and value by a tab character.

A Python Example:

To demonstrate how the Hadoop streaming utility can run Python as a MapReduce application on a Hadoop cluster, the WordCount application can be implemented as two Python programs: *mapper.py* and *reducer.py*. The code in mapper.py is the Python program that implements the logic in the map phase of WordCount. It reads data from stdin, splits the lines into words, and outputs each word with its intermediate count to stdout. The code below implements the logic in mapper.py.

Example mapper.py:

```
#!/usr/bin/env python
#!/usr/bin/ python
import sys
# Read each line from stdin
for line in sys.stdin:
    # Get the words in each line
    words = line.split()
    # Generate the count for each word
    for word in words:
```

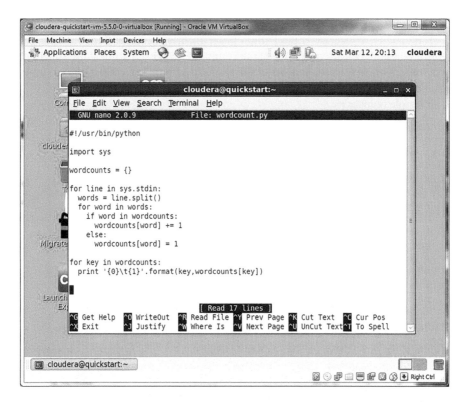

Fig. 4.13 Program to count words in a text file

```
# Write the key-value pair to stdout to be processed by
# the reducer.
# The key is anything before the first tab character and the
#value is anything after the first tab character.
print '{0}\t{1}'.format(word, 1)
```

Once you have saved the above code in mapper.py, change the permissions of the file by issuing

```
chmod a+x mapper2.py
```

Finally, type the following. This will serve as our input. "echo" command below simply prints on screen what input has been provided to it. In this case, it will print "jack be nimble jack be quick"

```
echo "jack be nimble jack be quick"
```

In the next line, we pass this input to our mapper program

```
echo "jack be nimble jack be quick"|./mapper.py
```

Next, we issue the sort command to do sorting.

```
echo "jack be nimble jack be quick"|./mapper2.py|sort
```

This is the way Hadoop streaming gives output to reducer.

The code in reducer.py is the Python program that implements the logic in the reduce phase of WordCount. It reads the results of mapper.py from stdin, sums the occurrences of each word, and writes the result to stdout. The code in the example implements the logic in reducer.py.

Example reducer.py:

```
#!/usr/bin/ python
import sys
curr_word = None
curr_count = 0
# Process each key-value pair from the mapper
for line in sys.stdin:
   # Get the key and value from the current line
   word, count = line.split('\t')
   # Convert the count to an int
   count = int(count)
   # If the current word is the same as the previous word,
   # increment its count, otherwise print the words count
   # to stdout
   if word == curr_word:
      curr_count += count
   else:
      # Write word and its number of occurrences as a key-value
      # pair to stdout
      if curr_word:
         print '{0}\t{1}'.format(curr_word, curr_count)
      curr_word = word
      curr_count = count
# Output the count for the last word
if curr_word == word:
   print '{0}\t{1}'.format(curr_word, curr_count)
```

Finally, to mimic overall functionality of MapReduce program, issue the following command:

```
echo "jack be nimble jack be quick"|./mapper2.py|sort|reducer2.py
```

Before attempting to execute the code, ensure that the *mapper.py* and *reducer.py* files have execution permission. The following command will enable this for both files:

```
chmod a+x mapper.py reducer.py
```

Also ensure that the first line of each file contains the proper path to Python. This line enables *mapper.py* and *reducer.py* to execute as stand-alone executables. The value #! /usr/bin/env python should work for most systems, but if it does not, replace /usr/bin/env python with the path to the Python executable on your system.

To test the Python programs locally before running them as a MapReduce job, they can be run from within the shell using the echo and sort commands. It is highly recommended to test all programs locally before running them across a Hadoop cluster.

```
$ echo 'jack be nimble jack be quick' | ./mapper.py
 | sort -t 1 | ./reducer.py
be 2
jack 2
nimble 1
quick 1
echo "jack be nimble jack be quick" | python mapper.py
 | sort | python reducer.py
```

Once the mapper and reducer programs are executing successfully against tests, they can be run as a MapReduce application using the Hadoop streaming utility. The command to run the Python programs *mapper.py* and *reducer.py* on a Hadoop cluster is as follows:

```
/usr/bin/hadoop jar /usr/lib/hadoop-mapreduce/
  hadoop-streaming.jar -files mapper.py,reducer.py -mapper
  mapper.py -reducer reducer.py -input /frost.txt -output /output
```

You can observe the output using the following command:

```
hdfs dfs -ls /output
hdfs dfs -cat /output/part-0000
The output above can be interpreted as follows:
'part-oooo' file mentions that this is the text output of first
    reducer in the MapReduce system. If there were multiple
    reducers in action, then we would see output such as
    'part-0000', 'part-0001', 'part-0002' and so on. Each such
    file is simply a text file, where each observation in the text
    file contains a key, value pair.
```

The options used with the Hadoop streaming utility are listed in Table 4.2.

A key challenge in MapReduce programming is thinking about a problem in terms of map and reduce steps. Most of us are not trained to think naturally in terms of MapReduce problems. In order to gain more familiarity with MapReduce programming, exercises provided at the end of the chapter help in developing the discipline.

Now that we have covered MapReduce programming, let us now move to the final core component of Hadoop—the resource manager or YARN.

Table 4.2 Hadoop stream utility options

Option	Description
-files	A command-separated list of files to be copied to the MapReduce cluster
-mapper	The command to be run as the mapper
-reducer	The command to be run as the reducer
-input	The DFS input path for the Map step
-output	The DFS output directory for the Reduce step

3.6 YARN

YARN (Yet Another Resource Negotiator) is a cluster resource management system for Hadoop. While it shipped as an integrated module in the original version of Hadoop, it was introduced as a separate module starting with Hadoop 2.0 to provide modularity and better resource management. It creates link between high-level applications (Spark, HBase, Hive, etc.) and HDFS environment and ensuring proper allocation of resources. It allows for running several different frameworks on the same hardware where Hadoop is deployed.

The main components in YARN are the following:

- Resource manager (one per cluster): Responsible for tracking the resources in a cluster, and scheduling applications.
- Node manager (one per every node): Responsible to monitor nodes and containers (slot analogue in MapReduce 1) resources such as CPU, memory, disk space, and network. It also collects log data and reports that information to the Resource Manager.
- Application master: It runs as a separate process on each slave node. It is responsible for sending heartbeats—short pings after a certain time period—to the resource manager. The heartbeats notify the resource manager about the status of each data node (Fig. 4.14).

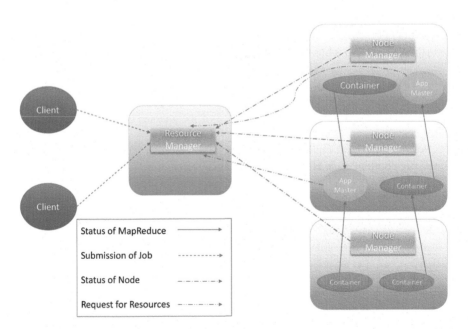

Fig. 4.14 Apache Hadoop YARN Architecture. (Source: https://data-flair.training/blogs/hadoop-yarn-tutorial/ (accessed on Aug 10, 2018))

Whenever a client requests the Resource Manager to run application, the Resource Manager in turn requests Node Managers to allocate a container for creating Application Master Instance on available (which has enough resources) node. When the Application Master Instance runs, it itself sends messages to Resource Manager and manages application.

3.7 Spark

In the previous sections, we have focused solely on Hadoop, one of the first and most widely used big data solutions. Hadoop was introduced to the world in 2005 and it quickly captured attention of organizations and individuals because of the relative simplicity it offered for doing big data processing. Hadoop was primarily designed for batch processing applications, and it performed a great job at that. While Hadoop was very good with sequential data processing, users quickly started realizing some of the key limitations of Hadoop. Two primary limitations were difficulty in programming, and limited ability to do anything other than sequential data processing.

In terms of programming, although Hadoop greatly simplified the process of allocating resources and monitoring a cluster, somebody still had to write programs in the MapReduce framework that would contain business logic and would be executed on the cluster. This posed a couple of challenges: first, a user had to know good programming skills (such as Java programming language) to be able to write code; and second, business logic had to be uniquely broken down in the MapReduce way of programming (i.e., thinking of a programming problem in terms of mappers and reducers). This quickly became a limiting factor for nontrivial applications.

Second, because of the sequential processing nature of Hadoop, every time a user would query for a certain portion of data, Hadoop would go through entire dataset in a sequential manner to query the data. This in turn implied large waiting times for the results of even simple queries. Database users are habituated to ad hoc data querying and this was a limiting factor for many of their business needs.

Additionally, there was a growing demand for big data applications that would leverage concepts of real-time data streaming and machine learning on data at a big scale. Since MapReduce was primarily not designed for such applications, the alternative was to use other technologies such as Mahout, Storm for any specialized processing needs. The need to learn individual systems for specialized needs was a limitation for organizations and developers alike.

Recognizing these limitations, researchers at the University of California, Berkeley's AMP Lab came up with a new project in 2012 that was later named as Spark. The idea of Spark was in many ways similar to what Hadoop offered, that is, a stable, fast, and easy-to-use big data computational framework, but with several key features that would make it better suited to overcome limitations of Hadoop and to also leverage features that Hadoop had previously ignored such as usage of memory for storing data and performing computations.

Spark has since then become one of the hottest big data technologies and has quickly become one of the mainstream projects in big data computing. An increasing number of organizations are either using or planning to use Spark for their project needs. At the same point of time, while Hadoop continues to enjoy the leader's position in big data deployments, Spark is quickly replacing MapReduce as the computational engine of choice. Let us now look at some of the key features of Spark that make it a versatile and powerful big data computing engine:

Ease of Usage: One of the key limitations of MapReduce was the requirement to break down a programming problem in terms of mappers and reducers. While it was fine to use MapReduce for trivial applications, it was not a very easy task to implement mappers and reducers for nontrivial programming needs.

Spark overcomes this limitation by providing an easy-to-use programming interface that in many ways is similar to what we would experience in any programming language such as R or Python. The manner in which this is achieved is by abstracting away requirements of mappers and reducers, and by replacing it with a number of operators (or functions) that are available to users through an API (application programming interface). There are currently 80 plus functions that Spark provides. These functions make writing code simple as users have to simply call these functions to get the programming done.

Another side effect of a simple API is that users do not have to write a lot of boilerplate code as was necessary with mappers and reducers. This makes program concise and requires less lines to code as compared to MapReduce. Such programs are also easier to understand and maintain.

In-memory computing: Perhaps one of the most talked about features of Spark is in-memory computing. It is largely due to this feature that Spark is considered up to 100 times faster than MapReduce (although this depends on several factors such as computational resources, data size, and type of algorithm). Because of the tremendous improvements in execution speed, Spark can handle the types of applications where turnaround time needs to be small and speed of execution is important. This implies that big data processing can be done on a near real-time scale as well as for interactive data analysis.

This increase in speed is primarily made possible due to two reasons. The first one is known as in-memory computing, and the second one is use of an advanced execution engine. Let us discuss each of these features in more detail.

In-memory computing is one of the most talked about features of Spark that sets it apart from Hadoop in terms of execution speed. While Hadoop primarily uses hard disk for data reads and write, Spark makes use of the main memory (RAM) of each individual computer to store intermediate data and computations. So while data resides primarily on hard disks and is read from the hard disk for the first time, for any subsequent data access it is stored in computer's RAM. Accessing data from RAM is 100 times faster than accessing it from hard disk, and for large data processing this results in a lot of time saving in terms of data access. While this difference would not be noticeable for small datasets (ranging from a few KB to few MBs), as soon as we start moving into the realm of big data process (tera bytes or more), the speed differences are visibly apparent. This allows for data processing to be done in a matter of minutes or hours for something that used to take days on MapReduce.

The second feature of Spark that makes it fast is implementation of an advanced execution engine. The execution engine is responsible for dividing an application into individual stages of execution such that the application executes in a time-efficient manner. In the case of MapReduce, every application is divided into a sequence of mappers and reducers that are executed in sequence. Because of the sequential nature, optimization that can be done in terms of code execution is very limited. Spark, on the other hand, does not impose any restriction of writing code in terms of mappers and reducers. This essentially means that the execution engine of Spark can divide a job into multiple stages and can run in a more optimized manner, hence resulting in faster execution speeds.

Scalability: Similar to Hadoop, Spark is highly scalable. Computation resources such as CPU, memory, and hard disks can be added to existing Spark clusters at any time as data needs grow and Spark can scale itself very easily. This fits in well with organizations that they do not have to pre-commit to an infrastructure and can rather increase or decrease it dynamically depending on their business needs. From a developer's perspective, this is also one of the important features as they do not have to make any changes to their code as the cluster scales; essentially their code is independent of the cluster size.

Fault Tolerance: This feature of Spark is also similar to Hadoop. When a large number of machines are connected together in a network, it is likely that some of the machines will fail. Because of the fault tolerance feature of Spark, however, it does not have any impact on code execution. If certain machines fail during code execution, Spark simply allocates those tasks to be run on another set of machines in the cluster. This way, an application developer never has to worry about the state of machines in a cluster and can rather focus on writing business logic for their organization's needs.

Overall, Spark is a general purpose in-memory computing framework that can easily handle programming needs such as batch processing of data, iterative algorithms, real-time data analysis, and ad hoc data querying. Spark is easy to use because of a simple API that it provides and is orders of magnitude faster than traditional Hadoop applications because of in-memory computing capabilities.

Spark Applications

Since Spark is a general purpose big data computing framework, Spark can be used for all of the types of applications that MapReduce environment is currently suited for. Most of these applications are of the batch processing type. However, because of the unique features of Spark such as speed and ability to handle iterative processing and real-time data, Spark can also be used for a range of applications that were not well suited for MapReduce framework. Additionally, because of the speed of execution, Spark can also be used for ad hoc querying or interactive data analysis. Let us briefly look at each of these application categories.

Iterative Data Processing

Iterative applications are those types of applications where the program has to loop or iterate through the same dataset multiple times in a recursive fashion (sometimes it requires maintaining previous states as well). Iterative data processing lies at the heart of many machine learning, graph processing, and artificial intelligence algorithms. Because of the widespread use of machine learning and graph processing these days, Spark is well suited to cater to such applications on a big data scale.

A primary reason why Spark is very good at iterative applications is because of the in-memory capability that Spark provides. A critical factor in iterative applications is the ability to access data very quickly in short intervals. Since Spark makes use of in-memory resources such as RAM and cache, Spark can store the necessary data for computation in memory and then quickly access it multiple times. This provides a power boost to such applications that can often run thousands of iterations in one go.

Ad Hoc Querying and Data Analysis

While business users need to do data summarization and grouping (batch or sequential processing tasks), they often need to interactively query the data in an ad hoc manner. Ad hoc querying with MapReduce is very limited because of the sequential processing nature in which it is designed to work for. This leaves a lot to be desired when it comes to interactively querying the data. With Spark it is possible to interactively query data in an ad hoc manner because of in-memory capabilities of Spark. For the first time when data is queried, it is read from the hard disk but, for all subsequent operations, data can be stored in memory where it is much faster to access the data. This makes the turnaround time of an ad hoc query an order of magnitude faster.

High-level Architecture

At a broad level, the architecture of Spark program execution is very similar to that of Hadoop. Whenever a user submits a program to be run on Spark cluster, it invokes a chain of execution that involves five elements: driver that would take user's code and submit to Spark; cluster manager like YARN; workers (individual machines responsible for providing computing resources), executor, and task. Let us explore the roles of each of these in more detail.

Driver Program

A driver is nothing but a program that is responsible for taking user's program and submitting it to Spark. The driver is the interface between a user's program and Spark library. A driver program can be launched from command prompt such as REPL (Read–Eval–Print Loop), interactive computer programming environment, or it could be instantiated from the code itself.

Cluster Manager

A cluster manager is very similar to what YARN is for Hadoop, and is responsible for resource allocation and monitoring. Whenever user request for code execution comes in, cluster manager is responsible for acquiring and allocating resources such as computational power, memory, and RAM from the pool of available resources in the cluster. This way a cluster manager is responsible and knows the overall state of resource usage in the cluster, and helps in efficient utilization of resources across the cluster.

In terms of the types of cluster managers that Spark can work with, there are currently three cluster manager modes. Spark can either work in stand-alone mode (or single machine mode in which a cluster is nothing but a single machine), or it can work with YARN (the resource manager that ships with Hadoop), or it can work with another resource manager known as Mesos. YARN and Mesos are capable of working with multiple nodes together compared to a single machine stand-alone mode.

Worker

A worker is similar to a node in Hadoop. It is the actual machine/computer that provides computing resources such as CPU, RAM, storage for execution of Spark programs. A Spark program is run among a number of workers in a distributed fashion.

Executor

These days, a typical computer comes with advanced configurations such as multi core processors, 1 TB storage, and several GB of memory as standard. Since any application at a point of time might not require all of the resources, potentially many applications can be run on the same worker machine by properly allocating resources for individual application needs. This is done through a JVM (Java Virtual Machine) that is referred to as an executor. An executor is nothing but a dedicated virtual machine within each worker machine on which an application executes. These JVMs are created as the application need arises, and are terminated once the application has finished processing. A worker machine can have potentially many executors.

Task

As the name refers, task is an individual unit of work that is performed. This work request would be specified by the executor. From a user's perspective, they do not have to worry about number of executors and division of program code into tasks. That is taken care of by Spark during execution. The idea of having these components is to essentially be able to optimally execute the application by distributing it across multiple threads and by dividing the application logically into small tasks (Fig. 4.15).

3.8 Spark Ecosystem

Similar to Hadoop, Spark ecosystem also has a number of components, and some of them are actively developed and improved. In the next section, we discuss six components that empower the Spark ecosystem (Fig. 4.16).

SPARK Core Engine

- Spark Core Engine performs I/O operations, Job Schedule, and monitoring on Spark clusters.
- Its basic functionality includes task dispatching, interacting with storage systems, efficient memory management, and fault recovery.

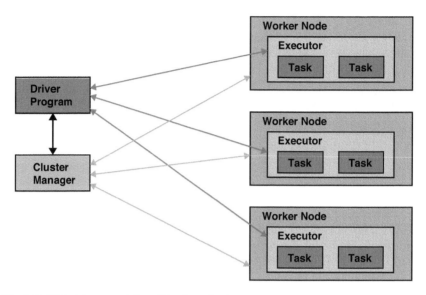

Fig. 4.15 Pictorial representation of Spark application

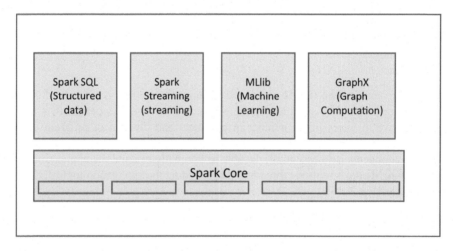

Fig. 4.16 Spark ecosystem

- Spark Core uses a fundamental data structure called RDD (Resilient Data Distribution) that handles partitioning data across all the nodes in a cluster and holds as a unit in memory for computations.
- RDD is an abstraction and exposes through a language integrated API written in either Python, Scala, SQL, Java, or R.

SPARK SQL

- Similar to Hive, Spark SQL allows users to write SQL queries that are then translated into Spark programs and executed. While Spark is easy to program, not everyone would be comfortable with programming. SQL, on the other hand, is an easier language to learn and write commands. Spark SQL brings the power of SQL programming to Spark.
- The SQL queries can be run on Spark datasets known as DataFrame. DataFrames are similar to a spreadsheet structure or a relational database table, and users can provide schema and structure to DataFrame.
- Users can also interface their query outputs to visualization tools such as Tableau for ad hoc and interactive visualizations.

SPARK Streaming

- While the initial use of big data systems was thought for batch processing of data, the need for real-time processing of data on a big scale rose quickly. Hadoop as a platform is great for sequential or batch processing of data but is not designed for real-time data processing.
- Spark streaming is the component of Spark that allows users to take real-time streams of data and analyze them in a near real-time fashion. Latency is as low as 1sec.
- Real-time data can be taken from a multitude of sources such as TCP/IP connections, sockets, Kafka, Flume, and other message systems.
- Data once processed can be then output to users interactively or stored in HDFS and other storage systems for later use.
- Spark streaming is based on Spark core, so the features of fault tolerance and scalability are also available for Spark Streaming.
- On top of it even machine learning and graph processing algorithms can be applied on real-time streams of data.

SPARK MLib

- Spark's MLLib library provides a range of machine learning and statistical functions that can be applied on big data. Some of the common applications provided by MLLib are functions for regression, classification, clustering, dimensionality reduction, and collaborative filtering.

GraphX

- GraphX is a unique Spark component that allows users to by-pass complex SQL queries and rather use GraphX for those graphs and connected dataset computations.
- A GraphFrame is the data structure that contains a graph and is an extension of DataFrame discussed above. It relates datasets with vertices and edges that produce clear and expressive computational data collections for analysis.

Here are a few sample programs using python for better understanding of Spark RDD and Data frames.

A sample RDD Program:
Consider a word count example.
Source file: input.txt
Content in file:
Business of Apple continues to grow with success of iPhone X. Apple is poised to become a trillion-dollar company on the back of strong business growth of iPhone X and also its cloud business.

Create a rddtest.py and place the below code and run from the unix prompt.

In the code below, we first read a text file named "input.txt" using textFile function of Spark, and create an RDD named text. We then run flatMap function on text RDD to generate a count of all of the words, where the output would be of the type (output,1). Finally, we run reduceByKey function to aggregate all observations with a common key, and produce a final (key, sum of all 1's) for each unique key. The output is then stored in a text file "output.txt" using the function saveAsTextFile.

```
text= sc.textFile("hdfs://input.txt")
wordcount = text.flatMap(lambda line: line.split(" ")).map
            (lambda word: (word, 1))\
            .reduceByKey(lambda a, b: a + b)
wordcount.saveAsTextFile("hdfs://output.txt")
```

$python rddtest.py

A sample Dataframe Program:
Consider a word count example.
Source file: input.txt
Content in file:
Business of Apple continues to grow with success of iPhone X. Apple is poised to become a trillion-dollar company on the back of strong business growth of iPhone X and also its cloud business.

Create a dftest.py and place the code shown below and run from the unix prompt.

In the code below, we first read a text file named "input.txt" using textFile function of Spark, and create an RDD named text. We then run the map function that takes each row of "text" RDD and converts to a DataFrame that can then be processed using Spark SQL. Next, we make use of the function filter() to consider only those observations that contain the term "business." Finally, search_word.count() function prints the count of all observations in search_word DataFrame.

```
text = sc.textFile("hdfs://input.txt")
  # Creates a DataFrame with a single column"
df = text.map(lambda r: Row(r)).toDF(["line"])
search_word = df.filter(col("line").like("%business%"))
# Counts all  words
search_word.count()
# Counts the word "business" mentioning MySQL
search_word.filter(col("line").like("%business%")).count()
# Gets   business word
search_word.filter(col("line").like("%business%")).collect()
```

3.9 Cloud Computing for Big Data

From the earlier sections of this chapter we understand that data is growing at an exponential rate, and organizations can benefit from the analysis of big data in making right decisions that positively impact their business. One of the most common issues that organizations face is with the storage of big data as it requires quite a bit of investment in hardware, software packages, and personnel to manage the infrastructure for big data environment. This is exactly where cloud computing helps solve the problem by providing a set of computing resources that can be shared though the Internet. The shared resources in a cloud computing platform include storage solutions, computational units, networking solutions, development and deployment tools, software applications, and business processes.

Cloud computing environment saves costs for an organization especially with costs related to infrastructure, platform, and software applications by providing a framework that can be optimized and expanded horizontally. Think of a scenario when an organization needs additional computational resources/storage capability/software has to pay only for those additional acquired resources and for the time of usage. This feature of cloud computing environment is known as elasticity. This helps organizations not to worry about overutilization or underutilization of infrastructure, platform, and software resources.

Figure 4.17 depicts how various services that are typically used in an organization setup can be hosted in cloud and users can connect to those services using various devices.

Features of Cloud Computing Environments

The salient features of cloud computing environment are the following:

- *Scalability*: It is very easy for organizations to add an additional resource to an existing infrastructure.
- *Elasticity*: Organizations can hire additional resources on demand and pay only for the usage of those resources.
- *Resource Pooling*: Multiple departments of an organization or organizations working on similar problems can share resources as opposed to hiring the resources individually.
- *Self-service*: Most of the cloud computing environment vendors provide simple and easy to use interfaces that help users to request services they want without help from differently skilled resources.
- *Low Cost*: Careful planning, usage, and management of resources help reduce the costs significantly compared to organizations owning the hardware. Remember, organizations owning the hardware will also need to worry about maintenance and depreciation, which is not the case with cloud computing.
- *Fault Tolerance*: Most of the cloud computing environments provide a feature of shifting the workload to other components in case of failure so the service is not interrupted.

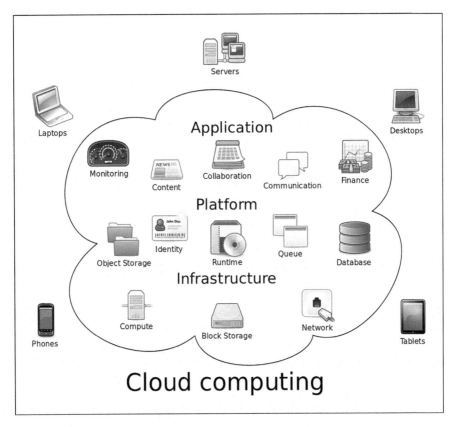

Fig. 4.17 An example of cloud computing setup in an organization. (Source: https://en.wikipedia.org/wiki/Cloud_computing#/media/File:Cloud_computing.svg (accessed on Aug 10, 2018))

Cloud Delivery Models

There are a number of on-demand cloud services, and the different terms can seem confusing at first. At the very basic, the purpose of all of the different services is to provide the capabilities of configuration, deployment, and maintenance of the infrastructure and applications that these services provide, but in different manners. Let us now have a look at some of the important terms that one could expect to hear in the context of cloud services.

- *Infrastructure as a Service (IaaS)*: IAAS, at the very core, is a service provided by a cloud computing vendor that involves ability to rent out computing, networking, data storage, and other hardware resources. Additionally, the service providers also give features such as load balancing, optimized routing, and operating systems on top of the hardware. Key users of these services are small and medium businesses and those organizations and individuals that do not want to make expensive investments in IT infrastructure but rather want to use IT infrastructure on a need basis.

- *Platform as a Service (PasS)*: PaaS is provided by a cloud computing environment provider that entails providing clients with a platform using which clients can develop software applications without worrying about the underlying infrastructure. Although there are multiple platforms to develop any kind of software application, most of the times the applications to be developed are web based. An organization that has subscribed for PaaS service will enable its programmers to create and deploy applications for their requirements.
- *Software as a Service (SaaS)*: One of the most common cloud computing services is SaaS. SaaS provides clients with a software solution such as Office365. This helps organizations avoid buying licenses for individual users and installing those licenses in the devices of those individual users.

Figure 4.18 depicts how responsibilities are shared between cloud computing environment providers and customers for different cloud delivery models.

Cloud Computing Environment Providers in Big Data Market

There are a number of service providers that provide various cloud computing services to organizations and individuals. Some of the most widely known service providers are Amazon, Google, and Microsoft. Below, we briefly discuss their solution offerings:

Amazon Web Services: Amazon Web Services (AWS) is one of the most comprehensive and widely known cloud services offering by Amazon. Considered as one of the market leaders, AWS provides a plethora of services such as hardware (computational power, storage, networking, and content delivery), software (database storage, analytics, operating systems, Hadoop), and a host of other services such as IoT. Some of the most important components of AWS are EC2 (for computing infrastructure), EMR (provides computing infrastructure and big data software such as Hadoop and Spark installed out of the box), Amazon S3 (Simple Storage Service) that provides low cost storage for data needs of all types), Redshift (a large-scale

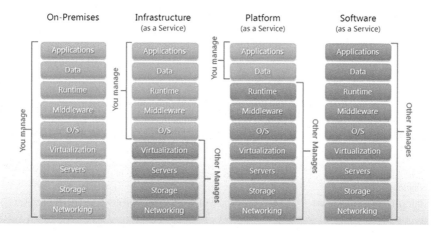

Fig. 4.18 Sharing of responsibilities based on different cloud delivery models. (Source: https://www.hostingadvice.com/how-to/iaas-vs-paas-vs-saas/ (accessed on Aug 10, 2018))

data storage and data warehouse solution), and RDS (a relational database service that provides instances of relational databases such as MySQL running out of the box).

Google Cloud Platform: Google Cloud Platform (GCP) is a cloud services offering by Google that competes directly with the services provided by AWS. Although the current offerings are not as diverse as compared to AWS, GCP is quickly closing the gap by providing bulk of the services that users can get on AWS. A key benefit of using GCP is that the services are provided on the same platform that Google uses for its own product development and service offerings. Thus, it is very robust. At the same point of time, Google provides many of the services for either free or at a very low cost, making GCP a very attracting alternative. The main components provided by GCP are Google Compute Engine (providing similar services as Amazon EC2, EMR, and other computing solutions offered by Amazon), Google Big Query, and Google Prediction API (that provides machine learning algorithms for end-user usage).

Microsoft Azure: Microsoft Azure is similar in terms of offerings to AWS listed above. It offers a plethora of services in terms of both hardware (computational power, storage, networking, and content delivery) and software (software as a service, and platform as a service) infrastructure for organizations and individuals to build, test, and deploy their offerings. Additionally, Azure has also made available some of the pioneering machine learning APIs from Microsoft under the umbrella of Microsoft Cognitive Toolkit. This makes very easy for organizations to leverage power of machine and deep learning and integrate it easily for their datasets.

Hadoop Useful Commands for Quick Reference

Caution: While operating any of these commands, make sure that the user has to create their own folders and practice as it may impact on the Hadoop system directly using the following paths.

Command	Description
hdfs dfs -ls/	List all the files/directories for the given hdfs destination path
hdfs dfs -ls -d /hadoop	Directories are listed as plain files. In this case, this command will list the details of hadoop folder
hdfs dfs -ls -h /data	Provides human readable file format (e.g., 64.0m instead of 67108864)
hdfs dfs -ls -R /hadoop	Lists recursively all files in hadoop directory and all subdirectories in hadoop directory
hdfs dfs -ls /hadoop/dat*	List all the files matching the pattern. In this case, it will list all the files inside hadoop directory that start with 'dat' hdfs dfs -ls/list all the files/directories for the given hdfs destination path
hdfs dfs -text /hadoop/derby.log	Takes a file as input and outputs file in text format on the terminal
hdfs dfs -cat /hadoop/test	This command will display the content of the HDFS file test on your stdout

(continued)

Command	Description
hdfs dfs -appendToFile /home/ubuntu/test1 /hadoop/text2	Appends the content of a local file test1 to a hdfs file test2
hdfs dfs -cp /hadoop/file1 /hadoop1	Copies file from source to destination on HDFS. In this case, copying file1 from hadoop directory to hadoop1 directory
hdfs dfs -cp -p /hadoop/file1 /hadoop1	Copies file from source to destination on HDFS. Passing -p preserves access and modification times, ownership, and the mode
hdfs dfs -cp -f /hadoop/file1 /hadoop1	Copies file from source to destination on HDFS. Passing -f overwrites the destination if it already exists
hdfs dfs -mv /hadoop/file1 /hadoop1	A file movement operation. Moves all files matching a specified pattern to a destination. Destination location must be a directory in case of multiple file moves
hdfs dfs -rm /hadoop/file1	Deletes the file (sends it to the trash)
hdfs dfs -rmr /hadoop	Similar to the above command but deletes files and directory in a recursive fashion
hdfs dfs -rm -skipTrash /hadoop	Similar to above command but deletes the file immediately
hdfs dfs -rm -f /hadoop	If the file does not exist, does not show a diagnostic message or modifies the exit status to reflect an error
hdfs dfs -rmdir /hadoop1	Delete a directory
hdfs dfs -mkdir /hadoop2	Create a directory in specified HDFS location
hdfs dfs -mkdir -f /hadoop2	Create a directory in specified HDFS location. This command does not fail even if the directory already exists
hdfs dfs -touchz /hadoop3	Creates a file of zero length at <path> with current time as the timestamp of that <path>
hdfs dfs -df /hadoop	Computes overall capacity, available space, and used space of the filesystem
hdfs dfs -df -h /hadoop	Computes overall capacity, available space, and used space of the filesystem. -h parameter formats the sizes of files in a human-readable fashion
hdfs dfs -du /hadoop/file	Shows the amount of space, in bytes, used by the files that match the specified file pattern
hdfs dfs -du -s /hadoop/file	Rather than showing the size of each individual file that matches the pattern, shows the total (summary) size
hdfs dfs -du -h /hadoop/file	Shows the amount of space, in bytes, used by the files that match the specified file

Source: https://linoxide.com/linux-how-to/hadoop-commands-cheat-sheet/ (accessed on Aug 10, 2018)

Electronic Supplementary Material

All the datasets, code, and other material referred in this section are available in www.allaboutanalytics.net.

- Data 4.1: Apple_discussion.txt
- Data 4.2: Children_names.txt
- Data 4.3: Numbers_dataset.csv
- Data 4.4: UserProfile.csv
- Data 4.5: Users.csv

Exercises

These exercises are based on either MapReduce or Spark and would ask users to code the programming logic using any of the programming languages they are comfortable with (preferably, but not limited to Python) in order to run the code on a Hadoop cluster:

Ex. 4.1 Write code that takes a text file as input, and provides word count of each unique word as output using MapReduce. Once you have done so, repeat the same exercise using Spark. The text file is provided with the name apple_discussion.txt.

Ex. 4.2 Now, extend the above program to output only the most frequently occurring word and its count (rather than all words and their counts). Attempt this first using MapReduce and then using Spark. Compare the differences in programming effort required to solve the exercise. (Hint: for MapReduce, you might have to think of multiple mappers and reducers.)

Ex. 4.3 Consider the text file children_names.txt. This file contains three columns—name, gender, and count, and provides data on how many kids were given a specific name in a given period. Using this text file, count the number of births by alphabet (not the word). Next, repeat the same process but only for females (exclude all males from the count).

Ex. 4.4 A common problem in data analysis is to do statistical analysis on datasets. In order to do so using a programming language such as R or Python, we simply use the in-built functions provided by those languages. However, MapReduce provides no such functions and so a user has to write the programming logic using mappers and reducers. In this problem, consider the dataset numbers_dataset.csv. It contains 5000 randomly generated numbers. Using this dataset, write code in MapReduce to compute five point summary (i.e., Mean, Median, Minimum, Maximum, and Standard Deviation).

Ex. 4.5 Once you have solved the above problem using MapReduce, attempt to do the same using Spark. For this you can make use of the in-built functions that Spark provides.

Ex. 4.6 Consider the two csv files: Users.csv, and UserProfile.csv. Each file contains information about users belonging to a company, and there are 5000 such records in each file. Users.csv contains following columns: FirstName, Surname, Gender, ID. UserProfile.csv has following columns: City, ZipCode, State, EmailAddress, Username, Birthday, Age, CCNumber, Occupation, ID. The common field in both files is ID, which is unique for each user. Using the two datasets, merge them into a single data file using MapReduce. Please remember that the column on which merge would be done is ID.

Ex. 4.7 Once you have completed the above exercise using MapReduce, repeat the same using Spark and compare the differences between the two platforms.

Further Reading

5 Reasons Spark is Swiss Army Knife of Data Analytics. Retrieved August 10, 2018, from https://datafloq.com/read/5-ways-apache-spark-drastically-improves-business/1191.

A secret connection between Big Data and Internet of Things. Retrieved August 10, 2018, from https://channels.theinnovationenterprise.com/articles/a-secret-connection-between-big-data-and-the-internet-of-things.

Big Data: Are you ready for blast off. Retrieved August 10, 2018, from http://www.bbc.com/news/business-26383058.

Big Data: Why CEOs should care about it. Retrieved August 10, 2018, from https://www.forbes.com/sites/davefeinleib/2012/07/10/big-data-why-you-should-care-about-it-but-probably-dont/#6c29f11c160b.

Hadoop and Spark Enterprise Adoption. Retrieved August 10, 2018, from https://insidebigdata.com/2016/02/01/hadoop-and-spark-enterprise-adoption-powering-big-data-applications-to-drive-business-value/.

How companies are using Big Data and Analytics. Retrieved August 10, 2018, from https://www.mckinsey.com/business-functions/mckinsey-analytics/our-insights/how-companies-are-using-big-data-and-analytics.

How Uber uses Spark and Hadoop to optimize customer experience. Retrieved August 10, 2018, from https://www.datanami.com/2015/10/05/how-uber-uses-spark-and-hadoop-to-optimize-customer-experience/.

Karau, H., Konwinski, A., Wendell, P., & Zaharia, M. (2015). *Learning spark: Lightning-fast big data analysis*. Sebastopol, CA: O'Reilly Media, Inc..

Reinsel, D., Gantz, J., & Rydning, J. (April 2017). Data age 2025: The evolution of data to life-critical- don't focus on big data; focus on the data that's big. IDC White Paper URL: https://www.seagate.com/www-content/our-story/trends/files/Seagate-WP-DataAge2025-March-2017.pdf.

The Story of Spark Adoption. Retrieved August 10, 2018, from https://tdwi.org/articles/2016/10/27/state-of-spark-adoption-carefully-considered.aspx.

White, T. (2015). *Hadoop: The definitive guide* (4th ed.). Sebastopol, CA: O'Reilly Media, Inc..

Why Big Data is the new competitive advantage. Retrieved August 10, 2018, from https://iveybusinessjournal.com/publication/why-big-data-is-the-new-competitive-advantage/.

Why Apache Spark is a crossover hit for data scientists. Retrieved August 10, 2018, from https://blog.cloudera.com/blog/2014/03/why-apache-spark-is-a-crossover-hit-for-data-scientists/.

Chapter 5
Data Visualization

John F. Tripp

1 Introduction

Data analytics is a burgeoning field—with methods emerging quickly to explore and make sense of the huge amount of information that is being created every day. However, with any data set or analysis result, the primary concern is in communicating the results to the reader. Unfortunately, human perception is not optimized to understand interrelationships between large (or even moderately sized) sets of numbers. However, human perception is excellent at understanding interrelationships between sets of data, such as series, deviations, and the like, through the use of visual representations.

In this chapter, we will present an overview of the fundamentals of data visualization and associated concepts of human perception. While this chapter cannot be exhaustive, the reader will be exposed to a sufficient amount of content that will allow them to consume and create quantitative data visualizations critically and accurately.

J. F. Tripp (✉)
Clemson University, Clemson, SC, USA
e-mail: jftripp@clemson.edu

© Springer Nature Switzerland AG 2019
B. Pochiraju, S. Seshadri (eds.), *Essentials of Business Analytics*, International Series in Operations Research & Management Science 264,
https://doi.org/10.1007/978-3-319-68837-4_5

2 Motivating Example

A Vice President of Sales wishes to communicate his division's performance to the executive team. His division has performed very well in year-over-year sales, but is still rather small compared to other divisions. He is concerned that when his current year numbers are presented alongside other divisions, it will not accurately reflect the excellent performance of his team.

He wishes to provide the sales figures as well as show the growth in sales, year over year, and compare his rate of growth to other divisions' rates. Providing this information using only tales of numbers would be difficult and time consuming. However, this information can be provided in one or two simple but intuitive graphs.

3 Methods of Data Visualization

3.1 Working with (and Not Against) Human Perception

Consider Fig. 5.1. When you see this graph, what do you believe is true about the "level" of the variable represented by the line? Is the level greater or less at point 2 compared with point 1?

If you are like most people, you assume that the level of the variable at point 2 is greater than the level at point 1. Why? Because it has been ingrained in you from childhood that when you stack something (blocks, rocks, etc.), the more you stack, the higher the stack becomes. From a very early age, you learn that "up" means "more."

Now consider Fig. 5.2. Based on this graph, what happened to gun deaths after 2005?[1]

Upon initial viewing, the reader may be led to believe that the number of gun deaths went down after 2005. However, look more closely, is this really what happened? If you observe the axes, you will notice that the graph designer inverted

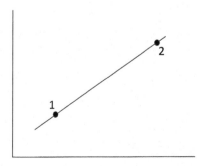

Fig. 5.1 A graph

---

[1] The "Stand Your Ground" law in Florida enabled people to shoot attackers in self-defense without first having to attempt to flee.

5 Data Visualization

Gun Deaths in Florida

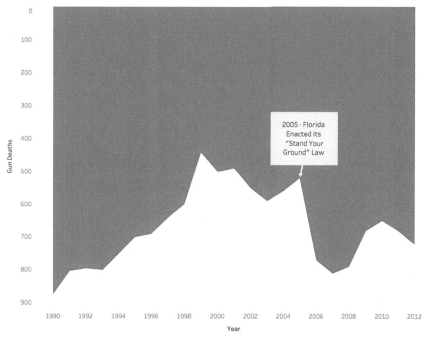

Fig. 5.2 A graph representing gun deaths in Florida over time

the y-axis, making larger values appear at lower points in the graph than smaller values. While many readers may be able to perceive the shift in the y-axis, not all will. For all readers, this is a fundamental violation of primal perceptive processes. It is not merely a violation of an established convention; it is a violation of the principles that drove the establishment of that convention.[2]

This example is simple, but it illustrates the need for data visualizers to be well trained in understanding human perception and to work with the natural understanding of visual stimuli.

3.2 Why We Visualize Data

Visual and cognitive processing in humans is not optimized to process relationships between large sets of numbers. While we are good at comparing several values against each other, we are simply not good enough at drawing inferences using sets or tables in order to communicate and compare trends for large groups of numbers. Let us look at a famous example.

---

[2]This example was strongly debated across the Internet when it appeared. For more information about the reaction to the graph, see the Reddit thread at http://bit.ly/2ggVV7V.

Table 5.1 Anscombe's Quartet

Set 1		Set 2		Set 3		Set 4	
X	Y	X	Y	X	Y	X	Y
10.0	8.04	10.0	9.14	10.0	7.46	8.0	6.58
8.0	6.95	8.0	8.14	8.0	6.77	8.0	5.76
13.0	7.58	13.0	8.74	13.0	12.74	8.0	7.71
9.0	8.81	9.0	8.77	9.0	7.11	8.0	8.84
11.0	8.33	11.0	9.26	11.0	7.81	8.0	8.47
14.0	9.96	14.0	8.10	14.0	8.84	8.0	7.04
6.0	7.24	6.0	6.13	6.0	6.08	8.0	5.25
4.0	4.26	4.0	3.10	4.0	5.39	19.0	12.50
12.0	10.84	12.0	9.13	12.0	8.15	8.0	5.56
7.0	4.82	7.0	7.26	7.0	6.42	8.0	7.91
5.0	5.68	5.0	4.74	5.0	5.73	8.0	6.89

In Table 5.1, the classic example of "Anscombe's Quartet" is presented. In 1973, Anscombe created these sets of data to illustrate the importance of visualizing data. When you consider these four data sets, what sense can you make of them? How much are the sets the same? How much are they different? These data sets are also referred to in Chap. 7 on linear regression.

> *Information, that is imperfectly acquired, is generally as imperfectly retained; and a man who has carefully investigated a printed table, finds, when done, that he has only a very faint and partial idea of what he had read; and that like a figure imprinted on sand, is soon totally erased and defaced*—William Playfair (1801, p. 3).

As Playfair notes in the quote above, even after a great deal of study, humans cannot make clear sense of the data provided in tables, nor can they retain it well. Turning back to Table 5.1, most people note that sets 1–3 are similar, primarily because the X values are the same, and appear in the same order. However, for all intents and purposes, these four sets of numbers are statistically equivalent. If a typical regression analysis was performed on the four sets of data in Table 5.1, the results would be *identical*. Among other statistical properties, the following are valid for all four sets of numbers in Table 5.1:

- $N = 11$.
- Mean of $X = 9.0$.
- Mean of $Y = 7.5$.
- Equation of regression line: $Y = 3 + 0.5X$.
- Standard error of regression slope $= 0.118$.
- Correlation coefficient $= 0.82$.
- $r^2 = 0.67$.

However, if one compares the sets of data visually, the differences between the data sets become immediately obvious, as illustrated in Fig. 5.3.

Tables of data are excellent when the purpose of the table is to allow the user to look up specific values, and when the relationships between the values

5 Data Visualization

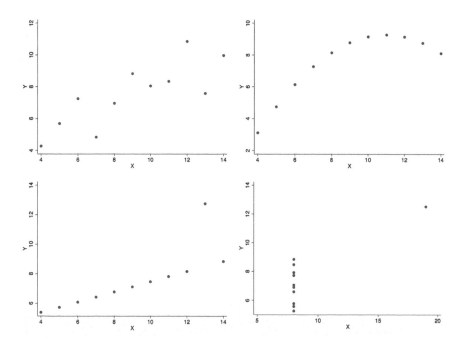

Fig. 5.3 Anscombe's Quartet

are direct. However, when relationships between sets or groups of numbers are intended for presentation, the human perception is much better served through graphical representations. The goal of the remainder of this chapter is to provide the reader with enough background into human visual perception and graphical data representation to become an excellent consumer and creator of graphical data visualizations.

3.3 Visualization as a Cognitive Aid

Humans are bombarded by information from multiple sources and through multiple channels. This information is gathered by our five senses, and processed by the brain. However, the brain is highly selective about what it processes and humans are only aware of the smallest fraction of sensory input. Much of sensory input is simply ignored by the brain, while other input is dealt with based on heuristic rules and categorization mechanisms; these processes reduce cognitive load. Data visualizations, when executed well, aid in the reduction of cognitive load, and assist viewers in the processing of cognitive evaluations.

When dealing with data visualization, likely the most important cognitive concept to understand is working memory. Working memory is the part of short-term memory that is concerned with immediate perceptual processing. Working memory is extremely limited, with the average human having the capacity to hold at most four "chunks" of information at once (see, e.g., Cowan 2010). For this reason, visualizers must build graphs that do not rely on the user to hold multiple chunks of information in order to "generate" understanding. Instead, visualizers should "do the work" for the user and visualize the relationships in question directly.

Practically, this means that visualizations are not effective when built for general purposes, for a generalized audience. For instance, in my experience with real-world corporate "dashboards," they tend to be designed in a manner that presents a generous amount of data that users might be able to leverage for different cognitive tasks. However, users would have to extrapolate the useful data themselves and perform additional steps to generate a result that assists in achieving their cognitive task. Instead, if visualizations were created with particular cognitive tasks in mind and constructed in a manner that would assist the viewer in the completion of the task, then the user would not have to do much in facilitating the understanding of the data.

Designing and building visualizations that are fit for purpose—for use in the processing of a defined cognitive task (communicating the status of sales in a region or the need for managerial intervention in a process, etc.)—is the key responsibility of any data visualizer. This will usually result in the need to create visualizations that are more tightly focused than those that already exist in businesses today. In doing so, we re-emphasize the need to reduce cognitive load for a particular cognitive task. A good visualization that is focused on a task, highlights key information, and provides computational assistance for viewers will allow them to build an understanding more quickly, and is of significantly more value than creating a single, multipurpose visualization that fails to assist in any of the viewers' cognitive tasks.

3.4 Six Meta-Rules for Data Visualization

For the remainder of this chapter, we will review and discuss six meta-rules for data visualization. These are based upon the work of many researchers and writers, and provide a short and concise summary of the most important practices regarding data visualization. However, it is important to note that these rules are intended to describe how we can attempt to represent data visually with the highest fidelity and integrity—not argue that all are immutable, or that trade-offs between the rules may not have to be made. The meta-rules are presented in Table 5.2.

By following these meta-rules, data visualizers will be more likely to graphically display the actual effect shown in the data. However, there are specific times and reasons when the visualizer may choose to violate these rules. Some of the reasons may be, for example, the need to make the visualization "eye-catching," such as for an advertisement. In these cases, knowing the effect on perceptive ability of breaking the rules is important so that the visualizer understands what is being

Table 5.2 Six meta-rules for data visualization

1. The simplest chart is usually the one that communicates most clearly. Use the "not wrong" chart—not the "cool" chart
2. Always directly represent the relationship you are trying to communicate. Do not leave it to the viewer to derive the relationship from other information
3. In general, do not ask viewers to compare in two dimensions. Comparing differences in length is easier than comparing differences in area
4. Never use color on top of color—color is not absolute
5. Do not violate the primal perceptions of your viewers. Remember, up means more
6. Chart with graphical and ethical integrity. Do not lie, either by mistake or intentionally

lost. However, in some cases, the reasons for violating the rules may be because the visualizer wishes to intentionally mislead. Examples of this kind of lack of visualization integrity are common in political contexts.

These rules are made to be understood, and then followed to the extent that the context requires. If the context requires an accurate understanding of the data, with high fidelity, the visualizer should follow the rules as much as possible. If the context requires other criteria to be weighed more heavily, then understanding the rules allows the visualizer to understand how these other criteria are biasing the visual perception of the audience.

Simplicity Over Complexity

Meta-Rule #1: The simplest chart is usually the one that communicates most clearly. Use the "not wrong" chart—not the "cool" chart.

When attempting to visualize data, our concern should be, as noted above, to reduce the cognitive load of the viewer. This means that we should eliminate sources of confusion. While several of the other meta-rules are related to this first rule, the concept of simplicity itself deserves discussion.

Many data visualizers focus on the aesthetic components of a visualization, much to the expense of clearly communicating the message that is present in the data. When the artistic concerns of the visualizer (e.g., the Florida dripping blood example above) overwhelm the message in the data, confusion occurs. Aside from artistic concerns, visualizers often choose to use multiple kinds of graphs to add "variety," especially when visualizing multiple relationships. For instance, instead of using three stacked column charts to represent different part to whole relationships, the visualizer might use one stacked column chart, one pie chart, and one bubble chart. So instead of comparing three relationships represented in one way, the viewer must attempt to interpret different graph types, as well as try to compare relationships. This increases cognitive load, as the viewer has to keep track of both the data values, as well as the various manners in which the data has been encoded.

Fig. 5.4 Example of unnecessary visualization

Instead, we should focus on selecting a "not wrong" graph.[3] To do this, one must understand both the nature of the data that is available, as well as the nature of the relationship being visualized. Data is generally considered to be of two types, quantitative and qualitative (or categorical). At its simplest, data that is quantitative is data that is (1) numeric, and (2) it is appropriate to use for mathematical operations (unit price, total sales, etc.). Qualitative or categorical data is data that is (1) numeric or text, and (2) is not appropriate (or possible) to use for mathematical operations (e.g., Customer ID #, City, State, Country, Department).

Is a Visualization Needed?

An initial analysis of the data is required to determine the necessity of a visual representation. For instance, in the graphic shown in Fig. 5.4, a text statement is redundantly presented as a graph.

When a simple statement is enough to communicate the message, a visualization may not be needed at all.

Table or Graph?

Once you decide that you need a visual representation of the data to communicate your message, you need to choose between two primary categories of visualization—tables vs. graphs. The choice between the two is somewhat subjective and, in many cases, you may choose to use a combination of tables and graphs to tell your data story. However, use the following heuristic to decide whether to use a table or a graph:

> *If the information being represented in the visualization needs to display precise and specific individual values, with the intention to allow the user to look up specific values, and compare to other specific values, choose a table. If the information in the visualization must display sets or groups of values for comparison, choose a graph.*

Tables are best used when:

1. The display will be used as a lookup for *particular values*.
2. It will be used to compare *individual values* not groups or series of values to one another.
3. Precision is required.
4. Quantitative information to be provided involves more than one unit of measure.
5. Both summary and detail values are included.

---

[3]By using the term "not wrong" instead of "right" or "correct," we attempt to communicate the fact that in many cases there is not a single "correct" visualization. Instead, there are visualizations that are more or less "not wrong" along a continuum. In contrast, there are almost always multiple "wrong" visualization choices.

If a Graph, Which Graph?

The exploration of choosing which graph to use is a topic that requires more space than is available in this chapter. However, the process of choosing the correct graph is fundamentally linked to the relationship being communicated. For each kind of relationship, there are multiple kinds of graphs that might be used. However, for many relationships, there are a small group of "best practice" graphs that fit that relationship best.

How Many Dimensions to Represent?

When representing data visually, we must decide how many dimensions to represent in a single graph. The maximum number of data dimensions that can be represented in a static graph is five and in an interactive graph is six. Table 5.3 provides a list of the most likely graphs to be used for various relationships and numbers of dimensions.

Table 5.3 Data relationships and graphs

Relationship	Most likely graph(s)	Keywords	Max. # of dimensions
Time series	Trend line Column chart Heat map Sparklines	Change Rise Increase Fluctuation Growth Decline/decrease Trend	4
Part to whole	Stacked column chart Stacked area chart Pareto chart (for two simultaneous parts to whole)	Rate or rate of total Percent or percentage of total Share "Accounts for X percent"	4
Ranking	Sorted bar/column chart	Larger than Smaller than Equal to Greater than Less than	4
Deviation	Line chart Column/bar chart Bullet graph	Plus or minus Variance Difference Relative to	4
Distribution	Box/whisker plot Histogram	Frequency Distribution Range Concentration Normal curve, bell curve	4
Correlation	Scatterplot Table pane	Increases with Decreases with Changes with Varies with	6
Geospatial	Choropleth (filled gradient) map	N/A	2

1. X-axis placement
2. Y-axis placement
3. Size
4. Shape
5. Color
6. Animation (interactive only, often used to display time)

However, many graph types, because of their nature, reduce the number of possible dimensions that can be displayed (Table 5.3). For instance, while a scatterplot can display all six dimensions, a filled map can only display two: the use of the map automatically eliminates the ability to modify dimensions 1–4. As such, we are left with the ability to use color to show different levels of one data dimension and animation to how the level of that one dimension changes over time.

While the maximum possible dimensions to represent is six, it is unlikely that most visualization (or any) should/would use all six. Shape especially is difficult to use as a dimensional variable and should never be used with size. Notice in Fig. 5.5 that it is difficult to compare the relative sizes of differently shaped objects.

One of the biggest issues in many visualizations is that visualizers attempt to encode too much information into a graph, attempting to tell multiple stories with a single graph. However, this added complexity leads to the viewer having a reduced ability to interpret any of the stories in the graph. Instead, visualizers should create

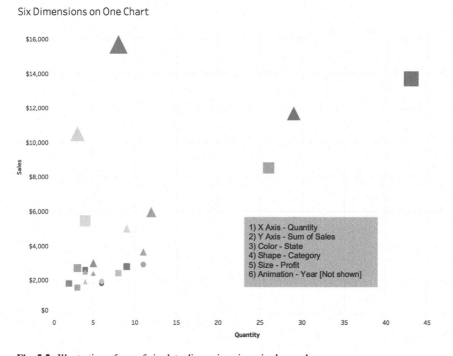

Fig. 5.5 Illustration of use of six data dimensions in a single graph

simpler, single story graphs, using the fewest dimensions possible to represent the story that they wish to tell.

Direct Representation

Meta-Rule #2: *Always directly represent the relationship you are trying to communicate. Don't leave it to the viewer to derive the relationship from other information.*

It is imperative that the data visualizer provide the information that the data story requires directly—and not rely on the viewer to have to interpret what relationships we intend to communicate. For instance, we often wish to tell a story of differences, such as deviations from plan, and budgets vs. actual. When telling a story of differences, do not rely on the viewer to calculate the differences themselves.

Figure 5.6 illustrates a visualization that relies on the viewer to calculate differences. Figure 5.7 presents the actual deviation through the profit margin, allowing the viewer to focus on the essence of the data set.

Again, the goal of the visualizer is to tell a story while minimizing the cognitive load on the viewer. By directly representing the relationship in question, we assist the viewer in making the cognitive assessment we wish to illustrate. When we "leave it to the viewer" to determine what the association or relationship is that we are trying to communicate, not only do we increase cognitive load, but we also potentially lose consistency in the way that the viewers approach the visualization.

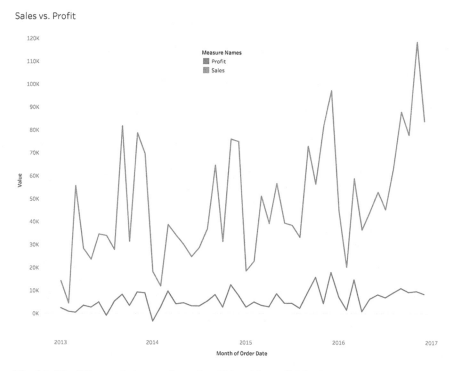

Fig. 5.6 The difference between sales and profit is not immediately clear

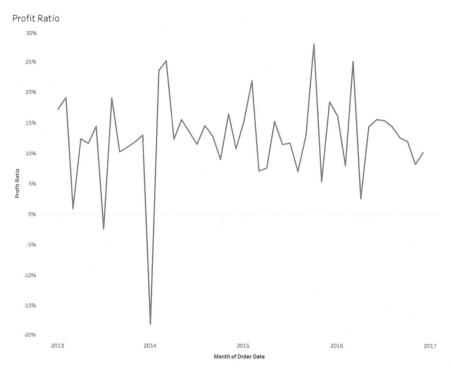

Fig. 5.7 The profit ratio is clear and intuitive

By properly visualizing relationships directly, we not only reduce cognitive load, we also reduce the potential for the viewer to misunderstand the story we are trying to tell.

Single Dimensionality

Meta-Rule #3: In general, do not ask viewers to compare in two dimensions. Comparing differences in length is easier than comparing differences in area.

When representing differences in a single data variable (e.g., sales by month, profit by product), visualizers should generally use visual comparisons on a single visual dimension—it is much easier for viewers to perceive differences in length or height than differences in area. By doing this we also avoid some common issues in the *encoding* of data.

While most modern software packages manage the encoding of data very well, whenever a visualizer chooses to use differences in two dimensions to represent changes in a single data dimension, visualizations tend to become distorted. Figure 5.8 illustrates the issue when visualizers directly represent data differences in two dimensions. In this example, the visualization is attempting to depict the ratio between two levels of a variable. However, the ratio of the circles being 1:2 is not

5 Data Visualization

Fig. 5.8 Difference between circles and lines

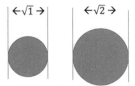
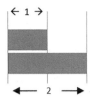

Table 5.4 Sales of fictitious salespeople

Salesperson	YTD sales (in $)	Share of sales
Deepika Padukone	1,140,000	37%
George Clooney	750,000	25%
Jennifer Garner	740,000	24%
Danny Glover	430,000	14%

immediately apparent from the visual. Using $\pi\left(\frac{d}{2}\right)^2$ we see that the "smaller" circle has an area of $\frac{\pi}{4}$ and the "larger" circle has an area of $\frac{\pi}{2}$, hence 1:2.

While both examples are properly encoded, the area of each exactly maintains the proportion of 1:2 that is found in the data, which of the visualizations more clearly communicates what is present in the data? The simple bars—these only vary on one dimension and therefore more clearly illustrate the relationship that is present in the data.

To 3D or Not to 3D

One of the software features that many people enjoy using, because it looks "cool," is 3D effects in graphs. However, like the example above, 3D effects create cognitive load for the viewer, and create distortion in perception. Let us look at a simple example. Table 5.4 presents some YTD sales data for four fictitious salespersons.

In this data set, Deepika is well above average, and Danny is well below average. This same info is presented in 2D and 3D, respectively, in Fig. 5.9a, b. However, when viewing the 2D representation, the pie chart is less clear than the table (from the chart, can you identify the #2 salesperson?). Moreover, the 3D chart greatly distorts the % share of Deepika due to perspective.

In fact, with 3D charts, the placement of the data points has the ability to change the perception of the values in the data. Figure 5.10a, b illustrate this point. Note that when Deepika is rotated to the back of the graph, the *perception* of her % share of sales is reduced.

Pie Charts?

Although the previous example uses pie charts, this was simply to illustrate the impact of 3D effect on placement. Even though its use is nearly universal, the pie chart is not usually the best choice to represent the part-to-whole relationship. This is due to the requirement it places on the viewer to compare differences on area instead of on a single visual dimension, and the difficulty that this causes in making comparisons.

Going back to the 2D example in Fig. 5.9a, it is very difficult to compare the differences between George and Jennifer. The typical response to this in practice is

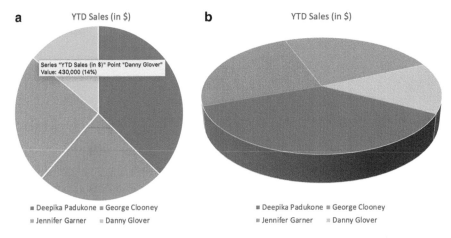

Fig. 5.9 (**a**) 2D pie chart of sales. (**b**) 3D pie chart of sales

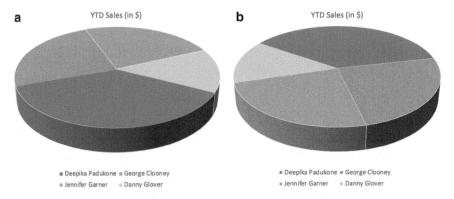

Fig. 5.10 (**a**) Chart with Deepika at front. (**b**) Chart with Jennifer at front

to add the % values to the chart, as Fig. 5.11 illustrates. However, at this point, what cognitive value does the pie chart add that the viewer would not have gained from Table 5.4?

Building a Fit-for-Purpose Visualization

When considering the visualization that is to be created, the visualizer *must* focus on the purpose of the viewer. Creating a stacked column, sorted bar chart, table, or even a pie chart, could all be "not wrong" decisions. Remember, if the user is interested in looking up precise values, the table might be the best choice. If the user is interested in understanding the parts-to-whole relationship, a stacked column or pie chart may be the best choice. Finally, if the viewer needs to understand rank order of values, the sorted bar chart (Fig. 5.12) may be the best option.

5 Data Visualization

Fig. 5.11 Annotated pie chart

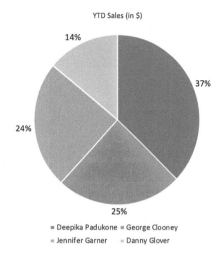

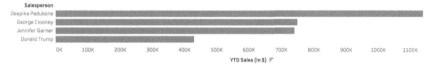

Fig. 5.12 Bar chart of sales

Fig. 5.13 Heat map of sales

Comparing Levels of a Dimension over Multiple Categories

While a stacked bar graph is excellent at showing levels of one variable over one set of categories, often we wish to compare levels of a variable over multiple categories. While we might choose a number of visualizations for this (including possibly a table, if precision is required), one visualization that is optimized for looking at the levels for a single variable at the intersection of two categorical dimensions is the heat map.

Heat maps use a color gradient (see next section for a discussion on the proper use of gradients), within a grid of cells that represent the possible intersections of two categories—perhaps sales by region by quarter (Fig. 5.13).

The heat map illustrates that the fourth quarter tends to be the strongest in all regions, and when compared to the other regions, the East region seems to perform

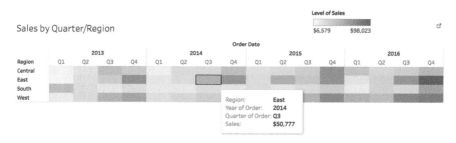

Fig. 5.14 Heat map of sales with interactive tool tip

Fig. 5.15 Perception of color on color

consistently highest in the fourth quarter. This type of data presentation is good for illustrating comparative performance, but only in a general way, as an overview. To add more value to the heat map, one might decide to add the actual number to the cells or, as a better choice, add a tool tip in an interactive visualization (Fig. 5.14).

In almost every case, there are multiple choices for visualizations that are "not wrong." The visualization that is "fit for purpose" is the one that properly illustrates the data story, in the fashion that is most compatible with the cognitive task that the viewer will use the visualization to complete.

Use Color Properly

Meta-Rule #4: Never use color on top of color—color is not absolute.

One of the most important concepts in visualization is the appropriate use of color. As with shape, color should be used to provide the viewer meaning. This means that color should be used consistently, and within several rules for human perception. In order to understand these rules, we must spend a few paragraphs on how humans perceive color.

First, color is not perceived absolutely by the human eye. Look at the example in Fig. 5.15. Which of the five small rectangles does your visual perception tell you is the lightest in color?

Most people (if they are honest) will quickly answer that the rightmost rectangle in Fig. 5.15 is the lightest in color. However, as Fig. 5.16 illustrates, all five small rectangles are actually the same color.

The optical illusion presented in Fig. 5.15 is caused by the human visual perception characteristic that colors, and differences in colors, are evaluated relatively, rather than absolutely. What this means in practice is that colors are perceived differently depending upon what colors are around (or "behind") them. The gradient

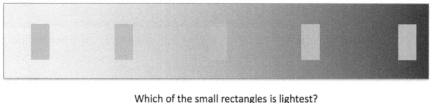

Which of the small rectangles is lightest?

Fig. 5.16 True colors of all bars

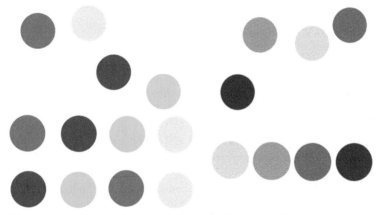

When rainbow colors are used, establishing order is inconsistent, even when the same person tries to order the colors in multiple tests. However, a gradient (of any color) is much more easily and consistently ordered.

Fig. 5.17 Ordering colored chips

behind the small rectangles in Fig. 5.15 causes changes in perception as to the color of the rectangle, which, in the context of data visualization, changes the *meaning* of the color of the rectangles.

Second, color is not perceived as having an order. Although the spectrum has a true order (e.g., Red, Orange, Yellow, Green, Blue, Indigo, Violet: "ROY G. BIV"), when viewing rainbow colors, violet is not perceived as "more" or "less" than red. Rainbow colors are simply perceived as different from one another, without having a particular "order." However, variation in the level of *intensity* of a single color is perceived as having an order. This is illustrated in Fig. 5.17, and in the following quote:

> *"If people are given a series of gray paint chips and asked to put them in order, they will consistently place them in either a dark-to-light or light-to-dark order. However, if people are given paint chips colored red, green, yellow, and blue and asked to put them in order, the results vary,"* according to researchers David Borland *and* Russell M. Taylor II, *professor of computer science at the University of North Carolina at Chapel Hill.*

Using Color with Meaning

Based upon the previous discussion, we now turn to the use of color and its meaning in data visualization. In general, use the following heuristic when deciding on how to use color:

When representing levels of a single variable, use a single-color gradient,[4] when representing categories of a variable, use rainbow colors.

Levels of a single variable (continuous data) are best represented using a gradient of a single color. This representation of a single color, with different levels of saturation, visually cues the user that, while the levels of the variable may be different, the color represents levels of the *same concept*. When dealing with categories (or categorical data), the use of different colors cues the users that the different categories represent *different concepts*.

Best Practices for Color

When building a visualization, it is easy to create something that is information-rich, but that does not always allow users to quickly zero in on what is the most important information in the graph. One way to do this is through the choice of colors. In general, colors that are lower in saturation, and are further from the primary colors on the color wheel are considered more "natural" colors, because they are those most commonly found in nature. These are also more soothing colors than the brighter, more primary colors.

For this reason, when designing a visualization, use "natural" colors as the standard color palette, and brighter, more primary colors for emphasis (Fig. 5.18).

By using more natural colors in general, viewers will more calmly be able to interpret the visualization. Using more primary colors for emphasis allows the visualizer to control when and where to drive the attention of the viewer. When this is done well, it allows the viewer to find important data more immediately, limiting the need for the viewer to search the entire visualization to interpret where the important information is. This helps to achieve the visualizer's goal of reducing the cognitive load on the viewer.

Use Viewers' Experience to Your Advantage

Meta-Rule #5: *Do not violate the primal perceptions of your viewers. Remember, up means more.*

In the example provided in Fig. 5.2, we reviewed the disorientation that can occur when the viewers' primal perceptive instincts are violated. When viewing the graph with the reversed Y-axis, the initial perception is that when the line moves down, it should have the meaning of "less." This is violated in Fig. 5.2. However, why

---

[4] Based upon the discussion of the concept of non-order in the perception of rainbow colors, the use of a two-color gradient will not have meaning outside of a particular context. For instance, a red–green gradient may be interpreted as having meaning in the case of profit numbers that are both positive and negative, but that same scale would not have an intuitive meaning in another context. As such, it is better to avoid multicolor gradients unless the context has a meaning already established for the colors.

5 Data Visualization

Fig. 5.18 Differences between standard colors and emphasis colors

is it that humans perceive "up," even in a 2D line graph, as meaning "more"? The answer lies in the experiences of viewers, and the way that the brain uses experience to drive perception.

For example, most children, when they are very young, play with blocks, or stones, or some other group of objects. They sort them, stack them, line them up, etc., and as they do so, they begin the process of wiring their brains' perceptive processes. This leads to the beginning of the brain's ability to develop categories—round is different from square, rough is different from smooth, large is different than small, etc. At the same time, the brain learns that "more" takes up more space, and "more," when stacked, grows higher. It is from these and other childhood experiences that the brain is taught "how the world works," and these primal perceptions drive understanding for the remainder of our lives.

When a visualizer violates these perceptions, it can cause cognitive confusion in the viewer (e.g., Fig. 5.2). It can also create a negative emotional reaction in the viewer, because the visualization conflicts with "how the world works." The visualizer who created Fig. 5.2 was more interested in creating a "dripping blood" effect than in communicating clearly to her viewer, and the Internet firestorm that erupted from the publication of that graph is evident in the emotional reaction of many of the viewers.

Another common viewer reaction is toward visualizations that present percentage representations. Viewers understand the concept of percent when approaching a graph, and they know that the maximum percent level is 100%. However, Fig. 5.19 illustrates that some graphs may be produced to add up to more than 100%.

This is because the graph is usually representing multiple part-to-whole relationships at the same time, but not giving the viewer this insight. The solution to this perception problem is to always represent the whole when presenting a percentage, so that viewers can understand to which whole each of the parts is being compared (Fig. 5.20).

Fig. 5.19 Annotations sum to 243%

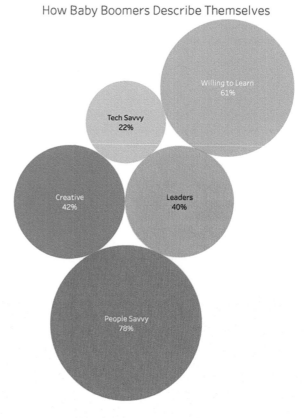

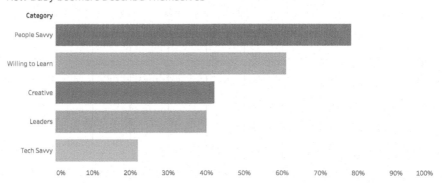

Fig. 5.20 Representing multiple parts to whole

There are obviously many more of these primal perceptions that viewers hold, and modern software packages make it rather difficult to accidentally violate them. In almost every case, these kinds of violations occur when the visualizer attempts to

5 Data Visualization

add some additional artistic dimension to a graph, instead of focusing on presenting the data with the highest possible fidelity.

Represent the Data Story with Integrity

Meta-Rule #6: Chart with graphical and ethical integrity. Do not lie, either by mistake or intentionally.

Finally, it is important that, at all times, visualizers work to accurately represent the story that is in their data. This means that the effect in the data must be accurately reflected in the visualization. Edward Tufte, in his classic book *The visual display of quantitative information*, provides a number of rules for charting graphical integrity. While we cannot cover them in detail here, we provide a summary.

First, Tufte introduces the "Lie Factor," which is the ratio between the effect in the visualization and the effect in the data. So if the effect in the data is 1, but in the visualization it is 2, the lie factor would be 2/1 or 2. In order for the visualization to accurately represent the data, the Lie Factor should be as close to 1 as possible. Figure 5.21 illustrates the Lie Factor.

Tufte's other rules for graphical integrity, when broken, create Lie Factor ratios that are higher or lower than 1. These other rules for graphical integrity are summarized below:

Use Consistent Scales. "A scale moving in regular intervals, for example, is expected to continue its march to the very end in a consistent fashion, without the muddling or trickery of non-uniform changes" (Tufte, p. 50). What this means is that when building axes in visualizations, the meaning of a distance should not

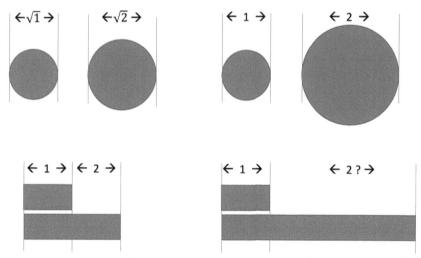

Fig. 5.21 Lie factors in proportional differences

change, so if 15 pixels represents a year at one point in the axis, 15 pixels should not represent 3 years at another point in the axis.

Standardize (Monetary) Units. "In time-series displays of money, deflated and standardized units [...] are almost always better than nominal units." This means that when comparing numbers, they should be standardized. For currency, this means using a constant unit (e.g., 2014 Euros, or 1950 USD). For other units, standardization requires consideration of the comparison being communicated. For instance, in a comparison between deaths by shooting in two states, absolute numbers may be distorted due to differences in population. Or in a comparison of military spending, it may be appropriate to use standardization by GDP, land mass, or population. In any case, standardization of data is an important concept in making comparisons, and should be carefully considered in order to properly communicate the relationship in question.

Present Data in Context. "Graphics must not quote data out of context" (Tufte, p. 60). When telling any data story, no data has meaning until it is compared with other data. This other data can be a standard value, such as the "normal" human body temperature, or some other comparison of interest such as year-over-year sales. For instance, while today's stock price for Ford might be $30, does that single data point provide you with any understanding as to whether that price is high, low, good, or bad? Only when data is provided within an appropriate context can it have meaning.

Further, as Tufte illustrates, visualizers choose what they wish to show, *and* what they choose to omit. By intentionally presenting data out of context, it is possible to change the story of that data completely. Figure 5.22 illustrates the data from Fig. 5.2 presented out of context, showing the year after the enactment of the "Stand Your Ground" law.

From the graph in Fig. 5.22 it is not possible to understand if the trend in gun deaths in Florida is any different than it was before the law was enacted. However, by presenting this data in this manner, it would be possible for the visualizer to drive public opinion that the law greatly increased gun deaths. Figure 5.23 adds context to this data set by illustrating that the trajectory of gun deaths was different after the law was enacted—it is clearly a more honest graph.

Show the Data. "Above all else show the data" (Tufte, p. 92). Tufte argues that visualizers often fill significant portions of a graph with "non-data" ink. He argues that as much as possible, show data to the viewer, in the form of the actual data, annotations that call attention to particular "causality" in the data, and drive viewers to generate understanding.

When visualizers graph with low integrity, it reduces the fidelity of the representation of the story that is in the data. I often have a student ask, "But what if we WANT to distort the data?". If this is the case in your mind, check your motives. If you are intentionally choosing to mislead your viewer, to lie about what the data say, stop. You should learn the rules of visualization so that (1) you don't break them and *unintentionally lie*, and (2) you can more quickly perceive when a visualization is lying to you.

Fig. 5.22 Lack of illustration of context

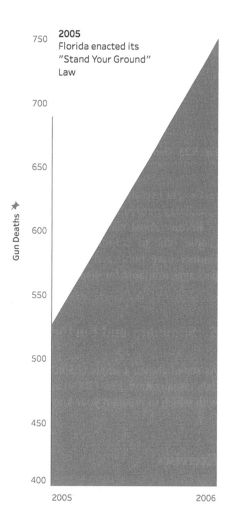

4 Software and Data Visualization

Based upon the widespread recognition of the power of the visual representation of data, and the emergence of sufficiently inexpensive computing power, many modern software packages have emerged that are designed specifically for data visualization. Even the more generalized software packages are now adding and upgrading their visualization features more frequently. This chapter was not intended to "endorse" a particular software package. It stands to illustrate some of the rules

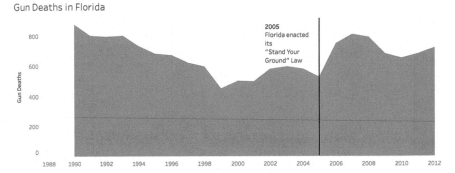

Fig. 5.23 Gun deaths in Florida, presented intuitively

that might explain why some software packages behave in a certain manner when visualizing particular relationships.

Because most visualizers do not have the freedom to select any software they choose (due to corporate standards), and because research companies such as Gartner have published extensive comparative analyses of the various software packages available for visualization, we do not recreate that here.

5 Summary and Further Reading

As stated above, a single chapter is far too little space to describe the intricacies of data visualization. Few (2009) and Tufte and Graves-Morris (1983) are good sources with which to broaden your knowledge of the "whys" of data visualization.

Exercises

Ex. 5.1 Answer the following conceptual questions:

(a) What is the key issue with using 3D graphs?
(b) When displaying differences in a single data dimension, explain why using differences in object area is suboptimal.
(c) How may data dimensions can you represent on a scatterplot graph?
(d) Which kind of color should you use when representing different levels of a single variable?
(e) What are some problems with gradients?
(f) Find an example of a quantitative graph in the media. Evaluate whether or not the graph is properly conforming to Tufte's principles described above.

Ex. 5.2 Scenarios

(a) You wish to provide a visualization that illustrates the rank and relative differences in various salespersons' results. Which graph would be the most "not wrong"?
(b) You wish to denote categories of customers using color. Should you use rainbow colors or gradient colors?
(c) You wish to illustrate the percentage of donations coming from a particular percentage of donors. What kind of relationship(s) are you attempting to illustrate? Which visualization would be the most "not wrong"?
(d) You wish to illustrate the amount that each of your product lines contributes to the percentage of total sales over a 5-year period. What would be your choice for the most "not wrong" graph?

Ex. 5.3 Two views of the same data

Go to the New York Times website, and view this interactive visualization.

http://www.nytimes.com/interactive/2012/10/05/business/economy/one-report-diverging-perspectives.html?_r=1&. Accessed on Feb 23, 2019.

While this visualization provides two views of the same data, critique its success at providing alternative viewpoints. Notice how the different visualizations utilize/break Tufte's rules in order to shift perception.

References

Anscombe, F. J. (1973). Graphs in statistical analysis. *American Statistician, 27*(1), 17–21.

Cowan, N. (2010). The magical mystery four: How is working memory capacity limited, and why? *Current Directions in Psychological Science, 19*(1), 51–57.

Few, S. (2009). *Now you see it: Simple visualization techniques for quantitative analysis*. Oakland, CA: Analytics Press.

Playfair, W. (1801). "The statistical breviary; shewing, on a principle entirely new, the resources for every state and kingdom in Europe; illustrated with stained copper-plate charts, representing the physical powers of each distinct nation with ease and perspicuity. By William Playair".

Tufte, E. R., & Graves-Morris, P. R. (1983). *The visual display of quantitative information* (Vol. 2, No. 9). Cheshire, CT: Graphics Press.

Chapter 6
Statistical Methods: Basic Inferences

Vishnuprasad Nagadevara

1 Introduction

The purpose of statistical analysis is to glean meaning from sets of data in order to describe data patterns and relationships between variables. In order to do this, we must be able to summarize and lay out datasets in a manner that allows us to use more advanced methods of examination. This chapter introduces fundamental methods of statistics, such as the central limit theorem, confidence intervals, hypothesis testing and analysis of variance (ANOVA).

2 Motivating Examples

A real estate agent would like to estimate the average price per square foot of residential property in the city of Hyderabad, India. Since it is not possible to collect data for an entire city's population in order to calculate the average value (μ), the agent will have to obtain data for a random sample of residential properties. Based on the average of this sample, they will have to infer, or draw conclusions, about the possible value of the population average.

Electronic supplementary material The online version of this chapter (https://doi.org/10.1007/978-3-319-68837-4_6) contains supplementary material, which is available to authorized users.

V. Nagadevara (✉)
IIM-Bangalore, Bengaluru, Karnataka, India
e-mail: nagadevara_v@isb.edu

© Springer Nature Switzerland AG 2019
B. Pochiraju, S. Seshadri (eds.), *Essentials of Business Analytics*, International Series in Operations Research & Management Science 264,
https://doi.org/10.1007/978-3-319-68837-4_6

A battery manufacturer desires to estimate the average life of batteries produced by his company. In order to collect data about the lifetime of each battery, he will have to use each one until it has run its course (employing, of course, a consistent and scientific method of battery use). Thus, the data collection process is destructive in nature. Consequently, it is not possible to collect the lifetime data for the entire population of batteries produced by the company. The manufacturer will have to resort to data collected from a random sample of batteries and use the sample average to draw conclusions about the population average.

The Commissioner of a city municipal corporation needs to estimate the strength of a recently constructed flyover before making payments to the contractor. In order to estimate the strength of the entire flyover, the engineers will have to take measurements of its strength at different locations. Based on the data, they can then make inferences about the strength of the entire bridge.

A retailer wants to estimate the proportion of its customer base who make electronic payments so that he can decide whether or not to install a credit card POS (point-of-sale) machine. To do so, he can collect the data of customers who would like to make electronic payments, and based on this sample data, he can estimate the population proportion.

All of the above examples require the collection of sample data and calculation of the sample mean ($\overline{X}$) or sample proportion (p). Inferences with respect to the population mean (μ) and population proportion (π) are drawn based on the values of the sample estimates $\overline{X}$ and p.

3 Methods of Basic Inference

3.1 Sample and Population

A population can be considered as a collection of entire set of measurements of any characteristic that we are interested in. A sample can be defined as a smaller set of measurements collected from the population. The science of statistics involves drawing conclusions or inferring about the population based on the measurements obtained from the sample. We need to employ random sampling methods in order to draw reliable inferences. A random sample is defined as a set of n elements selected from a population in such a way that the probability of selecting a set of n elements is same as that of any other set of n elements.

The elements of the population have a probability distribution and have a mean and standard deviation associated with the distribution. These values are called parameters of the population. In other words, numerical measures associated with the population mean, standard deviation, proportion etc. are called parameters. On the other hand, the corresponding numerical measures associated with the sample are called the statistics. Inferences about the population parameters are drawn based on the sample statistics. An example of some of the parameters and corresponding statistics is shown in Table 6.1.

The value of the sample statistic $\overline{X}$ can be used to draw inferences about the population mean μ. In other words, the value of $\overline{X}$ can be used to estimate the possible value of μ. Thus, $\overline{X}$ is considered as the estimate of μ. Since $\overline{X}$ provides a single value estimate corresponding to μ, it is referred to as a "point estimate." We will discuss about "interval estimates," which deal with a range estimation, later in this chapter. The sample statistics such as $\overline{X}$, S, and p are usually referred to as estimators, while the numerical values associated with them are referred to as estimates.

Let us consider a very small population of five companies. The names of the companies and their profit before tax (PBT) in millions are given in Table 6.2.

The population mean (μ) is 2.60 million, and the population standard deviation (σ) is 1.0198 million. Let us take all possible samples of size 3 from this population. These samples are ABC, ABD, ABE, etc. There are only ten such samples possible. Several samples have the same mean. For example, {A,C,E} and {C,D,E} have sample mean equal to 2, and thus the frequency of the sample mean equal to 2 is 2. The sample means ($\overline{X}$s) are calculated and summarized in Table 6.3.

It can be seen from Table 6.3 that the sample mean $\overline{X}$ is a discrete random variable in the sense that it can take any of the five possible values and that there is a probability associated with each of these values. Let $\overline{X}_i$ be the ith value taken

Table 6.1 Statistical measures and their notations

Numerical measure	Population (notation)	Sample (notation)
Mean	μ	$\overline{X}$
Variance	σ^2	S^2
Standard deviation	σ	S
Proportion	π	p

Table 6.2 Breweries and their profit before tax

Company name	Code	PBT (million)
Champion India Breweries	A	3
Charmy Breweries	B	4
Indian Breweries	C	2
Weka Breweries	D	3
Stoll Breweries	E	1

Table 6.3 Sample means, frequencies, and relative frequencies

Sample mean	Frequency	Relative frequency
2.00	2	0.2
2.33	2	0.2
2.67	3	0.3
3.00	2	0.2
3.33	1	0.1

by the sample mean and $P\left(\overline{X}_i\right)$ be the probability of observing the ith value. For example, $\overline{X}_1 = 2.00$, and $P\left(\overline{X}_1\right) = 0.2$. It can be seen that the expected value of the variable $\overline{X}$ is exactly equal to the population mean μ.

$$\sum_{i=1}^{5} \overline{X}_i . P\left(\overline{X}_i\right) = 2.60 = \mu.$$

The standard deviation of $\overline{X}$ (also referred to as $\sigma_{\overline{X}}$) is

$$\sigma_{\overline{X}} = \sum_{i=1}^{5} \left(\overline{X}_i - \mu\right)^2 . P\left(\overline{X}_i\right) = 0.4163.$$

There is a relationship between σ and $\sigma_{\overline{X}}$. The relationship is given by

$$\sigma_{\overline{X}} = \frac{\sigma}{\sqrt{n}} \sqrt{\left(\frac{N-n}{N-1}\right)},$$

where σ is the population standard deviation, N is the size of the population (5 in this particular case), and n is the sample size (3 in this particular case). This implies that the sample mean $\overline{X}$ has a probability distribution with mean μ and standard deviation $\sigma_{\overline{X}}$ as given above. The expression $\sqrt{\left(\frac{N-n}{N-1}\right)}$ used in the calculation of $\sigma_{\overline{X}}$ is called the finite population multiplier (FPM). It can be seen that the FPM approaches a value of 1 as the population size (N) becomes very large and sample size (n) becomes small. Hence, this is applicable to small or finite populations. When population sizes become large or infinite, the standard deviation of $\overline{X}$ reduces to $\frac{\sigma}{\sqrt{n}}$. The standard deviation of $\overline{X}$ is also called standard error. This is a measure of the variability of $\overline{X}'s$.

Just as $\overline{X}$ is a random variable, each of the sample statistics (*sample* standard deviation S, computed as the standard deviation of the values in a sample, *sample* proportion p, etc.) is also a random variable, having its own probability distributions. These distributions, in general, are called sampling distributions.

3.2 Central Limit Theorem

The fact that the expected value of $\overline{X}$ is equal to μ implies that on average, the sample mean is actually equal to the population mean. Thus, it makes $\overline{X}$ a very good estimator of μ. The standard error ($\sigma_{\overline{X}}$) which is equal to $\frac{\sigma}{\sqrt{n}}$ indicates that as the sample size (n) increases, the standard error decreases, making the estimator, $\overline{X}$, approach closer and closer to μ.

In addition to the above properties, the distribution of $\overline{X}$ tends to the normal distribution as the sample size increases. This is known as the central limit theorem (CLT). The central limit theorem states that when the population has a mean μ and standard deviation σ, and when a sample of sufficiently large size (n) is drawn from it, the sample mean $\overline{X}$ tends to a normal distribution with mean μ and standard deviation $\sigma_{\overline{X}}$.

The central limit theorem is very important in statistical inference because it states that the sample mean $\overline{X}$ (approximately) follows a normal distribution irrespective of the distribution of the population. It allows us to use the normal distribution to draw inferences about the population mean. We will be making extensive use of the CLT in the following sections.

3.3 Confidence Interval

We can see from the earlier section that the sample mean, $\overline{X}$, is a point estimate of the population mean. A point estimate is a specific numerical value estimate of a parameter. The best point estimate of the population mean μ is the sample mean $\overline{X}$, and since a point estimator cannot provide the exact value of the parameter, we compute an "interval estimate" for the parameter which covers the true population mean sufficiently large percentage of times. We have also learned in the previous section that $\overline{X}$ is a random variable and that it follows a normal distribution with mean μ and standard deviation $\sigma_{\overline{X}}$ or an estimate of it. The standard deviation is also referred to as "standard error" and can be shown to be equal to $\frac{\sigma}{\sqrt{n}}$ (or its estimate) where n is the sample size. We can use this property to calculate a confidence interval for the population mean μ as described next.

3.4 Confidence Interval for the Population Mean When the Population Standard Deviation Is Known

In order to develop a confidence interval for the population mean, μ, it is necessary that the population standard deviation, σ, is either known or estimated from the sample data. First, let us consider a scenario where σ is known. Figure 6.1 shows the sampling distribution of $\overline{X}$. The distribution is centered at μ, and the standard deviation (standard error) is given by $\sigma_{\overline{X}}$.

Let us identify two limits, L_L and U_L, each at a distance of $\pm 1.96\ \sigma_{\overline{X}}$ from μ. Note that in a normal distribution, a distance of 1.96 standard deviations from the mean will cover an area of 0.475. Given this, the area between these two limits is equal to 95%. What this implies is that if we take a large number of samples, each of size n, and calculate the sample average for each sample, 95% of those sample means will lie between the two limits, L_L and U_L. In other words,

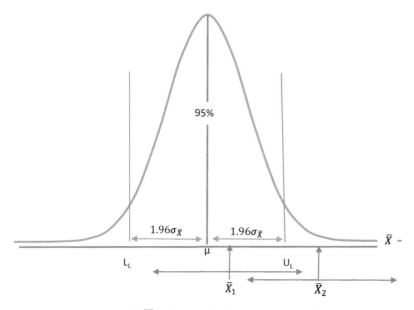

Fig. 6.1 Sampling distribution of $\overline{X}$

$$P\left(\mu - 1.96\sigma_{\overline{X}} \leq \overline{X} \leq \mu + 1.96\sigma_{\overline{X}}\right) = 0.95 \tag{6.1}$$

By rearranging the terms μ and $\overline{X}$ in the above expression, we get

$$P\left(\overline{X} - 1.96\sigma_{\overline{X}} \leq \mu \leq \overline{X} + 1.96\sigma_{\overline{X}}\right) = 0.95 \tag{6.2}$$

The way to interpret Expression (6.2) is that if we take a large number of samples of size n and calculate the sample means and then create an interval around each $\overline{X}$ with a distance of $\pm 1.96\ \sigma_{\overline{X}}$, then 95% of these intervals will be such that they contain the population mean μ. Let us consider one such sample mean, $\overline{X}_1$, in Fig. 6.1. We have built an interval around $\overline{X}_1$ whose width is the same as the width between L_L and U_L. This interval is such that it contains μ. It can be seen that an interval built around any $\overline{X}$ lying between L_L and U_L will always contain μ. It is obvious from Expression (6.1) that 95% of all $\overline{X}s$ will fall within this range. On the other hand, an interval built around any $\overline{X}$ which is outside this range (such as $\overline{X}_2$ in Fig. 6.1) will be such that it does not contain μ. In Fig. 6.1, 5% of all $\overline{X}s$ fall in this category.

It should be noted that because μ is a parameter and not a random variable, we cannot associate any probabilities with it. The random variable in Expression (6.2) above is $\overline{X}$. Consequently, the two values, $\overline{X} - 1.96\sigma_{\overline{X}}$ and $\overline{X} + 1.96\sigma_{\overline{X}}$, are also random variables. The appropriate way of interpreting the probability statement (Expression 6.2) is that there is a large number of $\overline{X}s$ possible and there are just as many intervals, each having a width of $2\times 1.96\ \sigma_{\overline{X}}$. Moreover, there is a 95%

probability that these intervals contain μ. The above statement can be rephrased like this: "We are 95% confident that the interval built around an $\overline{X}$ contains μ." In other words, we are converting the probability into a confidence associated with the interval. This interval is referred to as the "confidence interval," and its associated probability as the "confidence level." We can decide on the desired confidence level and calculate the corresponding confidence interval by using the appropriate z-value drawn from the standard normal distribution. For example, if we are interested in a 90% confidence interval, the corresponding z-value is 1.645, and this value is to be substituted in the place of 1.96 in Expression (6.2).

Let us now apply this concept to a business example. Swetha is running a number of small-sized restaurants in Hyderabad and wants to estimate the mean value of APC (average per customer check) in such restaurants. She knows that the population standard deviation σ = ₹ 520 (₹ stands for Indian Rupee). She has collected data from 64 customers, and the sample average $\overline{X}$ = ₹ 1586. The standard deviation of $\overline{X}$ is $\sigma_{\overline{X}} = \frac{520}{\sqrt{64}} = 65$. By substituting these values in Expression (6.2), we get

$$P(1586 - 1.96 \times 65 \leq \mu \leq 1586 + 1.96 \times 65) = 0.95$$

$$P(1458.6 \leq \mu \leq 1713.4) = 0.95$$

The above probability statement indicates that we are 95% confident that the population mean μ (which we do not know) will lie between ₹ 1458.6 and ₹ 1713.4. By using the sample mean (₹ 1586), we are able to calculate the possible interval that contains μ and associate a probability (or confidence level) with it. Table 6.4 presents a few more confidence intervals and their corresponding confidence levels as applied to this example.

Notice that the width of the confidence interval increases as we increase the confidence level. It may be noted that a larger confidence level means a higher z-value, and an automatic increase in width. The width of the confidence interval can be considered as a measure of the precision of the statement regarding the true value of the population mean. That is, the larger the width of the interval, the lower the precision associated with it. In other words, there is a trade-off between the confidence level and the precision of the interval estimate.

It will be good if we can achieve higher confidence levels with increased interval precision. The width of the confidence interval depends not only on the z-value

Table 6.4 Confidence intervals with different confidence levels

Confidence level	Lower limit	Upper limit	Width of the interval
90%	₹1479.075	₹1692.925	₹213.85
95%	₹1458.60	₹1713.40	₹254.80
99%	₹1418.625	₹1753.375	₹334.75

Table 6.5 Widths of confidence intervals with different sample sizes

Confidence level	Width of the interval	
	Sample size = 64	Sample size = 256
90%	₹213.85	₹106.925
95%	₹254.80	₹127.40
99%	₹334.75	₹167.375

as determined by the confidence level but also on the standard error. The standard error depends on the sample size. As the sample size, n, increases, the standard error decreases since $\sigma_{\overline{X}} = \frac{\sigma}{\sqrt{n}}$. By increasing the sample size (n), we can decrease the standard error ($\sigma_{\overline{X}}$) with a consequent reduction in the width of the confidence interval, effectively increasing the precision. Table 6.5 presents the width of the confidence interval for two sample sizes, 64 (earlier sample size) and 256 (new sample size), as applied to the earlier example. These widths are calculated for three confidence levels as shown in Table 6.5.

It should be noted that the precision of the 99% confidence interval with $n = 256$ is more profound than that of the 90% confidence interval with $n = 64$.

We can consider the confidence interval as the point estimate, $\overline{X}$, adjusted for the margin of error measured by the standard error, $\sigma_{\overline{X}}$. Obviously, the margin of error is augmented by the confidence level as measured by the z-value.

3.5 Sample Size Determination

We can use the concept of confidence intervals to determine the ideal sample size. Is the sample size of 64 ideal in the above example? Suppose Swetha wants to calculate the ideal sample size, given that she desires a 95% confidence interval with a width of $\pm$₹ 100. We already know that the margin of error is determined by the z-value and the standard error. Given that the confidence level is 95%, the margin of error or the width of the confidence interval is

$\pm 1.96 \sigma_{\overline{X}} = \pm 1.96 \left(\frac{\sigma}{\sqrt{n}}\right) = 100$. Thus, $n = (1.96 \times 520/100)^2 = 103.88 \approx 104$.

The formula for calculation of sample size is

$$n = \left(\frac{z * \sigma}{w}\right)^2 \quad (6.3)$$

where z = the standard normal variate corresponding to the confidence level, σ = the standard deviation of the population, and w = the width of the confidence interval (on one side of the mean).

It can be seen from Expression (6.3) that the sample size increases with a corresponding increase in the desired confidence level and the desired precision (the smaller the width, the higher the precision). Also, the larger the standard deviation, the larger the size of the required sample.

3.6 Confidence Intervals for μ When σ Is Unknown: The T Distribution

In many situations, the population standard deviation, σ, is not known. In such a situation, we can use s, the point estimate of σ, to calculate the confidence interval. The sample standard deviation s, the best estimate of σ, is calculated using the following formula:

$$s = \sqrt{\frac{\sum_{i=1}^{n}(X_i - \overline{X})^2}{n-1}} \qquad (6.4)$$

where:

- s is the sample standard deviation.
- X_i is the ith sample observation.
- $\overline{X}$ is the sample mean.
- n is the sample size.

The denominator, $(n - 1)$, is referred to as the degree of freedom (df). Abstractly, it represents the effect that estimates have on the overall outcome. For example, say we need to seat 100 people in a theater that has 100 seats. After the 99th person is seated, you will have only one seat and one person left, leaving you no choice but to pair the two. In this way, you had 99 instances where you could make a choice between at least two options. So, for a sample size of $n = 100$, you had $n - 1 = 99$ degrees of freedom. Similarly, since we are using $\overline{X}$ as a proxy for the population mean μ, we lose one degree of freedom. We cannot randomly choose all n samples since their mean is bounded by $\overline{X}$. As we begin using more proxy values, we will see how this affects a parameter's variability, hence losing more degrees of freedom.

Since we are using the sample standard deviation s, instead of σ in our calculations, there is more uncertainty associated with the confidence interval. In order to account for this additional uncertainty, we use a t-distribution instead of a standard normal distribution ("z"). In addition, we will need to consider the standard error of the sample mean $s_{\overline{X}} = \frac{s}{\sqrt{n}}$ when calculating the confidence intervals using t-distribution.

Figure 6.2 presents the t-distribution with different degrees of freedom. As the degrees of freedom increase, the t-distribution approaches a standard normal distribution and will actually become a standard normal distribution when the degrees of freedom reach infinity.

It can be seen from Fig. 6.2 that the t-distribution is symmetric and centered at 0. As the degrees of freedom increase, the distribution becomes taller and narrower, finally becoming a standard normal distribution itself, as the degrees of freedom become infinity.

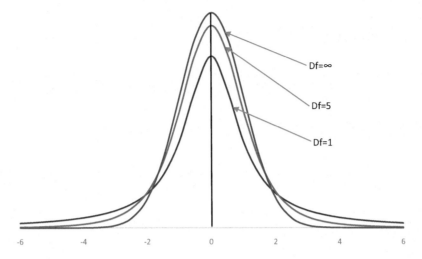

Fig. 6.2 t-Distribution with different degrees of freedom. When df = ∞, we achieve a standard normal distribution

$$\text{The t-value is defined as } t = \frac{\overline{X} - \mu}{s_{\overline{X}}} \text{ where } s_{\overline{X}} = \frac{s}{\sqrt{n}} \quad (6.5)$$

The calculation of the confidence interval for a t-distribution is very similar to that of a standard normal distribution except that we use a t-value instead of a z-value and $s_{\overline{X}}$ instead of $\sigma_{\overline{X}}$.

The formula for a 90% confidence interval when σ is not known is

$$P\left(\overline{X} - t_{0.9} * s_{\overline{X}} \leq \mu \leq \overline{X} + t_{0.9} * s_{\overline{X}}\right) = 0.90 \quad (6.6)$$

Let us now apply this to an example. Swetha is interested in estimating the average time spent by the customers in eating lunch in the restaurant. She collected data from 25 customers (*n*) and calculated the sample mean and sample standard deviation using this data. The sample mean ($\overline{X}$) was 28 min, and the sample standard deviation (*s*) was 12 min. The standard error of the sample mean is $s_{\overline{X}} = \frac{12}{\sqrt{25}} = 2.4$ min.

Figure 6.3 presents the t-distribution with 24 degrees of freedom (24 = *n* − *1* since we are using a proxy for the population mean μ). If we are interested in a 90% confidence interval, the corresponding t-value can be obtained from Table 6.12. Note that the combined area under the two tails is 0.1 (since the confidence level is 90% or 0.9), making the area under the right tail 0.05. In Table 6.12, we need to look under the column with right tail area = 0.05 and row Df = 24. This gives us a corresponding t-value of 1.711. The 90% confidence interval can now be calculated as

6 Statistical Methods: Basic Inferences

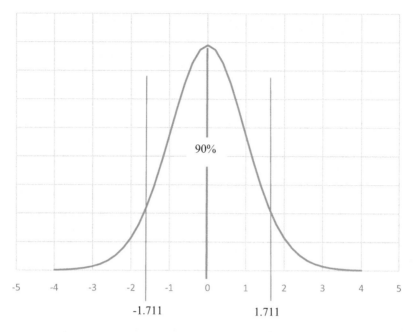

Fig. 6.3 t-Distribution with 24 degrees of freedom

$$P(28 - 1.711 \times 2.4 \leq \mu \leq 28 + 1.711 \times 2.4) = P(23.8936 \leq \mu \leq 32.1064) = 90\%.$$

It is apparent from this calculation that the interval of a t-distribution is somewhat less precise than that of a standard normal distribution since the width of the confidence interval obtained using t-distribution is larger than that obtained using z-distribution.

3.7 Large-Sample Confidence Intervals for the Population Proportion

Estimating the confidence interval for the population proportion π is very similar to that of the population mean, μ. The point estimate for π is the sample proportion p. The sampling distribution of p can be derived from binomial distribution. It is known that the binomial random variable is defined as the number of successes, given the sample size n and probability of success, π. The expected value of a binomial random variable is $n\pi$, and the variance is $n\pi(1-\pi)$. Consider a variable D_i which denotes the outcome of a Bernoulli process. Let us map each success of the Bernoulli process as $D_i = 1$ and failure as $D_i = 0$. Then the number of successes

X is nothing but ΣD_i and the sample proportion, $p = \frac{X}{n} = \frac{\Sigma D_i}{n}$. In other words, p is similar to the sample mean $\overline{D}$, and by virtue of the central limit theorem, p (as analogous to the sample mean) is distributed normally (when n is sufficiently large). The expected value and the variance of the estimator p are given below:

$$E(p) = E\left(\frac{X}{n}\right) = \frac{1}{n}E(X) = \frac{1}{n}n\pi = \pi \tag{6.7}$$

and

$$V(p) = V\left(\frac{X}{n}\right) = \frac{1}{n^2}V(X) = \frac{1}{n^2}.n\pi(1-\pi) = \frac{\pi(1-\pi)}{n} \tag{6.8}$$

In other words, the sample proportion p is distributed normally with the $E(p) = \pi$ (population proportion) and $V(p) = \pi(1-\pi)/n$. We use this relationship to estimate the confidence interval for π.

Swetha wants to estimate a 99% confidence interval for the proportion (π) of the families who come for dinner to the restaurant with more than four members in the family. This is important for her because the tables in the restaurant are "four seaters," and when a group consisting of more than four members arrive, the restaurant will have to join two tables together. She summarized the data for 160 families, and 56 of them had more than four members. The sample proportion is $56/160 = 0.35$. The standard error of p, σ_p, is the square root of V(p) which is equal to $\left(\frac{\pi(1-\pi)}{n}\right)^{0.5}$. In order to estimate the confidence interval for π, we need the σ_p which in turn depends on the value of π! We know that p is a point estimate of π, and hence we can use the value of p in coming up with a point estimate for σ_p. Thus, this point estimate $\widehat{\sigma}_p = \left(\frac{p(1-p)}{n}\right)^{0.5} = \left(\frac{0.35(1-0.35)}{160}\right)^{0.5} = 0.0377$. The z-value corresponding to the 99% confidence level is 2.575.

The 99% confidence level for the population proportion π is

$$P(0.35 - 2.575 \times 0.0377 \leq \pi \leq 0.35 + 2.575 \times 0.0377)$$

$$= P(0.2529 \leq \pi \leq 0.4471) = 0.99.$$

The above confidence interval indicates that Swetha should be prepared to join the tables for 25–45% of the families visiting the restaurant.

3.8 Sample Size Determination

We can determine the ideal sample size for estimating the confidence interval for π with a given confidence level and precision. The process is the same as that for the mean as shown in Expression (6.3)—which is reproduced below for ready reference.

$$n = \left(\frac{z.\sigma}{w}\right)^2$$

In the case of the population mean μ, we have assumed that the population standard deviation, σ, is known. Unfortunately, in the present case, the standard deviation itself is a function of π, and hence we need to approach it differently. Substituting the formula for σ in Expression (6.3), we get

$$n = \pi(1-\pi)\left(\frac{z}{w}\right)^2 \tag{6.9}$$

Needless to say, π(1 − π) is maximum when π = 0.5. Thus we can calculate the maximum required sample size (rather than ideal sample size) by substituting the value of 0.5 for π. Let us calculate the maximum required sample size for this particular case. Suppose Swetha wants to estimate the 99% confidence interval with a precision of ±0.06. Using the z-value of 2.575 corresponding to 99% confidence level, we get

$$n = (0.5)(0.5)\left(\frac{2.575}{0.06}\right)^2 = 460.46 \approx 461$$

In other words, Swetha needs to collect data from 461 randomly selected families visiting her restaurant. Based on Expression (6.9) above, it is obvious that the more the desired precision, the larger is the sample size and the higher the confidence level, the larger is the sample size. Sometimes, it is possible to get a "best guess" value for a proportion (other than 0.5) to be used in Expression (6.9). It could be based on a previous sample study or a pilot study or just a "best judgment" by the analyst.

3.9 Confidence Intervals for the Population Variance

The sample variance, s^2, calculated as shown in Expression (6.4) is the best estimator of the population variance, σ^2. Thus, we can use s^2 to make inferences about σ^2 by identifying the sampling distribution of $\frac{(n-1)s^2}{\sigma^2}$. The random variable $\frac{(n-1)s^2}{\sigma^2}$ has a chi-square (χ^2) distribution with (n − 1) degrees of freedom. The χ^2 distribution is defined only for non-negative values, and its shape depends on the degrees of freedom. Notice that unlike the normal distribution, the χ^2 distribution is not symmetric but is skewed to the right. Figure 6.4 presents the χ^2 distribution for three different values of degrees of freedom.

The χ^2 values for the different degrees of freedom and various confidence levels can be obtained from Table 6.13. In Table 6.13, consider the row corresponding to degrees of freedom 8 and columns corresponding to 0.95 and 0.05 under the right tail. These two values are 2.7826 and 15.5073. These two values are shown in Fig. 6.4, under the chi-square distribution curve with df = 8. This implies that the area

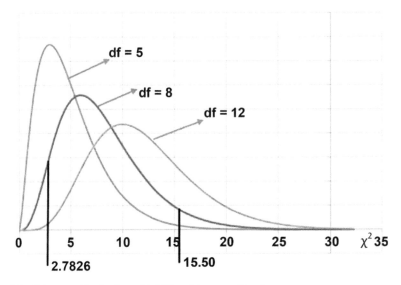

Fig. 6.4 Chi-square distribution with different degrees of freedom

to the right of 2.7826 is equal to 0.95 (i.e., the area to the left of 2.7826 is equal to $1 - 0.95 = 0.05$), and the area to the right of 15.5073 is equal to 0.05. Thus, the area enclosed between these two χ^2 values is equal to 0.90.

Swetha had earlier calculated the sample standard deviation, s, as 12 min. The variance S^2 is 144 min^2, and the degrees of freedom is 24. She wants to estimate a 95% confidence interval for the population variance, σ^2. The formula for this confidence interval is

$$P\left(\chi^2_{L,0.975} \leq \frac{(n-1)s^2}{\sigma^2} \leq \chi^2_{U,0.025}\right) = 0.95 \tag{6.10}$$

By rearranging the terms in Expression (6.10), we get

$$P\left(\frac{(n-1)s^2}{\chi^2_{U,0.025}} \leq \sigma^2 \leq \frac{(n-1)s^2}{\chi^2_{L,0.975}}\right) = 0.95$$

where $\chi^2_{L,0.975}$ is the "lower" or left-hand side value of χ^2 from Table 6.13, corresponding to column 0.975 and $\chi^2_{U,0.025}$ is the "upper" or right-side value corresponding to column 0.025. The two values, drawn from Table 6.13, are 12.4012 and 39.3641. The 95% confidence interval for σ^2 is

$$P\left(\frac{24 \times 144}{39.3641} \leq \sigma^2 \leq \frac{24 \times 144}{12.4012}\right) = P\left(87.7957 \leq \sigma^2 \leq 278.6827\right) = 0.95$$

In some situations, we want to confidently state that the parameter being estimated is far from zero or less than some given value. Such statements can be made using one-sided confidence intervals. One-sided confidence intervals can be similarly calculated as two-sided ones for the population mean, population proportion, or population standard deviation. For example, in the last calculation, $P(87.7957 \leq \sigma^2) = 0.975$. Sample size calculations are also similar and omitted. Also see one-sided hypothesis testing below.

3.10 Hypothesis Testing

When estimating the confidence interval for μ, π, or σ^2, we do not make any prior assumptions about the possible value of the parameter. The possible range within which the parameter value is likely to fall is estimated based on the sample statistic. However, sometimes the decision maker has enough reason to presume that the parameter takes on a particular value and would like to test whether this presumption is tenable or not based on the evidence provided by the sample data. This process is referred to as hypothesis testing.

Let us consider a simple example of hypothesis testing. In the field of electronics, some parts are coated with a specific material at a thickness of 50 μm. If the coating is too thin, the insulation does not work properly, and if it is too thick, the part will not fit properly with other components. The coating process is calibrated to achieve 50 μm thickness on average with a standard deviation of 8 μm. If the outcome is as expected (50 μm thickness), then the process is said to be in control, and if not, it is said to be out of control. Let us take a sample of 64 parts whose thickness is measured and sample average, $\overline{X}$, is calculated. Based on the value of $\overline{X}$, we need to infer whether the process is in control or not.

In order to effectively understand this, we create a decision table (refer to Table 6.6).

H_0 represents the null hypothesis, or the assumption that the parameter being observed is actually equal to its expected value, in this case that is $\mu = 50$ μm.

It is obvious from Table 6.6 that there are only four possibilities in hypothesis testing. The first column in Table 6.6 is based on the assumption that the "process is in control." The second column is that the "process is not in control." Obviously, these two are mutually exclusive. Similarly, the two rows representing the conclusions based on $\overline{X}$ are also mutually exclusive.

Situation 1: Null Hypothesis Is True

Let us analyze the situation under the assumption for the null hypothesis wherein the process is in control (first column) and μ is exactly equal to 50 μm. We denote this as follows:

H_0: $\mu_0 = 50$ μm.

Table 6.6 Decision table for hypothesis testing

		Null hypothesis is true The process is in control H_0: $\mu = 50$ μm	Null hypothesis is false The process is out of control H_1: $\mu \neq 50$ μm
Our conclusion based on the value of $\overline{X}$	Conclude that process is in control (accept H_0)	Q1 The conclusion is GOOD No error	Q2 Type II error probability is β
	Conclude that process out of control (reject H_0)	Q3 Type I error probability is α	Q4 The conclusion is GOOD No error

μ_0 is a notation used to imply that we do not know the actual value of μ, but its hypothesized value is 50 μm.

If our conclusion based on evidence provided by $\overline{X}$ puts us in the first cell of column 1, then the decision is correct, and there is no error involved with the decision. On the other hand, if the value of $\overline{X}$ is such that it warrants a conclusion that the process is not in control, then the null hypothesis is rejected, resulting in a Type I error. The probability of making a Type I error is represented by α. This is actually a conditional probability, the condition being "the process is actually in control." By evaluating the consequences of such an error and coming up with a value for α, it allows us to indicate the extent to which the decision maker is willing to tolerate a Type I error. Based on the value of α, we can evolve a decision rule to either accept or reject the proposition that the "process is in control."

The selection of α is subjective, and in our example we will pick 0.05 as the tolerable limit for α.

Suppose we take a sample of size 64 and calculate the sample mean $\overline{X}$. The next step is to calculate the limits of the possible values of $\overline{X}$ within which the null hypothesis cannot be rejected. Considering that α is set at 0.05, the probability of accepting the null hypothesis when it is actually true is $1 - \alpha = 0.95$. The sample mean is distributed normally with mean 50 and standard error $\frac{\sigma}{\sqrt{64}} = 1$. The limits of $\overline{X}$ within which we accept the null hypothesis is given by

$$P\left(\mu_0 - z \cdot \frac{\sigma}{\sqrt{n}} \leq \overline{X} \leq \mu_0 + z \cdot \frac{\sigma}{\sqrt{n}}\right) = P\left(50 - 1.96(1) \leq \overline{X} \leq 50 + 1.96(1)\right) = 0.95$$

In other words, our decision rule states:

If the sample mean $\overline{X}$ lies between 48.04 and 51.96 μm, then we can conclude that the process is in control. This range is referred to as the acceptance region. On the other hand, if $\overline{X}$ falls outside these two limits, there is enough evidence to conclude that the process is out of control, rejecting the null hypothesis and resulting in an error, the probability of which is not more than 0.05.

This is illustrated in Fig. 6.5. The shaded area under the two tails combined is equal to 0.05 and is usually referred to as the rejection region, since the null

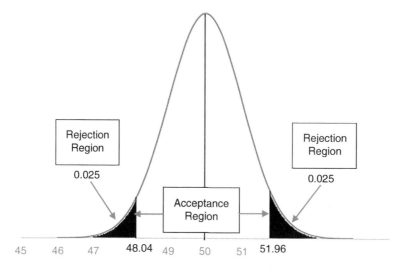

Fig. 6.5 Acceptance and rejection regions

hypothesis is rejected when $\overline{X}$ falls in this region. The probability of a Type I error (α) is equally divided between the two tail regions since such an equal division leads to the smallest range of an acceptance region, thereby maximizing the precision. It should be noted that the selection of α is completely within the control of the decision maker.

Situation 2: Null Hypothesis Is False

Now that we have a decision rule that says that we accept the null hypothesis as long as $\overline{X}$ is between 48.04 and 51.96, we can develop the procedure to calculate the probability of a Type II error. A Type II error is committed when the reality is that the process is out of control and we erroneously accept the null hypothesis (that the process is in control) based on evidence given by $\overline{X}$ and our decision rule. If the process is out of control, we are actually operating on the second column of Table 6.6. This implies that $\mu \neq 50$ μm. This also implies that μ can be any value other than 50 μm. Let us pick one such value, 49.5 μm. This value is usually denoted by μ_A because this is the alternate value for μ. This is referred to as the "alternate hypothesis" and expressed as

$H_1: \mu_0 \neq 50$

The distribution of $\overline{X}$ with $\mu_A = 49.5$ is presented in Fig. 6.6. As per our decision rule created earlier, we would still accept the null hypothesis that the process is under control, if $\overline{X}$ falls between 48.04 and 51.96 μm. This conclusion is erroneous because, with $\mu_A = 49.5$, the process is actually out of control. This is nothing but a Type II error. The probability of committing this error (β) is given by the area under the curve (Fig. 6.6) between the values 48.04 and 51.96. This area is 0.9209.

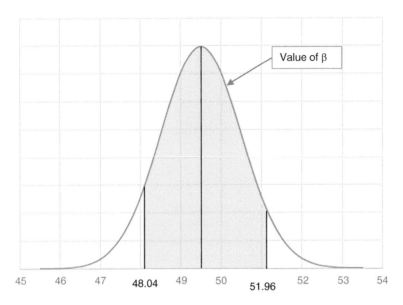

Fig. 6.6 Calculation of β when $\mu_A = 49.5$ μm

Table 6.7 Values of β and $1 - \beta$ for different values of μ

Alternate value for μ	β	$1 - \beta$
47.00	0.1492	0.8508
47.50	0.2946	0.7054
48.00	0.4840	0.5160
48.50	0.6770	0.3230
49.00	0.8299	0.1701
49.50	0.9209	0.0791
50.00	0.9500	0.0500
50.50	0.9209	0.0791
51.00	0.8299	0.1701
51.50	0.6770	0.3230
52.00	0.4840	0.5160
52.50	0.2946	0.7054
53.00	0.1492	0.8508

As a matter of fact, the alternate value for μ can be any value other than 50 μm. Table 6.7 presents values of β as well as $1 - \beta$ for different possible values of μ. $1 - \beta$, called the power of the test, is interpreted as the probability of rejecting the null hypothesis when it is not true (Table 6.6 Q4), which is the correct thing to do.

It can be seen from Table 6.7 that as we move farther away in either direction from the hypothesized value of μ (50 μm in this case), β values decrease. At $\mu_A = \mu_0$ (the hypothesized value of μ), $1 - \beta$ is nothing but α and β is nothing but $1 - \alpha$. This is clear when we realize that at the point $\mu_A = \mu_0$, we are no longer in quadrants Q2 and Q4 of Table 6.6 but have actually "shifted" to the first column.

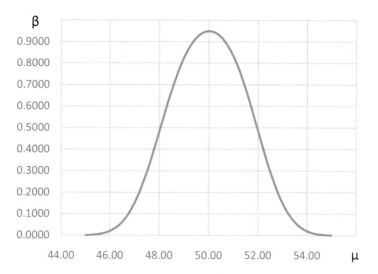

Fig. 6.7 Operating characteristic curve

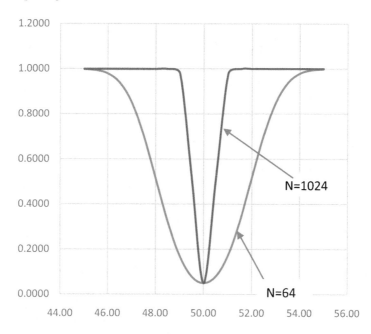

Fig. 6.8 Power curve

Figure 6.7 presents the values of β plotted against different values for μ. This curve is referred to as the "operating characteristic" curve. What is interesting is the plot of 1 − β against different values of μ, which is presented in Fig. 6.8.

As mentioned earlier, $1 - \beta$ is the probability of rejecting the null hypothesis when it is not true. Ideally, this probability should be as close to 1 as possible when μ is different from μ_0 and as close to zero when $\mu = \mu_0$. This indicates how powerful our test is with respect to the hypothesis. Hence, the curve presented in Fig. 6.8 is referred to as the "power curve." Figure 6.8 presents two power curves, one corresponding to a sample size of 64 and the other to 1024. It can be seen that the power curve approaches the ideal curve of a straight line at the mean when the sample size increases.

The sample mean, $\overline{X}$, based on the 64 observations turned out to be 47.8 μm. The conclusion we draw is that the process is out of control—the manufacturing process needs to be stopped and the entire process recalibrated. The error, if any, has a probability of not more than 0.05 associated with this conclusion.

It is interesting to note that it is possible to commit either a Type I error or a Type II error, but not both. It is important to understand that when we commit a Type I error, we not only have a clearly specified probability (α), but also the decision maker can select its value based on the implications of the Type I error. On the other hand, when we commit a Type II error, there is no single value for β. All we can do is calculate possible values of β corresponding to different, alternate values of μ. In other words, when we reject the null hypothesis, we commit a Type I error, which is completely in the control of the decision maker. Hence, the decision maker would like to set up the null hypothesis such that it could be rejected. This is the reason why we always say that "we do not have sufficient evidence to reject the null hypothesis" or "we fail to reject the null hypothesis" rather than saying "we accept the null hypothesis."

3.11 Steps Involved in the Hypothesis Test

The hypothesis test is a very important aspect of statistical inference. The following steps summarize the process involved in the hypothesis test:

1. State the null and alternate hypotheses with respect to the desired parameter (in this case, the parameter is the population mean, μ).
   Eg. H_0: $\mu_0 = 50\,\mu$m
   H_A: $\mu_0 \neq 50\,\mu$m
2. Identify the sample statistic that is to be used for testing the null hypothesis. (In this case, the sample statistic is $\overline{X}$.)
3. Identify the sampling distribution of the sample statistic (normal distribution in this case).
4. Decide on the value of α (0.05 in the above example).
5. Decide on the sample size (64 in the example).
6. Evolve the decision rule by calculating the acceptance/rejection region based on α and the sampling distribution (e.g., accept if $\overline{X}$ falls between 48.04 and 51.96 μm, reject otherwise).

(Notice that the above steps can be completed even before the data is collected.)
7. Draw the conclusions regarding the null hypothesis based on the decision rule.

3.12 Different Methods for Testing the Null Hypothesis

There are four different ways to test the null hypothesis, and all the four are conceptually similar. The methods described below use numbers drawn from an earlier example in our discussion. We also reference the interval of ±1.96 in relation to a confidence level of 95%.

H_0: $\mu_0 = 50$ μm
H_A: $\mu_0 \neq 50$ μm
$\sigma = 8$
$n = 64$
$\sigma_{\overline{X}} = \frac{8}{\sqrt{64}} = 1$ μm
$\overline{X} = 47.8$ μm

1. The first method is to build an interval around μ (based on α) and test whether the sample statistic, $\overline{X}$, falls within this interval.

    Example: The interval is $50 \pm 1.96\,(1) = 48.04$ to 51.96, and since $\overline{X}$ falls outside these limits, we reject the null hypothesis.

2. The second method is similar to estimating the confidence interval. Recall that when α is set at 0.05, the probability of accepting the null hypothesis when it is actually true is $1 - \alpha = 0.95$. Using the value of $\overline{X}$, we can build a confidence interval corresponding to $(1 - \alpha)$ and test whether the hypothesized value of μ falls within these limits.

    Example: The confidence interval corresponding to $(1-0.05) = 0.95$ is
    $47.8 \pm 1.96\,(1) = 45.84$ to 49.76. Since the hypothesized value of μ, 50 μm, falls outside these limits, we reject the null hypothesis.

3. Since the sampling distribution of $\overline{X}$ is a normal distribution, convert the values into standard normal, and compare with the z-value associated with α. It is important to note that the comparison is done with absolute values, since this is a two-sided hypothesis test.

    Example: The standard normal variate corresponding to $\overline{X}$ is $\frac{(\overline{X} - \mu)}{\sigma_{\overline{X}}} = \frac{47.8 - 50}{1} = -2.2$. The absolute value 2.2 is greater than 1.96 and hence we reject the null hypothesis. This process is shown in Fig. 6.9.

4. The fourth method involves calculating the probability (p-value) of committing a Type I error if we reject the null hypothesis based on the value of $\overline{X}$.

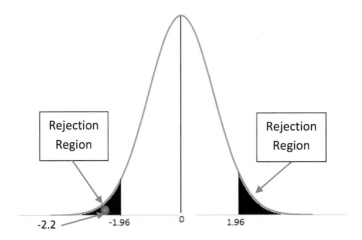

Fig. 6.9 Hypothesis test using standard normal transformation

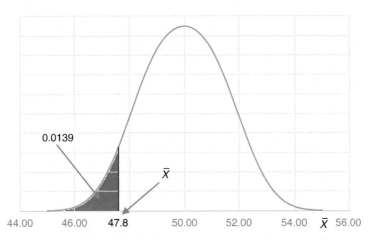

Fig. 6.10 Calculating p-value based on $\overline{X}$ value

3.13 Calculation of p-Value

Let us consider the sampling distribution of $\overline{X}$ shown in Fig. 6.10. It should be noted that this distribution is drawn based on the assumption that the null hypothesis is true.

(a) First, locate the $\overline{X}$ value of 47.8 μm on the x-axis, and note that its value is to the left of $\mu = 50$.

(b) Calculate the area under the curve to the left of 47.8 μm (if the $\overline{X}$ value happened to the right of μ, we would have calculated the area under the curve to the right). This area, highlighted in red, turns out to be 0.0139. This is the probability of committing an error, if we reject the null hypothesis based on any value of $\overline{X}$ less than or equal to 47.8.
(c) Since this is a two-sided test, we should be willing to reject the null hypothesis for any value of $\overline{X}$ which is greater than or equal to 52.2 μm (which is equidistant to μ on the right side). Because of the symmetry of a normal distribution, that area also happens to be exactly equal to 0.0139.
(d) In other words, if we reject the null hypothesis based on the $\overline{X}$ value of 47.8, the probability of a type I error is $0.0139 + 0.0139 = 0.0278$. This value is called the p-value, and it is different from α because α is determined by the decision maker, while p-value is calculated based on the value of $\overline{X}$.
(e) In this example, we have taken the value of α (which is the probability of committing a type I error) as 0.05, indicating that we are willing to tolerate an error level up to 0.05. The p-value is much less than 0.05, which implies that this error level, while rejecting the null hypothesis, is less than what we are willing to tolerate. In other words, we are willing to commit an error level up to 0.05 while the actual error level being committed is only 0.0278 (the p-value). Hence the null hypothesis can be rejected. In general, the null hypothesis can be rejected if the p-value is less than α.
(f) Since $p = 0.0278 < \alpha = 0.05$, we reject the null hypothesis.

Note that the conclusion is the same for all four methods.

3.14 One-Tailed Hypothesis Test

The example discussed earlier involves a rejection region under both the tails of the distribution and the α value to be distributed equally between the two tail regions. Sometimes, the rejection region is only under one of the two tails and is referred to as a one-tailed test. Consider the carrying capacity of a bridge. If the bridge is designed to withstand a weight of 1000 tons but its actual capacity is more than 1000 tons, there is no reason for alarm. On the other hand, if the actual capacity of the bridge is less than 1000 tons, then there is cause for alarm. The null and alternate hypotheses in this case are set up as follows:

H_0: $\mu_0 \leq 1000$ tons (*bridge designed to withstand weight of 1000 tons or less*).
H_A: $\mu_0 > 1000$ tons (*bridge designed to withstand weight > 1000 tons*).

The bridge is subjected to a stress test at 25 different locations, and the average ($\overline{X}$) turned out to be 1026.4 tons. There is no reason to reject the null hypothesis for any value of $\overline{X}$ less than 1000 tons. On the other hand, the null hypothesis can be rejected only if $\overline{X}$ is significantly larger than 1000 tons.

The rejection region for this null hypothesis is shown in Fig. 6.11. The entire probability (α) is under the right tail only. C is the critical value such that if the value of $\overline{X}$ is less than C, the null hypothesis is not rejected, and the contractor is asked

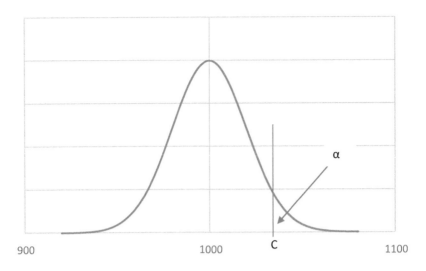

Fig. 6.11 Rejection region in a one-tailed test

to strengthen the bridge. If the value of $\overline{X}$ is greater than C, the null hypothesis is rejected, traffic is allowed on the bridge, and the contractor is paid. The consequence of a Type 1 error here is that traffic is allowed on an unsafe bridge leading to loss of life. The decision maker in this case would like to select a very small value for α. Let us consider a value of 0.005 for α.

If the population standard deviation is known, we can calculate C by using the formula

$$C = \mu_0 + z\frac{\sigma}{\sqrt{n}} \text{ where } z = 2.575 \text{ corresponding to } \alpha = .005.$$

The population standard deviation is not known in this particular case, and hence we calculate C using the sample standard deviation s and t-distribution. The formula is $C = \mu_0 + 2.787\frac{s}{\sqrt{25}}$ where s is the sample standard deviation and 2.787 is the t-value corresponding to 24 degrees of freedom and $\alpha = 0.005$. The sample standard deviation $s = 60$ tons, and the standard error is $60/\sqrt{25} = 12$ tons.

$$C = 1000 + 2.787(12) = 1033.444 \text{ tons.}$$

Two aspects need to be noted here. First, traffic is allowed on the bridge only if $\overline{X}$ is greater than 1033.44 tons, even though the required capacity is only 1000 tons. The additional margin of 33.444 tons is to account for the uncertainty associated with the sample statistics and playing it safe. If we reduce α further, the safety margin that is required will correspondingly increase. In the results mentioned earlier, $\overline{X} = 1026.4$ tons. Based on our value of C, the null hypothesis is not rejected, and the contractor is asked to strengthen the bridge. We should also note that the null hypothesis is set to $\mu_0 \leq 1000$ tons in favor of a stronger bridge and erring on the safer side. In fact, it would actually be favoring the contractor (albeit at the cost of safety) if the null hypothesis was set to $\mu_0 \geq 1000$ tons.

3.15 Hypothesis Test for Population Proportion π

We can extend the concept of the hypothesis test for μ to a similar hypothesis test for the population proportion, π. Swetha (see the section on confidence intervals) is concerned about the occupancy of her restaurant. She noticed that sometimes, the tables are empty, while other times there is a crowd of people in the lobby area outside, waiting to be seated. Obviously, empty tables do not contribute to revenue, but she is also worried that some of the people who are made to wait might go off to other restaurants nearby. She believes it is ideal that people only wait to be seated 20% of the time. Note that this will be her null hypothesis. She decides to divide the lunch period into four half-hour slots and observe how many customers, if any, wait to be seated during each time slot. She collected this data for 25 working days, thus totalling 100 observations (25 days × 4 lunch slots/day). She decides on the value of $\alpha = 0.05$.

The null hypothesis is as follows:

H_0: $\pi_0 = 0.2$ *(people wait to be seated 20% of the time)*.
H_A: $\pi_0 \neq 0.2$.

The corresponding sample statistic is the sample proportion, p. Swetha found that 30 out of the 100 time slots had people waiting to be seated. Since we develop the decision rule based on the assumption that the null hypothesis is actually true, the standard error of p is

$$\sigma_p = \sqrt{\frac{0.2(1-0.2)}{100}} = 0.04.$$

The p-value for this example is shown in Fig. 6.12. Note that the probability that p is greater than 0.3 is 0.0062 (and consequently the probability that p is less than 0.1 is also 0.0062, by symmetry). Thus, the p-value is $0.0062 \times 2 = 0.0124$. This is considerably lower than the 5% that Swetha had decided. Hence, the null hypothesis is rejected, and Swetha will have to reorganize the interior to create more seating space so that customers do not leave for the competition.

3.16 Comparison of Two Populations

There are many situations where one has to make comparisons between different populations with respect to their means, proportions, or variances. This is where the

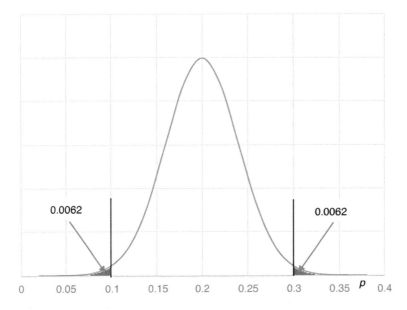

Fig. 6.12 Hypothesis test for π

real usefulness of the statistical inference becomes apparent. First, we will look at a very simple situation where the data is collected as a pair of observations from the same respondents. This is called paired-observation comparisons.

3.17 Paired-Observation Comparisons

Consider a situation where 16 workers from an automobile manufacturer are sent for 2 weeks of training for additional skill acquisition. The productivity of each of the workers is measured by the time taken to produce a particular component, before and after the training program. It was desired to test whether the training program was effective or not. In order to find the real effectiveness, the improvement with respect to each worker is calculated. This is simply the difference between the time taken to produce the same component, before and after the training program for each worker. This difference, D_i (for worker i), itself becomes the random variable. The null and alternate hypotheses are:

H_0: $\mu_{D0} \leq 0$ *(the time taken to produce the component before the training program is less than the time taken after the training program indicating that the training program is ineffective).*

$H_A: \mu_{D0} > 0$.

The average of D_i, $\overline{D}$, was 0.52 h, and the standard deviation was 1.8 h. The standard error is $1.8/\sqrt{16} = 0.45$ h. The p-value is calculated as $P(\overline{D} > 0.52 | \mu_{D0} = 0 \text{ and } s_{\overline{D}} = 0.45)$. Since we are using the sample standard deviation instead of the population standard deviation (σ), we need to calculate the probability using the t-distribution.

$$t_{calc} = \frac{\overline{D} - \mu_{D0}}{s_{\overline{D}}} = \frac{0.52 - 0}{0.45} = 1.1556 \text{ and df} = (16-1) = 15.$$

Using the t-distribution, the p-value is calculated as 0.1330 which is large enough for us not to reject the null hypothesis. Thus, the conclusion is that the training program was not effective.

3.18 Difference Between Independent Samples

Let us consider two samples which are drawn independently from two neighbourhoods in Hyderabad, India. The first sample refers to the daily profit from a restaurant in Jubilee Hills and the second refers to the daily profit for the same restaurant in Hitech City. Swetha, who owns both these restaurants, wants to test if there is a difference between the two population means. The null and alternate hypotheses are:

$H_0: \mu_1 = \mu_2 \rightarrow \mu_1 - \mu_2 = 0$.
$H_A: \mu_1 \neq \mu_2 \rightarrow \mu_1 - \mu_2 \neq 0$.

The original null hypothesis of $H_0: \mu_1 = \mu_2$ is restated as $\mu_1 - \mu_2 = 0$, because $(\mu_1 - \mu_2)$ can be estimated by $(\overline{X}_1 - \overline{X}_2)$ and the sampling distribution of $(\overline{X}_1 - \overline{X}_2)$ is normal with mean $(\mu_1 - \mu_2)$ and standard error $\sigma_{(\overline{X}_1 - \overline{X}_2)}$. When the standard deviations of the populations are known, $\sigma_{(\overline{X}_1 - \overline{X}_2)}$ is calculated as $\sqrt{\frac{\sigma_1^2}{n_1} + \frac{\sigma_2^2}{n_2}}$. The hypothesis can be simply tested by using standard normal distribution, as in the case of single sample, that is, $z_{calc} = \frac{(\overline{X}_1 - \overline{X}_2) - (\mu_1 - \mu_2)}{\sigma_{(\overline{X}_1 - \overline{X}_2)}}$, and comparing it with the value of the z-value corresponding to α, obtained from the standard normal tables.

When σ_1 and σ_2 are not known, we need to use the sample standard deviations s_1 and s_2 and calculate $s_{(\overline{X}_1 - \overline{X}_2)}$ and further use t-distribution, instead of z. Since σ_1 and σ_2 are not known, calculating $s_{(\overline{X}_1 - \overline{X}_2)}$ becomes somewhat involved. The calculation of $\sigma_{(\overline{X}_1 - \overline{X}_2)}$ depends on whether $\sigma_1 = \sigma_2$ or not. The formulae are presented below:

1. If $\sigma_1 \neq \sigma_2$, then $s_{(\overline{X}_1-\overline{X}_2)}$ is calculated as $\sqrt{\frac{s_1^2}{n_1} + \frac{s_2^2}{n_2}}$, and the associated degrees of freedom is calculated using the formula

$$df = \frac{\left(\frac{s_1^2}{n_1} + \frac{s_2^2}{n_2}\right)^2}{\frac{1}{n_1-1}\left(\frac{s_1^2}{n_1}\right)^2 + \frac{1}{n_2-1}\left(\frac{s_2^2}{n_2}\right)^2}.$$

2. If $\sigma_1 = \sigma_2 = \sigma$, then it implies that both s_1 and s_2 are estimating the same σ. In other words, we have two different estimates for the same population parameter, σ. Obviously, both these estimates have to be pooled to obtain a single estimate s_p (which is an estimate of σ) using the formula (the degrees of freedom is $n_1+n_2 - 2$).

$$s_p^2 = \frac{(n_1 - 1) s_1^2 + (n_2 - 1) s_2^2}{n_1 + n_2 - 2}$$

3. Now, $s_{(\overline{X}_1-\overline{X}_2)}$ is calculated using the formula

$$s_{(\overline{X}_1-\overline{X}_2)} = s_p \sqrt{\frac{1}{n_1} + \frac{1}{n_2}}$$

At this point it is necessary to test a hypothesis regarding the equality of variances between the two populations before proceeding with the hypothesis test regarding the population means. The null and alternate hypotheses are set up as follows:

$$H_0 : \sigma_1^2 = \sigma_2^2 \rightarrow \frac{\sigma_1^2}{\sigma_2^2} = 1$$

$$H_A : \sigma_1^2 \neq \sigma_2^2 \rightarrow \frac{\sigma_1^2}{\sigma_2^2} \neq 1$$

As we had done earlier, the original null hypothesis, $\sigma_1^2 = \sigma_2^2$, is transformed as $\frac{\sigma_1^2}{\sigma_2^2} = 1$ so that the ratio of sample standard deviations can be used as the test

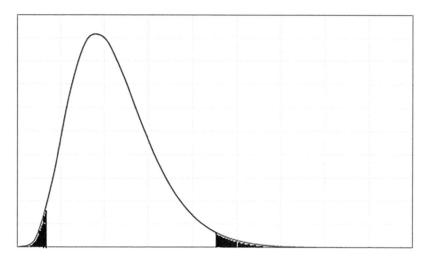

Fig. 6.13 F Distribution for df$_1$ = 17 and df$_2$ = 11

statistic. This null hypothesis can be tested using an "F" distribution. The random variable "F" is defined as

$$F = \frac{\chi_1^2/(n_1-1)}{\chi_2^2/(n_2-1)} = \frac{s_1^2/\sigma_1^2}{s_2^2/\sigma_2^2}.$$

The F distribution is based on two separate degrees of freedom, normally referred to as k$_1$, the numerator degrees of freedom, and k$_2$, the denominator degrees of freedom. Needless to say, the F distribution is defined only for non-negative values. If the null hypothesis is true, then $F = \frac{s_1^2}{s_2^2}$.

Let us consider the situation where Swetha wants to compare the average daily profit from two of her restaurants. She has collected data for 18 days from the Jubilee Hills restaurant (sample 1) and for 12 days from the Hitech City restaurant (sample 2) and calculated the average daily profit. The data is summarized below:

$$\overline{X}_1 = ₹\,13,750;\ \overline{X}_2 = ₹\,14,110;\ s_1^2 = 183,184\ ₹^2;\ s_2^2 = 97,344\ ₹^2$$

Assuming that the null hypothesis is true, the F statistic is $\frac{183,184}{97,344} = 1.882$. Figure 6.13 presents the F distribution for 17 and 11 degrees of freedom for the numerator and denominator respectively.

The p-value corresponding to the calculated F, 1.882, is 0.2881 and hence the null hypothesis cannot be rejected. Now, Swetha has to proceed under the condition of equal variances. First we calculate

$$s_p^2 = \frac{(n_1 - 1) s_1^2 + (n_2 - 1) s_2^2}{n_1 + n_2 - 2}.$$

It can be seen that the above expression for s_p^2 is nothing but the weighted average of the two sample variances, with the respective degrees of freedom as weights. The degrees of freedom associated with s_p^2 is the total number of observations minus 2 (because we are using two proxies, namely, $\overline{X}_1$ and $\overline{X}_2$).

The standard error of $(\overline{X}_1 - \overline{X}_2)$ is calculated as

$$s_{(\overline{X}_1 - \overline{X}_2)} = \sqrt{s_p^2 \left(\frac{1}{n_1} + \frac{1}{n_2}\right)} = 144.0781$$

$$t_{calc} = \frac{(\overline{X}_1 - \overline{X}_2) - 0}{s_{(\overline{X}_1 - \overline{X}_2)}} = \frac{-360}{144.0781} = -2.4987.$$

The p-value corresponding to the above calculated t is 0.0186 which is low enough for us to reject the null hypothesis of equality of population means. The average profits from the Hitech city restaurant are significantly more than those of Jubilee Hills.

3.19 Analysis of Variance (ANOVA)

We have just seen the procedure to compare the means of the two populations. In many situations, we need to compare the means of a number of populations. For example, Swetha has been receiving complaints that the time taken to collect the order from the customers is inordinately long in some of her restaurants. She wants to test whether there is any difference in the average time taken for collecting the orders in four of her restaurants. In such a situation, we use a technique called "analysis of variance (ANOVA)" instead of carrying out pairwise comparisons. Swetha has set up the null and alternate hypotheses as

H_0: $\mu_1 = \mu_2 = \mu_3 = \mu_4$.
H_A: At least one μ is different.

The following assumptions are required for ANOVA:

1. All observations are independent of one another and randomly selected from the respective population.
2. Each of the populations is approximately normal.
3. The variances for each of the populations are approximately equal to one another.

Table 6.8 Summary data for four restaurants

	Sample 1 Jubilee Hills	Sample 2 Hitech City	Sample 3 Banjara Hills	Sample 4 Kavuri Hills
Sample size	15	12	18	16
Sample mean (minutes)	6.8	4.2	7.6	5.4
Sample standard deviation	2.8	3.6	3.1	3.3

The ANOVA technique estimates the variances through two different routes and compares them in order to draw conclusions about the population means. She has collected data from each of the four restaurants and summarized it in Table 6.8.

First we estimate the pooled variance of all the samples, just the way we had done in the case of comparing two population means. Extending the formula to four samples, s_p^2 is defined as

$$s_p^2 = \frac{\sum (n_i - 1) s_i^2}{\sum (n_i - 1)}, (i = 1, .., 4)$$

Since one of the assumptions is that $\sigma_1^2 = \sigma_2^2 = \sigma_3^2 = \sigma_4^2 = \sigma^2$, the s_p^2 is an estimate of σ^2. It may be remembered that the variance of the sample means ($\overline{X}s$), $\sigma_{\overline{X}}^2 = \frac{\sigma^2}{n}$ (consequently, $\sigma^2 = n\sigma_{\overline{X}}^2$), and it can be estimated by calculating the variance of the $\overline{X}s$. The variance of the four sample means can be calculated by the formula $s_{\overline{X}}^2 = \frac{\sum \left(\overline{X}_i - \overline{\overline{X}}\right)^2}{k-1}$ where k is the number of samples and $\overline{\overline{X}}$ is the overall mean for all the observations. We can get an estimate of σ^2 by multiplying $s_{\overline{X}}^2$ with sample size. Unfortunately, the sample sizes are different and hence the formula needs to be modified as

$\hat{\sigma}^2 = \frac{\sum \left(\overline{X}_i - \overline{\overline{X}}\right)^2 n_i}{k-1}$ where $\hat{\sigma}^2$ is an estimate of σ^2 obtained through the "means" route. If the null hypothesis is true, these two estimates ($\hat{\sigma}^2$ and s_p^2) are estimating the same σ^2. On the other hand, if the null hypothesis is not true, $\hat{\sigma}^2$ will be significantly larger than s_p^2. The entire hypothesis test boils down to testing to see if $\hat{\sigma}^2$ is significantly larger than s_p^2, using the F distribution (testing for equality of variances). We always test to see if $\hat{\sigma}^2$ is larger than s_p^2 and not the other way, and hence this F test is always one-sided (one-tailed test on the right side). This is because $\hat{\sigma}^2$ is a combination of variation arising out of σ^2 as well as difference between means, if any.

Using the summary data provided in Table 6.8, we can calculate s_p^2 as 10.16 and $\hat{\sigma}^2$ as 32.93. The F value is $32.93/10.16 = 3.24$. The corresponding degrees

Table 6.9 ANOVA table

Source of variation	Degrees of freedom	Sum of squares	Mean squares	F
Between samples	$k-1$	$SSB = \sum_i \left(\overline{X}_i - \overline{\overline{X}}\right)^2 . n_i$	$MSB = \frac{SSB}{k-1}$	$F = \frac{MSB}{MSW}$
Within samples	$\sum_i (n_i - 1)$	$SSW = \sum_i \sum_j \left(X_{ij} - \overline{X}_i\right)^2$	$MSW = \frac{SSW}{\sum_i (n_i - 1)}$	
Total	$\sum_i n_i - 1$	$SST = \sum_i \sum_j \left(X_{ij} - \overline{\overline{X}}\right)^2$		

of freedom are as follows: Numerator degrees of freedom = (number of samples − 1) = (4 − 1) = 3 and denominator degrees of freedom = $\sum(n_i - 1)$ = (14 + 11 + 17 + 15) = 57. The p-value corresponding to the calculated F value is 0.0285. Since this value is small enough, Swetha can conclude that at least one of the restaurants is taking unduly long time in collecting the orders from the customers.

The calculations shown above are not only involved, but also amount to unnecessary duplication. Consequently, the calculation process is simplified by using what is called the "ANOVA Table," presented in Table 6.9. The basic concept is that the total variation in all the samples put together can be decomposed into two components, namely, "variation between samples" and "variation within samples."

The calculation formulae are as follows:

n_i = size of ith sample
X_{ij} = jth observation in ith sample (j = 1, ..., n_i)
$\overline{X}_i$ = mean of ith sample (i = 1, ..., k)
$\overline{\overline{X}}$ = overall mean (weighted average of all the $\overline{X}_i$s)

The total variation is measured by the sum of squares, total (SST) = $\sum_i \sum_j \left(X_{ij} - \overline{\overline{X}}\right)^2$.

Variation between samples is measured by the sum of squares, between (SSB) = $\sum_i \left(\overline{X}_i - \overline{\overline{X}}\right)^2 . n_i$.

Variation within samples is measured by the sum of squares, within (SSW) = $\sum_i \sum_j \left(X_{ij} - \overline{X}_i\right)^2$.

These values are put into the ANOVA table as shown in Table 6.9.

Table 6.10 presents the ANOVA table depicting the comparison between restaurants with respect to order taking time.

It can be seen from the above tables that SST = SSB + SSW. Hence, it is not necessary to calculate all the three from the formulae, even though individual formulae are given. Any one of them can be obtained after calculating the other

Table 6.10 ANOVA table for order taking time

Source of variation	Degrees of freedom	Sum of squares	Mean squares	F
Between samples	4−1 = 3	98.81	$\frac{98.81}{3} = 32.94$	$\frac{32.94}{10.16} = 3.2422$
Within samples	14 + 11 + 17 + 15 = 57	579.04	$\frac{579.04}{57} = 10.16$	
Total	61−1 = 60	677.85		

two, using the relationship between the three of them. Similar relationship exists between the degrees of freedom corresponding to each source. There are several excellent books that introduce the reader to statistical inference. We have provided two such references at the end of this chapter (Aczel & Sounderpandian 2009; Stine & Foster 2014).

Electronic Supplementary Material

All the datasets, code, and other material referred in this section are available in www.allaboutanalytics.net.

- Data 6.1: ANOVA.csv
- Data 6.2: Confidence_Interval.csv
- Data 6.3: Critical_Value_Sample_Prop.csv
- Data 6.4: Hypothesis_Testing.csv
- Data 6.5: Independent_Samples.csv
- Data 6.6: Paired_Observations.csv

Exercises

Ex. 6.1 Nancy is running a restaurant in New York. She had collected the data from 100 customers related to average expenditure per customer in US dollars, per customer eating time in minutes, number of family members, and average waiting time to be seated in minutes. *[Use the data given in Confidence_Interval.csv]*

(a) Given the population standard deviation as US$ 480, estimate the average expenditure per family along with confidence interval at 90%, 95%, and 99% levels of confidence.
(b) Show that increased sample size can reduce the width of the confidence interval with a given level of confidence, while, with the given sample size, the width of the confidence interval increases as we increase the confidence level.
(c) For a width of 150, determine the optimal sample size at 95% level of confidence.

Ex. 6.2 Nancy wants to estimate the average time spent by the customers and their variability. *[Use the data given in Confidence_Interval.csv]*

(a) Using appropriate test-statistic at 90% confidence level, construct the confidence interval for the sample estimate of average time spent by the customer.
(b) Using appropriate test-statistic at 95% confidence level, construct the confidence interval for the sample estimate of variance of time spent by the customer.

Ex. 6.3 The dining tables in the restaurant are only four seaters, and Nancy has to join two tables together whenever a group or family consisting more than four members arrive in the restaurant. *[Use the data given in Confidence_Interval.csv]*

(a) Using appropriate test-statistic at 99% confidence level, provide the confidence interval for the proportion of families visiting the restaurant with more than four members.
(b) Provide the estimate of the maximum sample size for the data to be collected to have the confidence interval with a precision of ±0.7.
(c) Suppose it is ideal for Nancy to have 22% families waiting for a table to be seated. Test the hypothesis that the restaurant is managing the operation as per this norm of waiting time. *[Use the data given in Critical_Value_Sample_Prop.csv]*

Ex. 6.4 In an electronic instrument making factory, some parts have been coated with a specific material with a thickness of 51 μm. The company collected data from 50 parts to measure the thickness of the coated material. Using various methods of hypothesis testing, show that null hypothesis that the mean thickness of coated material is equal to 51 μm is not rejected. *[Use the data given in Hypothesis_Testing.csv]*

Ex. 6.5 A multinational bridge construction company was contracted to design and construct bridges across various locations in London and New York with a carrying capacity of 1000 tons. The bridge was subjected to stress test at 50 different locations. Using appropriate test statistics at 1% level of confidence, test whether the bridge can withstand a weight of at least 1000 tons. *[Use the data given in Critical_Value_Sample_Prop.csv]*

Ex. 6.6 A multinational company provided training to its workers to perform certain production task more efficiently. To test the effectiveness of the training, a sample of 50 workers was taken and their time taken in performing the same production task before and after the training. Using appropriate test statistic, test the effectiveness of the training program. *[Use the data given in Paired_Observations.csv]*

Ex. 6.7 Bob is running two restaurants in Washington and New York. He wants to compare the daily profits from the two restaurants. Using the appropriate test statistic, test whether the average daily profits from the two restaurants are equal. *[Use the data given in Independent_Samples.csv]*

Ex. 6.8 Paul is running four restaurants in Washington, New York, Ohio, and Michigan. He has been receiving complaints that the time taken to collect the order from the customer is inordinately long in some of his restaurants. Test whether there is any difference in the average time taken for collecting orders in four of his restaurants. *[Use the data given in ANOVA.csv]*

Ex. 6.9 Explain the following terms along with their relevance and uses in analysis and hypothesis testing: *[Use data used in Exercises 1–8 as above.]*

(a) Standard normal variate
(b) Type I error
(c) Power of test
(d) Null hypothesis
(e) Degree of freedom
(f) Level of confidence
(g) Region of acceptance
(h) Operating characteristics curve

Appendix 1

See Tables 6.11, 6.12, 6.13, and 6.14.

Table 6.11 Standard normal distribution

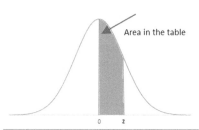

Z	0	0.01	0.02	0.03	0.04	0.05	0.06	0.07	0.08	0.09
0	0.0000	0.0040	0.0080	0.0120	0.0160	0.0199	0.0239	0.0279	0.0319	0.0359
0.1	0.0398	0.0438	0.0478	0.0517	0.0557	0.0596	0.0636	0.0675	0.0714	0.0753
0.2	0.0793	0.0832	0.0871	0.0910	0.0948	0.0987	0.1026	0.1064	0.1103	0.1141
0.3	0.1179	0.1217	0.1255	0.1293	0.1331	0.1368	0.1406	0.1443	0.1480	0.1517
0.4	0.1554	0.1591	0.1628	0.1664	0.1700	0.1736	0.1772	0.1808	0.1844	0.1879
0.5	0.1915	0.1950	0.1985	0.2019	0.2054	0.2088	0.2123	0.2157	0.2190	0.2224
0.6	0.2257	0.2291	0.2324	0.2357	0.2389	0.2422	0.2454	0.2486	0.2517	0.2549
0.7	0.2580	0.2611	0.2642	0.2673	0.2704	0.2734	0.2764	0.2794	0.2823	0.2852

(continued)

Table 6.11 (continued)

Z	0	0.01	0.02	0.03	0.04	0.05	0.06	0.07	0.08	0.09
0.8	0.2881	0.2910	0.2939	0.2967	0.2995	0.3023	0.3051	0.3078	0.3106	0.3133
0.9	0.3159	0.3186	0.3212	0.3238	0.3264	0.3289	0.3315	0.3340	0.3365	0.3389
1	0.3413	0.3438	0.3461	0.3485	0.3508	0.3531	0.3554	0.3577	0.3599	0.3621
1.1	0.3643	0.3665	0.3686	0.3708	0.3729	0.3749	0.3770	0.3790	0.3810	0.3830
1.2	0.3849	0.3869	0.3888	0.3907	0.3925	0.3944	0.3962	0.3980	0.3997	0.4015
1.3	0.4032	0.4049	0.4066	0.4082	0.4099	0.4115	0.4131	0.4147	0.4162	0.4177
1.4	0.4192	0.4207	0.4222	0.4236	0.4251	0.4265	0.4279	0.4292	0.4306	0.4319
1.5	0.4332	0.4345	0.4357	0.4370	0.4382	0.4394	0.4406	0.4418	0.4429	0.4441
1.6	0.4452	0.4463	0.4474	0.4484	0.4495	0.4505	0.4515	0.4525	0.4535	0.4545
1.7	0.4554	0.4564	0.4573	0.4582	0.4591	0.4599	0.4608	0.4616	0.4625	0.4633
1.8	0.4641	0.4649	0.4656	0.4664	0.4671	0.4678	0.4686	0.4693	0.4699	0.4706
1.9	0.4713	0.4719	0.4726	0.4732	0.4738	0.4744	0.4750	0.4756	0.4761	0.4767
2	0.4772	0.4778	0.4783	0.4788	0.4793	0.4798	0.4803	0.4808	0.4812	0.4817
2.1	0.4821	0.4826	0.4830	0.4834	0.4838	0.4842	0.4846	0.4850	0.4854	0.4857
2.2	0.4861	0.4864	0.4868	0.4871	0.4875	0.4878	0.4881	0.4884	0.4887	0.4890
2.3	0.4893	0.4896	0.4898	0.4901	0.4904	0.4906	0.4909	0.4911	0.4913	0.4916
2.4	0.4918	0.4920	0.4922	0.4925	0.4927	0.4929	0.4931	0.4932	0.4934	0.4936
2.5	0.4938	0.4940	0.4941	0.4943	0.4945	0.4946	0.4948	0.4949	0.4951	0.4952
2.6	0.4953	0.4955	0.4956	0.4957	0.4959	0.4960	0.4961	0.4962	0.4963	0.4964
2.7	0.4965	0.4966	0.4967	0.4968	0.4969	0.4970	0.4971	0.4972	0.4973	0.4974
2.8	0.4974	0.4975	0.4976	0.4977	0.4977	0.4978	0.4979	0.4979	0.4980	0.4981
2.9	0.4981	0.4982	0.4982	0.4983	0.4984	0.4984	0.4985	0.4985	0.4986	0.4986
3	0.4987	0.4987	0.4987	0.4988	0.4988	0.4989	0.4989	0.4989	0.4990	0.4990
3.1	0.4990	0.4991	0.4991	0.4991	0.4992	0.4992	0.4992	0.4992	0.4993	0.4993
3.2	0.4993	0.4993	0.4994	0.4994	0.4994	0.4994	0.4994	0.4995	0.4995	0.4995
3.3	0.4995	0.4995	0.4995	0.4996	0.4996	0.4996	0.4996	0.4996	0.4996	0.4997
3.4	0.4997	0.4997	0.4997	0.4997	0.4997	0.4997	0.4997	0.4997	0.4997	0.4998

Table 6.12 Cumulative probabilities for t-distribution

df	Right tail area				
	0.1	0.05	0.025	0.01	0.005
1	3.078	6.314	12.706	31.821	63.657
2	1.886	2.920	4.303	6.965	9.925
3	1.638	2.353	3.182	4.541	5.841
4	1.533	2.132	2.776	3.747	4.604
5	1.476	2.015	2.571	3.365	4.032
6	1.440	1.943	2.447	3.143	3.707
7	1.415	1.895	2.365	2.998	3.499
8	1.397	1.860	2.306	2.896	3.355
9	1.383	1.833	2.262	2.821	3.250
10	1.372	1.812	2.228	2.764	3.169
11	1.363	1.796	2.201	2.718	3.106
12	1.356	1.782	2.179	2.681	3.055
13	1.350	1.771	2.160	2.650	3.012
14	1.345	1.761	2.145	2.624	2.977
15	1.341	1.753	2.131	2.602	2.947
16	1.337	1.746	2.120	2.583	2.921
17	1.333	1.740	2.110	2.567	2.898
18	1.330	1.734	2.101	2.552	2.878
19	1.328	1.729	2.093	2.539	2.861
20	1.325	1.725	2.086	2.528	2.845
21	1.323	1.721	2.080	2.518	2.831
22	1.321	1.717	2.074	2.508	2.819
23	1.319	1.714	2.069	2.500	2.807
24	1.318	1.711	2.064	2.492	2.797
25	1.316	1.708	2.060	2.485	2.787
26	1.315	1.706	2.056	2.479	2.779
27	1.314	1.703	2.052	2.473	2.771
28	1.313	1.701	2.048	2.467	2.763
29	1.311	1.699	2.045	2.462	2.756
30	1.310	1.697	2.042	2.457	2.750
120	1.289	1.658	1.980	2.358	2.617

Table 6.13 Cumulative probabilities for chi-square distribution

df	Right tail area										
	0.995	0.990	0.975	0.950	0.900	0.100	0.050	0.025	0.010	0.005
1	0.000039	0.000157	0.000982	0.003932	0.0158	2.7055	3.8415	5.0239	6.6349	7.8794
2	0.010025	0.020101	0.050636	0.102587	0.2107	4.6052	5.9915	7.3778	9.2103	10.5966
3	0.071722	0.114832	0.215795	0.351846	0.5844	6.2514	7.8147	9.3484	11.3449	12.8382
4	0.2070	0.2971	0.4844	0.7107	1.0636	7.7794	9.4877	11.1433	13.2767	14.8603
5	0.4117	0.5543	0.8312	1.1455	1.6103	9.2364	11.0705	12.8325	15.0863	16.7496
6	0.6757	0.8721	1.2373	1.6354	2.2041	10.6446	12.5916	14.4494	16.8119	18.5476
7	0.9893	1.2390	1.6899	2.1673	2.8331	12.0170	14.0671	16.0128	18.4753	20.2777
8	1.3444	1.6465	2.1797	2.7326	3.4895	13.3616	15.5073	17.5345	20.0902	21.9550
9	1.7349	2.0879	2.7004	3.3251	4.1682	14.6837	16.9190	19.0228	21.6660	23.5894
10	2.1559	2.5582	3.2470	3.9403	4.8652	15.9872	18.3070	20.4832	23.2093	25.1882
11	2.6032	3.0535	3.8157	4.5748	5.5778	17.2750	19.6751	21.9200	24.7250	26.7568
12	3.0738	3.5706	4.4038	5.2260	6.3038	18.5493	21.0261	23.3367	26.2170	28.2995
13	3.5650	4.1069	5.0088	5.8919	7.0415	19.8119	22.3620	24.7356	27.6882	29.8195
14	4.0747	4.6604	5.6287	6.5706	7.7895	21.0641	23.6848	26.1189	29.1412	31.3193
15	4.6009	5.2293	6.2621	7.2609	8.5468	22.3071	24.9958	27.4884	30.5779	32.8013
16	5.1422	5.8122	6.9077	7.9616	9.3122	23.5418	26.2962	28.8454	31.9999	34.2672
17	5.6972	6.4078	7.5642	8.6718	10.0852	24.7690	27.5871	30.1910	33.4087	35.7185
18	6.2648	7.0149	8.2307	9.3905	10.8649	25.9894	28.8693	31.5264	34.8053	37.1565

19	6.8440	7.6327	8.9065	10.1170	11.6509	27.2036	30.1435	32.8523	36.1909	38.5823
20	7.4338	8.2604	9.5908	10.8508	12.4426	28.4120	31.4104	34.1696	37.5662	39.9968
21	8.0337	8.8972	10.2829	11.5913	13.2396	29.6151	32.6706	35.4789	38.9322	41.4011
22	8.6427	9.5425	10.9823	12.3380	14.0415	30.8133	33.9244	36.7807	40.2894	42.7957
23	9.2604	10.1957	11.6886	13.0905	14.8480	32.0069	35.1725	38.0756	41.6384	44.1813
24	9.8862	10.8564	12.4012	13.8484	15.6587	33.1962	36.4150	39.3641	42.9798	45.5585
25	10.5197	11.5240	13.1197	14.6114	16.4734	34.3816	37.6525	40.6465	44.3141	46.9279
26	11.1602	12.1981	13.8439	15.3792	17.2919	35.5632	38.8851	41.9232	45.6417	48.2899
27	11.8076	12.8785	14.5734	16.1514	18.1139	36.7412	40.1133	43.1945	46.9629	49.6449
28	12.4613	13.5647	15.3079	16.9279	18.9392	37.9159	41.3371	44.4608	48.2782	50.9934
29	13.1211	14.2565	16.0471	17.7084	19.7677	39.0875	42.5570	45.7223	49.5879	52.3356
30	13.7867	14.9535	16.7908	18.4927	20.5992	40.2560	43.7730	46.9792	50.8922	53.6720
40	20.7065	22.1643	24.4330	26.5093	29.0505	51.8051	55.7585	59.3417	63.6907	66.7660
50	27.9907	29.7067	32.3574	34.7643	37.6886	63.1671	67.5048	71.4202	76.1539	79.4900
60	35.5345	37.4849	40.4817	43.1880	46.4589	74.3970	79.0819	83.2977	88.3794	91.9517
70	43.2752	45.4417	48.7576	51.7393	55.3289	85.5270	90.5312	95.0232	100.4252	104.2149
80	51.1719	53.5401	57.1532	60.3915	64.2778	96.5782	101.8795	106.6286	112.3288	116.3211
90	59.1963	61.7541	65.6466	69.1260	73.2911	107.5650	113.1453	118.1359	124.1163	128.2989
100	67.3276	70.0649	74.2219	77.9295	82.3581	118.4980	124.3421	129.5612	135.8067	140.1695

Table 6.14 F distribution (F values for right tail area = 0.05)

α = 0.05

Denominator df	Numerator degrees of freedom																
	1	2	3	4	5	6	7	8	9	10	11	12	13	14	16	20	24	30
1	161	200	216	225	230	234	237	239	241	242	243	244	245	245	246	248	249	250
2	18.51	19.00	19.16	19.25	19.30	19.33	19.35	19.37	19.38	19.40	19.40	19.41	19.42	19.42	19.43	19.45	19.45	19.46
3	10.13	9.55	9.28	9.12	9.01	8.94	8.89	8.85	8.81	8.79	8.76	8.74	8.73	8.71	8.69	8.66	8.64	8.62
4	7.71	6.94	6.59	6.39	6.26	6.16	6.09	6.04	6.00	5.96	5.94	5.91	5.89	5.87	5.84	5.80	5.77	5.75
5	6.61	5.79	5.41	5.19	5.05	4.95	4.88	4.82	4.77	4.74	4.70	4.68	4.66	4.64	4.60	4.56	4.53	4.50
6	5.99	5.14	4.76	4.53	4.39	4.28	4.21	4.15	4.10	4.06	4.03	4.00	3.98	3.96	3.92	3.87	3.84	3.81
7	5.59	4.74	4.35	4.12	3.97	3.87	3.79	3.73	3.68	3.64	3.60	3.57	3.55	3.53	3.49	3.44	3.41	3.38
8	5.32	4.46	4.07	3.84	3.69	3.58	3.50	3.44	3.39	3.35	3.31	3.28	3.26	3.24	3.20	3.15	3.12	3.08
9	5.12	4.26	3.86	3.63	3.48	3.37	3.29	3.23	3.18	3.14	3.10	3.07	3.05	3.03	2.99	2.94	2.90	2.86
10	4.96	4.10	3.71	3.48	3.33	3.22	3.14	3.07	3.02	2.98	2.94	2.91	2.89	2.86	2.83	2.77	2.74	2.70
11	4.84	3.98	3.59	3.36	3.20	3.09	3.01	2.95	2.90	2.85	2.82	2.79	2.76	2.74	2.70	2.65	2.61	2.57
12	4.75	3.89	3.49	3.26	3.11	3.00	2.91	2.85	2.80	2.75	2.72	2.69	2.66	2.64	2.60	2.54	2.51	2.47
13	4.67	3.81	3.41	3.18	3.03	2.92	2.83	2.77	2.71	2.67	2.63	2.60	2.58	2.55	2.51	2.46	2.42	2.38
14	4.60	3.74	3.34	3.11	2.96	2.85	2.76	2.70	2.65	2.60	2.57	2.53	2.51	2.48	2.44	2.39	2.35	2.31

15	4.54	3.68	3.29	3.06	2.90	2.79	2.71	2.64	2.59	2.54	2.51	2.48	2.45	2.42	2.38	2.33	2.29	2.25
16	4.49	3.63	3.24	3.01	2.85	2.74	2.66	2.59	2.54	2.49	2.46	2.42	2.40	2.37	2.33	2.28	2.24	2.19
17	4.45	3.59	3.20	2.96	2.81	2.70	2.61	2.55	2.49	2.45	2.41	2.38	2.35	2.33	2.29	2.23	2.19	2.15
18	4.41	3.55	3.16	2.93	2.77	2.66	2.58	2.51	2.46	2.41	2.37	2.34	2.31	2.29	2.25	2.19	2.15	2.11
19	4.38	3.52	3.13	2.90	2.74	2.63	2.54	2.48	2.42	2.38	2.34	2.31	2.28	2.26	2.21	2.16	2.11	2.07
20	4.35	3.49	3.10	2.87	2.71	2.60	2.51	2.45	2.39	2.35	2.31	2.28	2.25	2.22	2.18	2.12	2.08	2.04
21	4.32	3.47	3.07	2.84	2.68	2.57	2.49	2.42	2.37	2.32	2.28	2.25	2.22	2.20	2.16	2.10	2.05	2.01
22	4.30	3.44	3.05	2.82	2.66	2.55	2.46	2.40	2.34	2.30	2.26	2.23	2.20	2.17	2.13	2.07	2.03	1.98
23	4.28	3.42	3.03	2.80	2.64	2.53	2.44	2.37	2.32	2.27	2.24	2.20	2.18	2.15	2.11	2.05	2.01	1.96
24	4.26	3.40	3.01	2.78	2.62	2.51	2.42	2.36	2.30	2.25	2.22	2.18	2.15	2.13	2.09	2.03	1.98	1.94
25	4.24	3.39	2.99	2.76	2.60	2.49	2.40	2.34	2.28	2.24	2.20	2.16	2.14	2.11	2.07	2.01	1.96	1.92
26	4.23	3.37	2.98	2.74	2.59	2.47	2.39	2.32	2.27	2.22	2.18	2.15	2.12	2.09	2.05	1.99	1.95	1.90
27	4.21	3.35	2.96	2.73	2.57	2.46	2.37	2.31	2.25	2.20	2.17	2.13	2.10	2.08	2.04	1.97	1.93	1.88
28	4.20	3.34	2.95	2.71	2.56	2.45	2.36	2.29	2.24	2.19	2.15	2.12	2.09	2.06	2.02	1.96	1.91	1.87
29	4.18	3.33	2.93	2.70	2.55	2.43	2.35	2.28	2.22	2.18	2.14	2.10	2.08	2.05	2.01	1.94	1.90	1.85
30	4.17	3.32	2.92	2.69	2.53	2.42	2.33	2.27	2.21	2.16	2.13	2.09	2.06	2.04	1.99	1.93	1.89	1.84

References

Aczel, & Sounderpandian. (2009). *Complete business statistics*. New York: McGraw-Hill.
Stine, R. A., & Foster, D. (2014). *Statistics for business: Decision making and analysis (SFB)* (2nd ed.). London: Pearson Education Inc..

Chapter 7
Statistical Methods: Regression Analysis

Bhimasankaram Pochiraju and Hema Sri Sai Kollipara

1 Introduction

Regression analysis is arguably one of the most commonly used and misused statistical techniques in business and other disciplines. In this chapter, we systematically develop a linear regression modeling of data. Chapter 6 on basic inference is the only prerequisite for this chapter. We start with a few motivating examples in Sect. 2. Section 3 deals with the methods and diagnostics for linear regression. Section 3.1 is a discussion on what is regression and linear regression, in particular, and why it is important. In Sect. 3.2, we elaborate on the descriptive statistics and the basic exploratory analysis for a data set. We are now ready to describe the linear regression model and the assumptions made to get good estimates and tests related to the parameters in the model (Sect. 3.3). Sections 3.4 and 3.5 are devoted to the development of basic inference and interpretations of the regression with single and multiple regressors. In Sect. 3.6, we take the help of the famous Anscombe (1973) data sets to demonstrate the need for further analysis. In Sect. 3.7, we develop the basic building blocks to be used in constructing the diagnostics. In Sect. 3.8, we use various residual plots to check whether there are basic departures from the assumptions and to see if some transformations on the regressors are warranted.

Electronic supplementary material The online version of this chapter (https://doi.org/10.1007/978-3-319-68837-4_7) contains supplementary material, which is available to authorized users.

B. Pochiraju
Applied Statistics and Computing Lab, Indian School of Business, Hyderabad, Telangana, India

H. S. S. Kollipara (✉)
Indian School of Business, Hyderabad, Telangana, India
e-mail: hemasri.kollipara@gmail.com

© Springer Nature Switzerland AG 2019
B. Pochiraju, S. Seshadri (eds.), *Essentials of Business Analytics*, International Series in Operations Research & Management Science 264,
https://doi.org/10.1007/978-3-319-68837-4_7

Suppose we have developed a linear regression model using some regressors. We find that we have data on one more possible regressor. Should we bring in this variable as an additional regressor, given that the other regressors are already included? This is what is explored through the added variable plot in Sect. 3.9.

In Sect. 3.10, we develop deletion diagnostics to examine whether the presence and absence of a few observations can make a large difference to the quantities of interest in regression, such as regression coefficient estimates, standard errors, and fit of some observations. Approximate linear relationships among regressors are called collinear relationships among regressors. Such collinear relationships can cause insignificance of important regressors, wrong signs for regression coefficient estimates, and instability of the estimates to slight changes in the data. Section 3.11 is devoted to the detection of collinearity and correction through remedial measures such as stepwise regression, best subset regression, ridge regression, and lasso regression. There we note that subset selection deserves special interest. How do we represent categorical regressors? This leads us to the study of dummy variables, which is done in Sect. 3.12. We also consider the case of several categories and ordinal categories. We mention the use of interactions to identify certain effects. Section 3.13 deals with the transformation of response variables and a specific famous method known as Box–Cox transformation. One of the assumptions in linear regression model estimation through least squares is that the errors have equal variance. What if this assumption is violated? What are the ill-effects? How does one detect this violation which is often called heteroscedasticity? Once detected, what remedial measures are available? These questions are dealt with in Sect. 3.14. Another assumption in the estimation of the linear regression model by the method of least squares is that of independence of the errors. We do not deal with this problem in this chapter as this is addressed in greater detail in the Forecasting Analytics chapter (Chap. 12). Section 3.15 deals with the validation of a developed linear regression model. Section 3.16 provides broad guidelines useful for performing linear regression modeling. Finally, Sect. 3.17 addresses some FAQs (frequently asked questions). R codes are provided at the end.

Chatterjee and Hadi (2012) and Draper and Smith (1998) are very good references for linear regression modeling.

2 Motivating Examples

2.1 Adipose Tissue Problem

We all have a body fat called adipose tissue. If abdominal adipose tissue area (AT) is large, it is a potential risk factor for cardiovascular diseases (Despres et al. 1991). The accurate way of determining AT is computed tomography (CT scan). There are three issues with CT scan: (1) it involves irradiation, which in itself can be harmful to the body; (2) it is expensive; and (3) good CT equipment are not available in smaller towns, which may result in a grossly inaccurate measurement of the area. Is there a way to predict AT using one or more anthropological measurements? Despres et al. (1991) surmised that obesity at waist is possibly a good indicator of large *AT*. So they thought of using waist circumference (*WC*) to predict *AT*. Notice that one requires only a measuring tape for measuring WC. This does not have any of the issues mentioned above for the CT scan. In order to examine their surmise, they got a random sample of 109 healthy-looking adult males and measured WC in cm and *AT* in cm^2 using a measuring tape and CT scan, respectively, for each. The data are available on our data website.

Can we find a suitable formula for predicting *AT* of an individual using their WC? How reliable is this prediction? For which group of individuals is this prediction valid? Is there one formula which is the best (in some acceptable sense)? Are there competing formulae, and if so, how to choose among them?

The dataset "wc_at.csv" is available on the book's website.

2.2 Newspaper Problem

Mr. Warner Sr. owns a newspaper publishing house which publishes the daily newspaper *Newsexpress*, having an average daily circulation of 500,000 copies. His son, Mr. Warner Jr. came up with an idea of publishing a special Sunday edition of *Newsexpress*. The father is somewhat conservative and said that they can do so if it is almost certain that the average Sunday circulation (circulation of the Sunday edition) is at least 600,000.

Mr. Warner Jr. has a friend Ms. Janelia, who is an analytics expert whom he approached and expressed his problem. He wanted a fairly quick solution. Ms. Janelia said that one quick way to examine this is to look at data on other newspapers in that locality which have both daily edition and Sunday edition. "Based on these data," said Ms. Janelia, "we can fairly accurately determine the lower bound for the circulation of your proposed Sunday edition." Ms. Janelia exclaimed, "However, there is no way to pronounce a meaningful lower bound with certainty."

What does Ms. Janelia propose to do in order to get an approximate lower bound to the Sunday circulation based on the daily circulation? Are there any assumptions that she makes? Is it possible to check them?

2.3 Gasoline Consumption

One of the important considerations both for the customer and manufacturer of vehicles is the average mileage (miles per gallon of the fuel) that it gives. How does one predict the mileage? Horsepower, top speed, age of the vehicle, volume, and percentage of freeway running are some of the factors that influence mileage. We have data on MPG (miles per gallon), HP (horsepower), and VOL (volume of cab-space in cubic feet) for 81 vehicles. Do HP and VOL (often called explanatory variables or regressors) adequately explain the variation in MPG? Does VOL have explaining capacity of variation in MPG over and above HP? If so, for a fixed HP, what would be the impact on the MPG if the VOL is decreased by 50 cubic feet? Are some other important explanatory variables correlated with HP and VOL missing? Once we have HP as an explanatory variable, is it really necessary to have VOL also as another explanatory variable?

The sample dataset[1] was inspired by an example in the book *Basic Econometrics* by Gujarati and Sangeetha. The dataset "cars.csv" is available on the book's website.

2.4 Wage Balance Problem

Gender discrimination with respect to wages is a hotly debated topic. It is hypothesized that men get higher wages than women with the same characteristics, such as educational qualification and age.

We have data on wage, age, and years of education on 100 men and 100 women with comparable distributions of age and years of education. Is it possible to find a reasonable formula to predict wage based on age, years of education, and gender? Once such a formula is found, one can try to examine the hypothesis mentioned above. If it is found that, indeed, there is a gender discrimination, it may be also of interest to examine whether women catch up with men with an increase in educational qualification. After accounting for gender difference and age, is it worthwhile to have higher educational qualification to get a higher wage? Can one quantify result in such a gain?

The dataset "wage.csv" is available on the book's website.

2.5 Medicinal Value in a Leaf

The leaf of a certain species of trees is known to be of medicinal value proportional to its surface area. The leaf is of irregular shape and hence it is cumbersome to determine the area explicitly. One scientist has thought of two measurements (which are fairly easy to obtain), the length and breadth of the leaf which are defined as follows: the length is the distance between the two farthest points in the leaf and the breadth is the distance between the two farthest points in the leaf perpendicular to the direction of the length. The scientist obtained the length, breadth, and area (area measured in the laborious way of tracing the leaf on a graph paper and counting the squares in the traced diagram) on 100 randomly selected leaves. Is it possible to find

---

[1]Original datasource is US Environmental Pollution Agency (1991), Report EPA/AA/CTAB/91-02 and referred to in the book "Basic Econometrics" by Gujarati and Sangeetha.

an approximate formula for the area of the leaf based on its length and breadth which are relatively easy to measure? The dataset "leaf.csv" is available on the book's website.

3 Methods of Regression

3.1 What Is Regression?

Let us consider the examples in Sect. 2. We have to find an approximate formula—for AT in terms of WC in Example 2.1, for the circulation of Sunday edition in terms of the circulation of daily edition in Example 2.2, for MPG in terms of HP, VOL, and WT in Example 2.3, for wage in terms of age, years of education, and gender in Example 2.4, and for the area of the leaf in terms of its length and width in Example 2.5. We call the variable of interest for which we want to get an approximate formula as the *response variable*. In the literature, the response variable is synonymously referred to as the *regressand*, and the *dependent variable* also. The variables used to predict the response variable are called *regressors*, *explanatory variables*, *independent variables*, or *covariates*. In Example 2.4, wage is the response variable and age, years of education, and gender are the regressors.

In each of these examples, even if the data are available on all the units in the population, the formula cannot be exact. For example, AT is not completely determined by WC. It also depends on gender, weight, triceps, etc. (Brundavani et al. 2006). Similarly, in other examples it is easy to notice that regressors do not completely determine the response variable. There are omitted variables on which we do not have data. We have a further limitation that the data are available on a sample and hence we can only estimate the prediction formula. Thus, there are two stages in arriving at a formula: (a) to postulate a functional form for the approximation which involves some parameters and (b) to estimate the parameters in the postulated functional form based on the sample data.

If we denote the response variable by Y, the regressors by $X_1, \ldots, X_k$, the parameters by $\theta_1, \ldots, \theta_r$, and the combined unobserved variables, called the error, ε, then we attempt to estimate an equation:

$$Y = f(X_1, \ldots, X_k, \theta_1, \ldots, \theta_r, \varepsilon). \tag{7.1}$$

In Example 2.2, we may postulate the formula as

$$Sunday\ circulation = \alpha + \beta\ daily\ circulation + \varepsilon. \tag{7.2}$$

Here the functional form f is linear in the regressor, namely, daily circulation. It is also linear in the parameters, α and β. The error ε has also come into the equation as an additive term.

In Example 2.4, notice that a natural functional form of area in terms of the length and breadth is multiplicative. Thus we may postulate

$$area = \alpha . length^{\beta_1} . breadth^{\beta_2} . \varepsilon \qquad (7.3)$$

Here the functional form is multiplicative in powers of length and breadth and the error. It is not linear in the parameters either.

Alternatively, one may postulate

$$area = \alpha . length^{\beta_1} . breadth^{\beta_2} + \varepsilon \qquad (7.4)$$

In this specification, the functional form is multiplicative in powers of length and breadth but additive in the error. It is not linear in the parameters either.

How does one know which of the two specifications (7.3) and (7.4) is appropriate? Or is it that neither of them is appropriate? There are ways to examine this based on the data. We shall deal with this in detail subsequently.

3.2 What Is Linear Regression?

If the functional form, f, in Eq. (7.1), is linear in the parameters, then the regression is called linear regression. As noted earlier, (7.2) is a linear regression equation. What about Eq. (7.3)? As already noted, this is not a linear regression equation. However, if we make a log transformation on both sides, we get

$$\log(area) = \log \alpha + \beta_1 \log(length) + \beta_2 \log(breadth) + \log \varepsilon \qquad (7.5)$$

which is linear in the parameters: $\log \alpha$, β_1, and β_2. However, this model is not linear in length and breadth.

Such a model is called an *intrinsically linear regression model*.

However, we cannot find any transformation of the model in (7.4) yielding a linear regression model. Such models are called *intrinsically nonlinear regression models*.

3.3 Why Regression and Why Linear Regression?

Regression is performed for one or more of the following reasons:

1. To predict the response variable for a new case based on the data on the regressors for that case.

2. To study the impact of one regressor on the response variable keeping other regressors fixed. For example, one may be interested in the impact of one additional year of education on the average wage for a person aged 40 years.
3. To verify whether the data support certain beliefs (hypotheses)—for example, whether $\beta_1 = \beta_2 = 1$ in the Eq. (7.3) which, if upheld, would mean that the leaf is more or less rectangular.
4. To use as an intermediate result in further analysis.
5. To calibrate an instrument.

Linear regression has become popular for the following reasons:

1. The methodology for linear regression is easily understood, as we shall see in the following sections.
2. If the response variable and the regressors have a joint normal distribution, then the regression (as we shall identify in Sect. 3.5, regression is the expectation of the response variable conditional on the regressors) is a linear function of the regressors.
3. Even though the model is not linear in the regressors, sometimes suitable transformations on the regressors or response variable or both may lead to a linear regression model.
4. The regression may not be a linear function in general, but a linear function may be a good approximation in a small focused strip of the regressor surface.
5. The methodology developed for the linear regression may also act as a good first approximation for the methodology for a nonlinear model.

We shall illustrate each of these as we go along.

3.4 Basic Descriptive Statistics and Box Plots

Analysis of the data starts with the basic descriptive summary of each of the variables in the data (the Appendix on Probability and Statistics provides the background for the following discussion). The descriptive summary helps in understanding the basic features of a variable such as the central tendency, the variation, and a broad empirical distribution. More precisely, the basic summary includes the minimum, the maximum, the first and third quartiles, the median, the mean, and the standard deviation. The minimum and the maximum give us the range. The mean and the median are measures of central tendency. The range, the standard deviation, and the interquartile range are measures of dispersion. The box plot depicting the five-point summary, namely, the minimum, the first quartile, the median, the third quartile, and the maximum, gives us an idea of the empirical distribution. We give below these measures for the (WC, AT) data.

From Table 7.1 and Fig. 7.1, it is clear that the distribution of WC is fairly symmetric, about 91 cm, and the distribution of AT is skewed to the right. We shall see later how this information is useful.

Table 7.1 Descriptive statistics for (WC, AT) data

Variable	min	1st Qu	Median	Mean	3rd Qu	Max	Std. dev.
WC (in cm)	63.5	80.0	90.8	91.9	104.0	121.0	13.55912
AT (in sq. cm)	11.44	50.88	96.54	101.89	137.00	253.00	57.29476

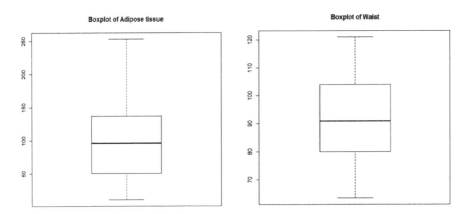

Fig. 7.1 Box plots of adipose tissue area and waist circumference

3.5 Linear Regression Model and Assumptions

The linear regression model can be written as follows:

$$\left.\begin{array}{l} Y = \beta_0 + \beta_1 X_1 + \cdots + \beta_k X_k + \varepsilon \\ E(\varepsilon | X_1, \ldots, X_k) = 0, \\ Var(\varepsilon | X_1, \ldots, X_k) = \sigma^2 (> 0) \end{array}\right\} \quad (7.6)$$

Note on notation: The notation $E(Y|X)$ stands for the expected value of Y given the value of X. It is computed using $P(Y|X)$, which stands for the conditional probability given X. For example, we are given the following information: when $X = 1$, Y is normally distributed with mean 4 and standard deviation of 1; whereas, when $X = 2$, Y is normally distributed with mean 5 and standard deviation of 1.1. $E(Y|X = 1) = 4$ and $E(Y|X = 2) = 5$. Similarly, the notation $Var(Y|X)$ is interpreted as the variance of Y given X. In this case $Var(Y|X = 1) = 1$ and $Var(Y|X = 2) = 1.21$.

The objective is to draw inferences (estimation and testing) related to the parameters $\beta_0, \ldots, \beta_k, \sigma^2$ in the model (7.6) based on the data $(y_i, x_{i1}, \ldots, x_{ik}), i = 1, \ldots, N$ on a sample with N observations from $(Y, X_1, \ldots, X_k)$. Note that Y is a column vector of size N × 1. The transpose of a vector Y (or matrix M) is written as Y^t (M^t), the transpose of a column vector is a row vector and vice versa.

7 Statistical Methods: Regression Analysis

We now write down the observational equations as

$$\left.\begin{array}{c} y_1 = \beta_0 + \beta_1 x_{11} + \cdots + \beta_k x_{1k} + \varepsilon_1 \\ \cdot \\ \cdot \\ \cdot \\ y_N = \beta_0 + \beta_1 x_{N1} + \cdots + \beta_k x_{Nk} + \varepsilon_N \\ \cdot \end{array}\right\} \quad (7.7)$$

The above model can be written compactly as

$$\left.\begin{array}{l} Y = Z\beta + \varepsilon \\ where \\ Y = (y_1, \ldots, y_N)^t, \; X = \left((x_{ij})\right), \; Z = (1 : X), \; \varepsilon = (\varepsilon_1, \ldots, \varepsilon_N)^t \end{array}\right\} \quad (7.8)$$

In (7.8), by $X = ((x_{ij}))$, we mean that X is a matrix, the element in the junction of the ith row and j^{th} column of which is x_{ij}. The size of X is $N \times k$. In the matrix Z, 1 denotes the column vector, each component of which is the number 1. The matrix 1 is appended as the first column and the rest of the columns are taken from X—this is the notation (1:X). The matrices X and Z are of orders $N \times k$ and $N \times (k+1)$, respectively.

We make the following assumption regarding the errors $\varepsilon_1, \ldots, \varepsilon_N$:

$$\varepsilon \mid X \sim N_N\left(0, \sigma^2 I\right) \quad (7.9)$$

that is, the distribution of the errors given X is an N-dimensional normal distribution with mean zero and covariance matrix equal to σ^2 times the identity matrix. In (7.9), I denotes the identity matrix of order $N \times N$. The identity matrix comes about because the errors are independent of one another, therefore, covariance of one error with another error equals zero.

From (7.8) and (7.9), we have

$$Y \mid X \sim N_N\left(Z\beta, \sigma^2 I\right).$$

In other words, the regression model $Z\beta$ represents the mean value of Y given X. The errors are around the mean values.

What does the model (7.7) (or equivalently 7.8) together with the assumption (7.9) mean? It translates into the following:

1. The model is linear in the parameters $\beta_0, \ldots, \beta_k$ (L).
2. Errors conditional on the data on the regressors are independent (I).
3. Errors conditional on the data on the regressors have a joint normal distribution (N).

4. The variance of each of the errors conditional on the data on the regressors is σ^2(E).
5. Each of the errors conditional on the data on the regressors has 0 mean.

In (4) above, E stands for equal variance. (5) above is usually called the exogeneity condition. This actually implies that the observed covariates are uncorrelated with the unobserved covariates. The first four assumptions can be remembered through an acronym: LINE.

Why are we talking about the distribution of $\varepsilon_i \mid X$? Is it not a single number? Let us consider the adipose tissue example. We commonly notice that different people having the same waist circumference do not necessarily have the same adipose tissue area. Thus there is a distribution of the adipose tissue area for people with a waist circumference of 70 cm. Likewise in the wage example, people with the same age and education level need not get exactly the same wage.

The above assumptions will be used in drawing the inference on the parameters. However, we have to check whether the data on hand support the above assumptions. How does one draw the inferences and how does one check for the validity of the assumptions? A good part of this chapter will be devoted to this and the interpretations.

3.6 Single Regressor Case

Let us consider Examples 2.1 and 2.2. Each of them has one regressor, namely, WC in Example 2.1 and daily circulation in Example 2.2. Thus, we have bivariate data in each of these examples. What type of relationship does the response variable have with the regressor? We have seen in the previous chapter that covariance or correlation coefficient between the variables is one measure of the relationship. We shall explore this later where we shall examine the interpretation of the correlation coefficient. But we clearly understand that it is just one number indicating the relationship. However, we do note that each individual (WC, AT) is an ordered pair and can be plotted as a point in the plane. This plot in the plane with WC as the X-axis and AT as the Y-axis is called the scatterplot of the data. We plot with the response variable on the Y-axis and the regressor on the X-axis. For the (WC, AT) data the plot is given below:

What do we notice from this plot?

1. The adipose tissue area is by and large increasing with increase in waist circumference.
2. The variation in the adipose tissue area is also increasing with increase in waist circumference.

The correlation coefficient for this data is approximately 0.82. This tells us the same thing as (1) above. It also tells that the strength of (linear) relationship between the two variables is strong, which prompts us to fit a straight line to the data. But by looking at the plot, we see that a straight line does not do justice for large values of waist circumference as they are highly dispersed. (More details on this later.) So the first lesson to be learnt is: *If you have a single regressor, first look at the scatterplot.* This will give you an idea of the form of relationship between the response variable and the regressor. If the graph suggests a linear relationship, one can then check the correlation coefficient to assess the strength of the linear relationship between the response variable and the regressor.

We have the following linear regression model for the adipose tissue problem:

$$\text{Model}: \quad AT = \beta_0 + \beta_1 WC + \varepsilon \tag{7.10}$$

$$\text{Data}: \quad (AT_i, WC_i), i = 1, \ldots 109 \tag{7.11}$$

$$\text{Model adapted to data}: \quad AT_i = \beta_0 + \beta_1 WC_i + \varepsilon_i, i = 1, \ldots, 109 \tag{7.12}$$

Assumptions: $\varepsilon_i \mid WC_i, i = 1, \ldots, 109$ are independently and identically distributed as normal with mean 0 and variance σ^2 often written in brief as

$$\varepsilon_i \mid WC_i, i = 1, \ldots, 109 \text{ are } iid \ N\left(0, \sigma^2\right) \text{ variables.} \tag{7.13}$$

Model described by (7.12) and (7.13) is a special case of the model described by (7.7) and (7.9) where k = 1 (For a single regressor case, k = 1, the number of regressors.) and N = 109.

A linear regression model with one regressor is often referred to as a *Simple Linear Regression* model.

Estimation of Parameters

From the assumptions it is clear that all the errors (conditional on the Waist Circumference values) are of equal importance.

If we want to fit a straight line for the scatterplot in Fig. 7.2, we look for that line for which some reasonable measure of the magnitude of the errors is small. A straight line is completely determined by its intercept and slope, which we denote by β_0 and β_1, respectively. With this straight line approximation, from (7.12), the error for the ith observation is $\varepsilon_i = AT_i - \beta_0 - \beta_1 WC_i, i = 1, \ldots, 109$.

A commonly used measure for the magnitude of the errors is the sum of their squares, namely, $\sum_{i=1}^{109} \varepsilon_i^2$. So if we want this measure of the magnitude of the errors to be small, we should pick up the values of β_0 and β_1, which will minimize $\sum_{i=1}^{109} \varepsilon_i^2$. This method is called the *Method of Least Squares*. This is achieved by solving the equations (often called *normal equations*):

$$\left. \begin{array}{l} \sum_{i=1}^{109} AT_i = 109\beta_0 + \beta_1 \sum_{i=1}^{109} WC_i \\ \sum_{i=1}^{109} AT_i WC_i = \beta_0 \sum_{i=1}^{109} WC_i + \beta_1 \sum_{i=1}^{109} WC_i^2 \end{array} \right\} \tag{7.14}$$

Fig. 7.2 Lo and behold! A picture is worth a thousand words

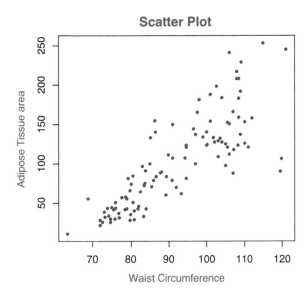

Solving the two equations in (7.14) simultaneously, we get the estimators $\widehat{\beta}_0, \widehat{\beta}_1$ of β_0, β_1, respectively, as

$$\left.\begin{array}{l}\widehat{\beta}_1 = \dfrac{\text{cov}(AT,WC)}{V(WC)} = \dfrac{\sum_{i=1}^{109} AT_i WC_i - \sum_{i=1}^{109} AT_i \sum_{i=1}^{109} WC_i/109}{\sum_{i=1}^{109} WC_i^2 - \left(\sum_{i=1}^{109} WC_i\right)^2/109} \\ \widehat{\beta}_0 = \overline{AT} - \widehat{\beta}_1 \overline{WC}\end{array}\right\} \quad (7.15)$$

(Let $v_1, \ldots, v_r$ be a sample from a variable V. By $\overline{V}$, we mean the average of the sample.)

Thus the estimated regression line, often called the *fitted model*, is

$$\widehat{AT} = \widehat{\beta}_0 + \widehat{\beta}_1 WC. \quad (7.16)$$

The predicted value (often called the *fitted value*) of the ith observation is given by

$$\widehat{AT_i} = \widehat{\beta}_0 + \widehat{\beta}_1 WC_i. \quad (7.17)$$

Notice that this is the part of the adipose tissue area for the ith observation explained by our fitted model.

The part of the adipose tissue area of the ith observation, not explained by our fitted model, is called the *Pearson residual*, henceforth referred to as *residual* corresponding to the ith observation, and is given by

$$e_i = AT_i - \widehat{AT_i}. \quad (7.18)$$

7 Statistical Methods: Regression Analysis

Thus, $AT_i = \widehat{AT_i} + e_i = \widehat{\beta}_0 + \widehat{\beta}_1 WC_i + e_i$. Compare this with (7.12). We notice that $\widehat{\beta}_0, \widehat{\beta}_1, e_i$ are the sample analogues of $\beta_0, \beta_1, \varepsilon_i$ respectively. We know that the errors, ε_i are unobservable. Therefore, we use their sample representatives e_i to check the assumptions (7.9) on the errors.

The sum of squares of residuals, $R_0^2 = \sum_{i=1}^{109} e_i^2$ is the part of the variation in the adipose tissue area that is not explained by the fitted model (7.16) estimated by the method of least squares. The estimator of σ^2 is obtained as

$$\widehat{\sigma}^2 = \frac{R_0^2}{107}. \tag{7.19}$$

The data has 109 degrees of freedom. Since two parameters $\widehat{\beta}_0, \widehat{\beta}_1$ are estimated, 2 degrees of freedom are lost and hence the effective sample size is 107. That is the reason for the denominator in (7.19). In the regression output produced by R (shown below), the square root of $\widehat{\sigma}^2$ is called the residual standard error, denoted as s_e.

Coefficient of Determination

How good is the fitted model in explaining the variation in the response variable, the adipose tissue area? The variation in the adipose tissue area can be represented by $\sum_{i=1}^{109} (AT_i - \overline{AT})^2$. As we have seen above, the variation in adipose tissue area not explained by our model is given by R_0^2. Hence the part of variation in the adipose tissue area that is explained by our model is given by

$$\sum_{i=1}^{109} (AT_i - \overline{AT})^2 - R_0^2. \tag{7.20}$$

Thus the proportion of the variation in the adipose tissue area that is explained by our model is

$$\frac{\sum_{i=1}^{109} (AT_i - \overline{AT})^2 - R_0^2}{\sum_{i=1}^{109} (AT_i - \overline{AT})^2} = 1 - \frac{R_0^2}{\sum_{i=1}^{109} (AT_i - \overline{AT})^2} \tag{7.21}$$

This expression is called the coefficient of determination corresponding to the model and is denoted by R^2. Formally, R^2 is the ratio of the variation explained by the model to total variation in the response variable.

It is easy to see that $0 \leq R^2 \leq 1$ always. How do we interpret the extreme values for R^2?

If R^2 is 0, it means that there is no reduction in the variation in the response variable achieved by our model and thus this model is useless. (Caution: This however does not mean that the regressor is useless in explaining the variation in the response variable. It only means that the function, namely, the linear function in this case, is not useful. Some other function of the same regressor may be quite useful. See Exercise 7.1.)

On the other hand, if $R^2 = 1$, it means that $R_0^2 = 0$, which in turn means that each residual is 0. So the model fits perfectly to the data.

Let us recall that R^2 is *the proportion of variation in the response variable that is explained by the model*.

In the case of a *single regressor*, one can show that R^2 is the square of the correlation coefficient between the response variable and the regressor (see Exercise 7.2). This is the reason for saying that the correlation coefficient is a measure of the strength of a linear relationship between the response variable and the regressor (in the single regressor case).

However, the above two are the extreme cases. For almost all practical data sets, $0 < R^2 < 1$. Should we be elated when R^2 is large or should we be necessarily depressed when it is small? Fact is, R^2 is but just one measure of fit. We shall come back to this discussion later (see also Exercise 7.2).

Prediction for a New Observation

For a new individual whose waist circumference is available, say, $WC = x_0$ cm, how do we predict his abdominal adipose tissue? This is done by using the formula (7.16). Thus the predicted value of the adipose tissue for this person is

$$\widehat{AT} = \widehat{\beta}_0 + \widehat{\beta}_1 x_0 \quad \text{sq.cm.} \tag{7.22}$$

with the standard error given by

$$s_1 = s_e \sqrt{1 + \frac{1}{107} + \frac{(x_0 - \overline{WC})^2}{\sum_{i=1}^{109}(WC_i - \overline{WC})^2}}. \tag{7.23}$$

The average value of the adipose tissue area for all the individuals with $WC = x_0$ cm is also estimated by the formula (7.22), with the standard error given by.

$$s_2 = s_e \sqrt{\frac{1}{107} + \frac{(x_0 - \overline{WC})^2}{\sum_{i=1}^{109}(WC_i - \overline{WC})^2}} \tag{7.24}$$

Notice the difference between (7.23) and (7.24). Clearly (7.23) is larger than (7.24). This is not surprising because the variance of an observation is larger than that of the average as seen in the Chap. 6, Statistical Methods—Basic Inferences. Why does one need the standard-error-formulae in (7.23) and (7.24)? As we see in Chap. 6, these are useful in obtaining the prediction and confidence intervals. Also note that the confidence interval is a statement of confidence about the true line—because we only have an estimate of the line. See (7.28) and (7.29) for details.

Testing of Hypotheses and Confidence Intervals

Consider the model (7.10). If $\beta_1 = 0$, then there is no linear relationship between AT and WC. Thus testing

$$H_0 : \beta_1 = 0 \quad against \quad H_1 : \beta_1 \neq 0 \tag{7.25}$$

is equivalent to testing for the usefulness of WC in predicting AT through a linear relationship. It can be shown that under the assumptions (7.13), $\widehat{\beta_1}$ as obtained in (7.15) has a normal distribution with mean β_1 and a suitable variance.

Thus, the test statistic to perform the test (7.25) is given by

$$\frac{\widehat{\beta_1}}{S.E.\left(\widehat{\beta_1}\right)} \qquad (7.26)$$

Where $S.E.(.)$ stands for the standard error of $(.)$.

As seen in Chap. 6, under the null hypothesis, (7.26) has a student's t distribution with 107 (109 minus 2) degrees of freedom. The corresponding p-value can be obtained as in Chap. 6. Testing $H_0 : \beta_1 = 0$ *against* $H_1 : \beta_1 \neq 0$ can be performed in a similar manner.

Again, as seen in Chap. 6, a 95% confidence interval for β_1 is given by

$$\left(\widehat{\beta_1} - t_{107, 0.025} S.E.\left(\widehat{\beta_1}\right), \widehat{\beta_1} + t_{107, 0.025} S.E.\left(\widehat{\beta_1}\right)\right) \qquad (7.27)$$

where $t_{107, 0.025}$ stands for the 97.5 percentile value of the student's t distribution with 107 degrees of freedom.

Similarly, a 95% confidence interval for the average value of the adipose tissue for all individuals having waist circumference equal to x_0 cm is given by

$$\left(\widehat{AT} - t_{107, 0.025} s_2, \widehat{AT} + t_{107, 0.025} s_2\right). \qquad (7.28)$$

Also, a 95% *prediction interval* for the adipose tissue for an individual having waist circumference equal to x_0 cm is given by

$$\left(\widehat{AT} - t_{107, 0.025} s_1, \widehat{AT} + t_{107, 0.025} s_1\right). \qquad (7.29)$$

From the expressions for s_i, $i = 1, 2$ (as in 7.23 and 7.24), it is clear that the widths of the confidence and prediction intervals is the least when $x_0 = \overline{WC}$ and gets larger and larger as x_0 moves farther away from $\overline{WC}$. This is the reason why it is said that the prediction becomes unreliable if you try to predict the response variable value for a regressor value outside the range of the regressor values. The same goes for the estimation of the average response variable value.

Let us now consider the linear regression output in R for regressing AT on WC and interpret the same.

```
> model<-lm(AT ~ Waist, data=wc_at)
> summary(model)

Call:
lm(formula = AT ~ Waist, data = wc_at)

Residuals:
     Min       1Q   Median       3Q      Max
-107.288  -19.143   -2.939   16.376   90.342
```

```
Coefficients:
            Estimate Std. Error t value Pr(>|t|)
(Intercept) -215.9815   21.7963  -9.909   <2e-16 ***
Waist          3.4589    0.2347  14.740   <2e-16 ***
---
Signif. codes:  0 '***' 0.001 '**' 0.01 '*' 0.05 '.' 0.1 ' ' 1

Residual standard error: 33.06 on 107 degrees of freedom
Multiple R-squared:  0.67,    Adjusted R-squared:  0.667
F-statistic: 217.3 on 1 and 107 DF,  p-value: < 2.2e-16
```

Interpretation of the Regression Output

From the output, the following points emerge.

1. From the five-point summary (min, 1Q, median, etc.) and box plot of the residuals, it appears that the distribution of the residuals is skewed to the left. It also shows that there are four residuals which are too far from the center of the data. The normality of the residuals needs to be examined more closely.
2. The estimated regression equation is

$$\widehat{AT} = -215.9815 + 3.4589 WC. \tag{7.30}$$

From here we conclude that one cm increase in WC leads to an increase of 3.4589 sq.cm. in AT on average.

3. The t-value in each row is the ratio of the coefficient estimate and the standard error (see 7.26). The corresponding p-values are given in the next column of the table. The estimated coefficient estimates for both intercept and WC are highly significant. (The hypothesis $\beta_0 = 0$ against $\beta_0 \neq 0$ and also, the hypothesis $\beta_1 = 0$ against $\beta_1 \neq 0$ are both rejected since each of the p-values is smaller than 0.05.) This means that, based on this model, waist circumference does contribute to the variation in the adipose tissue area.
4. The estimate of the error-standard deviation, namely, the residual standard error is 33.06 cm.
5. The coefficient of determination R^2 is 0.67. This means that 67% of the variation in the adipose tissue area is explained by this model. When there is only one regressor in the model, this also means that the square of the correlation coefficient (0.8186) between the adipose tissue area and the waist circumference is 0.67. The sign of the correlation coefficient is the same as the sign of the regression coefficient estimate of WC which is positive. Hence, we conclude that AT, on average, increases with an increase in WC.
6. What is the interpretation of the F-statistic in the output? This is the ratio of the explained variation in the response variable to the residual or "still unexplained" variation based on the fitted model (after adjusting for the respective degrees of freedom). Intuitively, the larger this value, the better the fit because a large

part of the variation in the response variable is explained by the fitted model if the value is large. How does one judge how large is large? Statistically this statistic has an F distribution with the numerator and denominator degrees of freedom under the null hypothesis of the ineffectiveness of the model. In the single regressor case, the F-statistic is the square of the t-statistic. In this case, both the t-test for the significance of the regression coefficient estimate of *WC* and the F-statistic test for the same thing, namely, whether *WC* is useful for predicting *AT* (or, equivalently, in explaining the variation in *AT*) through the model under consideration.

7. Consider the adult males in the considered population with a waist circumference of 100 cm. We want to estimate the average abdominal adipose tissue area of these people. Using (7.30), the point estimate is $-215.9815 + 3.4589 \times 100 = 129.9085$ square centimeters.

8. Consider the same situation as in point 7 above. Suppose we want a 95% confidence interval for the average abdominal adipose tissue area of all individuals with waist circumference equal to 100 cm. Using the formula (7.28), we have the interval [122.5827, 137.2262]. Now consider a specific individual whose waist circumference is 100 cm. Using the formula (7.29), we have the 95% prediction interval for this individual's abdominal adipose tissue as [63.94943, 195.8595].

9. In point 7 above, if the waist circumference is taken as 50 cm, then using (7.30), the estimated average adipose tissue area turns out to be $-215.9815 + 3.4589 \times 50 = -42.9365$ square centimeters, which is absurd. Where is the problem? The model is constructed for the waist circumference in the range (63.5 cm, 119.90 cm). The formula (7.30) is applicable for estimating the average adipose tissue area when the waist circumference is in the range of waist circumference used in the estimation of the regression equation. If one goes much beyond this range, then the confidence intervals and the prediction intervals as constructed in the point above will be too large for the estimation or prediction to be useful.

10. Testing whether *WC* is useful in predicting the abdominal adipose tissue area through our model is equivalent to testing the null hypothesis, $\beta_1 = 0$. This can be done using (7.26) and this is already available in the output before Fig. 7.3. The corresponding p-value is 2×10^{-16}. This means that if we reject the hypothesis $\beta_1 = 0$, based on our data, then we reject wrongly only 2×10^{-16} proportion of times. Thus we can safely reject the null hypothesis and declare that *WC* is useful for predicting *AT* as per this model.

While we used this regression model of AT on Waist, this is not an appropriate model since the equal variance assumption is violated. For a suitable model for this problem, see Sect. 3.16.

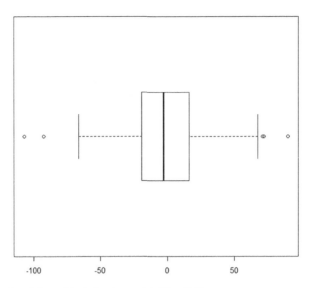

Fig. 7.3 Box plot of the residuals for the model AT vs WC

Table 7.2 Summary statistics of HP, MPG, and VOL

Variable	Min	1st quartile	Median	Mean	3rd quartile	Max.	Standard deviation
HP	49.0	84.0	100.0	117.5	140.0	322.0	57.1135
MPG	12.10	27.86	35.15	34.42	39.53	53.70	9.1315
VOL	50.00	89.00	101.00	98.77	113.00	160.00	22.3015

3.7 Multiple Regressors Case

Let us consider the example in Sect. 2.3. Here we have two regressors, namely, HP and VOL. As in the single regressor case, we can first look at the summary statistics and the box plots to understand the distribution of each variable (Table 7.2 and Fig. 7.4).

From the above summary statistics and plots, we notice the following:

(a) The distribution of MPG is slightly left skewed.
(b) The distribution of VOL is right skewed and there are two points which are far away from the center of the data (on this variable), one to the left and the other to the right.
(c) The distribution of HP is heavily right skewed.

Does point (a) above indicate a violation of the normality assumption (7.9)? Not necessarily, since the assumption (7.9) talks about the conditional distribution of MPG given VOL and HP whereas the box plot of MPG relates to the unconditional distribution of MPG. As we shall see later, this assumption is examined using residuals which are the representatives of the errors. Point (b) can be helpful when

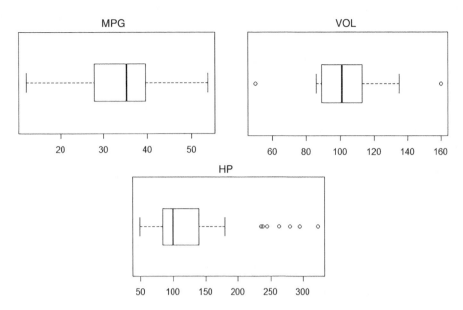

Fig. 7.4 Box plots of MPG, VOL, and HP

we identify, interpret and deal with unusual observations which will be discussed in the section. Point (c) will be helpful when we will look at the residual plots to identify suitable transformations which will be discussed in Sect. 3.10.

We can then look at the scatterplots for every pair of the variables: MPG, VOL, and HP. This can be put in the form of a matrix of scatterplots.

Scatterplots Matrix

The matrix of the scatterplots in Fig. 7.5, is called the scatterplot matrix of all the variables, namely, the response variable and the regressors. Unlike the scatterplot in the single regressor case, the scatterplot matrix in the multiple regressors case is of limited importance. In the multiple regressors case, we are interested in the influence of a regressor over and above that of other regressors. We shall elaborate upon this further as we go along. However, the scatterplots in the scatterplot matrix ignore the influence of the other regressors. For example, from the scatterplot of HP vs MPG (second row first column element in Fig. 7.2), it appears that MPG has a quadratic relationship with HP. But this ignores the impact of the other regressor on both MPG and HP. How do we take this into consideration? After accounting for these impacts, will the quadratic relationship still hold? We shall study these aspects in Sect. 3.10. The scatterplot matrix is useful in finding out whether there is almost perfect linear relationship between a pair of regressors. Why is this important? We shall study this in more detail in Sect. 3.13.

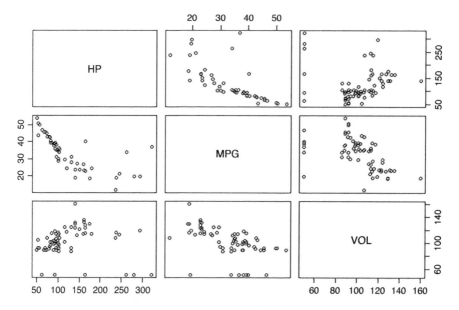

Fig. 7.5 Scatterplot matrix of variables

A linear regression model with two or more regressors is often referred to as a *multiple linear regression* model. Let us start with the following multiple linear regression model for the MPG problem.

$$\text{Model: } MPG = \beta_0 + \beta_1 HP + \beta_2 VOL + \varepsilon \tag{7.31}$$

Where does the error, ε, come from? We know that MPG is not completely determined by HP and VOL. For example, the age of the vehicle, the type of the road (smooth, bumpy, etc.) and so on impact the MPG. Moreover, as we have seen in the scatterplot: MPG vs HP, a quadratic term in HP may be warranted too. (We shall deal with this later.) All these are absorbed in ε.

$$\text{Data: } (MPG_i, HP_i, VOL_i), i = 1, \ldots, 81. \tag{7.32}$$

$$\text{Model adapted to data: } MPG_i = \beta_0 + \beta_1 HP_i + \beta_2 VOL_i + \varepsilon_i, i = 1, \ldots, 81 \tag{7.33}$$

Assumptions: $\varepsilon_i \mid HP_i, VOL_i, i = 1, \ldots, 81$ are independently and identically distributed as normal with mean 0 and variance σ^2 often written in brief as

$$\varepsilon_i \mid HP_i, VOL_i, i = 1, \ldots, 81 \text{ are } iid \ N\left(0, \sigma^2\right) \text{ variables.} \tag{7.34}$$

The model described by (7.33) and (7.34) is a special case of the model described by (7.7) and (7.9) where k = 2 and N = 81.

7 Statistical Methods: Regression Analysis

In the formula (7.32), β_1 and β_2 are the rates of change in MPG with respect to HP and VOL when the other regressor is kept fixed. Thus, these are partial rates of change. So strictly speaking, β_1 and β_2 should be called *partial regression coefficients*. When there is no confusion we shall refer to them as just regression coefficients.

Estimation of Parameters

As in the single regressor case, we estimate the parameters by *the method of least squares* (least sum of squares of the errors). Thus, we are led to the normal equations:

$$\left.\begin{array}{l}\sum_{i=1}^{81} MPG_i = 81\beta_0 + \beta_1 \sum_{i=1}^{81} HP_i + \beta_2 \sum_{i=1}^{81} VOL_i \\ \sum_{i=1}^{81} MPG_i HP_i = \beta_0 \sum_{i=1}^{81} HP_i + \beta_1 \sum_{i=1}^{81} HP_i^2 + \beta_2 \sum_{i=1}^{81} HP_i VOL_i \\ \sum_{i=1}^{81} MPG_i VOL_i = \beta_0 \sum_{i=1}^{81} VOL_i + \beta_1 \sum_{i=1}^{81} HP_i VOL_i + \beta_2 \sum_{i=1}^{81} VOL_i^2\end{array}\right\}$$
(7.35)

The solution of the system of equations in (7.35) yields the estimates $\widehat{\beta}_0, \widehat{\beta}_1, \widehat{\beta}_2$ of the parameters $\beta_0, \beta_1, \beta_2$ respectively.

Thus, the estimated regression equation or the *fitted model* is

$$\widehat{MPG} = \widehat{\beta}_0 + \widehat{\beta}_1 HP + \widehat{\beta}_2 VOL \quad (7.36)$$

Along similar lines to the single regressor case, the *fitted value* and the *residual* corresponding to the ith observation are, respectively,

$$\left.\begin{array}{l}\widehat{MPG}_i = \widehat{\beta}_0 + \widehat{\beta}_1 HP_i + \widehat{\beta}_2 VOL_i \\ e_i = MPG_i - \widehat{MPG}_i\end{array}\right\}. \quad (7.37)$$

We recall that the fitted value and the residual for the ith observation are the explained and the unexplained parts of the observation based on the fitted model.

Residual Sum of Squares

As before, we use their sample representatives e_i to check the assumptions (7.9) on the errors.

The sum of squares of residuals, $R_0^2 = \sum_{i=1}^{81} e_i^2$ is the part of the variation in MPG that is not explained by the fitted model (7.36) obtained by the method of least squares.

The estimator of σ^2 is obtained as

$$\widehat{\sigma}^2 = \frac{R_0^2}{78}. \quad (7.38)$$

The data has 81 degrees of freedom. Since three parameters $\widehat{\beta}_0, \widehat{\beta}_1, \widehat{\beta}_2$ are estimated, three degrees of freedom are lost and hence the effective sample size is $81 - 3 = 78$. That is the reason for the denominator in (7.38). As mentioned earlier, the square root of $\widehat{\sigma}^2$ is called the residual standard error in the R package.

Interpretation of the Regression Coefficient Estimates and Tests of Significance

From (7.37), we see that $\widehat{\beta}_1$ is the estimated change in MPG per unit increase in HP when VOL is kept constant. (It is easy to see that $\widehat{\beta}_1$ is the partial derivative, $\frac{\partial \widehat{MPG}}{\partial HP}$.) Notice the change in the interpretation of $\widehat{\beta}_1$ from that of the corresponding coefficient estimate in the single regressor case. The significance of $\widehat{\beta}_1$ is tested in the same way as in the single regressor case, that is, using the statistic as in (7.26). The degrees of freedom for the *t*-statistic are the same as those of the residual sum of squares, namely, 78 (the total number of observations minus the number of β parameters estimated).

Coefficient of Multiple Determination, R^2

As in the single regressor case, we ask the question: How good is the fitted model in explaining the variation in the response variable, MPG? The variation in MPG can be represented by $\sum_{i=1}^{81}(MPG_i - \overline{MPG})^2$. As we saw, the variation in MPG not explained by our fitted model is given by R_0^2. Hence the part of variation in MPG that is explained by our model is given by

$$\sum_{i=1}^{81}(MPG_i - \overline{MPG})^2 - R_0^2. \tag{7.39}$$

Thus the proportion of the variation in MPG that is explained by our fitted model is

$$\frac{\sum_{i=1}^{81}(MPG_i - \overline{MPG})^2 - R_0^2}{\sum_{i=1}^{81}(MPG_i - \overline{MPG})^2} = 1 - \frac{R_0^2}{\sum_{i=1}^{81}(MPG_i - \overline{MPG})^2} \tag{7.40}$$

This expression, similar to that in (7.21) is called the *coefficient of multiple determination* corresponding to the fitted model and is also denoted by R^2.

It is easy to see that $0 \leq R^2 \leq 1$ always. The interpretation of the extreme values for R^2 is the same as in the single regressor case.

Adjusted R^2

It can be shown that R^2 almost always increases with more regressors. (It never decreases when more regressors are introduced.) So R^2 may not be a very good criterion to judge whether a new regressor should be included. So it is meaningful to look for criteria which impose a penalty for unduly bringing in a new regressor into the model. One such criterion is Adjusted R^2 defined below. (When we deal with subset selection, we shall introduce more criteria for choosing a good subset of regressors.) We know that both R_0^2 and $\sum_{i=1}^{81}(MPG_i - \overline{MPG})^2$ are representatives of the error variance σ^2. But the degrees of freedom for the former is $81 - 3 = 78$ and for the latter is $81 - 1 = 80$. When we are comparing both of them as in (7.40), some feel that we should compare the measures per unit degree of freedom as then they will be true representatives of the error variance. Accordingly, adjusted R^2 is defined as

$$\text{Adj } R^2 = 1 - \frac{R_0^2/(81-3)}{\sum_{i=1}^{81}(MPG_i - \overline{MPG})^2/(81-1)}. \tag{7.41}$$

The adjusted R^2 can be written as $1 - \frac{n-1}{n-K}(1 - R^2)$, where n is the number of observations and K is the number of parameters of our model that are being estimated. From this it follows, that unlike R^2, adjusted R^2 may decrease with the introduction of a new regressor. However, adjusted R^2 may become negative and does not have the same intuitive interpretation as that of R^2.

As we can see, adjusted R^2 is always smaller than the value of R^2. A practical thumb rule is to examine whether R^2 and adjusted R^2 are quite far apart. One naïve way to judge this is to see if the relative change, $\frac{R^2 - \text{adj } R^2}{R^2}$, is more than 10%. If it is not, go ahead and interpret R^2. However, if it is, then it is an indication that there is some issue with the model—either there is an unnecessary regressor in the model or there is some unusual observation. We shall talk about the unusual observations in Sect. 3.12.

Let us now consider Example 2.3, the gasoline consumption problem. We give the R-output of the linear regression of MPG on HP and VOL in Table 7.3.

Table 7.3 Regression output

```
> model1<-lm(MPG ~ HP + VOL, data=Cars)
> summary(model1)

Call:
lm(formula = MPG ~ HP + VOL, data = Cars)

Residuals:
    Min      1Q  Median      3Q     Max
-8.3128 -3.3714 -0.1482  2.8260 15.4828

Coefficients:
             Estimate Std. Error t value Pr(>|t|)
(Intercept) 66.586385   2.505708  26.574  < 2e-16 ***
HP          -0.110029   0.009067 -12.135  < 2e-16 ***
VOL         -0.194798   0.023220  -8.389 1.65e-12 ***
---
Signif. codes:  0 '***' 0.001 '**' 0.01 '*' 0.05 '.' 0.1 ' ' 1

Residual standard error: 4.618 on 78 degrees of freedom
Multiple R-squared:  0.7507,    Adjusted R-squared:  0.7443
F-statistic: 117.4 on 2 and 78 DF,  p-value: < 2.2e-16
```

Fig. 7.6 The box plot of the residuals for the model MPG vs VOL and HP

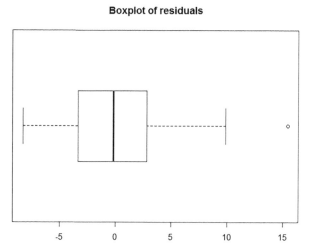

Interpretation of the Regression Output

From the output and the plot above, we notice the following:

1. From the five-point summary of the residuals and the box-plot in Fig. 7.6, we observe that the distribution of the residuals is skewed to the right. Also, there appears to be one unusually large residual which is marked by a small circle towards the right in the box plot. This may have an implication towards the deviation from normality of the residuals. We shall examine this more closely using the QQ plot of suitable residuals.
2. The estimated regression equation is:

$$\widehat{MPG} = 66.586385 - 0.110029 HP - 0.194798 VOL. \tag{7.42}$$

3. Do the signs of the regression coefficient estimates conform to the intuition? Higher horse power cars consume more petrol and hence give lower MPG. The negative sign of the regression coefficient conforms to this intuition. Similarly large cars consume more petrol justifying the negative coefficient estimate of VOL.
4. From (7.42) we infer that a 10 cubic feet increase in the volume of the vehicle, with no change in horse power, will lead to a decrease of approximately 2 miles (to be precise, $10 \times 0.194798 = 1.94798$ miles) per gallon. Notice the difference in the interpretation of the (partial) regression coefficient estimate of VOL from the interpretation of a regression coefficient estimate in a single regressor case. In the present case it is the rate of change in the response variable with respect to the regressor under consideration, keeping the other regressors fixed. From (7.42), a unit increase in HP, keeping the volume unchanged, will lead to a reduction in the mileage, that is, a reduction in MPG by 0.110029. This is based on the data on the sample collected, if another sample is used, then the coefficient estimate

of HP is unlikely to be exactly the same as 0.110029. So, can we give some realistic approximate bounds for the coefficient β_1 of HP in the population? In other words, we are seeking an interval estimate of the coefficient. A 95% confidence interval for the coefficient of HP is given by

$$\widehat{\beta}_1 \pm t_{78, 0.025} S.E.\left(\widehat{\beta}_1\right).$$

The computed 95% confidence interval for β_1 is $(-0.1280797, -0.09197832)$. Thus in the population of the vehicles, a unit increase in the horse power with no change in volume can lead to as high a decrease in the MPG as 0.128 or as low as 0.092. Such a statement can be made with 95% confidence.

5. The tests for $H_0 : \beta_1 = 0$ against $H_1 : \beta_1 \neq 0$ and $H_2 : \beta_2 = 0$ against $H_3 : \beta_2 \neq 0$ can be performed in the same way as in the single regressor case (see point 3 in the interpretation of the output in Sect. 3.6). What do these tests mean and why are they important? The estimated coefficients of HP and VOL are different from 0. But these are estimated from the sample under consideration. If we conclude that the coefficient of HP is different from 0, what are the chances that we are wrong? Since the p-value for rejecting H_0 against H_1 is less than 2×10^{-16}, the probability that we are wrong is less than 2×10^{-16} which is very small. So we can conclude safely that the coefficient of HP in the population is different from 0. Likewise, we can conclude that the coefficient of VOL is also different from 0. If the (partial) regression coefficient of HP were 0, it would mean that HP cannot at all explain the variation in MPG over and above whatever is explained by VOL. Thus in the presence of VOL, HP is not useful in explaining the variation in MPG. But that is not the case in this instance. From our results, we find that both HP (respectively VOL) is useful in explaining the variation in MPG in the presence of VOL (respectively HP).

6. The usual point estimate of the average value of MPG for all vehicles with HP = 150 and VOL = 100 cubic feet can be obtained using (7.12) and is given by

$$M\widehat{P}G = 66.586385 - 0.110029 \times 150 - 0.194798 \times 100 = 30.602235.$$

7. A 95% confidence interval for the average value of MPG for all vehicles with HP = 150 and VOL = 100 cubic feet can be obtained as follows.

Let Z be the matrix of order 81×3 where each element in the first column is 1, the second and third columns, respectively, are data on HP and VOL on vehicles 1 to 81 in that order. Let $u^t = \begin{pmatrix} 1 & 150 & 100 \end{pmatrix}$. Then the required confidence interval is given by the formula

$$\left(u^t \widehat{\beta} - \sqrt{u^t (Z^t Z)^{-1} u \widehat{\sigma}^2} \, t_{78, 0.025}, \, u^t \widehat{\beta} + \sqrt{u^t (Z^t Z)^{-1} u \widehat{\sigma}^2} \, t_{78, 0.025} \right)$$

where $\hat{\sigma}^2$, the estimate of the error variance, is the square of the residual standard error in the output. The computed interval is (29.42484, 31.77965).

A 95% prediction interval for MPG of a vehicle with HP = 150 and VOL = 100 cubic feet can be obtained as follows. Let Z, u, and $\hat{\sigma}^2$ be as specified in point 8 above. Then the required prediction interval is given by the formula:

$$\left(u^t\hat{\beta} - \sqrt{\left(1+u^t(Z^tZ)^{-1}u\right)\hat{\sigma}^2} t_{78,0.025},\ u^t\hat{\beta} + \sqrt{\left(1+u^t(Z^tZ)^{-1}u\right)\hat{\sigma}^2} t_{78,0.025}\right).$$

The computed interval is (21.33388, 39.87062). Notice the difference in the confidence and prediction intervals. The prediction interval is always larger than the corresponding confidence interval. Do you know why? The variation in the average is always smaller than the individual variation.

8. The coefficient of multiple determination, R^2 is 75.07%. This means that HP and VOL together explain 75.07% of the variation in MPG through the current model. The adjusted R square is 74.43%. Since the relative change $\frac{R^2 - \text{adj } R^2}{R^2}$ is much smaller than 10%, we go on to interpret the R^2 as described above.
9. What does the F-test do here? It tests whether the regressors used, namely, HP and VOL together, have any explaining capacity regarding the variance in MPG, through the model under consideration. If the F-test turns out to be insignificant, then this model is not worth considering. However, it is wrong to conclude that the regressors themselves are not useful for predicting MPG. Perhaps some other model using the same regressors may yield a different result.

3.8 Why Probe Further?

We have so far considered the cases of single regressor and multiple regressors in Linear Regression and estimated the model which is linear in parameters and also in regressors. We interpreted the outputs and obtained the relevant confidence and prediction intervals. We also performed some basic tests of importance. Are we done? First, let us consider the data sets of Anscombe (1973) as given in Table 5.1 and Fig. 5.3 of the data visualization chapter (Chap. 5). Look at Table 5.1. There are four data sets, each having data on two variables. The plan is to regress y_i on x_i, $i = 1, \ldots, 4$. From the summary statistics, we notice the means and standard deviations of the x's are the same across the four data sets and the same is true for the y's. Further the correlation coefficient in each of the four data sets are the same. Based on the formulae (7.15) adapted to these data sets, the estimated regression lines are the same. Moreover, the correlation coefficient is 0.82 which is substantial. So it appears that linear regression is equally meaningful in all the four cases and gives the same regression line.

Now let us look at the plots in the Fig. 5.3. For the first data set, a linear regression seems reasonable. From the scatterplot of the second data set, a parabola is more appropriate. For the third data set, barring the third observation (13, 12.74), the remaining data points lie almost perfectly on a straight line, different from the line in the plot. In the fourth data set, the x values are all equal to 8, except for the eighth observation where the x value is 19. The slope is influenced by this observation. Otherwise we would never use these data (in the fourth data set) to predict y based on x.

This reiterates the observation made earlier: One should examine suitable plots and other diagnostics to be discussed in the following sections before being satisfied with a regression. In the single regressor case, the scatterplot would reveal a lot of useful information as seen above. However, if there are several regressors, scatterplots alone will not be sufficient since we want to assess the influence of a regressor on the response variable after controlling for the other regressors. Scatterplots ignore the information on the other regressors. As we noticed earlier, the residuals are the sample representatives of the errors. This leads to the examination of some suitable residual plots to check the validity of the assumptions in Eq. (7.9). In the next section, we shall develop some basic building blocks for constructing the diagnostics.

3.9 Leverage Values, Residuals, and Outliers

The Hat Matrix: First, we digress a little bit to discuss a useful matrix. Consider the linear regression model (7.8) with the assumptions (7.9). There is a matrix H, called the *hat matrix* which transforms the vector of response values Y to the vector of the fitted values $\widehat{Y}$. In other words, $\widehat{Y} = HY$. The hat matrix is given by $Z(Z^tZ)^{-1}Z^t$. It is called the hat matrix because when applied to Y it gives the estimated (or hat) value of Y. The hat matrix has some good properties, some of which are listed below

(a) $Var\left(\widehat{Y}\right) = \sigma^2 H$.
(b) $Var(residuals) = \sigma^2(I - H)$, where I is the identity matrix.

Interpretation of Leverage Values and Residuals Using the Hat Matrix

Diagonal elements h_{ii} of the hat matrix have a good interpretation. If h_{ii} is large, then the regressor part of data for the ith observation is far from the center of the regressor data. If there are N observations and k regressors, then h_{ii} is considered to be large if it is $\frac{2(k+1)}{N}$ or higher.

Let us now look at the residuals. If the residual corresponding to the ith residual is large, then the fit of the ith observation is poor. To examine whether a residual is large, we look at a standardized version of the residual. It can be shown that the mean of each residual is 0 and the variance of the ith residual is given by $Var(e_i) = (1 - h_{ii})\sigma^2$. We recall that σ^2 is the variance of the error of an observation (of course conditional on the regressors). The estimate of σ^2 obtained from the model after dropping the ith observation, denoted by $\widehat{\sigma}^2_{(i)}$ is preferred (since large i^{th} residual also has a large contribution to the residual sum of squares and hence to the estimate of σ^2).

An observation is called an *Outlier* if its fit in the estimated model is poor, or, equivalently, if its residual is large.

The statistic that is used to check whether the ith observation is an outlier is

$$r_i = \frac{e_i - 0}{\sqrt{(1 - h_{ii})\widehat{\sigma}^2_{(i)}}}, \tag{7.43}$$

often referred to as the ith studentized residual, which has a t distribution with $N - k - 1$ degrees of freedom under the null hypothesis that the ith observation is not an outlier.

When one says that an observation is an outlier, it is in the context of the estimated model under consideration. The same observation may be an outlier with respect to one model and may not be an outlier in a different model (see Exercise 7.3).

As we shall see, the residuals and the leverage values form the basic building blocks for the deletion diagnostics, to be discussed in Sect. 3.12.

3.10 Residual Plots

We shall now proceed to check whether the fitted model is adequate. This involves the checking of the assumptions (7.9). If the fitted model is appropriate then the residuals are uncorrelated with the fitted values and also the regressors. We examine these by looking at the residual plots:

(a) Fitted values vs residuals
(b) Regressors vs residuals

In each case the residuals are plotted on the Y-axis. If the model is appropriate, then each of the above plots should yield a random scatter. The deviation from the random scatter can be tested, and an R package command for the above plots gives the test statistics along with the p-values.

Let us look at the residual plots and test statistics for the fitted model for the gasoline consumption problem given in Sect. 3.7.

How do we interpret Fig. 7.7 and Table 7.4? Do they also suggest a suitable corrective action in case one such is warranted?

We need to interpret the figure and the table together.

We notice the following.

1. The plot HP vs residual does not appear to be a random scatter. Table 7.4 also confirms that the p-value corresponding to HP is very small. Furthermore, if we

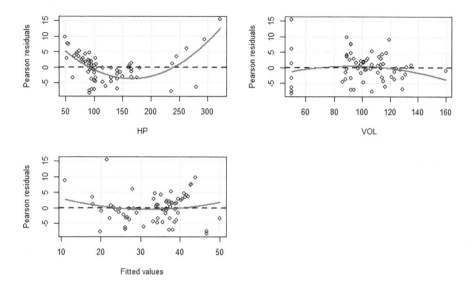

Fig. 7.7 The residual plots for gasoline consumption model in Sect. 3.7

Table 7.4 The tests for deviation from random scatter

```
> residualPlots(model1)
              Test stat  Pr(>|t|)
HP              10.042     0.000
VOL             -1.575     0.119
Tukey test       1.212     0.226
```

look at the plot more closely we can see a parabolic pattern. Thus the present model is not adequate. We can try to additionally introduce the square term for HP to take care of this parabolic nature in the plot.

2. The plot VOL vs residual is not quite conclusive whether it is really a random scatter. Read this in conjunction with Table 7.4. The p-value corresponding to VOL is not small. So we conclude that there is not a significant deviation from a random scatter.
3. Same is the case with the fitted values vs residual. For fitted values vs residual, the corresponding test is the Tukey test.

A note of caution is in order. The R package command for the residual plots leads to plots where a parabolic curve is drawn to notify a deviation from random scatter. First, the deviation from random scatter must be confirmed from the table of tests for deviation from random scatter by looking at the p-value. Second, even if the p-value is small, it does not automatically mean a square term is warranted. Your judgment of the plot is important to decide whether a square term is warranted or something else is to be done. One may wonder why it is important to have the plots if, anyway, we need to get the confirmation from the table of tests for random scatter. The test only tells us whether there is a deviation from random scatter but it does not guide us to what transformation is appropriate in case of deviation. See, for example, the residual plot of HP vs Residual in Fig. 7.7. Here we clearly see a parabolic relationship indicating that we need to bring in HP square. We shall see more examples in the following sections.

Let us add the new regressor which is the square of HP and look at the regression output and the residual plots.

The following points emerge:

1. From Table 7.5, we notice that HP_Sq is highly significant (p-value: 1.2×10^{-15}) and is positive.
2. R square and Adj. R square are pretty close. So we can interpret R square. The present model explains 89.2% of the variation in MPG. (Recall that the model in Sect. 3.7 explained only 75% of the variation in MPG.)
3. Figure 7.8 and Table 7.6 indicate that the residuals are uncorrelated with the fitted values and regressors.
4. Based on Table 7.5, how do we assess the impact of HP on MPG? The partial derivative of $M\widehat{P}G$ with respect to HP is $-0.4117 + 0.001808$ HP. Unlike in the model in Sect. 3.7 (see point 4), in the present model, the impact of unit increase in HP on the estimated MPG, keeping the VOL constant, depends on the level of HP. At the median value of HP (which is equal to 100—see Table 7.2), one unit increase in HP will lead to a reduction in $M\widehat{P}G$ by 0.2309 when VOL is kept constant. From the partial derivative, it is clear that as long as HP is smaller than 227.710177, one unit increase in HP will lead to a reduction in $M\widehat{P}G$, keeping VOL constant. If HP is greater than this threshold, then based on this model, $M\widehat{P}G$ will increase (happily!) with increasing HP when the VOL is held constant.

The question is: Are we done? No, we still need to check a few other assumptions, like normality of the errors. Are there some observations which are driving the results? Are there some important variables omitted?

Suppose we have performed a linear regression with some regressors. If the plot of fitted values vs residuals shows a linear trend, then it is an indication that there is an omitted regressor. However, it does not give any clue as to what this regressor is. This has to come from domain knowledge. It may also happen that, after the regression is done, we found another variable which, we suspect, has an influence on the response variable. In the next section we study the issue of bringing in a new regressor.

3.11 Added Variable Plot and Partial Correlation

Let us consider again the gasoline consumption problem. Suppose we have run a regression of MPG on VOL. We feel that HP also has an explaining power of the variation in MPG. Should we bring in HP? A part of MPG is already explained by VOL. So the unexplained part of MPG is the residual e (unexplained part of MPG) after regressing MPG on VOL. There is a residual value corresponding to

Table 7.5 Regression output of MPG vs HP, HP_SQ (Square of HP), and VOL

```
> model2<-lm(MPG ~ HP + VOL + HP_sq, data=Cars)
> summary(model2)

Call:
lm(formula = MPG ~ HP + VOL + HP_sq, data = Cars)

Residuals:
    Min      1Q  Median      3Q     Max
 -8.288  -2.037   0.561   1.786  11.008

Coefficients:
              Estimate Std. Error t value Pr(>|t|)
(Intercept)  7.744e+01  1.981e+00  39.100  < 2e-16 ***
HP          -4.117e-01  3.063e-02 -13.438  < 2e-16 ***
VOL         -1.018e-01  1.795e-02  -5.668  2.4e-07 ***
HP_sq        9.041e-04  9.004e-05  10.042  1.2e-15 ***
---
Signif. codes:  0 '***' 0.001 '**' 0.01 '*' 0.05 '.' 0.1 ' ' 1

Residual standard error: 3.058 on 77 degrees of freedom
Multiple R-squared: 0.892,    Adjusted R-squared: 0.8878
F-statistic: 212.1 on 3 and 77 DF,  p-value: < 2.2e-16
```

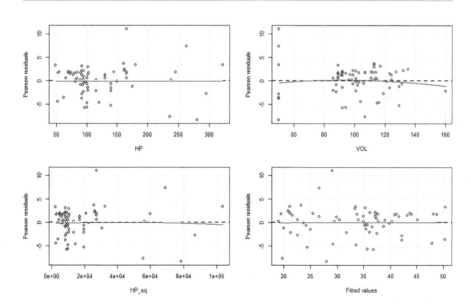

Fig. 7.8 Residual plots corresponding to the fitted model in Table 7.5

each observation. Let us call these residual values e_i, $i = 1, \ldots, 81$. So we will bring in HP if it has an explaining power of this residual. At the same time, a part of the explaining power of HP may already have been contained in VOL. Therefore, the

Table 7.6 The tests for deviation from uncorrelatedness for the fitted model in Table 7.5

```
> residualPlots(model2)

             Test stat  Pr(>|t|)
HP              0.371     0.712
VOL            -0.703     0.484
HP_sq          -0.528     0.599
Tukey test      0.089     0.929
```

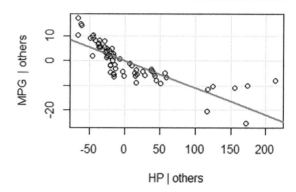

Fig. 7.9 Added variable plot

part of HP that is useful for us is the residual f after regressing HP on VOL. Let us call these residuals $f_i, i = 1, \ldots, 81$. It now boils down to the single regressor case of regressing e on f. It is natural to look at the scatterplot $(f_i, e_i), i = 1, \ldots, 81$. This scatterplot is called the *Added Variable Plot*. The correlation coefficient is called the *partial correlation coefficient* between MPG and HP fixing (or equivalently, eliminating the effect of) VOL. Let us look at this added variable plot (Fig. 7.9).

The following points emerge from the above added variable plot.

(a) As the residual of HP increases, the residual of MPG decreases, by and large. Thus it is useful to bring in HP into the regression in addition to VOL.

(b) We can notice a parabolic trend in the plot, suggesting that HP be brought in using a quadratic function. Thus, it reinforces our earlier decision (see Sect. 3.10.) to use HP and HP_SQ. Barring exceptions, in general, the added variable plot suggests a suitable functional form in which the new regressor can be added.

(c) The partial correlation coefficient is the strength of a linear relationship between the two residuals under consideration. For the gasoline consumption example, the partial correlation coefficient between MPG and HP keeping VOL fixed is -0.808. The correlation coefficient between MPG and HP is -0.72. (As we know the negative sign indicates that as HP increases, MPG decreases.) Is it surprising?

It is a common misconception that the correlation coefficient between two variables is always larger than or equal to the partial correlation coefficient between these variables after taking away the effect of other variables. Notice that the correlation coefficient between MPG and HP represents the strength of the linear relationship between these variables ignoring the effect of VOL. The partial correlation coefficient between MPG and HP after taking away the effect of VOL is the simple correlation coefficient between the residuals e and f which are quite different from MPG and HP, respectively. *So a partial correlation coefficient can be larger than or equal to or less than the corresponding correlation coefficient.*

We give an interesting formula for computing the partial correlation coefficient using just the basic regression output. Let T denote the t-statistic value for a regressor. Let there be N observations and $k(\geq 2)$ regressors plus one intercept. Then the partial correlation coefficient between this regressor and the response variable (Greene 2012, p. 77) is given by

$$\frac{T^2}{T^2 + (N - k - 1)}. \qquad (7.44)$$

Recall (see 7.38) that $N - k - 1$ is the degree of freedom for the residual sum of squares.

3.12 Deletion Diagnostics

Look at the third data set of Anscombe (1973) described in Sect. 3.8 and its scatterplot (Fig. 5.3) in the Data Visualization chapter (Chap. 5). But for the third observation (13, 12.74), the other observations fall almost perfectly on a straight line. Just this observation is influencing the slope and the correlation coefficient. In this section, we shall explain what we mean by an influential observation, give methods to identify such observations, and finally discuss what one can do with an influential observation.

We list below some of the quantities of interest in the estimation of a linear regression model.

1. Regression coefficient estimates
2. Fit of an observation
3. Standard errors of the regression coefficient estimates
4. The error variance
5. Coefficient of multiple determination

In a linear regression model (7.8) with the assumptions as specified in (7.9), no single observation has any special status. If the presence or absence of a particular observation can make a large (to be specified) difference to some or all of the quantities above, we call such an observation an *influential observation*.

Let us describe some notation before we give the diagnostic measures. Consider the model specified by (7.8) and (7.9) and its estimation in Sect. 3.7. Recall the definitions of the fitted values, $\widehat{y}_i$, the residuals, e_i, the residual sum of squares, R_0^2, and the coefficient of multiple determination, R^2, given in Sect. 3.7. Let $\widehat{\beta}$ denote the vector of the estimated regression coefficients.

In order to assess the impact of an observation on the quantities (1)–(5) mentioned above, we set aside an observation, say the ith observation and estimate the model with the remaining observations. We denote the vector of estimated regression coefficients, fitted value of the jth observation, residual of the jth observation, residual sum of squares and the coefficient of multiple determination after dropping the ith observation by $\widehat{\beta}_{(i)}, \widehat{y}_{j(i)}, e_{j(i)}, R^2_{0(i)}, R^2_{(i)}$ respectively.

We give below a few diagnostic measures that are commonly used to detect influential observations.

(a) *Cook's distance*: This is an overall measure of scaled difference in the fit of the observations due to dropping an observation. This is also a scaled measure of the difference between the vectors of regression coefficient estimates before and after dropping an observation. More specifically, Cook's distance after dropping the ith observation, denoted by $Cookd_i$, is proportional to $\sum_{j=1}^{N}(\widehat{y}_j - \widehat{y}_{j(i)})^2$ where N is the number of observations. ($Cookd_i$ is actually a squared distance.) The ith observation is said to be influential if $Cookd_i$ is large. If $Cookd_i$ is larger than a cutoff value (usually 80th or 90th percentile value of F distribution with parameters k and $N - k - 1$ where N is the number of observations and k is the number of regressors), then the ith observation is considered to be influential. In practice, a graph is drawn with an observation number in the X-axis and Cook's distance in the Y-axis, called the index plot of Cook's distance, and a few observations with conspicuously large Cook's distance values are treated as influential observations.

(b) $DFFITS_i$: This is a scaled absolute difference in the fits of the ith observation before and after the deletion of the ith observation. More specifically, $DFFITS_i$ is proportional to $|\widehat{y}_i - \widehat{y}_{i(i)}|$. Observations with $DFFITS$ larger than $2\sqrt{\frac{k+1}{N}}$ are flagged as influential observations.

(c) $COVRATIO_i$: $COVRATIO_i$ measures the change in the overall variability of the regression coefficient estimates due to the deletion of the ith observation. More specifically it is the ratio of the determinants of the covariance matrices of the regression coefficient estimates after and before dropping the ith observation. If $|COVRATIO_i - 1| > \frac{3(k+1)}{N}$, then the ith observation is flagged as an influential observation in connection with the standard errors of the estimates. It is instructive to also look at the index plot of $COVRATIO$.

(d) The scaled residual sum of squares estimates the error variance as we have seen in (7.38). The difference in the residual sum of squares R^2_0 and $R^2_{0(i)}$ before and after deletion of the ith observation, respectively, is given by

$$R^2_0 - R^2_{0(i)} = \frac{e_i^2}{1 - h_{ii}}.$$

Thus the ith observation is flagged as influential in connection with error variance if it is an outlier (see 7.43).

Two points are worth noting:
(a) If an observation is found to be influential, it does not automatically suggest "off with the head." The diagnostics above are only markers suggesting that an influential observation has to be carefully examined to find out whether there is an explanation from the domain knowledge and the data collection process why it looks different from the rest of the data. Any deletion should be contemplated only after there is a satisfactory explanation for dropping, from the domain knowledge.
(b) The diagnostics are based on the model developed. If the model under consideration is found to be inappropriate otherwise, then these diagnostics are not applicable.

We shall illustrate the use of Cook's distance using the following example on cigarette consumption. The data set "CigaretteConsumption.csv" is available on the book's website.

Example 7.1. A national insurance organization in USA wanted to study the consumption pattern of cigarettes in all 50 states and the District of Columbia. The variables chosen for the study are given in Fig. 7.10.

Variable	Definition
Age	Median age of a person living in a state
HS	% of people over 25 years of age in a state who completed high school
Income	Per capita personal income in a state (in dollars)
Black	% of blacks living in a state
Female	% of females living in a state
Price	Weighted average price (in cents) of a pack of cigarettes in a state
Sales	Number of packs of cigarettes sold in a state on a per capita basis

The R output of the regression of Sales on the other variables is in Table 7.7.

The index plots of Cook's distance and studentized residuals are given in Fig. 7.10.

From the Cook's distance plot, observations 9, 29, and 30 appear to be influential. Observations 29 and 30 are also outliers. (These also are influential with respect to error variance.) On scrutiny, it turns out that observations 9, 29, and 30 correspond to Washington DC, Nevada, and New Hampshire, respectively. Washington DC is the capital city and has a vast floating population due to tourism and otherwise. Nevada is different from a standard state because of Las Vegas. New Hampshire does not impose sales tax. It does not impose income tax at state level. Thus these three states behave differently from other states with respect to cigarette consumption. So it is meaningful to consider regression after dropping these observations.

The corresponding output is provided in Table 7.8.

Table 7.7 Regression results for cigarette consumption data

```
> model3<-lm(Sales ~ ., data=CigaretteConsumption[,-1])
> summary(model3)

Call:
lm(formula = Sales ~ ., data = CigaretteConsumption[, -1])

Residuals:
    Min      1Q  Median      3Q     Max
-48.398 -12.388  -5.367   6.270 133.213

Coefficients:
             Estimate Std. Error t value Pr(>|t|)
(Intercept) 103.34485  245.60719   0.421  0.67597
Age           4.52045    3.21977   1.404  0.16735
HS           -0.06159    0.81468  -0.076  0.94008
Income        0.01895    0.01022   1.855  0.07036 .
Black         0.35754    0.48722   0.734  0.46695
Female       -1.05286    5.56101  -0.189  0.85071
Price        -3.25492    1.03141  -3.156  0.00289 **
---
Signif. codes:  0 '***' 0.001 '**' 0.01 '*' 0.05 '.' 0.1 ' ' 1

Residual standard error: 28.17 on 44 degrees of freedom
Multiple R-squared:  0.3208,    Adjusted R-squared:  0.2282
F-statistic: 3.464 on 6 and 44 DF,  p-value: 0.006857
```

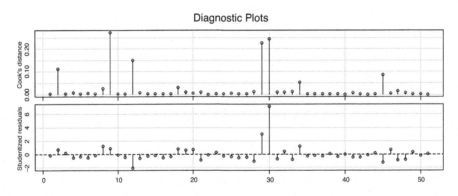

Fig. 7.10 Index plots for cigarette consumption model in Table 7.7

Notice the appreciable changes in the regression coefficient estimates and standard errors. The coefficients of HS and Price changed from -0.062 and -3.255 to -1.172 and -2.782, respectively. While the income coefficient estimate has not changed very much (from 0.019 to 0.021), the standard error got almost halved from 0.010 to 0.005, thereby Income became highly significant from being insignificant at 5% level. There are also changes in other coefficient estimates (including changes in sign), but we are not emphasizing them since they are significant in both the

Table 7.8 Regression results for cigarette consumption data after dropping observations 9, 29, and 30

```
> model4<- lm (Sales ~ ., data = CigaretteConsumption[-c(9,29,30),-1])
> summary(model4)

Call:
lm(formula = Sales ~ ., data = CigaretteConsumption[-c(9, 29,
    30), -1])

Residuals:
    Min      1Q  Median      3Q     Max
-40.155  -8.663  -2.194   6.301  36.043

Coefficients:
             Estimate Std. Error t value Pr(>|t|)
(Intercept) 100.68317  136.24526   0.739 0.464126
Age           1.84871    1.79266   1.031 0.308462
HS           -1.17246    0.52712  -2.224 0.031696 *
Income        0.02084    0.00546   3.817 0.000448 ***
Black        -0.30346    0.40567  -0.748 0.458702
Female        1.12460    3.07908   0.365 0.716810
Price        -2.78195    0.57818  -4.812 2.05e-05 ***
---
Signif. codes:  0 '***' 0.001 '**' 0.01 '*' 0.05 '.' 0.1 ' ' 1

Residual standard error: 15.02 on 41 degrees of freedom
Multiple R-squared:  0.4871,    Adjusted R-squared:  0.412
F-statistic: 6.489 on 6 and 41 DF,  p-value: 7.195e-05
```

models. There is also significant reduction in the residual standard error (from 28.17 to 15.02). Furthermore, R^2 has improved from 0.32 to 0.49.

This is not to say that we are done with the analysis. There are more checks, such as checks for normality, heteroscedasticity, etc., that are pending.

3.13 Collinearity

Let us revisit Example 2.3 (cars) which we analyzed in Sects. 3.7 and 3.10. We incorporate data on an additional variable, namely, the weights (WT) of these cars. A linear regression of MPG on HP, HP_SQ, VOL, and WT is performed. The output is given below.

Compare the output in Tables 7.5 and 7.9. The following points emerge:

(a) WT is insignificant, as noticed in Table 7.9.
(b) VOL which is highly significant in Table 7.5 turns out to be highly insignificant in Table 7.9. Thus, once we introduce WT, both VOL and WT become insignificant which looks very surprising.
(c) The coefficient estimates of VOL in Tables 7.5 and 7.9 (corresponding to the models without and with WT, respectively) are −0.1018 and −0.0049 which are

Table 7.9 Output for the regression of MPG on HP, HP_SQ, VOL, and WT

```
> model5<-lm(MPG ~ HP + HP_sq + VOL + WT, data = Cars)
> summary(model5)

Call:
lm(formula = MPG ~ HP + HP_sq + VOL + WT, data = cars)

Residuals:
    Min      1Q   Median      3Q      Max
-8.3218  -2.0723  0.5592   1.7386  10.9699

Coefficients:
              Estimate Std. Error  t value  Pr(>|t|)
(Intercept)  7.734e+01  2.126e+00   36.374   < 2e-16 ***
HP          -4.121e-01  3.099e-02  -13.299   < 2e-16 ***
HP_sq        9.053e-04  9.105e-05    9.943  2.13e-15 ***
VOL         -4.881e-02  3.896e-01   -0.125     0.901
WT          -1.573e-01  1.156e+00   -0.136     0.892
---
Signif. codes:  0 '***' 0.001 '**' 0.01 '*' 0.05 '.' 0.1 ' ' 1

Residual standard error: 3.078 on 76 degrees of freedom
Multiple R-squared:  0.8921,    Adjusted R-squared:  0.8864
F-statistic:   157 on 4 and 76 DF,  p-value: < 2.2e-16
```

quite far apart. Also, the corresponding standard errors are 0.01795 and 0.3896. Once we introduce WT, the standard error of the coefficient for VOL changes by more than 20 times. Furthermore, the value of the coefficient is halved. Thus, we notice that the standard error has gone up by as high as 20 times and the magnitude of the coefficient is halved once we introduce the variable WT.

(d) There is virtually no change in R square.

Since there is virtually no change in R square, it is understandable why WT is insignificant. But why did VOL, which was highly significant before WT was introduced, became highly insignificant once WT is introduced? Let us explore. Let us look at the scatterplot matrix (Fig. 7.11).

One thing that is striking is that VOL and WT are almost perfectly linearly related. So in the presence of WT, VOL has virtually no additional explaining capacity for the variation in the residual part of MPG not already explained by WT. The same is the situation with WT that it has no additional explaining capacity in the presence of VOL. If both of them are in the list of regressors, both of them become insignificant for this reason. Let us look at the added variable plots for VOL and WT in the model corresponding to Table 7.9 which confirm the same thing (Fig. 7.12).

It is said that there is a collinear relationship among some of the regressors if one of them has an almost perfect linear relationship with others. If there are collinear relationships among regressors, then we say that there is the problem of *Collinearity*.

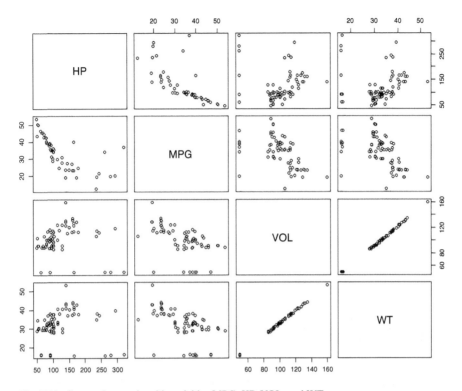

Fig. 7.11 Scatterplot matrix with variables MPG, HP, VOL, and WT

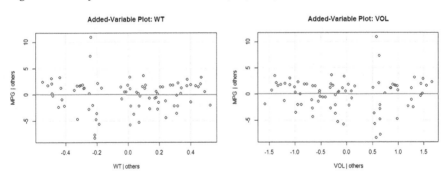

Fig. 7.12 Added variable plots for WT and VOL

Why should we care if there is collinearity? What are the symptoms? How do we detect collinearity and if detected, what remedial measures can be taken?

Some of the symptoms of collinearity are as follows:

(a) R square is high, but almost all the regressors are insignificant.
(b) Standard errors of important regressors are large, and important regressors become insignificant.

(c) Very small changes in the data produce large changes in the regression coefficient estimates.

We noticed (b) above in our MPG example.

How does one detect collinearity? If the collinear relation is between a pair of regressors, it can be detected using the scatterplot matrix and the correlation matrix. In the MPG example, we detected collinearity between VOL and WT this way. Suppose that more than two variables are involved in a collinear relationship. In order to check whether a regressor is involved in a collinear relationship, one can use *variance inflation factors (VIF)*.

The variance inflation factor for the ith regressor, denoted by VIF_i is defined as the factor by which the variance of a single observation, σ^2, is multiplied to get the variance of the regression coefficient estimate of the ith regressor. It can be shown that VIF_i is the reciprocal of $1 - R_i^2$ where R_i^2 is the coefficient of multiple determination of the ith regressor with the other regressors. The ith regressor is said to be involved in a collinear relationship if R_i^2 is large. There is no unanimity on how large is considered to be large, but 95% is an accepted norm. If R_i^2 is 95% or larger, then VIF_i is at least 20.

Variance decomposition proportions (VP): Table 7.10 (see Belsley et al. 2005) is sometimes used to identify the regressors involved in collinear relationships. We shall explain below how to use the VP table operationally for the case where there are four regressors and an intercept. The general case will follow similarly. For more details and the theoretical basis, one can refer to BKW.

By construction, the sum of all the elements in each column starting from column 3 (columns corresponding to the intercept and the regressors), namely, $\sum_{j=0}^{4} \pi_{ji}$, is 1 for $i = 0, 1, \ldots, 4$.

Algorithm

Step 0: Set $i = 0$.

Step 1: Check whether the condition index c_{5-i} is not more than 30. If yes, declare that there is no serious issue of collinearity to be dealt with and stop. If no, go to step 2.

Step 2: Note down the variables for which the π values in the $(5-i)^{th}$ row are at least 0.5. Declare these variables as variables involved in a collinear relationship. Go to Step 3.

Table 7.10 Variance decomposition proportions table

S. No.	Condition index	Intercept	X_1	X_2	X_3	X_4
1	c_1	π_{00}	π_{01}	π_{02}	π_{03}	π_{04}
2	c_2	π_{10}	π_{11}	π_{12}	π_{13}	π_{14}
3	c_3	π_{20}	π_{21}	π_{22}	π_{23}	π_{24}
4	c_4	π_{30}	π_{31}	π_{32}	π_{33}	π_{34}
5	c_5	π_{40}	π_{41}	π_{42}	π_{43}	π_{44}

Table 7.11 Variance inflation factors

```
> vif(model5)
      HP     HP_sq       VOL        WT
26.45225  26.32768 637.51477 634.06751
```

Table 7.12 Variance decomposition proportions

```
> colldiag(model5)
Condition
Index            Variance Decomposition Proportions
         intercept      HP        HP_sq    VOL    WT
1   1.000    0.001    0.000    0.001   0.000  0.000
2   2.819    0.004    0.001    0.024   0.000  0.000
3  11.652    0.511    0.002    0.008   0.000  0.000
4  29.627    0.358    0.983    0.956   0.000  0.000
5 338.128    0.127    0.013    0.012   1.000  0.999
```

Step 3: Delete the row $5 - i$ from the table and calibrate the π values in each column corresponding to the intercept and the regressors so that the corresponding columns is 1.

Step 4: Replace i by $i+1$ and go to step 1,

When the algorithm comes to a stop, say, at $i = 3$, you have 2 (i.e., $(i - 1)$) collinear relationships with you.

Let us return to the MPG example and the model that led us to the output in Table 13.1. The VIFs and the variance decomposition proportions table are given in Tables 7.11 and 7.12:

The VIFs of VOL and WT are very high (our cutoff value is about 20), and thereby imply that each of VOL and WT is involved in collinear relationships. This also explains what we already observed, namely, VOL and WT became insignificant (due to large standard errors). The VIFs of HP and HP_SQ are marginally higher than the cutoff.

From the variance decompositions proportions table, we see that there is a collinear relationship between VOL and WT (condition index is 338.128 and the relevant π values corresponding to VOL and WT are 1 and 0.999, respectively). The next largest condition index is 29.667 which is just about 30. Hence, we can conclude that we have only one mode of collinearity.

What remedial measures can be taken once the collinear relationships are discovered?

Let us start with the easiest. If the intercept is involved in a collinear relationship, subtract an easily interpretable value close to the mean of each of the other regressors involved in that relationship and run the regression again. You will notice that the regression coefficient estimates and their standard errors remain the same as in the earlier regression. Only the intercept coefficient and its standard error will change. The intercept will no longer be involved in the collinear relationship.

Consider one collinear relationship involving some regressors. One can delete the regressor that has the smallest partial correlation with the response variable given the other regressors. This takes care of this collinear relationship. One can repeat this procedure with the other collinear relationships.

We describe below a few other procedures, stepwise regression, best subset regression, ridge regression, and lasso regression, which are commonly employed to combat collinearity in a blanket manner.

It may be noted that subset selection in the pursuit of an appropriate model is of independent interest for various reasons, some of which we mention below.

(a) The "kitchen sink" approach of keeping many regressors may lead to collinear relationships.
(b) Cost can be a consideration, and each regressor may add to the cost. Some balancing may be needed.
(c) Ideally there should be at least ten observations per estimated parameter. Otherwise one may find significances by chance. When the number of observations is not large, one has to restrict the number of regressors also.

The criteria that we describe below for selecting a good subset are based on the residual sum of squares. Let us assume that we have one response variable and k regressors in our regression problem. Suppose we have already included r regressors ($r < k$) into the model. We now want to introduce one more regressor from among the remaining $k - r$ regressors into the model. The following criteria place a penalty for bringing in a new regressor. The coefficient of multiple determination, R^2 unfortunately never decreases when a new regressor is introduced. However, adjusted R^2 can decrease when a new regressor is introduced if it is not sufficiently valuable. We introduce a few other criteria here for which the value of the criterion increases unless the residual sum of squares decreases sufficiently by introducing the new regressor, indicating that it is not worth introducing the regressor under consideration. The current trend in stepwise regression is to start with the model in which all the k regressors are introduced into the model and drop the regressors as long as the criterion value decreases and stop at a stage where dropping a regressor increases the criterion value. The object is to get to a subset of regressors for which the criterion has the least value. We use the following notation:

N = The number of observations
k = The total number of regressors
r = The number of regressors used in the current model
$\sum_{i=1}^{N}(y_i - \bar{y})^2$ = The sum of squared deviations of the response variable from its mean
$\left(R_0^2\right)_r$ = The sum of squared residuals when the specific subset of r regressors is used in the model
$\left(R_0^2\right)_k$ = The sum of squared residuals when all the k regressors are used in the model

The major criteria used in this connection are given below:

(a) Adjusted R^2: $1 - \frac{N-1}{N-r} \frac{\left(R_0^2\right)_r}{\sum_{i=1}^{N}(y_i - \bar{y})^2}$ (see also Eq. 7.11)

(b) AIC : $\log\left(\frac{(R_0^2)_r}{N}\right) + \frac{2r}{N}$

(c) BIC : $\log\left(\frac{(R_0^2)_r}{N}\right) + \frac{r \log N}{N}$

(d) Mallow's C_p: $\dfrac{(R_0^2)_r}{(R_0^2)_k/(N k-1)} - n + 2r$

Note: The AIC, or Akaike Information Criterion, equals twice the negative of the log-likelihood penalized by twice the number of regressors. This criterion has general applicability in model selection. The BIC or Bayes Information Criterion is similar but has a larger penalty than AIC and like AIC has wider application than regression.

Among (a), (b), and (c) above, the penalty for introducing a new regressor is in the ascending order. The criterion (d) compares the residual sum of squares of the reduced model with that of the full model. One considers the subset models for which C_p is close to r and chooses the model with the least number of regressors from among these models.

We illustrate the stepwise procedure with the cigarette consumption example (Example 7.1) using the criterion AIC. We give the R output as in Table 7.13.

The first model includes all the six regressors, and the corresponding AIC is 266.56. In the next stage one of the six regressors is dropped at a time, keeping all other regressors, and the AIC value is noted. When Female is dropped, keeping all other regressors intact, the AIC is 264.71. Likewise, when age is dropped, the AIC is 265.79, and so on. We notice the least AIC corresponds to the model dropping Female. In the next stage, the model with the remaining five regressors is considered. Again, the procedure of dropping one regressor from this model is considered and the corresponding AIC is noted. (The dropped regressor is brought back and its AIC is also noted. In the case of AIC this is not necessary because this AIC value is already available in a previous model. However, for the case of Adj. R^2 this need not necessarily be the case.) We find that the least AIC equaling 263.23 now corresponds to the model which drops Black from the current model with the five regressors. The procedure is repeated with the model with the four regressors. In this case dropping any variable from this model yields an increased AIC. The stepwise procedure stops here. Thus, the stepwise method yields the model with the four regressors, age, HS, income, and price. The corresponding estimated model is given below.

Compare Tables 7.14 and 7.8. We notice that the significance levels of HS, Income, and Price have remained the same (in fact, the p-values are slightly smaller in the subset model). Age, which was insignificant in the full model (Table 7.8), is now (Table 7.14) significant at 5% level (p-value is 0.039). So when some undue regressors are dropped, some of the insignificant regressors may become significant. It may also be noted that while R^2 dropped marginally from 0.4871 to 0.4799 corresponding to full model and the subset model, respectively, there is a substantial increase in adjusted R^2 from 0.412 in the full model to 0.4315 in the subset model.

Table 7.13 Stepwise regression

```
> step <-stepAIC (model1, direction="both")

Start:  AIC=266.56
Sales ~ Age + HS + Income + Black + Female + Price

          Df Sum of Sq     RSS     AIC
- Female   1      30.1  9284.4  264.71
- Black    1     126.3  9380.6  265.21
- Age      1     240.0  9494.3  265.79
<none>                  9254.3  266.56
- HS       1    1116.7 10370.9  270.03
- Income   1    3288.6 12542.8  279.15
- Price    1    5225.4 14479.7  286.05

Step:  AIC=264.71
Sales ~ Age + HS + Income + Black + Price

          Df Sum of Sq     RSS     AIC
- Black    1      99.4  9383.8  263.23
<none>                  9284.4  264.71
- Age      1     629.1  9913.4  265.86
+ Female   1      30.1  9254.3  266.56
- HS       1    1099.8 10384.1  268.09
- Income   1    3366.8 12651.2  277.57
- Price    1    5198.8 14483.2  284.06

Step:  AIC=263.23
Sales ~ Age + HS + Income + Price

          Df Sum of Sq     RSS     AIC
<none>                  9383.8  263.23
+ Black    1      99.4  9284.4  264.71
+ Female   1       3.2  9380.6  265.21
- Age      1     991.8 10375.6  266.05
- HS       1    1573.0 10956.8  268.67
- Income   1    3337.8 12721.6  275.83
- Price    1    5216.3 14600.1  282.44
```

Best Subset Regression

In stepwise regression, we considered 15 subsets, as can be seen from Table 7.14. But there are $2^6 - 1 = 63$ subsets with at least one regressor. The Best Subset regression (using AIC or BIC) considers all these subsets in a systematic manner and delivers that subset for which the AIC (or BIC) is the least. In case of Adjusted R^2, it delivers that subset for which the adjusted R^2 is the largest. However, if the number of regressors is large, it generally gets unwieldy to search for the best subset.

7 Statistical Methods: Regression Analysis

Table 7.14 Best Stepwise model from stepAIC

```
> best_model_step <-lm(Sales ~ Age + HS + Income + Price, data=Cig_data)
> summary(best_model_step)

Call:
lm(formula = Sales ~ Age + HS + Income + Price, data = Cig_data)

Residuals:
    Min      1Q  Median      3Q     Max
-40.196  -8.968  -1.563   8.525  36.117

Coefficients:
             Estimate Std. Error t value Pr(>|t|)
(Intercept) 124.966923  37.849699   3.302 0.001940 **
Age           2.628064   1.232730   2.132 0.038768 *
HS           -0.894433   0.333147  -2.685 0.010267 *
Income        0.019223   0.004915   3.911 0.000322 ***
Price        -2.775861   0.567766  -4.889 1.45e-05 ***
---
Signif. codes:  0 '***' 0.001 '**' 0.01 '*' 0.05 '.' 0.1 ' ' 1

Residual standard error: 14.77 on 43 degrees of freedom
Multiple R-squared:  0.4799,    Adjusted R-squared:  0.4315
F-statistic:  9.92 on 4 and 43 DF,  p-value: 8.901e-06
```

In such a case, stepwise regression can be employed. For the cigarette consumption data, the best subset regression using AIC leads to the same regressors as in the stepwise regression using AIC.

Ridge Regression and Lasso Regression

In the least squares method (see estimation of parameters in Sects. 3.6 and 3.7), we estimate the parameters by minimizing the error sum of squares. Ridge regression and lasso regression minimize the error sum of squares subject to constraints that place an upper bound on the magnitude of the regression coefficients. Ridge regression minimizes the error sum of squares subject to $\sum \beta_i^2 \leq c$, where c is a constant. Lasso regression (Tibshirani 1996) minimizes the error sum of squares subject to $\sum |\beta_i| \leq d$, where d is a constant. Both these methods are based on the idea that the regression coefficients are bounded in practice. In ridge regression all the regressors are included in the regression and the coefficient estimates are nonzero for all the regressors. However, in lasso regression it is possible that some regressors may get omitted. It may be noted that both these methods yield biased estimators which have some interesting optimal properties under certain conditions.

The estimates for the regression coefficients for the Cigarette consumption data are given in Table 7.15.

In ridge regression, all the regressors have nonzero coefficient estimates which are quite different from those obtained in Table 7.8 or Table 7.14. The signs match, however. Lasso regression drops the same variables as in stepwise regression and best subset regression. The coefficient estimates are also not too far off.

In practice, it is better to consider stepwise/best subset regression and lasso regression and compare the results before a final subset is selected.

Table 7.15 Coefficients using ridge and lasso regressions

Ridge regression

```
> Cig_data<-CigaretteConsumption[-c(9,29,30),-1]
> lambda <- 10^seq(10, -2, length = 100)
> ridge.mod <- glmnet(as.matrix(Cig_data[,-7]),as.matrix(Cig_data[,7]), alpha
 = 0, lambda = lambda)
> predict(ridge.mod, s = 0, Cig_data[,-7], type = 'coefficients')[1:6,]

   (Intercept)           Age            HS        Income         Black        Female
   100.59166959    1.85775082   -1.16705566    0.02079292   -0.30019556    1.11722448


> cv.out <- cv.glmnet(as.matrix(Cig_data[,-7]),as.matrix(Cig_data[,7]), alpha
 = 0)
> bestlam <- 13.1
> coef(cv.out, s=bestlam)

7 x 1 sparse Matrix of class "dgCMatrix"
                       1
(Intercept) 62.407535361
Age          1.602311458
HS          -0.258625440
Income       0.007815691
Black        0.038966928
Female       0.892841343
Price       -1.366561732
```

Lasso regression

```
> lasso.mod <- glmnet(as.matrix(Cig_data[,-7]),as.matrix(Cig_data[,7]), alpha
 = 1, lambda = lambda)
> predict(lasso.mod, s = 0, Cig_data[,-7], type = 'coefficients')[1:6,]

   (Intercept)           Age            HS        Income         Black        Female
   101.17866407    1.86717911   -1.16229575    0.02075092   -0.29588646    1.09634187


> cv.out1 <- cv.glmnet(as.matrix(Cig_data[,-7]),as.matrix(Cig_data[,7]), alph
a = 1)
> bestlam1 <- 0.7
> coef(cv.out1, s=bestlam1)

7 x 1 sparse Matrix of class "dgCMatrix"
                       1
(Intercept) 120.81017762
Age           2.40000527
HS           -0.70182973
Income        0.01614734
Black         .
Female        .
Price        -2.46939620
```

3.14 Dummy Variables

Gender discrimination in wages is a highly debated topic. Does a man having the same educational qualification as a woman earn a higher wage on average? In order to study the effect of gender on age controlling for the educational level, we use data

7 Statistical Methods: Regression Analysis

Table 7.16 Wage equation for males and females

```
> dummy_reg <-lm(Wage ~ Education + Female, data=female)
> summary(dummy_reg)

Call:
lm(formula = Wage ~ Education + Female, data = female)

Residuals:
    Min      1Q  Median      3Q     Max
-12.440  -3.603  -1.353   1.897  91.603

Coefficients:
             Estimate Std. Error t value Pr(>|t|)
(Intercept) -0.85760    0.51279  -1.672   0.0945 .
Education    0.95613    0.04045  23.640   <2e-16 ***
Female      -2.26291    0.15198 -14.889   <2e-16 ***
---
Signif. codes:  0 '***' 0.001 '**' 0.01 '*' 0.05 '.' 0.1 ' ' 1

Residual standard error: 7.006 on 8543 degrees of freedom
Multiple R-squared:  0.07949,  Adjusted R-squared:  0.07928
F-statistic: 368.9 on 2 and 8543 DF,  p-value: < 2.2e-16
```

on hourly average wage, years of education, and gender on 8546 adults collected by Current Population Survey (1994), USA. Gender is a qualitative attribute. How does one estimate the effect of gender on hourly wages? An individual can either be male or female with no quantitative dimension. We need a quantifiable variable that can be incorporated in the multiple regression framework, indicating gender. One way to "quantify" such attributes is by constructing an artificial variable that takes on values 1 or 0, indicating the presence or absence of the attribute. We can use 1 to denote that the person is a female and 0 to represent a male. Such a variable is called a Dummy Variable. A *Dummy Variable* is an indicator variable that reveals (indicates) whether an observation possesses a certain characteristic or not. In other words, it is a *device to classify data into mutually exclusive categories such as male and female.*

We create the dummy variable called "female," where female = 1, if gender is female and female = 0, if gender is male. Let us write our regression equation:

$$Y = \beta_0 + \beta_1 X + \beta_2 \text{female} + \varepsilon,$$

where Y = Wage and X = Years of education.

The regression output is shown in Table 7.16.

Education is significant. One year of additional schooling will lead to an increase in wage by $0.956. How do we interpret the intercept and the coefficient of Female? The estimated regression equation is

$$\widehat{Y} = -0.8576 + 0.9561 \text{ Education} - 2.2629 \text{ Female}.$$

Fig. 7.13 Parallel regression lines for the wages of males and females

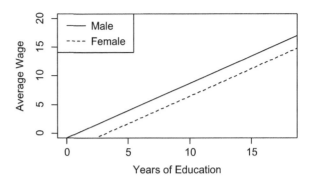

When Female = 1 (that is for females), intercept = $\beta_0 + \beta_2$. When Female = 0 (that is for males), intercept = β_0.

Thus, for the same level of education, the difference in the average wage between a female and a male is the difference in intercept, β_2 (−2.2629). In other words, a female receives $2.2629 less average hourly wage as compared to a male. A male with 10 years of education earns $(−0.8576 + 0.9561 * 10) = $8.7034 of hourly wage. A female with 10 years of education earns $(−0.8576 + 0.9561 * 10- 2.2629) = $6.4405 of hourly wage. Thus $\beta_2 = -2.2629$ is the additional effect of female on wages. Here, male is called the *base category* because the effect of the gender female on wages is measured over and above that of being male.

Here, the dummy variable acts as an Intercept Shifter. Notice that the regression line for the male is −0.8576 + 0.9561 * Education and that for female is −3.1205 + 0.9561 * Education. Thus, the two regression lines differing only by intercept are parallel shifts of each other. Such a dummy is sometimes called intercept dummy (Fig. 7.13).

At this point one may wonder: what if I take the dummy Male = 1 for male and 0 for female? Will the regression results for the same data set change? No, they will remain the same.

Intercept Dummy fits a different intercept for each qualitative characteristic. What if the relationship is different—the effect of female is not just a uniform negative number added to the base wage but instead depends on the level of education? In fact, discrimination may work in many ways: a uniform lower wage for women across all education levels (modeled by intercept dummy), or lower wages for women versus men as education increases (incremental return from education is more for men than for women) or the gap in the wages of men and women may reduce with more education. Clearly, the first model specification is inadequate if the second or the third case occurs! This leads us to a new specification.

Relax the assumption of parallel slopes or equal returns to education for men and women by introduction of new a variable "interaction term" defined as:

$$\text{Education}^*\text{Female} = \begin{cases} Education & Female = 1 \\ 0 & Female = 0 \end{cases}.$$

7 Statistical Methods: Regression Analysis

Table 7.17 Wage equation with interaction term between female and education

```
> dummy_reg2 <-lm(Wage ~ Education + Female + Female.Education, data=female)
> summary(dummy_reg2)

Call:
lm(formula = Wage ~ Education + Female + Female.Education, data = female)

Residuals:
    Min      1Q  Median      3Q     Max
-12.168  -3.630  -1.340   1.904  91.743

Coefficients:
                  Estimate Std. Error t value Pr(>|t|)
(Intercept)        0.07569    0.69543   0.109    0.913
Education          0.88080    0.05544  15.887  < 2e-16 ***
Female            -4.27827    1.02592  -4.170 3.07e-05 ***
Female.Education   0.16098    0.08104   1.986    0.047 *
---
Signif. codes:  0 '***' 0.001 '**' 0.01 '*' 0.05 '.' 0.1 ' ' 1

Residual standard error: 7.005 on 8542 degrees of freedom
Multiple R-squared:  0.07992,   Adjusted R-squared:  0.07959
F-statistic: 247.3 on 3 and 8542 DF,  p-value: < 2.2e-16
```

Thus, the interaction term is the product of the dummy variable and education. Female and Education interact to produce a new variable Female * Education. For the returns to education data, here is the regression output when we include the interaction term (Table 7.17).

We notice that the interaction effect is positive and significant at the 5% level. What does this mean?

The predicted wage for Male is given by

$$0.07569 + 0.88080*\text{Education} - 4.27827*(\text{Female}=0) + 0.16098*\{(\text{Female}=0)*\text{Education}\}$$
$$= 0.07569 + 0.88080*\text{Education}$$

The predicted Wage for Female is given by

$$0.07569 + 0.88080*\text{Education} - 4.27827*(\text{Female}=1) + 0.16098*\{(\text{Female}=1)*\text{Education}\}$$
$$= 0.07569 + 0.88080*\text{Education} - 4.27827 + 0.16098*\text{Education}$$
$$= -4.20258 + 1.04178*\text{Education}$$

Notice that for the two regression equations, both the slope and the intercepts are different!

Good news for the feminist school! An additional year of education is worth more to females because $\beta_3 = 0.16098 > 0$. An additional year of education is worth about $ 0.88 extra hourly wage for men. An additional year of education is worth about $1.04178 extra hourly wage for women. A man with 10 years of education earns: $ (0.07569 + 0.88080 * 10) = $ 8.88369 average hourly wage. A woman with 10 years of education earns: $ (−4.20258 + 1.04178 * 10) = $ 6.21522 average hourly wage.

Fig. 7.14 Regression equations for males and females

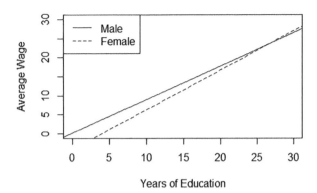

Thus, we see that there are two effects in work: (a) The female wage-dampening effect (through lower intercept for women) across education and (b) narrowing of gap in wage with years of education. This is depicted visually in Fig. 7.14.

It appears from the above figure that women start earning more than men starting from 27 years of education. Unfortunately, this conclusion cannot be drawn from the data on hand as the maximum level of education in our data is 18 years.

So far we have considered the case of two categories. In the returns to education data set considered in this section, the dataset refer to the variable "PERACE." An individual can come from five different races—WHITE, BLACK, AMERICAN INDIAN, ASIAN, OR PACIFIC ISLANDER, OTHER. The question under consideration is: Is there also racial discrimination in wages? How to model race as a regressor in the wage determination equation? Clearly, one dummy variable taking values 0, 1 will not work! One possibility is that we assign five different dummy variables for the five races. $D_1 = 1$, if white and $= 0$, otherwise; $D_2 = 1$, if Black and $= 0$, otherwise; $D_3 = 1$, if American Indian and $= 0$, otherwise; $D_4 = 1$, if Asian or Pacific Islander and $= 0$, otherwise; and $D_5 = 1$, if other and $= 0$, otherwise.

The regression output after introducing these dummies into the model is as given in Table 7.18.

What are the NAs corresponding to the others? Clearly, there is some problem with our method. R did not compute estimates for the "Other" dummy citing the reason "1 not defined because of singularities." The issue actually is the following: for every individual one and only one D1–D5 is 1 and the rest are all 0. Hence the sum of D1–D5 is 1 for each individual. Hence there is perfect collinearity between the intercept and the dummies D1–D5.

What is the solution? When you have n categories, assign either (a) n Dummies and no intercept OR b) (n − 1) Dummies and an intercept. In our example, for the five races, either assign four dummies and an intercept or just assign five dummies but no intercept.

R automatically inputted four dummies to denote the five race categories and one dummy to denote gender. If female $= 0$, and all four race dummies (D1, D2, D3, D4) $= 0$, then estimated regression equation is

Table 7.18 Wage equation for different ethnic groups

```
> dummy_reg3 <-lm(Wage ~ Education + Female + D1+D2+D3+D4+D5, data=female)
> summary(dummy_reg3)

Call:
lm(formula = Wage ~ Education + Female + D1 + D2 + D3 + D4 + D5, data = femal
e)

Residuals:
    Min      1Q  Median      3Q     Max
-12.527  -3.615  -1.415   1.963  92.124

Coefficients: (1 not defined because of singularities)
            Estimate Std. Error t value Pr(>|t|)
(Intercept) -2.04604    0.87495  -2.338   0.0194 *
Education    0.95173    0.04055  23.470   <2e-16 ***
Female      -2.25709    0.15205 -14.845   <2e-16 ***
D1           1.34520    0.74842   1.797   0.0723 .
D2           0.71014    0.77995   0.911   0.3626
D3           1.01328    0.98776   1.026   0.3050
D4           0.72720    0.85495   0.851   0.3950
D5                NA         NA      NA       NA
---
Signif. codes:  0 '***' 0.001 '**' 0.01 '*' 0.05 '.' 0.1 ' ' 1

Residual standard error: 7.003 on 8539 degrees of freedom
Multiple R-squared:  0.08069,   Adjusted R-squared:  0.08004
F-statistic: 124.9 on 6 and 8539 DF,  p-value: < 2.2e-16
```

Table 7.19 Transforming education into categories

Education years	Category
≤ 12	School
13–16	College
≥ 17	Higher education

$$\widehat{y} = -2.04604 + 0.95173\,{}^*\text{Education}$$

Thus, the intercept here denotes the effect of all the excluded categories—that is, the effect of the base category "Other male." All the dummies measure the effect over and above the base category "Other male." Looking at the R-output in Table 7.18, we infer that none of the race dummies is significant. (White is just about significant at the 10% level.) Whether you are white or black or any other race does not affect your wages. No racial discrimination in wages! But since the coefficient female is negative in estimate and highly significant, there is gender discrimination in wages!

Consider the "Education" variable. Till now, we were estimating the incremental effect of an additional year of education on wages. Moving education level from class 5 to class 6 is not so much likely to make a difference to wages. Rather, going from school to college or college to higher education may make a difference. Perhaps a more sensible way to model education is to group it into categories as show in Table 7.19.

The categories defined here (school, college, and higher education) are also qualitative but they involve an ordering—a college graduate is higher up the

Table 7.20 Transforming an ordinal variable into dummy variables

Observation	Category	College (D1)	Higher_Ed (D2)
1	School	0	0
2	College	1	0
3	Higher education	1	1

Table 7.21 Wage equation with education dummies

```
> dummy_reg4 <-lm(Wage ~ Female + College + Higher_Ed + Female.College + Fema
le.Higher_Ed, data=female)
> summary(dummy_reg4)

Call:
lm(formula = Wage ~ Female + College + Higher_Ed + Female.College +
    Female.Higher_Ed, data = female)

Residuals:
    Min      1Q  Median      3Q     Max
-13.694  -3.688  -1.539   2.068  91.112

Coefficients:
                 Estimate Std. Error t value Pr(>|t|)
(Intercept)      10.1322     0.1440  70.356  < 2e-16 ***
Female           -2.2438     0.2049 -10.950  < 2e-16 ***
College           1.7267     0.2281   7.570 4.12e-14 ***
Higher_Ed         6.0848     0.7416   8.205 2.65e-16 ***
Female.College    0.3164     0.3144   1.006    0.314
Female.Higher_Ed  0.6511     1.0940   0.595    0.552
---
Signif. codes:  0 '***' 0.001 '**' 0.01 '*' 0.05 '.' 0.1 ' ' 1

Residual standard error: 7.093 on 8540 degrees of freedom
Multiple R-squared:  0.05676,   Adjusted R-squared:  0.05621
F-statistic: 102.8 on 5 and 8540 DF,  p-value: < 2.2e-16
```

education ladder than a school pass out and one with a degree higher than a college degree is still higher up. To incorporate the effect of education categories, ordinary dummy variable is not enough. The effect of College on wages is over and above the effect of schooling on wages. The effect of "Higher Education" on wages will be some notches higher than that of college education on wages. Assign dummy variables as shown in Table 7.20.

The output after incorporating these dummies usually called *ordinal dummies* is shown in Table 7.21.

How do we interpret this output? The increment in the hourly wage for completing college education over high school is $1.7267 and that for completing higher education over college degree is $6.0848. Both these are highly significant (based on the p-values).

For identifying that there is a dummy in play using residual plots see Exercise 7.4.

For an interesting application of dummy variables and interactions among them, see Exercise 7.5.

For the use of interaction between two continuous variables in linear regression, see the chapter on marketing analytics (Chap. 19).

3.15 Normality and Transformation on the Response Variable

One of the assumptions we made in the linear regression model (Sect. 3.5) is that the errors are normally distributed. Do the data on hand and the model proposed support this assumption? How does one check this? As we mentioned several times in Sect. 3.7 and later, the residuals of different types are the representatives of the errors. We describe below one visual way of examining the normality of errors, known as Q-Q plot of the studentized residuals (see 7.43) for the definition of a studentized residual). In Q-Q plot, Q stands for quantile. First, order the observations of a variable of interest in the ascending order. We recall that the first quartile is that value of the variable below which 25% of the ordered observations lie and above which 75% of the ordered observations lie. Let p be a fraction such that $0 < p < 1$. Then the pth quantile is defined as that value of the variable below which a proportion p of the ordered observations lie and above which a proportion $1 - p$ of the ordered observations lie. In the normal Q-Q plot of the studentized residuals, the quantiles of the studentized residuals are plotted against the corresponding quantiles of the standard normal distribution. This plot is called the normal Q-Q plot of the studentized residuals. Let the ith quantile of studentized residuals (often referred to as sample quantile) be denoted by q_i and the corresponding quantile of the standard normal distribution (often referred to as theoretical quantile) be denoted by t_i. If the normality assumption holds, in the ideal case, $(t_i, q_i), i = 1, \ldots, N$ fall on a straight line. Since we are dealing with a sample, it is not feasible that all the points fall on a straight line. So a confidence band around the ideal straight line is also plotted. If the points go out of the band, then there is a concern regarding normality. For more details regarding the Q-Q plots, you can read stackexchange.[2]

We give below the Q-Q plots of the studentized residuals for different models related to Example 2.1. For Example 2.1, we consider the models AT vs WC, $\sqrt{AT}$ vs WC, and log AT vs WC and WC^2. (We shall explain in the case study later why the latter two models are of importance.) The Q_Q plots are given in Fig. 7.15.

Compare the three plots in Fig. 7.15. We find that AT vs WC is not satisfactory as several points are outside the band. The plot of $\sqrt{AT} vs WC$ is somewhat better and that of $\log AT vs WC, WC^2$ is highly satisfactory.

There are also formal tests for testing for normality. We shall not describe them here. The interested reader may consult Thode (2002). A test tells us whether the null hypothesis of normality of errors is rejected. However, the plots are helpful in taking the remedial measures, if needed.

If the residuals support normality, then do not make any transformation as any nonlinear transformation of a normal distribution never yields a normal distribution. On the other hand, if Q-Q plot shows significant deviation from normality, it may be due to one or more of several factors, some of which are given below:

---

[2]https://stats.stackexchange.com/questions/101274/how-to-interpret-a-qq-plot (Accessed on Feb 5, 2018).

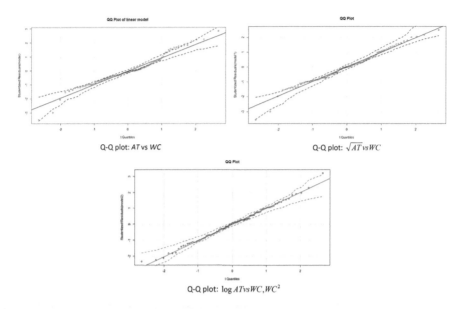

Fig. 7.15 Q-Q plots for the adipose tissue example

(a) Presence of influential observations.
(b) Omission of some important regressors which are correlated with the regressors included in the model.
(c) The response variable requires a transformation.

When you find nonnormality of studentized residuals, first check (a) and (b) above and if one of them is the case, then take care of them by the techniques we already developed. If the nonlinearity still persists, then contemplate a transformation on the response variable. Power transformations are what are commonly used. There is an oriented way of determining the power popularly known as the Box–Cox transformation. We describe the same hereunder.

Box–Cox Transformation

$$Y^{(\lambda)} = \begin{cases} \frac{Y^\lambda - 1}{\lambda \tilde{Y}^{\lambda-1}} & \lambda \neq 0 \\ \tilde{Y} \log Y & \lambda = 0 \end{cases} \tag{7.45}$$

where $\tilde{Y} = (y_1 \times y_2 \times \ldots y_N)^{\frac{1}{N}}$ is the geometric mean of the data on the response variable.

With this transformation, the model will now be $Y^{(\lambda)} = Z\beta + \varepsilon$. This model has an additional parameter λ over and above the parametric vector β. Here the errors are minimized over β and λ by the method of least squares. In practice, for various values of λ in $(-2, 2)$, the parametric vector β is estimated and the corresponding estimated log-likelihood is computed. The values of the estimated log-likelihood (y-

Fig. 7.16 Box–Cox plot

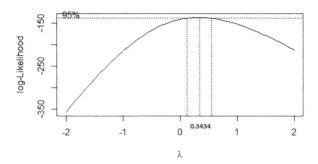

axis) are plotted against the corresponding value of λ. The value of λ at which the estimated log-likelihood is maximum is used in (7.45) to compute the transformed response variable. Since the value of λ is estimated from a sample, a confidence interval for λ is also obtained. In practice, that value of λ in the confidence interval is selected which is easily interpretable. We shall illustrate this with Example 2.1 where the response variable is chosen as *AT* and the regressor as *WC*. The plot of log-likelihood against λ is given below (Fig. 7.16).

Notice that the value of λ at which the log-likelihood is the maximum is 0.3434. This is close to $1/3$. Still it is not easily interpretable. But 0.5 is also in the confidence interval and this corresponds to the square-root which is more easily interpretable. One can use this and make the square-root transformation on the *AT* variable.

3.16 Heteroscedasticity

One of the assumptions in the least squares estimation and testing was that of equal variance of the errors (E in LINE). If this assumption is violated, then the errors do not have equal status and the standard least squares method is not quite appropriate. The unequal variance situation is called heteroscedasticity. We shall talk about the sources for heteroscedasticity, the methods for detecting the same, and finally the remedial measures that are available.

Consider AT–Waist problem (Example 2.1). Look at Fig. 7.2 and observation (ii) following the figure. We noticed that the variation in adipose tissue area increases with increasing waist circumference. This is a typical case of heteroscedasticity.

We shall now describe some possible sources for heteroscedasticity as given in Gujarati et al. (2013).

(a) Error-learning models: As the number of hours put in typing practice increases, the average number of typing errors as well as their variance decreases.
(b) As income increases, not only savings increases but the variability in savings also increases—people have more choices with their income than to just save!
(c) Error variance changes with values of X due to some secondary issue.

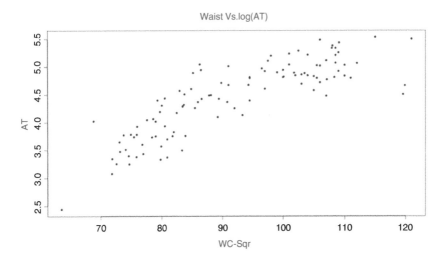

Fig. 7.17 Scatterplot of Waist vs Log AT

(d) Omitted variable: Due to the omission of some relevant regressor which is correlated with a regressor in the model, the omitted regressor remains in the error part and hence the error demonstrates a pattern when plotted against X—for example, in the demand function of a commodity, if you specify its own price but not the price of its substitutes and complement goods available in the market.
(e) Skewness in distribution: Distribution of income and education—bulk of income and wealth concentrated in the hands of a few.
(f) Incorrect data transformation, incorrect functional form specification (say linear X instead of Quadratic X, the true relation).

How does one detect heteroscedasticity? If you have a single regressor as in the case of Example 2.1 one can examine the scatterplot. If there are more regressors, one can plot the squared residuals against the fitted values and/or the regressors. If the plot is a random scatter, then you are fine with respect to the heteroscedasticity problem. Otherwise the pattern in the plot can give a clue regarding the nature of heteroscedasticity. For details regarding this and for formal tests for heteroscedasticity, we refer the reader to Gujarati et al. (2013).

Coming back to Example 2.1, one way to reduce the variation among adipose tissue values is by transforming AT to Log AT. Let us look at the scatterplot of Log AT vs Waist (Fig. 7.17).

We notice that the variation is more or less uniform across the range of Waist. However, we notice that there is a parabolic relationship between Log AT and Waist. So we fit a linear regression of Log AT on Waist and the square of Waist. The output is given below in Table 7.22.

The corresponding normal Q-Q plot is shown in Fig. 7.18.

7 Statistical Methods: Regression Analysis

Table 7.22 Regression output for the linear regression of Log AT on Waist and Square of Waist

```
> modellog<-lm(log(AT)~ Waist + Waist_sq, data=wc_at)
> summary(modellog)


Call:
lm(formula = log(AT) ~ Waist + Waist_sq, data = wc_at)

Residuals:
     Min       1Q   Median       3Q      Max
-0.69843 -0.20915  0.01436  0.20993  0.90573

Coefficients:
              Estimate Std. Error t value Pr(>|t|)
(Intercept) -7.8240714  1.4729616  -5.312 6.03e-07 ***
Waist        0.2288644  0.0322008   7.107 1.43e-10 ***
Waist_sq    -0.0010163  0.0001731  -5.871 5.03e-08 ***
---
Signif. codes:  0 '***' 0.001 '**' 0.01 '*' 0.05 '.' 0.1 ' ' 1

Residual standard error: 0.308 on 106 degrees of freedom
Multiple R-squared:  0.779,     Adjusted R-squared:  0.7748
F-statistic: 186.8 on 2 and 106 DF,  p-value: < 2.2e-16
```

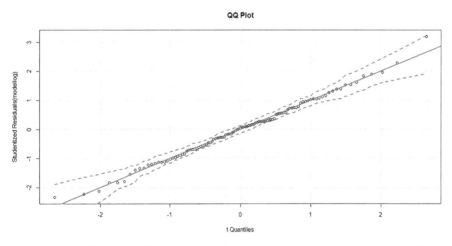

Fig. 7.18 Normal Q-Q plot of the standardized residuals in the linear regression of Log AT on Waist and Square of Waist

The corresponding plot of fitted values against residuals is given in Fig. 7.19.
Thus, both the plots tell us that the model is reasonable.
Alternatively, one can look at the linear regression of AT on Waist and look at the plot of squared residuals on Waist which is given in Fig. 7.20.

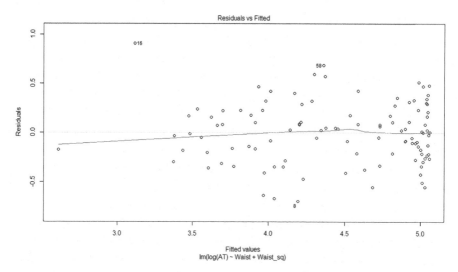

Fig. 7.19 Plot of fitted values vs residuals in the linear regression of log AT on Waist and Square of Waist

Fig. 7.20 Plot of Waist vs squared residuals in the regression of AT vs Waist

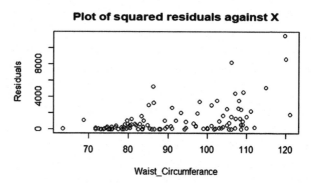

This shows a quadratic relationship and hence it suggests a regression of AT/Waist on 1/Waist (see Gujarati et al. 2013). We give the output of this regression in Table 7.23.

However, it can be checked that the corresponding normal Q-Q plot is not satisfactory.

Even though the usual least squares method, often called the ordinary least squares (OLS), is not appropriate, sometimes people use OLS with robust standard errors adjusted for heteroscedasticity. People also perform generalized least squares which can be performed as follows.

First run OLS. Get the residuals, $e_1, \ldots, e_N$. Divide the ith row of the data by $\frac{|e_i|}{1-h_{ii}}$. Now run OLS on the new data. For details, refer to Gujarati et al. (2013).

Table 7.23 Regression output of AT/Waist on 1/Waist

```
> modelinv<-lm(ATdWaist ~ Waistin ,data=wc_at)
> summary(modelinv)

Call:
lm(formula = ATdWaist ~ Waistin, data = wc_at)

Residuals:
     Min       1Q   Median       3Q      Max
-0.91355 -0.21062 -0.02604  0.17590  0.84371

Coefficients:
             Estimate Std. Error t value Pr(>|t|)
(Intercept)    3.5280     0.2158   16.35   <2e-16 ***
Waistin     -222.2786    19.1967  -11.58   <2e-16 ***
---
Signif. codes:  0 '***' 0.001 '**' 0.01 '*' 0.05 '.' 0.1 ' ' 1

Residual standard error: 0.3305 on 107 degrees of freedom
Multiple R-squared:  0.5562,    Adjusted R-squared:  0.552
F-statistic: 134.1 on 1 and 107 DF,  p-value: < 2.2e-16
```

3.17 Model Validation

The linear regression model is developed using the sample on hand. Using the residual sum of squares, R^2, and other measures that we discussed, we can examine the performance of the model on the sample. But how does it perform on the population in general? How can we get an idea regarding this? For this purpose, if we have a reasonable size data, say of about 100 observations or more, about 20–30% of randomly selected observations from the sample are kept aside. These data are called *validation data*. The remaining data are called *training data*. The model is developed using the training data. The developed model's performance on the validation data is then checked using one or more of the following measures:

(a) Root Mean Square Error (RMSE)

We have the residual sum of squares, R_0^2 for the training data set. Using the developed model, predict the response variable for each of the observations in the validation data and compute the residual. Look at the residual sum of squares, $R_0^2(V)$ for the validation data set. Let N_1 and N_2 denote the number of observations in the training and validation data sets, respectively. The residual sum of squares per observation is called the RMSE. Let RMSE(T) and RMSE(V) denote the RMSE for training and validation data sets. Compare the RMSE for the training and validation data sets by computing

|RMSE (V) – RMSE(T)|/RMSE(T). If this is large, then the model does not fit well to the population. A thumb rule is to use a cutoff of 10%.

(b) Comparison of R^2

Define achieved R^2 for the validation data set as $R^2(V) = 1 - \frac{R_0^2(V)}{s^2}$, where s^2 is the sum of squared deviations of the response variable part of the validation data from their mean. Compare the achieved R^2 with R^2 of the training data set in the same manner as the RMSE above.

(c) Compare the box plots of the residuals in the training and validation data sets.

If we have a small data set, we can use the cross-validation method described as follows. Keep one observation aside and predict the response variable of this observation based on the model developed from the remaining observations. Get the residual. Repeat this process for all the observations. Let the sum of squared residuals be denoted by $R_0^2(C)$. Compute the cross-validation R^2 as

$$R^2(C) = 1 - \frac{R_0^2(C)}{\sum (y_i - \bar{y})^2}.$$

Interpret this the same way as R^2.

3.18 Broad Guidelines for Performing Linear Regression Modeling

There is no clear-cut algorithm for linear regression modeling. It is a combination of science, art, and technology. We give below broad guidelines one may follow in the pursuit of linear regression modeling for cross-sectional data (data collected on several units at one time point). We assume that we are dealing with data of reasonable size (as specified in Sect. 3.17).

Step 1. Keep aside a randomly chosen subset of about 20–30% observations. This will form the validation data. The remaining subset will be the training data. We shall develop the model based on the training data.

In steps 2–12, we work with the training data.

Step 2. Obtain the summary statistics of all the variables. This will help in understanding the measures of central tendency and dispersion of the individual variables.

Step 3. Obtain the box plots of all the variables. These will help in understanding the symmetry and skewness. Even if there is skewness, do not try to transform the variables to achieve symmetry at this stage. Remember that we need normality of the response variable conditional on the regressors and not necessarily the unconditional normality of the response variable. These plots may become helpful in making transformations at a later stage (see step 7).

Step 4. Obtain scatterplot matrix and correlation matrix of all the variables. We understand that the scatterplots and the correlations give us an idea of the linear relationship between two variables ignoring the impact of the other variables.

However, in the case of several regressors, we seek the impact of a regressor on the response variable after taking away the impact of other regressors on both the response variable and this regressor. A scatterplot matrix helps in understanding if there is collinearity between two regressors. It may give some broad idea of the relationship between pairs of the variables.

Step 5. Check whether there are categorical variables among the regressors. If so, following the guidelines given in Sect. 3.14, transform the categorical variables into appropriate dummy variables.

Step 6. Run a linear regression of the response variable on all the regressors. Check whether R square and adjusted R square are quite apart. If so, you know that there is issue with this model. Probably some unnecessary regressors are in the model or some observations are influential. These may become apparent in the later steps. Check whether some important regressors are insignificant or their coefficient estimates are of a wrong sign. If so, there may be a collinearity issue which will become clear when collinearity is examined in step 9.

Step 7. Obtain the residual plots—Fitted Values vs Residuals and Regressors vs Residuals. Follow the instructions in Sect. 3.10 for the necessary action. There can be occasions when the plots are not conclusive regarding the transformations. In such a case, one can look at the box plots for guidance. Sometimes these plots may indicate that some important variable correlated with the present set of regressors is missing. If so try to get data on a candidate variable based on the domain knowledge and follow the instructions in Sect. 3.11 to examine its usefulness and the form in which it should be included, if found suitable.

Step 8. Check for influential observations. Follow the instructions in Sect. 3.12.

Step 9. Check for collinearity among regressors. VIFs and variance decomposition proportions will help in detecting the collinear relationships. Follow the instructions in Sect. 3.13 for identifying collinear relationships and for remedial measures.

Step 10. Check for heteroscedasticity of the errors and take remedial actions as suggested in Sect. 3.16.

Step 12. Check for normality of the residuals and make a transformation on the response variable, if necessary, following the instructions in Sect. 3.15.

Step 13. Perform the validation analysis as in Sect. 3.17.

Step 14. If the model validation is successful, then fit a final regression following steps 6–12. Interpret the regression summary and translate your technological solution to the business problem into a business solution and prepare a report accordingly. If your model validation is not successful, then you are back to modeling and try fitting a suitable alternative model following steps 7–12.

One might wonder: where is the art in all this? Everything seems sequential. Often there is no unique transformation that the residual plots suggest. Again, there is no unique way of taking care of collinearity. Also, there is no unique way of tackling heteroscedasticity or nonnormality. In all such cases, several alternative models are indicated. The data scientist has to take a call based on his or her experience. Further, when collinearity is taken care of, one may find a new observation is influential. The reverse also is possible. So some iterations may be required.

4 FAQs

1. I have fitted a linear regression model to my data and found that $R^2 = 0.07$. Should I abandon performing regression analysis on this data set?

   Answer: There are several points to consider.

   (a) If R^2 is significantly different from 0 (based on the F-test), and if the assumptions are not violated, then one can use the model to study the impact of a significant regressor on the response variable when other regressors are kept constant.
   (b) If R^2 is not significantly different from 0, then this model is not useful. This is not to say that you should abandon the data set.
   (c) If the object is to predict the response variable based on the regressor knowledge of a new observation, this model performs poorly.
   (d) Suitable transformations may improve the model performance including R^2. (See the Exercise 7.1.) The basic principle is to look for simple models, as more complicated models tend to over-fit to the data on hand.

2. I have data for a sample on a response variable and one regressor. Why should I bother about regression which may not pass through any point when I can do a polynomial interpolation which passes through all the points leading to a perfect fit to the data?

   Answer: The data is related to a sample and our object is to develop a model for the population. While it is true that the polynomial interpolation formula is an exact fit to the data on hand, the formulae will be quite different if we bring in another observation or drop an observation. Thus, the model that we develop is not stable and is thus not suitable for the problem on hand. Moreover, the regressor considered is unlikely to be the only variable which impacts the response variable. Thus, there is error intrinsically present in the model. Interpolation ignores this. This is one of the reasons why over-fitting leads to problems.

3. I have performed the linear regression and found two of the regressors are insignificant. Can I drop them?

   Answer: There can be several reasons for a regressor to be insignificant:

   (a) This regressor may be involved in a collinear relationship. If some other regressor, which is also insignificant, is involved in this linear relationship, you may find the regressor insignificant. (See Table 7.9 where both *VOL* and *WT* are insignificant. Drop the variable *WT* and you will find that *VOL* is significant.)

(b) A variable in the current regression may not be significant. It is possible that if you bring in a new regressor, then in the presence of the new regressor, this variable may become significant. This is due to the fact that the correlation coefficient between a regressor and response variable can be smaller than the partial regression coefficient between the same two variables fixing another variable. (See Models 3 and 4 in Exercise 7.4.)

(c) Sometimes some unusual observations may make a regressor insignificant. If there are reasons to omit such observations based on domain knowledge, you may find the regressor significant.

(d) Sometimes moderately insignificant regressors are retained in the model for parsimony.

(e) Dropping a regressor may be contemplated as a last resort after exhausting all possibilities mentioned above.

4. I ran a regression of sales on an advertisement and some other regressors and found the advertisement's effect to be insignificant and this is not intuitive. What should I do?

Answer: Examine points (a)–(c) in (3) above. Sometimes model misspecification can also lead to such a problem. After exhausting all these possibilities, if you still find the problem then you should examine whether your advertisement is highly uninspiring.

5. I have performed a linear regression. The residual plots suggested that I need to bring in the square term of a regressor also. I did that and once I included the square term and found from the variance decompositions proportions table that there is a collinearity among the intercept, the regressor, and its square. What should I do?

Answer: Subtract an interpretable value close to the mean from the regressor and repeat the same with its square term. Usually this problem gets solved.

6. I have performed a linear regression and got the following estimated equation:
$\widehat{Y} = -4.32 + 542.7X_1 - 3.84X_2 + 0.043X_3$. Can I conclude that the relative importance of X_1, X_2, X_3 is in that order?

Answer: The value of the regression coefficient estimate depends on the scale of measurement of that regressor. It is possible that X_1, X_2, X_3 are measured in centimeters, meters, and kilometers, respectively. The way to assess is by looking at their t-values. There is also some literature on relative importance of regressors. One may see Kruskal (1987) and Gromping (2006).

7. How many observations are needed to perform a meaningful linear regression?

Answer: The thumb rule is at least ten observations per estimated parameter. At least 30 observations are needed to estimate a linear regression model with two regressors and an intercept. Otherwise you may get significance by chance.

8. I made two separate regressions for studying the returns to education—one for men and the other for women. Can I compare the regression coefficients of both the models for education?

Answer: If you want to test the equality of coefficients in the two models, you need an estimate of the covariance between the two coefficient estimates, which cannot be obtained from separate regressions.

9. How do I interpret the intercept in a linear regression equation?
   Answer: We shall answer through a couple of examples.
   Consider the wage equation $\widehat{wage} = 14.3 + 2.83^*edu$, where *edu* stands for years of education. If you also have illiterates in the data used to develop the model, then the intercept 14.3 is the average wage of an illiterate (obtained from the wage equation by taking $edu = 0$). However, if your data is on individuals who have at least 7 years of education, do not interpret the intercept.
   Consider again the wage equation in Table 7.21. Here the intercept is the average wage of males having education of at most 12 years.

Electronic Supplementary Material

All the datasets, code, and other material referred in this section are available in www.allaboutanalytics.net.

- Data 7.1: AnscombesQuarter.csv
- Data 7.2: cars.csv
- Code 7.1: cars.R
- Data 7.3: cigarette_consumption.csv
- Code 7.2: cigarette_consumption.R
- Data 7.4: female.csv
- Code 7.3: female.R
- Data 7.5: healthcare1997.csv
- Data 7.6: leaf.csv
- Data 7.7: US_Dept_of_Commerce.csv
- Data 7.8: wage.csv
- Data 7.9: wc-at.csv
- Code 7.4: wc-at.R

Exercises

Ex. 7.1 Let the probability distribution of X be as follows:

Value	−2	−1	0	1	2
Probability	0.2	0.2	0.2	0.2	0.2

Define $Y = 3X^2 + 2$ and $Z = X^2$. Show that X and Y are uncorrelated. Show also that the correlation coefficient between Y and $Z = X^2$ is 1.

7 Statistical Methods: Regression Analysis

Ex. 7.2 Consider the following data on the variables X and Y.

Y	0.6	0.2	0.2	0.2	0.1	0.1	0.1	0.05	0.05	0.05
X	2.01	2.0	2.0	2.0	2.0	2.0	2.0	2.0	2.0	2.0

(a) Do you use this X to predict Y?
(b) Can you guess the value of the correlation coefficient between X and Y?
(c) Now compute the correlation coefficient. Are you surprised?
(d) What happens to the correlation coefficient if 0.6 for Y is changed to 2.0 keeping the rest of the data unchanged? What can you conclude from this?
(e) What happens to the correlation coefficient if you change the value 2.01 for X to 50.0? How do you justify the answer?

Ex. 7.3 Consider data set 3 in Anscombe's quartet. Show that the observation 3 and 6 are influential but only 3rd observation is an outlier in the model $Y = \beta_0 + \beta_1 X + error$.

Ex. 7.4 Based on data on 100 adult males and 100 adult females on wage, years of education (YOE), age and sex (1 for male and 0 for female), perform the following linear regressions:

1. Model 1: Wage on age
2. Model 2: Wage on age and sex
3. Model 3: Wage on age and years of education
4. Model 4: Wage on age, sex, and years of education

In each case obtain also the residual plots including the normal Q-Q plot.
Based on the analysis, answer the questions (a)–(j) given below.

(a) Notice that in models 1 and 3, age is insignificant. But it is significant in models 2 and 4. What is the reason for this?
(b) Interpret the residual plots: Normal Q-Q plot, age vs residual and fitted values vs residual in model 1.
(c) Based on your examination of the plots mentioned in (b), what action would you take and why?
(d) Compare the residual plots of models 1 and 3. What differences do you notice? How do you explain these differences?
(e) Consider the residual plots in model 4. You notice two clusters in the plot of fitted values vs residuals. However, there are no clusters in the residual plots of age vs residuals and YOE vs residuals. Is it strange? Give reasons for your answer.
(f) Does the output of model 3 indicate that there is collinearity between age and YOE? Give reasons for your answer.
(g) Compare the residual plots of YOE vs residuals in models 3 and 4. What do you think is the real reason for the difference? Do you believe that adding a square term of YOE in model 3 will improve the fit?

(h) In model 4, consider a male and a female of same age, 28, and the same education, of 10 years. Who earns more and by how much? Does it change if the age is 40 and the education is 5 years?
(i) If one wishes to test whether the females catch up with males with increasing level of education, how would you modify model 4 and what test do you perform?
(j) Notice that the Q-Q plots in models 1 and 3 are similar. Are they satisfactory? What about those in models 2 and 4? What is the reason for this dramatic change?

Ex. 7.5 In order to study the impact of bank deregulation on income inequality, yearly data was collected on the following for two states, say 1 and 0 during the years 1976 to 2006. Bank deregulation was enacted in state 1 and not in state 0. Gini index is used to measure income inequality. To control for time-varying changes in a state's economy, we use the US Department of Commerce data ("US_Dept_of_Commerce.csv") to calculate the growth rate of per capita Gross State Product (GSP). We also control for the unemployment rate, obtained from the Bureau of Labor Statistics, and a number of state-specific, time-varying sociodemographic characteristics, including the percentage of high school dropouts, the proportion of blacks, and the proportion of female-headed households.

Name	Description
Log_gini	Logarithm of Gini index of income inequality
Gsp_pc_growth	Growth rate of per capita Gross State Product (2000 dollars)
Prop_blacks	Proportion blacks
Prop_dropouts	Proportion of dropouts
Prop_female_headed	Proportion female-headed households
Unemployment	Unemployment
Post	Bank deregulation dummy
Treatment	Denoting two different states 1 and 0
Interaction	Post*treatment
Wrkyr	Year of study

Perform the linear regression of Log_gini on the rest of variables mentioned in the above table. Report your findings. How do you interpret the regression coefficient of the interaction? Are there any assumptions you are making over and above the usual linear regression assumptions?

Ex. 7.6 From the given dataset on health care outcomes ("healthcare1997.csv") in 1997 for many countries, you are required to develop a relationship between composite health care attainment measure (response variable) and the potential drivers of health care attainment. Develop the model on a training data set and examine the validity on the validation data.

The variables along with their descriptions are as follows:

COMP = Composite measure of health care attainment
DALE = Disability adjusted life expectancy (other measure)
HEXP = Per capita health expenditure
HC3 = Educational attainment
OECD = Dummy variable for OECD country (30 countries)
GINI = Gini coefficient for income inequality
GEFF = World Bank measure of government effectiveness*
VOICE = World Bank measure of democratization of the political process*
TROPICS = Dummy variable for tropical location
POPDEN = Population density*
PUBTHE = Proportion of health expenditure paid by public authorities
GDPC = Normalized per capita GDP

References

Anscombe, F. J. (1973). Graphs in statistical analysis. *American Statistician, 27*, 17–21.
Belsley, D. A., Kuh, E., & Welsch, R. E. (2005). *Regression diagnostics*. New York: John Wiley and Sons.
Brundavani, V., Murthy, S. R., & Kurpad, A. V. (2006). Estimation of deep-abdominal-adipose-tissue (DAAT) accumulation from simple anthropometric measurements in Indian men and women. *European Journal of Clinical Nutrition, 60*, 658–666.
Chatterjee, S., & Hadi, A. S. (2012). *Regression analysis by example* (5th ed.). New York: John Wiley and Sons.
Current Population Survey. (1994) United States Department of Commerce. Bureau of the Census.
Despres, J. P., Prud'homme, D., Pouliot, A. T., & Bouchard, C. (1991). Estimation of deep abdominal adipose-tissue accumulation from simple anthropometric measurements in men. *American Journal of Clinical Nutrition, 54*, 471–477.
Draper, N., & Smith, H. (1998). *Applied regression analysis* (3rd ed.). New York: John Wiley and Sons.
Greene, W. H. (2012). *Econometric analysis* (7th ed.). London: Pearson Education.
Gromping, U. (2006). Relative importance for linear regression in R. *Journal of Statistical Software, 17*, 1–27.
Gujarati, D. N., Porter, D. C., & Gunasekar, S. (2013). *Basic econometrics* (5th ed.). New Delhi: Tata McGrawHill.
Kruskal, W. (1987). Relative importance by averaging over orderings. *American Statistician, 41*, 6–10.
Thode, H. C., Jr. (2002). *Testing for normality*. New York: Marcel Dekker.
Tibshirani, R. (1996). Regression shrinkage and selection via lasso. *Journal of Royal Statistical Society, Series B, 58*, 267–288.

Chapter 8
Advanced Regression Analysis

Vishnuprasad Nagadevara

Three topics are covered in this chapter. In the main body of the chapter, the tools for estimating the parameters of regression models when the response variable is binary or categorical are presented. The appendices cover two other important techniques, namely, maximum likelihood estimate (MLE) and how to deal with missing data.

1 Introduction

In general, regression analysis requires that the response variable or the dependent variable is a continuous and quantifiable variable, while the independent or explanatory variables can be either quantifiable or indicator (nominal or categorical) variables. The indicator variables are managed using dummy variables as discussed in Chap. 7 (Statistical Methods: Linear Regression Analysis). Sometimes it becomes necessary to use indicator or categorical variables as the response variable. Some of such situations are discussed in the next section.

Electronic supplementary material The online version of this chapter (https://doi.org/10.1007/978-3-319-68837-4_8) contains supplementary material, which is available to authorized users.

V. Nagadevara (✉)
IIM-Bangalore, Bengaluru, Karnataka, India
e-mail: nagadevara_v@isb.edu

2 Motivation

A software firm is interested in predicting employee attrition. The response variable is categorical with only two possible values, namely, "whether the employee had left the company" or "not." The firm would like to build a mathematical model which can predict the attrition based on various demographic and behavioral variables as the explanatory variables.

A financial institution would like to predict whether a particular customer is creditworthy or not. The response variable is again categorical with only two possibilities: "a customer has good credit rating" or "a customer has bad credit rating." Here again, the company would like to build a model that can predict the credit rating (good or bad) using various explanatory variables such as occupation, annual income, family size, and past repayment history.

Another example of a categorical response variable is where a pharmaceutical company would like to predict which of the patients is likely to stop using their product. A good example would be a diabetic patient using a special drug having a special ingredient which makes it more effective. Such a drug is likely to be more expensive because of the patent involved with the special ingredient which enables the body to absorb the medicine more effectively. There is a tendency among many diabetic patients to worry about the expensive nature of the drug, once the problem is under control, and many of them might switch to a less expensive drug without realizing that the special ingredient is missing in that drug. The company would like to predict those patients who are likely to switch to some other drug so that they can create a personalized campaign to retain them.

Similar is the situation with respect to telecom companies. Customer churn is a major problem in the telecom sector. It is very important for any telecom company to be able to predict customer churn so that it can create special campaigns, even at individual level, in order to retain them.

In all the above examples, the response variable is categorical and also binary in nature. As in the case of categorical explanatory variables, we can use dummy variables to represent a categorical response variable. Let us take the example of employee attrition. We can code Y = 1 if the employee had left the company and Y = 0 if the employee has not left the company. Since linear regression requires the response variable to be a quantifiable continuous variable, we modify the model to predict the probabilities of Y (the binary variable) taking one of the two values. Thus, the model can be described as

$$p = P(Y = 1|X = x) = \alpha + \beta x + \epsilon$$

where p is the probability that Y takes on a value of 1 given the explanatory variable x. α and β are the regression coefficients and ε is the error term.

There are two problems associated with the above model. The first is that the above model is un-bounded, implying p can take on any value depending on the values of X and the regression coefficients. In reality, p cannot be less than 0 or greater than 1 (since p is probability). The second problem is that it brings in heteroscedasticity. An additional problem is that the values of p are not known. We cannot use ordinary least squares (OLS) because of these two problems. Hence, we modify our model to take care of these issues.

Since Y takes only two values, namely, 0 and 1, the error term ε is given by

$$\varepsilon = \begin{Bmatrix} 0 - (\alpha + \beta x) \, if \, Y = 0 \\ 1 - (\alpha + \beta x) \, if \, Y = 1 \end{Bmatrix}$$

The error term ε does not follow normal distribution since it can take on only two values for a given x.

3 Methods in Advanced Regression: Logistic Response Function

The relationship between p and x can be represented by a logistic regression model. This model results in a "S-shaped" curve as shown in Fig. 8.1. The functional form of the model is shown below:

$$p = P\,(Y = 1|X = x) = \frac{e^{\alpha+\beta x}}{1 + e^{\alpha+\beta x}}$$

If $\alpha + \beta x$ is very small (a large negative value), then p is equal to 0. When $\alpha + \beta x$ is very large (a large positive value), then p is equal to 1. p can never be less than 0 or greater than 1. Thus, this functional form ensures that the response variable p will always remain within the required bounds. If $\alpha + \beta x$ is equal to 0, then p is equal to 0.5 implying that the explanatory variable has no impact on the response variable. This equation is usually referred to as the "logistic response function." Since Y can take on only two values in the present case, this function is generally referred to as the "Binary" logistic response function. This function is nonlinear in parameters α and β; hence the parameters cannot be estimated using ordinary least squares. A simple modification of the logistic response function can lead to a linear transformation. It can be easily seen that

$$(1 - p) = \frac{1}{1 + e^{\alpha+\beta x}}$$

Then the odds ratio

$$\frac{p}{(1 - p)} = e^{\alpha+\beta x}$$

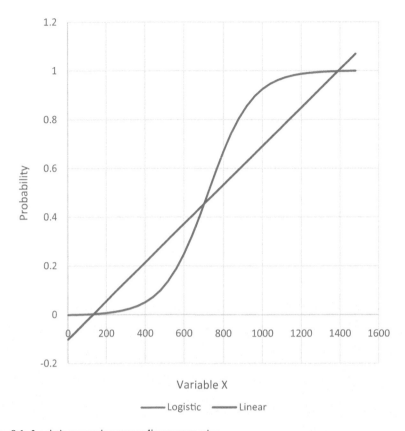

Fig. 8.1 Logistic regression versus linear regression

Taking log on both sides of the equation, we get

$$\log\left(\frac{p}{(1-p)}\right) = \alpha + \beta x$$

The above model is usually referred to as the "logit (*log*istic probability un*it*) model."

3.1 Estimation of Parameters in Logistic Regression

The parameters of the logistic regression are estimated using the *maximum likelihood method*. A brief description of the maximum likelihood method is given below. (Please see Appendix 2 for more details on the maximum likelihood method.)

8 Advanced Regression Analysis

Consider the binary logistic response function

$$p = P(Y = 1|X = x) = \frac{e^{\alpha+\beta x}}{1 + e^{\alpha+\beta x}} = f(x).$$

The probability function of a binary logistic response function for a specific observation of the response variable Y_i for a given value of x_i is given by

$$P(Y) = f(x_i)^{Y_i}(1 - f(x_i))^{(1-Y_i)}.$$

If there are n observations, the likelihood function is given by

$$L = P(Y_1, Y_2, \ldots, Y_n) = \prod_{i=1}^{n} f(x_i)^{Y_i}(1 - f(x_i))^{(1-Y_i)}.$$

The log of the above likelihood function is

$$\ln(L) = LL = \sum_{i=1}^{n} Y_i \ln\{f(x_i)\} + \sum_{i=1}^{n} (1 - Y_i) \ln\{1 - f(x_i)\}.$$

We can take the partial derivatives of the above log likelihood function (LL) with respect to α and β and equate the resultant equations to 0. Solving these two equations for the two unknowns α and β, we can obtain their estimates. These two equations do not have a closed form solution and hence the solution is obtained numerically through an iterative process.

3.2 Interpretation of β

Unlike in regression analysis, the slope coefficient β needs to be interpreted differently in logistic regression. The sign of β is important because it determines whether *f(x)* is increasing or decreasing when the value of the explanatory variable *x* is increasing. If β is 0, the curve becomes horizontal and parallel to the X-axis indicating that the response variable is independent of X. The absolute value of β indicates the rate of increase or decrease in the odds ratio. In other words, "The slope coefficient (β) can be interpreted as the rate of change in the 'log odds' as X changes." This is not very intuitive. We can exponentiate (take anti-log) both sides of the logit function and then try to interpret the slope coefficient. When we exponentiate, the odds ratio becomes an exponential function of the explanatory variable x. Now, e^β is the effect of the independent variable on the "odds ratio." In other words, the odds of Y increase multiplicatively by e^β for every unit increase in x.

Let us consider an example where we know the existing odds and would like to estimate the new odds corresponding to an increase in the value of x. Let us assume that the response variable is purchasing a washing machine and the independent variable is monthly income. Let us say, the current value of x is $ 12 (income measured in $ '000s); corresponding odds is 1 (implying that the probability is 0.5) and the exponentiated β is 1.16. We are interested in finding the increase in the odds and probability of purchasing the washing machine when the monthly income increases by $ 2 (in '000). The increase in odds can be calculated by using the following formula:

New Odds = Old Odds × Exponentiate β×Change in income=1×1.16×2 = 2.32.

The corresponding probability is 2.32/(1+2.32) = 0.6988. Thus, the probability of purchasing a washing machine increases from 0.5 to 0.6988 when the monthly income increases from $ 12,000 to $ 14,000.

The logistic regression can be extended from a single explanatory variable to multiple explanatory variables easily. If we have three explanatory variables, the logistic response function becomes

$$p = P(Y = 1|X = x) = \frac{e^{\alpha+\beta_1 x_1+\beta_2 x_2+\beta_3 x_3}}{1+e^{\alpha+\beta_1 x_1+\beta_2 x_2+\beta_3 x_3}}.$$

Similarly, the logit function becomes

$$\frac{p}{(1-p)} = e^{\alpha+\beta_1 x_1+\beta_2 x_2+\beta_3 x_3}.$$

3.3 Logistic Regression Model Diagnostics

There are three diagnostic tests available for testing the logistic regression model. They are as follows:

- Likelihood ratio test
- Wald's test
- Hosmer and Lemeshow test

Likelihood Ratio

The likelihood ratio test compares two different models, one without any independent variables and the other with independent variables. This is a chi-square test with degrees of freedom equal to the number of independent variables. This test is used to check whether the variance explained by the model is more than the unexplained variance. In other words, this is an omnibus test which can be used to test the entire model and comparable to the F test in multiple regression. The chi-square value is calculated based on two models, the first without any explanatory

8 Advanced Regression Analysis

variables and the second with the explanatory variables. The null hypothesis with respect to the likelihood ratio test is as follows:

$H_0: \beta_1 = \beta_2 = \ldots = \beta_k = 0$.

H_A: at least one β_j is not equal to zero.

Let us start with the log likelihood function given below and also the data presented in Table 8.1.

Table 8.1 Data for employee attrition model

S.no.	Employed	Experience	S.no.	Employed	Experience	S.no.	Employed	Experience
1	Y	7	51	Y	48	101	N	73
2	Y	6	52	Y	2	102	N	73
3	Y	19	53	Y	24	103	N	79
4	Y	7	54	Y	36	104	Y	66
5	Y	19	55	Y	68	105	Y	72
6	Y	2	56	N	36	106	Y	84
7	Y	6	57	Y	42	107	Y	54
8	Y	6	58	Y	30	108	Y	66
9	Y	7	59	N	48	109	Y	72
10	Y	12	60	Y	49	110	Y	3
11	Y	19	61	N	48	111	N	84
12	Y	19	62	Y	48	112	Y	66
13	Y	4	63	Y	51	113	Y	91
14	Y	19	64	Y	2	114	N	93
15	Y	7	65	Y	54	115	N	81
16	Y	3	66	Y	21	116	Y	96
17	Y	4	67	Y	51	117	Y	102
18	Y	16	68	N	48	118	N	78
19	Y	3	69	N	48	119	Y	72
20	Y	6	70	Y	30	120	N	84
21	Y	18	71	Y	42	121	N	96
22	Y	32	72	Y	66	122	Y	72
23	Y	24	73	Y	48	123	N	102
24	Y	6	74	Y	30	124	Y	85
25	Y	32	75	N	63	125	N	91
26	N	24	76	Y	66	126	Y	108
27	Y	24	77	N	55	127	Y	102
28	Y	27	78	Y	7	128	Y	108
29	Y	36	79	N	54	129	N	120
30	Y	24	80	N	60	130	Y	96
31	Y	4	81	Y	54	131	N	72
32	Y	24	82	N	60	132	Y	144
33	Y	16	83	N	60	133	Y	129
34	N	24	84	N	60	134	Y	144
35	Y	15	85	N	60	135	Y	120
36	Y	30	86	Y	60	136	N	152

(continued)

Table 8.1 (continued)

S.no.	Employed	Experience	S.no.	Employed	Experience	S.no.	Employed	Experience
37	Y	6	87	N	60	137	N	132
38	Y	19	88	N	61	138	Y	37
39	Y	19	89	N	64	139	Y	132
40	N	36	90	Y	72	140	Y	132
41	Y	36	91	N	72	141	N	228
42	Y	54	92	N	78	142	Y	192
43	Y	36	93	Y	60	143	N	240
44	Y	27	94	Y	15	144	Y	204
45	Y	36	95	Y	84	145	Y	201
46	Y	36	96	N	72	146	Y	117
47	Y	51	97	N	72	147	Y	288
48	Y	8	98	Y	78			
49	Y	24	99	Y	61			
50	Y	36	100	N	72			

$$\ln(L) = LL = \sum_{i=1}^{n} Y_i \ln\{f(x_i)\} + \sum_{i=1}^{n} (1 - Y_i) \ln\{1 - f(x_i)\}.$$

Table 8.1 has 147 observations corresponding to employees of a particular company. The data is also provided in csv format (refer to employee_attrition.csv on the website). It contains data on employee attrition. The table has details about who had left the company and also those employees continuing with the company (Variable name: "Employed" with "Y" implying that the employee is continuing with the company and "N" implying that the employee had left) and the months of work experience (Variable name: "Experience") within the company. Out of the 147 (denoted by N) employees, 41 (denoted by N_L) had left the company, and the remaining 106 (denoted by N_C) are continuing with the company. The objective of the exercise is to predict those who are likely to leave the company.

When there are no independent variables in the logistic regression, the estimates of probability are given by

$$P\left(Y = "N"\right) = \frac{N_L}{N} = \frac{41}{147} = 0.2789.$$

$$P\left(Y = "Y"\right) = \frac{N_C}{N} = \frac{106}{147} = 0.7211.$$

In order to convert the log likelihood function (LL) to a chi-square distribution, we multiply it with −2. Substituting $\frac{N_L}{N}$ for $f(x_i)$ and $\frac{N_C}{N}$ for $\{1 - f(x_i)\}$ in the log likelihood function, we get

8 Advanced Regression Analysis

Table 8.2 Logit function with experience as explanatory variable

Iteration history[a,b,c,d]

Iteration	−2 Log likelihood	Coefficients	
		Constant	Variance
1	165.970	1.70738	−0.01073
2	165.786	1.60188	−0.01038
3	165.786	1.60486	−0.01040
4	165.786	1.60487	−0.01040

[a]Method: Enter
[b]Constant is included in the model
[c]Initial −2 Log likelihood: 174.025
[d]Estimation terminated at iteration number 4 because parameter estimates changed by less than .001

$$-2LL_0 = -2\left[N_L \ln\left(\frac{N_L}{N}\right) + N_C \ln\left(\frac{N_C}{N}\right)\right] = 174.025.$$

We denote the above as $-2LL_0$ because there are 0 explanatory variables. The logit function is fitted using "Experience" as the explanatory variable, and the result is reproduced in Table 8.2. It can be seen from the table that it required four iterations to obtain the estimates for α and β. The corresponding $-2LL$ is 165.786. Let us refer to this as $-2LL_1$ because there is only one explanatory variable. The chi-square value is calculated as the difference between $-2LL_0$ and $-2LL_1$ (i.e., 174.025 − 165.786 = 8.239). The degree of freedom corresponding to this chi-square is 1 because there is one independent variable.

The R code ("Advanced_Regression_Analysis.R") to generate tables and charts referred in the chapter is provided on the website. Table 8.23 (Relevant R functions) in Appendix 3 contains a summary of the R commands used in this chapter.

The p-value corresponding to chi-square value of 8.239 with 1 degree of freedom is 0.004. Considering the small p-value, the null hypothesis can be rejected, concluding that the explanatory variable "Experience" has a significant impact on the probability of attrition.

Wald's Test

While the likelihood ratio test is used for testing the significance of the overall model, Wald's test is used to test the significance of each of the individual β_js. The null hypothesis in this case is:

H_0: $\beta_j = 0$.
H_A: $\beta_j \neq 0$.

The statistic for Wald's test follows chi-square distribution with 1 degree of freedom. The statistic for each β_j is calculated as

$$W = \left(\frac{\widehat{\beta}_j - 0}{S_e\left(\widehat{\beta}_j\right)}\right)^2,$$

Table 8.3 Logistic regression coefficients for the employee attrition model

	β	S.E.	Wald	df	Sig.	Exp(β)	95% C.I. for Exp(β)	
							Lower	Upper
Constant	1.605	.311	26.596	1	.000	4.977		
Experience	−.010	.004	7.567	1	.006	.990	.982	.997

where $S_e\left(\widehat{\beta}_j\right)$ is the standard error of the estimate $\widehat{\beta}_j$. The logistic regression coefficients along with the standard errors and Wald's statistics and significance levels are presented in Table 8.3.

The null hypothesis that β = 0 is rejected based on the p-value (Sig.) of 0.006. Table 8.3 also presents the 95% confidence interval for β. The confidence interval is calculated based on the formula below:

A two-sided confidence interval of (1 − α) level for β is $\widehat{\beta} \pm Z_{\alpha/2} S_e\left(\widehat{\beta}\right)$ where Z is the standard normal value corresponding to a confidence level of (1 − α) × 100%.

Hosmer–Lemeshow Test

The Hosmer–Lemeshow test is also a chi-square test which is used as a goodness of fit test. It checks how well the logistic regression model fits the data. The test divides the dataset into deciles and calculates observed and expected frequencies for each decile. Then it calculates the "H" statistic using the formula

$$H = \sum_{j=1}^{2} \sum_{i=1}^{g} \frac{\left(O_{ij} - E_{ij}\right)^2}{E_{ij}} = \sum_{i=1}^{g} \frac{(O_{1i} - E_{1i})^2}{n_i \pi_i (1 - \pi_i)}$$

where O_{ij} is the observed frequency of ith group and jth category, E_{ij} is the expected frequency of *ith* group and *jth* category, n_i is the number of observations in *ith* category, g is the number of groups, and π_i is the predicted risk for *ith* group. The statistic H is somewhat similar to the chi-square used in the goodness of fit tests.

The cases are grouped together according to their predicted values from the logistic regression model. Specifically, the predicted values are arrayed from lowest to highest and then separated into several groups of approximately equal size. While the standard recommendation by Hosmer and Lemeshow is ten groups, the actual number of groups can be varied based on the distribution of the predicted values.

The null and alternate hypotheses for Hosmer–Lemeshow test are:

H_0: the logistic regression model fits the data.

H_A: the logistic regression model does not fit the data.

Table 8.4 presents the contingency table for Hosmer–Lemeshow test, while Table 8.5 presents the Hosmer–Lemeshow statistic with the significance levels. The contingency table had only nine classes owing to the number of observed and expected frequencies. Consequently, the chi-square of the final model had only seven degrees of freedom.

Table 8.4 Contingency table for the Hosmer–Lemeshow test

Group	Employed = N		Employed = Y		Total
	Observed	Expected	Observed	Expected	
1	5	9.225	13	8.77	18
2	7	5.132	8	9.86	15
3	10	4.829	6	11.17	16
4	9	5.006	9	12.993	18
5	6	4.319	11	12.680	17
6	2	3.171	12	10.825	14
7	2	4.498	20	17.501	22
8	0	2.212	12	9.788	12
9	0	2.602	15	12.397	15

Table 8.5 Hosmer and Lemeshow test

Chi-square	df	Sig.
26.389	7	.000429

Based on the p-value (Sig. = .000429) corresponding to chi-square value 26.389 with 7 degrees of freedom, we reject the null hypothesis that the logistic regression fits the data. In other words, the logistic regression model with only one explanatory variable does not fit the data, and the model needs to be revised. This will be done at a later part of this chapter by adding additional explanatory variables.

3.4 Cox and Snell R^2, Nagelkerke R^2, and McFadden R^2

When the dependent variable is continuous in nature, we can use R^2 as a measure of the percentage of variation explained by the explanatory variables. On the other hand, such calculation of R^2 is not possible with logistic regression where the dependent variable is binary. It is possible to calculate pseudo R^2 values. These pseudo R^2 values compare the likelihood functions with and without the explanatory variables. Three popular pseudo R^2 values that are used with logistic regression are Cox and Snell R^2, Nagelkerke R^2, and McFadden R^2.

Cox and Snell R^2 is defined as

$$\text{Cox and Snell } R^2 = 1 - \left(\frac{L\ (Intercept\ only\ model)}{L\ (Full\ model)}\right)^{2/n}$$

where n is the sample size. The full model is the one which has the explanatory variables included in the model and used for further analysis such as calculating class probability or prediction.

Table 8.6 Pseudo R^2 values

−2 Log likelihood	Cox and Snell R^2	Nagelkerke R^2	McFadden R^2
165.786	.0473	.0785	0.4734

It is not necessary that the maximum possible value of Cox and Snell R^2 is 1. Nagelkerke R^2 adjusts Cox and Snell R^2 such that the maximum possible value is equal to 1.

The Nagelkerke R^2 is calculated using the formula

$$\text{Nagelkerke } R^2 = \frac{1 - \left(\frac{L(Intercept\ only\ model)}{L(Full\ model)}\right)^{2/n}}{1 - \left(L(Intercept\ only\ model)\right)^{\frac{2}{n}}}$$

McFadden R^2 is defined as

$$\text{McFadden } R^2 = 1 - \frac{Ln\left(L_{Full\ model}\right)}{Ln\left(L_{Intercept\ only\ model}\right)}$$

where $L_{Intercept\ only\ model}$ (L_0) is the value of the likelihood function for a model with no predictors and $L_{Full\ model}$ (L_M) is the likelihood for the model being estimated using predictors.

The rationale for this formula is that $\ln(L_0)$ plays a role analogous to the residual sum of squares in linear regression. Consequently, this formula corresponds to a proportional reduction in "error variance." Hence, this is also referred to as "Deviance R^2."

These Pseudo R^2 values are intended to mimic the R^2 values of linear regression. But the interpretation is not the same. Nevertheless, they can be interpreted as a metric for the amount of variation explained by the model, and in general, these values can be used to gauge the goodness of fit. It is expected that the larger the value of pseudo R^2, the better is the fit. It should also be noted that only Nagelkerke R^2 can reach a maximum possible value of 1.0.

The three pseudo R^2 values for the logistic regression model using the attrition data (data from Table 8.1) are presented in Table 8.6.

3.5 Selection of Cutoff Probability

Normally the logistic model sets the cutoff value at 0.5 by default. It implies that those observations which have a predicted probability of less than or equal to 0.5 are classified as belonging to one category and those with a probability of above 0.5

are classified as belonging to the other category. This cutoff value can be tweaked in order to improve the prediction accuracy. The tweaking can be based on either the classification plot or Youden's index.

Classification Plot

The classification plot is created by plotting the frequencies of observations on the Y-axis and the predicted probability on the X-axis. It also presents the cutoff value of probability on the X-axis. The classification plot for the attrition model obtained from SPSS is presented in Fig. 8.2.[1]

The classification table with the cutoff value of 0.5 is presented in Table 8.7. There are a series of "N"s and "Y"s below the X-Axis. These indicate the cutoff of 0.5 (i.e., the point where Ns change to Ys). Each N and Y in the graph above the X-Axis represents 5 data points. There are two Ns and four Ys to the left of 0.5 but not shown because of the granularity. There are a few Ns to the right of 0.5,

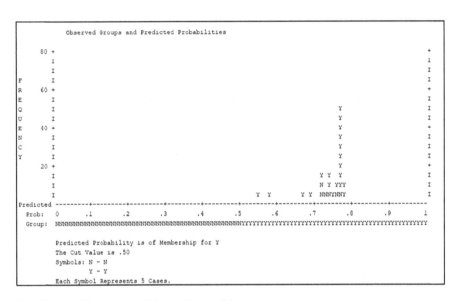

Fig. 8.2 Classification plot of the attrition model

Table 8.7 Classification table with a cutoff value of 0.5

		Observed		
		Continuing or not		
Predicted		N	Y	Overall percentage
Continuing or not	N	2	4	
	Y	39	102	
Percentage correct		4.9	96.2	70.7

---

[1] R may produce different looking output for the same chart.

Table 8.8 Classification table with a cutoff value of 0.7

Predicted		Observed		Overall percentage
		Continuing or not		
		N	Y	
Continuing or not	N	17	22	
	Y	24	84	
Percentage correct		41.5	79.2	68.7

and these are misclassifications. If the cutoff is shifted beyond 0.5, some of these will be correctly classified. A cutoff of 0.7 is also tried and found to be better in classifying the Ns more correctly as shown in Table 8.8. The downside is that more Ys are misclassified by the shift in the cutoff value.

It can be seen from Fig. 8.2 and Table 8.7 that many of those who have left the company are misclassified as "continuing with the company." The overall prediction accuracy of 70.7% is of no use because the model is completely ineffective with respect to those who have left the company. The very objective of the exercise is to be able to predict who are likely to leave the company so that appropriate strategy (even at an individual level) can be created to retain them. It is important to predict accurately, as many of those who are likely to leave the company, even at the cost of misclassifying those who are going to stay with the company. The prediction accuracy of those who are likely to leave the company can be increased by tweaking the cutoff value. By analyzing Fig. 8.2, it can be deduced that a cutoff value of 0.70 can give better results in terms of predicting those who are not continuing with the company. The prediction accuracies with 0.7 as the cutoff value are presented in Table 8.8. The new cutoff value of 0.7 has resulted in a significant increase in the prediction accuracy of those who are likely to leave the company. The overall prediction accuracy has marginally gone down to 68.7%.

Youden's Index

As in the case of any predictive technique, the sensitivity and specificity of statistics can be calculated for the logistic regression. The formulae for sensitivity and specificity are reproduced below for quick reference. The values of sensitivity and specificity corresponding to the predictions shown in Table 8.7 are also given below.

$$Sensitivity = \frac{True\ Positive\ (TP)}{True\ Positive\ (TP) + False\ Negative\ (TN)} = 41.5\%$$

$$Specificity = \frac{True\ Negative\ (TN)}{True\ Negative\ (TN) + False\ Positive\ (FP)} = 79.2\%$$

Youden's index enables us to calculate the cutoff probability such that the following function is maximized.

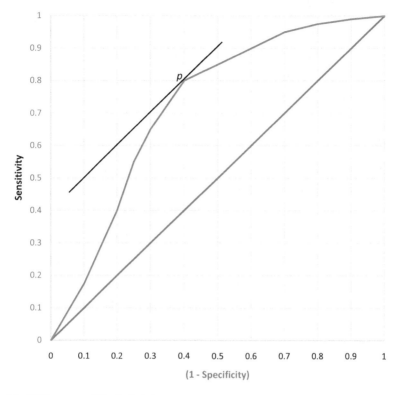

Fig. 8.3 ROC curve and Youden's index

$$Youden's\ Index = \max_{p} \{Sensitivity(p) + Specificity(p) - 1\}$$

Youden's index can be shown graphically in order to understand its interpretation better. Let us start with a receiver operating characteristics (ROC) curve. The ROC curve plots the false positive rate (1– specificity) on the X-axis and true positive rate (sensitivity) on the Y-axis. Consider the ROC curve shown in Fig. 8.3. The cutoff probability for which the distance to the ROC curve from the diagonal is maximum. In other words, it is the point where a line which is parallel to the diagonal is tangent to the ROC curve. Figure 8.3 presents the ROC curve for a hypothetical prediction model and is meant for illustration purposes only (this does not correspond to the data presented in Table 8.1). In Fig. 8.3, this point is denoted by p (the corresponding probability is 0.4).

Discordant and Concordant Pairs

The performance of a logistic regression model can be evaluated using discordant pairs and concordant pairs. Once the logistic regression model is built, the predictions and the associated probabilities can be estimated using the model. Table 8.9 presents the predications along with the other details for four selected observations.

Table 8.9 Predictions along with other details for four selected observations

Employee Sl.no.	Employed	Experience	Probability
142	Y	192	0.4034
125	N	91	0.6590
136	N	152	0.5061
90	Y	72	0.7019

Let us consider the first two observations (Employee numbers 142 and 125) of Table 8.7. The probabilities estimated by the model are 0.4034 and 0.6590, respectively. There is no cutoff probability that can predict these two observations correctly. If we set the cutoff at a point less than 0.4034, Employee number 125 will be misclassified as "Yes." If we set the cutoff above 0.6590, Employee number 142 will be misclassified. If we set the cutoff between 0.4034 and 0.6590, both observations will be misclassified. This type of pairs is called "discordant" pairs. Now, let us consider the last two observations (Employee numbers 136 and 90) of Table 8.7. The estimated probabilities are 0.5061 and 0.7019. If we set the cutoff at any point between these two probabilities, both the observations are classified correctly. Such pairs are called "concordant" pairs. Needless to say, a logistic regression model with a large number of concordant pairs is preferred.

4 Logistic Regression with Multiple Explanatory Variables

The attrition dataset has many more explanatory variables in addition to Experience (Refer to dataset "employee_attrition_nvars.csv" on the website). These variables are:

- TotExp_Binned: Total Experience (Binned)
- Age
- Gender
- ExpinTeam_Binned: Experience in the Team (Binned)
- Pos_Binned: Position in the company (Categorical)
- ExpCrnt_Binned: Experience in the current position (Binned)
- Tech_Binned: Technology specialization (Categorical)
- TotalJobHops: Total Job Hops
- CL_binned: Change in Use of Casual Leave (CL)
- PL_binned: Change in Use of Privilege Leave (PL)
- LC_Binned: Late coming to work

The logit function with k explanatory variables is

$$\frac{p}{(1-p)} = e^{\alpha + \beta_1 x_1 + \beta_2 x_2 + \beta_3 x_3 + \cdots + \beta_k x_k}$$

Table 8.10 Variables in the final equation

Coefficients:	Estimate	Std.Error	z value	Pr(>|z|)	
(Intercept)	10.8953	3.4421	3.165	0.00155	**
Age	0.1915	0.1244	1.539	0.12385	
GenderM	1.7867	0.9476	1.885	0.05937	.
TotExp_Binned	-3.8473	0.9676	-3.976	7.00E-05	***
ExpinTeam_Binned	-3.8797	0.8983	-4.319	1.57E-05	***
Pos_Binned1	4.7522	1.6286	2.918	0.00352	**
Pos_Binned2	1.6765	1.053	1.592	0.11134	
Pos_Binned3	1.4696	1.4107	1.042	0.29755	
Pos_Binned4	3.9062	1.6072	2.43	0.01508	*
TotalJobHops	-1.0511	0.3978	-2.642	0.00824	**

Significant codes: 0 '***' 0.001 '**' 0.01 '*' 0.05 '·' 0.1

A logistic regression model is built using these variables with "Continuing or Not" as the dependent variable. Both forward and backward methods with Akaike information criterion (AIC) value as the selection criterion is used. The results are discussed below.

The forward method starts with no explanatory variables (null model) and goes on adding one explanatory variable at a time, based on the AIC values until it reaches the full model. On the other hand, the backward method starts with the full model and goes on dropping explanatory variables, one at a time, based on the AIC values.

Only 134 observations out of 147 available in the dataset could be used because of some missing values in 13 observations. Of the 134 employees, 38 had left the company, and the remaining 96 are continuing with the company. The value of $-2LL_0$ for this model is 159.81. The variables that are included in the final model along with the estimated coefficients and other details are presented in Table 8.10. The variable "Position in the company" is categorical, and there are four dummy variables created to represent the five levels in the company. This variable as a whole is significant and hence the four dummy variables are included in the model. The value of $-2LL_M$ (-2Log likelihood function of the full model) is 62.96, and the chi-square value ($-2LL_0 + 2LL_M$) is 96.84. The degrees of freedom are 9 since there are nine variables in the final model. The p-value corresponding to this chi-square value is 0.000.

The stepwise method dropped Tech_Binned in the first step, LC_Binned in the second step, PL_binned in the third step, and ExpCrnt_Binned in the fourth step. CL_binned is removed in the fifth step because it turned out as statistically not significant in the fourth step.

The final model was obtained in five steps, and the model summary along with the pseudo R^2 values is presented in Table 8.11.

The Hosmer and Lemeshow test indicates that the model fits the data very well. The chi-square value corresponding to the final model is 8.5455, and the corresponding p-value is 0.287. The contingency table had only nine classes owing to the number of observed and expected frequencies. Consequently, the chi-square

Table 8.11 Model summary. Estimation terminated at iteration number 6 because parameter estimates changed by less than .001

Step	−2 Log likelihood	Cox and Snell R^2	Nagelkerke R^2	McFadden R^2
1	57.76	.533	.765	.638
2	58.34	.531	.762	.635
3	59.94	.525	.754	.625
4	61.39	.520	.747	.616
5	62.96	.515	.739	.606

Table 8.12 Contingency table for Hosmer and Lemeshow test

	Employed or not = N		Employed or not = Y		
Group	Observed	Expected	Observed	Expected	Total
1	13	14.3253	2	0.6747	15
2	11	11.6908	4	3.3092	15
3	12	8.0197	3	6.9803	15
4	2	3.1177	13	11.8822	15
5	0	0.6933	14	13.3067	14
6	0	0.1260	15	14.8740	15
7	0	0.0228	15	14.9771	15
8	0	.0040	17	16.9960	17
9	0	.0002	13	12.9999	13

Table 8.13 Hosmer and Lemeshow test

Chi-square	df	Sig.
8.5455	7	.287

of the final model had only 7 degrees of freedom. The contingency table and the Hosmer and Lemeshow test are presented in Tables 8.12 and 8.13, respectively.

Model Selection

When we have multiple explanatory variables, the model selection can be facilitated by Akaike information criterion (AIC) or Bayesian information criterion (BIC). The concepts of AIC and BIC were discussed in Chap. 7 on linear regression analysis. In the case of logistic regression, these criteria are defined as

$$AIC = -2\,(\log likelihood\ of\ the\ model) + 2r$$

$$BIC = -2\,(\log likelihood\ of\ the\ model) + r\,(\log n)$$

where n is the number of observations and r is the number of explanatory variables used in the model. As in the case of linear regression, lower values of AIC and BIC are preferred. The AIC and BIC values corresponding to each of the models (each step represents a specific model) are presented in Table 8.14.

In addition to AIC and BIC, prediction accuracies can also be used for model selection. Table 8.15 presents the classification matrix for the final model. The

Table 8.14 AIC and BIC values of different models

Step (Model)	−2 Log likelihood	AIC	BIC
1	57.758	85.76	126.33
2	58.347	84.35	122.02
3	59.949	83.95	118.72
4	61.386	83.38	115.26
5	62.96	82.96	111.94

Table 8.15 Classification Matrix of the final model

			Observed		
			Employed or not		
Predicted			N	Y	Overall percentage
Continuing or not	N		29	9	
	Y		9	87	
Percentage correct			76.3	90.6	86.6

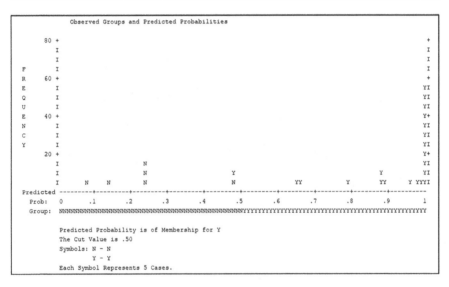

Fig. 8.4 Classification table of the final model with a cutoff value of 0.5

model is able to predict those who left the company with an accuracy level of 76.3%. The prediction accuracy with respect to those who are continuing with the company is 90.6%. The overall prediction accuracy of the final model is 86.6%.

The classification plot of the final model is presented in Fig. 8.4. It can be seen from the classification plot that a cutoff value of 0.5 is appropriate.

The final model is significant and fits the data the best. The variables that are significant in determining the probability of attrition are total experience (binned), experience in the team, and the position in the company (a categorical variable). This model can predict those who are likely to leave the company with an accuracy level of about 76.3%. In summary, this model can be effectively used to predict those who are likely to attrite at the individual level and take necessary action.

5 Multinomial Logistic Regression (MNL)

The discussion above on logistic regression was with respect to dependent variables which are qualitative and binary in nature. In other words, the dependent variable can take on only two possible values. There are many situations where the dependent variable is qualitative in nature but has more than two categories, that is, multinomial. When these categories are nominal, then such models are called polytomous models. These categories can also be ordinal in nature. A typical example would be ratings of employees into five categories, namely, (1) Excellent, (2) Very Good, (3) Average, (4) Below Average, and (5) Poor. We would like to predict the performance of the employee based on different explanatory variables such as age, education, and years of work experience. Such models where the dependent variable is ordinal are called ordinal logistic regression models. The concept of binary logistic regression can be easily extended to multinomial logistic regression.

Consider a situation where there is a qualitative dependent variable with k categories. We select one of the k categories as the base category and build logit functions relative to it. If there is no ordinality in the categories, any of the k categories can be selected as the base category. The logit function with m independent variables and k^{th} category as the base is defined as

$\ln\left[\frac{p_j(X_i)}{p_k(X_i)}\right] = \alpha_j + \beta_{1j}X_{1i} + \beta_{2j}X_{2i} + \cdots + \beta_{mj}X_{mi}$ where $j = 1, 2, \ldots, (k-1)$ and $i = 1$ to n and $\sum_{j=1}^{k} p_j = 1$

$$p_j(X_i) = \frac{e^{\alpha_j + \beta_{1j}X_{1i} + \beta_{2j}X_{2i} + \cdots + \beta_{mj}X_{mi}}}{1 + \sum_{j=1}^{k-1} e^{\alpha_j + \beta_{1j}X_{1i} + \beta_{2j}X_{2i} + \cdots + \beta_{mj}X_{mi}}}$$

As in the case of binomial logistic regression, the parameters of the multinomial regression are also estimated by using the maximum likelihood method.

Application of Multinomial Logistic Function

The following application demonstrates the prediction of quality of mulberry cocoons (refer to dataset "Quality_Index.csv" on the website) based on various characteristics. Mulberry cocoons are divided into three categories, namely, Low (1), Medium (2), and High (3), based on a quality index (QINDEX). A multinomial logistic regression is built to predict the quality categories. The third category, namely, High, is selected as the base category. There are several characteristics that determine the quality of cocoons. These characteristics are described below:

1. Single cocoon weight (COCWT): This is simply the average weight of a cocoon. This is usually calculated by selecting 25 cocoons at random, taking the total weight, and then calculating the average of a single cocoon weight. This is measured in grams or centigrams.
2. Shell weight (SHELLWT): This is the average of the single shell weight. The shell is the portion of the cocoon after the pupae is removed. This is calculated by taking the same 25 cocoons that are used for calculating the single cocoon weight. The pupae are removed from these 25 cocoons, and then the average weight of the shells is calculated. The shell yields the raw silk and hence the higher the shell weight, the higher the yield of the raw silk. This is also measured in grams or centigrams.
3. Shell ratio (SR): This is defined as the ratio of average shell weight to the average single cocoon weight and expressed as a percentage. This ratio actually estimates the raw silk content of each cocoon. Thus, the higher the shell ratio, the better is the quality.
4. Filament length (FILENGTH): This is the total length of the silk filament reeled from the cocoon. This is measured in *meters*.
5. Filament size (FILSIZE): This is the thickness of the silk filament. This is also expressed as the denier. The denier is expressed as the weight of the silk filament measured in grams for 9,000 m of the filament. A lower denier implies finer silk filament and hence is more desirable.
6. Reelability (REEBLTY): This is a measure of the re-reelability of the silk filament. It is the ratio of the cocoon reeled without break and the total number of cocoons casted, and it is measured as a percentage. This ratio is calculated from the number of times of casting filaments and the number of cocoons reeled. This characteristic actually measures the frequency of breakages of the filament during reeling.
7. Raw silk (RAWSILK): This is a measure of the raw silk expressed as a percentage. It is the ratio of the number of kilograms of cocoons required to produce 1 kilogram of raw silk and expressed as a percentage.
8. Neatness (NEATNESS): This measures the neatness of the silk filament. This is expressed as a percentage. The number of small knots and loops and the frequency of distribution on raw silk are represented as percentage by comparing a sample of 20 panels taken on a seriplane board, with the standard photographs for neatness defects. This characteristic has an impact on the quality of the fabrics woven from the silk.
9. Boil loss (BOILLOSS): Boil loss or degumming loss is the loss of sericin that is used as the gum for binding the silk filaments together in the form of a cocoon. Cocoons selected for reeling are boiled in soap solution for removing the gum or sericin. This is the ratio of the weight of cocoons after degumming to the original weight of cocoons (green cocoons) and expressed as a percentage.

There are 18,127 observations of which 5903 are of low quality, 6052 are of medium quality, and the remaining 6172 are of high quality.

Table 8.16 Estimated coefficients for multinomial logit model

Coefficients					
	(Intercept)	COCWT	SHELLWT	SR	FILENGTH
1	183.77739	20.03949	−1.07946	−0.4463738	−0.03074
2	87.55541	18.80128	−0.961	0.3470156	−0.01544
	FILSIZE	REEBLTY	RAWSILK	NEATNESS	BOILLOSS
1	−13.037409	−1.0407	−0.86267	−0.5900141	1.806278
2	−7.259055	−0.59009	−0.30789	−0.2680136	0.819668
Std. errors					
	(Intercept)	COCWT	SHELLWT	SR	FILENGTH
1	0.012280555	0.019688	0.015999	0.03656516	0.000623
2	0.007261254	0.015588	0.011393	0.02861125	0.000447
	FILSIZE	REEBLTY	RAWSILK	NEATNESS	BOILLOSS
1	0.2044784	0.010252	0.018907	0.007368245	0.029541
2	0.13739	0.00743	0.012727	0.005826297	0.02001
Residual deviance			8895.627		
AIC			8935.627		

Table 8.17 Pseudo R-square values

Cox and Snell	.818
Nagelkerke	.921
McFadden	.777

Table 8.18 Model fitting information

	Model fitting criteria			Likelihood ratio tests		
Model	AIC	BIC	−2 Log likelihood	Chi-square	df	Sig.
Intercept only	39827.07	39842.684	39823.07			
Final	8935.627	9091.73	8895.627	30927.446	18	.000

Since there are three categories, two different functions were built for prediction. These were Low vs. High and Medium vs. High. The estimated coefficients along with the standard errors are presented in Table 8.16. All the coefficients are statistically significant at the 5% level.

Pseudo R^2 values are presented in Table 8.17. All the three R^2s, namely, Cox and Snell R^2, Nagelkerke R^2, and McFadden R^2, are fairly high.

The model fitting information values (AIC, BIC, and −2LL) along with the chi-square value are presented in Table 8.18. The degrees of freedom are 18 (since there are nine explanatory variables and two functions). The p-value of 0.000 suggests that the models are statistically significant.

The prediction accuracies are presented in Table 8.19. The two models predict the quality with an accuracy level of 88.7%.

Table 8.19 Classification

Observed	Predicted			
	Low	Medium	High	Percent correct
Low	5407	496	0	91.6%
Medium	596	5036	420	83.2%
High	0	536	5636	91.3%
Overall percentage	33.1%	33.5%	33.4%	88.7%

6 Conclusion

The flexibility that logistic regression offers makes it a very useful technique in predicting categorical dependent variables (binary as well as multinomial). Logistic regression predicts the probability rather than the event itself. Unlike linear regression, it ensures that the probability remains between the limits, namely, 0 and 1. Logistic regression is more suitable for prediction when majority of the explanatory variables are metric in nature. While binary logit function can be used for predicting categorical variables which can take on only two values, the technique can be easily extended to variables that take on more than two values. The multinomial logistic model can also be used to predict variables which are ordinal in nature. The ordinal logistic model is beyond the scope of this chapter. Interested readers may look at the additional readings given at the end of this chapter.

Further Reading

- Agresti, A. (2002). *Categorical data analysis* (2nd ed.). Hoboken, NJ: John Wiley and Sons.
- Chatterjee, S., & Hadi, A. S. (2012). *Regression analysis by example* (5th ed.). Hoboken, NJ: Wiley-Interscience.
- DeMaris, A. (1995). A tutorial on logistic regression. *Journal of Marriage and Family, 57*(4), 956–968.
- Kleinbaum, D. G., & Klein, M. (2010). *Logistic regression, a self-learning text* (3rd ed.). New York: Springer.
- Menard, S. (2002). Applied logistic regression analysis. Sage University paper (Vol. 106).

Electronic Supplementary Material

All the datasets, code, and other material referred in this section are available in www.allaboutanalytics.net.

- Code 8.1: Advanced_Regression_Analysis.R
- Data 8.1: Employee_attrition.csv
- Data 8.2: Employee_attrition_nvars.csv
- Data 8.3: Quality_Index.csv

Exercises

Global TV Week is a weekly magazine providing news and information about different television programs. The magazine tried to predict who is likely to buy a subscription. They carried out a campaign and collected data from 100 respondents. The magazine tried to build logistic regression models using age as the predictor variable to make the prediction. Out of the 100 respondents, only 72 actually subscribed to the magazine.

Ex. 8.1 Based on the above information, calculate the -2 log likelihood ($-2LL_0$) function without any explanatory variables.

Ex. 8.2 The $-2LL$ function values for the four iterations are given in the table below. Based on these values, calculate the χ^2 value. Carry out the omnibus test based on this χ^2 value. What are the degrees of freedom for this test? What is the p-value corresponding to this test? What is the conclusion regarding the validation of the model?

Iteration history[a,b,c,d]

Iteration	−2 Log likelihood	Coefficients	
		Constant	Age
1	114.451	3.057	−.080
2	114.205	3.488	−.092
3	114.204	3.504	−.092
4	114.204	3.504	−.092

[a]Method: Enter
[b]Constant is included in the model
[c]Initial −2 Log Likelihood: 114.451
[d]Estimation terminated at iteration number 4 because parameter estimates changed by less than .001

Ex. 8.3 The beta coefficients and the standard errors are given in the table below. Calculate Wald's statistic and test the two coefficients for statistical significance.

8 Advanced Regression Analysis

What are the degrees of freedom associated with Wald's statistic? What are the p-values?

	β	S.E.
Age	−.092	.045
Constant	3.504	1.276

Ex. 8.4 Calculate a 95% confidence interval for the β coefficient for the variable "Age." What is the interpretation of the coefficient for Age, −0.092?

Ex. 8.5 The model summary is given in the table below. Comment on the model based on the pseudo R^2s and −2LL.

−2 Log likelihood	Cox & Snell R^2	Nagelkerke R^2
114.204	.043	.062

Ex. 8.6 Hosmer and Lemeshow chi-square is given below. What is the conclusion with respect to the model based on Hosmer and Lemeshow test?

Chi-square	Df	Sig.
23.437	8	.001

Ex. 8.7 The prediction matrix for the above logistic regression is given below. Calculate Youden's index based on this matrix.

		Predicted		
		Subscribed or not		Percentage correct
Observed		N	Y	
Subscribed or not	N	1	27	3.6
	Y	3	69	95.8
Overall percentage				70.0

Appendix 1: Missing Value Imputation

Missing data is a common issue in almost all analyses. There are a number of ways for handling missing data, but each of them has its own advantages and disadvantages. This section discusses some of the methods of missing value imputation and the associated advantages and disadvantages.

The type of missing values can be classified into different categories. These categories are described below:

(a) *Missing at random*: when the probability of non-response to a question depends only on the other items where the response is complete, then it is categorized as "Missing at Random."
(b) *Missing completely at random*: if the probability of a value missing is the same for all observations, then it is categorized as "Missing Completely at Random."
(c) *Missing value that depends on unobserved predictors*: when the value that is missing depends on information that has not been recorded and the same information also predicts the missing values. For example, a discomfort associated with a particular treatment might lead to patients dropping out of a treatment leading to missing values.
(d) *Missing values depending on the variable itself*: this occurs when the probability of missing values in a variable depends on the variable itself. Persons belonging to very-high-income groups may not want to report their income which leads to missing values in the income variable.

Handling Missing Data by Deletion

Many times, missing data problem can be handled simply by discarding the data. One method is to exclude all observations where the values are missing. For example, in regression analysis, any observation which has either the values of the dependent variable or any independent variable is missing, such observation is excluded from the analysis. There are two disadvantages of this method. The first is that the exclusion of observations may introduce bias, especially if those excluded differ significantly from those which are included in the analysis. The second is that there may be only a few observations left for analysis after deletion of observations with missing values.

This method is often referred to as "Complete Case Analysis." This method is also called "List-wise Deletion". This method is most suited when there are only a few observations with missing values.

The next method of discarding the data is called "Available Case Analysis" or "Pair-wise Deletion." The analysis is carried out with respect to only those observations where the values are available for a particular variable. For example, let us say, out of 1000 observations, information on income is available only for 870 and information on age is available for 960 observations. The analysis with respect to age is carried out using 960 observations, whereas the analysis of income is carried out for 870 observations. The disadvantage of this method is that the analysis of different variables is based on different subsets and hence these are neither consistent nor comparable.

Handling Missing Data by Imputation

These methods involve imputing the missing values. The advantage is that the observations with missing values need not be excluded from the analysis. The missing values are replaced by the best possible estimate.

- **Mean Imputation**

This method involves substituting missing values with the mean of the observed values of the variable. Even though it is the simplest method, mean imputation reduces the variance and pulls the correlations between variables toward "zero." Imputation by median or mode instead of mean can also be done. In order to maintain certain amount of variation, the missing values are replaced by "group mean." The variable with missing values is grouped into different bins and mean values of each group (bin) are calculated. The missing values in any particular group are replaced by the corresponding group mean, instead of overall mean.

- **Imputation by Linear Regression**

The missing value can be imputed by using simple linear regression. The first step is to treat the variable with missing values as the dependent variable and identify several predictor variables in the dataset. The identification of the predictor variables can be done using correlation. These variables are used to create a regression equation, and the missing values of the dependent variable are estimated using the regression equation. Sometimes, an iterative process is used where all the missing values are first imputed and using the completed set of observations, the regression coefficients are re-estimated, and the missing values are recalculated. The process is repeated until the difference in the imputed values between successive iterations is below a predetermined threshold. While this method provides "good estimates" for the missing values, the disadvantage is that the variables tend to fit too well because the missing values themselves are estimated using the other variables of the dataset itself.

- **Imputation with Time Series Data**

Certain imputation methods are specific to time series data. One method is to carry forward the last observation or carry backward the next observation. Another method is to linearly interpolate the missing value using the adjacent values. This method works better where the time series data exhibits trend. Wherever seasonality is involved, linear interpolation can be carried out after adjusting for seasonality.

- **Imputation of Categorical Variables**

Categorical variables, by their nature, require different methods for imputation. Imputation by mode is the simplest method. Yet it will introduce bias, just the same way as imputation by mean. Missing values of categorical variables can be treated as a separate category in itself. Different prediction models such as classification trees, k-nearest neighbors, logistic regression, and clustering can be used to estimate the missing values. The disadvantage of these methods is that it requires building high-level predictive models which in itself can be expensive and time consuming.

As mentioned earlier, missing data is a major issue in data analysis. While there are number of methods available for imputing the missing values, there is no such method that is the "best." The method that needs to be used depends on the type of dataset, the type of variables, and the type of analysis.

Further Reading

- Enders, C. K. (2010). *Applied missing data analysis*. New York: The Guilford Press.
- Little, R. J. A., & Rubin, D. B. (2002). *Statistical analysis with missing data* (2nd ed.). Hoboken, NJ: John Wiley & Sons, Inc.
- Yadav, M. L. (2018). Handling missing values: A study of popular imputation packages in R. *Knowledge-Based Systems*. Retrieved July 24, 2018, from https://doi.org/10.1016/j.knosys.2018.06.012.

Appendix 2: Maximum Likelihood Estimation

One of the most commonly used techniques for estimating the parameters of a mathematical model is the least squares estimation, which is commonly used in linear regression. Maximum likelihood estimation (MLE) is another approach developed for the estimation of parameters where the least squares method is not applicable, especially when the estimation involves complex nonlinear models. MLE involves an iterative process, and the availability of computing power has made MLE more popular recently. Since MLE does not impose any restrictions on the distribution or characteristics of independent variables, it is becoming a preferred approach for estimation.

Let us consider a scenario where a designer boutique is trying to determine the probability, π, of a purchase made by using a credit card. The boutique is interested in calculating the value of p which is a maximum likelihood estimate of π. The boutique had collected the data of 50 purchases and found that 32 out of 50 were credit card purchases and the remaining are cash purchases.

The maximum likelihood estimation process starts with the definition of a likelihood function, $L(\beta)$, where β is the vector of unknown parameters. The elements of the vector β are the individual parameters $\beta_0, \beta_1, \beta_2, \ldots, \beta_k$. The likelihood function, $L(\beta)$, is the joint probability or likelihood of obtaining the data that was observed. The data of the boutique mentioned above can be described by binomial distribution with 32 successes observed out of 50 trials. The likelihood function for this example can be expressed as

$$P(X = 32 | N = 50 \text{ and } \pi) = L(\pi) = K\pi^{32}(1-\pi)^{18}$$

where K is constant, N is the number of trials, and π is the probability of success, which is to be estimated. We can take the first derivative of the above function, equate it to zero, and solve for the unknown parameter. Alternatively, we can substitute different values for π (i.e., possible estimated values of π; let us call it p) and calculate the corresponding value of the likelihood function. Usually, these values turn out to be extremely small. Consequently, we can take the log values of the likelihood function. Since there is a one-to-one mapping of the actual values of the likelihood function and its log value, we can pick the value of π which will maximize the log likelihood function. The values of the log likelihood function (also known as log likelihood function) are plotted against the possible values of π in Fig. 8.5. The value of the log likelihood function is maximum when the value of p is equal to 0.64. This 0.64 is the maximum likelihood estimate of π.

The above example is extremely simple, and the objective here is to introduce the concept of likelihood function and log likelihood function and the idea of maximization of the likelihood function/log likelihood function. Even though in reality most applications of MLE involve multivariate distributions, the following example deals with univariate normal distribution. The concepts of an MLE with univariate normal distribution can be easily extended to any other distribution with multiple variables.

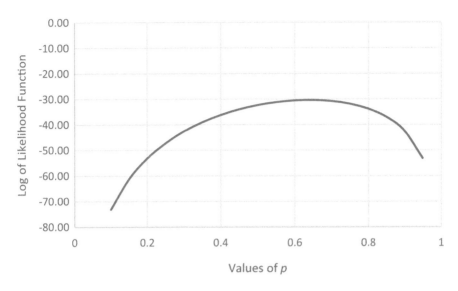

Fig. 8.5 Log of likelihood function and possible values of p

Table 8.20 Sample data

Sl. no.	Income ('000)	Sl. no.	Income ('000)
1	131	11	80
2	107	12	97
3	88	13	98
4	75	14	92
5	83	15	76
6	136	16	103
7	72	17	84
8	113	18	91
9	109	19	124
10	130	20	82

Consider the data on income levels presented in Table 8.20. There are 20 observations, and we would like to estimate the two parameters, mean and standard deviation of the population from which these observations are drawn.

The estimation process involves the probability density function of the distribution involved. Assuming that the above observations are drawn from a univariate normal distribution, the density function is

$$L_i = \frac{1}{\sqrt{2\pi}\sigma} e^{-\frac{(x_i-\mu)^2}{2\sigma^2}}$$

where X_i is the value of ith observation, μ is the population mean, σ is the population standard deviation, and L_i is the likelihood function corresponding to the ith observation. Here, L_i is the height of the density function, that is, the value of the density function, f(x).

8 Advanced Regression Analysis

Table 8.21 Likelihood values of the sample values with $\mu = 100$ and $\sigma^2 = 380$

Sl. no.	X_i	L_i	Sl. no.	X_i	L_i
1	131	0.0058	11	80	0.0121
2	107	0.0192	12	97	0.0202
3	88	0.0169	13	98	0.0204
4	75	0.0090	14	92	0.0188
5	83	0.0140	15	76	0.0096
6	136	0.0037	16	103	0.0202
7	72	0.0073	17	84	0.0146
8	113	0.0164	18	91	0.0184
9	109	0.0184	19	124	0.0096
10	130	0.0063	20	82	0.0134

The joint probability of two events, E_i and E_j, occurring is the product of the two probabilities, considering that the two events are independent of each other. Even though L_i and L_j, the two likelihood functions associated with observations i and j, are not exactly probabilities, the same rule applies. There are 20 observations in the given sample (Table 8.20). Thus, the likelihood of the sample is given by the product of the corresponding likelihood values.

The sample likelihood is given by

$$L = \prod_{i=1}^{20}\left[\frac{1}{\sqrt{2\pi}\sigma}e^{-\frac{(X_i-\mu)^2}{2\sigma^2}}\right]$$

The likelihood values of the 20 observations are presented in Table 8.21. These values are calculated based on $\mu = 100$ and $\sigma^2 = 380$ (these two values are taken from a set of all possible values).

To get the sample likelihood, the above likelihood values are to be multiplied. Since the likelihood values are small, the sample likelihood will have an extremely small value (in this case, it happens to be 7.5707×10^{-39}). It will be much better to convert the individual L_i values to their log values, and the log likelihood of the sample can be obtained simply by adding the log values. The log likelihood of the sample is obtained by

$$\log L = \sum_{i=1}^{20}\log\left[\frac{1}{\sqrt{2\pi}\sigma}e^{-\frac{(X_i-\mu)^2}{2\sigma^2}}\right]$$

Table 8.22 presents the values of L_i along with their log values. The log likelihood value (log L value) of the entire sample obtained from Table 8.22 is –87.7765. The Log L_i values in the above table are based on assumed values of $\mu = 100$ and $\sigma^2 = 380$. The exercise is repeated with different values of μ. The log L values of the entire sample are plotted against the possible values of μ in Fig. 8.6.

Table 8.22 Likelihood values along with the log values

Sl. no.	X_i	L_i	$Log(L_i)$	Sl. no.	X_i	L_i	$Log(L_i)$
1	131	0.0058	−5.1535	11	80	0.0121	−4.4153
2	107	0.0192	−3.9535	12	97	0.0202	−3.9009
3	88	0.0169	−4.0785	13	98	0.0204	−3.8943
4	75	0.0090	−4.7114	14	92	0.0188	−3.9732
5	83	0.0140	−4.2693	15	76	0.0096	−4.6469
6	136	0.0037	−5.5943	16	103	0.0202	−3.9009
7	72	0.0073	−4.9206	17	84	0.0146	−4.2259
8	113	0.0164	−4.1114	18	91	0.0184	−3.9956
9	109	0.0184	−3.9956	19	124	0.0096	−4.6469
10	130	0.0063	−5.0732	20	82	0.0134	−4.3153

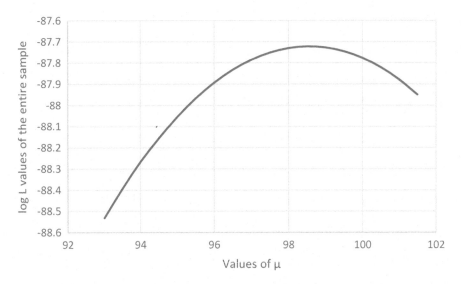

Fig. 8.6 Log L values of the entire sample and possible values of μ

The maximum likelihood estimate for σ^2 based on Fig. 8.7 is 378. It can be concluded that the sample data has a univariate normal distribution with the maximum likelihood estimates of $\widehat{\mu} = 98.5$ and $\sigma^2 = 378$.

It can be seen from Fig. 8.6 that the maximum value of log L is obtained when value of μ = 98.5. This is the maximum likelihood estimate ($\widehat{\mu}$) for the population parameter, μ. The entire exercise is repeated with different possible values of σ^2 while keeping the value of $\widehat{\mu}$ at 98.5. The values of log L corresponding to different values of σ^2 are plotted in Fig. 8.7.

This appendix provides a brief description of the maximum likelihood estimation method. It demonstrated the method by using a univariate normal distribution. The same technique can be extended to multivariate distribution functions. It is obvious that the solution requires many iterations and considerable computing power. It may

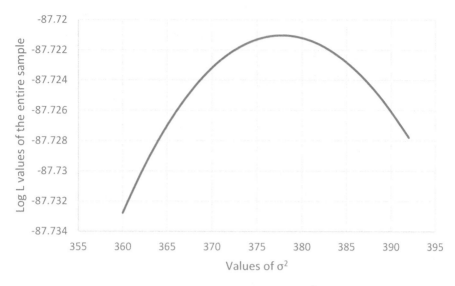

Fig. 8.7 Log L values of the entire sample and possible values of σ^2

also be noted that the values of likelihood function are very small, and consequently, log likelihood function values tend to be negative. It is a general practice to use $-\log$ likelihood function (negative of log likelihood function) and correspondingly minimize it instead of maximizing the log likelihood function.

Further Reading

- Eliason, S. R. (2015). Maximum likelihood estimation: Logic and practice. In *Quantitative applications in the social sciences book* (p. 96). Thousand Oaks, CA: Sage Publications.

Appendix 3

We provide the R functions and command syntax that are used to build various tables and charts referred in the chapter. Refer to Table 8.23. It can be helpful for practice purpose.

Table 8.23 Relevant R functions

Table/Figure/ Section	R function	Required package

Table 8.2	# to estimate logistic model parameters:
	glm(<formulae>, family=binomial("logit"), data=<dataset name>)
	# likelihood ratio test
	pchisq(<chisq value>, <degree of freedom>)
	# log likelihood value
	logLik(<model name>)
		"stats"

Table 8.3	# Wald test
	wald.test(<coefficients>, <variance covariance matrix>, Terms='n')
	where 'n' represents Wald test of nth regressor
		"aod"

Table 8.4	#Hosmer-Lemeshow Test
	hoslem.test(<response variable>, fitted(<model name>), g=<n>)
	where "n" = number of groups
		"ResourceSelection"

Table 8.6	# Pseudo R Squares
	PseudoR2(<model name>, c("pseudo R-square name1", "pseudo R-square name2", ...))
		"DescTools"

Table 8.7	# Predict probability value for response variable
	predict(<model name>,<type="response">)
		–

Section 2	# Relevel factors (define base value)
	relevel(<variable name>,ref="base category")
	# optimal cut off value
	optimalCutoff(as.factor(<response variable>), <predicted variable>)
	# Youden index
	youdensIndex(as.numeric(<response variable>), <predicted variable>, threshold=<selected cut off>)
		"InformationValue"

(continuted)

Table 8.23 Relevant R functions

Table/Figure/Section	R function	Required package

Section 2	# Sensitivity
	sensitivity(<classification table>)
	# Specificity
	specificity(<classification table>)
		"caret"

Figure 8.3	# Performance of the model at different cut off values
	performance(prediction(<predicted variable>, <response variable>), 'tpr', 'fpr')
	# Plot ROC
	plot(performance)
		"ROCR"

Table 8.10	# Model selection by AIC value
	stepAIC(<logistic model name>)
		"MASS"

Table 8.14	# AIC value
	AIC(<model name>)
	#BIC value
	BIC(<model name>)
		–

Table 8.16	multinom(<formulae>, data=<dataset name>)	"nnet"

Chapter 9
Text Analytics

Sudhir Voleti

1 Introduction

The main focus of this textbook thus far has been the analysis of numerical data. Text analytics, introduced in this chapter, concerns itself with understanding and examining data in word formats, which tend to be more unstructured and therefore more complex. Text analytics uses tools such as those embedded in R in order to extract meaning from large amounts of word-based data. Two methods are described in this chapter: bag-of-words and natural language processing (NLP). This chapter is focused on the bag-of-words approach. The bag-of-words approach does not attribute meaning to the sequence of words. Its applications include clustering or segmentation of documents and sentiment analysis. Natural language processing uses the order and "type" of words to infer the meaning. Hence, NLP deals more with issues such as parts of speech.

2 Motivating Text Analysis

Consider the following scenarios: A manager wants to know the broad contours of what customers are saying when calling into the company's call center. A firm wants to know if there are persistent patterns in the content of their customer feedback

---

Electronic supplementary material The online version of this chapter (https://doi.org/10.1007/978-3-319-68837-4_9) contains supplementary material, which is available to authorized users.

S. Voleti (✉)
Indian School of Business, Hyderabad, Telangana, India
e-mail: sudhir_voleti@isb.edu

© Springer Nature Switzerland AG 2019
B. Pochiraju, S. Seshadri (eds.), *Essentials of Business Analytics*, International Series in Operations Research & Management Science 264,
https://doi.org/10.1007/978-3-319-68837-4_9

records, customer complaints, or customer service calls/emails. An investor wants to know what major topics surround press coverage of a particular company. All these cases entail analyzing unstructured, open-ended text data, which may not be amenable to measurement and scaling along any of the four primary data scales. By several estimates, the vast majority of data flooding into organizations is unstructured, and much of this unstructured data is textual in form. Multiple customer touch-points in firms today, such as the call transcripts of a call center; email to customer service departments; social media outreach (Facebook comments, tweets, blog entries); speech transcripts, conference proceedings, press articles, statutory filings by friendly and rival firms; notes by field agents, salespeople, insurance inspectors, auditors; open-ended questions in direct interviews, surveys, typically yield unstructured data. Thus, there appears to be no getting away from the analysis of text data, especially in the field of business analytics. There is vast potential to unlocking sizeable value and competitive advantage.

2.1 An Illustrative Example: Google Flu Detector

Consider an illustrative example of text analysis from a non-business scenario. In 2009, a new flu virus was discovered, named H1N1. This virus spread quickly and there was no ready vaccine to deploy. Since it was potentially contagious, the best-case scenario was that the identified H1N1 cases would restrict their movements to stay at home avoiding contact with others. The US government had limited options—the primary agency dealing with the H1N1 threats, the Center for Disease Control (CDC) in Atlanta, had an information deficit. CDC's numbers came from doctors nationwide at the county level and were only updated every 1–2 weeks. But in that time, uninformed and unaware patients could accidentally spread H1N1 to even more people. It was at this stage that Google approached the CDC. Google reasoned that anyone with flu-like symptoms would likely search for it online and would probably use Google to do so. Moreover, they could pinpoint the origination of these searches for symptoms corresponding to H1N1 with great accuracy (right down to the street, house, hour and minute level). But to sort the wheat from the chaff, Google would need data on symptoms unique to H1N1, symptoms common to the flu, and most importantly—confirmed cases of flu in years past. The idea was to train a machine using data comprising years 2003–2008 to know which of the searched symptoms corresponded to those found in actual confirmed flu cases, and thereafter predict who might have the flu and/or H1N1 in the year going forward. The famous Google Flu detector was well received but later got embroiled in controversy and challenges regarding its prediction claims (par for the course in academia).

2.2 Data Sources for Text Mining

So what text data sources are individuals and organizations finding most common or useful for their various purposes? Figure 9.1 shows the results of a survey conducted at the 2014 Journal of Data Analysis Techniques (JDAT) conference in Europe, attended by academia and industry alike. The results are unsurprising. Social media and other user-generated content sources top the list—microblogs, full-length blogs, online forums, Facebook postings, etc. followed by more "traditional" text content sources such as mainstream media news articles, surveys (of customers, employees), reports, and medical records and compliance filings.

Data collection and extraction from these sources is now possible at scale by the development of new tools, standards, protocols, and techniques (e.g., Application Programming Interfaces (APIs)).

What are the building blocks of text analysis? What is the basic unit of analysis? How similar or different is text analysis from the analysis of more structured, metric data? The next section offers a briefing.

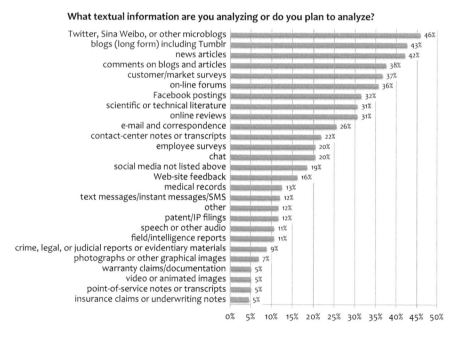

Fig. 9.1 Common sources for text data

3 Methods of Text Analysis

There are two broad approaches to handling text analysis. The first, bag-of-words, assumes that words in the text are "exchangeable," that is, their order does not matter in conveying meaning. While this assumption vastly oversimplifies the quirks of text as a means of communication, it vastly reduces the dimensionality of the text object and makes the analysis problem very tractable. Thus, if the order did not matter then two occurrences of the same word in any text document could be clustered together and their higher-level summaries (such as counts and frequencies) could be used for analysis. Indeed, the starting point of most text analysis is a data object called the term-document matrix (TDM) which lists the counts for each term (word/phrase token) in each document within a given set of text documents. The second approach, natural language processing (NLP), attempts to interpret "natural language" and assumes that content (as also context) depends on the order and "type" of words used. Hence, NLP deals more with issues such as parts of speech and named entity recognition.

Consider a passage from Shakespeare's *As You Like It*, "All the world's a stage, and all the men and women merely players: they have their exits and their entrances; and one man in his time plays many parts..." While some may see profound meaning in this (admittedly nonrandom) collection of words, Fig. 9.2 displays what happens when we process this text input in a computer's text analysis program (in this case, the open source R platform, in particular its *"tm"* and *"stringr"* packages). The code that runs the above processing uses two user-defined functions, *"Clean_String()"* and *"Clean_Text_Block()"*, for which a tutorial can be found on Matt Denny's Academic Website[1] or you may refer to the code (Fig_9.2_Shakespeare_quote.R) available on the book's website. The breaking up of text into "words" (or more technically, word "tokens") is called *tokenization*. While what we see in this simple routine are single-word tokens (or unigrams), it is possible to define, identify, and extract phrases of two or more words that "go

```
>clean_sentence <- Clean_String(sentence)
>print(clean_sentence)
 [1] "all"    "the"    "world"  "s"      "a"         "stage"    "and"    "all"
 [9] "the"    "men"    "and"    "women"  "merely"    "players"  "they"   "have"
[17] "their"  "exits"  "and"    "their"  "entrances" "and"      "one"    "man"
[25] "in"     "his"    "time"   "plays"  "many"      "parts"
>clean_speech <- Clean_Text_Block(sentence)
>str(clean_speech)    #unlist and view output
List of 3
 $ num_tokens    : int 30
 $ unique_tokens: int 24
 $ text          : chr [1:30] "all" "the" "world" "s" ...
>
```

Fig. 9.2 Basic text processing and unigram tokenization

---

[1] Tutorial link—http://www.mjdenny.com/Text_Processing_In_R.html (accessed on Dec 27, 2017).

together" in the text (bigrams, trigrams, etc.). However, that would require us to consider the order in which the words first occurred in the text and would form a part of NLP.

3.1 Some Terminology

To introduce some terminology commonly associated with text analysis, let us consider a sample dataset called the "ice-cream"[2] dataset. It contains survey responses from over 5000 consumers to an online survey run by a mid-sized, regional US retail chain that wanted feedback and opinion on its soon-to-be-launched line of "light" store-branded ice-creams. Let us suppose the store-brand name is *"Wows."* Download the dataset from the website and save as a text file on your local machine in the working directory.

The responses are to one particular survey question of interest, "If *Wows* offered a line of light ice-creams, what flavors would you want to see? Please be as specific as possible." Each row (including the empty rows) in the dataset is a document, and the stack of documents is a text corpus. The following are some observations made after a cursory run through the first few sets of responses. Notice the blank rows (i.e., empty documents)—these are consumers who chose not to answer this open-ended question. Also notice the quirks of language and grammar. There are terms with typos in them (chocolate spelled without the second "o" or vanilla spelled with a single "l," etc.). Sometimes, the same terms occur both in lower and uppercase format in different documents. There are filler words—grammar's scaffolding—such as connectors, pronouns ("all," "for," "of," "my," "to," etc.) which, in a bag-of-words world, may not make much sense and would likely swamp other words with their relatively high frequency. And quite often, punctuations are all over the place.

It may help to mitigate the effects of such quirks of language by "standardizing" words and word-forms using some general rules that apply to the vast majority of situations in everyday text (though by no means, all of them). Thus, we could consider dropping the filler words, converting all the remaining terms to lowercase to avoid case-sensitivity effects, remove all special characters (numbers, punctuations, etc.), and "stem" the remaining words. To understand the process of stemming, think of each word as a tree and different variations of that word (borne by prefixes and suffixes) as branches. Stemming cuts out the branches and retains only the core "stem" of the word. Thus, for instance, the stem-word "run" replaces "runs," "running," etc.

As mentioned previously, tokenization is the process of breaking up a cleaned corpus into individual terms composed of 1-, 2-, or more word phrases. For example, "ice-cream" is a 2-word token (bigram), whereas ice and cream are 1-word tokens

---

[2]The dataset "icecream.csv" can be downloaded from the book's website.

```
># View a sample of the DTM, sorted from most to least frequent token count
>dtm_clean <- dtm_clean[,order(apply(dtm_clean,2,sum),decreasing=T)]
>inspect(dtm_clean[1:5,1:5])
<<DocumentTermMatrix (documents: 5, terms: 5)>>
Non-/sparse entries: 11/14
Sparsity            : 56%
Maximal term length: 7
Weighting           : term frequency (tf)
Sample              :
    Terms
Docs butter chip chocol mint vanilla
   1      0    0      1    0       1
   3      0    0      1    0       1
   4      1    0      1    0       0
   5      0    1      1    0       0
   6      1    0      1    0       1
>
```

Fig. 9.3 First few cells in a document term matrix in R

(unigrams). After a corpus is tokenized, a simple frequency table can be built to determine how many times each token occurred in each document. This frequency table is called the term-document matrix (TDM). It is one of the basic objects of analysis in text analytics. Each cell in a TDM records the frequency of a particular token in a particular document. TDMs tend to be large, sparse matrices. The TDM for the ice-cream dataset has a dimension of 907 (tokens) × 2213 (docs). Figure 9.3 shows the first few cells for the first few documents of a TDM for the ice-cream dataset, in order of the most frequently occurring tokens (listed left to right). Despite this, we find 56% sparsity (or, proportion of empty cells). Notice also that empty rows have been dropped. Notice that "chocolate-chip" and "chocolate" are two distinct tokens; the first is a bigram and the second a unigram. The code "icre" is available on the book's website for reference.

Having obtained a TDM, the next logical step is to display the simplest yet meaningful parts of the information contained in the TDM in a form that people can easily read and digest. One such output display avenue is the "wordcloud," which displays the most frequently occurring tokens in the corpus as a mass or cloud of words. The higher a word's frequency in the corpus, the larger its font size. Aside from font size, no other feature of the wordcloud—color, orientation, etc.—is informative. Figure 9.4a displays the wordcloud of the most frequently occurring tokens in the ice-cream dataset. The prevalence of bigrams can be seen. One can control how many tokens to show here, to avoid clutter.

A document-term matrix (DTM) is the transpose of a TDM, with documents as rows and tokens as columns, but is sometimes used interchangeably with TDM. There are generally two ways one can "weigh" a DTM. The first uses simple term frequency (or TF in short) wherein each column merely records token frequency per document. The issue with this method is that the token frequency may not necessarily imply token importance. For instance, in a dataset about opinions on ice-cream flavors, the bigram "ice-cream" is likely to occur with high frequency since it is expected to show up in most documents. However, its importance in relation to understanding customers' flavor preferences is relatively low. Hence, a

9 Text Analytics

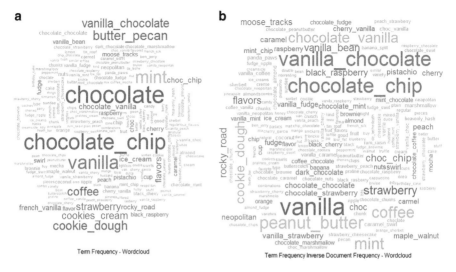

Fig. 9.4 (**a, b**) Wordclouds under the TF and TFIDF weighing schemes

second weighing scheme for tokens across documents, labeled TFIDF for "term frequency–inverse document frequency," has gained popularity. The basic idea is that a token's frequency should be normalized by the average number of times that token occurs in a document (i.e., document frequency). Thus, tokens with very high frequency spread evenly across documents tend to get discounted whereas those which occur more in some documents rather than all over the corpus gain in importance. Thus, "butter-scotch" would get a higher TFIDF score than "ice-cream" because only a subset of people would say the former, thereby raising its inverse document frequency. Various TFIDF weighing schemes have been proposed and implemented in different text analysis packages. It is prudent to test a few to see if they make more sense in the context of a given corpus than the simple TF scheme. Figure 9.4b displays a TFIDF wordcloud. The differences between that and the TF wordcloud in Fig. 9.4a are visibly apparent.

3.2 Co-occurrence Graphs (COG)

Wordclouds are very basic in that they can only say so much. Beyond simple term frequency, one might also want to know which tokens occur most frequently *together* within a document. For instance, do "vanilla" and "chocolate" go together? Or do "vanilla" and "peanut butter" go together more? More generally, does someone who uses the term "big data" also say "analytics"? A co-occurrence graph (COG) highlights token-pairs that tend to co-occur the most within documents across the corpus. The idea behind the COG is straightforward. If two tokens co-occur in documents more often than by random chance, we can use a network graph

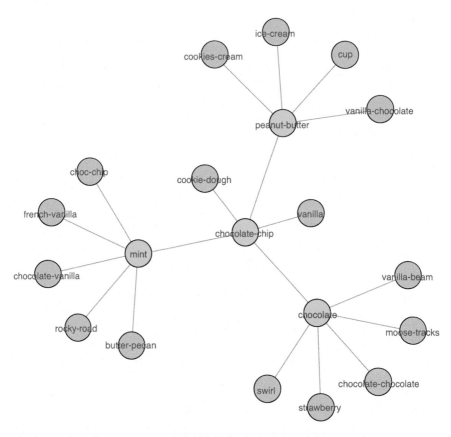

Fig. 9.5 A cleaned co-occurrence graph (COG) for the ice-cream dataset

framework to "connect" the two tokens as nodes with a link or an "edge." Because the odds are high that any two tokens will co-occur at least once in some document in the corpus, we introduce connection "thresholds" that ensure two node tokens are linked only if they co-occur more than the threshold number of times. Even so, we often see too many connections across too many nodes. Hence, in "cleaned" COGs, we designate nodes as "primary" (central to the graph) and "secondary" (peripheral nodes which only connect to one central node at a time), and we suppress interconnections between peripheral nodes for visual clarity. The R *shiny* app (discussed in the next section) can be used to generate a COG for the ice-cream dataset as shown in Fig. 9.5. Figure 9.5 displays one such cleaned COG that arises from the top tokens in the ice-cream dataset. We can interpret the COG in Fig. 9.5 as follows. We assume the peripheral (pink) nodes connect to one another only through the central (green) nodes, thus taking away much of the clutter in the earlier COG. Again, the links or edges between nodes appear only if the co-occurrence score crosses a predefined threshold.

3.3 Running Elementary Text Analysis

Today, a wide variety of platforms and tools are available to run standard text analysis routines. To run the analysis and obtain outputs similar to the ones shown in the figures, do the following.

- Open RStudio and copy-paste the code for *"shiny"* example[3] from author's github account into a script file.
  - Now, select lines 37–39 (#Basic Text-An-app) in RStudio and click "Run." RStudio first installs required libraries from the source file (line 38). Then it launches the *shiny* app for Factor An (line 39).
- Examine what the app is like—the Input Sidebar and the Output Tabs. This entire process has been described in the YouTube video tutorial[4] for a sample dataset and set of *shiny* apps in R.
- Now use the app and read in the ice-cream data, either as a text or a csv file.
- Once the app has run (may take up to several minutes to finish processing depending on the size of the dataset and the specification of the local machine), explore the output tabs to see what shows up. The video tutorial provides some assistance in this regard.

3.4 Elementary Text Analysis Applications

A number of interesting downstream applications become available once we have reduced the text corpus to an analysis object—the TDM. In the rest of this section, we discuss two such applications: (1) clustering or segmentation of documents based on text content and (2) elementary sentiment analysis.

Clustering Documents Using Text Bases

Clustering has had a long and interesting history in analytics, particularly in business where marketers are keen users of clustering methods under the broad rubric of "segmentation." However, while many applications in the past have relied on metric variables in datasets, we now examine a scenario where text data could become the basis for clustering documents. For our illustrative example, consider a marketing problem. Conceptually, segmentation in marketing implies the process of grouping together customers who share certain characteristics of interest, into homogenous segments. The rationale for segmentation is that it is more effective and efficient to pitch a value proposition to a relatively homogeneous segment than

---

[3]https://raw.githubusercontent.com/sudhir-voleti/profile-script/master/sudhir%20shiny%20app%20run%20lists.txt (accessed on Dec 27, 2017).

[4]https://www.youtube.com/watch?v=tN6FYIOe0bs (accessed on Dec 27, 2017) Sudhir Voleti is the creator of video.

to a heterogeneous mass. So, how do marketers segment and on what basis? Well, several bases can be considered: demographic (age, income, family size, zip code, etc.) and psychographic (brand conscious, price sensitive, etc.) are two examples. The ideal basis for marketers specifically is "customer need," which is often latent and hard to observe. In such instances, text offers a way to access qualitative insights that would otherwise not be reachable by traditional means. After segmentation, marketers would typically evaluate which segments are worth pursuing and how to target them.

Let us run a segmentation of the ice-cream survey's respondents based on the ice-cream flavor preferences they have freely stated in unstructured text form. The idea is to simply take the DTM and run the k-means clustering algorithm (see the Chap. 15 on Unsupervised Learning) on it. Thus, documents that are "similar" based on the tokens used would group together. To proceed with using the k-means algorithm, we must first input the number of clusters we want to work with.

To run segmentation, select codes (#Segmentation-discriminant-targeting App) in RStudio and click "Run." The line 27 launches the *shiny* app for segmentation, discriminant, and classification. Upload the ice-cream data for segmentation (refer Fig 9.6). We can estimate the number of clusters through the *scree plot* on the "Summary—Segmentation" tab. Looking at where the "elbow" is sharpest (refer Fig. 9.7), we can arrive at what a reasonable number of clusters would be. The current example suggests that 2, 4, or 7 clusters might be optimal. Let us go with 7 (since we have over a thousand respondents).

We can interpret the output as follows. The "Summary—Segmentation" tab yields segment sizes, centroid profiles, and other classic descriptive data about the obtained clusters. The tab "Segmentation—Data" yields a downloadable version of segment assignment to each document. The two output tabs "Segmentation—Wordcloud" and "Segmentation co-occurrence" display segment wordclouds and COGs alongside segment sizes. We can observe that the rather large Segment 1 (at 44% of the sample) seems to be those who prefer vanilla followed by those who prefer butter-pecan at Segment 2 (8.4% of the sample) and so on.

Sentiment Analysis

Sentiment mining is an attempt to detect, extract, and assess value judgments, subjective opinion, and emotional content in text data. This raises the next question, "How is sentiment measured?" Valence is the technical term for the subjective inclination of a document—measured along a positive/neutral/negative continuum. Valence can be measured and scored.

Machines cannot determine the existence of sentiment content in a given word or phrase without human involvement. To aid machines in detecting and evaluating sentiment in large text corpora, we build word-lists of words and phrases that are likely to carry sentiment content. In addition, we attach a "score" to each word or phrase to represent the amount of sentiment content found in that text value. Since sentiment is typically measured along a positive/neutral/negative continuum, the steps needed to perform an analysis are fairly straightforward. First, we start with a processed and cleaned corpus. Second, we match the stem of each document in the corpus with positive (or negative) word-lists. Third, we score each document in the

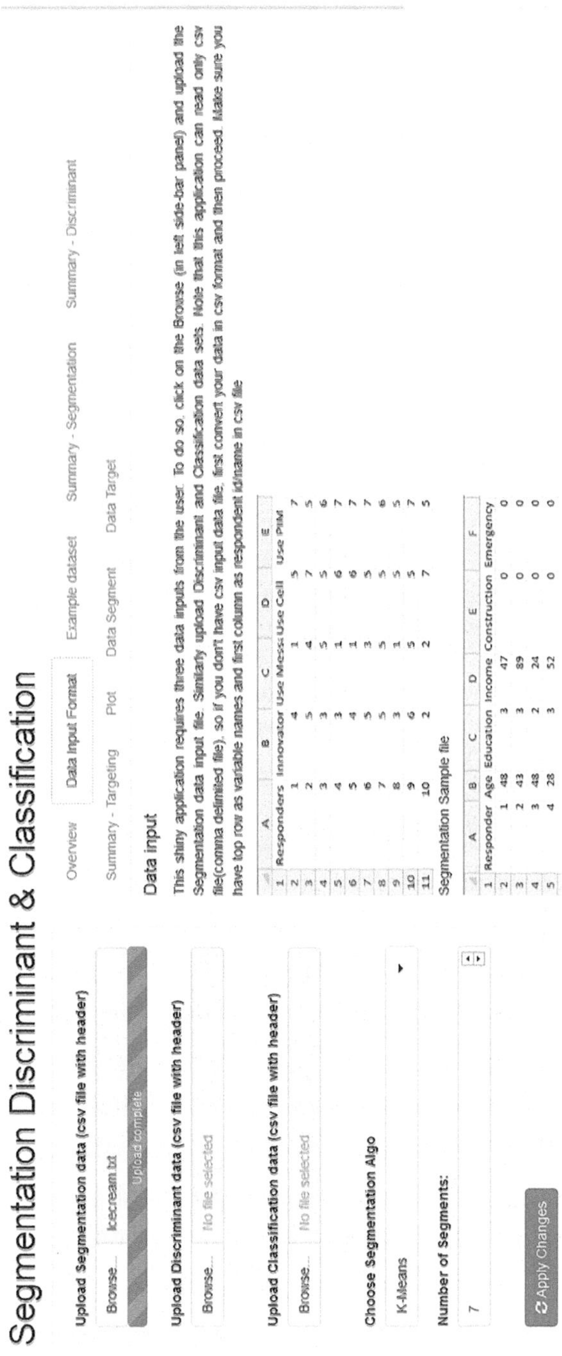

9 Text Analytics

Fig. 9.6 Shiny app for segmentation and classification

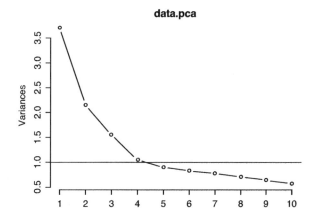

Fig. 9.7 K-means segmentation

corpus with a positive (or negative) "polarity." We can then plot the distribution of sentiment scores for each document as well as for the corpus as a whole. The positive and negative word-lists we use in the App are from the Princeton dictionary[5]. Since they are general, their effectiveness is somewhat limited in detecting sentiment in a given content. To customize sentiment analysis, one can and probably should build one's own context-specific sentiment scoring scheme giving valence weights for the most common phrases that occur in the domain of interest. Ideally, businesses and organizations would make domain-specific wordlists (and corresponding sentiment scores) for greater accuracy.

Let us see the result of running sentiment analysis on the ice-cream dataset. In the R app that you previously launched, look at the last two output tabs, namely "sentiment analysis" and "sentiment score data." One reason why the sentiment-laden wordclouds in the first output tab are so small is that most sentiment-laden tokens are adjectives while the ice-cream dataset consists of mostly nouns. Once sentiment scores by document are obtained, a host of downstream analysis becomes possible. The documents can now be sorted based on their "polarity" toward the topic at hand (i.e., ice-cream flavors). Such analyses are useful to businesses studying brand mentions on social media, for instance. Running trackers of sentiment over time can also provide useful information to managers. What we saw in this app was elementary sentiment mining. More advanced versions of sentiment mining and polarity-scoring schemes, leveraging natural language processing or NLP (discussed in detail in the next section), can be performed.

Let us step back and look at what we have covered with elementary text analysis thus far. In a nutshell, we have been able to rapidly crunch through raw text input on a scalable level, reduce open-ended text to a finite dimensional object (TDM), apply

---

[5]http://wordnet.princeton.edu/ (accessed on Feb 7, 2018).

9 Text Analytics 295

standard analysis techniques (k-means segmentation) to this object, and sense what might be the major preference groups and important attributes that emerged through our analysis. These techniques help us to achieve a state in which a survey method can be leveraged for business insight. Next, we leave behind "basic" text analysis and head into a somewhat more advanced version wherein we will uncover hidden structure beneath text data, that is, latent topic mining and modeling.

3.5 Topic Mining Text Corpora

We know that organizations have huge data stored in text form—and it is likely that people are looking for ways to extract "meaning" from these texts. Imagine files piled up on a desk—the pile represents a corpus and each individual file a "document." One may ask, can we "summarize" the text content of those files? If so, what might a summary look like? One answer is that it should contain broad commonalities in coherent, inter-related text content patterns that occur across the files in the pile. Thus, one way to view meaning is in the form of coherent, condensed "topics" or "themes" that underlie a body of text. In the past, automated analysis has relied on simple models that do not directly address themes or topics, leaving us to derive meaning through a manual analysis of the text corpus. An approach to detect and extract coherent "themes" in the text data has become popular. It is the basis for topic mining of text corpus described in detail below.

To illustrate what underlying themes might look like or mean in a text corpus context, we take a simple example. Consider a text corpus of 20 product reviews. Suppose there are two broad themes or topics in the structure of the corpus, "price" and "brand." Further, suppose that each document is a mixture of these two topics in different proportions. We can then argue that a document that talks about "price" 90% of the time, and "brand" 10%, should have nine times more "price" terms than "brand" terms. A topic model formalizes this intuition mathematically. The algorithm yields which tokens belong to which topics with what probability, and which documents "load" (have strong association, see below for an example) on which topics with what proportion. Given this, we can sort, order, plot, analyze, etc. the tokens and the documents.

So, how can we directly mine for these latent topics or themes in text? The basic idea is analogous to the well-known Factor Analytic procedures (see the Chap. 15 on Unsupervised Learning). Recall what happens in traditional factor analysis. A dataset with R rows and C columns is factorized into two components— an R × F scores matrix and an F × C loadings matrix (R stands for the number of observations, C stands for the attributes of each observation, F then stands for the number of factors into which the data is decomposed). These factors are then labeled and interpreted in terms of the composite combinations of variables that load on them. For example, we can characterize each factor by observing which variables "load" highest onto it (Variable 1 positively, Variable 4 negatively, etc.). Using this we interpret what this means and give each factor an informative label.

Now let us see what would happen if instead of a traditional metric variable dataset, our dataset was a document term matrix (DTM) with D documents and T terms? And instead of conventional factors, we use the term "Topic Factors" for the factor. This would result in a D × F scores matrix of documents-on-factors, and an F × T loadings matrix of terms-on-factors. What are these matrices? What do these scores and loadings mean in a text context? Let us find out through a hands-on example.

The next dataset we will use is a comma-separated values (CSV) file containing the mission statements from a subset of Fortune 1000 firms. It is currently stored on the author's github page. You can run the following lines of R code in RStudio to save the file on your local machine.

```
> # saving the data file as a .csv in your local machine
> mission.stments.data = read.csv("https://raw.githubusercontent.
  com/sudhir-voleti/sample-data-
  sets/master/Mission%20Statements%20v1.csv")
> # save file as data1.csv on your local machine
> write.csv(mission.stments.data, "data1.csv")
> ### Topic Mining App
> source("https://raw.githubusercontent.com/sudhir-voleti/text-
  topic-analysis-shinyapp/master/dependency-text-topic-analysis-
  shinyapp.R")
> runGitHub("text-topic-analysis-shinyapp", "sudhir-voleti")
```

Now invoke the *Topic mining* app in the *shiny* apps list (code above) and explore the app's input fields and output tabs. Now read in the saved .csv file on mission statements into the app. Similar to the process used in our basic text analysis app, this too tokenizes the corpus, creates a TDM as its basic unit of analysis, and applies a latent topic model upon it. It calls upon the user (in this case, us) to tell it the optimal number of topics we think there are in the corpus. The default in the app is 2, but this can be manually changed.

The "TDM & Word Cloud" tab shows the corpus level wordcloud and the "Topic Model—Summary" tab shows the top few phrases loading onto each topic factor. The next three tabs display topic model output. At first glance, the output in the "Topics Wordcloud" and "Topics Co-occurrence" tabs may seem no different from those in the Basic Text An app. However, while the Basic Text An app represented TDM-based clusters (wherein entire documents are hard-allocated to one cluster or another), what we see now are topic factors that can co-exist within the same document. To recap, this model says in a nutshell that every document in the corpus is a mixture of topics in some proportion and that each topic is a distribution over word or phrase tokens with some topic membership probability. Both the topic membership probability vector per token and topic proportions vector per document can be simultaneously estimated in a Bayesian manner.

Returning to the example at hand, Fig. 9.8a, b show the wordclouds from Topics 1 and 2 respectively. Topic 1 seems to emphasize "corporation" and allied terms (services, systems, provide), whereas Topic 2 emphasizes "Customers" with allied terms (solutions, employees, product-services). This suggests that Topic 1 is perhaps a measure of company (and perhaps, product) centricity whereas Topic 2 is that of customer centricity in firms' mission statements. As before, the co-occurrence graphs are used in conjunction with the wordclouds for better interpretation.

9 Text Analytics

Fig. 9.8 (**a**, **b**) Wordclouds for topic factors 1 and 2 for the mission statements dataset

Finally, the last tab in the app, "Data with topic proportions" yields a downloadable file that shows the proportion of each document that is dominated by tokens loading on a particular topic. Thus, in this example, Microsoft corporation seems to have a 70–30 split between the company and customer centricity of its mission statement while that of toymaker Mattel, Inc. is 15–85.

4 Natural Language Processing (NLP)

While the "bag-of-words" approach we have seen thus far is useful, it has severe limitations. Human language or "natural language" approach is much more complex than a bag-of-words approach. The same set of words used in a different order could produce a different meaning. Even the same set of words uttered in a different tone could have a different meaning. Often, what precedes or succeeds particular sentences or paragraphs can impact the contextual meaning of any set of words. This brings our discussion to the natural language processing or NLP. By definition, NLP is a set of techniques that enable computers to detect nuances in human language that humans are able to detect automatically. Here, "nuances" refers to entities, relationships, context, and meaning among other things. So, what are some things we humans process automatically when reading or writing "natural" text? We parse text out into paragraphs and sentences—while we may not explicitly label the parts-of-speech (nouns, verbs, etc.), we can certainly understand and identify them. We notice names of people, places, dates, etc. ("entities") as they come up. And we can infer whether a sentence or paragraph portrays a happy, angry, or sad tone. What even human children appear to do effortlessly presents some tall challenges to the machine. Human language is too rich and subtle for computer languages to capture anywhere near the total amount of information "encoded" in it.

Between R and Python, the two main open-source data science alternatives, which is better for NLP? Python's natural language toolkit[6] (NLTK) is a clear winner here, but R is steadily closing the gaps with each passing quarter. The Apache OpenNLP[7] package in R provides access from the local machine to some trained NLP models. In what follows, let us very briefly see some of the main functions NLP of written text involves.

The simplest NLP functions involve recognizing sentences (and words) as distinct data forms. Tokenization does achieve some of this for words but sentence annotations come well within the ambit of NLP. So how does a machine identify that one sentence has ended and another has begun? Typically, large corpora of preclassified and annotated text content are fed to the machine and the machine trains off this data. The output of applying this trained algorithm on new or "virgin" data is an annotated document that delineates every word and sentence in the text separately. How can we use these sentence level annotations? Theoretically, sentences could now take the place of documents to act as our "rows" in the TDM. A simple sentence expresses a single idea, typically (unlike compound sentences or paragraphs). So, instead of doing sentiment analysis at the document level—we can do it at the sentence level. We can then see which associations (adjectives) appear with what nouns (brands, people, places, etc.). Co-occurrence graphs (COGs) can be built giving more weight to words co-occurring in sentences than in the document as a whole. We now have a building block that we could scale up and apply to documents and to corpora, in principle.

A popular application of NLP is in the Named Entity Recognition (NER) space. To illustrate, imagine a large pile of files on your work desk constituting of documents stacked into a corpus. Your task is to identify and extract every instance of a person's name, or an organization's name or phone-numbers or some combination of the above that occur in that corpus. This would then become an NER problem. An entity is basically a proper noun, such as the name of a person or place. In R, OpenNLP's NER annotator identifies and extracts entities of interest from an annotated document that we saw previously. OpenNLP can find dates, locations, money, organizations, percentages, people, and times (corresponding to "date," "location," "money," "organization," "percentage," "person," "misc"). The quality of the data recovery and the results that we get depend hugely on the type and training of the NER algorithm being employed. OpenNLP's NER annotator is fairly basic but does well on western entities. A more detailed exposition of NLP can be found in several excellent books on the subject, see for example Bird et al. (2009) and Robinson and Silge (2017). Also, Exercise 9.2 provides step-by-step guidance to NER problem-solving using OpenNLP package in R.

---

[6]NLTK package and documentation are available on http://www.nltk.org/ (accessed on Feb 10, 2018).

[7]Apache OpenNLP package and documentation are available on https://opennlp.apache.org/ (accessed on Feb 10, 2018).

5 Summary and Conclusion

Advances in information technology have enabled us to capture an ever-increasing amount of data both within and outside organizations, of which a significant portion is textual in form. For analysis, text data are organized into distinct "documents" that are stacked into a "corpus." The analysis of text data proceeds by first breaking down the text content into atomic form called "tokens"—basic units of analysis—that can easily be read into machines. Once a corpus has been tokenized, basic analysis structures such as term-document matrices (TDM) emerge which can form the building blocks of downstream analyses using linear algebra and econometric principles. We saw an overview of the kinds of basic analyses possible with text data: TDMs, display output through wordclouds and COGs, the use of clustering algorithms such as k-means upon a TDM to yield groups of documents that use similar token sets, the use of external wordlists to match tokens and give rise to weighted tokens (as happens in sentiment mining and analysis), etc.

We went further and explored latent topic mining in text corpora. The basic idea there was that TDMs could be factorized to yield latent text structure—coherent semantic themes that underlie and span the corpus—wherein each document is a composite of latent topics in some unknown but estimable proportion and each topic itself is a probability distribution over the token set. Finally, we saw a brief introduction to the quirks and challenges associated with directly mining spoken and written language as is, that is, natural language processing (NLP). The aim of this chapter is to provide an overview to business applications of such text data and we discuss through many examples as well as hands-on exercises with datasets and shiny apps in R (containing automated and interactivity enabled workflows) a way to do so.

Electronic Supplementary Material

All the datasets, code, and other material referred in this section are available in http://www.allaboutanalytics.net.

- Data 9.1: Generate_Document_Word_Matrix.cpp
- Code 9.1: Github_shiny_code.R
- Code 9.2: Icecream.R
- Data 9.2: Icecream.txt

Exercises

Ex. 9.1 Analyzing a simple set of documents.

Imagine you are a consultant for a movie studio. Your brief is to recommend the top 2–3 movie aspects or attributes the studio should focus on in making a sequel.

Go to IMDB and extract 100 reviews (50 positive and 50 negative) for your favorite movie.

(a) Preprocess the data like removing punctuation marks, numbers, ASCII characters, converting whole text to lowercase/uppercase, and removing stop-words and stemming.
    Do elementary text analysis
(b) Create document term matrix.
(c) Check word-clouds and COGs under both TF (Term Frequency) and TFIDF (term frequency—inverse document frequency) weighing schemes for which configurations appear most meaningful/informative.
(d) Iterate by updating the stop-words list, etc.
(e) Compare each review's polarity score with its star rating. You can choose to use a simple cor() function to check correlation between the two data columns.
(f) Now, make a recommendation. What movie attributes or aspects (plot? star cast? length? etc.) worked well, which the studio should retain? Which ones did not work well and which the studio should change?

Explore with trial-and-error different configurations of possibilities (what stop-words to use for maximum meaning? TF or IDF? etc.) in the text analytics of a simple corpus. You may also use topic modeling if you wish.

Ex. 9.2 NLP for Entity recognition.

(a) Select one well-known firm from the list of the fortune 500 firms.
(b) For the selected firm, scrape its Wikipedia page.
(c) Using openNLP, find all the locations and persons mentioned in the Wikipedia page.
    Note: You can use either openNLPs NER functionality, or, alternately, use the noun-phrase home-brewed chunker. If using the latter, manually separate persons and locations of interest.
(d) Plot all the extracted locations from the Wikipedia page on a map.
(e) Extract all references to numbers (dollar amounts, number of employees, etc.) using Regex.

Algorithm:

Step 1: Web scraping to choose one company among the list of Fortune 500 firms. For example, I have chosen Walmart.

Step 2: Navigate to the wiki page of the selected firm. For example: Copy-paste the top wiki paragraphs into a string in R.

Step 3: Install OpenNLP in R

1. Load the required packages in R and perform basic tokenization.
2. Now generate an annotator which will compute sentence and word annotations in openNLP.
3. Now, Annotate Persons and locations using the entity annotator. Example: Maxent_Entity_Annotator(Kind = "location") and Maxent_Entity_Annotator(Kind = "person").

Step 4: Load ggmap, rworldmap package to plot the locations in the map.
Step 5: Using regular expressions in R, match the patterns of numbers and display the results.

Ex. 9.3 Empirical topic modeling.

(a) Choose three completely different subjects. For example, choose "cricket, "macroeconomics," and "astronomy."
(b) Scrape Wikipedia pages related to the given subjects. Make each paragraph as a document and annotate each document for its respective category (approx. 50 paragraph should be analyzed for each category). For example on subject cricket we can search One Day International, test cricket, IPL etc.
(c) Now create a simulated corpus of 50 documents thus: The first of the 50 documents is a simple concatenation of the first document from subject 1, from subject 2 and from subject 3. Likewise, for the other 49 documents.

   Thus, our simulated corpus now has "composite" documents, that is, documents composed of three distinct subjects each.
(d) Run the latent topic model code for k = 3 topics on this simulated corpus of 50 composite documents.
(e) Analyze the topic model results—Word clouds, COGs, topic proportions in documents.
(f) See

- Whether the topic model is able to separate each subject from other subjects. To what extent is it able to do so?
- Are there mixed tokens (with high lift in more than one topic)? Are the highest LIFT tokens and the document topic proportions (ETA scores) clear and able to identify each topic?

References

Bird, S., Klein, E., & Loper, E. (2009). *Natural language processing with Python*. Sebastopol, CA: O'Reilly Media.

Robinson, D., & Silge, J. (2017). *Text mining with R: A tidy approach*. Sebastopol, CA: O'Reilly Media.

Part II
Modeling Methods

Chapter 10
Simulation

Sumit Kunnumkal

1 Introduction

The goal of this chapter is to provide an understanding of how simulation can be an effective business analytics technique for informed decision making. Our focus will be on applications and to understand the steps in building a simulation model and interpreting the results of the model; the theoretical background can be found in the reference textbooks described at the end of the chapter. Simulation is a practical approach to decision making under uncertainty in different situations. For example: (1) We have an analytical model and we would like to compare its output against a simulation of the system. (2) We do not have an analytical model for the entire system but understand the various parts of the system and their dynamics well enough to model them. In this case, simulation is useful in putting together the various well-understood parts to examine the results. In all these cases, the underlying uncertainty is described, the model developed in a systematic way to model the decision variables, when necessary describe the dynamics of the system, and use simulation to capture values of the relevant outcomes. This chapter sets out the steps necessary to do all the above in a systematic manner.

Electronic supplementary material The online version of this chapter (https://doi.org/10.1007/978-3-319-68837-4_10) contains supplementary material, which is available to authorized users.

S. Kunnumkal (✉)
Indian School of Business, Hyderabad, Telangana, India
e-mail: sumit.kunnumkal@queensu.ca

2 Motivating Examples

We will use the following example throughout the chapter: Consider a fashion retailer who has to place an order for a fashion product well in advance of the selling season, when there is considerable uncertainty in the demand for the product. The fashion item is manufactured in its factories overseas and so the lead time to obtain the product is fairly long. If the retailer orders too large a quantity, then it is possible that the retailer is left with unsold items at the end of the selling season, and this being a fashion product loses a significant portion of its value at the end of the season. On the other hand, if the retailer orders too little, then it is possible that the product may be stocked out during the selling season. Since the lead time of the product is long compared to the length of the selling season (typically 12 weeks), the retailer is unable to replenish inventory of the product during the selling season and stock outs represent a missed sales opportunity. The retailer would like to understand how much to order factoring in these trade-offs.

The example described above is an example of a business problem where decisions have to be made in the presence of uncertainty, considering a number of different trade-offs. In this chapter, we will study Monte Carlo simulation, which is an effective technique to make decisions in the presence of uncertainty.

What Is Simulation?

At a high level, a simulation is a "virtual mirror" where we build a model of our business on a computer. A simulation model has three broad purposes. The first is to model an existing business situation. Then, to understand the impact of making a change to the business and finally to understand what may be the optimal intervention. For example, a manufacturing company may be interested in building a simulation model to describe its current production system and understand where the bottlenecks are and where material may be piling up. Next, it may build on this model to understand the impact of making a process change or augmenting the capacity of a particular machine in its production line. Finally, it may use this simulation model to understand the optimal process flow or capacity expansion plan in its production facility. As another example, a manager of a customer service center may want to build a simulation model of its existing system to understand the average on-hold times for its customers. Next, she may want to use the simulation model to understand the impact of an intervention to bring down the on-hold times, say, by augmenting the number of agents during peak hours. Finally, she may be considering a range of interventions from augmenting capacity to better agent training, and may like to use the model to understand the effectiveness of the different intervention strategies in bringing down the on-hold times. She can then use a cost–benefit analysis to decide how to best reduce the on-hold times of the customers.

Computer Simulation vs Physical Simulation

Throughout this chapter, we use simulation to refer to a computer simulation where we model a business problem on a computer. We distinguish this from a physical simulation, which typically involves building a small-scale replica or running a small pilot study. As an example of a physical simulation, consider testing a scale model of a new aircraft wing design in a wind tunnel to understand its performance. As another example, a fast-food chain may want to carry out a small pilot study at one of its restaurants to understand the impact of process redesign on the customer wait times. An advantage of a physical simulation compared to a computer simulation is that it requires fewer assumptions as to how the system functions. For example, in the case of the fast-food chain, the customer wait times can be obtained by simply measuring how long customers wait in the pilot study. It requires minimal assumptions on the nature of the arrival and service processes and how the system functions.

In a computer simulation, on the other hand, we model the business problem on a computer and this naturally requires a greater understanding of the process, the constraints, and the relations between the different variables. Therefore, in a computer simulation the modeler has to capture the most relevant aspects of the business problem and think critically about the set of assumptions underpinning the model. Some advantages of a computer simulation over a physical simulation are the shorter implementation times and the reduced costs.

Applications

While simulation has its origins in World War II, it continues to find new and interesting applications: HARVEY is a biomedical engineering software that simulates the flow of blood throughout the human body based on medical images of a patient. This can be a useful tool to inform surgical planning or to design new drug delivery systems (Technology Review 2017). GE uses a simulation model of a wind farm to inform the configuration of each wind turbine before the actual construction (GE Look ahead 2015). UPS has developed a simulation software called ORION ride that it uses to simulate the effectiveness of new package delivery routes before actually rolling them out (Holland et al. 2017). Jian et al. (2016) describe an application to bike-sharing systems, while Davison (2014) describes applications in finance.

The continued interest in building simulation models to answer business questions stems from a number of reasons. For one, the greater availability of data allows for models that are able to describe the underlying uncertainty more accurately. A second reason is the growing complexity of business problems in terms of the volume and frequency of the transactions as well as the nonlinear nature of the relationships between the variables. Nonlinear models tend to quickly become challenging to analyze mathematically and a simulation model is a particularly effective technique in such situations. Furthermore, the performance metrics obtained by simulation models, such as expected values, are also more appropriate for business situations involving a large volume of transactions. Expected values can be interpreted as long run averages and are more meaningful when applied to a large number of repeated transactions. Finally, advances in computing hardware and software also make it possible to run large and complex simulation models in practice.

Advantages and Disadvantages

One of the main advantages of a simulation model is that it is easy to build and follow. A simulation model is a virtual mirror of the real-world business problem. It is therefore easy to communicate what the simulation model is doing, since there is a one-to-one correspondence between elements in the real-world problem and those in the simulation model. A simulation model is also flexible and can be used to model complex business processes, where the mathematical analysis becomes difficult. It is also possible to obtain a range of performance metrics from the simulation model and this can be particularly helpful when there are multiple factors to be considered when making decisions.

One disadvantage of a simulation model is that it often tends to be a "black-box" model. It describes what is happening, but does not provide insight into the underlying reasons. Since simulation models are easy to build, there is a tendency to add a number of extraneous features to the model. This drives up the complexity, which makes it difficult to derive insight from the model. On the other hand, it usually does not improve the quality of the solution.

3 Simulation Modeling Method

Simulation modeling comprises of a number of logically separate steps. The first step in building a simulation model is to identify the input and the output random variables and the relationship between the two. We think of the input random variables as the underlying drivers of uncertainty. We think of the output random variables as the performance metrics of interest. The output depends on the input and since the input is random, so is the output. Mathematically, if X is the input random variable and Y is the output random variable, then we let $Y = f(X)$ to denote the dependence of Y on X. The function f(.) relates the output to the input and we say that the output is a function of the input. We note that simulation is most helpful in situations where the function f(.) is nonlinear. In the fashion retail example described previously, the demand for the product is the input random variable. The performance metric of interest is the profit, which is a function of the input since profits depend on sales. On the other hand, in the customer service center example, the random inputs are the arrival times and service times of the customers. The performance metric of interest is the on-hold time, which depends on the number of customers requesting service and the amount of time it takes to serve them.

The next step is to generate a random sample of the input and use the relation between the input and the output to generate a random sample of the output.

Let $X_1, \ldots, X_n$ be a random sample of size n of the input. The random sample $X_1, \ldots, X_n$ is drawn from the distribution of X. In the appendix, we describe how this can be done on a computer. Given the random sample of the input, we obtain a random sample of size n of the output: $Y_1, \ldots, Y_n$ by using the relation $Y_i = f(X_i)$. The function *f(.) will depend on the application.*

The third step is to interpret and make sense of the output of the simulation. Here we use the sample of the output to estimate the performance metrics of interest. We note that since we use a random sample of the input, the sample of the output is random as well and the estimates of the output that we obtain from the simulation change with the input sample. Moreover, there is sampling error since we work with a finite sample. Therefore, the results of the simulation are both random and approximate and we use concepts from statistical sampling theory to interpret the results.

We illustrate the steps using the fashion retail example.

Example 10.1 (Fashion Retailer): A fashion retailer purchases a fashion product for Rs (Indian rupees) 250 and sells it for Rs 500. The retailer is not sure what the demand for this product will be. Based on experience (past sales for similar products, market judgment etc.), she thinks that the demand for the product over the selling season will be normally distributed with a mean of 980 and a standard deviation of 300. (In the appendix, we describe how input distributions can be obtained from historical data.) The retailer also incurs a fixed cost of Rs 150,000 (administrative salaries, fixed overhead charges, etc.), which is independent of the sales volume. Assume for the sake of simplicity that this is the only product that the retailer sells and that the retailer can meet all of the demand for the product. Further, assume that any unsold product can be returned to the manufacturer and the cost recouped. Based on these assumptions, we would like to answer the following questions:

(a) What is the retailer's expected profit over the selling season?
(b) What is the standard deviation of the retailer's profit?
(c) What is the probability that the retailer will not break even at the end of the selling season?
(d) What is the probability that the retailer's profits will exceed Rs. 100,000?

Solution: We first determine the input and the output random variables and the relation between the two.

- Step 1: Specify input—demand. We denote demand by the random variable X. We are given that X is normally distributed with a mean of 980 and a standard deviation of 300.
- Step 2: Specify output—profit. We denote profit by the random variable Y.
- Step 3: Relate the input to the output: The unit profit margin is Rs. 250. We multiply this by the demand and subtract off the fixed costs to obtain the profit. Therefore, *Y = 250X − 150,000.*

Before we go and build a simulation model, we note that we can answer the questions exactly in this case since the relation between the input and the output is linear. We therefore do the exact analysis first so that we have a benchmark to compare the simulation results later.

Exact analysis: Since Y is a linear function of X, we can use linearity of expectations and conclude that $E[Y] = 250 E[X] - 150{,}000 = Rs\ 95{,}000$. Therefore, the expected profit is Rs. 95,000.

We have $V[Y] = 250^2\ V[X] = 250^2 * 300^2$. The standard deviation of Y is therefore $250 * 300 = 75{,}000$. Therefore, the standard deviation of the retailer's profit is Rs. 75,000.

Since X is normally distributed and Y is a linear function of X, Y is also normally distributed with a mean of 95,000 and a standard deviation of 75,000. We can use z-tables or Excel functions to determine the answer. For example, using the Excel function NORM.DIST(.), we get $P(Y <= 0) = 0.10$ and $P(Y >= 100{,}000) = 0.47$. Therefore, there is about a 10% chance that the retailer would not break even and there is a 47% chance that the retailer's profits would exceed Rs. 100,000.

Simulation model: We next build a simulation model to answer the same questions and compare the results from the simulation model to those obtained from the exact analysis. We have already described the input and the output random variables and the relationship between the two. In a simulation model, we first generate a random sample of size n of the input. So let $X_1, \ldots, X_n$ be a random sample of size n of the demand drawn from a normal distribution with a mean of 980 and a standard deviation of 300. Given the sample of the input, we then obtain a sample of the output by using the relation between the input and the output. Let $Y_1, \ldots, Y_n$ be the sample of size n of the output (profit) where $Y_i = 250 X_i - 150{,}000$. Finally, we use the sample of the output to estimate the performance measures of interest. In particular, we use the sample mean $\overline{Y} = \frac{Y_1 + \cdots + Y_n}{n}$ to estimate $E[Y]$. We use the sample variance $S_Y^2 = \frac{(Y_1 - \overline{Y})^2 + \cdots + (Y_n - \overline{Y})^2}{n-1}$ to estimate $V[Y]$ (refer to the Chap. 6 on basic inferences). We use the fraction of the output sample that is smaller than zero to estimate the probability $P(Y <= 0)$ and the fraction of the output sample that is larger than 100,000 to estimate the probability $P(Y >= 100{,}000)$. That is, letting $I_i = 1$ if $Y_i <= 0$ and $I_i = 0$ otherwise, we estimate $P(Y <= 0)$ using $\frac{I_1 + \cdots + I_n}{n}$. We estimate the probability $P(Y >= 100{,}000)$ in a similar way.

Implementing the simulation model: We describe how to implement the simulation model in Excel using @Risk. @Risk is an Excel add-in that is part of the Palisade DecisionTools Suite.[1] We note that there are a number of other software packages and programming languages that can be used to build simulation models. While some of the implementation details vary, the modeling concepts remain the same. We also note that Excel has some basic simulation capabilities and we describe this in the appendix.

---

[1] http://www.palisade.com/decisiontools_suite (accessed on Jan 25, 2018).

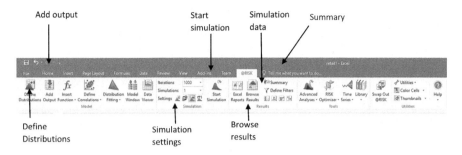

Fig. 10.1 @Risk toolbar

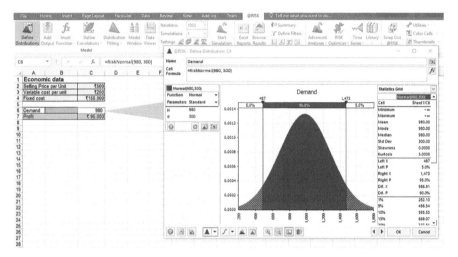

Fig. 10.2 Defining the input distribution

Implementation in @Risk

Once the Palisade DecisionTools Suite is installed, @Risk can be launched from the Start Menu in Windows. The *@Risk* toolbar is shown in Fig. 10.1. *@Risk* has many features and we will only cover the basic ones in this chapter. The more advanced features can be understood from the *@Risk* documentation.[2]

- Step 1: We specify the input cell corresponding to demand using the "Define Distributions" command from the *@Risk* toolbar (see Fig. 10.2). @Risk has a number of in-built distributions including the binomial, exponential, normal, Poisson, and uniform. It also allows for custom distributions through the "Discrete Distribution" option. In our example, demand is normally distributed. So we select the normal distribution from the list of options and specify the mean

---
[2]http://www.palisade.com/risk/ (accessed on Jan 25, 2018).

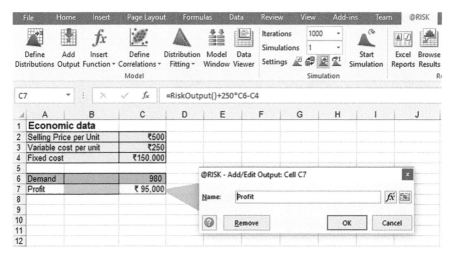

Fig. 10.3 Defining the output cell

(980) and the standard deviation (300) of the demand random variable. Note that alternatively, the demand distribution can be directly specified in the input cell by using the *@Risk* function *RISKNORMAL(.)*.

- Step 2: Next, we specify the relation between the input and the output in the output cell ($Y = 250X - 150,000$).
- Step 3: We use the "Add Output" command from the *@Risk* toolbar to indicate that the cell corresponding to profit is the output cell (see Fig. 10.3).
- Step 4: Before we run the simulation, we click the "Simulation Settings" button. Under the "General" tab, we set the number of iterations to be 1000 (see Fig. 10.4). The number of iterations corresponds to the sample size. Here, the sample size $n = 1000$. We later describe how we can determine the appropriate sample size for our simulation model. Under the "Sampling" tab, we set the sampling type to "Monte Carlo," we use the default option for the generator (Mersenne Twister) and fix the initial seed to 1 (see Fig. 10.5). We briefly comment on the generator as well as the seed. There are different algorithms to generate random numbers on a computer. *@Risk* has a number of such built-in algorithms and Mersenne twister is one such algorithm. We provide more details in the appendix. The seed provides the starting key for the random number generator. Briefly, the seed controls for the random sample of the input that is generated and fixing the seed fixes the random sample of size n (here 1000) that is drawn from the input distribution. It is useful to fix the seed initially so that it becomes easier to test and debug the simulation model. We provide more details in the appendix.
- Step 5: We are now ready to run the simulation. We do so by clicking "Start Simulation."

10 Simulation

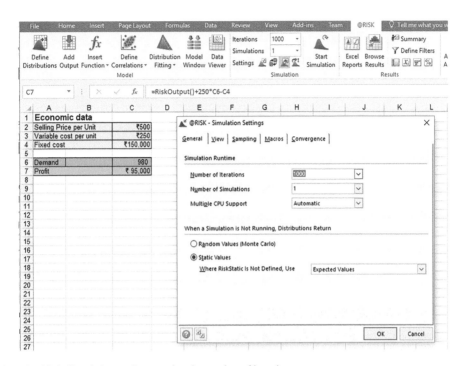

Fig. 10.4 Simulation settings: setting the number of iterations

- Step 6: After *@Risk* runs the simulation, the raw data generated by *@Risk* can viewed by clicking on "Simulation data" (see Fig. 10.6). The column corresponding to the input (demand) shows the 1000 values drawn from the normal distribution with a mean of 980 and a standard deviation of 300. The column corresponding to output (profit) shows the output sample, which is obtained by applying the relation between the input and the output ($Y = 250X - 150,000$) to the input sample. The simulation data can be exported to an Excel file for further analysis. *@Risk* also provides summary statistics and this can be accessed, for example, through "Browse Results" (see Fig. 10.7).

Simulation Results

The summary statistics indicate that the sample mean of profit is Rs 94,448 and the sample standard deviation of profit is Rs 75,457. Therefore, our estimate of the retailer's expected profit from the simulation model is Rs 94,448 and that of the standard deviation is Rs 75,457. Interpreting the area under the histogram between an interval as an estimate of the probability that the random variable is contained in that range, we estimate $P(Y <= 0)$ as 9.9% and $P(Y => 100,000) = 47.8\%$. Therefore, based on the simulation model we think there is roughly a 10% chance of not breaking even and a 48% chance of the profits exceeding Rs 100,000.

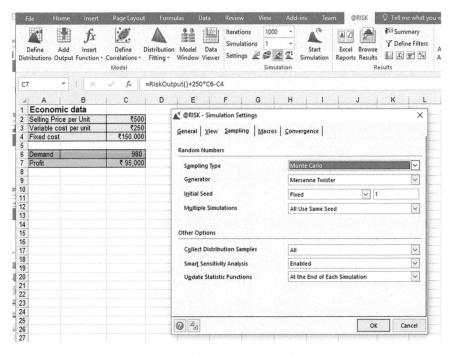

Fig. 10.5 Simulation settings: sampling type, generator, and initial seed

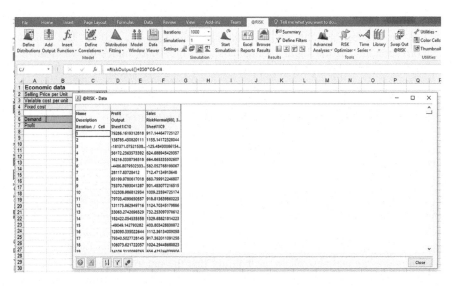

Fig. 10.6 Simulation data

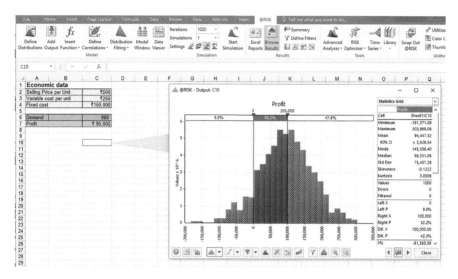

Fig. 10.7 Simulation statistics

Table 10.1 Comparison of simulation results with the results from the exact analysis

	Simulation model	Exact analysis
Expected profit (in Rs)	94,448	95,000
Standard deviation of profit (in Rs)	75,457	75,000
Sample size (no. of iterations)	1000	
95% confidence interval	[89771, 99124]	
P(Y <= 0)	9.9%	10.3%
P(Y >= 100,000)	47.8%	47.3%

Comparison of the Simulation Results to the Exact Analysis

Table 10.1 summarizes the results obtained from the simulation model and compares it to the exact analysis. We observe that the simulation results are close but not exact. The error comes from finite sampling. Moreover, the results are random in that if we had taken a different random sample of size 1000 of the input, the results would have been slightly different (see Table 10.2). Therefore, we assess the accuracy of the simulation results by building a confidence interval. We use the formula $\bar{x} \pm 1.96\, s/\sqrt{n}$ to construct a 95% confidence interval for the expected profit (see, e.g., Stine and Foster (2014)).

Table 10.2 compares the simulation results obtained using different seeds. The second column shows the simulation results when the initial seed is set to 1, and the last column shows the simulation results when the initial seed is set to 2. Note that the estimates change a little when we change the seed since we change the sample of the input random variable.

Table 10.2 Comparison of simulation results for different values of the initial seed

	Simulation model Initial seed = 1	Simulation model Initial seed = 2
Expected profit (in Rs)	94,448	92,575
Standard deviation of profit (in Rs)	75,457	73,367
Sample size (no. of iterations)	1000	1000
95% confidence interval	[89771, 99124]	[88027, 97122]
P(Y <= 0)	9.9%	10.0%
P(Y >= 100,000)	47.8%	44.0%

Table 10.3 Comparison of simulation results for different sample sizes

Sample size (no. of iterations)	Sample mean (in Rs)	Sample standard deviation (in Rs)	Confidence interval
10	37,560	87,536	[−16696, 91816]
100	89,151	79,561	[73557, 104745]
1000	94,448	75,457	[89771, 99124]
10,000	94,392	74,536	[92931, 95853]

The initial seed is fixed to 1

We note that the question as to which is the "right" seed is not meaningful since the estimates always have sampling error associated with them. The more relevant question to ask is regarding the sample size (or the number of iterations in @Risk) since this determines the accuracy of the simulation results. Table 10.3 shows how the accuracy of the simulation estimates changes with the sample size. We observe that as the sample size increases, we obtain progressively more accurate estimates; the sample mean of profit we obtain from the simulation model is closer to the population mean (95,000) and the confidence intervals for the expected profit are also narrower. This is a natural consequence of the central limit theorem, which states that as the sample size increases, the sample mean gets more and more concentrated around the population mean (see, e.g., Ross (2013)). As a result, we get more accurate estimates as the sample size increases.

The natural question that then arises is: what should be the appropriate sample size? This in general depends on the nature of the business question being answered by the simulation model and so tends to be quite subjective. The main ideas come from statistical sampling theory. We first fix a margin of error, that is, a target width of the confidence interval that we are comfortable with. We then determine the sample size so that the actual width of the confidence interval matches with the target width. Therefore, if we let e be the target margin of error (for a 95% confidence interval), then we determine the sample size n as

$$n = \left[\frac{1.96\, s_n}{e}\right]^2.$$

In the above equation, s_n is the sample standard deviation, which in turn depends on the sample size n. We break this dependence by running the simulation model for a small number of iterations (say, 100 or 200) to obtain an estimate of the sample standard deviation s_n and use this estimate in the above formula to determine the appropriate value of n. For example, we might use the sample standard deviation corresponding to 100 iterations ($s_{100} = 79,561$) from Table 10.3 as an approximation to s_n in the above equation. So if we have a margin of error $e = 1000$, the required sample size n is

$$n = \left[\frac{1.96 * 79561}{1000} \right]^2 \approx 24317.$$

3.1 Decision Making Under Uncertainty

We have used simulation so far simply as an evaluation tool to obtain the summary statistics associated with a performance metric. We now build on this and add a decision component to our model. In particular, we consider business settings where we have to make decisions under uncertainty and see how we can use simulation to inform decision making.

Suppose that we have to pick an action "a" from a range of possible alternatives $\{a_1, \ldots, a_K\}$. In making this decision, we have a performance metric (output) in mind and we would like to choose the action that optimizes this performance metric. The output variable is affected not only by the action that we take (decisions) but also by uncertain events (inputs). That is, if we let Y denote the output variable, we have $Y = f(X, a)$, where X represents the random input and $f(.,.)$ is a function that relates the output to our decisions as well as the random inputs. If the function $f(.,.)$ is nonlinear, then the mathematical analysis quickly becomes challenging. In such cases, a simulation model can be a useful alternative.

The simulation approach remains quite similar to what we have discussed in the previous section. The first step is to identify the decision variable, as well as the input and the output random variables. We specify the range of the decision variable, the distribution of the input random variable and the relation between the output and the decision and the input variables. Once we do this, we evaluate the outcomes associated with each possible action using simulation and then pick the action that optimizes the output metric. That is, we pick an action a_k from the list of possible actions. We then generate a random sample of the input $X_1, \ldots, X_n$. Given the action a_k and the random sample of the input, we obtain a random sample of the output using $Y_i = f(X_i, a_k)$. We use the random sample of the output to estimate the performance metric of interest when we choose action a_k. We repeat this process for each action in the list and pick the action that optimizes the performance metric. Note that since the results of the simulation are random and approximate, the previous statements regarding the interpretation of the simulation results continue to hold.

We next build on the fashion retailer example to illustrate these steps.

Example 10.2 (Fashion Retailer, Continued)

A fashion retailer purchases a fashion product for Rs 250 and sells it for Rs 500. The retailer has to place an order for the product before the start of the selling season when the demand is uncertain. The demand for the product is normally distributed with a mean of 980 and a standard deviation of 300. If the retailer is left with unsold items at the end of the selling season, it disposes them off at a salvage value of Rs 100. The retailer also incurs a fixed cost of Rs 150,000, which is independent of the sales volume. How many units of the product should the retailer order? Assume that the retailer's objective is to maximize its expected profit.

Solution

We first specify the decision variable, as well as the input and the output random variables. Then we relate the output to the input and the decision variables.

- Step 1: Specify decision—stocking quantity. We let q denote the stocking decision. The theoretical range for the stocking quantity is $[0, \infty)$. However, for practical reasons (minimum batch size, budget constraints, etc.) we may want to impose lower and upper limits on the stocking quantity. Here, given that the demand is normally distributed, we consider order quantities that are within one standard deviation above and below the mean. So we will consider $q \in \{680, 730, \ldots, 1280\}$ (q increases in step size $= 50$). We note that we could have considered an extended range and even a more refined range for the decision variable. The trade-off is the increased solution time stemming from evaluating the profit (output) at each possible value of the decision.
- Step 2: Specify input—demand. We denote demand by the random variable X. X is normally distributed with a mean of 980 and a standard deviation of 300.
- Step 3: Specify output—profit. We denote profit by the random variable Y.
- Step 4: Relate the input and the decision to the output. The unit profit margin is Rs. 250. We have sales $= min(X, q)$ since we cannot sell more than what is demanded (X) and what we have in stock (q). It follows that the number of unsold items is the difference between the stocking quantity and the sales. That is, *Unsold = q − Sales*. We have *Revenue = 250 Sales + 100 Unsold*, where the first term captures the revenues from sales at the full price while the second term captures the salvage value of the unsold items. On the other hand, *Cost = 250 q + 150,000*. Therefore, we have that the profit $Y = Revenue - Cost = 250$ *Sales + 100 Unsold − 250q − 150,000*.

Now, we proceed to build a simulation model to determine the optimal stocking quantity. We note that the problem we are considering is an example of a newsvendor problem, which is a very well-studied model in the operations management literature and it is possible to obtain an analytical expression for the optimal ordering quantity (see, e.g., Porteus (2002)). We do not dwell on the exact mathematical analysis here and instead focus on the simulation approach. An advantage of the simulation model is that it can be easily adapted to the case where the retailer may have to manage a portfolio of products as well as accommodate a number of other business constraints.

Simulation Model

We have already described how the decision variable together with the random input affects the output. We evaluate the expected profit corresponding to each stocking level in the list {680, 730, ..., 1280} and pick the stocking level that achieves the highest expected profit. So we pick a stocking level q_k from the above list. Then we generate a random sample of size n of the input. So, let $X_1, \ldots, X_n$ be a random sample of size n of the demand drawn from a normal distribution with a mean of 980 and a standard deviation of 300. Given the sample of the input, we then obtain a sample of the output by using the relation between the decision, the input, and the output. Let $Y_1, \ldots, Y_n$ be the sample of size n the output (profit) where

$$Y_i = 250\, Sales_i + 100\, Unsold_i - 250 q_k - 150000$$
$$= 250 \min(X_i, q_k) + 100\, (q_k - \min(X_i, q_k)) - 250 q_k - 150000.$$

Finally, we use the sample of the output to estimate the expected profit corresponding to ordering q_k units. We repeat this process for all the possible stocking levels in the list to determine the optimal decision.

Implementing the simulation model in @Risk: We specify the decision cell corresponding to the stocking quantity using the RISKSIMTABLE(.) function. The argument to the RISKSIMTABLE(.) function is the list of stocking quantities that we are considering (see Fig. 10.8). The remaining steps are similar to the previous simulation model: we specify the input cell corresponding to demand using the RISKNORMAL(.) function and specify the output cell corresponding to profit using the RISKOUTPUT(.) function. We link the decision variable and the input to the output using the relation described above.

Before we run the simulation model, we specify the "Number of Iterations" in the "Simulation Settings" tab. We set the "Number of Iterations" to be 1000 as before. Now we would like to generate an input sample of size 1000 for each possible value of the decision variable. There are 13 possible stocking decisions that we are considering ({680, ..., 1280}) and so we would like an input sample of size 1000 to be generated 13 times, one for each stocking decision in the list. We specify this by setting the "Number of Simulations" to be 13 (see Fig. 10.9). In the "Sampling" tab, we set the "Sampling Type," "Generator," and "Initial Seed" as before. We also set "Multiple Simulations" to use the same seed (see Fig. 10.10). This ensures that the same input sample of size 1000 is used to evaluate the profits corresponding to all of the stocking decisions. That is, the same underlying uncertainty drives the outcomes associated with the different actions. As a result, it is more likely that any differences in the output (expected profit) associated with the different decisions (stocking quantity) are statistically significant. The benefit of using common random numbers can be formalized mathematically (see, e.g., Ross (2013)).

	A	B	C	D	E	F	G	H
1	**Economic data**					List of possible stocking quantities		
2	Selling Price per Unit		₹500			680		
3	Variable cost per unit		₹250			730		
4	Salvage value per unit		₹100			780		
5	Fixed cost		₹150,000			830		
6						880		
7	Stocking quantity		680			930		
8						980		
9						1030		
10	Demand		980			1080		
11	Sales		680			1130		
12	Unsold		0			1180		
13	Revenue		₹ 340,000			1230		
14	Cost		₹ 320,000			1280		
15	Profit		₹ 20,000					

Fig. 10.8 Determining the optimal stocking quantity using simulation. The decision cell is specified using the RISKSIMTABLE(.) function and the argument to the function is the list of possible stocking quantities described in the list F2:F14 in the Excel spreadsheet

After running the simulation, we view the simulation results by clicking on the "Summary" button. The table summarizes the simulation statistics for the different values of the ordering quantity (see Fig. 10.11). It indicates that the expected profit is maximized by ordering 1080 units.

We make two observations regarding the simulation results. First, the expected profit for each stocking level is estimated as the sample average of a sample of size 1000. It remains to be verified that the results are statistically significant. That is, are the differences we see in the samples representative of a corresponding difference in the populations? This question can be answered using standard statistical tests (see, e.g., Stine and Foster (2014)). Second, we note that ordering 1080 units is the best choice from the list $\{680, \ldots, 1280\}$. It is possible that we could further increase profits if we were not restricted to these ordering levels. We can evaluate the profits on a more refined grid of ordering levels to check if this is indeed the case. The trade-off is the increased computation time stemming from evaluating the profit for a larger number of stocking levels.

Our analysis so far was based on the assumption that the retailer was interested in maximizing its expected profit for a single product. We now extend the model in a couple of directions.

10 Simulation

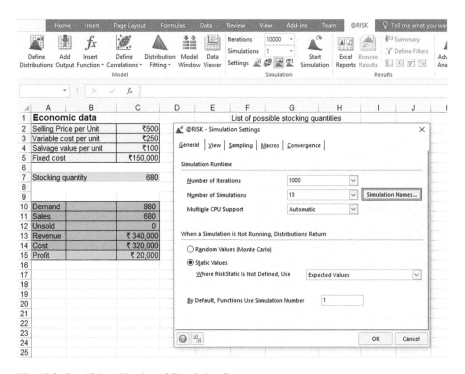

Fig. 10.9 Specifying "Number of Simulations"

Example 10.3 (Fashion Retailer, Stockouts)

Suppose that the retailer also cares for opportunities lost. That is, if the retailer runs out of inventory of the product then some customers who visit the store would be unable to purchase the product. If stockouts occur frequently and a large number of customers do not find the product available, the customers may take their business elsewhere and this would impact the profits of the retailer in the long run. Consequently, the retailer would also like to factor in lost demand when making the stocking decisions. Suppose that the retailer would like to ensure that the expected percentage unmet demand is no larger than 5%. How many units of the product should the retailer order in order to maximize expected profits while ensuring that the stockout constraint is satisfied?

Solution

We can easily modify our simulation model to answer this question. The decision variable and the input random variable remain the same. In addition to the profit metric, we add the expected percentage unmet demand as a second metric in our simulation model. We define the percentage unmet demand as $max(X - q, 0)/X$, where X is the demand and q is the stocking level. Note that there is unmet demand

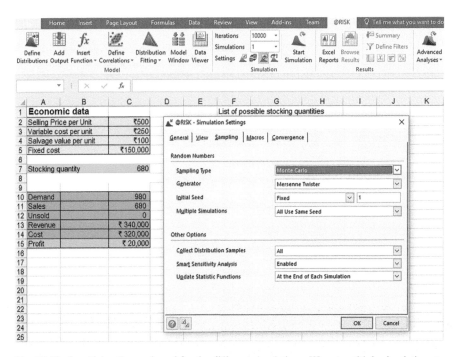

Fig. 10.10 Specifying the seed used for the different simulations. We set multiple simulations to use the same seed

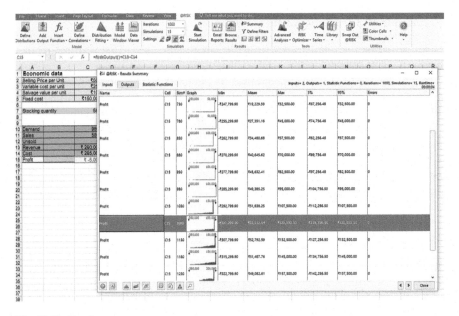

Fig. 10.11 Results summary

Table 10.4 Expected profit and expected % unmet demand as a function of the ordering quantity

Stocking quantity	Expected profit	Exp. % unmet demand
680	₹10,102	28.94
730	₹19,229	24.95
780	₹27,351	21.31
830	₹34,480	17.98
880	₹40,645	14.94
930	₹45,632	12.24
980	₹49,385	9.87
1030	₹51,838	7.83
1080	₹53,012	6.11
1130	₹52,792	4.70
1180	₹51,487	3.54
1230	₹49,082	2.61
1280	₹45,836	1.85

only when $X > q$ and so $X - q > 0$. We add the percentage unmet demand as a second output in our spreadsheet model (using the RISKOUTPUT(.) function).

Table 10.4 describes the summary results we obtain after running the simulation model again. The first column gives the stocking level, the second column gives the corresponding expected profit while the last column gives the expected percent unmet demand. We notice that the previous stocking level of 1080 which maximized the expected profit results in the expected percent unmet demand being around 6%, violating the 5% threshold. The feasible stocking levels, that is, those which satisfy the 5% stockout constraint, are {1130, 1180, 1230, 1280}. Among these stocking levels, the one which maximizes the expected profit is 1130. Therefore, based on our simulation model we should order 1130 units of the product, which results in an expected profit of around Rs 52,792 and the expected percent unmet demand being around 4.7%.

Example 10.4 (Fashion Retailer, Multiple Products)

We now consider the case where the retailer sells multiple products that are potential substitutes. That is, if a customer does not find the product that she is looking for on the shelf, she may switch and buy a product that is a close substitute. To keep things simple, let us assume that the retailer sells two products during the selling season. The base (primary) demands for the two products are independent, normally distributed random variables. The cost and demand characteristics of the two products are described in Table 10.5. In addition, the retailer incurs a fixed cost of Rs 150,000 that is independent of the sales volumes. The two products are substitutes in that if one product is stocked out then some of the customers interested in purchasing that product might switch over and purchase the other product (provided it is in stock). If product 1 is stocked out, then 70% of the unmet demand for that product shifts to product 2. On the other hand, if product 2 is stocked out, then 30% of the demand for that product shifts to product 1. Therefore, there is

Table 10.5 Cost and demand characteristics of the two products

	Product 1	Product 2
Unit selling price	Rs 500	Rs 300
Unit cost price	Rs 250	Rs 200
Unit salvage value	Rs 100	Rs 50
Demand	Normal with a mean of 980 and a standard deviation of 300	Normal with a mean of 2000 and a standard deviation of 500

a secondary demand stream for each product that is created when the other product is stocked out. What should be the stocking levels of the two products for the retailer to maximize its expected profits?

Solution

We determine the decision variable, as well as the input and the output random variables. We then describe the relation between the output and the decision and the input variables.

- Step 1: Specify decisions—stocking quantities of the two products. We let q_1 denote the stocking level of product 1 and q_2 denote the stocking level of product 2. For each product, we consider stocking levels that are within one standard deviation of the mean demand. Therefore, we consider $q_1 \in \{680, 730, \ldots, 1280\}$ and $q_2 \in \{1500, 1550, \ldots, 2500\}$. We again note that it is possible to work with an expanded range of values and also consider a more refined set of grid points, at the expense of greater computational effort.
- Step 2: Specify input—demands for the two products. We let X_1 denote the primary demand random variable for product 1 and X_2 denote the primary demand random variable for product 2. X_1 is normally distributed with a mean of 980 and a standard deviation of 300, while X_2 is normally distributed with a mean of 2000 and a standard deviation of 500. Furthermore, X_1 and X_2 are independent random variables.
- Step 3: Specify output—profit. We denote profit by the random variable Y.
- Step 4: Relate the input and the decision to the output. We have primary $Sales_i = min(X_i, q_i)$ for $i = 1, 2$, where primary sales refer to the sales generated from the primary demand for that product. The remaining inventory of product i is therefore $Inventory_i = q_i - $ Primary sales$_i$. On the other hand, the portion of the demand that cannot be satisfied from the on-hand inventory, unmet $Demand_i = X_i - $ Primary sales$_i$. Now, if there is unmet demand for product 2, then 30% of that is channeled to product 1. Therefore, the secondary sales of product 1 is the smaller of the remaining inventory of product 1 (which remains after satisfying the primary demand for product 1) and the secondary demand for product 2. That is, $Secondary\ Sales_1 = min(Inventory_1, 0.3 * Unmet\ Demand_2)$. By following a similar line of reasoning, the secondary sales of product 2, $Secondary\ Sales_2 = min(Inventory_2, 0.7 * Unmet\ Demand_1)$. The number of unsold items of product i is, therefore, $Unsold_i = q_i - $ Primary $Sales_i - $ Secondary $Sales_i$.

Tallying up the revenues and costs, we have that the *total revenue* $= 250 *$ *Primary Sales*$_1$ $+ 250 *$ *Secondary Sales*$_1$ $+ 100 *$ *Unsold*$_1$ $+ 200 *$ *Primary Sales*$_2$ $+ 200 *$ *Secondary Sales*$_2$ $+ 50 *$ *Unsold*$_2$. The total cost $= 250q_1 + 200q_2 + 150{,}000$. We obtain the profit, Y, as the difference between the total revenue and the total cost.

Implementing the Simulation Model in @Risk

Now we have two decision variables q_1 and q_2, where q_1 takes values in the range $\{680, 730, \ldots, 1280\}$ and q_2 takes values in the range $\{1500, 1550, .., 2500\}$. So there are 273 (13 * 21) possible combinations of q_1 and q_2 that we have to consider. While it is possible to implement the model using the RISKSIMTABLE(.) function, this would involve creating a list of all the possible combinations of the two decision variables and can be cumbersome. An alternative way to implement the simulation model is using the RISK Optimizer function (see Fig. 10.12). Under "Model Definition" we specify the optimization goal (maximize) and the metric that is optimized (mean value of profit). Under "Adjustable Cell Ranges," we specify the decision variables and the range of values that they can take (between the minimum and the maximum stocking levels, in increments of 50, see Fig. 10.13). We start the optimization routine by clicking on the "Start" button under the RISK Optimizer tab (see Fig. 10.14) and after @Risk has finished, we obtain the results from the "Optimization Summary" report (see Figs. 10.14 and 10.15). From the optimization summary report, we see that the optimal stocking level for product 1 is 1230 units and the optimal stocking level for product 2 is 1600 units. The corresponding expected profit is Rs 228,724. It is interesting to note that the optimal stocking levels of the two products when considered together are different from the

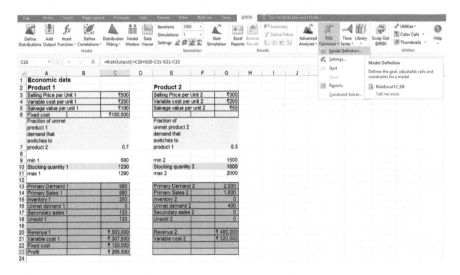

Fig. 10.12 Using RISK Optimizer

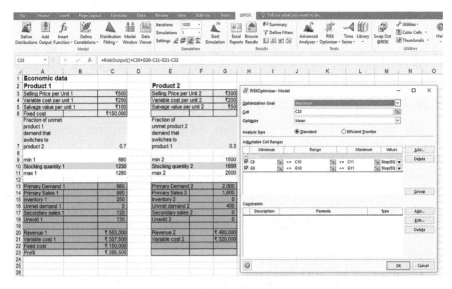

Fig. 10.13 Specifying the optimization objective and the decision variables in RISK Optimizer

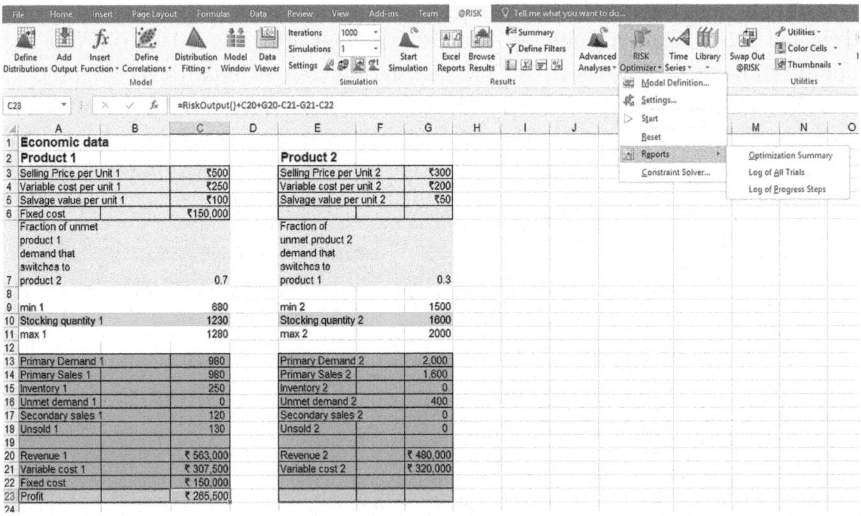

Fig. 10.14 Obtaining the results of the optimization

optimal stockings levels when considered independently. If product 1 is analyzed independently, we have from the earlier analysis (Example 10.2) that its optimal stocking level is 1080 units. On the other hand, if product 2 is considered in isolation, then its optimal stocking level turns out to be 1900 units. When both products are considered together, factoring in the substitution effects, the optimal stocking level of product 1 increases to 1230 units, while that of product 2 decreases to 1600 units.

Goal	
Cell to Optimize	Sheet1!C23
Statistic to Optimize	Mean
Type of Goal	Maximum
Results	
Valid Trials	143
Total Trials	143
Original Value	₹ 183,316
+ soft constraint penalties	0.00
= result	₹ 183,316
Best Value Found	₹ 228,724
+ soft constraint penalties	0.00
= result	₹ 228,724
Best Trial Number	37
Time to Find Best Value	0:00:36
Reason Optimization Stopped	Best solution found
Time Optimization Started	11/15/2017 15:26
Time Optimization Finished	11/15/2017 15:27
Total Optimization Time	0:00:41
Adjustable Cell Values	Sheet1!C10
Original	680
Best	1230
Adjustable Cell Values	Sheet1!G10
Original	2000
Best	1600

Fig. 10.15 Optimization summary report

4 Conclusion

In this chapter, we have given an overview of how simulation can be an effective technique to make decisions in the presence of uncertainty. Simulation is a particularly effective tool when there is a nonlinear relation between the input and the output random variables. In order to build a simulation model, we have to (1) specify the decision variables and their ranges, (2) the input random variables and their distributions, and (3) describe how the output depends on the decisions as well as the inputs. The output of a simulation is (1) random since we work with a random sample of the input and (2) approximate since we work with a finite sample of input. The results of the simulation model therefore have to be carefully interpreted using concepts from statistical sampling theory.

5 Solved Case Study "Coloring the World"

Daniel operates a successful civil work firm, *Coloring the World*. He has been running this decades-old family business for the last 10 years. *Coloring the World* provides painting and civil work services to commercial buildings and large apartments. Daniel has a dedicated sales team which is very active in identifying construction projects that may require the firm's services. He has another sales team which keeps following up with the businesses from established customers who may require refurbishment or painting work. Both the teams are quite active in generating leads. He also has a robust customer relationship system in place to engage with existing clients that helps in building long term association with the customers.

An existing customer, *Feel Good Fabrics*, a maker of cotton and linen cloth, has sent *Coloring the World* an RFP (Request for Proposal) to paint its entire manufacturing plant. Though Daniel has provided various small refurbishment work support to this client, he knows that this is a big requirement that can fetch good recognition and lead to a long-term association with the customer. With his earlier experience, Daniel knows that Brian Painters has also been approached for this offer and suspects that Whitney-White colors (W&W) is also trying hard to get empaneled on *Feel Good Fabrics'* vendors list. Daniel does not want to lose this opportunity to create an impactful relationship with *Feel Good's* commercial and operation team.

Daniel has competed with Brian and W&W for many other projects and believes he can more or less estimate the bidding strategies of these competitors. Assuming that these competing firms are bidding for this contract, Daniel would like to develop a bid that offers him a good shot at winning, but also does not result in a loss on the contract since the firm has many expenses including labor, paints, and materials.

Daniel estimates that *Brian's painters* bid could be anywhere between $450,000 and $575,000. As for *Whitney-White colors*, Daniel predicts the bid to be as low as $425,000 or as high as $625,000, but he thinks $550,000 is most likely. If Daniel bids too high, one of the competitors is likely to win the contract and Daniel's company will get nothing. If, on the other hand, he bids too low, he will probably win the contract but may have to settle for little or no profit, even a possible loss.

Due to the complexity in the plant structure of *Feel Good Fabrics*, Daniel in consultation with his service department estimates the direct cost to service the client at $300,000. Realizing that the costs are actually uncertain, Daniel takes this number to be expected value of the costs and thinks that the actual cost will be normally distributed around this mean value with a standard deviation of $25,000.

The preparation and budget estimation cost $10,000 to Daniel since that includes in-person visits and technical test clearance by the *Feel Good Fabrics* team. How much should Daniel bid in order to maximize his expected profit?

Solution

The underlying source of uncertainty is in the bids of the two competitors and in the direct costs. The decision is how much to bid for the project. Note that *Coloring the World* wins the project if they bid the lowest. Given the bids of the competitors we also observe that *Coloring the World* will win the project for sure if they bid lower than $425,000. So it does not make sense to bid anything lower than $425,000 since it does not further improve their chances of winning the contract. On the other hand, bidding lower shrinks their profit margins. *Coloring the World* will also never win the project if they bid more than $575,000. Therefore, the optimal bidding strategy must lie between $425,000 and $575,000.

Simulation Model:

Input random variables
Direct cost (D) = Normal (300000, 25000)
Brian painters' bid (B) = Uniform (450000, 575000)
W&W bid (W) = Triangular(425000, 550000, 625000)
Decision variable
Coloring the World's bid C, Range is between $425,000 and $575,000.
Output
Profit = IF (C < min(B, W), C − D, 0) − 10,000

From the simulation model, we find that a bid of $450,000 is optimal. The expected profits will be $137,000 on average. The 95% confidence interval is [$135,000, $140,000].

The number of simulations is number of times we would like to repeat the simulation. Here we would like to evaluate profits corresponding to the 13 different ordering quantities and so we set "Number of Simulations" to be 13.

Electronic Supplementary Material

All the datasets, code, and other material referred in this section are available in www.allaboutanalytics.net.

- Data 10.1: Priceline_Hotelbids.xlsx
- Data 10.2: Watch_bids.xlsx

Exercise Caselets

Caselet 10.1: Watch the Time

Christopher Luis is a consultant for an international architectural firm that specializes in building skyscrapers. He has consulted for various projects in China, India, USA, and other countries. Because of his profession, Chris travels to different countries to better understand the project location and local culture. All his travels help him customize and personalize designs based on the geography of the site and the local availability of material and labor.

Through his travel, Chris has developed a passion toward collecting antique watches. So much so, he is more famous as a watch collector among his family and friends than as a famous architect. He has a collection of more than a hundred antique watches representing different cultures and manufacturing styles. He keeps looking for a new variety of mechanical watch to add to his collection during his travel across the globe. He has even subscribed to bidding websites that regularly add a variety of new watches to their product pages. He has been successful many a time in winning auctions but unfortunately at very high prices. Chris believes in paying the right price for the product and an added extra based on his emotional attachment to the product.

Over time, Chris has developed a bidding strategy that combines bidding at selective time points and at a specific price ratio considering the existing bids. He has started winning bids at lower prices in recent times, but he is a little disappointed as his new strategy did not work out with a few watches that he wanted to win desperately. Though, thankfully due to the new strategy he has not paid high prices on the winning deals.

Chris also notices that he is investing a lot of time and effort following up on multiple websites and placing the bid. Some websites even limit the number of bids for each customer and he is running out of number of bids very quickly. The other websites even charge per bid to the customers in order to restrain the customers from placing multiple bids of small differences. His experience tells that the winning bid ranges between $1.5\times$ and $3\times$ of the first few bid prices and follows a "regular" price trajectory that can help in estimating the final price. The price trajectories of the 100 products are provided on the book's website "Watch_bids.xlsx". The table contains the number of bids and the highest bid so far at 36 h before closing, 24 h, 12 h, and so on.

He wishes to develop a new strategy to place a single bid or a maximum of two bids at specific time points rather than following up multiple times and wasting time and effort monitoring the outcome.

- When should Chris bid in order to secure the deal?
- At what price ratio should Chris bid to secure the deal?
- How can he add extra value of emotional quotient to the bid in terms of timing and price?

Caselet 10.2: Winter Jackets

Monika Galore is excited as she graduates as a textile engineer from a premier engineering school in the USA. She received the best student award because of her outstanding innovation and contribution during her internship with a world-famous researcher, *Akihiko Karamorita*. Akihiko is a distinguished professor in the textile and chemical engineering department at her university. She mentored Monika in developing a fabric that not only can protect the skin from a harsh winter but also is light to wear. Alongside completing her dissertation, Monika and Akihiko patented the idea.

Many manufacturers offered them a huge price and life-time employment in research and development teams to sell their idea. *Monika* and *Akihiko* discussed commercializing the product rather than selling the idea to established manufacturers. They decided that they will contract the cloth and garment manufacturing to a quality and trusted supplier and sell the product online through e-commerce websites.

Under the women entrepreneurship scheme, Monika decided to set up her own plant to manufacture the fabric. Professor Akihiko approached her ex-students who had expertise in winter-wear manufacturing and designing. A few of them showed interest and also bid for the contract. Monika and Akihiko decided to contract the manufacturing of jackets to a third party. They outsourced the product manufacturing to *Vimal Jain*, a close friend of Monika and trusted student of Professor Akihiko, considering his family experience in textile manufacturing. *Vimal* proposed two designs—a Sports jacket and a Trendy jacket using the special fabric. Though he also proposed to mix this specific thread with another material in order to cater to different geographic needs, Monika and Akihiko rejected the idea and decided to target the niche segment of woolen cloths. Vimal agreed to design, manufacture, and supply the two types of jackets and in different sizes small, medium, large, and extra-large to the retail locations.

The product was initially sold only through an online channel. However, looking at the increasing demand, Monika and Akihiko decided to go offline and partnered with a large retail chain that had store presence across the world. The retailer negotiated a 20% profit margin on the items sold with a condition to return unsold products at 80% of the purchase price. Since the manufacturing of fabric and garment are done at different locations and by different manufacturers, it is essential to estimate the demand in advance to optimize inventory at various stages. Also, as this is a seasonal and/or fashion product, the excess inventory of unsold products may lead to deep discounts. The product demand also depends on the severity of weather.

They requested *Chris Alfo*, head of operations with the partner retailer to estimate the demand based on his experience with comparable products. Note: The numbers in demand estimate table are in thousands (Table 10.6).

Monika estimated the manufacturing cost of one jacket at $125 and fixed 40% markup when selling to the retailer. She thought that retailers would add 20% profit margin. Monika also found that unsold products can be sold at 50% of the cost in deep-discount outlets. She knew that all the stock has to be cleared within the same year considering the contemporary nature of fashion. Looking at historical weather reports, she estimated the probability of mild winter at 0.7 and cold at 0.3. The customers may switch between the product designs based on availability of the product. In case of extreme demand and unavailability, if winter is mild, the probability to switch from one design to another is 0.5 while in case of cold winter it is 0.9. Monika is planning how much to manufacture every year in order to procure

Table 10.6 Demand estimates for sizes and types of jackets

Type of jacket	Small	Medium	Large	Extra-large
Sports	150	300	180	90
Trendy	210	450	365	120

raw material and finalize the manufacturing contract with Vimal. She has to estimate demand in such a way that she does not end up with too much unsold product, as well as, does not lose the opportunity to sell more.

Caselet 10.3: Priceline—A Reverse Auction

Priceline popularized the name your own price (NYOP). For example, in its website,[3] it advertises "For Deeper Discounts Name Your Own Price®." In this model, called a reverse auction, the buyer specifies the product or service and names a price at which the buyer is willing to purchase the product. On the other side, sellers offer products or services at two or more prices and also the number of products that are available. For example, a room in a three-star hotel on September 21st in downtown Manhattan for a one-night stay could be a product. Several hotels would offer, say, anywhere from 1 to 5 rooms at rates such as $240 and $325. When a bid is made, the market-maker (in this case Priceline) picks a seller (a hotel that has made rooms available) at random and sees if there is a room that is available at a rate that is just lower than the bidder's price. If no such price is available, the market-maker chooses another seller, etc. More details about the model are found in Anderson and Wilson's article.[4]

The NYOP mechanism is fairly complex because it involves buyers, sellers, and the intermediary who are acting with limited information about the actual market conditions. For example, even the intermediary is not aware of the exact supply situation. How does this model benefit everyone? The buyer benefits because it creates a haggle-free environment: In case the bid fails the buyer cannot bid for a day, thus forcing the bidder to either reveal the true reservation price or be willing to forgo an opportunity to purchase in order to learn more about the model and act with delay. The seller benefits because the model avoids direct price competition. The intermediary benefits because it gets to keep the difference between the bidder's and the seller's price.

Barsing is a boutique hotel situated in downtown Manhattan. It has 100 rooms, out of which 65 are more or less indistinguishable with regard to size and amenities. It often finds some of its rooms remain unsold even during peak seasons due to the relative newness of the hotel and its small size. The variable cost of a room-night is around $75. Sometimes this cost may increase or decrease by 10% depending on the amount of cleaning and preparation necessary. Richard Foster who manages Barsing started offering rooms on an NYOP program called HotelsManhattan. He classifies Barsing's as a mid-range hotel with a family atmosphere. He feels that the program was intriguing and requires constant tinkering to get the price right.

Foster's assistant, Sarah, was tasked with reviewing the data and recommending an automatic approach to making rooms available on HotelsManhattan. Historical data on the number of bids made on the past 40 weekdays are available to her, see "Priceline_Hotelbids.xlsx" on book's website. Typically, 4–5 rooms are available to sell using the NYOP program.

---

[3] www.priceline.com (accessed on Aug 17, 2017).

[4] Anderson and Wilson (2011).

1. Assume that the bidders are fixed in number. If Sarah uses one price, what price maximizes the expected profit if four rooms are available?
2. Assume that the number of bidders is random. What is the optimal single price?
3. Assume that the number of bidders is random and Sarah can specify two prices. How should she set those prices to maximize her expected profit from four rooms?

Appendix 1: Generating Random Numbers on a Computer

Virtually all computer simulations use mathematical algorithms to generate random variables. These are potentially very fast, and can be replicated at will. However, it should be clear that no such sequence can be truly random. Consequently, we refer to such a sequence as a pseudo-random sequence. This is because one can in fact predict the sequence of numbers generated provided one had a sophisticated knowledge of the way the algorithm is designed to operate. This is not a huge concern from a practical standpoint as most of the popular algorithms generate sequences that are virtually indistinguishable from a truly random sequence.

A commonly used generator is the linear congruential generator (LCG), which obtains a sequence $x_0, x_1, \ldots$ of integers via the recursion

$$x_{n+1} = (ax_n + c) \bmod m,$$

where a, c, and m are integers also known, respectively, as the multiplier, increment, and modulus. The number x_0 is called the initial seed. The mod operator applied to two numbers returns the remainder when the first number is divided by the second. So 5 mod 3 = 2 and 6 mod 3 = 0. Since each number in the sequence generated lies between 0 and $m - 1$, x_n/m is a number that lies between 0 and 1. Therefore, the sequence $x_1/m, x_2/m, \ldots$ has the appearance of a sequence of random numbers generated from the uniform distribution between 0 and 1. However, note that since each x_n lies between 0 and $m - 1$, the sequence must repeat itself after a finite number of values. Therefore, we would like to choose the values of a, c, and m so that a large number of values can be generated before the sequence repeats itself. Moreover, once the numbers a, c, m, and x_0 are known, the sequence of numbers generated by the LCG is completely deterministic. However, if these numbers were unknown to us and we were only observing the sequence generated by the LCG, it would be very hard for us to distinguish this sequence from a truly random sequence. For example, Fig. 10.16 shows the frequency histogram of the first 100 numbers of the sequence $x_1/m, x_2/m, \ldots$ obtained by setting $a = 16,807$, $m = 2,147,483,647$, $c = 0$, and $x_0 = 33,554,432$. It has the appearance of being uniformly distributed between 0 and 1, and it can be verified that the uniform distribution indeed is the distribution that best fits the data. By setting the seed of a random number generator, we fix the sequence of numbers that is generated by the algorithm. Thus, we are also able to easily replicate the simulation.

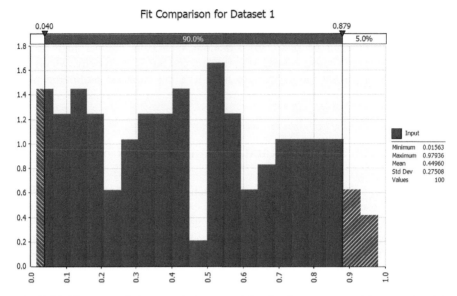

Fig. 10.16 Histogram of the sequence generated by the linear congruential generator with $a = 16,807$, $m = 2,147,483,647$, $c = 0$, and $x_0 = 33,554,432$

The LCG algorithm generates a sequence of numbers that have the appearance of coming from a uniform distribution. It is possible to build on this to generate pseudo-random numbers from other probability distributions (both discrete as well as continuous distributions). We refer the reader to Ross (2013) and Law (2014) for more details on the algorithms and their properties.

Appendix 2: Fitting Distributions to Data

When we build a simulation model, a key assumption is that we know the distributions of the input random variables. In practice, the distributions of the input random variables come from statistically testing the observed data to find out the distribution that best fits the data. The statistical tests are referred to as goodness-of-fit tests. The underlying idea is to compare the observed distribution with the hypothesized distribution and measure the discrepancy between the two. If the discrepancy is small, then it suggests that the hypothesized distribution is a good fit to the observed data. Otherwise, the hypothesized distribution is a poor fit. The error metrics and the hypothesis tests can be formalized; see Ross (2013) and Law (2014) for details. Here we only focus on how we can use @Risk to run the goodness-of-fit tests.

Given historical data, we can use the "Distribution Fitting" tool in @Risk to find out the distribution that best matches the data (see Fig. 10.17). After describing the

10 Simulation

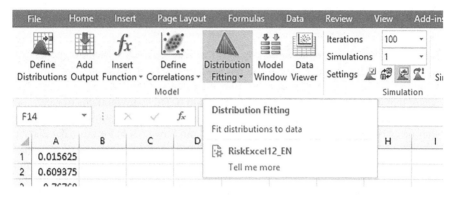

Fig. 10.17 Distribution Fitting tool in @Risk

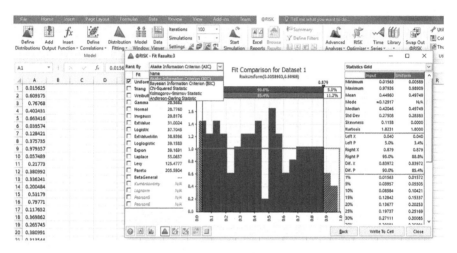

Fig. 10.18 Goodness-of-fit test results

nature of the data set (discrete or continuous) as well as additional details regarding the range of values it can take (minimum, maximum values), we run the Distribution Fitting tool. This gives us a range of distributions and their corresponding fit values. We can broadly think of the fit values as measuring the error between the observed data and the hypothesized distribution and a smaller fit value indicates a better fit in general. There are different goodness-of-fit tests available. The fit values as well as the relative ranking of the distributions in terms of their fit can vary depending on the test that is used. We refer the reader to Ross (2013) and Law (2014) for more details regarding the different goodness-of-fit tests and when a given test is more applicable (Fig. 10.18).

Appendix 3: Simulation in Excel

Excel has some basic simulation capabilities. The RAND(.) function generates a (pseudo) random number from the uniform distribution between 0 and 1. The RANDBETWEEN(.,.) function takes two arguments a and b, and generates an integer that is uniformly distributed between a and b. There are methods that can use this sequence of uniform random numbers as an input to generate sequences from other probability distributions (see Ross (2013) and Law (2014) for more details). A limitation of using the RAND(.) and RANDBETWEEN(.,.) functions is that it is not possible to fix the initial seed and so it is not possible to easily replicate the results of the simulation.

As mentioned, there are a number of other packages and programming languages that can be used to build simulation models. For example, the code snippet below implements the fashion retailer simulation described in Example 10.1 in R:

Sales = rnorm(1000, 980, 300)

Profit = 250 * Sales − 150,000.

The first line generates a random input sample of size 1000 from the normal distribution with a mean of 980 and a standard deviation of 300. The second line generates a random sample of the output (profit) by using the random sample of the input and the relation between the profit and the sales. Note that the seed as well as the random number generator can be specified in R; refer the R documentation.

References

Anderson, C. K., & Wilson, J. G. (2011). Name-your-own price auction mechanisms – Modeling and future implications. *Journal of Revenue and Pricing Management, 10*(1), 32–39. https://doi.org/10.1057/rpm.2010.46.

Davison, M. (2014). *Quantitative finance: A simulation-based approach using excel*. London: Chapman and Hall.

GE Look ahead. (2015). *The digital twin: Could this be the 21st-century approach to productivity enhancements*. Retrieved May 21, 2017, from http://gelookahead.economist.com/digital-twin/.

Holland, C., Levis, J., Nuggehalli, R., Santilli, B., & Winters, J. (2017). UPS optimizes delivery routes. *Interfaces, 47*(1), 8–23.

Jian, N., Freund, D., Wiberg, H., & Henderson, S. (2016). Simulation optimization for a large-scale bike-sharing system. In T. Roeder, P. Frazier, R. Szechtman, E. Zhou, T. Hushchka, & S. Chick (Eds.), *Proceedings of the 2016 winter simulation conference*.

Law, A. (2014). *Simulation modeling and analysis*. McGraw-Hill Series in Industrial Engineering and Management.

Porteus, E. (2002). *Foundations of stochastic inventory theory*. Stanford business books, Stanford California.

Ross, S. (2013). *Simulation*. Amsterdam: Elsevier.

Stine, R., & Foster, D. (2014). *Statistics for business decision making and analysis*. London: Pearson.

Technology Review. (2017). September 2017 edition.

Chapter 11
Introduction to Optimization

Milind G. Sohoni

1 Introduction

Broadly, one may describe **management science** as an interdisciplinary study of problem solving and decision making in human organizations. Management science uses a combination of analytical models and behavioral sciences to address complex business and societal problems. Often, finding a solution involves recognizing the optimization model at the core of the business application, formulating it appropriately, and solving it to gain managerial insights. The classic problems of management science include finding the right mix of inputs to minimize the cost of producing gasoline (blending problem), production planning in a manufacturing setup, inventory and workforce optimization to minimize the expected cost of meeting operating plans for matching supply to demand (aggregate planning), deciding optimal routes to offer for an airline (network planning), assigning crew to manage a schedule (crew scheduling), and maximizing the expected return subject to acceptable levels of variance (portfolio planning). Several textbooks are available that describe (Bazaraa et al., 2011; Bertsimas and Tsitsiklis, 1997; Chvátal, 1983; Wagner, 1969) these and several of other applications. Moreover, almost every chapter in this book includes examples of optimization, such as service level optimization in healthcare analytics and supply chain analytics, portfolio selection in Financial Analytics, inventory optimization in supply chain analytics as well as retail analytics, and price and revenue optimization in pricing analytics. The chapter on simulation provides a glimpse at combining simulation and optimization, for example, when the objective function is difficult to evaluate but can be specified using a simulation model. The case studies in this book illustrate the use of

M. G. Sohoni (✉)
Indian School of Business, Hyderabad, Telangana, India
e-mail: milind_sohoni@isb.edu

optimization in diverse settings: insurance fraud detection, media planning, and airline route planning. Optimization is also embedded in many predictive analytics methods such as forecasting, least square, lasso and logistics regression, maximum likelihood estimation, and backpropagation in neural networks.

It is natural to ask: What is meant by optimality? What is an algorithm? Is there a step-by-step approach that can be used to set up a model? What is the difference between constrained and unconstrained optimization? What makes these problems easy or hard to solve? How to use software to model and solve these problems? There are several books that cover some of these concepts in detail. For example, Bazaraa et al. (2011), Bertsimas and Tsitsiklis (1997), and Luenberger and Ye (1984) contain precise definitions and detailed descriptions of these concepts. In this chapter, we shall illustrate some of the basic ideas using one broad class of optimization problems called linear optimization. Linear optimization covers the most widely used models in business. In addition, because linear models are easy to visualize in two dimensions, it offers a visual introduction to the basic concepts in optimization. Additionally, we also provide a brief introduction to other optimization models and techniques such as integer/discrete optimization, non-linear optimization, search methods, and the use of optimization software.

But before we continue further we briefly touch upon the need to build mathematical models. Representing real-world systems as abstract models, particularly mathematical models, has several advantages. By analyzing the model in a virtual setting, the modeler is able to decide on actions to follow or policies to implement. The modeler is able to gain insights into the complex relationships among many decision variables and inform his/her judgment before selecting/formulating a policy to bring into action.

Figure 11.1 schematically represents the basic idea of model building. A good model abstracts away from reality without losing the essence of the trade-off

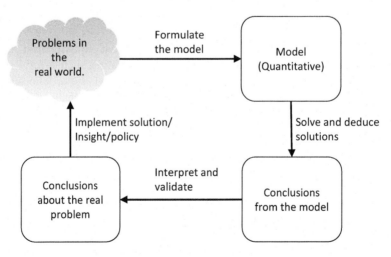

Fig. 11.1 Abstracting from a real-world problem

that needs to be considered. Once the model is analyzed the output needs to be interpreted appropriately and implemented in the real world with suitable modifications.

2 Methods in Optimization: Linear Programming

In this section we focus on a class of analytical models, and their solution techniques, that capture a real-world problem as a mathematical model (linear program) and understand how to interpret their analytical solutions and develop insights. Such mathematical programs are often referred to as (linear) optimization problems (models). The basic idea behind any optimization model is to find the "best" (optimal) solution, with respect to some objective, among all possible (feasible) solutions. While the **objective** depends on the real-world problem that is being modeled, it is represented as a mathematical function capturing the trade-off between the **decisions** that need to be made. The feasible solutions depend on the **constraints** specified in the real-world problem and are also represented by mathematical functions. A general mathematical program then tries to identify an extreme point (i.e., minimum or maximum) of a mathematical function that satisfies a set of constraints. Linear programming (LP) is the specialization of mathematical programming where both, the function—called the **objective function**—and the problem constraints are linear. We explain the notion of linearity in Sect. 2.3. The type of decision variables and constraints depends on the techno-socio-economic nature of the real-world application. However, irrespective of the domain of the application, an important factor for the applicability of optimization methodology is **computational tractability**. With modern computing technology, tractability requirements imply the existence of effective, and efficient, algorithmic procedures that are able to provide a fast solution to these models in a systematic manner.

The Simplex algorithm is one such powerful computational procedure for LPs that is readily applicable to very large-scale applications, sometimes including hundreds of thousands of decision variables. George Dantzig is credited with the development of the Simplex algorithm in 1947, as one of the first mathematical programming algorithms. Today, there are numerous successful implementations[1] of the Simplex algorithm that routinely solve complex techno-socio-economic problems. It is noteworthy that the success of the Simplex method, and the wide application of LP, has led to the broader field of **operations research/management science** being accepted as a scientific approach to decision making.

Over the past few decades, techniques to solve linear optimization problems have evolved significantly. While implementation (and computational speed) of the Simplex algorithm has improved dramatically in most commercially avail-

---

[1]The reader is referred to http://pubsonline.informs.org/journal/inte (accessed on Jul 22, 2018) to read about several industrial applications of optimization problems.

able solvers, newer mathematical algorithms and implementations have also been developed that compete with the Simplex algorithm effectively. From a business analytics standpoint, however, understanding the models being built to address the optimization problem, the underlying assumptions, and pertinent interpretation of the obtained analytical solutions are equally important. In this chapter, we discuss these details of the linear modeling. We will try to build our understanding using a prototypical LP example in Sect. 2.1 and two-dimensional geometry in Sect. 2.4. The insights gained are valid for higher-dimensional problems too and also reveal how the Simplex algorithm works. For a detailed description of the Simplex algorithm and other solution algorithms the reader is referred to Bazaraa et al. (2011), Bertsimas and Tsitsiklis (1997), and Chvátal (1983).

2.1 A Prototype LP Problem: Glass Manufacturer's Profit

Consider the following problem[2] for a manufacturer who produces two types of glasses, P_1 and P_2. Suppose that it takes the manufacturer 6 h to produce 100 cases of P_1 and 5 h to produce 100 cases of P_2. The production facility is operational for 60 h per week. The manufacturer stores the week's production in her own stockroom where she has an effective capacity of 15,000 ft^3. Hundred cases of P_1 occupy 1000 ft^3 of storage space, while 100 cases of P_2 require 2000 ft^3 due to special packaging. The contribution margin of P_1 is $5 per case; however, the only customer available will not accept more than 800 cases per week. The contribution of P_2 is $4.5 per case and there is no limit on the amount that can be sold in the market. The question we seek to answer is the following: *How many cases of each product should the glass manufacturer produce per week in order to maximize the total weekly contribution/profit?*

2.2 The LP Formulation and Solution

It is important to first notice that this is an optimization problem. The objective is to maximize the weekly profit. Furthermore, we are going to maximize the glass manufacturer's weekly profit by adjusting the weekly production levels for P_1 and P_2. Therefore, these weekly production levels are our **control/decision variables**. For ease of representation, we denote our decision variables in hundreds of cases of P_1 and P_2. Let these decision variables be represented as follows:

---

[2]This example is based on the example described in Chapter 1 of "Applied Mathematical Programming", by Bradley et al. (1977).

11 Optimization

x_1:	Number of units (in hundreds of cases) of product P_1 produced weekly, and
x_2:	Number of units (in hundreds of cases) of product P_2 produced weekly

Using these decision variables, we can now represent the manufacturer's objective function analytically as:

$$\max \quad f(x_1, x_2) \equiv 500x_1 + 450x_2. \qquad (11.1)$$

Equation (11.1) is called the **objective function**, and the coefficients 500 and 450 are called the **objective function coefficients**.

In our problem description, however, the manufacturer is resource constrained, i.e., the manufacturer has limited weekly production and storage capacity. Additionally, the demand for P_1 in the market is limited. Hence, we need to represent these **technological constraints** in our analytical formulation of the problem. First, let's focus on the **production constraint**, which states that the manufacturer has 60 h of production capacity available for weekly production. As mentioned in the problem statement, 100 cases of P_1 require 6 h of production time and that of P_2 require 5 h of production time. The technological constraint imposing this production limitation that our total weekly production doesn't exceed the available weekly production capacity is analytically expressed by:

$$6x_1 + 5x_2 \leq 60. \qquad (11.2)$$

Notice that in (11.2) time is measured in hours. Following a similar line of reasoning, the **storage capacity constraint** is analytically represented as:

$$10x_1 + 20x_2 \leq 150. \qquad (11.3)$$

From our problem statement, we know that the weekly demand for P_1 does not exceed 800 cases. So we need not produce more than 800 cases of P_1 in the week. Thus, we add a maximum demand constraint as follows:

$$x_1 \leq 8. \qquad (11.4)$$

Constraints (11.2), (11.3), and (11.4) are known as the **technological constraints** of the problem. In particular, the coefficients of the variables $x_i, i = 1, 2$, are known as the **technological coefficients** while the values on the **right-hand side** of the three inequalities are referred to as the right-hand side (**rhs**) vector of the constraints.

Finally, we recognize that the permissible value for variables $x_i, i = 1, 2$, must be nonnegative, i.e.,

$$x_i \geq 0; \ i = 1, 2, \qquad (11.5)$$

since these values express production levels. These constraints are known as the **variable sign restrictions**. Combining (11.1)–(11.5), the LP formulation of our problem is as follows:

$$\max \quad 500x_1 + 450x_2 \tag{11.6}$$

$$\begin{aligned}
\text{s.t.} \quad & 6x_1 + 5x_2 \leq 60 && : \text{Production constraint,} \\
& 10x_1 + 20x_2 \leq 150 && : \text{Storage capacity constraint,} \\
& x_1 \leq 8 && : \text{Max demand for } P_1, \\
& x_1, x_2 \geq 0 && : \text{Non-negativity constraint.}
\end{aligned}$$

2.3 The General Form of a LP Formulation

In general, a maximization linear programming problem (LPP) can be represented analytically as follows:

Objective function:

$$\max \quad f(x_1, x_2, \ldots, x_n) \equiv c_1 x_1 + c_2 x_2 + \cdots + c_n x_n \tag{11.7}$$

s.t. Technological constraints:

$$a_{i1}x_1 + a_{i2}x_2 + \cdots + a_{in}x_n \begin{pmatrix} \leq \\ = \\ \geq \end{pmatrix} b_i, \quad i = 1, \ldots, m, \tag{11.8}$$

Sign restrictions:

$$(x_j \geq 0) \text{ or } (x_j \leq 0) \text{ or } (x_j \text{ is unrestricted}), \quad j = 1, \ldots, n. \tag{11.9}$$

The formulation (11.7)–(11.9) has the general structure of a mathematical programming problem. Moreover, there is a specific structure to this formulation, i.e., the functions involved in the problem objective and the left-hand side (lhs) of the technological constraints are **linear**. It is the assumptions implied by linearity that to a large extent determine the applicability of the above model in real-world applications. To understand this concept of linearity a bit better, assume that the different decision variables $x_1, \ldots, x_n$ correspond to various activities from which any solution is eventually constructed. Essentially, the assigned values in a solution indicate the activity level in the plan considered. Each technological constraint of (11.8) imposes some restriction on the consumption of a particular resource (similar to the production and storage resources described in the prototype example (11.6).) Under this interpretation, the linearity property implies the following:

Additivity assumption: The total consumption of each resource and the overall objective value are the aggregates of the resource consumptions and the contributions to the problem objective, resulting by carrying out each activity independently.

Proportionality assumption: The consumptions and contributions for each activity are proportional to the actual activity level.

Divisibility assumption: Each variable is allowed to have fractional values (continuous variables).

Certainty assumption: Each coefficient of the objective vector and constraint matrix is known with certainty (not a random variable).

It is informative to understand how we implicitly applied this logic when we derived the technological constraints of the prototype example: (1) Our assumption that the processing of each case of P_1 and P_2 required constant amounts of time, respectively, implies proportionality, and (2) the assumption that the total production time consumed in the week is the aggregate of the manufacturing times required for the production of each type of glass, if the corresponding activity took place independently, implies additivity.

It is important to note how the linearity assumption restricts our modeling capabilities in the LP framework: For example, we cannot immediately model effects like economies of scale in the problem structure, and/or situations in which resource consumption of resources by complementary activities takes place. In some cases, one can approach these more complicated problems by applying some **linearization** scheme—but that requires additional modeling effort.

Another approximation, implicit in many LPPs, is the so-called **divisibility assumption**. This assumption refers to the fact that for LP theory and algorithms to work, the decision variables must be **real** valued. However, in many business problems, we may want to restrict values of the decision variables to be integers. For example, this may be the case with the production of glass types, P_1 and P_2, in our prototype example or production of aircraft. On the other hand, continuous quantities, such as tons of steel to produce and gallons of gasoline to consume, are divisible. That is, if we solved a LPP whose optimal solution included the consumption of 3.27 gallons of gasoline, the answer would make sense to us; we are able to consume fractions of gallons of gasoline. On the contrary, if the optimal solution called for the production of 3.27 aircraft, however, the solution probably would not make sense to anyone.

Imposing integrality constraints for some, or all, variables in a LPP turns the problem into a **(mixed) integer programming** (MIP or IP) problem. The computational complexity of solving an MIP problem is much higher than that of a LP. Actually, MIP problems belong to the notorious class of *NP-complete* problems, i.e., those problems for which there is no known/guaranteed polynomial bound on the solution time to find an optimal solution. We will briefly discuss the challenge of solving MIPs later in Sect. 3.

Finally, before we conclude this discussion, we define the **feasible region of the LP** of (11.7)–(11.9), as the entire set of vectors $\langle x_1, x_2, \ldots, x_n \rangle$ (notice that each variable is a coordinate of an n-dimensional vector) that satisfy the technological constraint (11.8) and the sign restrictions (11.9). An **optimal solution** to the problem is any feasible vector that further satisfies the optimality requirement defined by (11.7).

Before we conclude this section, it is worth mentioning that there are several commercially available software packages such as Microsoft Excel's Solver or XLMiner add-ins, Python, SAS, GAMS, etc. to solve LPPs. However, many business applications can be easily modeled using Microsoft Excel's **Solver** add-in program because of the convenient spreadsheet interface available. There are a few online tutorials available to understand how to input a LP model in Solver. We provide a few weblinks in Appendix. We do not describe the steps involved in building a spreadsheet-based LP model in Excel Solver. However, in Appendix section "Spreadsheet Models and Excel Solver Reports" we describe the output generated from Excel Solver, which is closely linked to the discussion provided in the rest of this chapter.

2.4 Understanding LP Geometry

In this section, we develop a solution approach for LP problems, which is based on a geometrical representation of the feasible region and the objective function. While we will work with our prototype example with two decision variables, the insights we gain through this exercise will readily carry over to LPPs with n variables. The number of decision variables in a LP determines the problems **dimensionality**. A two-dimensional (2-D) problem can be represented in a Cartesian coordinate system (two-dimensional space with axes perpendicular to each other) and problems with n variables can be represented by n-dimensional spaces based on a set of n mutually perpendicular axes. In particular, the n-dimensional space (think coordinates) to be considered has each dimension defined by one of the LP variables x_j. While a 2-D problem is easy to represent and visualize, to maintain sanity, it is advisable not to visualize higher-dimensional spaces beyond three dimensions.

As will be shown later with our prototype example, the objective function, in an n-dimensional space, is represented by its **contour plots**, i.e., the sets of points that correspond to the same objective value. As mentioned earlier, to facilitate the visualization of the concepts involved, we shall restrict ourselves to the two-dimensional case. To the extent that the proposed approach requires the visualization of the underlying geometry, it is applicable only for LPs with up to three variables.

2.4.1 Feasible Region for Two-Variable LPs

The primary idea behind the geometrical representation is to correspond every vector $\langle x_1, x_2 \rangle$, denoting the decision variables of a two-variable LP, to the point with coordinates (x_1, x_2) in a two-dimensional (planar) Cartesian coordinate system. Remember that the set of constraints determine the feasible region of the LP. Thus, under aforementioned correspondence, the feasible region is depicted by the set of points that satisfy the LP constraints and the sign restrictions simultaneously. Since all constraints in a LPP are expressed by linear inequalities, we must first characterize the set of points that constitute the solution space of each linear inequality. The intersection of the solution spaces corresponding to each technological constraint and/or sign restriction will represent the LP feasible region. Notice that a constraint can either be an equality or an inequality in LPP. We first consider the feasible region corresponding to a single equality constraint.

The Feasible Space of a Single Equality Constraint Consider an equality constraint of the type

$$a_1 x_1 + a_2 x_2 = b \tag{11.10}$$

Assuming $a_2 \neq 0$, this equation corresponds to a **straight line** with **slope** $s = \frac{a_1}{a_2}$ and **intercept** $d = \frac{b}{a_2}$. In the special case where $a_2 = 0$, the feasible space *locus* of (11.10) is still a straight line perpendicular to the x_1-axis, intersecting it at the point $\left(\frac{b}{a_1}, 0\right)$. It is noteworthy that an equality constraint restricts the dimensionality of the feasible space by *one degree of freedom*, i.e., in the case of a 2-D problem, it turns the feasible space from a planar area to a line segment.

The Feasible Space of a Single Inequality Constraint Consider the constraint:

$$a_1 x_1 + a_2 x_2 \left(\begin{array}{c}\leq\\ \geq\end{array}\right) b \tag{11.11}$$

The feasible space is one of the closed **half-planes** defined by the equation of the line corresponding to this inequality: $a_1 x_1 + a_2 x_2 = b$. Recollect that a line divides a 2-D plane into two halves (half-planes), i.e., the portion of the plane lying on each side of the line. One simple technique to determine the half-plane comprising the feasible space of a linear inequality is to test whether the point $(0, 0)$ satisfies the inequality. In case of a positive answer, the **feasible space** is the half-space containing the origin. Otherwise, it is the half-space lying on the other side which does not contain the origin.

Consider our prototype LP Sect. 2.1 described earlier. Figure 11.2 shows the feasible regions corresponding to the individual technological and nonnegativity constraints. In particular, Fig. 11.2c shows the entire feasible region as the intersection of the half-spaces of the individual constraints. Note that, for our prototype problem, the feasible region is bounded on all sides (the region doesn't extend to infinity in any direction) and nonempty (has at least one feasible solution).

Infeasibility and Unboundedness Sometimes, the constraint set can lead to an **infeasible** or **unbounded** feasible region.

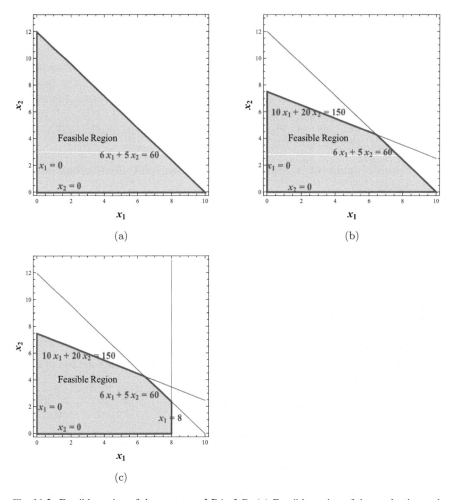

Fig. 11.2 Feasible region of the prototype LP in 2-D. (**a**) Feasible region of the production and nonnegative constraints. (**b**) Feasible region of the storage and production constraint. (**c**) The entire feasible region

An infeasible region implies the constraints are "contradictory" and hence the intersection set of the half-spaces is empty. An unbounded feasible region may mean that the optimal solution could go off to $-\infty$ or $+\infty$ if the objective function "improves" in the direction in which the feasible region is unbounded.

Consider again our original prototype example. Suppose there is no demand restriction on the number of cases of P_1 and the manufacturer requires that at least 1050 cases of P_1 are produced every week. These requirements introduce two new constraints into the problem formulation, i.e.,

$$x_1 \geq 10.5.$$

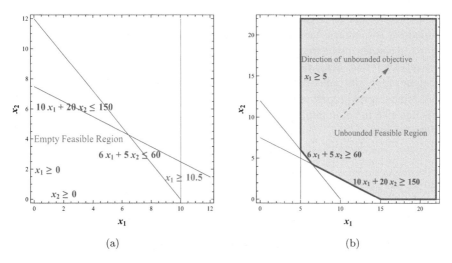

Fig. 11.3 Infeasible and unbounded feasible regions in 2-D. (**a**) Feasible region is empty (Infeasible). (**b**) Feasible region is unbounded

Figure 11.3a shows the feasible region for this new problem which is empty, i.e., there are no points on the (x_1, x_2)-plane that satisfy all constraints, and therefore our problem is infeasible (over-constrained).

To understand unbounded feasible regions visually, consider a situation wherein we change our prototype LP such that the manufacturer must use at least 60 h of production, must produce at least 500 cases of P_1, and must use at least 15,000 units of storage capacity. In this case the constraint set changes to

$$x_1 \geq 5,$$
$$6x_1 + 5x_2 \geq 60,$$
$$10x_1 + 20x_2 \geq 150,$$

and the feasible looks like the region depicted in Fig. 11.3b. It is easy to see that the feasible region of this problem is unbounded, Furthermore, in this case our objective function, $500x_1 + 450x_2$ can take arbitrarily large values and there will always be a feasible production decision corresponding to that arbitrarily large profit. Such a LP is characterized as **unbounded**. It is noteworthy, however, that even though an unbounded feasible region is a necessary condition for a LP to be unbounded, it is not sufficient (e.g., if we were to minimize our objective function, we would get a finite value).

Representing the Objective Function A function of two variables $f(x_1, x_2)$ is typically represented as a surface in an (orthogonal) three-dimensional space, where two of the dimensions correspond to the independent variables x_1 and x_2, while the third dimension provides the objective function value for any pair (x_1, x_2). In the context of our discussion, however, we will use the concept of **contour plots**. Suppose α is some constant value of the objective function, then for any given range of α's, a contour plot depicts the objective function by identifying the set of points (x_1, x_2) such that $f(x_1, x_2) = \alpha$. The plot obtained for any fixed value of α is a contour of the function. Studying the structure of a contour identifies some patterns that depict useful properties of the function. In the case of 2-D LPPs, the linearity of the objective function implies that any contour can be represented as a straight line of the form:

$$c_1 x_1 + c_2 x_2 = \alpha. \qquad (11.12)$$

It is noteworthy that for a maximization (minimization) problem, this starting line is sometimes referred to as an **isoprofit (isocost)** line. Assuming that $c_2 \neq 0$ (o.w., work with c_1), (11.12) can be rewritten as:

$$x_2 = -\frac{c_1}{c_2} x_1 + \frac{\alpha}{c_2}. \qquad (11.13)$$

Consider the objective function $500 x_1 + 450 x_2$ in our prototype example. Let us draw the first isoprofit line as $500 x_1 + 450 x_2 = \alpha$ (the dashed red line in Fig. 11.4), where $\alpha = 1000$ and superimpose it over our feasible region. Notice that the intersection of this line with the feasible region provides all those production decisions that would result in a profit of exactly \$1000.

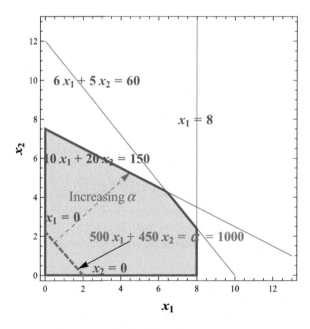

Fig. 11.4 Drawing isoprofit lines over the feasible region

11 Optimization

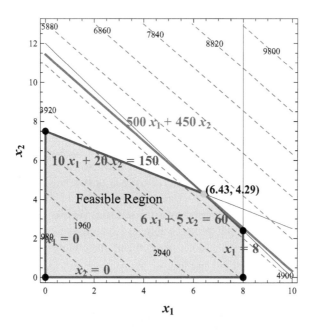

Fig. 11.5 Sweeping the isoprofit line across the feasible region until it is about to exit. An optimal solution exists at a corner point (vertex)

As we change the value of α, the resulting isoprofit lines have constant slope and varying intercept, i.e., they are parallel to each other (since by definition isoprofit/isocost lines cannot intersect). Hence, if we continuously increase α from some initial value α_0, the corresponding isoprofit lines can be obtained by "sliding" the isoprofit line corresponding to $f(x_1, x_2) = \alpha_0$ parallel to itself, in the direction of increasing (decreasing) intercepts, if c_2 is positive (negative.) This "improving direction" of the isoprofit line is denoted by the dashed magenta arrow in Fig. 11.4.

Figure 11.5 shows several isoprofit lines, superimposed over the feasible region, for our prototype problem.

Finding the Optimal Solution It is easy to argue that an optimal solution to a LPP will never lie in the interior of the feasible region. To understand why this must be true, consider the prototype example and let us assume that an optimal solution exists in the interior. It is easy to verify that by simply increasing the value of either x_1 or x_2, or both—as long as we remain feasible—we can improve the objective value. But this would contradict the fact that the point in the interior is an optimal solution. Thus, we can rule out the possibility of finding an optimal solution in the interior of the feasible region. So then, if an optimal solution exists, it must lie somewhere on the boundary of the feasible region. The "sliding motion" described earlier suggests a way for finding the optimal solution to a LPP. The basic idea is to keep sliding the isoprofit line in the direction of increasing α's, until we cross (or are just about to slide beyond) the boundary of the LP feasible region. For our prototype

LPP, this idea is demonstrated in Fig. 11.5. The dashed red lines are the contour lines and the solid red line is the contour line corresponding to that value of α such that any further increase would result in the objective line crossing the feasible region, i.e., an infinitesimal increase in α would result in the contour line moving parallel to itself but not intersecting the feasible region. Thus, the objective value is maximized at that point on the boundary beyond which the objective function crosses out of the feasible region. In this case that point happens to be defined by the intersection of the constraint lines for the production capacity and storage capacity, i.e., $6x_1 + 5x_2 = 60$ and $10x_1 + 20x_2 = 150$. The coordinates of the optimal point are $x_1^\star = 6.43$ and $x_2^\star = 4.29$. The maximal profit is $f\left(x_1^\star, x_2^\star\right) = 5142.86$.

In fact, notice that the optimal point (the green dot) is one of the **corner** points (the black dots) of the feasible region depicted in Fig. 11.5 and is **unique**. The optimal corner point is also referred to as the **optimal vertex**.

In summary, if the optimal vertex is uniquely determined by a set of intersecting constraints and the optimal solution only exists at that unique corner point (vertex), then we have a unique optimal solution to our problem. See Fig. 11.6.

LPs with Many Optimal Solutions A natural question to ask is the following: Is the optimal solution, if one exists, always unique? To analyze this graphically, suppose the objective function of our prototype problem is changed to

$$225x_1 + 450x_2.$$

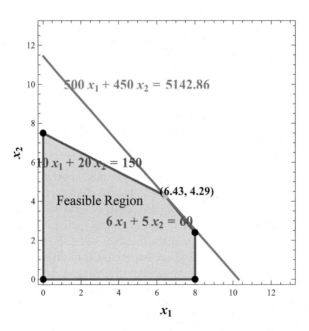

Fig. 11.6 A unique optimal solution exists at a single corner point (vertex)

11 Optimization

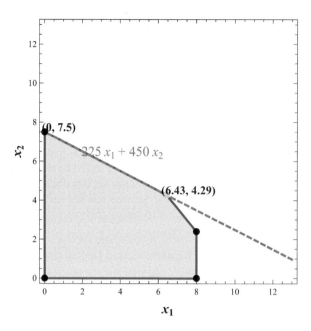

Fig. 11.7 Multiple optimal solutions along a face (includes corner points)

Notice that any isoprofit line corresponding to the new objective function is parallel to the line corresponding to the storage constraint:

$$10x_1 + 20x_2 = 150.$$

Therefore, if we try to apply the graphical optimizing technique described earlier, we get the situation depicted in Fig. 11.7, i.e., every point in the line segment between points (0,7.5) and (6.43, 4.29), along the storage constraint line, is an optimal point, providing the optimal objective value of $5142.86.

It is worth noticing that even in this case of many optimal solutions, we have two of them corresponding to "corner" points of the feasible region, namely, points (0,7.5) and (6.43, 4.29).

Summarizing the above discussion, we have shown that a two-variable LP can either have a unique optimal solution that corresponds to a "corner" point of the feasible region or have many optimal solutions that correspond to an entire "edge" of the feasible region, or be unbounded, or be infeasible. **This is true for general n-dimensional LPPs too**. There is a famous theorem, called the **fundamental theorem of linear programming**, which states that if a LP has a bounded optimal solution, then it must have one that is an extreme point of its feasible region. The **Simplex algorithm** (a solution algorithm embedded in most software) essentially exploits this fundamental result to reduce the space to be searched for an optimal solution.

The fundamental theorem of linear programming states the following:

Theorem 1 (The Fundamental Theorem of Linear Programming) *If a LP has a bounded optimal solution, then there exists an extreme point of the feasible region that is optimal.*

Another important fact about the feasible region of a LP is that it is **convex**. A convex set is defined as follows: Let y_1 and y_2 be any two points belonging to a set S. Then S is a convex set if and only if all points y, belonging to the line segment joining y_1 and y_2, also belong to the set S. In mathematical terms y can be expressed as $y = \alpha y_1 + (1-\alpha)y_2$, for all values of $\alpha \in [0, 1]$, and the set S is a convex set if y also belongs to the set S. An example of a convex set is a circle in two dimensions. The feasible region of a LP is **polyhedral** because it is defined by linear equalities.

2.4.2 Binding Constraints, LP Relaxation, and Degeneracy

A **constraint is binding** if it passes through the optimal vertex, and **nonbinding** if it does not pass through the optimal vertex. If we increase the rhs value of a $\leq$ constraint (see Fig. 11.8a) or decrease the rhs value of a $\geq$ constraint, we "relax" the constraint, i.e., we enlarge the feasible region to include additional points that simultaneously satisfy all the constraints of the original LP. **Relaxing** the LP can only "improve" optimal objective value, i.e., the inclusion of additional feasible points in the feasible region does not remove any of the original feasible points (including the original optimal solution). All that can happen is that one of the new feasible points may provide a better objective function value. On the other hand,

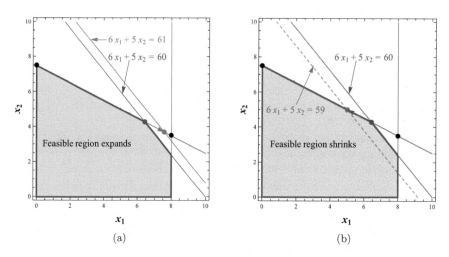

Fig. 11.8 Relaxing and tightening the feasible region. (**a**) Relaxing (expanding) the feasible region. (**b**) Tightening (shrinking) the feasible region

11 Optimization

if we **tighten the LP** by decreasing the rhs value of a $\leq$ constraint or increase the rhs value of a $\geq$ constraint (see Fig. 11.8b), the optimal objective value can only deteriorate. In Fig. 11.8 we demonstrate how the feasible region expands (shrinks) as we increase (decrease) the rhs value of the production constraint in our prototype LP.

If a constraint is binding, changing its rhs will change the optimal solution and the objective value. In general, as the rhs of the binding constraint is changed, the optimal vertex *slides along* the intersection of changing constraint and, as the optimal vertex moves, the optimal objective value changes (unless the corner point or vertex is degenerate, which we will discuss later). If a constraint is not binding, then tightening it (a bit) or relaxing it (as much as you please) will not change the optimal solution or the optimal objective value.

A **slack variable** for nonbinding $\leq$ constraint is defined to be the difference between its rhs and the value of the left-hand side of the constraint evaluated at the optimal vertex (in the n-dimensional space). Suppose the optimal vertex is represented by $\langle x_1^*, x_2^*, \ldots, x_n^* \rangle$. Formally, the slack s_i of the ith $\leq$ constraint is defined as

$$s_i = b_i - \sum_{j=1}^{n} a_{ij} x_j^*.$$

Similarly, the **surplus** associated with a nonbinding $\geq$ constraint is the extra value that may be reduced from the constraint's left-hand side function before the constraint becomes binding, i.e., the left-hand side equals the rhs. The formal definition of the surplus variable of an ith $\geq$ constraint is:

$$\text{surplus}_i = \sum_{j=1}^{n} a_{ij} x_j^* - b_i.$$

Any LP involving inequality constraints can be converted into an equivalent LP involving just equality constraints (simply add slack and surplus variables). After such a conversion, the LP formulation can be written as (here we consider only a maximization problem and assume that the constraints are of $\leq$ type):

$$\max \quad \sum_{i=1}^{n} c_i x_i$$

s.t.

$$\begin{pmatrix} a_{11}x_1 + \cdots + a_{1n}x_n & +s_1 & +0 & +0 & +\cdots & +0 & = b_1 \\ a_{21}x_1 + \cdots + a_{2n}x_n & +0 & +s_2 & +0 & +\cdots & +0 & = b_2 \\ \vdots & \vdots & \vdots & \vdots & +0 & +\vdots & +s_i & +\vdots & \vdots & \vdots \\ a_{m1}x_1 + \cdots + a_{mn}x_n & +0 & +0 & +0 & \cdots & +s_m & = b_m \end{pmatrix},$$

$$x_j \geq 0 \quad j = 1, \ldots, n$$

with m equality constraints and n variables (where we can assume $n > m$). Then, theory tells us that each vertex of the feasible region of this LP can be found by: choosing m of the n variables (these m variables are collectively known as the **basis** and the corresponding variables are called **basic variables**); setting the remaining $(n - m)$ variables to zero; and solving a set of simultaneous linear equations to determine values for the m variables we have selected. Not every selection of m variables will give a nonnegative solution. Also, enumerating all possible solutions can be very tedious, though, there are problems where, if m were small, the enumeration can be done very quickly. Therefore, the Simplex algorithm tries to find an "adjacent" vertex that improves the value of the objective function. There is one problem to be solved before doing that: if these values for the m variables are all >0, then the vertex is **nondegenerate**. If one or more of these variables is zero, then the vertex is **degenerate**. This may sometimes mean that the vertex is **over-defined**, i.e., there are more than necessary binding constraints at the vertex. An example of a degenerate vertex in three dimensions is

$$x_1 + 4x_3 \leq 4$$
$$x_2 + 4x_3 \leq 4$$
$$x_1, x_2, x_3 \geq 0$$

The three-dimensional feasible region looks like the region in Fig. 11.9. Notice that the vertex $(0,0,1)$ has four planes defining it.

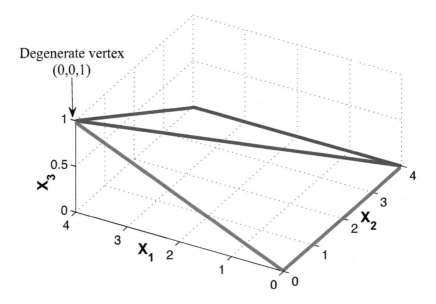

Fig. 11.9 An example of a degenerate vertex

For two-dimensional problems degeneracy is **not** an issue—if there are three constraints binding at a vertex, then one of them is redundant, i.e., one of the constraints can be expressed as a linear combination of the other two and hence removed (not considered). However, in more than two dimensions it may not be possible to eliminate any of the constraints because they may be linearly independent. Consequently, the Simplex algorithm may get stuck at a degenerate vertex (cycling) and may not move to an adjacent vertex that improves the objective value during execution. There are special methods to overcome this problem with degeneracy allowing the Simplex algorithm to break out of cycling at a degenerate vertex.

2.5 Shadow Prices, Reduced Costs, and Sensitivity Analysis

The **shadow price**, associated with a particular constraint, is the **change in the optimal value of the objective function per unit increase in the rhs value for the constraint, all other problem data remaining unchanged**. Equivalently, the shadow price is also the rate of deterioration in the objective value obtained by restricting that constraint. Shadow prices are also called **dual values**. Shadow price is discussed in detail in Chap. 23 on Pricing Analytics.

The **reduced cost** associated with the nonnegativity constraint for each variable is the shadow price of that constraint, i.e., the corresponding change in the objective function per unit increase in the lower bound of the variable. Algebraically, we can express the reduced cost of activity j as

$$\bar{c}_j = c_j - \sum_{i=1}^{m} a_{ij} y_i$$

where c_j is the objective coefficient of activity j, y_i is the shadow price (dual value) associated with constraint i, and a_{ij} is the amount of resource i (corresponds to constraint i) used per unit of activity j. The operation of determining the reduced cost of an activity, j, from the shadow prices of the constraints and the objective function is generally referred to as **pricing out an activity**. To understand these computations, consider the prototype LP described earlier. Suppose the manufacturer decides to add another set of glasses, P_3, to his product mix. Let us assume that P_3 requires 8 h of production time per 100 cases and occupies 1000 cubic units of storage space. Further, let the marginal profit from a case of P_3 be $6. If x_3 represents the decision of how many hundreds of cases of P_3 to produce, then the new LP can be rewritten as:

max	500	x_1	+450	x_2	+600	x_3			Shadow prices at optimality ↓
Production:	6	x_1	+5	x_2	+8	x_3	≤	$b_1 = 60$	: $y_1^\star = 78.57$
Storage:	10	x_1	+20	x_2	+10	x_3	≤	150,	: $y_2^\star = 2.86$
Demand:	1	x_1					≤	8,	: $y_3^\star = 0$
Sign restrictions:		x_1,		x_2,		x_3	≥	0.	

Suppose, we solve this LP to find the following optimal solution:

Optimal obj. value	= 5142.86		
Decision variables ↓		Shadow prices ↓	
$x_1^\star$	= 6.43	First constraint: $y_1^\star$ (**binding**)	= 78.57
$x_2^\star$	= 4.29	Second constraint: $y_2^\star$(**binding**)	= 2.86
$x_3^\star$	= 0	Third constraint: $y_3^\star$ (**non-binding**)	= 0

The **reduced cost of variable** x_1 (here $j = 1$) **at optimality**, i.e., $\bar{c}_1$, is computed as follows, where $c_1 = 500$, $a_{11} = 6$, $a_{21} = 10$, $a_{31} = 1$.

$$\bar{c}_1 = c_1 - \left(a_{11}y_1^\star + a_{21}y_2^\star + a_{31}y_3^\star\right),$$
$$= 500 - (6 \times 78.57 + 10 \times 2.86 + 1 \times 0),$$
$$= 0.$$

Similarly, for variable x_3 ($j = 3$) the reduced cost is (where $c_3 = 600$, $a_{13} = 8$, $a_{23} = 10$, $a_{33} = 0$)

$$\bar{c}_3 = c_3 - \left(a_{13}y_1^\star + a_{23}y_2^\star + a_{33}y_3^\star\right),$$
$$= 600 - (8 \times 78.57 + 10 \times 2.86 + 0 \times 0),$$
$$= -57.14.$$

Now, suppose we want to compute the **shadow price** of the production constraint. Let b_1 denote the rhs of the production constraint (C1). Currently, $b_1 = 60$ as stated in the formulation above. Notice that the current optimal objective value is 5142.86 when $b_1 = 60$. Let us define the optimal value as a function of rhs of the production constraint, i.e., b_1 and denote it as $Z^\star(b_1)$. Thus, $Z^\star(60) = 5142.86$. Now suppose we keep all other values the same (as mentioned in the formulation) but change b_1 to 61 and recompute the optimal objective value. Upon solving the LP we get the new optimal objective value of $Z^\star(61) = 5221.43$. Then, using the definition of the shadow price of a constraint, the shadow price of the production constraint is computed as follows:

$$\text{Shadow price of C1} = \frac{Z^\star(61) - Z^\star(60)}{61 - 60}$$
$$= \frac{5221.43 - 5142.86}{1}$$
$$= 78.57.$$

Notice that the shadow price is the **rate at which the optimal objective changes with respect to the rhs of a particular constraint all else remaining equal.** It **should not** be interpreted as the absolute change in the optimal objective value.

Notice two important facts: (1) The *reduced cost of basic variables is 0*, i.e., $\bar{c}_j$ equals 0 for all basic x_j (see Sect. 2.4.2 for the definition), and (2) Since c_j equals zero for slack and surplus variables (see Sect. 2.4.2 for definition) the reduced cost of these variables is always the negative of the shadow price corresponding to the respective constraints. The *economic interpretation of a shadow price*, y_i (associated with resource i), is the *imputed value* of resource i. The term $\sum_{i=1}^{m} a_{ij} y_i$ is interpreted as the total value of the resource used per unit activity j. It is thus the *marginal resource cost* for using that activity. If we think of the objective coefficients c_j as being the marginal revenues, the reduced costs, $\bar{c}_j$, are simply the *net marginal revenues*.

An intuitive way to think about reduced costs is as follows: If the optimal solution to a LP indicates that the optimal level of a particular decision variable is zero, it must be because the objective function coefficient of this variable (e.g., its unit contribution to profits or unit cost) is not beneficial enough to justify its "inclusion" in the decision. The reduced cost of that decision variable tells us the amount by which the objective function coefficients must improve for the decision variable to become "attractive enough to include" and take on a nonzero value in the optimal solution. Hence the reduced costs of all decision variables that take nonzero values in the optimal solution are, by definition, zero ⇒ no further enhancement to their attractiveness is needed to get the LP to use them, since they are already "included." In economic terms, the values imputed to the resources (x_j) are such that the *net marginal revenue* is zero on those activities operated at a positive level, i.e., *marginal revenue = marginal cost* (MR = MC).

Shadow prices are only locally accurate (shadow prices are valid over a particular range, i.e., as long as the *set of binding constraints does not change* the shadow price of a constraint remains the same.); if we make dramatic changes in the constraint, naively multiplying the shadow price by the magnitude of the change may mislead us. In particular, the shadow price holds only within an allowable range of changes to the constraints rhs; outside of this allowable range the shadow price may change. This allowable range is composed of two components. The **allowable increase** is the amount by which the rhs may be increased before the shadow price can change; similarly, the **allowable decrease** is the corresponding reduction that may be applied to the rhs before a change in the shadow price can take place (whether this increase or decrease corresponds to a tightening or a relaxation of the constraint depends on the direction of the constraints inequality). **A constraint is binding** if it passes through the optimal vertex, and **nonbinding** if it does not pass through the optimal vertex (constraint C3 in the example above). For a binding constraint, the geometric intuition behind the definition of a shadow price is as follows: By changing the rhs of a binding constraint, we change the optimal solution as it slides along the *other binding constraints*. Within the allowable range of changes to the rhs, the optimal vertex slides in a straight line, and the optimal objective value changes at a constant rate (which is the shadow price). Once we cross the limits indicated by the allowable increase or decrease, however, the optimal vertex's slide changes because **the set of binding constraints change**. At some point the constraint, whose rhs is being modified, may become nonbinding and a new vertex is optimal. For a nonbinding constraint the shadow price (or dual value) is always zero.

Consider the prototype LP described earlier where the rhs value of production constraint is 60. In Fig. 11.10 we show how the feasible region changes and when the set of binding constraints change as we perturb the rhs value of the production constraint. Notice that in Fig. 11.10a the storage constraint drops out of the set of binding constraints and in Fig. 11.10c the demand constraint becomes binding. In between these two extremes, the set of binding constraints, as shown in Fig. 11.10b, remains unchanged. The range over which the current optimal shadow price of 78.57 remains unchanged is from 37.5 to 65.5 (allowable increase is 5.5 and allowable decrease is 22.5). That is, if the rhs of the production constraint were to vary in the range from 37.5 to 65.5 (values of $b_1 \in [37.5, 65.5]$) the shadow price would be constant at 78.57.

Currently, the value of $b_1 = 60$. In Fig. 11.11 we plot the optimal objective value $Z^\star(b_1)$ as a function of b_1, the rhs of production constraint, when b_1 is in the range [37.5, 65.5]. All other values are kept the same. Notice, as we vary b_1, the optimal objective value changes linearly at the rate of the shadow price, i.e., 78.57.

When the reduced cost of a decision variable is nonzero (implying that the value of that decision variable is zero in the optimal solution), the reduced cost is also reflected in the allowable range of its objective coefficient. In this case, one of the allowable limits is always infinite (because making the objective coefficient less attractive will never cause the optimal solution to include the decision variable in the optimal solution); and the other limit, by definition, is the reduced cost (for it is the amount by which the objective coefficient must improve before the optimal solution changes).

2.5.1 One Hundred Percent Rule

While performing sensitivity analysis changes to the objective coefficients, rhs values, or consumption levels are analyzed one at a time. Changing these objective coefficients or rhs values simultaneously does not guarantee that the optimal

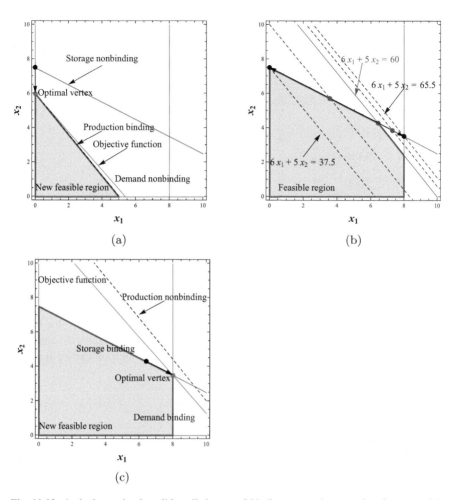

Fig. 11.10 A shadow price is valid until the set of binding constraints remains the same. (**a**) Decreasing the rhs beyond the range. (**b**) The range of the rhs for which shadow price remains constant. (**c**) Increasing the rhs beyond the range

solution is conserved. Simultaneous changes can be implemented and a conservative bound on these simultaneous changes can be computed using the **100% rule**. First, we compute the ranges for the rhs values assuming changes are made one at a time. The 100% rule implies the following: If simultaneous changes are made to the rhs (or the objective coefficients) values of more than one constraint (variable) in such a way that the sum of the fractions of allowable range utilized by these changes is less than or equal to one, the optimal basis (variables that are included in the optimal decision) remains unchanged. Consider the example described earlier where the rhs value of constraint C1 is 60. If we solve the LP, it turns out that at optimality the allowable range for the shadow price of 78.57, in the current optimal solution, is 37.5–65.5. That is, if the rhs of constraint C1 were to be in the range 37.5–65.5

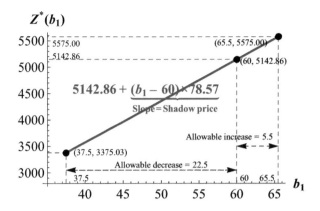

Fig. 11.11 Plot of $Z^*(b_1)$ vs. b_1 for the production constraint, when $b_1 \in [37.5, 65.5]$

(values of $b_1 \in [37.5, 65.5]$), the shadow price of C1 at optimality would be 78.57 for all these values of the rhs. Further, notice that the current rhs value of constraint C2 is 150 and the range for its shadow price (2.86, corresponding to the current optimal solution) is between 128 and 240. Suppose we reduce the rhs value of the first constraint to $b_1^{new} \leq 60$ and increase the rhs value of the second constraint to $b_2^{new} \geq 150$. The 100% rule suggests that the current solution (basis) remains optimal as long as

$$\frac{60 - b_1^{new}}{60 - 37.5} + \frac{b_2^{new} - 150}{240 - 150} \leq 1.$$

2.6 A Quick Note About LP Optimality

It is evident from the earlier discussion that any optimal solution to a LPP has a very specific structure. We reiterate the optimal structure of any LPP below:

1. The shadow price of nonbinding constraint is always 0. A binding constraint may have a nonzero shadow price. Together, this implies

   Slack (or surplus) on a constraint × shadow price of the constraint = 0.

2. Every decision variable has a reduced cost associated with it. Basic variables, at optimality, have a zero reduced cost and nonbasic variables may have a nonzero reduced cost. This implies that

   Reduced cost of a variable × the optimal value of the variable = 0.

3. Finally, it is easy to verify that for a LP, at optimality

> The optimal objective value = Product of the rhs value of a constraint
> × the shadow price of the constraint,
> summed over all the constraints, i.e.,

$$\sum_{j=1}^{n} c_j x_j^\star = \sum_{i=1}^{m} b_i y_i^\star,$$

where $y_i^\star$ is the shadow price of the ith constraint at optimality, b_i is the value of the rhs of constraint i, c_j is the objective coefficient of the jth decision variable, and $x_j^\star$ is the optimal value of the jth decision variable. For the prototype problem described earlier,

$$\sum_{j=1}^{n} c_j x_j^\star = (500 \times 6.429) + (450 \times 4.285) = 5142.8.$$

$$\sum_{i=1}^{m} b_i y_i^\star = (60 \times 78.571) + (150 \times 2.857) = 5142.8.$$

Conditions (1) and (2) together are called the **complementary slackness conditions of optimality**. All the three conditions, (1), (2), and (3) provide an easily **verifiable certificate of optimality** for any LPP. This is one of the fascinating features of any LP optimal solution—the certificate of optimality comes with the solution. Thus, combining the search from vertex to vertex and examining the solution for optimality gives an algorithm (the Simplex algorithm) to solve LPs very efficiently!

3 Methods in Optimization: Integer Programming—Enforcing Integrality Restrictions on Decision Variables

Introducing integrality constraints on decision variables has its advantages. First, we can model more realistic business requirements (economic indivisibility, for example) and second, the model allows us more flexibility such as modeling business logic using binary variables. However, the main disadvantage lies in the

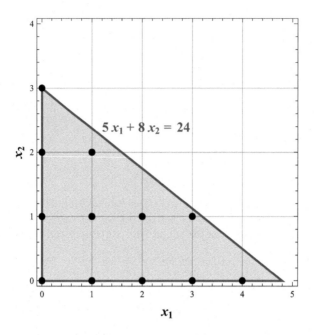

Fig. 11.12 Integer feasible region for (P)

difficulty of solving the model and guaranteeing optimality of the solution. Let us consider a simple example to understand where the computational challenge arises. Consider the following example:

$$(P) \quad \max \quad 3x_1 + 4x_2$$
$$\text{s.t.} \quad 5x_1 + 8x_2 \leq 24,$$
$$x_1, x_2 \geq 0 \quad \text{and integer.}$$

What is the optimal solution for this problem?

Notice that the mathematical representation is very similar to the corresponding LP with the added constraint that both x_1 and x_2 must be restricted to integral values. In Fig. 11.12 we represent the feasible region of this problem.

It is noteworthy that the LP relaxation, i.e., when we ignore the integrality restriction on both the decision variables, is the entire gray region included in the triangle. However, the integer formulations must restrict any solution to the lattice points within the LP feasible region. This "smallest" polyhedral set including all the lattice points is sometimes referred to as the **convex hull** of the integer programming problem. It is readily observable that the LP relaxation includes the integer programming problem's (IP) feasible region (**convex hull**) as a strict subset. One may argue that it should be possible to solve the LP relaxation of (P) and then simply round up or round down the optimal decision variables appropriately.

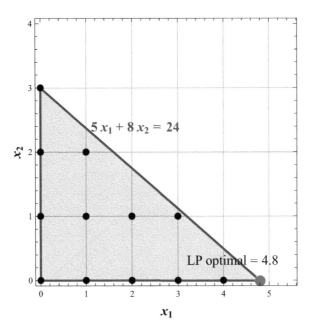

Fig. 11.13 LP solution to (P)—it's not integral valued

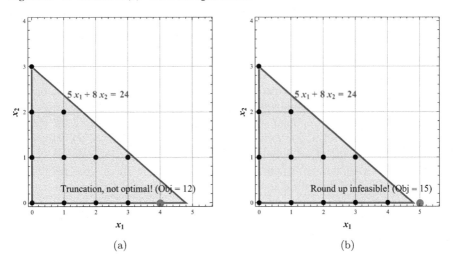

Fig. 11.14 Finding an integer solution by truncating or rounding-up a LP solution may not work. (**a**) Truncating (not optimal). (**b**) Rounding-up (infeasible)

But as Fig. 11.13 illustrates the corner point, at which the LP will always find its optimal solution, need not be integer valued. In this examples the LP relaxation optimal value is found at the vertex (4.8,0). As Fig. 11.14 shows, truncating or rounding-up the optimal LP solution doesn't provide the integer optimal solution.

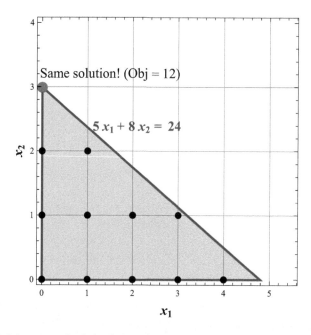

Fig. 11.15 Had the truncated solution been optimal, the LP would have found it at another corner point! That's why it is not optimal

While rounding-up renders the solution infeasible, had the truncated solution been optimal, the LP would have found at another corner point as shown in Fig. 11.15.

In this simple example, it turns out that the IP optimal solution is in the interior of the LP feasible region as shown in Fig. 11.16.

Thus, finding the IP optimal solution is much harder than looking for optimal solution of the LP relaxation (which is guaranteed to be found at a corner point of the LP polyhedra if an optimal solution exists) because the solution can lie in the interior of the corresponding LP feasible region. If it were possible to get an accurate mathematical (polyhedral) description of the **convex hull** using linear constraints, then one could solve the resulting problem (after including these additional constraints) as a LP and guarantee that the LP corner point solution would indeed be optimal to the IP too. However, there is no known standard technique to develop these constraints systematically for any IP and get an accurate mathematical description of the convex hull. Developing such constraints are largely problem specific and tend to exploit the specific mathematical structure underlying the formulation.

So what carries over from LPs to IPs (or MILPs)? The idea of feasibility is unchanged. One can define and compute shadow price in an analogous fashion. The linear relaxation to an integer problem provides a bound on the attainable solution but does not say anything about the feasibility of the problem. There is no way to verify optimality from the solution—instead one must rely on other methods to

11 Optimization

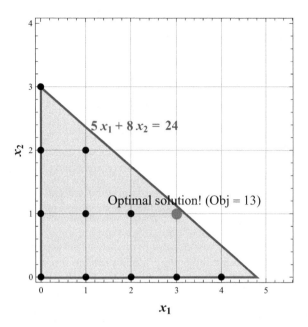

Fig. 11.16 Unfortunately, the IP optimal solution is not at a "corner" point of the original LP feasible region. It is much harder to find

verify optimality. Search methods are used but not going from vertex to vertex! That's why IPs are so hard to solve. Of course, this is not to imply that problems shouldn't be modeled and solved as IPs. Today's state-of-the-art algorithms are very efficient in solving large instances of IPs to optimality but, unlike the case of LPs, guaranteeing optimality is not possible in general. A detailed study of the theory and practice of integer programming can be found in Bertsimas and Tsitsiklis (1997), Nemhauser and Wolsey (1988), and Schrijver (1998).

Next we briefly illustrate a basic **branch-and-bound** solution technique to solve IPs.

3.1 The Branch-and-Bound Method

The basic idea behind the naive branch-and-bound (B&B) method is that of *divide and conquer*. Notice that the feasible region of the LP relaxation of an IP, i.e., when we ignore the integrality constraints, is always larger than that of the feasible region of the IP. Consequently, any optimal solution to the LP relaxation provides a bound on the optimal IP value. In particular, for a minimization problem the LP relaxation will result in a lower bound and for a maximization problem it will result in a upper bound. If $Z^\star_{LP}$ denotes the optimal objective value of the LP relaxation and $Z^\star_{IP}$ denotes the optimal solution to the IP, then

$$Z^\star_{LP} \geq Z^\star_{IP} \quad \text{for a maximization problem, and}$$
$$Z^\star_{LP} \leq Z^\star_{IP} \quad \text{for a minimization problem.}$$

The B&B method divides the feasible region (partitions it) and solves for the optimal solution over each partition separately. Suppose F is the feasible region of the IP and we wish to solve $\min_{x \in F} \mathbf{c}'\mathbf{x}$. Consider a partition $F_1, \ldots F_k$ of F. Recollect, a partition implies that the subsets are collectively exhaustive and mutually exclusive, i.e.,

$$F_i \bigcap F_j = \emptyset \quad \text{and} \quad \bigcup_{i=1}^{k} F_i = F.$$

Then, for a minimization problem (equivalently for a maximization problem),

$$\min_{x \in F} \mathbf{c}'\mathbf{x} = \min_{1 \leq i \leq k} \left\{ \min_{x \in F_i} \mathbf{c}'\mathbf{x} \right\}.$$

In other words, we optimize over each subset separately. The idea hinges on the fact that if we can't solve the original problem directly, we might be able to solve the smaller subproblems recursively. Dividing the original problem into subproblems is the idea of **branching**. As is readily observable, a naive implementation of the B&B is equivalent to complete enumeration and can take a arbitrarily long time to solve.

To reduce the computational time most B&B procedures employ an idea called **pruning**. Suppose we assume that each of our decision variables have finite upper and lower bounds (not an unreasonable assumption for most business problems). Then, any feasible solution to our minimization problem provides an upper bound $u(F)$ on the optimal IP objective value.[3] Now, after branching, we obtain a lower bound $b(F_i)$ on the optimal solution for each of the subproblems. If $b(F_i) \geq u(F)$, then we don't need to consider solving the subproblem i any further. This is because we already have a solution better than any that can be found in partition F_i. One typical way to find the lower bound $b(F_i)$ is by solving the LP relaxation. Eliminating exploring solution in a partition by creating an appropriate bound is called **pruning**. The process of iteratively finding better values of $b(F_i)$ and $u(F)$ is called **bounding**. Thus, the basic steps in a **LP-based B&B procedure** involve:

LP relaxation: first solve the LP relaxation of the original IP problem. The result is one of the following:

1. The LP is infeasible $\implies$ IP is infeasible.
2. The LP is feasible with an integer solution $\implies$ Optimal solution to the IP.
3. LP is feasible but has a fraction solution $\implies$ Lower bound for the IP.

In the first two cases of step 1, we are done. In the third case, we must branch and recursively solve the resulting subproblems.

Branching: The most common way to branch is as follows: Select a variable i whose value x_i is fractional in the LP solution. Create two subproblems: in

---

[3] Typically, we could employ a heuristic procedure to obtain an upper bound to our problem.

one subproblem, impose the constraint $x_i \geq \lceil x_i \rceil$. In the other subproblem, impose the constraint $x_i \leq \lfloor x_i \rfloor$. This is called a **branching rule** (it is the simplest branching rule). Notice that doing so creates two subproblems yet does not eliminate any integer feasible solutions to the original problem. Hence, this branching rule is **valid**, i.e., the constraints generated are **valid inequalities**.

Pruning: After branching we solve the subproblems recursively. Now we consider the following: if the optimal objective value of the LP relaxation is greater than the current upper bound, we need not consider the current subproblem further (pruning), that is, if $Z_{LP}^{i\star} > Z_{IP}$, then prune subproblem i. This is the key to the potential efficiency of the problem.

Before we summarize the steps of the B&B algorithm, we describe some implementation terminology. If we picture the subproblems graphically, they form a **search tree**. Each subproblem is linked to its **parent** and eventually to its **children**. Eliminating a problem from further consideration is called pruning. The act of bounding and then branching is called **processing**. A subproblem that has not yet been considered is called a **candidate** for processing. The set of candidates for processing is called the **candidate list**. Using this terminology, the LP-based B&B procedure (for a minimization problem) can be summarized as follows:

1. To begin, we find an upper bound U using a preprocessing/heuristic routine.
2. We start with the original problem on the candidate list.
3. Select problem S from the candidate list and solve the LP relaxation to obtain the lower bound $b(S)$.

    (a) If LP is infeasible $\implies$ candidate is pruned.
    (b) Otherwise, if $b(S) \geq U \implies$ candidate is pruned.
    (c) Otherwise, if $b(S) < U$ and the solution is feasible for the IP $\implies$ update $U \leftarrow b(S)$.
    (d) Otherwise, *branch* and add the new subproblem to the candidate list.

4. If the candidate list is nonempty, go to step 2. Otherwise, the algorithm is done.

There are several ways to select a candidate in step 2. The **best-first** technique chooses a candidate with the lowest lower bound. Other possibilities are to use a **depth-first** or **breadth-first** technique. The depth-first technique is most common. The depth-first and breadth-first techniques differ in the way the B&B search tree is traversed to select the next candidate to the explored. The reader is referred to Nemhauser and Wolsey (1988) for details on the search procedures. Detail of the B&B algorithm and other procedures such as the cutting plane algorithm can also be found in Wolsey (1998). It is noteworthy that most commercial solvers build on the basic B&B procedure described here and combine it with generating constraints automatically, reducing the number of binary/integer variables used, and using pre- and postprocessing heuristics to generate "good" (from a business implementation standpoint) and feasible IP solutions quickly. For a partial list of LP and IP solvers available to solve large IPs, the reader is referred to https://neos-server.org/neos/solvers/index.html (accessed on Jul 22, 2018).

3.2 A B&B Solution Tree Example

To illustrate the implementation of the LP-based B&B algorithm, consider the following binary variable problem:

$$(P1) \max \quad 8x_1 + 11x_2 + 6x_3 + 4x_4,$$
$$\text{s.t.} \quad 5x_1 + 7x_2 + 4x_3 + 3x_4 \leq 14,$$
$$x_i \in \{0, 1\} \quad \forall \, i = 1, \ldots, 4.$$

The linear relaxation of optimal solution is $\mathbf{x}^\star = \{1, 1, , 0.5, 0\}$ with an objective value of 22. Notice that this solution is not integral. So we choose a fractional variable to branch on, which is x_3. Essentially, we generate two subproblems: one with the constraints $x_3 = 0$ and the other with the constraint $x_3 = 1$. We illustrate the entire B&B solution tree in Fig. 11.17.

The solution tree shows the LP-relaxation upper bounds (since this is maximization problem) at each node (subproblem) and the variables that were branched on at each iteration (these are the fractional valued variables in the LP solution at that node). The integer valued solutions are marked in red, which provide the lower bounds. We employ the depth-first search process to select candidates to solve

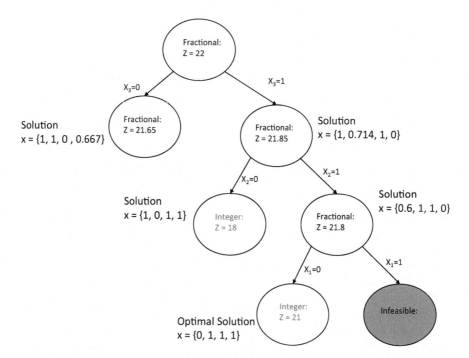

Fig. 11.17 LP-based B&B solution tree for problem P1

iteratively and always choose the "right-side" child candidate. Thus, we begin by branching on x_3 followed by x_2 and then x_1. Notice that by fixing $x_1 = x_2 = x_3 = 1$ we arrive at an infeasible solution at the rightmost node of the tree (fourth level). However, the left child candidate at the same level, i.e., when $x_3 = x_2 = 1$ and $x_1 = 0$, gives us an integer feasible solution with objective value $Z_{IP} = 21$. This is the best IP lower bound solution we have so far—our incumbent IP solution (0,1,1). Now, when we step one level higher to explore the node when $x_3 = 1$ and $x_2 = 0$, we get a LP solution with an objective value 18 (also happens to be integer valued), which is lesser than 21, our incumbent IP solution. Hence, we prune the sub-tree (not shown in the figure) rooted at that node (where the optimal objective value is 18). Similarly, we don't need to explore the sub-tree to the left of the root node, i.e., when we fix $x_3 = 0$ because that sub-tree can never get us a better integer solution than what we already have with our incumbent solution.

4 Methods in Optimization: Nonlinear Optimization Models

A vast majority of problems in real business applications are essentially nonlinear in nature. In fact, linear programming models are a subset of nonlinear models. One may also consider LP models to be an approximation of the real problem. In this section, we discuss a few examples of nonlinear optimization models in statistics, econometrics, and data analytics. However, we do not discuss the algorithmic details of the nonlinear solution techniques. For a detailed discussion on the theory and application of nonlinear programming we refer the readers to Bazaraa et al. (2013), Bertsekas (1999), and Boyd and Vandenberghe (2004). We begin by illustrating the use of optimization in simple linear regression.

4.1 Estimating Coefficients in an Ordinary Least Squares Regression

Optimization plays a very important role in the estimation of coefficients in linear regression models. Consider the method of ordinary least squares (OLS) where, using sample data, we wish to estimate the parameters of a linear relationship between the dependent (**response**) variable $\mathbf{Y} = \langle Y_1, \ldots, Y_n \rangle$ and the corresponding independent (**predictor**) variable $\mathbf{X} = \langle X_1, \ldots, X_n \rangle$, i.e., using the sample observations we try to fit a linear relationship:

$$y_i = \hat{\beta}_1 + \hat{\beta}_2 x_i + \hat{\varepsilon}_i \quad \forall \, i = 1, \ldots, n. \tag{11.14}$$

In (11.14), ε_i is the random error (residual) associated with the ith observation and $\left(\hat{\beta}_1, \hat{\beta}_2\right)$ are unbiased estimates of the (true) parameters of the linear function (β_1, β_2). Alternately, the relationship can be expressed as $Y_i = \mathbb{E}[Y \mid x_i] + \varepsilon_i$, where $\mathbb{E}[Y \mid x_i] = \beta_1 + \beta_2 x_i$ is the conditional expectation of all the responses, Y_i, observed when the predictor variable takes a value x_i. It is noteworthy that capital letters indicate random variables and small letters indicate specific values (instances). For example, suppose we are interested in computing the parameters of a linear relationship between a family's weekly income level and its weekly expenditure. In this case, the weekly income level is the predictor (x_i) and the weekly expense is the response (y_i). Figure 11.18a shows a sample of such data collected, i.e., sample of weekly expenses at various income levels. The scatterplot in Fig. 11.18b shows the fitted OLS regression line.

In order to construct the unbiased estimates $\left(\hat{\beta}_1, \hat{\beta}_2\right)$ OLS involves minimizing the sum of squared errors, i.e.,

$$\min \sum_{i=1}^{n} \varepsilon_i^2 = \sum_{i=1}^{n} \left(y_i - \hat{\beta}_1 - \hat{\beta}_2 x_i\right)^2. \quad (11.15)$$

	X (Income level)								
	80	100	120	140	160	180	200	220	240	260
Y (Expense)	55	65	79	80	102	110	120	135	137	150
	60	70	84	93	107	115	136	137	145	152
	65	74	90	95	110	120	140	140	155	175
	70	80	94	103	116	130	144	152	165	178
	75	85	98	108	118	135	145	157	175	180
		88		113	125	140		160	189	185
				115				162		191
Total	325	462	445	707	678	750	685	1043	966	1211
E[Y|X]	65	77	89	101	113	125	137	149	161	173

(a)

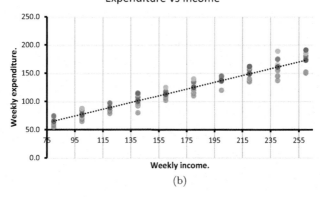

(b)

Fig. 11.18 Fitting an OLS regression line. (**a**) Weekly expenditure at various income levels. (**b**) Weekly expenditure as a function of weekly expense

For a give sample, notice that $\left[\sum_{i=1}^{n} y_i\right]$, $\left[\sum_{i=1}^{n} y_i x_i\right]$, $\left[\sum_{i=1}^{n} x_i^2\right]$, $\left[\sum_{i=1}^{n} y_i^2\right]$ are constants. Hence, (11.15) is a simple quadratic function of the parameters, i.e.,

$$-2\hat{\beta}_1 \left[\sum_{i=1}^{n} y_i\right] - 2\hat{\beta}_2 \left[\sum_{i=1}^{n} y_i x_i\right] + \left(\hat{\beta}_2\right)^2 \left[\sum_{i=1}^{n} x_i^2\right] + \left[\sum_{i=1}^{n} y_i^2\right]$$

and represents an unconstrained quadratic optimization problem. There are two ways of solving this problem. One, we can use differential calculus and solve it by setting derivatives to equal zero. We get what are known as normal equations (see Chap. 7 on linear regression). Gujarati (2009) also provides a detailed description of the analytical solution to this nonlinear parameter estimation optimization problem. Two, we can use a descent method as follows:

Step 1: We start with an initial solution (may be computed using a heuristic approach).
Step 2: We then find a value improving direction and move along that direction by changing $\hat{\beta}_1$ and $\hat{\beta}_2$, slightly.
Step 3: Repeat the steps 1 and 2, until the gain from such a move is very small (**stopping criteria**).

Like in LPPs, this problem does not have local optimal solutions—in other words, once we are unable to improve the solution we know we are at or close to the global optimal solution.

4.2 Estimating Coefficients Using Maximum Likelihood Estimation

As described in Sect. 4.1 we consider a two-variable model:

$$y_i = \hat{\beta}_1 + \hat{\beta}_2 x_i + \varepsilon_i$$

where Y_i is the response variable and x_i is the predictor variable. The method of maximum likelihood estimation (MLE), like the OLS method, helps us estimate the linear regression parameters $\left(\hat{\beta}_1, \hat{\beta}_2\right)$. In the MLE approach, we assume that the sample collected is made of independent and identically distributed observations (y_i, x_i) and that the error terms follow a normal distribution with mean zero and variance σ^2. This implies that Y_i are normally distributed with mean $\beta_1 + \beta_2 x_i$ and variance σ^2. Consequently, the joint probability density function of $Y_1, \ldots, Y_n$ can be written as

$$f\left(Y_1, \ldots, Y_n \mid \beta_1 + \beta_2 x_i, \sigma^2\right).$$

But given that the sample points are drawn independently, we express the joint probability density function as a product of the individual density functions as

$$f\left(Y_1, \ldots, Y_n \mid \beta_1 + \beta_2 x_i, \sigma^2\right)$$
$$= f\left(Y_1 \mid \beta_1 + \beta_2 x_i, \sigma^2\right) f\left(Y_2 \mid \beta_1 + \beta_2 x_i, \sigma^2\right) \cdots f\left(Y_n \mid \beta_1 + \beta_2 x_i, \sigma^2\right)$$

where

$$f(Y_i) = \frac{1}{\sigma\sqrt{2\pi}} e^{\left[-\frac{1}{2}\frac{(Y_i - \beta_1 - \beta_2 x_i)^2}{\sigma^2}\right]}$$

which is the density function of a normally distributed random variable. For given values of the response variable the likelihood function, $\text{LF}\left(\hat{\beta}_1, \hat{\beta}_2, \sigma^2\right)$, is written as

$$\text{LF}\left(\hat{\beta}_1, \hat{\beta}_2, \sigma^2\right) = \frac{1}{\sigma^n \left(\sqrt{2\pi}\right)^n} e^{\left[-\frac{1}{2}\sum_{i=1}^{n}\frac{(Y_i - \hat{\beta}_1 - \hat{\beta}_2 x_i)^2}{\sigma^2}\right]}.$$

The method of MLE computes $\left(\hat{\beta}_1, \hat{\beta}_2\right)$ such that the probability of observing the given $\mathbf{y} = \langle y_1, \ldots, y_n \rangle$ is maximum (as high as possible.) Notice that this is a nonlinear optimization problem that maximizes the likelihood function over $\hat{\beta}_1$ and $\hat{\beta}_2$. One natural way to solve this problem is to convert LF function into its log form, i.e.,

$$\ln \text{LF}\left(\hat{\beta}_1, \hat{\beta}_2, \sigma^2\right) = -n \ln \sigma - \frac{n}{2} \ln (2\pi) - \frac{1}{2} \sum_{i=1}^{n} \frac{\left(Y_i - \hat{\beta}_1 - \hat{\beta}_2 x_i\right)^2}{\sigma^2},$$

$$= -\frac{n}{2} \ln \sigma^2 - \frac{n}{2} \ln (2\pi) - \frac{1}{2} \sum_{i=1}^{n} \frac{\left(Y_i - \hat{\beta}_1 - \hat{\beta}_2 x_i\right)^2}{\sigma^2}.$$

Maximizing the log-likelihood function is a simple unconstrained quadratic optimization problem (just as seen in Sect. 4.1). Once again we refer the readers to Gujarati (2009) for details about the analytical solution to this optimization problem. MLE is also covered in detail in Chap. 8 (advanced regression and missing data) of this book.

4.3 The Logit Model for Classification

Unlike the case discussed in Sect. 4.1, sometimes we encounter situations wherein the response variables take binary outcomes. For example, consider a binary model, in which x_i (predictor) is the price of a product and y_i (response) is whether a customer purchased a product. In this case, the response variable $y_i \in \{0, 1\}$. Fitting an OLS regression model, in this case, may not be appropriate because the response variable must be restricted to the interval $[0, 1]$ and there exists no such restriction in the standard linear regression model. Instead, we use a binary outcome model that tries to estimate the conditional probability that $y_i = 1$ as a function of the independent variable, i.e., $\Pr\{Y_i = 1 \mid x_i\} = F\left(\hat{\beta}_1 + \hat{\beta}_2 x_i\right)$, where the function $F(\cdot)$ represents the cumulative density function of a probability distribution. One common model used is the **logit** model, where $F(\cdot)$ is the logistic distribution function, i.e.,

$$F\left(\hat{\beta}_1 + \hat{\beta}_2 x_i\right) = \frac{e^{\left[\hat{\beta}_1 + \hat{\beta}_2 x_i\right]}}{1 + e^{\left[\hat{\beta}_1 + \hat{\beta}_2 x_i\right]}}.$$

Assuming that the observations in the sample data are independent of each other, the conditional likelihood of seeing the n outcomes in our sample data is given by

$$\prod_{i=1}^{n} \Pr\{Y = y_i \mid x_i\} = \prod_{i=1}^{n} F\left(\hat{\beta}_1 + \hat{\beta}_2 x_i\right)^{y_i} \times \left[1 - F\left(\hat{\beta}_1 + \hat{\beta}_2 x_i\right)\right]^{(1-y_i)}$$

because $y_i \in \{0, 1\}$ and $\langle y_1, \ldots, y_n \rangle$ represents a sequence of Bernoulli trials. As described in Sect. 4.2, a natural way to solve this problem is to convert the likelihood function into its log form, i.e.,

$$\ln \mathrm{LF}\left(\hat{\beta}_1, \hat{\beta}_2\right) = \sum_{i=1}^{n} y_i \ln F\left(\hat{\beta}_1 + \hat{\beta}_2 x_i\right) + \sum_{i=1}^{n} (1 - y_i) \ln\left[1 - F\left(\hat{\beta}_1 + \hat{\beta}_2 x_i\right)\right],$$

$$= \sum_{i=1}^{n} y_i \ln\left[\frac{e^{\left[\hat{\beta}_1 + \hat{\beta}_2 x_i\right]}}{1 + e^{\left[\hat{\beta}_1 + \hat{\beta}_2 x_i\right]}}\right] + \sum_{i=1}^{n} (1 - y_i) \ln\left[\frac{1}{1 + e^{\left[\hat{\beta}_1 + \hat{\beta}_2 x_i\right]}}\right],$$

$$= \sum_{i=1}^{n} y_i \left[\hat{\beta}_1 + \hat{\beta}_2 x_i - \ln\left(1 + e^{\left[\hat{\beta}_1 + \hat{\beta}_2 x_i\right]}\right)\right]$$

$$- \sum_{i=1}^{n} (1 - y_i) \left(\ln\left(1 + e^{\left[\hat{\beta}_1 + \hat{\beta}_2 x_i\right]}\right)\right),$$

$$= \sum_{i=1}^{n} y_i \left[\hat{\beta}_1 + \hat{\beta}_2 x_i\right] - \sum_{i=1}^{n} \ln\left(1 + e^{\left[\hat{\beta}_1 + \hat{\beta}_2 x_i\right]}\right).$$

The optimization problem reduces to choosing the parameters (coefficients) $\hat{\beta}_1$ and $\hat{\beta}_2$ to maximize the log-likelihood function, $\ln \text{LF}\left(\hat{\beta}_1, \hat{\beta}_2\right)$. This is a nonlinear optimization problem but cannot be solved analytically using standard differential calculus. We may have to resort to approximately solving it numerically (e.g., see Newton's method in Bazaraa et al., 2013). It is noteworthy that this type of formulation can be used for making **multi-class predictions/classifications**, where Y can take on more than two values (not just binary). See Chaps. 15, 16, and 17 on machine learning techniques for discussion on these types of problems. Several specialized algorithms have been developed to solve this problem efficiently. Moreover, it is somewhat straightforward to connect this to a machine learning problem! The multi-class prediction can be seen to be equivalent to a single-layer neural network using softmax loss function (see Chaps. 16 and 17 on Supervised Learning and Deep Learning). The connection between learning and optimization is an advanced topic well worth pursuing.

As described in this section, the techniques and solution methodologies for solving nonlinear optimization problems can be varied. For a partial list of algorithmic procedures to solve nonlinear problems the reader is referred to https://neos-guide.org/algorithms (accessed on Jul 22, 2018).

5 Discussion

In this chapter, we touched upon the basics of optimization. In particular, we focused on formulating, solving, and interpreting solutions of LPPs. LPs have been used in a large number of business and scientific applications over the last few decades. It is important to understand that while the LP methodology is very efficient and easy to model, there are larger classes of optimization techniques that help model business and scientific applications even more closer to reality, integer programming being one of them. Finally, we briefly described nonlinear optimization models and showed a few examples that are closely related to often-used econometric models.

For a complete taxonomy of the types of mathematical programs/optimization techniques encountered in theory and practice, we refer the readers to NEOS guide.[4] With data sciences, machine learning, and analytics gaining importance, the use of LPs (and optimization methods in general) will continue to grow. In a sense, optimization models will eventually become ubiquitous.

---

[4]https://neos-guide.org/content/optimization-taxonomy (accessed on Jul 22, 2018).

Appendix

Spreadsheet Models and Excel Solver Reports

There are a few online tutorials available to understand how to input a LP model in Solver. The two commonly used websites are Solver[5] and Microsoft support[6] page. This section describes the various fields in the LP reports generated by Microsoft Solver and how to locate the information related to shadow prices, reduced costs, and their ranges after the model has been solved. We use the prototype example referred earlier to describe these reports.

The Answer Report

Figure 11.19 shows the **answer** report generated by Excel Solver for our prototype problem. We describe the entries in this report.

Target Cell The initial value of the objective function (to be maximized or minimized), and its final optimal objective value.

Adjustable Cells The initial and final values of the decision variables.

Constraints Maximum or minimum requirements that must be met, whether they are met just barely (binding) or easily (not binding), and the values of the slacks (excesses) leftover. Binding constraints have zero slacks and nonbinding ones have positive slacks.

Target Cell (Max)

Cell	Name	Original Value	Final Value
$K$34	PROFIT Z	0	5142.86

Adjustable Cells

Cell	Name	Original Value	Final Value
$G$33	Objective: max 500 x1 + 450 x2 X1	0.00	6.43
$H$33	Objective: max 500 x1 + 450 x2 X2	0.00	4.29

Constraints

Cell	Name	Cell Value	Formula	Status	Slack
$I$36	Production	60.000	$I$36<=$K$36	Binding	0
$I$37	Storage	150.000	$I$37<=$K$37	Binding	0
$I$38	Demand	6.429	$I$38<=$K$38	Not Binding	1.571428571

Fig. 11.19 Answer report

---

[5]https://www.solver.com/excel-solver-online-help (accessed on Jul 22, 2018).
[6]https://support.office.com/en-us/article/define-and-solve-a-problem-by-using-solver-5d1a388f-079d-43ac-a7eb-f63e45925040 (accessed on Jul 22, 2018).

Adjustable Cells

Cell	Name	Final Value	Reduced Cost	Objective Coefficient	Allowable Increase	Allowable Decrease
$G$33	Objective: max 500 x1 + 450 x2 X1	6.43	0.00	500	40	275
$H$33	Objective: max 500 x1 + 450 x2 X2	4.29	0.00	450	550	33.33

Constraints

Cell	Name	Final Value	Shadow Price	Constraint R.H. Side	Allowable Increase	Allowable Decrease
$I$36	Production	60.000	78.571	60	5.5	22.5
$I$37	Storage	150.000	2.857	150	90	22
$I$38	Demand	6.429	0.000	8	1E+30	1.57

Fig. 11.20 Sensitivity report (shadow prices and validity range)

The Sensitivity Report

Figure 11.20 shows the **sensitivity** report generated by Excel Solver for our prototype problem. Below we describe the entries in this report.

Adjustable Cells The decision variables, their cell addresses, names, and optimal values.

Reduced Cost This relates to decision variables that are bounded, from below (such as by zero in the nonnegativity requirement), or from above (such as by a maximum number of units that can be produced or sold). Recollect:

1. A variable's **reduced cost** is the amount by which the optimal objective value will change if that bound was relaxed or tightened.
2. If the optimal value of the decision variable is at its specified upper bound, the reduced cost is the amount by which optimal objective value will improve (go up in a maximization problem or go down in a minimization problem) if we relaxed the upper bound by increasing it by one unit.
3. If the optimal value of the decision variable is at its lower bound, its reduced cost is the amount by which the optimal objective value will be hurt (go down in a maximization problem or go up in a minimization problem) if we tightened the bound by increasing it by one unit.

Objective Coefficient The unit contribution of the decision variable to the objective function (unit profit or cost).

Allowable Increase and Decrease The amount by which the coefficient of the decision variable in the objective function can change (increase or decrease) before the optimal solution (the values of decision variables) changes. As long as an objective coefficient changes within this range, the current optimal solution (i.e., the values of decision variables) will remain optimal (although the value of the objective function optimal objective value will change as the objective coefficient changes, even within the allowable range).

Shadow Price Recollect:

1. The shadow price associated with each constraints measures the amount of change in the optimal objective value optimal objective value that would result from changing that constraint by a small amount.
2. In general, it is the increase in optimal objective value resulting from an increase in the right-hand side of that constraint.
3. Its absolute value measures the *marginal* (or incremental) *improvement* in optimal objective value (i.e., an increase in the maximum profit or a decrease in the minimum cost) if that constraint was *relaxed* (i.e., if the lower limit was reduced or the upper limit was increased) by one unit. Similarly, it is the *marginal degradation* in optimal objective value (i.e., if the lower limit was raised or the upper limit was reduced) by one unit. For example, if the constraint represents limited availability of a resource, its shadow price is the amount by which the optimal profit will increase if we had a little more of that resource and we used it in the best possible way. It is then the maximum price that we should be willing to pay to have more of this resource. Equivalently, it is the *opportunity cost* of not having more of that resource.

Allowable Increase and Decrease Recollect:

1. This is the amount by which the constraint can be relaxed or tightened before its shadow price changes. if the constraint imposes an upper limit, and it is relaxed by increasing this limit by more than the "allowable increase," the optimal objective value will still improve but at a lower rate, so the shadow price will go down below its current value. Similarly, if the upper limit on the constraint is decreased by more than the "allowable decrease," the optimal objective value will degrade at an even higher rate and its shadow price will go up.
2. If the constraint imposes a lower limit and that constraint is relaxed by decreasing the limit by more than the "allowable decrease," the optimal objective value will still improve but only at a lower rate and the shadow price will decrease. If, on the other hand, the lower limit is increased by more than the "allowable increase," the constraint becomes tighter, the optimal objective value will degrade faster, and the shadow price will increase. Thus, there are decreasing marginal benefits to relaxing a constraint, and increasing marginal costs of tightening a constraint.

It should be noted that all of the information in the sensitivity report assumes that only one parameter is changed at a time. Thus, the effects of relaxing or tightening two constraints or changing the objective coefficients of two decision variables cannot be determined from the sensitivity report. Often, however, if the changes are small enough to be within the respective allowable ranges, the total effect can be determined by simply adding the individual effects.

In an **Excel report** degeneracy can be spotted by looking at the rhs values of any of the constraints. **If the constraints (for the range over which the optimal shadow price is valid) have an allowable increase or allowable decrease of**

zero, then the LP is degenerate. One has to be careful while interpreting optimal solutions for degenerate LPs. For example:

1. When a solution is degenerate, the reduced costs may not be unique. Additionally, the objective function coefficients for the variable cells must change by at least as much (and possibly more than) their respective reduced costs before the optimal solution would change.
2. Shadow prices and their ranges can be interpreted in the usual way, but they are not unique. Different shadow prices and ranges may apply to the problem (even if the optimal solution is unique).

Exercises

Ex. 11.1 (LP Modeling)

Ex. 11.1.1 Retail Outlet Staffing

Consider a retail shop that is open 7 days a week. Based on past experience, the number of workers needed on a particular day is given as follows:

Day	Mon	Tue	Wed	Thu	Fri	Sat	Sun
Number	15	12	17	16	19	14	11

Every employee works five consecutive days and then takes off two days, repeating this pattern indefinitely. Our goal is to minimize the number of employees that staff the outlet. Define your variables, constraints, and objective function clearly.
Develop a Solver model and solve for the optimal staffing plan.

Ex. 11.1.2 Managing a Portfolio

We are going to manage an investment portfolio over a 6-year time horizon. We begin with ₹1,000,000, and at various times we can invest in one or more of the following:

(a) Savings account X, annual yield 6%
(b) Security Y, 2-year maturity, total yield 14% if bought now, 12% thereafter
(c) Security Z, 3-year maturity, total yield 17%
(d) Security W, 4-year maturity, total yield 22%

To keep things simple we will assume that each security can be bought in any denomination. We can make savings deposits or withdrawals anytime. We can buy Security Y any year but year 3. We can buy Security Z anytime after the first year. Security W, now available, is a one-time opportunity. Write down a LP model to maximize the final investment yield. Assume all investments must mature on or before year 6 and you cannot sell securities in between. Define your decision variables and constraints clearly.

Ex. 11.2 (Interpreting the Sensitivity Report)

AC manufactures two television models, Astros and Cosmos. Each Astro set sells for $300 and each Cosmo sells for $250 a set. AC purchases components for an Astro set for $260 and components for a Cosmo set cost $190.

Production of each model involves circuit board fabrication, picture tube construction, and chassis assembly. There are two separate and completely automated lines for circuit board fabrication, one for Astro and one for Cosmo. However, the picture tube and chassis assembly departments are shared in the production of both sets.

The capacity of the Astro circuit board fabrication line is 70 sets per day, while that of the Cosmo line is 50 sets per day. The picture tube department has 20 workstations, while the chassis department has 16 workstations. Each workstation can process one TV set at a time and is operated 6 h a day. Each Astro set takes 1 h for chassis assembly and 1 h for tube production. Each Cosmo set requires 2 h for picture tube production and 1 h for chassis assembly.

Workers in the picture tube and chassis assembly departments are paid $10 an hour. Heating, lighting, and other overhead charges amount to $1000 per day.

1. How many Astros and Cosmos should AC produce each day? What will be the maximum profit?
2. How should they allocate the available resources among the two models? Where are the bottlenecks?
3. Suppose that due to raw material shortage AC could make only 30 circuit boards for Cosmos each day. What will be the effect on their operation?
4. Suppose workers in the picture tube department are willing to work overtime for a premium of $21 an hour. How many hours of overtime, if any, should they employ? How will they use it?
5. If a workstation in the picture tube department breaks down, how will it affect AC's production and profit?
6. If a chassis assembly workstation breaks down, how will it affect AC's production plan and profit?
7. How much would you be willing to pay to increase Astro's circuit board capacity?
8. Suppose AC has developed a new model that uses same circuit boards as a Cosmos and requires 3 h of the picture tube time. If its profit margin is expected to be a high $42, should they produce it?
9. If the profit margin on Astro goes up to $40 a set, how would it affect the firm's production plan and the daily profit? What if it goes down by $10 a set?
10. How much must Cosmos's price increase before you will consider producing more Cosmos?

Ex. 11.3 (Modeling with Binary Variables)

1. Consider the knapsack set $X_1 = \{x_1, \} \ldots, x_5 \in \{0, 1\} : 3x_1 - 4x_2 + 2x_3 - 3x_4 + x_5 \leq 2$. Is the constraint $x_2 + x_4 \geq 1$ a valid inequality? Why or why not? *Note*: Suppose we formulate an integer program by specifying a rational

polyhedron $P = \{\mathbf{x} \in \mathbb{R}_+^n \,|\, \mathbf{Ax} \leq \mathbf{b}\}$ and define $S = \mathbb{Z}^n \cap P$, where $\mathbb{Z}_+^n$ is the n-dimensional set of nonnegative integers. Thus, $S = \{\mathbf{x} \in \mathbb{Z}_+^n \,|\, \mathbf{Ax} \leq \mathbf{b}\}$ and conv(S) is the convex hull of S, i.e., the set of points that are convex combinations of points in S. Note that conv$(S) \subseteq S$; "Ideal" if conv$(S) = S$. An inequality $\pi^T \mathbf{x} \leq \pi_0$ is called a *valid inequality* if it is satisfied by all points in S.
2. Solve using the branch-and-bound (B&B) algorithm. Draw the B&B tree, show your branches, LP solutions, lower and upper bounds. You may simply branch in sequence x_1 followed by x_2 and so on.

$$\max 9x_1 + 3x_2 + 5x_3 + 3x_4$$
$$\text{s.t. } 5x_1 + 2x_2 + 5x_3 + 4x_4 \leq 10$$
$$x_1, \ldots, x_4 \in \{0, 1\}.$$

References

Bazaraa, M. S., Jarvis, J. J., & Sherali, H. D. (2011). *Linear programming and network flows.* Hoboken, NJ: Wiley.
Bazaraa, M. S., Sherali, H. D., & Shetty, C. M. (2013). *Nonlinear programming: Theory and algorithms.* Hoboken, NJ: Wiley.
Bertsekas, D. P. (1999). *Nonlinear programming.* Belmont, MA: Athena Scientific.
Bertsimas, D., & Tsitsiklis, J. N. (1997). *Introduction to linear optimization* (Vol. 6). Belmont, MA: Athena Scientific.
Boyd, S., & Vandenberghe, L. (2004). *Convex optimization.* Cambridge: Cambridge University Press.
Bradley, S. P., Hax, A. C., & Magnanti, T. L. (1977). Applied mathematical programming.
Chvátal, V. (1983). *Linear programming.* New York: WH Freeman.
Gujarati, D. N. (2009). *Basic econometrics.* New York: Tata McGraw-Hill Education.
Luenberger, D. G., & Ye, Y. (1984). *Linear and nonlinear programming* (Vol. 2). Berlin: Springer.
Nemhauser, G. L., & Wolsey, L. A. (1988). *Interscience series in discrete mathematics and optimization: Integer and combinatorial optimization.* Hoboken, NJ: Wiley.
Schrijver, A. (1998). *Theory of linear and integer programming.* Chichester: Wiley.
Wagner, H. M. (1969). *Principles of operations research: With applications to managerial decisions.* Upper Saddle River, NJ: Prentice-Hall.
Wolsey, L. A. (1998). *Integer programming.* New York: Wiley.

Chapter 12
Forecasting Analytics

Konstantinos I. Nikolopoulos and Dimitrios D. Thomakos

*Of course, **there is no accurate forecast**, but at times this shifts the focus for ... If **there is no** perfect plan, is **there such thing** as a good enough plan? ...* [1]

1 Introduction

Forecasting analytics (FA) is a subset of predictive analytics focusing only on predictions about the future. This does not necessarily include predicting exercises typically in the likes of regression analysis aiming at the "holy grail" of causality! In forecasting analytics, we do not underestimate the importance of causality, but we can live without it, and as long as we can accurately predict elements of the future, we are good to go.

In a nutshell, *forecasting analytics* is the extensive use of data and quantitative models as well as evidence-based management and judgment so as to produce alternative point and density estimates, paths, predictions, and scenarios for the future.

Electronic supplementary material The online version of this chapter (https://doi.org/10.1007/978-3-319-68837-4_12) contains supplementary material, which is available to authorized users.

---

[1] By Kirk D. Zylstra, 2005, Business & Economics, John Wiley & Sons.

K. I. Nikolopoulos (✉)
Bangor Business School, Bangor, Gwynedd, UK
e-mail: k.nikolopoulos@bangor.ac.uk

D. D. Thomakos
University of Peloponnese, Tripoli, Greece

Forecasting analytics is probably the most difficult part of the analytics trio: descriptive, prescriptive, and predictive analytics. More challenging as it is about the future, and although everybody is right once the forecasts are set, only the very few (and brave!) will be right when the future is realized and the forecast accuracy is evaluated.... That is the judgment day for any predictive analytics professional (PAP).

Forecasting analytics is the key for an effective and efficient applied business and industrial forecasting process. *Applied* ... as the focus is primarily on evidence-based practical tools and algorithms for business, industrial and operational forecasting methods and applications, rather than upon problems from economics and finance. The rather more advanced techniques required for the latter are more of the core of a more focused chapter on financial predictive analytics (FPA). Similar is the case and the narrower focus of marketing analytics (MA).

Forecasting analytics is also the next big thing in the employment front, with millions of jobs on demand expected in the next few years.[2]

Forecasting analytics is a crucial function in any twenty-first century company and is the one that can truly give a competitive advantage to nowadays managers and entrepreneurs as a bad forecast can be translated into:

Either

> ... *lost sales,* thus poor service and unsatisfied customers!

Or

> ... *products left in shelves,* thus high inventory and logistics costs!

Wait a minute ... this sounds like a *lose-lose* situation! If you do not get it exactly right, you will lose money—one way or another. What's more, as you might have guessed, *you will not ever get it exactly right!* Even the most advanced forecasting system, only by pure chance, will give you a perfect forecast...

Thus, the angle of this chapter, and our sincere advice to the reader would be to:

> "...make sure you do your best to get an as-accurate-*forecast*-as-possible,[3] and learn to live with the *uncertainty* that will inevitably come with this forecast ...[4]" (exactly as the introductory quote wisely suggests).

In practice,[5] although forecasting is a key function in operations, it is usually very poorly performed. This chapter aims to shed light on the practical aspects of everyday business forecasting analytics, by adopting some well-informed, academically proven, and easily implemented processes, which in most cases are just simple heuristics. Given that this is just a chapter and not an entire book—as it may

---

[2]Fisher, Anne (May 21, 2013), Big Data could generate millions of new jobs, http://fortune.com/2013/05/21/big-data-could-generate-millions-of-new-jobs [Accessed on Oct 1, 2017].

[3]To take into account all available *Information* that is relevant to the specific forecasting task—usually referred to as *Marketing Intelligence*.

[4]This line is taken from Makridakis et al. (2009).

[5]Armstrong, S. (2001), *Principles of Forecasting*, Kluwer Academic Publishers.

well could be—focus is given on the process (that we abbreviate as AFA—applied forecasting analytics process) and some basic techniques but not all the specific techniques and algorithms used at each stage.

If you had googled ... "forecasting" (back in 2005) here is what you would have got:

> Forecasting is the process of estimation in unknown situations. Prediction is a similar, but more general term, and usually refers to estimation of time series, cross-sectional or longitudinal data. In more recent years, Forecasting has evolved into the practice of Demand Planning in everyday business forecasting for manufacturing companies. The discipline of demand planning, also sometimes referred to as supply chain forecasting, embraces both statistical forecasting and consensus process.[6]

The first sentence of the quote is the most important one: forecasting is more-or-less about *estimating in unknown situations*—thus your only weapon is the past and how much the latter resembles the former. This wiki-quote continues with nicely distinguishing between *forecasting* and *prediction*. In this chapter, we will use them interchangeably.[7] The next sentence fully aligns with the beliefs of the authors of this chapter: *Forecasting has evolved into the practice of Demand Planning in everyday business forecasting for manufacturing companies* ... sometimes referred to as supply chain forecasting.

The world of everyday business forecasting, comes with the assumption that some kind of regularly observed quantitative information will be available for the products under consideration. In other words, *time-series* data will be available.

A time series (Fig. 12.1) is just a series of observations over a long period of time; those observations are usually taken in equally distanced periods (months, week, quarters, etc.). That is typically how data look like in business and operational forecasting; in most cases, observations for more than 3 years per product are available, while these are recorded quite frequently (every month or less).

But this is not always the case as:

- You may have *cross-sectional* data, data referring to the same point of time but for different product/services, etc.—for example, sales for ten different car makes in a given day.
- You may have *no data* at all—so you end up using entirely your judgment as to make some forecasts.

In this chapter, focus is basically given to time-series forecasting[8] and how this integrates efficiently with judgmental adjustments. These adjustments are driven

---

[6]From Wikipedia, https://en.wikipedia.org/wiki/Forecasting (accessed on Feb 22, 2018).

[7]Among other similar terms like: projecting, extrapolating, foreseeing, etc. In a business context all these terms could be used.

[8]Due to limited space, other forecasting methods such as Bayesian forecasting technique, Artificial Neural Network, and special event forecasting are not included.

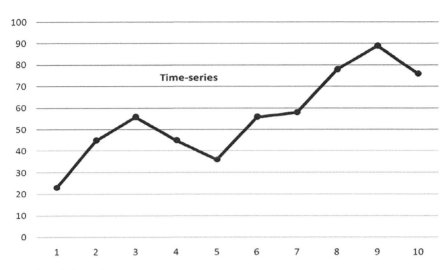

Fig. 12.1 A time series

from all sources of marketing intelligence.[9] Analysis and forecasting of cross-sectional data that is mainly the focus of regression analysis, but still remain out of the core scope of this chapter (please see Chap. 7 of this volume).

In Fig. 12.2, the aforementioned process would be to …:

… get the *thick-line* right in the first place …!

This thick-line stands for the history of the specific product you are interested to forecast. As explained in the data collection chapter (Chap. 2), data-related problems such as outliers, missing data, and sudden shifts need to be treated before forecasting. Also, data transformations, such as taking roots, logarithms, and differencing, might be required to conform to the requirements of the model.

The next logical step would be to project this thick-line into the future: forecasting the available time series. Time-series forecasting is based on the assumption that a particular variable will behave in the future in much the same way as it behaves in the past.[10] Thus, the dotted-line should be the "natural" extension of the thick-black-line. Natural … in the sense that history repeats itself. This is the basic assumption of statistical forecasting; thus *Statistics*—abbreviated Stats—is the second fundamental part of the forecasting process.

We will call this a *point-forecast*. In most of the cases you will usually be interested in many points of time in the future, so forecasts for the full forecasting *horizon* as it is usually termed, and not just a single-point forecast, are of interest.

---

[9]Marketing intelligence or market intelligence—http://en.wikipedia.org/wiki/Market_Intelligence (accessed on Feb 22, 2018).

[10]Keast, S., & Towler, M. (2009). Rational Decision-making for Managers: An Introduction (Chapter 2). John Wiley & Sons.

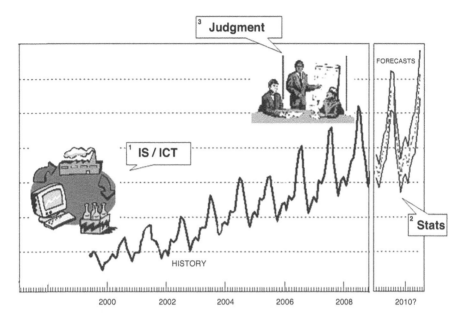

Fig. 12.2 What is needed in order to prepare a good set of forecasts?

In order to live with the aforementioned risk, we would like to have the *black-lines* as well, as shown in Fig. 12.2. Those lines are the forecast/*prediction intervals*—in this case symmetric over and under the point forecasts—and their very reason for existence is to give a sense of the *uncertainty* around point forecasts. In essence they tell you:

> ... if it's not going to be the *dotted-line*, then with great confidence it would be something from the *lower black-line* up to the *upper black-line*!

We would like to set these confidence levels around 95%, thus being 95% certain that the future unknown demand will appear somewhere between those solid-lines. But in real life this results in something that managers totally dislike: solid-lines being far out from the dotted line ... And as a result, managers go one step back and require only the point forecasts to be reported to them. This is the reason that most advanced forecasting software—FSS (stands for forecasting support systems) usually do not report the prediction intervals at all.

Another critical part of the forecasting process, as presented in Fig. 12.2, is *Human Judgment*! Humans don't really like machines... They're afraid of them! They think they will get their jobs and eventually they will get fired! As a result, they dislike ready-made solutions that do not require their intervention. They would like to have some ownership of the produced forecast. So ... they *Adjust*!

They basically adjust for two reasons:

(a) Because they think that they are *better forecasters* than the FSS system in front of them! They believe they're better at selecting and/or optimizing, as well as calibrating the available forecasting models provided to them by the FSS ... *that is obviously wrong*! FSS use advanced optimizers to select among thousands of values as to initialize, optimize, select, tune, and fine-tune the 100s of models available to their forecasting engine. So when an FSS suggests a model, usually termed as the *Expert* or *Auto* forecast, a very serious optimization procedure has taken place, and a challenge is more often than not futile.
(b) Because they think that they KNOW something that the FSS system in front of them does not! Now if they really know something, they would be correct to act. If the information is reliably sourced and they are confident they are doing the right thing, they should go for it. But, be aware, there are certain rules to how these adjustments should be made.
(c) What are the conceptual differences between forecasting and Statistics? Time-series Forecasting used to be a part of Statistics—nowadays (thanks to Makridakis[11] et al.) it is a far more generic and multidisciplinary scientific field. Multidisciplinary should be already obvious as we have clearly identified (a) Stats (so Math, Statistics), (b) IS/ICT (so Information and Computer Sciences), and (c) Human Judgment (so Psychology), being all essential parts of the discipline.

From a methodological point of view, *forecasters*—in contrast to *Statisticians*, quest for an optimal model in a different way, so:

How do you define the "best" forecasting method/model?

Statistics fundamentally makes the assumption that there is a *true underlying model* under the observed data series, that is, the black-dotted-line under the noisy time series in Fig. 12.3. If we identify that true underlying model, then all we have to do is project it into the future, in our case the grey-dotted line. However, is this the best possible forecast?

---

[11]The true birth of the Forecasting discipline dates back to late 1970s, early 1980s at the hands of Spyros Makridakis (at INSEAD), Robert Fildes (then at Manchester Business School, now in Lancaster University), and Scott Armstrong (Wharton). Benito Carbone also played a key role in the early stages. The result was to create two journals International Journal of Forecasting—IJF (Elsevier) and Journal of Forecasting—JoF (Wiley), a conference ISF (https://isf.forecasters.org/, accessed on Feb 22, 2018), an Institute IIF (www.forecasters.org, accessed on Feb 22, 2018), in a word ... a DISCIPLINE! Many have followed since then and are part of the forecasting community now, including the authors of these texts, but history was written by those 3–4 men and their close associates. More details can be found in the interview of Spyros for IJF: Fildes, R. and Nikolopoulos, K. (2006) "Spyros Makridakis: An Interview with the International Journal of Forecasting". International Journal of Forecasting, 22(3): 625–636.

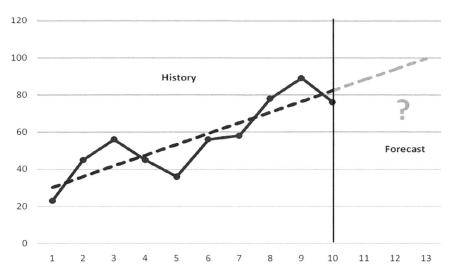

Fig. 12.3 Forecasting, extrapolation and ... statistics

In essence, in statistics we try to find the model that *best fits* the data. And since we expect history to repeat itself, we project it and we are happy. Thus, the statistical forecasting recipe is: Find the best fit → Get the job done → Sleep tight!

However, in time-series forecasting, history very rarely repeats itself!

Forecasters instead focus on which model *forecasts best* rather than which model fits best! Hold on, we have an *oxymoron* here? How can we know which model forecasts best since we do not know the future?

To resolve this, we do our first forecasting *trick*: we hide a part of the series, usually the very recent one: A 20% of the most recent part of the series is usually enough. Others suggest we have to hide as much as the forecasting horizon we are interested in—thus if we have to forecast 3 months ahead we should hide the last 3 months of the available data. We call this the *holdout* data (or sample) and we will use it to evaluate which model forecasts best. For example, we hide the last year of our time series, and we use the previous years to forecast this last hidden one, with a variety models, and the one model that goes "closer" to the hidden values is the model that ... forecasts "best." And this of course is not necessarily the one that fits the whole available dataset the best (the standard technique used in statistics).

Unfortunately, our approach is not bullet-proof either ... as:

There is no guarantee that the model that forecasts best, will keep on forecasting best ...

However, it still produces on average better forecasts than the model that fits best! At least, that is what most empirical investigations suggest. The next section discusses time-series techniques, process, and other applications.

2 Methods and Quantitative Approaches of Forecasting

2.1 Data and Statistics

Quantitative techniques mainly rely on data, statistical models, and estimation techniques for forecasting and are the basic elements of forecasting process as an aid to business decision-making, corporate planning, and management. Data can be in several forms: spoken and unspoken expression, alphanumeric written language, and other forms of communications and may consists of numbers, texts, signs, and images. These data can be time series, cross section, or combination of the two, that is, panel data. More often, in business and industry, real-time geographically distributed data on sales, orders, stocks, returns, failures, scheduling, logistics, budget, and information on competitors are very important and used in forecasting. Detecting abnormalities and irregularities, identifying outliers and missing observations, and cleaning and editing datasets for internal consistency are important aspects and one of the basic steps of setting up a quantitative forecasting model. One may in addition need to take into account known variations, such as holidays, calendar days correction (e.g., for leap year), special events, and changes in inflation rate.

Outliers are observations whose values are influenced by external factors and deviate markedly from other observations in the sample. They fall outside the 95% confidence interval around the mean of the dataset and affect forecast accuracy of the quantitative models (Hanke and Wichern 2005). There are numerous ways of identifying outliers and dealing with them including visualization and graphical presentation and newly developed methods like trimming and winsorizing (Jose and Winkler 2008). Trimmed means deleting the k smallest and the k largest observations from the sample, that is, observations are trimmed at each end. In case of winsorized means of N data points, the outer most k-values on either end are replaced with the $(k + 1)$st and $(N - k - 1)$th value at either end. Alternatively, the top and bottom values for trimming and winsorizing are determined by using some fixed percent criteria. For example, using 95% confidence interval criteria, top 2.5% and bottom 2.5% values are trimmed or winsorized.

Quantitative forecasting methods use historical patterns from time series in their prediction of future values (Makridakis et al. 1998). These historical patterns present in the data are broken down into various components using the methods of moving averages and autocorrelation analysis. The process is called decomposition of time series and pattern. The data is usually divided into seasonal, trend, cyclical, and irregular or error components. Each component is analyzed separately and the trend-cycle components are used for forecasting with the application of various statistical techniques and models. The method of moving average can use an additive or a multiplicative approach. The additive approach is used when the seasonal fluctuations do not change with the level of the series while the multiplicative approach is used when fluctuations change with the level.

Seasonal and Cyclical Adjustment: Seasonal fluctuations and changes can occur and repeat more or less regularly and periodically within a year and behavior of the data show predictability. The drivers of seasonal demands and supply are climate or festivals which repeat every year during a particular month. The most widely used tool to test and determine seasonality in time series is plotting the autocorrelation function (ACF).

Analysis of the autocorrelation coefficient or autocovariance function (ACF) which shows the relationship between current and lagged values of a time series is a way to decompose the data and investigate repeating patterns and presence of a periodic signal obscured by noise. The autocorrelation coefficient can be used to detect the presence of stationary, seasonality, trend, and random variability in the data. Specific aspects of autocorrelation processes such as unit root, trend stationary, autoregressive, and moving averages can be computed. The autocorrelation coefficient (r_k) is computed as:

$$r_k = \frac{\sum_{i=k+1}^{n} \left(Y_i - \overline{Y}\right)\left(Y_{i-k} - \overline{Y}\right)}{\sum_{i=1}^{n} \left(Y_i - \overline{Y}\right)^2}.$$

where k = time lag, n is the number of observations, and Y = observed value. Close to zero values of r_k indicate no autocorrelation—the series is not related to each other at any lag k and the variability in the values is random with zero mean and constant variance (Fig. 12.4a). If there is a trend the r_k value is high initially then drops off to zero (Fig. 12.4b). In the case there is a seasonal pattern in the data series, r_k reappears in cycles, for example, of 4 or 12 lags depending on quarterly or yearly series (Fig. 12.4c).

The autocorrelation between two observations at prior time steps, that is, correlations between observations at predetermined or specified time lags, in a data series consists of direct correlations among themselves, as well as indirect correlations with observation at intervening time steps, that is, correlations with observations in between specified time lag. The ACF comprises both direct and indirect correlations among observations and does not control for correlations of a particular observation with observations at other than the specified lag. An alternative to ACF is the partial autocorrelation function (PACF) in which indirect correlations with values at shorter lags are excluded and only direct correlations with its own lagged values are taken. (Under the assumption of stationarity, the jth PACF value is obtained by regressing the present values against the past j values and taking the coefficient of the jth value as the estimate of the PACF coefficient.) The ACF and the PACF both play important roles in identifying the extent of lags in autoregressive model such as the ARIMA model discussed in a subsequent section.

Cyclical fluctuations and data behavior indicate regular changes over a period of more than 1 year, at least 2 years, and can be analyzed and forecasted. The classical decomposition method provides ways to separate cyclical movements in the data.

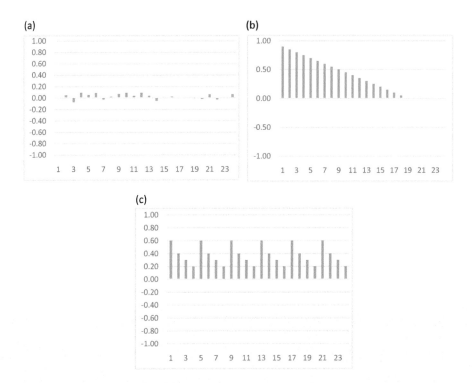

Fig. 12.4 Autocorrelation function

2.2 Time-Series Methods

A time-series data may consist of seasonal fluctuations, a trend, cyclical change, and irregular components. A simple divide and conquer approach could be to remove the seasonal and trend components by directly providing estimates of those using any number of simple techniques and using the smoothing techniques to forecast. After obtaining such a forecast, the trend and seasonal components can be added back. Below, we describe smoothing methods (different from the divide and conquer approach) for handling all three types of series, those without trend and seasonality, those without seasonality, and those with all the three.

2.2.1 NAÏVE Method (NF)

A naïve method assumes no seasonality and no trend-cycle in the data and simply sets the latest available actual observation to be the point forecast for periods in the future. Sometimes seasonally adjusted data is used and the forecasts are re-seasonalized. The naïve model is a kind of a random walk model. The NF is considered a simple benchmark against which the more advanced results may be compared. In some ways this is reasonable: the naïve method measures the volatility inherent in the data and in many systems nothing works better than what happened yesterday.

2.2.2 Average and Moving Average (MA) Methods

A simple average is obviously easy to compute but misses trends and recent changes in the series. Moving averages are a simple way of smoothing out the seasonality and noise in a series to reveal the underlying signal of trend used for forecasting. In the simplest version, the forecast for the future periods is set equal to the average of the observations of the past k periods. One may wish to optimize on the value of k. Variants of the simple moving average are weighted and exponentially weighted moving averages. In weighted moving average, different weights are assigned to various point observations within a seasonal period to be used for averaging, while in exponentially weighting, higher weight is assigned for the latest point observation of a season and lesser weights are assigned in a continuously decreasing manner to the earlier point observations.

One will notice that there will be no forecast for the first k periods unless fewer periods are used to produce the initial forecast. Also, the prediction after the last period of data will be same for every period thereafter.

2.2.3 Simple Exponential Smoothing (SES)

In this method one assumes the absence of trend and seasonality in the data. Brown (1956) is credited with the development of the single exponential smoothing methodology.

The following formula is used to forecast using the SES method:

$$F_{t+1} = \alpha * Y_t + (1 - \alpha) * F_t;$$

where Y_t is the actual observation in the period t and F_t is the forecast value from $(t - 1)$ period.

Also, $e_t = Y_t - F_t$, is the error between the observation and forecast value.

By substitution, one may also write:

$$F_t = \alpha * Y_{t-1} + (1 - \alpha) * F_{t-1}$$
$$= \alpha * Y_{t-1} + \alpha * (1 - \alpha) * Y_{t-2} + \alpha * (1 - \alpha)^2 * Y_{t-3} + \cdots + \alpha * (1 - \alpha)^{n-1}$$
$$* Y_{t-n} + (1 - \alpha)^n * F_{t-n}.$$

The best α can be found using an optimization approach or simply by trial and error. The forecast can be started in many ways. A popular method is to use the first actual value as the first forecast ($Y_1 = F_1$) or set the average of the first few values as the value of the first forecast.

In sheet "SES" of spreadsheet "FA-Excel Template.xlsx" we have provided sample data (which can be changed) and the value of the smoothing constant, α, that can be changed. As α changes from zero to 1, the forecast will be seen to follow the most recent value more closely. In the Appendix in the section "SES Method" we provide the R command for SES. The data is shown in Table 12.6 and the output in Table 12.10.

In the adaptive-response-rate single exponential smoothing (ARRSES) α can be modified as changes occur:

$$F_{t+1} = \alpha_t Y_t + (1 - \alpha_t) F_t$$

$$\alpha_{t+1} = ABS \left(\frac{A_t}{M_t} \right)$$

$$A_t = \beta e_t + (1 - \beta) A_{t-1}$$

$$M_t = \beta * ABS(e_t) + (1 - \beta) M_{t-1}$$

$$e_t = Y_t - F_t.$$

Here, β is a smoothing constant to change α. In this case the smoothing constant α changes over time. The idea behind the approach is that when A and M are close to one another, then the errors have the same sign and this might indicate bias. In that case adjusting the value of α closer to one might restart the forecasting process with the most recent observation.

Starting this method is somewhat more complicated. One may set $Y_1 = F_1$, $A_1 = M_1 = 0, \alpha_2 = \alpha_3 = \alpha_4$ equal to preset value, say 0.3. The last is done so that we have a few values to warm-up before changing the value of α. In sheet "ARRSES" of spreadsheet "FA-Excel Template.xlsx" we have provided sample data (which can be changed) and the value of the smoothing constant, α, set equal to 0.3 for the first three values. This value changes as time progresses. The value of β is set to 0.5. The data is shown in Table 12.6 and the output in Table 12.10.

2.2.4 Holt Exponential Smoothing (HES)

Brown's SES methodology (1956) was extended by Holt (1957) who added a parameter for smoothing the short-term trend. The current value is called the level of the series or L_t. The change in levels ($L_t - L_{t-1}$) is used to determine the trend during the period t. Then, the trend is smoothed using the previously forecast value, that is, $T_t = (1 - \beta)T_{t-1} + \beta*(L_t - L_{t-1})$. This method is also called Double Exponential Smoothing (DES). The formulae are:

$$L_t = \alpha Y_t + (1-\alpha)(L_{t-1} + T_{t-1})$$

$$T_t = \beta(L_t - L_{t-1}) + (1-\beta)T_{t-1}$$

$$F_{t+m} = L_t + mT_t, \quad m = 1, 2, \ldots$$

In order to start the forecast, one may set $L_1 = Y_1$ and the slope can be obtained by regressing initial values of the series against time. Search methods can be used to select the "optimal" values of the two smoothing constants, α and β. (The word optimal is in quotes because the criterion for optimization could be minimizing different types of errors, including errors one step or two steps ahead.) An example is shown in sheet "Holt" of spreadsheet "FA-Excel Template.xlsx". Appendix 1 section "Holt method" lists the R command. The data is shown in Table 12.6 and the output in Table 12.10.

2.2.5 Holt–Winters' trend and seasonality method

Holt–Winters' method is a smoothing method that takes both trend and seasonality into account known as Error, Trend, and Seasonality (ETS) or triple exponential smoothing as three components (viz., level, trend, and seasonality) in the data are used and smoothened to arrive at forecast values. It is a variant of Holt method of exponential smoothing in which a component of seasonality index along with trend and level is also added to arrive at forecast:

$$L_t = \alpha \frac{Y_t}{S_{t-s}} + (1-\alpha)(L_{t-1} + T_{t-1})$$

$$T_t = \beta(L_t - L_{t-1}) + (1-\beta)T_{t-1}$$

$$S_t = \gamma \frac{Y_t}{L_t} + (1-\gamma)S_{t-s}$$

$$F_{t+m} = (L_t + T_t m)S_{t-s+m}.$$

where S_t denotes the seasonal component, s is the length of a season, and γ is the seasonal smoothing factor. Note that after each step we need to renormalize the seasonal factors to add up to k ("Periods in Season"). The initial values for L_s, b_s, and S_s can be initially calculated as:

$$L_s = \frac{1}{s}(Y_1 + Y_2 + \cdots + Y_s)$$

$$b_s = \frac{1}{s}\left[\frac{Y_{s+1} - Y_1}{s} + \frac{Y_{s+2} - Y_2}{s} + \cdots + \frac{Y_{s+s} - Y_s}{s}\right]$$

$$S_1 = \frac{Y_1}{L_s}, S_2 = \frac{Y_2}{L_s}, \ldots, S_s = \frac{Y_s}{L_s}.$$

In the additive form seasonality is added to the forecast, instead of being multiplied.

$$L_t = \alpha (Y_t - S_{t-s}) + (1 - \alpha)(L_{t-1} + T_{t-1})$$

$$T_t = \beta (L_t - L_{t-1}) + (1 - \beta) T_{t-1}$$

$$S_t = \gamma (Y_t - L_t) + (1 - \gamma) S_{t-s}$$

$$F_{t+m} = L_t + T_t m + S_{t-s+m}.$$

The initial values for level and trend can be chosen like in the multiplicative method. The seasonality values can be estimated as below to start the forecast:

$$S_1 = Y_1 - L_1, \; S_2 = Y_2 - L_2, \ldots, \; S_s = Y_s - L_s.$$

The data is shown in Table 12.6 and the forecast output (produced by R) is shown in Table 12.10 for both methods. The R command is listed in the Appendix in the section "Holt–Winters Method."

2.2.6 Damped Exponential Smoothing for Holt's Method

When the trend in the observation has a nonlinear pattern, the damped method of exponential smoothing can be used. It is a variant of Holt's method in which only a fraction of trend forecast values of current and earlier periods are added to L_t to arrive at F_{t+1}:

$$L_t = \alpha Y_t + (1 - \alpha)(L_{t-1} + \phi T_{t-1})$$

$$T_t = \beta (L_t - L_{t-1}) + (1 - \beta) \phi T_{t-1}$$

$$F_{t+m} = L_t + \left(\phi + \phi^2 + \cdots + \phi^m\right) T_t.$$

where φ is the damped parameter for the trend coefficient T_t. The forecast can be started just as in the Holt's method for FIT. Usually, the damping parameter is set to be greater than 0.8 but less than 1.

The data is given in Table 12.7 and the output in Table 12.11. The same example is given in sheet "Damped Holt" of spreadsheet "Forecasting Analytics-Excel Template". The R command is listed in the Appendix in "Damped Holt Method" section.

2.2.7 The Theta model

This methodology provides a procedure to exploit the embedded useful data information components in the form of short-term behavior and long-term trend before applying a forecasting method. The idea is to modify the local curvature of the time series before forecasting.

In a simple version, the Theta model decomposes the seasonally adjusted series into two data series called Theta lines and the forecast is a combination of the values obtained from the two theta lines (Assimakopoulos and Nikolopoulos 2000; Thomakos and Nikolopoulos 2014). The forecast from the first Theta line provides the long-term trend of the data, and is obtained from a regression line $\widehat{Y}_t = \widehat{\beta}_0 + \widehat{\beta}_t t$, where $\widehat{Y}_t$ is forecast at time t. The second Theta line is computed by first setting a new time series equal to $2Y_t - \widehat{Y}_t$. The forecast value for the second line is obtained using SES which is discussed earlier. The point forecasts of the two Theta lines are combined using equal weight of ½.[12]

The data is given in Table 12.8 and output in Table 12.7. The same example is given in sheet "Theta" of spreadsheet "Forecasting Analytics-Excel Template". The R command is listed in the Appendix in the "Theta method" section.

2.2.8 Advances in Time-Series Processes

The Autoregressive Integrated Moving Average (ARIMA) Framework:
Extrapolation models are most frequently and widely used in forecasting with a large dataset, and among them, the exponential smoothing forecast approaches have been the most popular method (Petropoulos et al. 2014). The other advances in quantitative forecasting approaches and class of models are based on the ARMA framework (Box and Jenkins 1970), Bayesian method of forecasting (Harrison and Stevens 1976), state space models (Chatfield 2005), and application of neural networks (Andrawis et al. 2011; Tseng et al. 2002). Only the ARMA method is discussed below.

ARMA approach is warranted when there is evidence of autocorrelation. Usually, the first step in applying the ARMA framework would be to study the data for evidence of stationarity. This can be done by looking at the plots of ACF and PACF as described earlier. For example, the 95% critical values are approximately at $\pm 1.96/n^{0.5}$, where n is the number of data points (these are shown in the plots below as dotted lines). Other methods include *unit root* tests, such as the Dickey-Fuller test. One of the common techniques of removing nonstationarity is differencing and seasonal differencing. In seasonal differencing, values one season apart are differenced.

---

[12]https://link.springer.com/chapter/10.1007/978-3-642-25646-2_56 *accessed on Sep 11, 2017.*

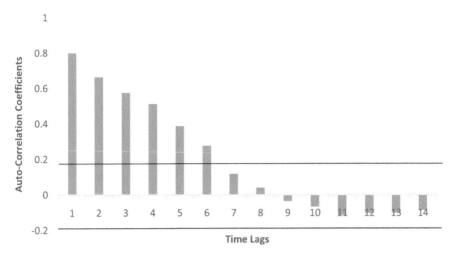

Fig. 12.5 Autocorrelation function

Box and Jenkins developed the ARMA model in 1970. The autoregressive part (or AR) of the ARMA model can be written as $y_t = c + \varphi_1 y_{t-1} + \varphi_2 y_{t-2} + \cdots + \varphi_p y_{t-p} + e_t$, where e_t is white noise and p lagged values are used. This is a multiple regression with lagged values of y_t as predictors. The lagged explanatory variable becomes stochastic and contemporaneously correlated with the error term, making the forecast stochastic and creating bias leading to loss of confidence that comes with estimator bias and variance (please see Chap. 7 on Regression for details). The moving average (or MA) part of the model includes $y_t = c + e_t + \omega_1 e_{t-1} + \omega_2 e_{t-2} + \cdots + \omega_p e_{t-q}$, which is a multiple regression with q past errors as predictors. (A common confusion is with the MA methods discussed earlier. There the data itself was averaged. Here, the errors are averaged.)

The ACF and PACF can be used to identify the lag structure of an ARMA model. ACF is used to estimate the MA-part and PACF is used to estimate the AR-part, for example, in Fig. 12.5 we show the ACF and PACF plots of difference in data. Both ACF and PACF are decaying, there is a drop off after the time-lag 6 in ACF (Fig. 12.5), and there is spike at the time-lag 1 in PACF (Fig. 12.6). Therefore, the appropriate lag structure could be ARMA (1, 6).

The ARIMA model: ARIMA forecasting is used when the condition of no-autocorrelation and homoscedasticity are violated. Then, it requires transformation of the data series to stabilize both variance as well as mean. A data series is said to be stationary when the mean and variance are constant over time. ARIMA was developed to handle nonstationary data by differencing d-times. Later, Engle (1982) introduced autoregressive conditional heteroscedastic (ARCH) models which "describe the dynamic changes in conditional variance as a deterministic function of past values." When "additional dependencies are permitted on lags of

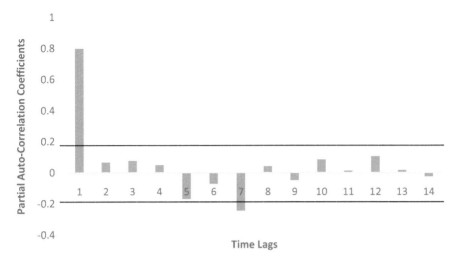

Fig. 12.6 Partial autocorrelation function

the conditional variance the model" is called generalized ARCH (GARCH) model and share many properties of ARMA (Bollerslev et al. 1994; Taylor 1997).

The data series is plotted against time to identify nonstationary, that is, changing means and variances over time. For a nonstationary series, the value of the autocorrelation coefficient, r_1, is often large and positive and the autocorrelation function (ACF) decreases slowly while it drops to zero relatively quickly for stationary data. To stabilize the varying mean due to seasonality and trend, data differencing is done, while AR and MA processes are used to incorporate autocorrelation in lagged values of the time series and the linear combination of error terms whose values change contemporaneously over time. Combining autoregressive and moving average models, the ARIMA $(p; d; q)$ model can be written as: $y_t = c + \varphi_1 y_{t-1} + \varphi_2 y_{t-2} + \cdots + \varphi_p y_{t-p} + e_t + \omega_1 e_{t-1} + \omega_2 e_{t-2} + \cdots + \omega_p e_{t-q}$, where, AR: p = order of the autoregressive part, I: d = degree of first differencing involved, and MA: q = order of the moving average part. While it appears that one has to search for a number of values, practically just values of 0, 1, and 2 for p, d, q suffice to generate a large number of models.

Maximum Likelihood Estimation (MLE) of ARIMA Model: Having specified the model order, after checking for stationarity, the ARIMA parameters are estimated using the MLE method and use of a nonlinear numerical optimization technique. One can minimize AIC $= -2 \log(L) + 2 (P + q + k + 1)$ or BIC $=$ AIC $+ \log(T)(p + q + k-1)$, where L is likelihood of the data, k $= 1$, if constant $\neq 0$ and k $= 0$, if constant $= 0$ to get a good model. An approximate estimate of $-2\log(L)$ is given by $n(1 + log(2\pi)) + n\, log(\sigma^2)$, where n is the number of data points and σ^2 is the variance of the residuals.

The R command is given in the Appendix in "ARIMA method" section. The data and summary output of R on an example is given in Table 12.1. The complete

Table 12.1 Forecasting Sofa demand using ARIMA

Months	Demand	Months	Demand	Months	Demand	Months	Demand
1	98	14	99	27	81	40	93
2	82	15	93	28	93	41	90
3	84	16	82	29	91	42	84
4	85	17	84	30	81	43	82
5	99	18	88	31	86	44	82
6	90	19	93	32	81	45	98
7	92	20	83	33	97	46	91
8	83	21	95	34	88	47	85
9	86	22	93	35	96	48	86
10	90	23	92	36	96	49	88
11	95	24	92	37	97	50	90
12	91	25	97	38	90		
13	87	26	88	39	88		
Three months ahead forecast values (See Table 12.2 for calculations)						51	92.84
						52	92.47
						53	91.39

Table 12.2 Estimated ARIMA(2,1,2) Model

Variables	Coefficients	Standard error
AR1	0.9298	0.1429
AR2	−0.2561	0.1471
MA1	−1.9932	0.1048
MA2	0.9999	0.1048
Log likelihood = −151.47, aic = 312.94		

output is in Table 12.13. The same data can be found in sheet "Data - ARIMA" in csv format.

EXAMPLE: The monthly demand of sofa (in thousands) by a company for the last 50 months is given below in Table 12.1. The problem is to provide the forecast of sofa demand for the company for the next 3 months using the ARIMA model.

Solution:
Assume that we want to fit the ARIMA model (2, 1, 2). Assume that the data is named as ARCV. The R command is: *fitted* ← arima(ARCV, order = c(2, 1, 2)). Here, fitted is where the output will be placed.

The ARIMA parameters of order p = 2, d = 1, q = 2 are estimated using MLE (maximum likelihood estimation) methods and automated nonlinear numerical optimization techniques. The coefficients are obtained by calling *fitted*. The output is given in Table 12.2.

The forecast equation is (to be written by the user): $Y_t = 0.929 \ast Y_{t-1} - 0.256 \ast Y_{t-2} - 1.993 \ast e_{t-1} + 0.999 \ast e_{t-2} + $ Error. The example reveals that after the estimate the equation has to be written in the forecast equation form to predict values in the future. The forecast values can be obtained by using the command forecast (*fitted*, h = 3).

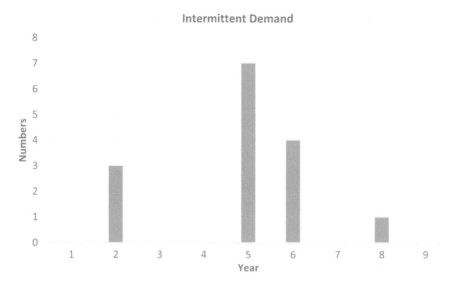

Fig. 12.7 Intermittent demand

2.3 Forecasting Intermittent Demand

The SES method assumes a constant probability for the occurrence of nonzero values which is often violated leading to count data or intermittent series (Lindsey and Pavur 2008). It is found that around 60% of the stock-keeping units in industrial settings can be characterized as intermittent (Johnston et al. 2003). Intermittent demand is characterized by infrequent demand arrivals and variable demand sizes when demand occurs. As Fig. 12.7 shows, there are "periods with demand followed by periods of no demand at all, and on top of this even the demand volume (when realized) comes with significant variation. There are two things to forecast: when the next demand period is going to be realized? And, whenever demand is realized, what will be the volume of this demand?" The basic technique is to combine different time block and different methods have been proposed to doing so. The SES method performs poorly in cases of stochastic intermittent demand.

Croston (1972) developed methodology for forecasting such cases and suggested decomposition of intermittent series into nonzero observations and the time intervals between successive nonzero values. The two series, namely, quantity and the intervals, are extrapolated separately. An updating is done for both quantity and interval series only after a nonzero value occurs in quantity series.

2.3.1 Croston's Approach (CR)

This approach applies SES independently to demand size y and inter-demand interval τ independently, where $\tau = 1$ for non-intermittent demand:

$$F_{t+1} = \frac{\widehat{y}_{t+1}}{\widehat{\tau}_{t+1}}.$$

where $\widehat{y}_{t+1}$ and $\widehat{\tau}_{t+1}$ are the forecast of the demand size and interval. Both are updated at each time t for which $y_t \neq 0$. An example is provided in sheet "Croston and SBA" in spreadsheet "Forecasting Analytics-Excel Template". The R command is given in the Appendix in the "Croston and SBA method" section. The data is in Table 12.9 and output is in Table 12.14 in the Appendix.

2.3.2 Syntetos and Boylan Approximation (SBA)

Syntetos and Boylan (2001) found that Croston's methodology provides upward biased forecast. Subsequently, they proposed an improved Croston's methodology in which the final forecasts are multiplied by a *debiasing factor* derived from the value of the smoothing parameter of intervals (Syntetos and Boylan 2005). Syntetos and Boylan (2005) found that Croston method is biased on stochastic intermittent demand and corrected the bias by modifying the forecasts to:

$$F_{t+1} = \left(1 - \frac{\beta}{2}\right) \frac{\widehat{y}_{t+1}}{\widehat{\tau}_{t+1}}.$$

SBA works well for intermittent demand but is biased for non-intermittent demand. Syntetos and Boylan (2001) avoided this problem by using a forecast:

$$F_t = \left(1 - \frac{\beta}{2}\right) \frac{\widehat{y}_{t+1}}{\widehat{\tau}_{t+1} - \frac{\beta}{2}}.$$

This removes the bias but it increases the variance of the forecast. Other variants include that of Leven and Segerstedt (2004).

None of these variants handle obsolescence well. When obsolescence occurs these methods continue to forecast a fixed nonzero demand forever. An example of SBA is provided in sheet "Croston and SBA" in spreadsheet "Forecasting Analytics – Excel Template". The R command is given in the Appendix in the "Croston and SBA Method" section. The data is shown in Table 12.9 and output is in Table 12.14 in the Appendix.

Recent development in the area of forecasting intermittent demand include the work of Babai et al. (2012), Kourentzes (2014), Kourentzes et al. (2014), Nikolopoulos et al. (2011a, b), Prestwich et al. (2014), Rostami-Tabar et al. (2013), Spithourakis et al. (2011), and Teunter et al. (2011).

2.4 Bootstrapping Method

Often, the forecasting task is to predict demand over a fixed leadtime. In this case bootstrapping might be used. Bootstrapping (Efron 1979) is a statistical method of inference that uses draws from sample to create an approximate distribution. Willemain et al. (2004) produce accurate forecasts of the demand of nine companies over a fixed lead time compared to exponential smoothing or Croston's method. We illustrate with an example.

EXAMPLE: Demand for an Automobile Part

Suppose, we would like to forecast the automobile part demand for the next 3 months. Historically, the 24 monthly demand for the part is given as follows (Table 12.3).

Solution:

Bootstrap scenarios of possible total demands for 3-month lead periods are created by taking random sample with replacement as follows:

1. Months: 3,17,21; demand: $7 + 0 + 0 = 7$.
2. Months: 1,20,8; demand: $0 + 13 + 0 = 13$.
3. Months: 6,14,19; demand: $2 + 9 + 5 = 16$.

Continuing this process, we can build the demand distribution for the given lead time.

3 AFA—Applied Forecasting Analytics Process

Forecasting in business is performed at an operational, tactical, and strategic level:

- At the *operational* level—where the focus of this chapter is, we are mostly interested in being "*roughly right within the limited available time,*" given in order to prepare the forecasts. This involves short-term forecasting tasks; real-life applications are usually a few weeks/months ahead. For some category of products (e.g., dairy) we may even need more frequent forecasts (every day or every other day).
- *Tactical* forecasting involves short- to mid-term forecasting, usually 3–12 months ahead. Cumulative and individual point forecasts are needed for this period, as well as, we would incorporate the effect of forthcoming events like promotions and supply interruptions.

Table 12.3 Demand data for automobile part

Month	1	2	3	4	5	6	7	8	9	10	11	12
Demand	0	0	7	4	0	2	0	0	0	11	0	0
Month	13	14	15	16	17	18	19	20	21	22	23	24
Demand	15	9	0	0	0	0	5	13	0	21	0	0

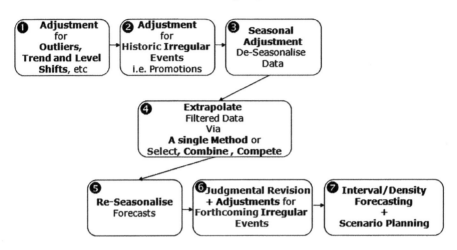

Fig. 12.8 AFA forecasting process

- At the *strategic* level we usually look into forecasting horizons that go beyond a year and involve the impact of rare events (like major international crises as the recent one regarding energy prices and the global credit system), new product development, product withdrawals, capacity amendments, and scenario planning.

The aforementioned *forecasting horizons* are only indicative, and often met in supply chain forecasting. There are many forecasting applications where a strategic forecast is just for a few months ahead! So, in order to avoid any confusion, we use the terms: "forecast for x steps ahead" or "forecast for x periods ahead", without specifying what steps/periods stand for. These steps could be anything from minutes to years depending on the application area. Typically *short-term* forecasting involves 1–3 steps ahead; *medium-term* or mid-term is for 4–12 steps ahead; and *long-term* anything over 12 steps ahead.[13]

This chapter proposes a simple *seven-step* forecasting process tailored for operational forecasting tasks. This process is illustrated in Fig. 12.8, and is abbreviated as "*AFA forecasting process*" or just *AFA* for short.

AFA provides detailed guidance on how to prepare operational forecasts for a *single* product. This process should be:

(a) *Repeated* for every product in your inventory, and
(b) *Rerun* each time new demand/sales data becomes available.

Thus, if we observe our inventory of 100 products on a monthly basis, we should run AFA every month for each of the hundred products we manage.

---

[13] For this and other types of forecast classifications, see Hibon and Makridakis (2000).

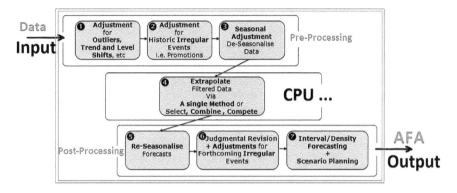

Fig. 12.9 I/O and CPU of the AFA forecasting process!

Let us start decoding what these boxes stand for; there is one box for each of the *seven steps* of the AFA process. The upper three form the *preprocessing* phase, the one in the middle the *main-forecasting-task*, and the latter three the *post-processing* phase (Fig. 12.9).

It looks like a typical Black-Box approach.[14] However, we believe it is more like a "Grey-Box" approach! A situation where you will be able to understand most of the things that are happening throughout AFA, however rely on automated tools to deliver for you!

Let us explore the AFA process as illustrated in Fig. 12.8:

- First Box: the BAD things...

  Each single time series comes with a number of problems. Some of these are dead-obvious but some are well hidden. To cut the long story short, we must deal with all these "bad things" and prepare a series with no missing values, no extremely low/high values (outliers), no level or trend shifts; this would involve automated detection algorithms for such problems and suggested solutions in order to adjust the original series into new series, filtered for all the aforementioned problems.

- Second Box: the GOOD things...

  In time-series forecasting it is often very difficult to tell good things from bad ones ... just like in real life! A 'good thing' in a time series is a *special event* (SE), often termed in literature as *irregular, infrequent* or *rare* event; it could be a promotion, a production interruption, news, regulation, etc., in general anything that could make demand deviate substantially from regular levels! But why is something irregular good? Simply because it is an information-rich period, a period with special interest where an external/exogenous event has driven excess or limited demand respectively. So it would look exactly like an outlier, but we will know what exactly happened. From a mathematics perspective, the way you

---

[14] A standard engineering expression, for a situation or a solution where something seems to work fine, but we are not sure why and definitely do not know for how long it will keep on working!!

detect and subsequently adjust periods with special events, is identical to the one used to treat outliers.
- Third Box: the REGULAR things...

In forecasting, finding regularities and patterns in a series is an essential task; usually termed *periodicity*, things that repeat themselves on a regular basis. If the regularity, the repetition happens within a year then we will call this phenomenon *seasonality*; for less than a year *mini-cycles*, while for more than a year, big *economic/financial-cycles*. In any case, removing the effect of these cycles at this stage of the AFA process (and reintroduce them later on) has been empirically proven to work very well, as argued in various empirical investigations.[15]

After successfully completing these three steps of the preprocessing phase of our time series, we should by now have a nice *filtered*[16]—*smoothed*—series, that will look almost[17] like a straight *line*, either entirely flat or with a constant-*ish* trend (upward or downward). Now, it is about time to extend this line into the future... Now we are ready to forecast!

- THE (fourth) BOX: FORECASTING...

This is where all the fun is .... Let us try forecasting: extrapolate the series in the future. We will not just choose a method—and that is it! (where would the fun be after all...?).

We basically employ three fundamental strategies ... the "three forecasting tricks" as I fancy calling them:

– "**Select**": my mother always says that "Experience is all that matters ...""; and she is probably right. Thus if a method worked well in the past, we should probably stick to it, and keep on selecting that same one for the task of extrapolation. Furthermore, some methods may have been proven to work better for some products while other methods better for other products; so there are "horses for courses" and once again we are better off sticking to them. In essence, we could build a nice table—a *selection protocol* (SP) as we will call it more formally, where in one column there is a list of our products, while in the other column the forecasting methods and models that have worked well in the past for the respective products. An illustrative example is shown in Table 12.4.

– "**Combine**":

When in doubt ... combine!

a very good piece of advice I dare say. When a method has worked well in the past for a certain product, but the new Statistician on the block... insists that method X is the new panacea in time-series forecasting, then why not combine those two? So get a set of forecasts from your trusty chosen method, get another set of forecasts from the new highly promising method, and then average these

---

[15] For more information please visit http://www.forecastingprinciples.com/, or read "Principles of Forecasting: A Handbook for Researchers and Practitioners, J. Scott Armstrong (ed.): Norwell, MA: Kluwer Academic Publishers, 2001".

[16] A term often used in engineering applications.

[17] Of course there would still be some *noise* over this line.

Table 12.4 Forecasting selection protocols

Company X	
'Best' FORECASTING methods over the last 2 years	
Product 1	Method A
Product 2	Method B
Product 3	Method A
...	...

to get a final set of forecasts. If you believe more in the former (or the later) you could easily differentiate the weights respectively as to express your belief, for example, via a 30% weight to the experience-based method and 70% to the new one. Rule of thumb: *"Combining always works!"* (In other words: Combining most of the times outperforms the individual performance of the methods being combined.)

- **"Compete,"** the true reason forecasters exist: (empirical) *Forecasting Competitions*! We do not trust anything, and from all the available methods and models, applied on all the available history, we will find the one that forecasts "best." These criteria typically include average or median error metrics like MAE, MdAE, MAPE, MDAPE, MASE, and MdASE.

Sometimes, we even apply these tricks simultaneously—for example, (a) we compete only among methods that have performed well in the past, or (b) we combine the winner of the competition and the top performing method in the past as described in a selection protocol, or ...

- Fifth Box: Superimpose regular patterns.

  By now, our forecasts should look like a straight line, either *flat* or with a certain *slope*. If we have identified regularities in step three, then we need to bring them back into the game, in other words we superimpose these patterns onto the extrapolation. Once this step completed, our forecasts will have *ups* and *downs*, and will look like a natural extension of the cycles and seasonality observed in the history of the series.

- Sixth Box: Human Judgment.

  This is where humans come into the game. No matter how sophisticated the process so far, people—usually referred to as *forecasters* or *experts*—want to intervene at this stage; primarily to introduce market intelligence? This is usually performed in two phases: (a) an initial phase, where experts roughly revise all provided forecasts by changing[18] them by a percentage x% (e.g., increase all monthly forecasts for the full next year by 10%), and (b) a more targeted one, where some specific forecasts in the future are adjusted for the potential impact of special events like promotions (e.g., increase by an extra 1000 units the sales forecast for next September due to an expected advertising campaign).

- Seventh (Last) Box: Density forecasting + SCENARIOS; living with Uncertainty!

---

[18]Usually increasing the forecasts, due to an optimism bias (more on this and other types of bias in Chap. 13).

Table 12.5 AFA output

– A list of *adjustments* made to the original data due to problems
– A list of *adjustments* made to the original data due to special events
– A set of *Seasonal Indices* if seasonality was identified
– Sets of *Cyclical indices* if min-cycles or major economic cycles were identified
– A set of statistical *point-forecasts* (each one for each respective forecasting horizon)
– A set of judgmentally *revised* forecasts plus *Notes* explaining the reasons for adjustment
– Two sets of *prediction intervals*, under and over the provided forecasts
– An estimation of the *Bias* of those forecasts: A tendency to consistently under-forecast or over-forecast
– An estimation of the *expected accuracy* of those forecasts: In the form of past errors
– An estimation of the *uncertainty* of those forecasts: In the form of the standards deviation of the forecasts
– An estimation of the *endogenous difficulty* of the forecasting task; in the form of the *noise*[19] existing in the original time series. In statistical and mathematical sciences, we believe that under any observed phenomena (time series in our case), there is an underlying signal where whatever is not explained and captured from it, is described as noise in the series
– *Statistical significance*[20] of the forecasts: In the form of comparisons with standard forecasting benchmarks
– *Economics significance*[21] of the forecasts: In the form of the financial implications of our forecasts as in stock holding costs [22], or trading financial results [23]

- In this final step, we try to cope with the *uncertainty* that comes with the produced forecasts. Firstly, we usually provide a set of *confidence* or *prediction* intervals, associated with the point forecasts for the full forecasting horizon, as shown in Fig. 12.2; this is also known as *density* forecasting. There are theoretical as well empirical ways so as to produce these intervals. The most popular way to deal with the uncertainty around the provided forecast is by building scenarios. These practically derivate from the produced forecasts, but we will treat them as an indispensable part of the AFA process.

We have seen the input; we have roughly seen the steps within the "grey-box"; let us stay a bit more on the output of AFA. When you started reading this chapter, you probably thought it would all be about a number or a few numbers—if forecasts for more periods ahead were required. By now, it should have become obvious that far more output—in numerical and narrative form—will be available. Practically every step of the AFA process is producing some output, which is consisted by-and-large of what is contained in Table 12.5. AFA Output (which is not exhaustive).

---

[19] Noise is a term met in many sciences. I prefer the electrical engineering definition of it where Noise can block, distort, or change/interfere with the meaning of a message in both human and electronic communication.

[20] Armstrong 2001.

[21] Timmerman and Granger 2004.

[22] Syntetos et al. 2010.

[23] Maris et al. 2007; Bozos et al. 2008.

4 A Few More Interesting FORECASTING Applications

So what are we really fascinated to forecast?

Whenever I say outside my inner academic circle that I am a forecaster or into forecasting/predictive analytics, I typically get three responses:

What's going to be the weather tomorrow?
Can you forecast the numbers for the lottery?
Can you forecast the stock market?

My answer to all these is: ... "Unfortunately NO". And this brings us back to the fundamental question: "what can realistically be forecasted, and what can not?" Maybe more interestingly the latter is what people are really interested in forecasting...

The aforementioned questions are clearly beyond the scope of this chapter... particularly the weather! The following list is not exhaustive, but we believe captures most of the things people are interested to forecast:

- Gambling / Individual and team performance in sports.
- Weather forecasting.
- Transportation forecasting.
- Economic forecasting/Major Economic shocks.
- Technology forecasting.
- Earthquake prediction/Major catastrophes.
- Land use/Real estate forecasting.
- Long term/Strategic forecasting/Foresight.

5 Evaluating Forecast Accuracy

The measures such as mean absolute error (MAE), mean squared error (MSE), root mean squared error (RMSE), and mean absolute percent error (MAPE) are used to evaluate accuracy of a particular forecast (Hyndman 2014).

Let, y_t denote the tth observation and $\widehat{y}_{t|t-1}$ denotes its forecast based on all previous data, where $t = 1, 2 \ldots T$. Then, the following measures, mean absolute error, means square error, root mean squared error, and mean absolute percentage error are useful.

$$MAE = T^{-1} \sum_{t=1}^{T} \left| \left(y_t - \widehat{y}_{t|t-1} \right) \right|$$

$$MSE = T^{-1} \sum_{t=1}^{T} \left(y_t - \widehat{y}_{t|t-1} \right)^2$$

$$RMSE = \sqrt{\left(T^{-1}\sum_{t=1}^{T}(y_t - \widehat{y}_{t|t-1})^2\right)}$$

$$MAPE = 100T^{-1}\sum_{t=1}^{T}\frac{|y_t - \widehat{y}_{t|t-1}|}{|y_t|}$$

Some practical considerations: It is not appropriate to compare these error measures across models that do and do not incorporate trend and seasonality. The reason is that even those accommodations can be somewhat ad hoc. Moreover, the errors could be correlated, and therefore adding them across periods might create a wrong notion of overall accuracy. ACF diagnoses of the errors might reveal patterns that can help identify seasonality and trends as mentioned in several places in the chapter. In addition, out of sample testing is also recommended.

The error estimates are for the forecast for the current period. However, they can also be computed two steps or three steps ahead. This is done by comparing the actual versus the forecast obtained for that period, but two or three periods ago.

Often, prediction intervals are necessary to the user. For example, one might like to know how the uncertainty in the forecast values changes for different forecast horizons. The inventory planner would like to use the prediction interval to source sufficient number of parts to ensure s/he does not run out of stock more than 10% of the time. A farmer might like to know what extent price can deviate if the produce is harvested and sold next week instead of right now. As one may anticipate, making these types of predictions involves making assumptions about the structure of error distribution. Forecasting packages often produce 1.25 times the MAD as an estimate of standard deviation. One may also compute a heuristic interval by using past data, generate a forecast, and determine the interval by trial and error.

6 Conclusion

This chapter aims to provide managers/executives as well as managers-to-be (PG/MBA students), with the necessary background knowledge and software tools to run a successful business forecasting analytics function. Ok ... but there at least 50 titles out there in business forecasting! ... Why do we need another one recapturing the analytics end of? This chapter is not about all the things you could possibly do when you are faced with a forecasting task. It is not or about guiding you through a methodology[24] tree, where all possible options are given, and it is up to you to decide where to go. If this is what you are looking for, then the best place to go is www.forecastingprinciples.com (accessed on Feb 22, 2018); led by

---

[24]www.forecastingprinciples.com [Accessed on Oct 1, 2017].

J. Scott Armstrong[25] and the International Institute of Forecasters,[26] where you get a gateway to the amazing world of forecasting free of cost.

Furthermore, this chapter is not about giving you all the underlying theory and mathematics of the discipline. In fact, mathematics and statistics, theorems, and axioms are kept to the absolute minimum. "Everything is kept as simple as possible, ... but not simpler!"[27] So there will be a few formulae, but expressed in a way that does not require a mathematical background to follow. If you were looking for the mathematics of forecasting then the leading textbook of the field "Forecasting Methods and Applications"—by Makridakis et al. (1998)—is your reference point. For engineers like me, that prefer the "do it yourself" approach, the second edition of the latter book is particularly useful as most of the forecasting algorithms are presented in such a way that their implementation is very straightforward in a standard programming language.

Now, if you need subjective approach then and judgmental forecasting is your weapon of choice when approaching forecasting tasks, asked then Goodwin (2006) and along with Wright and Goodwin (1998), are probably the way to go.

A *forecasting process* we strongly believe will significantly enhance the forecasting performance in your company/private or public organization; and is a process consisting roughly of two basic elements:

(a) A fairly accurate set of *forecasts*.
(b) A good estimate of the *uncertainty* around them.

Of course, it would be up to you, once faced with real-life problems, how to use these forecasts, and more importantly how to take countermeasures and back-up policies as to cope with the predicted uncertainty. Living with scenarios built around this uncertainty is the key to your business success.

Electronic Supplementary Material

All the datasets, code, and other material referred in this section are available in www.allaboutanalytics.net.

- Data 12.1: Data - ARIMA.csv
- Data 12.2: Data - Croston and SBA.csv
- Data 12.3: Data - Damped Holt.csv
- Data 12.4: Data - SES, ARRSES, Holt, HoltWinter.csv
- Data 12.5: Data - Theta.csv
- Data 12.6: FA - Excel Template.xlsx
- Code 12.1: Forecasting Analytics.R
- Data 12.7: Forecasting chapter - Consolidated Output.xlsx

---

[25]Professor J. Scott Armstrong, http://www.jscottarmstrong.com/ [Accessed on Oct 1, 2017].
[26]International Institute of Forecasters, https://forecasters.org/ [Accessed on Oct 1, 2017].
[27]A famous quote attributed to Albert Einstein.

Exercises

Ex. 12.1 Using time-series data on annual production of tractors in India from 1991 to 2016, provide forecast of the production of tractors (in millions) for the year 2017 using Theta model (combining regression and SES methods, $\alpha = 0.4$).

Year	1991	1992	1993	1994	1995	1996	1997	1998	1999	2000	2001	2002	2003
Production	14.5	14.8	15.1	15.4	15.7	16.0	16.3	16.7	17.0	17.3	17.7	18.0	18.4
Year	2004	2005	2006	2007	2008	2009	2010	2011	2012	2013	2014	2015	2016
Production	18.8	19.3	19.8	20.3	20.8	21.3	21.9	22.4	23.0	23.5	24.1	24.7	25.4

Ex. 12.2 Given the monthly production (in millions) of mobiles by a company for the last 20 months, provide the forecast of mobile production for the company for the next 3 months using the ARIMA (2,1,2) model.

Months	1	2	3	4	5	6	7	8	9	10
Production	3.40	3.43	3.47	3.50	3.54	3.57	3.61	3.65	3.68	3.72
Months	**11**	**12**	**13**	**14**	**15**	**16**	**17**	**18**	**19**	**20**
Production	3.77	3.82	3.77	3.82	3.77	3.82	3.77	3.82	3.77	3.82

Ex. 12.3 Monthly demand (in millions) for an automobile spare part is given as follows:

Months	1	2	3	4	5	6	7	8	9	10	11	12
Demand	3	0	1	0	0	8	0	0	0	2	0	5
Months	**13**	**14**	**15**	**16**	**17**	**18**	**19**	**20**	**21**	**22**	**23**	**24**
Demand	0	0	0	1	4	0	0	0	3	?	?	?

Based on the above time-series data, provide 3 months ahead forecast for the spare part using Croston and SBA methods.

Ex. 12.4 Using the provided Excel templates, create:

(a) A version of Holt exponential smoothing where both the level smoothing parameter and the trend smoothing parameter are equal,
(b) A version of damped Holt exponential smoothing where alpha (α) = a, beta (β) = a2, and phi (φ) = a3.

Appendix 1

Example: The monthly sales (in million USD) of Vodka is given for the period 1968–1970. We want to forecast the sales for the year 2016 using various forecasting methods—SES, ARRSES, Holt, Holt–Winters (Additive/Multiplicative).

Data: The data can be downloaded from the book's website and the dataset name is "*Data - SES, ARRSES, Holt, HoltWinter.csv*". You can also refer to Table 12.6 for data.

R Code (to read data)
```
read.csv ("filename.ext", header = TRUE)
```

SES Method
Install forecast package
```
install.packages("forecast")
```

R function
```
ses (<Univariate vector of observations>, h = <number of periods
   to forecast)
```

Note: The *ses* function in R by default optimizes both the value of alpha and the initial value.

In case you prefer the output for a specified alpha value then use parameter

```
<initial = "simple">
```

and set the alpha value in the parameters.

```
ses(<univariate vector of observations>, h = <number of periods
   to forecast>, alpha = < >, initial = "simple")
```

The above code will set the first forecast value equal to first observation. If alpha is omitted it will optimize for alpha.

Table 12.6 Data for SES, Holt, ARRSES, and Holt–Winters method

Period (t)	Vodka (Y_t)	Period (t)	Vodka (Y_t)	Period (t)	Vodka (Y_t)
Jan-68	42	Jan-69	21	Jan-70	47
Feb-68	40	Feb-69	31	Feb-70	38
Mar-68	43	Mar-69	33	Mar-70	91
Apr-68	40	Apr-69	39	Apr-70	107
May-68	41	May-69	70	May-70	89
Jun-68	39	Jun-69	79	Jun-70	116
Jul-68	46	Jul-69	86	Jul-70	117
Aug-68	44	Aug-69	125	Aug-70	274
Sep-68	45	Sep-69	55	Sep-70	137
Oct-68	38	Oct-69	66	Oct-70	171
Nov-68	40	Nov-69	93	Nov-70	155
Dec-68	49	Dec-69	99	Dec-70	143

Holt Method
Install forecast package

```
install.packages("forecast")
```

R function.

```
holt (<univariate vector of observations>, h = <number of
periods to forecast>)
```

Note: The *holt* function by default optimizes both the value of alpha and the initial value.

In case you prefer the output for a specified alpha and beta value then use

```
<initial = "simple">
```

parameter and set the alpha and beta values in the parameters.

```
holt (<univariate vector of observations>, h = <number of
periods to forecast>, alpha = < >, beta = < >, initial =
"simple")
```

The above code sets first level equal to first value and trend as difference of first two values. If alpha is omitted it will optimize for alpha.

Holt–Winters Method
Install stats package

```
install.packages("stats")
```

R function

```
HoltWinters (<name of dataset>, alpha = <>, beta = <>, gamma =
<>, seasonal = c("additive", "multiplicative"), start.periods =
2, l.start = NULL, b.start = NULL, s.start = NULL, optim.start
= c(alpha = 0.3, beta = 0.1, gamma = 0.1),
optim.control = list())
```

The value of *alpha*, *beta*, and *gamma* can be either initialized by specifying <alpha>, <beta>, <gamma> and if they are NULL it will optimize the values as specified in optim.start. You can also specify starting values of *alpha*, *beta*, and *gamma* to optimize using <*optim.start*> parameter. Seasonality can be considered *additive* or *multiplicative*. The <*start.periods*> is the initial data used to start the forecast (minimum 2 seasons of data). Starting values of level <*l.start*>, trend <*b.start*>, and seasonality <*s.start*> can be either be initialized or optimized by setting equal to NULL.

For the *HoltWinters* function, the dataset must be defined as a time-series (ts) type. A dataset can be converted to time-series type, using the below code:

```
ts (<name of dataset>, frequency = number of periods in a season)
```

Damped Holt Method

Data: The data can be downloaded from the book's website and the dataset name is "*Data - Damped Holt.csv*". You can also refer to Table 12.7 for data.

Install forecast package

```
install.packages("forecast")
```

Table 12.7 Data for damped exponential smoothing using Holt's method

Period (t)	Demand (Y_t)	Period (t)	Demand (Y_t)
1	818	8	805
2	833	9	808
3	817	10	817
4	818	11	836
5	805	12	855
6	801	13	853
7	803	14	851

Table 12.8 Data for Theta model

Time (t)	Cars (Y_t)	Time (t)	Cars (Y_t)
1	13.31	13	18.12
2	13.6	14	18.61
3	13.93	15	19.15
4	14.36	16	19.55
5	14.72	17	20.02
6	15.15	18	20.53
7	15.6	19	20.96
8	15.94	20	21.47
9	16.31	21	22.11
10	16.72	22	22.72
11	17.19	23	23.3
12	17.64	24	23.97

R function

```
holt (<univariate vector of observations>, h = <number of
 periods to forecast>, damped = TRUE)
```

Note: The *holt* function by default optimizes both the value of alpha and the initial value.

In case you prefer the output for a specified alpha, beta, and phi values then use *<initial = "simple">* parameter and set the alpha, beta, and phi values in the parameters.

```
holt (<univariate vector of observations>, h = <number of
 periods to forecast>, damped = TRUE, alpha = < >, beta = < >,
 phi =   < >)
```

If *alpha*, *beta*, and *phi* are omitted, it will optimize for these values.

Theta method

Data: The data can be downloaded from the book's website and the dataset name is "*Data - Theta.csv*". You can also refer to Table 12.8 for data.

Install forecTheta package

```
Install.packages("forectheta")
```

R function

```
stm (ts (<univariate vector of observations>), h = <number of
 periods to forecast>, par_ini = c (y[1]/2, 0.5,2))
```

Refer https://cran.r-project.org/web/packages/forecTheta/forecTheta.pdf for more details.

Note: You may try either "stm" or "stheta." There is a slight difference in the implementation of the original method.

ARIMA method

Data: The data can be downloaded from the book's website and the dataset name is "*Data - ARIMA.csv*".

Install forecast package

```
install.packages("forecast")
```

R function

```
arima (ts (<univariate vector of observations>, freq = <period
  of data>), order = c(<p>,<d>,<q>))
```

To view the fitted coefficients, store the output and call that array.

To forecast, use the command:

```
forecast (<name of output>, h = <number of periods to forecast>)
```

Croston and SBA method

Data: The data can be downloaded from the book's website and the dataset name is "*Data - Croston and SBA.csv*". You can also refer to Table 12.9 for data.

Install tsintermittent package

```
install.packages("tsintermittent")
```

R function

```
crost (ts(<univariate vector of observations>), h = <number of
  periods to forecast>, w = c(<>,<>), init = c(<>,<>), type =
  "croston", init.opt = FALSE)
```

Refer https://cran.r-project.org/web/packages/tsintermittent/tsintermittent.pdf for more details.

<*crost*> function operates on the time-series vector. Initial values can be either differently chosen or provided as a vector of demand and interval value. <*type*> refers to the model used. Cost to the optimization criterion. If <*init.opt*> is *TRUE*, it will optimize the initial values. If <*w*> is *NULL*, it will optimize the smoothing parameters.

Table 12.9 Data for Croston and SBA model

Months (t)	Actual Demand Number (Y_t)
1	5
2	0
3	7
4	28
5	0
6	0
7	11
8	0
9	4
10	19
11	0

Consolidated Forecast Output for Vodka Example

See Tables 12.10, 12.11, 12.12, 12.13, and 12.14.

Table 12.10 Consolidated output of SES, ARRSES, Holt, Holt–Winters methods (*R Output, ^Excel Output)

Period	Vodka	SES^	ARRSES^	Holt^	Holt–Winters Additive*	Holt–Winters Multiplicative*
t	Y_t	F_t	F_t	F_t	F_t	F_t
Jan-68	42	42.0000	42.0000	42.0000	–	–
Feb-68	40	42.0000	42.0000	42.0000	–	–
Mar-68	43	41.0000	41.4000	41.2474	–	–
Apr-68	40	42.0000	41.8800	43.7804	–	–
May-68	41	41.0000	41.3160	43.2701	–	–
Jun-68	39	41.0000	41.1590	43.5794	–	–
Jul-68	46	40.0000	39.9062	42.5508	–	–
Aug-68	44	43.0000	45.2167	46.1132	–	–
Sep-68	45	43.5000	44.4632	46.6752	–	–
Oct-68	38	44.2500	44.5858	47.4834	–	–
Nov-68	40	41.1250	42.1887	43.7545	–	–
Dec-68	49	40.5625	40.5625	43.0670	–	–
Jan-69	21	44.7813	47.5915	47.9867	13.3136	17.0694
Feb-69	31	32.8906	34.1680	34.0851	25.4230	24.3483
Mar-69	33	31.9453	31.8512	33.1256	30.4248	25.8316
Apr-69	39	32.4727	32.7419	33.9700	40.4559	31.7168
May-69	70	35.7363	36.4148	38.0008	71.4978	57.2978
Jun-69	79	52.8682	45.2818	58.0939	79.4149	64.2547
Jul-69	86	65.9341	75.3264	73.0838	73.0716	61.8609
Aug-69	125	75.9670	85.5710	84.1672	84.3076	65.0093
Sep-69	55	100.4835	123.8677	111.8418	117.0539	73.0378
Oct-69	66	77.7418	55.6361	84.0182	82.3369	65.9793
Nov-69	93	71.8709	59.8772	76.2076	75.8113	71.2471
Dec-69	99	82.4354	65.2778	87.9222	93.4888	86.3133
Jan-70	47	90.7177	80.4700	97.0133	69.5189	44.0442
Feb-70	38	90.7177	55.8669	97.0133	63.7863	63.7920
Mar-70	91	90.7177	53.2861	97.0133	46.3351	66.0826
Apr-70	107	90.7177	70.0641	97.0133	75.7723	77.0393
May-70	89	90.7177	85.5019	97.0133	126.1014	136.6830
Jun-70	116	90.7177	88.0616	97.0133	114.8522	150.6100
Jul-70	117	90.7177	109.1998	97.0133	109.6536	158.7803
Aug-70	274	90.7177	116.2344	97.0133	119.5669	219.8930
Sep-70	137	90.7177	262.3861	97.0133	208.5555	98.9493
Oct-70	171	90.7177	137.8308	97.0133	185.8300	114.7506
Nov-70	155	90.7177	143.9737	97.0133	195.4645	157.7524
Dec-70	143	90.7177	145.0455	97.0133	–	166.1216

Table 12.11 Forecast using Damped Holt method

Period	Demand	Forecast 1 period ahead^
t	Y_t	F_t
1	818	818.0000
2	833	824.3000
3	817	835.8020
4	818	829.1143
5	805	823.9495
6	801	811.7567
7	803	802.4828
8	805	799.7078
9	808	800.7737
10	817	804.2789
11	836	812.5885
12	855	829.5993
13	853	850.6078
14	851	858.8332

Table 12.12 Forecast using Theta method

Year	Time t	Cars Y_t	Forecast F_t
1993	1	13.31	
1994	2	13.6	
1995	3	13.93	13.848
1996	4	14.36	14.110
1997	5	14.72	14.440
1998	6	15.15	14.781
1999	7	15.6	15.158
2000	8	15.94	15.564
2001	9	16.31	15.944
2002	10	16.72	16.320
2003	11	17.19	16.709
2004	12	17.64	17.131
2005	13	18.12	17.564
2006	14	18.61	18.016
2007	15	19.15	18.483
2008	16	19.55	18.979
2009	17	20.02	19.437
2010	18	20.53	19.899
2011	19	20.96	20.381
2012	20	21.47	20.842
2013	21	22.11	21.322
2014	22	22.72	21.867
2015	23	23.3	22.437
2016	24	23.97	23.012
2017	**25**		**23.6244**

Table 12.13 Forecast using ARIMA method

Month	Production of Sofa (in Thousands)	Forecast* F_t
1	98	97.9020
2	82	92.7517
3	84	86.6899
4	85	87.2354
5	99	90.2576
6	90	92.3250
7	92	89.9272
8	83	89.0541
9	86	87.5629
10	90	89.2152
11	95	90.6584
12	91	90.7141
13	87	89.0086
14	99	89.5715
15	93	90.8333
16	82	87.9907
17	84	86.9563
18	88	89.0024
19	93	90.6564
20	83	90.5054
21	95	89.8928
22	93	91.6043
23	92	90.1840
24	92	89.4144
25	97	89.2052
26	88	88.5374
27	81	86.6061
28	93	87.6779
29	91	90.0032
30	81	88.9099
31	86	88.8512
32	81	90.9446
33	97	92.2831
34	88	93.9458
35	96	92.0745
36	96	92.2296
37	97	90.4160
38	90	88.5144
39	88	86.7300
40	93	86.9814
41	90	87.4583
42	84	86.7020
43	82	86.9679

(continued)

Table 12.13 (continued)

Month	Production of Sofa (in Thousands)	Forecast*F_t
44	82	88.7995
45	98	91.3190
46	91	92.8558
47	85	90.5459
48	86	90.3639
49	88	91.6516
50	90	92.5493
51		**92.84**
52		**92.47**
53		**91.39**

Table 12.14 Forecast using Croston and SBA methods

Months	Actual demand, Number	Croston Forecast^F_t	SBA Forecast^F_t
1	5	5.0000	4.0000
2	0	5.0000	4.0000
3	7	4.1429	3.3143
4	28	11.8387	9.4710
5	0	11.8387	9.4710
6	0	11.8387	9.4710
7	11	6.7942	5.4354
8	0	6.7942	5.4354
9	4	4.8438	3.8750
10	19	8.4280	6.7424
11	0	8.4280	6.7424

References

Andrawis, R. R., Atiya, A. F., & El-Shishiny, H. (2011). Forecast combinations of computational intelligence and linear models for the NN5 time series forecasting competition. *International Journal of Forecasting, 27*, 672–688.

Armstrong, J. S. (2001). *Principles of forecasting: A handbook for researchers and practitioners*. Dordrecht: Kluwer Academic Publishers.

Assimakopoulos, V., & Nikolopoulos, K. (2000). The theta model: A decomposition approach to forecasting. *International Journal of Forecasting, 16*, 521–530.

Babai, M. Z., Ali, M., & Nikolopoulos, K. (2012). Impact of temporal aggregation on stock control performance of intermittent demand estimators: Empirical analysis. *OMEGA: The International Journal of Management Science, 40*, 713–721.

Bollerslev, T., Engle, R. F., & Nelson, D. B. (1994). ARCH models. In R. F. Engle & D. L. McFadden (Eds.), *Handbook of econometrics* (Vol. 4, pp. 2959–3038). Amsterdam: North-Holland.

Box, G. E. P., & Jenkins, G. M. (1970). Time series analysis: Forecasting and control. San Francisco, Holden Day (revised ed. 1976).

Bozos, K., Nikolopoulos, K., & Bougioukos, N. (2008). Forecasting the value effect of seasoned equity offering announcements. In *28th international symposium on forecasting ISF 2008, June 22–25 2008*. France: Nice.

Brown, R. G. (1956). *Exponential smoothing for predicting demand.* Cambridge, MA: Arthur D. Little Inc.

Chatfield, C. (2005). Time-series forecasting. *Significance, 2*(3), 131–133.

Croston, J. D. (1972). Forecasting and stock control for intermittent demands. *Operational Research Quarterly, 23,* 289–303.

Efron, B. (1979). Bootstrap methods: Another look at the jackknife. *The Annals of Statistics, 7,* 126.

Engle, R. F. (1982). Autoregressive conditional heteroscedasticity with estimates of the variance of the United Kingdom inflation. *Econometrica, 50,* 987–1008.

Goodwin, P. (2006). Decision Analysis for Management Judgement, 3rd Edition Chichester: Wiley.

Hanke, J. E., & Wichern, D. W. (2005). *Business forecasting* (8th ed.). Upper Saddle River: Pearson.

Harrison, P. J., & Stevens, C. F. (1976). Bayesian forecasting. *Journal of the Royal Statistical Society (B), 38,* 205–247.

Hibon, M., & Makridakis, S. (2000). The M3 competition: Results, conclusions and implications. *International Journal of Forecasting, 16,* 451–476.

Holt, C. C. (1957). Forecasting seasonals and trends by exponentially weighted averages. O. N. R. Memorandum 52/1957. Pittsburgh: Carnegie Institute of Technology. Reprinted with discussion in 2004. *International Journal of Forecasting, 20,* 5–13.

Hyndman, R. J. (2014). *Forecasting – Principle and practices.* University of Western Australia. Retrieved July 24, 2017, from robjhyndman.com/uwa.

Johnston, F. R., Boylan, J. E., & Shale, E. A. (2003). An examination of the size of orders from customers, their characterization and the implications for inventory control of slow moving items. *Journal of the Operational Research Society, 54*(8), 833–837.

Jose, V. R. R., & Winkler, R. L. (2008). Simple robust averages of forecasts: Some empirical results. *International Journal of Forecasting, 24*(1), 163–169.

Keast, S., & Towler, M. (2009). *Rational decision-making for managers: An introduction.* Hoboken, NJ: John Wiley & Sons.

Kourentzes, N. (2014). Improving your forecast using multiple temporal aggregation. Retrieved August 7, 2017, from http://kourentzes.com/forecasting/2014/05/26/improving-forecasting-via-multiple-temporal-aggregation.

Kourentzes, N., Petropoulos, F., & Trapero, J. R. (2014). Improving forecasting by estimating time series structural components across multiple frequencies. *International Journal of Forecasting, 30,* 291–302.

Leven and Segerstedt. (2004). Referred to in Syntetos and Boylan approximation section.

Lindsey, M., & Pavur, R. (2008). A comparison of methods for forecasting intermittent demand with increasing or decreasing probability of demand occurrences. In K. D. Lawrence & M. D. Geurts (Eds.), *Advances in business and management forecasting (advances in business and management forecasting)* (Vol. 5, pp. 115–132). Bingley, UK: Emerald Group Publishing Limited.

Makridakis, S., Hogarth, R., & Gaba, A. (2009). *Dance with chance: Making luck work for you.* London, UK: Oneworld Publications.

Makridakis, S., Wheelwright, S. C., & Hyndman, R. J. (1998). *Forecasting: Methods and applications* (3rd ed.). New York: John Wiley and Sons.

Maris, K., Nikolopoulos, K., Giannelos, K., & Assimakopoulos, V. (2007). Options trading driven by volatility directional accuracy. *Applied Economics, 39*(2), 253–260.

Nikolopoulos, K., Assimakopoulos, V., Bougioukos, N., Litsa, A., & Petropoulos, F. (2011a). The theta model: An essential forecasting tool for supply chain planning. *Advances in Automation and Robotics, 2,* 431–437.

Nikolopoulos, K., Syntetos, A., Boylan, J., Petropoulos, F., & Assimakopoulos, V. (2011b). ADIDA: An aggregate/disaggregate approach for intermittent demand forecasting. *Journal of the Operational Research Society, 62,* 544–554.

Petropoulos, F., Makridakis, S., Assimakopoulos, V., & Nikolopoulos, K. (2014). 'Horses for Courses' in demand forecasting. *European Journal of Operational Research, 237,* 152–163.

Prestwich, S. D., Tarim, S. A., Rossi, R., & Hnich, B. (2014). Forecasting intermittent demand by hyperbolic-exponential smoothing. *International Journal of Forecasting, 30*(4), 928–933.

Rostami-Tabar, B., Babai, M. Z., Syntetos, A. A., & Ducq, Y. (2013). Demand forecasting by temporal aggregation. *Naval Research Logistics, 60*, 479–498.

Spithourakis, G. P., Petropoulos, F., Babai, M. Z., Nikolopoulos, K., & Assimakopoulos, V. (2011). Improving the performance of popular supply chain forecasting techniques: An empirical investigation. *Supply Chain Forum: An International Journal, 12*, 16–25.

Syntetos, A. A., & Boylan, J. E. (2001). On the bias of intermittent demand estimates. *International Journal of Production Economics, 71*, 457–466.

Syntetos, A. A., & Boylan, J. E. (2005). The accuracy of intermittent demand estimates. *International Journal of Forecasting, 21*, 303–314.

Syntetos, A. A., Nikolopoulos, K., & Boylan, J. E. (2010). Judging the judges through accuracy-implication metrics: The case of inventory forecasting. *International Journal of Forecasting, 26*, 134–143.

Taylor, A. R. (1997). On the practical problems of computing seasonal unit root tests. *International Journal of Forecasting, 13*(3), 307–318.

Teunter, R. H., Syntetos, A., & Babai, Z. (2011). Intermittent demand: Linking forecasting to inventory obsolescence. *European Journal of Operational Research, 214*, 606–615.

Thomakos, D. D., & Nikolopoulos, K. (2014). Fathoming the theta method for a unit root process. *IMA Journal of Management Mathematics, 25*, 105–124.

Timmerman, A., & Granger, C. W. J. (2004). Efficient market hypothesis and forecasting. *International Journal of Forecasting, 20*, 15–27.

Tseng, F., Yu, H., & Tzeng, G. (2002). Combining neural network model with seasonal time series ARIMA model. *Technological Forecasting and Social Change, 69*, 71–87.

Willemain, T. R., Smart, C. N., & Schwarz, H. F. (2004). A new approach to forecasting intermittent demand for service parts inventories. *International Journal of Forecasting, 20*, 375–387.

Wright, G., & Goodwin, P. (1998). *Forecasting with judgement.* Chichester and New York: John Wiley and Sons.

Chapter 13
Count Data Regression

Thriyambakam Krishnan

1 Introduction

Business analysts often encounter data on variables which take values 0, 1, 2, ... such as the number of claims made on an insurance policy; the number of visits of a patient to a particular physician; the number of visits of a customer to a store; etc. In such contexts, the analyst is interested in explaining and/or predicting such outcome variables on the basis of explanatory variables. In insurance, the expected number of claims per year in terms of the policy holder's characteristics helps to set premium rates for various insurer profiles. In a recreational example, how much to charge for particular facilities depending on the participants' profile can be determined from such regression exercises. The number of visits to a physician and such patient information can be modeled to optimize health-care resource uses. The number of customer arrivals, the number of new product launches, the number of items purchased in a grocery store and such phenomena can be modeled to determine business strategies in a retail context. In all these cases, a standard linear regression model is not suitable and models such as the Poisson regression model, negative

---

Electronic Supplementary Material The online version of this chapter (https://doi.org/10.1007/978-3-319-68837-4_13) contains supplementary material, which is available to authorized users.

T. Krishnan (✉)
Chennai Mathematical Institute, Chennai, India
e-mail: sridhar@illinois.edu

© Springer Nature Switzerland AG 2019
B. Pochiraju, S. Seshadri (eds.), *Essentials of Business Analytics*, International Series in Operations Research & Management Science 264,
https://doi.org/10.1007/978-3-319-68837-4_13

binomial regression model, etc. are more appropriate. These models help unravel the distributional effects of influencing factors rather than merely mean effects. Furthermore, extensions of these models, called zero-inflated models, help tackle high incidence of 0 counts in the data. This chapter covers the following:

- Understanding what a count variable is
- Getting familiar with standard models for count data like Poisson and negative binomial
- Understanding the difference between a linear regression model and a count data regression model
- Understanding the formulation of a count data regression model
- Becoming familiar with estimating count data regression parameters
- Learning to predict using a fitted count data regression
- Learning to validate a fitted model
- Learning to fit a model with an offset variable

2 Motivating Business Problems

The following are specific examples of business problems that involve count data analysis. We list the response and predictor variables below:

- In a study of the number of days of reduced activity in the past 2 weeks due to illness or injury, the following predictors are considered: gender, age, income, as well as the type of medical insurance of the patient.
- In an application of Poisson regression, the number of fish (remember "poisson" means fish in French) caught by visitors to a state park is analyzed in terms of the number of children in the group, camping one or more nights during stay (binary variable), and the number of persons in the group.
- In an example like the one above, there is scope for an excessive number of zero counts and so a zero-inflated model might turn out to be appropriate.
- In an insurance application, the issue is one of predicting the number of claims that an insurer will make in 1 year from third-party automobile insurance. The predictor variables are the amount insured, the area they live in, the make of the car, the no-claim bonus they received in the last year, the kilometers they drove last year, etc. A zero-inflated model called the Hurdle model has been found to be a reasonable model for the data.
- In another insurance example, we want to understand the determinants of the number of claims "Claim count": this is a count variable (discrete; integer values) where the possible explanatory variables are: the number of vehicles in the policy (an integer numeric variable) and age of the driver.
- In an experiment in AT & T Bell Laboratories, the number of defects per area of printed wiring boards by soldering their leads on the board was related to five possible influences on solderability.

3 Methods of Count Data

3.1 Poisson Regression Model

Let us consider the fifth problem in the list above. A part of the data is given below. The entire dataset named *"numclaims.csv"* is available on the book's website. The columns correspond to the number of claims, the number of vehicles insured, and the age of the insured (Table 13.1).

Is an ordinary least-squares linear regression an appropriate model and method for this problem? No, because this method assumes that the errors (and hence the conditional distributions of the claims count given the predictor values) are normal and also that the variances of the errors are the same (homoskedasticity). These assumptions are not tenable for claims count since it is a discrete variable. Moreover, the claims count is more likely distributed Poisson and hence for different values of the predictor variables the Poisson means and hence the Poisson variance will be different for different cases.

Poisson regression is appropriate when the conditional distributions of Y are expected to be Poisson distributions. This often happens when you are trying to regress on count data (for instance, the number of insurance claims in 5 years by a population of auto-insurers). Count data will be, by its very nature, discrete as opposed to continuous. When we look at a Poisson distribution, we see a spiked and stepped histogram at each value of X, as opposed to a smooth continuous curve. Moreover, the Poisson histogram is often skewed. Further, the distribution of Y, for a small value of Y, is not symmetric. In a Poisson regression the conditional distribution of Y changes shape and spreads as Y changes. However, a Poisson distribution becomes normal shaped, and wider, as the Poisson parameter (mean) increases.

The conditional distribution graphs for normal and Poisson are given below where the differences in the assumptions are apparent. Note that the model implies heteroskedasticy since for the Poisson distribution mean also equals variance (Fig. 13.1).

A Poisson distribution-based regression model could be stated as

$$log(\mu) = \beta_0 + \beta_1 x_1 + \beta_2 x_2 + \ldots + \beta_p x_p$$

Table 13.1 Insurance claims count table

numclaims	numveh	Age
1	3	41
1	3	42
1	3	46
1	3	46
0	1	39

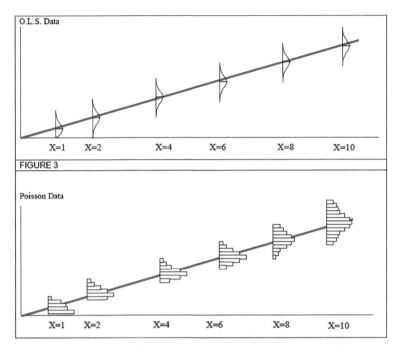

Fig. 13.1 Conditional distributional graphs for normal and Poisson distributions. Source: An Animated Guide: An Introduction To Poisson Regression Russ Lavery, NESUG 2010

where μ is the expected value of y an assumed Poisson distributed response (dependent) variable and $x = (x_1, x_2, \ldots, x_p)$ are predictor (independent) variables. In general the x_i variables need not appear in a linear form.

In general, the x_i variable may be a combination of variables in the form $x_j * x_k$ meaning interaction between these two variables (or factors). A reason for modeling the $\log(\mu)$ rather than μ itself is that it has been observed in practice that with count data effects of predictors are often multiplicative rather than additive—small counts produce small effects and large counts produce large effects. Whereas in a linear regression model the regression coefficients indicate the absolute change in the expected value of the response variable for a unit change in the predictor value, in the Poisson regression model these regression coefficients indicate the **relative** change in the expected value of the response variable. Also, note that deviation from the assumed linear form on the right-hand side will lead to a poor fit. Like in all regression models some trial and error is required to identify the "best" model. The case study in this chapter will illustrate the idea in the context of count data models.

Another reason for using log on the left side (called log link) is that the right-hand side can take any real value and the Poisson variable can take only non-negative values; this issue is resolved by taking the log of the expected value of Y.

13 Count Data Regression

This model is an example of what are called generalized linear models. Generally, maximum likelihood estimates of the regression parameters or their approximations are used. Once the expected value of the response Poisson variable is worked out, the probabilities of various possible values of the response are immediately worked out.

3.2 Poisson Regression Results for the Number of Claims Data

The R command and results are as follows. The interpretation is given in the next section (Table 13.2).

Table 13.2 Poisson regression R output

```
> pois<- glm(insurance ~ numveh + age, family="poisson",
data=numclaims)
> summary(pois)

Call:
glm(formula = numclaims ~ numveh + age, family = "poisson",
data=numclaims)

Deviance Residuals:
    Min       1Q   Median       3Q      Max
-2.1840  -0.9003  -0.5891   0.3948   2.9539

Coefficients   Estimate  Std. Error  z value  Pr(>|z|)
Intercept      -5.578057    0.676823   -8.242    <2e-16
numveh          0.123273    0.163261    0.755      0.45
age             0.086121    0.009586    8.984    <2e-16
---
Signif. codes:  0 '***' 0.001 '**' 0.01 '*' 0.05 '.' 0.1 ' ' 1

(Dispersion parameter for poisson family taken to be 1)

Null deviance: 287.67 on 199 degrees of freedom
Residual deviance: 203.45 on 197 degrees of freedom
AIC: 385.51
Number of Fisher Scoring iterations: 6

> logLik(pois)
'log Lik.' -189.753 (df=3)

> with(pois,cbind(res.deviance = deviance,df=df.residual,\\
p = pchisq(deviance, df.residual,lower.tail=FALSE)))

res.deviance    df         p
  203.4512     197   0.3612841
```

3.2.1 Regression Coefficients and Their Significance

The Poisson regression coefficients, their standard errors, z-scores for testing the hypothesis of the regression coefficient to be zero, and the p-values are given. The regression coefficient for age is 0.086121 which means that the expected log(count) for 1 year increase in age is 0.086121 and so the ratio of counts at age $x + 1$ to age x is $\exp(0.086121) \approx 1.09$. Similarly for the number of vehicles this ratio (often called incident rates) is $\exp(0.123273) = 1.131193$.

3.2.2 Deviance, Deviance Residuals, and Residual Deviance

A saturated model is one which contains a separate indicator parameter for each observation and so fits the data as closely as possible. Perfect fit means: $\mu_i = y_i$. This is not useful since there is no data reduction, since the number of parameters equals the number of observations. This model attains the maximum achievable log likelihood (equivalently the minimum of $-2 \log L_s$). This is used as a baseline for comparison to other model fits.[1]

The residual deviance is defined as

$$D_m \equiv 2(\log L_s - \log L_m)$$

where L_m is the maximized likelihood under the model in question and L_s is the maximized likelihood under a saturated model. In this case $\log(L_m)$ can be obtained using the *logLik* function of model in R, in this example we get -189.753. Thus, residual deviance

$$-2(\log L_m - log L_s)$$

is analogous to (and, indeed, is a generalization of) the residual sum of squares for a linear model. Here, residual deviance (also called deviance) is reported as 203.45. This implies $\log L_s = (2 * 189.753 - 203.45)/2 = 88.03$. The smaller the residual deviance, the better is the model since the regressors have captured a large part of the deviance. The chi-square test of 203.45 on 197 degrees of freedom produces a *p*-value of 0.3613 indicating a good fit. One may alternately say that we are testing the null hypothesis that the coefficients that are not in the model but in the saturated model are all equal to zero. The high *p*-value says we cannot reject this hypothesis. A thumb rule for a good fit is that the residual deviance is smaller than its degrees of freedom. The deviance residuals are the contributions of each case (observation) to the residual deviance and if the model fits well they should be approximately normally distributed around 0. In this case there is a bit of skewness

---

[1]The maximum log likelihood when $\mu_i = y_i$ is given by: $\sum_i (y_i log(y_i) - y_i - log(y_i!))$.

13 Count Data Regression

since the median is slightly different from 0. The deviance reduces as the model fit improves. If the model exactly fits the data, then the deviance is zero. As an approximation

$$D^* \sim \chi^2_{n-\dim(\boldsymbol{\beta})}$$

if the model is correct. The approximation can be good in some cases and is exact for the strictly linear model.

3.2.3 Analysis of Deviance

The residual deviance behaves similar to residual sum of squares of a linear model, therefore it can be used similar to residual variance in least square and is suitable for maximum likelihood estimates. For example, after the exploratory data analysis (EDA) identifies important covariates one can use the partial deviance test to test for significance of individual or groups of covariates. Example: The software reports null deviance, which is the deviance when only one parameter, the mean of all observations, is used to explain the number of claims. The deviance reported is 287.67 on 199 degrees of freedom. The difference in deviance between the null model and the model with three explanatory variables $= 287.67 - 203.45 = 84.22$. The chi-square test with 2 degrees of freedom (i.e., 199–197) yields a p-value close to zero. The two models can also be tested using a standard ANOVA method as shown below:

```
R Code and Output

%%% Model with zero parameters %%%
> pois0<-glm(numclaims~NULL, family="poisson", data=numclaims)
> pois0

Call: glm(formula = numclaims ~ NULL, family = "poisson",
data = numclaims)

Coefficients:
(Intercept)
 -0.462
Degrees of Freedom: 199 Total (i.e. Null);  199 Residual
Null Deviance:     287.7
Residual Deviance: 287.7   AIC: 465.7


 %%% Model with two parameters %%%
> pois<-glm(numclaims~numveh + age, family="poisson",
 data=numclaims)
> pois

Call:  glm(formula = numclaims ~ numveh + age,
 family = "poisson", data = numclaims)
Coefficients:
```

```
(Intercept)        numveh            age
-5.57806          0.12327         0.08612
Degrees of Freedom: 199 Total (i.e. Null);  197 Residual
Null Deviance:     287.7
Residual Deviance: 203.5   AIC: 385.5
> anova(pois, pois1, test="Chisq")
Analysis of Deviance Table
Model 1: numclaims ~ numveh + age
Model 2: numclaims ~ NULL
Resid. Df Resid. Dev Df Deviance  Pr(>Chi)
1        197    203.45
2        199    287.67 -2  -84.221 < 2.2e-16 ***
---
Signif. codes:  0 *** 0.001 ** 0.01 * 0.05 . 0.1   1
```

3.2.4 Residual Deviance and AIC

The residual deviance is used to check goodness of fit or significance of covariates. Another criterion is often used to compare different models. The Akaike information criterion (AIC) is a criterion to compare models. It is possible to reduce residual deviance by adding more parameters to the model even if it is not going to be useful for prediction. In order to control the number of parameters and achieve parsimony in the model, a penalty is introduced for the number of parameters. This penalized value of $-2\log L_m$ is the AIC criterion. AIC is defined as AIC $= -2\log L_m + 2\times$ the number of parameters. Thus in the example above: AIC $= 379.5061 + 2 * 3 = 385.5061$. Evidently the lower the value of AIC, the better is the model. For example, the model with just one parameter, the overall mean, has an AIC value of 465.7 which is higher than 385.5061.

3.2.5 Dispersion Parameter

The error distributions assumed in our models lead to a relationship between mean and variance. For the normal errors it is constant. In Poisson, the mean is equal to variance, hence the dispersion parameter is 1. The dispersion parameter is used to calculate standard errors. In other distributions, it is often considered a parameter and estimated from data and presented in the output.

3.2.6 Prediction Using the Model

We now use the model to predict the number of claims for two cases with the number of vehicles and age of driver as inputs: Case 1: (2, 48), Case 2: (3,50).

R command for creating predictions and confidence limits is
$predict(pois, newdata, interval = "confidence")$

Predictions of the expected value of the number of claims: Case 1: 0.3019, Case 2: 0.4057

with respective standard errors of 0.0448 and 0.0839. You can also calculate these by hand, for example, for Case 1: $-5.578 + 2 * 0.1233 + 48 * 0.0861 = 0.3019$.

3.2.7 Diagnostic Plots

Some of the basic diagnostic plots are illustrated in Fig. 13.2. These are similar to the plots in regression, see Chap. 7 (Linear Regression Analysis). However, brief descriptions are given below for a quick recap.

3.2.8 Residuals vs Fitted

For the model to be a good fit, residuals should lie around the central line like a set of random observations without a pattern. This plot has the predicted values of μ_i on the x-axis and $y_i - \mu_i$ on the y-axis. In this graph (Fig. 13.2), most of the residuals lie on one side of the central line showing unsatisfactory fit.

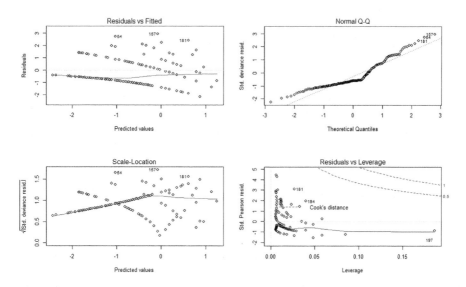

Fig. 13.2 Graphs of fit

3.2.9 Normal Q-Q

Although this is a Poisson model, residuals should behave approximately like a normal distribution for which the normal Q-Q plot should be a straight line, which is not the case here (Fig. 13.2).

3.2.10 Scale-Location Plot

In the Poisson model, the scale (spread) should vary as the location (mean). If the spread is larger than the mean on the whole, it is a sign of *overdispersion*. The graph (Fig. 13.2) shows the ID of cases that violate this phenomenon. This graph does not indicate the expected kind of relationship, showing lack of fit.

3.2.11 Residuals vs Leverage

This plot is meant to find influential cases, that is, those which by themselves change the regression parameters, in terms of a statistic known as Cook's distance. The graph (Fig. 13.2) indicates the IDs of such influential points. One needs to examine the reasons for this and if justified these points may be removed from the dataset.

3.2.12 Exposure Variable and Offset

The counts modeled as a Poisson distribution may depend on another variable which when used in the denominator may define a rate. For instance, in the insurance context the sum insured may be an exposure variable in which case one might like to model the rate: number of claims/sum insured. This situation is handled by multiplying both sides by the exposure variable and taking the log. This results in a term log(exposure) as an additive regressor. The term log(exposure) is often called an offset variable. Ideally the offset regressor should have a coefficient of 1 so that when moved to the left side a rate is defined.

3.3 *Possible Problems with Poisson Regression*

There are two potential problems with Poisson regression: *Overdispersion* and *excessive zeros*. We describe each below along with possible solutions. Poisson distribution has the property that the mean equals variance. However, not infrequently data display the phenomenon of **overdispersion** meaning that the (conditional) variance is greater than the (conditional) mean. One reason for this is omitted or unobserved heterogeneity in the data or an incorrect specification of the model not using the correct functional form of the predictors or not including interaction terms.

The second potential problem is the excess number of 0's in the counts which is more than what is expected from a Poisson distribution, called **zero inflation**. The implication of this situation is that standard errors of regression estimates and their p-values are small. There are statistical tests available for checking this. See the example of the overdispersion test in the next section. One way of dealing with this heterogeneity of data is to specify an alternative distribution model for the data. One such alternative distribution more general than the Poisson is the **negative binomial** distribution, which can be looked upon as a result of modeling the overdispersion as gamma distributed across means. Zero inflation is generally dealt with by modeling separately "true zeros" due to the Poisson process and "excess zeros" by a separate process.

3.4 Negative Binomial Regression

A way of modeling overdispersed count data is to assume a negative binomial (NB) distribution for $y_i|x_i$ which can arise as a gamma mixture of Poisson distributions. One parameterization of its probability density function is

$$f(y; \mu, \theta) = \frac{\Gamma(y+\theta)}{\Gamma(\theta)y!} \frac{\mu^y \theta^\theta}{(\mu+\theta)^{y+\theta}},$$

μ is mean and θ is shape parameter, Γ is the Gamma function and Variance, $V(\mu) = \mu + \frac{\mu^2}{\theta}$.

We illustrate the use of this model with a case study. First, we fit a Poisson model and then the negative binomial model.

4 A Case Study on Canadian Insurance Data: Poisson Model with Offset

The source of this data is: "Poisson regression" by Claudia Czado and TU München,[2] and An Actuarial Note by Bailey and Simon (1960).[3]

The data is provided for private passenger automobile liability for non-farmers for all of Canada excluding Saskatchewan. We have to fit the model to estimate the number of claims using the given data. The raw data *"canautoins.csv"* is available on the book's website.

---

[2]http://www.statistics.ma.tum.de/fileadmin/w00bdb/www/czado/lec6.pdf. Accessed on May 11, 2018.
[3]https://www.casact.org/pubs/proceed/proceed59/59159.pdf. Accessed on May 11, 2018.

Table 13.3 Private passenger automobile liability for non-farmers for all of Canada excluding Saskatchewan

Variable	Description
Merit	3 licensed and accident free ≥ 3 years
	2 licensed and accident free 2 years
	1 licensed and accident free 1 year
	0 all others
Class	1 pleasure, no male operator < 25
	2 pleasure, non-principal male operator < 25
	3 business use
	4 unmarried owner and principal operator < 25
	5 married owner and principal operator < 25
Insured	Number of years the car is insured
Premium	Earned premium in 1000s
Claims	Number of claims
Cost	Total cost of the claim in 1000s of dollars

The variable *Merit* measures the number of full years since the insured's most recent accident or since the insured became licensed. The variable *Class* is a concatenation of age, sex, use, and marital status. The variables *Insured* and *Premium* are two measures of the risk exposure of the insurance companies. The variable *premium* is the premium in 1000s for protection actually provided during the experience period. Please refer to Table 13.3 for the detailed description.

We should observe that we are given the count of claims for each Merit-Class combination. Thus, this data is aggregated over the same Merit-Claim class. First, observe that such aggregation of data over the same category does not change the MLE estimate. In other words, say we had claim data for every insured person. If we ran the MLE estimate for the disaggregated dataset, we would get the same estimate of coefficients and the same significance levels. Second, note that the fully saturated model will include all interaction terms between Merit and Class. Finally, the classification based on Merit and Class, as well as the definition of these categories is based on experience and data analysis done in the past. For further details, see the note by Bailey and Simon (1960).

Results with Log(Insured) as Offset

We estimate the Poisson model where log(Insured) is given as an offset variable. Recall the offset variable is equivalent to modeling a rate instead of a count, in this case, claims per insured years.

Table 13.4 provides the R code and model output.

The residual deviance shows a very poor fit with p-value $4.48229e-16$. The other elements of the output are to be interpreted as indicated in the earlier example. Since there is poor fit, can it be because of overdispersion or zero inflation? First, we test for overdispersion. This requires the additional package in R: *AER*.[4]

---

[4] ftp://cran.r-project.org/pub/R/web/packages/AER/AER.pdf. Accessed on May 11, 2018.

Table 13.4 Results with offset

```
> pois_ofs <- glm(Claims~ Merit + Class + Premium + Cost,\\
offset=log(Insured), data=canautoins, family="poisson")
> summary(pois_ofs)

Deviance Residuals:
   Min      1Q   Median      3Q      Max
-3.526   -1.505   0.196   1.204    4.423

Coefficients:
             Estimate Std. Error z value Pr(>|z|)
(Intercept) -2.081e+00  2.103e-02 -98.953  < 2e-16 ***
merit1      -9.306e-02  1.308e-02  -7.117 1.10e-12 ***
merit2      -1.652e-01  1.549e-02 -10.663  < 2e-16 ***
merit3      -4.067e-01  8.623e-03 -47.162  < 2e-16 ***
class2       2.523e-01  1.897e-02  13.300  < 2e-16 ***
class3       3.965e-01  1.280e-02  30.966  < 2e-16 ***
class4       4.440e-01  9.788e-03  45.356  < 2e-16 ***
class5       1.854e-01  2.414e-02   7.680 1.59e-14 ***
Premium     -8.537e-06  1.391e-06  -6.137 8.39e-10 ***
Cost         2.064e-05  3.902e-06   5.289 1.23e-07 ***
---
Signif. codes:  0 *** 0.001 ** 0.01 * 0.05 . 0.1  1

(Dispersion parameter for poisson family taken to be 1)

Null deviance: 33854.158  on 19  degrees of freedom
Residual deviance:    95.418  on 10  degrees of freedom
AIC: 325.83

> with (pois_ofs, cbind(res.deviance = deviance, df = df.residual,
p = pchisq(deviance, df.residual,lower.tail=FALSE)))
res.deviance df        p
[1,]    95.41849 10 4.48229e-16
```

The R command for the test that the mean is equal to the variance and the output are given below. The additional parameter in the R Command: *trafo=1* means that the ratio of mean and variance = 1.

Test for Overdispersion

```
> dispersiontest(pois_ofs)

Overdispersion test

data:  pois_ofs
z = 2.9394, p-value = 0.001644
alternative hypothesis: true dispersion is greater than 1
sample estimates:
dispersion
  4.784538
```

What does the test say? It rejects the null hypothesis and suggests a scaling of the variance by the factor of 4.784. Next, we estimate the negative binomial model for this data. In this model, we use log(Insured) as an independent variable (to match the use of it as an offset in the Poisson Regression).

This model is a much better fit as shown by the p-value of 0.0246. It is better than that of the Poisson model and also can be seen by the considerable reduction of AIC from 325.83 for the Poisson model to 277.5 for the negative binomial model. We can compute AIC by using definition $-2 \log L_m + 2 * (number of parameters)$. As we are also estimating theta, a shape parameter, it is added in the number of parameters while computing AIC. Thus, in the example above: $AIC = 253.493 + 2*(11+1) = 277.49$. However, at 5% level it is still not a good fit. In other words, we reject the hypothesis that the coefficients in the saturated model that are not in the negative binomial model are all equal to zero. Perhaps, the poor fit might be due to missing variables or due to non-linearity in the underlying variables. These issues can be further explored as detailed in the exercises.

4.1 Models for Zero Inflation: ZIP and ZINB Models and Hurdle Models

In ZIP (zero inflated poisson) and ZINB (zero inflated negative binomial) models, the count variable and the excess zero values are generated by two different processes both regressed on the predictors. The two processes are a Poisson or negative binomial count model (which could produce 0 counts) and a logit model for excess zeros. In contrast a hurdle model assumes all zeros are generated by a process and the positive counts are generated by a truncated Poisson or negative binomial process. Which model to use will depend on the structural way zeros arise and the design of the experiment. They may lead to quite different results and interpretations. We do not go into the details of these models here. The exercises in the chapter illustrate the use of these models (Table 13.5).

We can use *"zeroinfl"* function to run Zero Inflation models in *R*. For ZIP and ZINB model, use "poisson" and "negbin," respectively, in *family* parameter. The sample R command is:

```
zeroinfl(y~ x1+x2, data=inputdata, family="poisson",
  link="logit")
zeroinfl(y~ x1+x2, data=inputdata, family="negbin",
  link="logit")
```

Refer to *pscl* package documentation to know more about running zero inflation models (ZIP, ZINB) in R.

Table 13.5 Negative binomial model results

```
>neg_bin<-glm.nb(Claims~Merit+Class+Premium+Cost+log(Insured),
+ data=canautoins,init.theta=3557)
> summary(neg_bin)

Deviance Residuals:
Min        1Q      Median      3Q        Max
-1.8544   -0.7172  0.1949    0.5656    1.4852

Coefficients:
Estimate Std. Error z value Pr(>|z|)
(Intercept)  -7.163e-01  5.104e-01  -1.403  0.16053
merit1       -2.034e-01  4.031e-02  -5.046  4.50e-07 ***
merit2       -2.829e-01  5.075e-02  -5.575  2.48e-08 ***
merit3       -2.142e-01  7.114e-02  -3.010  0.00261 **
class2       -6.703e-02  1.183e-01  -0.567  0.57082
class3        1.851e-01  8.816e-02   2.099  0.03579 *
class4        2.180e-01  8.312e-02   2.623  0.00873 **
class5       -2.310e-01  1.510e-01  -1.529  0.12625
Premium      -2.435e-06  3.081e-06  -0.790  0.42941
Cost          4.824e-06  8.250e-06   0.585  0.55870
log(Insured)  8.969e-01  3.895e-02  23.025  < 2e-16 ***
---
Signif. codes:  0 *** 0.001 ** 0.01 * 0.05 . 0.1   1

(Dispersion parameter for Negative Binomial(3556.609)
 family taken to be 1)

Null deviance: 127831.875  on 19  degrees of freedom
Residual deviance:     19.072  on  9  degrees of freedom
AIC: 277.49

Number of Fisher Scoring iterations: 1

Theta:  3557
Std. Err.:  1783

2 x log-likelihood:  -253.493
> with(neg_bin, cbind(res.deviance = deviance,df = df.residual,
+ p = pchisq(deviance, df.residual,lower.tail=FALSE)))
res.deviance df         p
[1,]     19.07203  9 0.02458736
```

5 Summary and Further Reading

This chapter introduces count data regression where a response variable is a count (taking values 0, 1, 2, ...) which is regressed on a set of explanatory variables. The basic models for such a regression—the Poisson regression and the negative binomial regression—are introduced and discussed with examples. Methods of measuring goodness of fit and validating the models are also discussed. The problems of overdispersion in the Poisson model and of zero inflation are briefly discussed and solutions to these problems are mentioned. Several excellent texts are listed in the reference section for further reading, such as Cameron and Trivedi (2013), Jackman (2006), Winkelmann (2015), Zeileis et al. (2008), and Simonoff (2003).

Electronic Supplementary Material

All the datasets, code, and other material referred in this section are available in www.allaboutanalytics.net.

- Data 13.1: numclaim.csv
- Data 13.2: canautoins.csv
- Data 13.3: orsteindata.csv
- Code 13.1: count_data.R
- Data 13.4: Additional datasets are available on Jeff Simonoff's website.[5]

Exercises

Ex. 13.1 You are given a sample of subjects randomly selected for an Italian study on the relation between income and whether one possesses a travel credit card (such as American Express or Diner's Club). At each level of annual income in millions of Lira (the currency in Italy before euro), the table indicates the number of subjects sampled and the number of these subjects possessing at least one travel credit card. Please refer to the data *"creditcard.csv"* available on the book's website. The dataset is taken from Pennsylvania State University.[6]

This example has information on individuals grouped by their income, the number of individuals (cases) within that income group and number of credit cards. Notice that the number of individuals is the frequency of the data point and not a regressor.

(a) What is the estimated average rate of incidence, that is, the usage of credit cards given the income?
(b) Is income a significant predictor?
(c) Does the overall model fit?

---

[5]http://people.stern.nyu.edu/jsimonof/AnalCatData/Data/Comma_separated/. Accessed on May 11, 2018.
[6]https://onlinecourses.science.psu.edu/stat504/node/170. Accessed on Apr 15, 2018.

(d) How many credit cards do you expect a person with income of 120 million Lira to have?
(e) Also test for overdispersion and zero inflation.

Ex. 13.2 Ornstein's dataset (*"orsteindata.csv"*) is on interlocking directorates among 248 dominant Canadian firms. The number of "interlocks" for each firm is the number of ties that a firm maintained by virtue of its board members and top executives also serving as board members or executives of other firms in the dataset. This number is to be regressed on the firm's "assets" (billions of dollars), "nation" of control (Canada, the United States, the United Kingdom, or another country), and the principal "sector" of operation of the firm (ten categories, including banking, other financial institutions, heavy manufacturing, etc.) The asymmetrical nature of the response, a large number of 0s make the data not suitable for ordinary least-squares regression. The response is a count.

To understand coding of categorical variable, refer to examples in *Dummy Variable* section in *Chapter 7 (Linear Regression Analysis)*. In this exercise, you can consider "CAN" (Canada) as the reference category for "Nation" variable and "AGR" (Agriculture) for "Sector" variable.

(a) Fit a Poisson regression model for the number of interlocking director and executive positions shared with other major firms. Examine its goodness of fit.
(b) Discuss the results from an economic point of view. Which variables are most important in determining the number of interlocking director and executive positions shared with other major firms?
(c) Fit a negative binomial and compare with Poisson model.
(d) Examine whether adjusting for zero inflation improves the model by fitting ZIP and ZINB models.
(e) Compare the outputs of different models. Which metrics should we look at?
(f) Discuss which model is the best and why. Recommend further steps to improve the model.

Ex. 13.3 Introduce all interaction terms between "Merit" and "Class" in the Canadian Insurance model of Sect. 4. Run the Poisson regression with log(Insured) as offset.

(a) Which interaction terms are significant?
(b) Do you see that this is the fully saturated model because there is only one observation for every unique combination of Merit and Class?
(c) Rerun the model retaining only the significant interaction terms as well as all the original variables. What would you conclude based on this investigation? How does it help an insurance rating agency?

References

Bailey, R. A., & Simon, L. (1960). Two studies in automobile insurance rate-making. *ASTIN Bulletin, 1*, 192–217.

Cameron, A. C., & Trivedi, P. K. (2013). *Regression analysis of count data*. Econometric Society Monograph No. 53. Cambridge: Cambridge University Press.

Jackman, S. D. (2006). *Generalized linear models*. Thousand Oaks: Sage Publications.

Simonoff, J. S. (2003). *Analyzing categorical data*. New York: Springer. http://people.stern.nyu.edu/jsimonof/AnalCatData/.

Winkelmann, R. (May 2015). *Counting on count data models*. Bonn: IZA World of Labor. https://wol.iza.org.

Zeileis, A., Kleiber, C., & Jackman, S. (2008). Regression models for count data in R. *Journal of Statistical Software, 27*, 1–25. http://www.jstatsoft.org/.

Chapter 14
Survival Analysis

Thriyambakam Krishnan

1 Introduction

Survival analysis is a collection of statistical techniques for the analysis of data on "time-to-event" as a response variable and its relationships to other explanatory variables. The notion of "event" depends on the context and the applications. The event in question may be dealt as may happen in a biomedical context or churning in a business context or machine failure in an engineering context. Survival methods are characterized by "censoring" by which the event in question may not have happened (at the time observations end) for certain observational units (cases) in the data; yet, such censored data are useful and are judiciously used in survival analysis. In that sense, survival analysis methods differ from techniques such as regression analysis. The topics covered in this chapter are:

- Understanding time-to-event data and survival probabilities
- Understanding the notion of censoring
- Understanding the survival curve and other ways of representing survival distributions

---

Electronic Supplementary Material The online version of this chapter (https://doi.org/10.1007/978-3-319-68837-4_14) contains supplementary material, which is available to authorized users.

T. Krishnan (✉)
Chennai Mathematical Institute, Chennai, India
e-mail: sridhar@illinois.edu

© Springer Nature Switzerland AG 2019
B. Pochiraju, S. Seshadri (eds.), *Essentials of Business Analytics*, International Series in Operations Research & Management Science 264,
https://doi.org/10.1007/978-3-319-68837-4_14

- Learning to compute the Kaplan–Meier survival curve
- Learning to fit and validate a Cox proportional hazards model
- Learning to fit and validate parametric survival models

2 Motivating Business Problems

Survival analysis can provide tremendous insights and improved understanding into patterns of customer behavior depending upon their profiles and key performance indicators, especially in regard to churning, attrition, product purchase pattern, insurance claims, credit card default, etc. It can be used to compute customer lifetime values as a function of their past behaviors and contributions to a business, which in turn can be used to fine-tune campaigns. It can also be used to study organizational behaviors like bankruptcy, etc. The data required is a set of cases (suitably selected) where "lifetime" information (even if censored, but with censoring information) and information on possible drivers of such lifetimes is available. Some specific examples of survival analysis are given below:

- Business bankruptcy (time to bankruptcy) analysis on the basis of explanatory variables such as profitability, liquidity, leverage, efficiency, valuation ratio, etc. A firm not bankrupt at the time of end of data collection yields a censored observation that has to be interpreted in the analysis as samples of firms that have not yet failed (Lee 2014).
- Analysis of churn pattern in the telecom industry and impact of explanatory variables like the kind of plan, usage, subscriber profile like age, gender, household size, income, etc. on churn pattern. This information may be useful to reduce churn (Lu and Park 2003).
- Analysis of lifespan of car insurance contracts in terms of car's age, type of vehicle, age of primary driver, etc. may be carried out using survival analysis techniques to measure profitability of such contracts.
- Estimating a customer lifetime value (CLV) to a business on the basis of past revenue from the customer and an estimate of their survival probabilities based on their profile is a standard application of survival analysis techniques and results. This type of analysis is applicable to many types of business, this helps plan different campaign strategies depending on estimated lifetime value.

3 Methods of Survival Analysis

3.1 Time-to-Event Data and Censoring

Survival times are follow-up times from a defined starting point to the occurrence of a given event. Some typical examples are the time from the beginning of a customer-ship to churning; from issue of credit card to the first default; from beginning of an insurance to the first claim, etc. Standard statistical techniques do not apply because the underlying distribution is rarely normal; and the data are often "censored."

A survival time is called "censored" when there is a follow-up time but the defined event has not yet occurred or is not known to have occurred. In the examples above, the survival time is censored if the following happens: at the end of the study if the customer is still transacting; the credit card customer has not defaulted; the insurance policy holder has not made a claim. Concepts, terminology, and methodology of survival analysis originate in medical and engineering applications, where the prototype events are death and failure, respectively. Hence, terms such as lifetime, survival time, response time, death, and failure are current in the subject of survival analysis. The scope of applications is wider including in business, such as customer churn, employee attrition, etc. In Engineering these methods are called reliability analysis. In Sociology it is known as event-history analysis.

As opposed to survival analysis, regression analysis considers uncensored data (or simply ignores censoring). Logistic regression models proportion of events in groups for various values of predictors or covariates; it ignores time. Survival analysis accounts for censored observations as well as time to event. Survival models can handle time-varying covariates (TVCs) as well.

3.2 Types of Censoring

The most common form of censoring is right-censoring where a case is removed from the study during the study, or the observational part of the study is complete before the event occurs for a case. An example is where in an employee attrition study, an employee dies during the observational period (case removed) or may be still employed at the end of observations (event has not occurred). An observation is left-censored if its initial time at risk is unknown, like in a medical study in which the time of contracting the disease is unknown. The same observation may be both right- and left-censored, a circumstance termed interval-censoring. Censoring complicates the estimation of survival models and hence special techniques are required. If for a case (observational unit) the event of interest has not occurred then, all we know is that the time to event is greater than the observed time. In this chapter, we only consider right-censoring. One can consult Gomez et al. (1992) for left-censoring, Lagakos (1979) for right-censoring, and Sun (2006) for interval-censoring.

Observations that are censored give us no information about when the event occurs, but they do give us a bound on the length of their survival. For such observations, we know that they survived at least up to some observed time t^c and that their true lifetime is some $t^* \geq t^c$. In the dataset, for each observation, a censoring indicator c_i is created such that

$$c_i = \begin{cases} 1 \text{ if not censored} \\ 0 \text{ if censored.} \end{cases}$$

Censored observations are incorporated into the likelihood (or for that matter, in other approaches as well) as probability $t^* \geq t^c$, whereas uncensored observations are incorporated into the likelihood through the survivor density. This idea is illustrated below.

Suppose the lifetime (T) distribution is exponential (λ) with density function $f(t|\lambda) = \lambda e^{-\lambda t}$. Suppose an observation t is a censored observation. Then the contribution to the likelihood is $P(T \geq t) = e^{-\lambda t}$. Suppose an observation t is an uncensored observation. Then the contribution to the likelihood is $\lambda e^{-\lambda t}$. Suppose t_1, t_2 are censored, and u_1, u_2, u_3 are uncensored, then the likelihood function is

$$L(\lambda) = e^{-\lambda t_1} \times e^{-\lambda t_2} \times \lambda e^{-\lambda u_1} \times \lambda e^{-\lambda u_2} \times \lambda e^{-\lambda u_3},$$

$$\log(L(\lambda)) = -\lambda(t_1 + t_2 + u_1 + u_2 + u_3) + 3\log(\lambda),$$

maximizing which gives the maximum likelihood estimates of the parameters of the survival density.

3.3 Survival Analysis Functions

Survival time or lifetime T is regarded as a positive-valued continuous variable. Let $f(t)$: probability density function (pdf) of T.

Let $F(t)$: cumulative distribution function (CDF) of $T = P(T \leq t)$. $S(t)$: Survival function of T defined as $S(t) = 1 - F(t) = P(T > t)$.

The hazard function plays an important role in modeling exercises in survival analysis. It is defined below:

Let $h(t)$: hazard function or instantaneous risk (of death) function. It is defined as

$$h(t) = \lim_{dt \to 0} \frac{P(t \leq T \leq t + dt | T \geq t)}{dt} = \frac{f(t)}{S(t)} = -\frac{d}{dt}\log(S(t)).$$

It can be seen that

$$S(t) = e^{-\int_0^t h(x)dx}.$$

The function $H(t) = \int_0^t h(x)dx$ is called the cumulative hazard and is the aggregate of risks faced in the interval 0 to t. It can be shown that the mean (or expected) life $\int_0^\infty tf(t)dt$ is also $\int_0^\infty S(t)dt$. The hazard function has the following interpretation:

If a customer has been with a provider for 2 years, what is the probability he will attrite in the next year? Such questions are answered using the hazard rate. Answer: $H(1) = \int_2^3 h(t)dt$. The hazard rate is a function of time. Some simple types of hazard functions are:

Increasing hazard: A customer who has continued for 2 years is more likely to attrite than one that has stayed 1 year
Decreasing hazard: A customer who has continued for 2 years is less likely to attrite than one that has stayed 1 year
Flat hazard: A customer who has continued for 2 years is no more or less likely to attrite than one that has stayed 1 year

3.4 Parametric and Nonparametric Methods

Once we have collected time-to-event data, our first task is to describe it—usually this is done graphically using a survival curve. Visualization allows us to appreciate temporal patterns in the data. If the survival curve is sufficiently nice, it can help us identify an appropriate distributional form for the survival time. If the data are consistent with a parametric form of the distribution, then parameters can be derived to efficiently describe the survival pattern and statistical inference can be based on the chosen distribution by specifying a parametric model for $h(t)$ based on a particular density function $f(t)$ (parametric function). Otherwise, when no such parametric model can be conceived, an empirical estimate of the survival function can be developed (i.e., nonparametric estimation). Parametric models usually assume some shape for the hazard rate (i.e., flat, monotonic, etc.).

3.5 Nonparametric Methods for Survival Curves

Suppose there are no censored cases in the dataset. Then let $t_1, t_2, \ldots, t_n$ be the event-times (uncensored) observed on a random sample. The empirical estimate of the survival function, $\hat{S}(t)$, is the proportion of individuals with event-times greater than t.

$$\hat{S}(t) = \frac{\text{Number of event-times} > t}{n}. \tag{14.1}$$

When there is censoring $\hat{S}(t)$ is not a good estimate of the true $S(t)$; so other nonparametric methods must be used to account for censoring. Some of the standard methods are:

1. Kaplan–Meier method
2. Life table method, and
3. Nelson–Aalen method

We discuss only the Kaplan–Meier method in this chapter. For Life table method, one can consult Diener-West and Kanchanaraksa[1] and for Nelson–Aalen method, one may consult the notes provided by Ronghui (Lily) Xu.[2]

3.6 Kaplan–Meier (KM) method

This is also known as Product-Limit formula as will be evident when the method is described. This accounts for censoring. It generates the characteristic "stair case" survival curves. It produces an intuitive graphical representation of the survival curve. The method is based on individual event-times and censoring information. The survival curve is defined as the probability of surviving for a given length of time while considering time in intervals dictated by the data. The following assumptions are made in this analysis:

- At any time, cases that are censored have the same survival prospects as those who continue to be followed.
- Censoring is independent of event-time (i.e., the reason an observation is censored is unrelated to the time of censoring).
- The survival probabilities are the same for subjects recruited early and late in the study.
- The event happens at the time specified.

The method involves computing of probabilities of occurrence of events at certain points of time dictated by when events occur in the dataset. These are conditional probabilities of occurrence of events in certain intervals. We multiply these successive conditional probabilities to get the final estimate of the marginal probabilities of survival up to these points of time.

3.6.1 Kaplan–Meier Estimate as a Product-Limit Estimate

With censored data, Eq. (14.1) needs modifications since the number of event-times $> t$ will not be known exactly. Suppose out of the n event-times, there are k distinct times $t_1, t_2, \ldots, t_k$. Let event-time t_j repeat d_j times. Besides the event-times $t_1, t_2, \ldots, t_k$, there are also censoring times of cases whose event-times are not observed. The Kaplan–Meier or Product-Limit (PL) estimator of survival at time t is

$$\hat{S}(t) = \prod_{j:t_j \leq t} \frac{(r_j - d_j)}{r_j} \text{ for } 0 \leq t \leq t^+, \tag{14.2}$$

where $t_j, j = 1, 2, \ldots, n$ is the total set of event-times recorded (with t^+ as the maximum event-time), d_j is the number of events at time t_j, and r_j is the number

---

[1] http://ocw.jhsph.edu/courses/FundEpi/PDFs/Lecture8.pdf (accessed on Apr 27, 2018).
[2] http://www.math.ucsd.edu/~rxu/math284/slect2.pdf (accessed on Apr 27, 2018).

of individuals at risk at time t_j. Any case where a censoring time is a t_j is included in the r_j, as also cases whose event-time is t_j. This estimate can be considered a nonparametric maximum likelihood estimate.

3.6.2 Kaplan–Meier Method: An Example

The aim of the study in this example is to evaluate attrition rates of employees of a company. Data were collected over 30 years over $n = 23$ employees. Follow-up times are different for different employees due to different starting points of employment. The number of months with company is given below where + indicates still employed (censored):

6, 12, 21, 27, 32, 39, 43, 43, 46+, 89, 115+, 139+, 181+, 211+,

217+, 261, 263, 270, 295+, 311, 335+, 346+, 365+

The same data is named as "employ.csv" and available on the book's website. The following is the data dictionary.

Variable	Description
ID	The unique id of the employee
att	Represent 1 if uncensored and 0 if censored
months	No. of months the employee worked in the company

Survival rates are computed as follows:

P(surviving t days)=P(surviving day t | survived day $t-1$).P(surviving day $t-1$ | survived day $t-2$).P(surviving day $t-2$ | survived day $t-3$) $\vdots$ P(surviving day 3 | survived day 2).P(surviving day 2 | survived day 1).P(surviving day 1)

Standard errors of survival probabilities are computed using Greenwood's formula as follows:

$$\hat{V}(S(\hat{t})) = S(\hat{t})^2 \sum_{t_i \leq t} \frac{d_i}{n_i(n_i - d_i)}$$

Table 14.1 gives the survival probabilities computed by the Kaplan–Meier method. Notice that the probabilities are computed only at those time points where an event happens. In the table, n.risk is the r_j and n.event is the d_j in the formula, "survival" is the estimate of the survival probability $s(t)$ at time t. This table leads to the "stair-case" survival curve presented in the graph. The curve represents the probabilities of survival (y-axis) beyond the time points marked on the x-axis. Notice that we get revised estimates only at those points where an event is recorded in the data. The little vertical lines indicate the censored times in the data (Fig. 14.1).

Table 14.1 Survival probabilities using KM method

Time	n.risk	n.event	Survival	Std. err
6	23	1	0.957	0.0425
12	22	1	0.913	0.0588
21	21	1	0.870	0.0702
27	20	1	0.826	0.0790
32	19	1	0.783	0.0860
39	18	1	0.739	0.0916
43	17	2	0.652	0.0993
89	14	1	0.606	0.1026
261	8	1	0.530	0.1143
263	7	1	0.454	0.1205
270	6	1	0.378	0.1219
311	4	1	0.284	0.1228

Fig. 14.1 Kaplan–Meier Curve

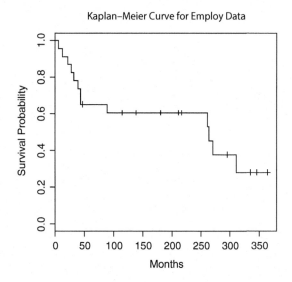

```
> employsurv <- survfit(Surv(months, att)~ 1, conf.
type="none",data = employ) summary(employsurv)
> plot(employsurv,mark.time = TRUE, xlab="Months",
ylab="Survival Probability",main="Kaplan-Meier Curve
for Employ Data")
```

3.7 Regression Models for Survival Data: Semiparametric Models

What happens when you have several covariates that you believe contribute to survival? For example, in job attrition data, gender, age, etc. may be such covariates. In that case, we can use stratified KM curves, that is, different survival curves for

14 Survival Analysis

different levels of a categorical covariate, possibly drawn in the same frame. Another approach is the Cox proportional hazards model.

Of all survival analysis functions, the hazard function captures the essence of the time process. Survival analysis uses a regression model-like structure into hazard function $h(t)$. The $h(t)$ being a rate should be positive with infinite range. To achieve this $h(t)$ is formulated as $h(t) = e^\beta$. Covariates (explanatory variables) x (a vector with components $(1, x_1, x_2, \ldots, x_p)$) is included by being additive in the log scale. Formulation:

$$\log[h(t, x)] = \boldsymbol{\beta}^T x = \beta_0 + \beta_1 x_1 + \beta_2 x_2 + \ldots + \beta_p x_p$$

or

$$h(t, x) = e^{\boldsymbol{\beta}^T x} \tag{14.3}$$

Equation (14.3) can be modified by introducing a function of time with a distribution model like a Weibull model. This will then be a fully parametric hazard function model, and describe a survival time distribution as an error component of regression, and describe how this distribution changes as a function of covariates (the systematic component). Such fully parametric models help predict survival time distributions for specific covariate conditions. If only relative survival experience is required under two or more conditions after adjusting for covariates, fully parametric models may be too unwieldy with too many restrictions. If we only need parameters in the systematic component of the model, then models with fully parametric regression leaving out the dependence on time unspecified may be useful. These are called semiparametric models.

A model of the form

$$h(t, x, \boldsymbol{\beta}) = h_0(t) r(x, \boldsymbol{\beta})$$

is such a formulation. $h_0(t)$ describes how the hazard function changes over time, $r(x, \boldsymbol{\beta})$ describes how the hazard function changes with the covariates. It is necessary that $h(t, x, \boldsymbol{\beta}) > 0$. Then $h(t, x, \boldsymbol{\beta}) = h_0(t)$ when $r(x, \boldsymbol{\beta}) = 1$. $h_0(t)$ is called the baseline hazard function—a generalization of intercept in regression.

The $h_0(t)$ which is the baseline hazard rate when $X = \underline{0} = (0, 0, \ldots, 0)$; this serves as a convenient reference point although an individual with $X = \underline{0}$ may not be a realistic one. Hazard ratio (HR) between two cases with x_1, x_2 is given by

$$\mathrm{HR}(t, x_1, x_2) = \frac{r(x_1, \boldsymbol{\beta})}{r(x_2, \boldsymbol{\beta})}$$

and does not depend on $h_0(t)$. Cox proposed the form $r(x, \boldsymbol{\beta}) = e^{(x^T \boldsymbol{\beta})}$ so that $h(t, x, \boldsymbol{\beta}) = h_0(t) e^{x^T \boldsymbol{\beta}}$. Then $\mathrm{HR}(t, x_1, x_2) = e^{(x_1 - x_2)^T \boldsymbol{\beta}}$. This is called Cox model, proportional hazards model, or Cox proportional hazards model.

3.8 Cox Proportional Hazards Model (Cox PH model)

This is a semiparametric model (part parametric, part nonparametric). It makes no assumptions about the form of $h(t)$ (nonparametric part). It assumes a parametric form for the effect of the explanatory variables on the hazard, but makes the assumption that the hazards are proportional over follow-up time. In most situations, we are more interested in studying how survival varies as a function of explanatory variables rather than the shape of the underlying hazard function. The Cox PH model is well suited for this purpose.

Let $X = (X_1, X_2, \ldots, X_p)$ be the explanatory variables. The model is

$$\log \frac{h(t|X)}{h(t)} = X^T \beta = \beta_1 X_1 + \beta_2 X_2 + \ldots + \beta_p X_p.$$

The model can also be written as $h(t|X) = h(t)e^{(X^T \beta)}$. The model can also be written as $S(t|X) = S(t|X) = 0)e^{(X^T \beta)}$. Predictor effects are the same for all t. No assumptions are made on the forms of S, h, f.

The hazard rate in PH models increases or decreases as a function of the covariates associated with each unit. The PH property implies that absolute differences in x imply proportionate differences in the hazard rate at each t. For some $t = \bar{t}$, the ratio of hazard rates for two units i and j with vectors of covariates x_i and x_j is:

$$\frac{h(\bar{t}, x_i)}{h(\bar{t}, x_j)} = e^{(x_i - x_j)\beta}.$$

Because the baseline hazards drop out in the equation it indicates that the baseline hazard rate for unit i is $e^{(X_i - X_j)\beta}$ times different from that of unit j. Importantly, the right-hand side of the equation does not depend on time, i.e., the proportional difference in the hazard rates of these two units is fixed across time. Put differently, the effects of the covariates in PH models are assumed to be fixed across time.

Estimates of the β's are generally obtained using the method of maximum partial likelihood, a variation of the maximum likelihood method. Partial likelihood is based on factoring the likelihood function using the multiplication rule of probability and discarding certain portions that involve nuisance parameters. If a particular regression coefficient β_j is zero, then the corresponding explanatory variable, X_j, is not associated with the hazard rate of the response; in that case, X_j may be omitted from any final model for the observed data. The statistical significance of explanatory variables is assessed using Wald tests or, preferably, likelihood ratio tests. The Wald test is an approximation to the likelihood ratio test. The likelihood is approximated by a quadratic function, an approximation which is generally quite good when the model fits the data.

In PH regression, the baseline hazard component, $h(t)$ vanishes from the partial likelihood. We only obtain estimates of the regression coefficients associated with the explanatory variables. Notice that $h(t) = h(t|x) = \beta_0$. Take the case of a single explanatory variable X. Then $\beta = \log \frac{h(t|x=1)}{h(t)}$. Thus β is the log of the relative hazard of group with $X = 1$ to the hazard of group with $X = 0$. $e^{(\beta)}$ is the relative risk of $X = 1$ to $X = 0$. So sometimes PH regression is called relative risk regression.

Concordance is a measure of goodness-of-fit of the model and defined as probability of agreement for any two randomly chosen observations. The large concordance value (possible maximum being 1) indicates a good fit.

3.9 Semiparametric vs Parametric Models

A parametric survival model completely specifies h(t) and S(t) and hence is more consistent with theoretical S(t). It enables time-quantile prediction possible. However, the specification of the underlying model S(t) makes this exercise a difficult one. On the other hand, the Cox PH model, a semiparametric one leaves the distribution of survival time unspecified and hence may be less consistent with a theoretical S(t); an advantage of the Cox model is that the baseline hazard is not necessary for estimation of hazard ratio.

A semiparametric model has only the regression coefficients as parameters and is useful if only the study of the role of the explanatory variables is of importance. In a full parametric model, besides the role of the explanatory variables, survival curves for each profile of explanatory variables can be obtained.

Some advantages of fully parameterized models are: maximum likelihood estimates (MLEs) can be computed. The estimated coefficients or their transforms may provide useful business information. The fitted values can provide survival time estimates. Residual analysis can be done for diagnosis.

Many theoretical specifications are used based on the form of S(t) (or f(t)) in survival analysis. Some of them are: Weibull, log-normal, log-logistic, generalized gamma, etc.

The regression outputs of a semiparametric and a full parametric are not directly comparable although one may compare the relative and absolute significance (p-values) of the various regressors. However, using the form of the parametric function's h(t) it is possible to strike a relationship between the parametric model's regression coefficients and Cox regression coefficients.

A parametric model is often called the accelerated failure time model (AFT model) because according to this model, the effect of an explanatory variable is to accelerate (or decelerate) the lifetime by a constant as opposed to say, the Cox proportional hazards model wherein the effect of an explanatory variable is to multiply hazard by a constant.

4 A Case Study

In this section, we discuss various methods of survival analysis through an example of a customer churn data of an online retail company. The observations are made up to a certain point of time only and if the customer is still there then it is censored and if the customer leaves it is denoted as uncensored. We also have many covariates which explain the activities of the customers. We are interested in analyzing the customer churn behavior with the help of survival time of a customer and *dead_flag* which indicates censored or uncensored along with 16 covariates. The dataset "churn.csv" and R code "Survival_Analysis.R" are available at the website. The variables chosen for the study are given in Table 14.2.

Table 14.2 Data dictionary

Variable	Definition
ptp_months	Profitable time period in months
dead_flag	Censor case or not: 0 indicates censored case
tenure_month	Tenure of user in months
unsub_flag	Email unsubscription status: 1 indicates unsubscribed
ce_score	Confidence score of user
items_home	No. of items purchased in home division
items_Kids	No. of items purchased in kids division
items_Men	No. of items purchased in men's division
items_Women	No. of items purchased in women's division
avg_ip_time	Average time between purchases
returns	No. of product returns
acq_sourcePaid	Has the user joined through paid channel or not
acq_sourceReferral	Has the user joined through referral channel or not
mobile_site_user	Does the user use mobile channel
business_name	First purchase division of user
redeemed_exposed	No. of offers redeemed or No. of offers given
refer_invite	No. of Referral joined or No. of invites sent
revenue_per_month	Revenue or tenure of user

4.1 Cox PH Model

We analyze the churn data to fit a Cox PH model (semiparametric model). The results are provided in Table 14.3. The output will be in two tables where the first table contains the regression coefficients, the exponentiated coefficients which are equivalent to estimated hazard ratios, standard errors, z tests, corresponding p-values and the second table contains exponentiated coefficients along with the reciprocal of exponentiated coefficients and values at 95% confidence intervals.

```
> churncoxph <- coxph(Surv(tenure_month, dead_flag) ~
ptp_months+unsub_flag+ce_score+items_Home+items_Kids+
items_Men+items_women
+avg_ip_time+returns +acq_sourcePaid+acq_
sourceReferral+mobile_site_user+business_name+redeemed
_exposed+refer_invite+avg_ip_time_sq+revenue_per_month,
data=churn)
> summary(churncoxph)
> predict(churncoxph, newdata=churn[1:6,], type="risk")
```

From the output, the estimated hazard ratio for *business_nameKids* vs *business_nameHome* is under column "exp(coef)" which is 1.8098 with 95% CI (1.7618, 1.8591). Similarly, exp(-coef) provides estimated hazard rate for *business_nameHome* vs *business_ nameKids* which is 0.5525 (the reciprocal of 1.8098). For continuous variables, exp(coef) is estimated hazard ratio for one unit increment in x, "(x+1)" vs "x" and exp(-coef) provides "x" vs 1 unit increment in x, "(x+1)". From the table the concordance is 0.814, which is large enough and thus indicating a good fit.

Besides interpreting the significance or otherwise of the explanatory variables and their relative use in predicting hazards, the output is useful in computing the relative risk of two explanatory variable profiles or relative risk with respect to the average profile, i.e., $e^{(X_i - X_j)'\beta}$, where X_i contains particular observation and X_j contains average values. The relative risks of the first six cases with respect to the average profile are: 3.10e-11, 0.60, 0.0389, 1.15, 0.196, and 0.182 (refer Table 14.3 for β values). We can compute the survival estimates of fitted model and obtain Cox adjusted survival curve.

```
> summary(survfit(churncoxph))
> plot(survfit(churncoxph),main= "Estimated Survival
Function by PH model", ylab="Proportion not churned")
```

Table 14.3 Cox PH model output

n= 214995, number of events= 117162

	coef	exp(coef)	se(coef)	z	Pr(>|z|)		exp(-coef)	lower .95	upper .95
ptp_months	−9.683e−02	9.077e−01	7.721e−04	−125.417	< 2e−16	***	1.1017	0.9063	0.9091
unsub_flag	3.524e−01	1.422e+00	6.529e−03	53.973	< 2e−16	***	0.7030	1.4044	1.4408
ce_score	−1.245e+00	2.879e−01	2.220e−02	−56.079	< 2e−16	***	3.4736	0.2756	0.3007
items_Home	−3.461e−02	9.660e−01	2.130e−03	−16.250	< 2e−16	***	1.0352	0.9620	0.9700
items_Kids	−7.456e−02	9.282e−01	2.521e−03	−29.570	< 2e−16	***	1.0774	0.9236	0.9328
items_Men	3.182e−03	1.003e+00	9.949e−04	3.198	0.00138	**	0.9968	1.0012	1.0051
items_Women	1.935e−03	1.002e+00	6.936e−04	2.790	0.00527	**	0.9981	1.0006	1.0033
avg_ip_time	1.427e−03	1.001e+00	9.936e−05	14.362	< 2e−16	***	0.9986	1.0012	1.0016
returns	−1.481e−01	8.624e−01	3.020e−03	−49.024	< 2e−16	***	1.1596	0.8573	0.8675
acq_sourcePaid	4.784e−02	1.049e+00	9.992e−03	4.788	1.69e−06	***	0.9533	1.0287	1.0697
acq_sourceReferral	−2.626e−01	7.690e−01	6.354e−03	−41.333	< 2e−16	***	1.3003	0.7595	0.7787
mobile_site_user	−3.644e−01	6.946e−01	2.278e−02	−15.998	< 2e−16	***	1.4396	0.6643	0.7263
business_nameKids	5.932e−01	1.810e+00	1.371e−02	43.264	< 2e−16	***	0.5525	1.7618	1.8591
business_nameMen	−9.704e−02	9.075e−01	1.220e−02	−7.951	1.89e−15	***	1.1019	0.8861	0.9295
business_nameWomen	−3.631e−01	6.955e−01	1.091e−02	−33.279	< 2e−16	***	1.4378	0.6808	0.7106
redeemed_exposed	−3.089e−01	7.342e−01	1.261e−02	−24.491	< 2e−16	***	1.3620	0.7163	0.7526
refer_invite	−3.870e−01	6.791e−01	8.996e−03	−43.014	< 2e−16	***	1.4725	0.6672	0.6912
avg_ip_time_sq	−5.027e−07	1.000e+00	1.970e−07	−2.552	0.01072	*	1.0000	1.0000	1.0000
revenue_per_month	1.712e−03	1.002e+00	2.555e−05	67.024	< 2e−16	***	0.9983	1.0017	1.0018

Signif. codes: 0 *** 0.001 ** 0.01 * 0.05 . 0.1 1
Concordance= 0.814 (se = 0.001)
Rsquare= 0.409 (max possible = 1)
Likelihood ratio test= 113002 on 19 df, p=0
Wald test = 75990 on 19 df, p=0
Score (logrank) test = 92819 on 19 df, p=0

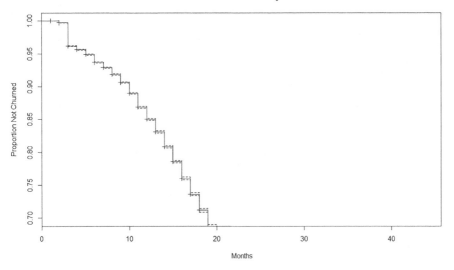

Fig. 14.2 Estimated survival function by PH method

4.2 Log-Logistic Model

Now we analyze the same data to fit the log-logistic parametric model. A simple way of stating the log-logistic model is by failure odds:

$$\frac{1 - S(t)}{S(t)} = \lambda t^p$$

where p is the shape parameter and λ is a function of predictor variables and regression parameters.

Following is the R code to fit the log-logistic model on the given data.

```
> aftloglogis<-survreg(formula = Surv(tenure_month,
dead_flag) ~ ptp_months + unsub_flag + +ce_score +
items_Home + items_Kids + items_Men +
+items_women + avg_ip_time, data = churn, dist =
"loglogistic")
> summary(aftloglogis)
> predict(aftloglogis, churn[1:10, ], type="quantile",
p=c(0.1,0.5,0.9))
```

The results are given in Table 14.4.

Coefficient estimates correspond to covariate coefficient estimates. Also of significant interest is the log-likelihood, which is used to find the Akaike information criterion (AIC), i.e., AIC $= -2 \log L + 2 \times$ number of parameters $= 917{,}817$. This is useful for comparison with any other model fitted on the same data (the lower the better).

Table 14.4 Output of parametric model log-logistic

	Value	Std. error	z	p
(Intercept)	2.484162	4.38e-03	566.79	0.00e+00
ptp_months	0.063756	4.26e-04	149.76	0.00e+00
unsub_flag	-0.269003	4.09e-03	-65.80	0.00e+00
ce_score	1.041445	1.27e-02	82.21	0.00e+00
items_Home	0.005020	8.80e-04	5.70	1.17e-08
items_Kids	-0.004644	5.57e-04	-8.34	7.73e-17
items_Men	0.002426	4.73e-04	5.12	2.99e-07
items_Women	0.013681	5.12e-04	26.74	1.67e-157
avg_ip_time	-0.000857	2.78e-05	-30.82	1.37e-208

```
Log logistic distribution
Loglik(model)= -458898.3 Loglik(intercept only)= -505430.2
Chisq= 93063.76 on 8 degrees of freedom, p= 0
Number of Newton-Raphson Iterations: 5
n= 214995
```

Table 14.5 Predicting survival time using the log-logistic model

Case	0.1	0.5	0.9
[1,]	1004.83620	2359.47444	5540.3255
[2,]	14.43473	33.89446	79.5882
[3,]	43.76790	102.77221	241.3213
[4,]	18.29105	42.94956	100.8506
[5,]	26.14241	61.38547	144.1404
[6,]	28.95115	67.98072	159.6268
[7,]	143.45923	336.85927	790.9855
[8,]	89.83391	210.94067	495.3137
[9,]	5.07855	11.92504	28.0014
[10,]	52.18694	122.54111	287.7410

The survival time difference for 1 month increase in tenure (ptp_months) is $\exp(0.063756) = 1.066$ increase, and from email unsub to sub (unsub_flag) is $\exp(-0.269003) = 0.764$ decrease.

For new data, any number of quantiles (importantly the 0.5 quantile, the median) of survival times can be predicted for input cases of regressors, effectively predicting the survival curves. The following is an example of 0.1, 0.5, 0.9 quantiles for the first ten cases in the dataset from the above model (Table 14.5). From the predicted values the median time is 2359.47 months for the first observation and for second observation it is only 33.89 months. You can similarly interpret other values.

4.3 Weibull Model

Next, we fit a Weibull parametric model on the same data. In the Weibull model,

$$S(t) = e^{(-\lambda * tp)}$$

where p is the shape parameter and λ is a function of predictor variables and regression parameters.

We can use the following R code to fit the Weibull model:

```
> aftweibull<-survreg(Surv(tenure_month, dead_flag) ~
ptp_months+unsub_flag+ce_score+items_Home+items_Kids+
items_Men++items_women+avg_ip_time, data=churn, dist
= "weibull")
> summary(aftweibull)
> predict(aftweibull,
coxfulla[1:10, ],type="quantile", p=c(0.1,0.5,0.9))
```

Coefficient estimates in Table 14.6 correspond to covariate coefficient estimates. Also of significant interest is the log-likelihood, which is used to find the Akaike information criterion (AIC), i.e., AIC $= -2 \log L + 2 \times$ number of parameters $= 909{,}304$. This is useful for comparison with any other model fitted on the same data (the lower the better).

The survival time difference for 1 month increase in tenure(ptp_months) is $\exp(0.056311) = 1.06$ increase, and from email unsub to sub (unsub_flag) is $\exp(-0.192530) = 0.825$ decrease (refer Table 14.6). Here, we observe that the Weibull model is predicting better than the log-logistic model as it has lower AIC value compared to the log-logistic model.

Table 14.6 Output of the Weibull parametric model

	Value	Std. error	z	p
(Intercept)	2.806480	3.88e-03	724.10	0.00e+00
ptp_months	0.056311	4.47e-04	126.09	0.00e+00
unsub_flag	-0.192530	3.32e-03	-57.96	0.00e+00
ce_score	0.746628	1.14e-02	65.52	0.00e+00
items_Home	0.008579	1.14e-04	9.07	1.23e-19
items_Kids	-0.001338	6.23e-04	-2.15	3.18e-02
items_Men	0.001414	4.75e-04	2.98	2.92e-03
items_Women	0.014788	5.44e-04	27.19	8.25e-163
avg_ip_time	-0.000858	2.68e-05	-32.05	2.52e-225

```
Weibull distribution
Loglik(model)= -454641.8  Loglik(intercept only)= -498568.4
Chisq= 87853.26 on 8 degrees of freedom, p= 0
Number of Newton-Raphson Iterations: 8
n= 214995
```

Table 14.7 Predicting survival time using the Weibull model

	[,1]	[,2]	[,3]
[1,]	1603.92	4180.3	7697.2
[2,]	12.66	33.0	60.8
[3,]	33.72	87.9	161.8
[4,]	17.01	44.3	81.7
[5,]	20.52	53.5	98.5
[6,]	23.04	60.0	110.6
[7,]	105.66	275.4	507.1
[8,]	77.17	201.1	370.3
[9,]	5.34	13.9	25.6
[10,]	38.51	100.4	184.8

For new data, any number of quantiles (importantly the 0.5 quantile, the median) of survival times can be predicted for input cases of regressors, effectively predicting the survival curves. The following is an example of 0.1, 0.5, 0.9 quantiles for the first ten cases in the dataset from the above model (Table 14.7). From the predicted values the median time for the first observation is 4180.3 months and for the second observation it is only 33.0 months. You can similarly interpret other values.

5 Summary

This chapter introduces the concepts and some of the basic techniques of survival analysis. It covers a nonparametric method of estimating a survival function called the Kaplan–Meier method, a semiparametric method of relating a hazard function to covariates in the Cox proportional hazards model, and a fully parametric method of relating survival time to covariates in terms of a regression as well as estimating quantiles of survival time distributions for various profiles of the covariate values. Survival analysis computations can be easily carried out in R with specialized packages such as *survival, KMsurv, survreg, RPub*, and innumerable other packages. Several textbooks provide the theory and explanations of the methods in detail. These include Gomez et al. (1992), Harrell (2001), Kleinbaum and Klein (2005), Hosmer et al. (2008), Klein and Moeschberger (2003), Lawless (2003), Sun (2006), Springate (2014), as well as websites given in the references.

Electronic Supplementary Material

All the datasets, code, and other material referred in this section are available in www.allaboutanalytics.net.

- Data 14.1: churn.csv
- Data 14.2: employ.csv
- Data 14.3: nextpurchase.csv
- Code 14.1: Survival_Analysis.R

Exercises

The data file *nextpurchase.csv* (refer website for dataset) relates to the purchase of fertilizers from a store by various customers. Each row relates to a customer. The study relates to an analysis of "time-to-next-purchase" starting from the previous purchase of fertilizers. "Censoring" is 0 if the customer has not returned for another purchase of a fertilizer since the first one. Censoring is 1 if he has returned for the purchase of a fertilizer since his earlier one. "Days" is the number of days since last purchase (could be a censored observation). "Visits" is the number of visits to the shop in the year not necessarily for the purchase of a fertilizer. "Purchase" is the amount of all purchases (in $'s) during the current year so far. "Age" is the customer's age in completed years. "Card" is 1 if they used a credit card; else 0.

Ex. 14.1 Without taking into account the covariates, use the Kaplan–Meier method to draw a survival curve for these customers.

Ex. 14.2 Fit the Weibull parametric model and predict the 0.1 (0.1) 0.9 quantiles of a customer aged 45, who uses a credit card, who spent $100 during the year so far and who has visited the shop four times in the year so far (not necessarily to purchase fertilizers).

Ex. 14.3 Rework the parametric Weibull exercise using the log-logistic parametric model.

Ex. 14.4 Rework the parametric Weibull exercise using the Cox PH model.

Useful functions for the Weibull distribution: (You need not know these to run this model.)

Density: $f(t) = k\lambda^k t^{k-1} e^{-(\lambda t)^k}$; Survival $S(t) = e^{(-\lambda t)^k}$; Hazard $h(t) = \lambda^k k t^{k-1}$; Cumulative Hazard: $H(t) = (\lambda t)^k$

References

Gomez, G., Julia, O., Utzet, F., & Moeschberger, M. L. (1992). Survival analysis for left censored data. In J. P. Klein & P. K. Goel (Eds.), *Survival analysis: State of the art* (pp. 269–288). Boston: Kluwer Academic Publishers.

Harrell, F. E. (2001). *Regression modeling strategies: With applications to linear models, logistic regression, and survival analysis* (2nd ed.). New York: Springer.

Hosmer, D. W., Jr., Lemeshow, S., & May, S. (2008). *Applied survival analysis: Regression modeling of time to event data* (2nd ed.). Hoboken, NJ: Wiley.

Klein, J. P., & Moeschberger, M. L. (2003). *Survival analysis: Techniques for censored and truncated data* (2nd ed.). New York: Springer.

Kleinbaum, D. G., & Klein, M. (2005). *Survival analysis: A self-learning text* (2nd ed.). New York: Springer.

Lagakos, S. W. (1979). General right censoring and its impact on the analysis of survival data. *Biometrics*, 139–156.

Lawless, J. F. (2003). *Statistical models and methods for lifetime data* (2nd ed.). Hoboken, NJ: Wiley.

Lee, M.-C. (2014). Business bankruptcy prediction based on survival analysis approach. *International Journal of Computer Science & Information Technology (IJCSIT)*, 6(2), 103. https://doi.org/10.5121/ijcsit.2014.6207.

Lu, J. & Park, O. (2003). *Modeling customer lifetime value using survival analysis—An application in the telecommunications industry*. Data Mining Techniques, 120–128 http://www2.sas.com/proceedings/sugi28/120-28.pdf.

Springate, D. (2014). *Survival analysis: Modeling the time taken for events to occur*. RPubs by RStudio. https://rpubs.com/daspringate/survival.

Sun, J. (2006). *The statistical analysis of interval censored failure time data*. New York: Springer.

Chapter 15
Machine Learning (Unsupervised)

Shailesh Kumar

We live in the age of data. This data is emanating from a variety of natural phenomena, captured by different types of sensors, generated by different business processes, or resulting from individual or collective behavior of people or systems. This observed sample data (e.g., the falling of the apple) contains a view of reality (e.g., the laws of gravity) that generates it. *In a way, reality does not know any other way to reveal itself but through the data we can perceive about it.*

The goal of unsupervised learning is essentially to "reverse engineer" as much of this reality from the data we can sample from it. In this chapter, we will explore unsupervised learning—an important paradigm in machine learning—that helps uncover the proverbial needle in the haystack, discover the grammar of the process that generated the data, and exaggerate the "signal" while ignoring the "noise" in it. In particular, we will explore methods of projection, clustering, density estimation, itemset mining, and network analysis—some of the core unsupervised learning frameworks that help us perceive the data in different ways and hear the stories it is telling about the reality it is sampled from. The examples, corresponding code, and exercises for the chapter are given in the online appendices.

1 Introduction

> The most elementary and valuable statement in Science, the beginning of Wisdom is—'I do not know'
> —Star Trek.

S. Kumar (✉)
Reliance Jio, Navi Mumbai, Maharashtra, India
e-mail: skumar.0127@gmail.com

Any scientific process begins with observation (data), formulating a hypothesis about the observation, testing the hypothesis through experimentation, and validating and evolving the hypothesis until it "fits the observation." Most scientific discoveries start with the ability to "observe" the data objectively followed by a pursuit to discover the "why" behind "what" we observe. The broad field of data *science* follows a similar scientific process by first trying to understand the nuances in the data, formulating a variety of hypotheses about, for example, what cause (e.g., a bad customer experience) might lead to what effect (e.g., customer churn), or which variables (e.g., education and age) might be correlated with others (e.g., income). It then provides algorithms to validate these hypotheses by building and interpreting models both descriptive and predictive, and finally, it enables us to take decisions to make the businesses, processes, applications, infrastructures, cities, traffic, economies, etc. more efficient.

The broad field of machine learning has evolved over the last several decades to generate a very large collection of modeling paradigms—including the supervised learning paradigm covered in the next chapter and the unsupervised learning paradigm, the subject of this chapter. Apart from these, there are a number of other paradigms such as the semi-supervised learning, active learning, and reinforcement learning. We will first understand the core differences between the supervised and unsupervised learning paradigms and then go into the various frameworks available within the unsupervised learning paradigm.

Supervised vs. Unsupervised Learning

Any intelligent system—including our own brain—does a variety of things with the data it observes:

- *Summarizes and organizes the data* (e.g., a business (retail or finance) might want to segment all its customers into a coherent group of similar customers based on their demographics and behavior).
- *Infers the grammar of the data* (e.g., typically what products in a retail market basket "go together" or what word will follow a sequence of words, say, "as soon as").
- *Interprets the data semantically* (e.g., a speech-enabled interface tries to first interpret the speech command of a user to text and from text to user intent).
- *Finds significant patterns in data* (e.g., which words typically occur before or after others, which sets of products are purchased together, what genes get activated together, or which neurons fire together).
- *Predicts what is about to happen* (e.g., in a bank or telecom businesses can predict that a certain customer is about to churn or an IoT system can predict that a certain part is about to fail).
- *Optimizes the best action given the prediction* (e.g., give a certain offer to the customer to prevent churn or preorder the part before it fails, to avert the unfavorable predicted future).

Some of these tasks require us to just observe the data in various ways and find structures and patterns in it. Here there is no "mapping" from some input to some output. Here we are just given a lot of data and asked to find something "interesting" in it, to reveal from data insights that we might not be aware of. For example, one might find product bundles that "go together" in a retail point of sale data or the fact that age, income, and education are correlated in a census data. The art and science of finding such structures in data without any particular end use-case in mind falls under unsupervised learning. Here we are just "reading the book" of the data and not "trying to answer a specific question" about the data. It is believed that in early childhood, most of what our brain does is unsupervised learning. For example:

- *Repeated Patterns*: when a baby hears the same set of sounds over and over again (e.g., "no"), it learns that this sound seems important and creates and stores a pattern in the brain to recognize that sound whenever it comes. It may not "understand" what the sound means but registers it as important because of repetition. The interpretation of this pattern might be learnt later as it grows.
- *Sequential patterns*: a child might register the fact that a certain event (e.g., ringing of a doorbell) is typically followed by another event (e.g., someone opens the door). This sequential pattern learning is key to how we pick up music, art, and language (mother tongue) even without understanding its grammar but by simply observing these sequential patterns over and over.
- *Co-occurrence patterns*: a child might recognize that two things always seem to co-occur together (e.g., whenever she sees eyes, she also sees nose, ear, and mouth). A repeated co-occurrence of same objects in the same juxtaposition leads to the recognition of a higher order object (e.g., the face).

In all these patterns, the grammar of the data is being learnt for no specific purpose except that it is there.

Supervised learning, on the other hand, is a mapping from a set of observed features to either a class label (classification paradigm) or a real value (regression paradigm) or a list of items (recommendation or retrieval paradigm), etc. Here we deliberately learn a mapping between one set of inputs (e.g., a visual pattern on a paper) and an output (e.g., this is letter "A"). This mapping is used both in interpreting and assigning names (or classes) to the patterns we have learnt (e.g., the sound for "dad" and "mom") as a baby in early childhood, which now are interpreted to mean certain people, or to the visual patterns one has picked up in childhood which are now given names (e.g., "this is a ball," "chair," "cat"), etc. This mapping is also used for learning cause (a disease) and effect (symptoms) relationships or observation (e.g., customer is not using my services as much as before) and prediction (e.g., customer is about to churn) relationships. A whole suite of supervised learning paradigms is discussed in the next chapter. In this chapter we will focus only on unsupervised learning paradigms.

Unsupervised Learning Paradigms

I don't know what I don't know—the Second Order of Ignorance

One of the most important frameworks in machine learning is unsupervised learning that lets us "observe" the data systematically, holistically, objectively, and often creatively to discover the nuances of the underlying process that generated the data, the grammar in the data, and insights that we didn't know existed in the data in the first place. In this chapter, we will cover five unsupervised learning paradigms:

- **Projections**—which is about taking a high dimensional data and finding lower dimensional projections that will help us both visualize the data and see if the data really belongs to a lower dimensional "manifolds" or is it inherently high dimensional. In particular, we will study various broad types of projection algorithms such as (a) principal components analysis (PCA) that try to minimize loss of variance, (b) self-organizing maps that try to smear the data on a predefined grid, and (c) multidimensional scaling (MDS) that try to preserve pairwise distances between data points after projection.
- **Clustering**—which is about taking the entire set of entities (customers, movies, stars, gene sequences, LinkedIn profiles, etc.) and finding "groups of similar entities" or hierarchies of entities. In a way our brain is a compression engine and it tries to map what we are observing into groups or quantization. Clustering ignores what might be noise or unimportant (e.g., accent when trying to recognize the word in a speech might not be important). It is also useful in organizing a very large amount of data into meaningful clusters that can then be interpreted and acted upon (e.g., segment-based marketing). In particular, we will study (a) partitional clustering, (b) hierarchical clustering, and (c) spectral clustering.
- **Density Estimation**—which is about quantifying whether a certain observation is even possible or not given the entire data. Density estimation is used in fraud detection scenarios where certain patterns in the data are considered normal (high probability) while certain other patterns might be considered outlier or abnormal (low probability). In particular, we will study both parametric and nonparametric approaches to learning how to compute the probability density of a record.
- **Pattern Recognition**—which is about finding the most frequent or significant repetitive patterns in the data (e.g., "people who buy milk also buy bread," or what words typically follow a given sequence of words). These patterns reveal the grammar of the data simply be relative frequency of patterns. High frequency patterns are deemed important or signal, while low frequency patterns are deemed noise. In particular, we will study (a) market-basket analysis, where patterns from sets are discovered, and (b) n-grams, where patterns from sequences are discovered.
- **Network Analysis**—which is about finding structures in what we call a network or graph data, for example, communities in social networks (e.g., terrorist cells, fraud syndicates), importance of certain nodes over others given the link structure of the graph (e.g., PageRank), and finding structures of interests (e.g., gene pathways, money laundering schemes, bridge structures). Graph theory and

network analysis algorithms when applied to real-word networks can generate tremendous insights that are otherwise hard to perceive.

Modeling and Optimization

Before we dive into the five paradigms, we will make another horizontal observation that will help us become a better "formulator" of a business problem into a machine learning problem—the key quality of a data scientist. Most machine learning algorithms—whether supervised or unsupervised—boil down to some form of an optimization problem. In this section, we will develop this *way of thinking* that what we really do in machine learning is a four stage optimization process:

- **Intuition**: We develop an intuition about how to approach the problem as an optimization problem.
- **Formulation**: We write the precise mathematical objective function in terms of data using intuition.
- **Modification**: We modify the objective function into something simpler or "more solvable."
- **Optimization**: We solve the modified objective function using traditional optimization approaches.

As we go through the various algorithms, we will see this common theme. Let us take two examples to highlight this process—as one of the goals of becoming a data scientist is to develop this systematic process of thinking about a business problem.

The Mean of a Set of Numbers

First we start with a very simple problem and formulate this as an objective function and apply the remaining steps. Consider a set of N numbers $\mathbf{X} = \{x_1, x_2, \ldots, x_N\}$. Now let us say we want to find the mean of these numbers. We know the answer already but that answer is not a formula we memorize, it is actually a result of an optimization problem. Let us first make an assumption (like in Algebra) that let m be the mean we are looking for. This is called the *parameter* we are trying to find.

Intuition: What makes m the *mean* of the set $\mathbf{X}$? The intuition says that mean is a point that is "closest to all the points." We now need to formulate this intuition into a mathematical objective function.

Formulation: Typically, in an objective function there are three parts: The unknown parameter we are optimizing (in this case m), the data (in this case $\mathbf{X}$), and the constraints (in this case there are no constraints). We can write the objective function as the sum of absolute distance between the point m and each data point that we must minimize to find m as a function of $\mathbf{X}$.

$$J(m \mid \mathbf{X}) = \sum_{n=1}^{N} |m - x_n|$$

Modification: Now the above objective function makes sense intuitive, but it is not easy to optimize it from a mathematical perspective. Hence, we come up with a more "solvable" or "cleaner" version of the same function. In this case, we want to make it "differentiable" and "convex." The following objective function is also known as sum of squared error (SSE).

$$J(m|\mathbf{X}) = \sum_{n=1}^{N} (m - x_n)^2$$

Now we have derived an objective function that matches our intuition as well as is mathematically easy to optimize using traditional approaches—in this case, simple calculus.

Optimization: The most basic optimization method is to set the derivative of the objective w.r.t. the parameter to zero:

$$\frac{\partial J(m|\mathbf{X})}{\partial m} = \sum_{n=1}^{N} \frac{\partial (m - x_n)^2}{\partial m} = 2\sum_{n=1}^{N}(m - x_n) = 0 \Rightarrow \widehat{m} = \frac{1}{N}\sum_{n}^{N} x_n$$

So we see that there is no *formula* for mean of a set of numbers. That formula is a *result* of an optimization problem. Let us see one more example of this process.

Probability of Heads

Let us say we have a two-sided (possibly biased) coin that we tossed a number of times and we know how many times we got heads (say H) and how many times we got tails (say T). We want to find the probability p of heads in the next coin toss. Again, we know the answer, but let us again go through the optimization process to find the answer. Here the data is (H, T) and parameter is p.

Intuition: The intuition says that we want to find that parameter value p that explains the data the most. In other words, if we knew p, what would be the likelihood of seeing the data (H, T)?

Formulation: Now we formulate this as an optimization problem by assuming that all the coin tosses are independent (i.e., outcome of previous coin tosses does not affect the outcome of the next coin toss) and the process (the probability p) is constant throughout the exercise. Now if p is the probability of seeing a head, then the joint probability of seeing H heads is p^H. Also since heads and tails are the only two options, probability of seeing a tail is $(1-p)$ and seeing T tails is $(1-p)^T$. The final *Likelihood* of seeing the data (H, T) is given by the product of the two:

$$J(p|H, T) = p^H (1-p)^T$$

Modification: The above objective function captures the intuition well but is not mathematically easy to solve. We modify this objective function by taking the log of it. This is typically called the *Log Likelihood* and is used commonly in both supervised and unsupervised learning.

$$J(p|H, T) = H \ln p + T \ln (1-p)$$

Optimization: Again we will use calculus—setting the derivative w.r.t. the parameter to zero.

$$\frac{\partial J(p|H,T)}{\partial p} = \frac{H}{p} - \frac{T}{1-p} = 0 \Rightarrow \widehat{p} = \frac{H}{H+T}$$

So we see how the intuitive answer that we remember for this problem is actually not a *formula* but a *solution* to an optimization problem.

Machine learning is full of these processes. In the above examples, we saw two types of objective functions (sum of squared error and log likelihood) which cover a wide variety of machine learning algorithms. Also, here we were lucky. The solutions to the objective functions we formulated were simple *closed-form* solutions where we could just write the parameters in terms of the data (or some statistics on the data). But, in general, the objective functions might become more complex and the solution might become more iterative or nonlinear. In any case, the process remains the same and we will follow pretty much the same process in developing machine learning (unsupervised in this chapter and supervised in next) models.

Visualizations

Machine learning is really the art of marrying our understanding of the data, our appreciation for domain knowledge, and our command on the algorithms that are available to us. Here we will explore the power of simple visualizations that reveal a lot about the nuances in the data. This is the essential first step in any data science journey. Through this process, we will also understand how to read the data (and not just ask questions), learn the art of listening to data (and not just testing our own hypotheses), and then apply the right transformations and prepare the data for subsequent stages of insight generation and modeling.

Histograms

One of the simplest and most powerful visualizations are the histograms of each dimension of the data. Figure 15.1 shows histograms of the Higgs-Boson data.[1] This reveals some interesting facts. None of the dimensions is actually

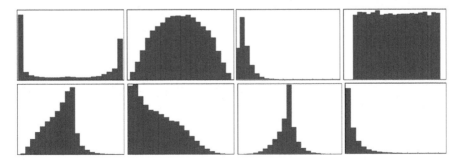

Fig. 15.1 Histograms of eight different dimensions of Higgs boson dataset (none is normally distributed)

---

[1] https://www.kaggle.com/c/Higgs-boson (Retrieved September 5, 2018).

normally distributed. The distributions we see here are exponential, log-normal, doubly exponential, a combination of linear and exponential, parabolic, and reverse normal distributions. Any transformation such as min-max or z-scoring on these dimensions that assumes normality will not yield the desired results. This insight is very important—when we choose certain algorithms (e.g., K-means later), we make certain implicit assumptions about the nature of the distance functions and feature distributions. Histograms help us validate or invalidate those assumptions and can therefore force us to transform some of the features (e.g., by taking their logs) to make them closer to the assumptions that are used by those techniques.

Log Transforms

One of the most common distributions in real-world data is not really a normal distribution but a log-normal, exponential, or Zip's law distribution. For example, the income distribution of any reasonably large population will be exponentially distributed. Frequency of words and their rank order have a Zip's law distribution and so on. One of the common and practical things is to try the log of a feature with such distributions instead of using the feature "as is." For example, PCA or K-means clustering that depends on normal distribution assumption and uses Euclidean distances performs better when we undo the effect of this exponential distribution by taking the log of the features. Figure 15.2 shows an example of the Higgs boson features with log-normal or exponential distribution (left) and their distribution after taking the log (right). In both cases, new insights emerge from this process. Hence, it is important to explore the histogram of each feature and determine whether log will help or not.

Scatter Plots

Another simple yet very effective tool for understanding data is to visualize the scatter plot between all pairs (or triples) of dimensions and even color code

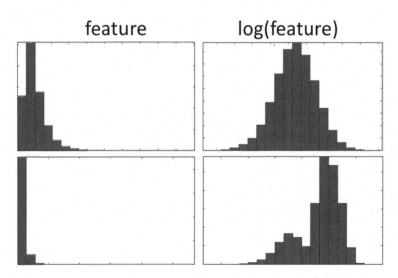

Fig. 15.2 Histogram of a feature (left) and its log (right). Taking log is a useful transformation

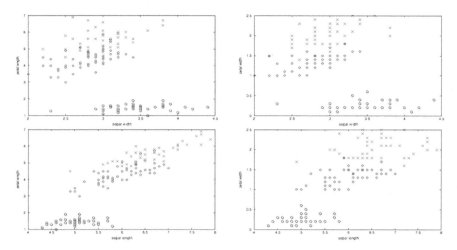

Fig. 15.3 A few scatter plots of IRIS data show that two classes are closer to each other than the third

each point by another property. Scatter plots reveal the structure of the data in the projected spaces and develop our intuition about what techniques might be best suited for this data, what kind of features we might want to extract, which features are more correlated to each other, which features are able to discriminate the classes better, etc.

Figure 15.3 shows the scatter plot of the IRIS dataset[2] between a few pairs of dimensions. The color/shape coding of a point is the class (type of Iris flower) the point represents. This immediately shows that two of the three classes of flowers are more similar to each other than the third class.

While histograms give us a one-dimensional view of the data and scatter plots give us two- or three-dimensional view of the data, they are limited in what they can do. We need more sophisticated methods to both visualize the data in lower dimensions and extract features for next stages. This is where we resort to a variety of projection methods discussed next.

2 Projections

One of the first things we do when we are faced with a lot of data is to get a grasp of it from both domain perspective and statistical perspective. More often than not, any real-world data is comprised of large number of features either because each record inherently is comprised of large number of input features (i.e., there are lots of sensors or the logs contain many aspects of each entry) or because we have engineered a large number of features on top of the input data. High dimensionality has its own problems.

---

[2]https://en.wikipedia.org/wiki/Iris_flower_data_set (Retrieved September 5, 2018).

- First, it becomes difficult to visualize the data and understand its structure. A number of methods help optimally project the data into two or three dimensions to exaggerate the signal and suppress the noise and make data visualization possible.
- Second, just because there are a large number of features does not mean that the data is inherently high dimensional. Many of the features might be correlated with other features (e.g., age, income, and education levels in census data). In other words, the data might lie in a lower dimensional "manifold" within the higher dimensional space. Projection methods uncorrelate these dimensions and discover the lower linear and nonlinear manifolds in the data.
- Finally, the *curse of dimensionality* starts to kick in with high dimensional data. The amount of data needed to build a model grows exponentially with the number of dimensions. For all these reasons a number of projection techniques have evolved in the past. The unsupervised projection techniques are discussed in this chapter. Some of the supervised projection techniques (e.g., Fisher Discriminant Analysis) are discussed in the Supervised Learning chapter.

In this section, we will introduce three different types of projection methods: principal components analysis, self-organizing maps, and multidimensional scaling.

2.1 Principal Components Analysis

Principal components analysis (PCA) is one of the oldest and most commonly used projection algorithms in machine learning. It linearly projects a high dimensional multivariate numeric data (with possibly correlated features) into a set of lower orthogonal (uncorrelated) dimensions where the first dimension captures most of the variance, next dimension—while being orthogonal to the first—captures the remaining variance, and so on. Before we go into the mathematical formulation of PCA, let us take a few examples to convey the basic intuition behind orthogonality and principalness of dimensions in PCA.

- **The Number System**: Let us take a number (e.g., 1974) in our base ten number system. It is represented as a weighted sum of powers of 10 (e.g., $1974 = 1 \times 10^3 + 9 \times 10^2 + 7 \times 10^1 + 4 \times 10^0$). Each place in the number is independent of another (hence orthogonal), and the digit at the ones place is least important, while the digit at the thousands' place is the most important. If we were forced to mask one of the digits to zero by *minimizing the loss of information*, it would be the ones place. So here thousands' place is the first principal component, hundreds' is the second, and so on.
- **Our Sensory System**: Another example of PCA concept is our sensory system. We have five senses—vision, hearing, smell, taste, and touch. There are two

15 Machine Learning (Unsupervised)

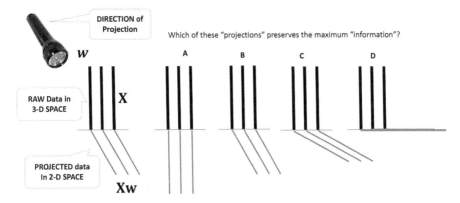

Fig. 15.4 The idea of a projection and loss of information as a result of projection

properties we want to highlight about our sensory system: First, all the five senses are orthogonal to each other, that is, they capture a very different perspective of reality. Second, the amount of information they capture about reality is not the same. The vision sense perhaps captures most of the information, followed by auditory, then taste, and so on. We might say that vision is the first principal component, auditory is the second, and so on. So PCA captures the notion of orthogonality and different amount of information in each of the dimensions.

Let us first understand the idea of "projection" and "loss of information." Figure 15.4 shows the idea of a projection in another way. Consider the stumps in a cricket game—this is the raw data. Now imagine we hold a torch light (or use the sun) to "project" this three-dimensional data on to the two-dimensional field. The shadow is the projection. The nature of the shadow depends on the angle of the light. In Fig. 15.4 we show four options. Among these options, projection A is "closest to reality," that is, loses minimal amount of information, while option D is farthest from reality, that is, loses all the information. Thus, there is a notion of the "optimal" projection w.r.t a certain objective called "loss of information."

We will use the above intuition to develop an understanding of principal components analysis. We first need to define the notion of "loss of information due to projection." Let $\mathbf{X} = \left[x_n^T \right]_{n=1}^{N}$ be the $N \times D$ data matrix with N rows where each row is a D-dimensional data point. Let $w^{(k)}$ be the kth principal component, constrained to be a unit vector, such that $k \leq \min \{D, N-1\}$. We will use the same four-stage process to develop PCA:

Intuition: The real information in the data from a statistical perspective is the *variability* in it. So any projection that maximizes the variance in the projected space is considered the first principal component (direction of projection), $w^{(1)}$. Note that since the input data $\mathbf{X}$ is (or is transformed to have) zero mean, its linear projection will also be zero mean (Fig. 15.5).

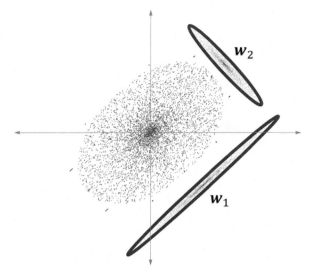

Fig. 15.5 The first and second principal components of a zero-mean data cloud

Formulation: Let $y_n = x_n^T w$ be the projection of the data point on w. The variance of the entire projected data is given by: $\sum_{n=1}^{N}(y_n)^2 = \sum_{n=1}^{N}(x_n^T w)^2$. Thus, the first principal component is found by solving:

$$w_1 = \arg\max_{\|w\|=1} \sum_{n=1}^{N} \left(x_n^T w\right)^2$$

Modification: The above is a "constrained optimization problem" where the constraint is that the projection is a unit vector, that is, $\|w\| = 1$. We can rewrite this constrained objective function as an unconstrained objective function as follows:

$$w_1 = \arg\max_{\|w\|=1} \|\mathbf{X}w\|^2 = \arg\max \left\{ \frac{w^T \mathbf{X}^T \mathbf{X} w}{w^T w} \right\}$$

Optimization: The standard solution to this problem is the first *eigenvector* of the positive semi-definite matrix $\mathbf{X}^T\mathbf{X}$, that is, $w_1 = eig_1(\mathbf{X}^T\mathbf{X})$, with the maximum value being the first **eigenvalue**, λ_1.

The kth principal component is derived by first removing the first $k-1$ principal components from $\mathbf{X}$ and then finding the first principal component in the residual ($\mathbf{X}_0 = \mathbf{X}$):

$$\mathbf{X}_k = \mathbf{X}_{k-1} - \mathbf{X} w_{k-1}(w_{k-1})^T$$

And from this, we iteratively find:

$$w_{k-1} = \arg \max_{\|w\|=1} \|X_k w\|^2 = \arg \max \left\{ \frac{w^T X_k^T X_k w}{w^T w} \right\}$$

In general, the first k principal components of the data correspond to the first k eigenvectors (that are both orthogonal and decreasing in the order of variance captured) of the covariance matrix of the data. The percent variance captured by the first d principal components out of D is given by the sum of squares of the first d eigenvalues of $X^T X$.

$$S(d|D) = \frac{\lambda_1^2 + \lambda_2^2 + \cdots + \lambda_d^2}{\lambda_1^2 + \lambda_2^2 + \cdots + \lambda_D^2}$$

Figure 15.6 shows the eigenvalues (above) and the fraction of variance captured (below) as a function of the number of principal components for MNIST data which is 28×28 images of handwritten data. Top 30 principal components capture more than 95% of variance in the data. The same can be seen in Figure 15.7 that shows

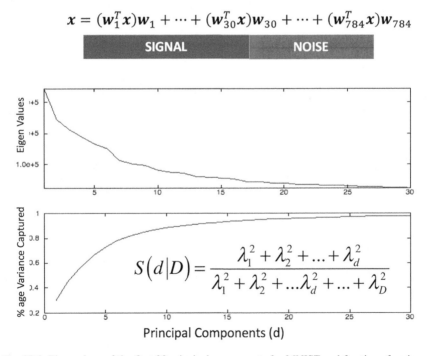

Fig. 15.6 Eigenvalues of the first 30 principal components for MNIST and fraction of variance captured

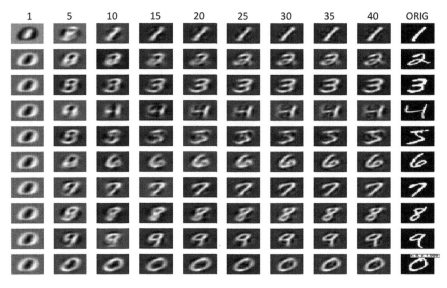

Fig. 15.7 Reconstruction of the digits by projecting them into k dimensions and back

the reconstruction of the ten digits when the data is projected to different number of principal components and reconstructed back. Again, we can see that although the original data is 784-dimensional (28×28), the top 30–40 dimensions capture the essence of the data (signal).

A number of other such linear (e.g., independent components analysis) and nonlinear (e.g., principal surfaces) projection methods with different variants of the loss of information objective function have been proposed. PCA is a "type" of projection method. Next we study another "type" of projection method which is very different in nature compared to the PCA-like methods.

To learn more about principal components analysis, refer to Chap. 3 (Sect. 3.4.3) in Han et al. (2011), Chap. 12 (Sect. 12.2) in Murphy (2012), and Chap. 14 (Sect. 14.5.1) in Friedman et al. (2001).

2.2 Self-Organizing Maps

Another classical approach to project data into 2–3 dimensions is self-organizing map (SOM) approach that uses competitive self-organization to smear the input data on a predefined grid structure.

Intuition: Figure 15.8 shows the basic intuition behind SOM. The left part shows the original data in a high dimensional space. Two points close to each other in this original space map to either the same or nearby grid points on the right grid also known as the "map."

15 Machine Learning (Unsupervised)

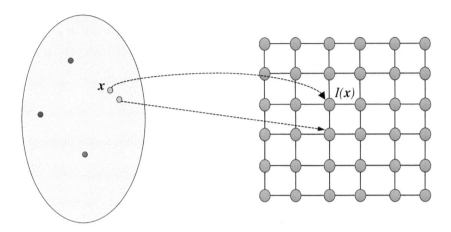

Fig. 15.8 Nearby points in the original space (left) map to nearby or same point in the SOM grid (right)

Formulation: A SOM is defined say M grid points organized typically in a rectangular (each grid point has four neighbors) or hexagonal (each grid point has three neighbors) grid. Each grid point is associated with a weight vector (the parameters) and a neighborhood structure.

Let $\mathbf{X} = \{x_n \in \mathbf{R}^D\}_{n=1}^{N}$ be the set of N data points each in a D dimensional space. Let $\mathbf{W} = \{w_m\}_{m=1}^{M}$ be the weights associated with the M grid points. The goal is to learn these weights so they "quantize" the input space in such a way that the weights associated with nearby grid points are similar to each other, that is, there is a smooth transition between weights on the grid.

Initially (in iteration $t = 0$) the weights are set to random. Then with each iteration the weights are updated through competitive learning, that is, (a) each data point is first associated with that grid point whose weights are closest to the data point itself, (b) then this grid point weights are updated to move closer to the data point, and (c) not only that, the weights of the "nearby grid points" are also moved toward this data point, albeit to a slightly lesser degree.

Optimization: SOM is learnt through an iterative algorithm where (a) each data point is first associated with the nearest grid point and (b) weights of all the grid points are updated depending on how far they are from the grid point associated with the data point. SOM starts with random initial weights: $\mathbf{W}(0) = \{w_m(0)\}_{m=1}^{M}$ and updates these weights iteratively as follows:

- Associate each data point with its nearest grid point (image of the data point) in iteration t

$$I_t(n) = \arg\min_{m=1...M} ||x_n - w_m(t)||$$

- Compute degree of association $\theta_t(n,m)$ between the nth data point and the mth grid point, such that it decreases with the distance between m and $I_t(n)$, $\delta(I_t(n), m)$:

$$\theta_t(n, m) = \exp\left(-\frac{\delta(I_t(n), m)}{\sigma(t)^2}\right), \forall m = 1 \ldots M$$

- Now each of the grid point weights w_m is updated in the direction of the input x_n with different degrees that depends on $\theta_t(n, m)$:

$$w_m(t+1) = w_m(t) + \eta(t)\theta_t(n, m)(x_n - w_m(t))$$

- Decrease the learning rate $\eta(t)$ and the variance $\sigma(t)$ as iterations progress.

Figure 15.9 shows a semantic space of a news corpus comprising of millions of news articles from the last ten years of a country. Here word embeddings (300-dimensional semantic representation of words such that two words with similar meaning are nearby in the embedding space) of the top 30K most frequent words (minus the stop words) are smeared over a 30 × 30 SOM grid. Two words close to each other in the embedding space are mapped to either the same or nearby grid

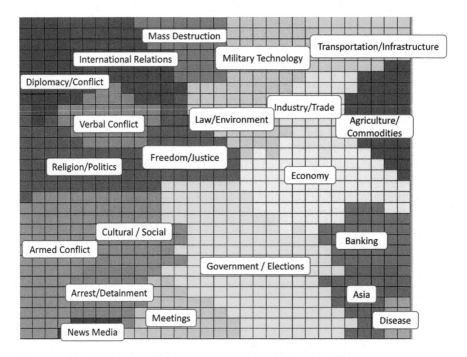

Fig. 15.9 Word embeddings of a large news corpus visualized on a 2D SOM

points. The grid vectors quantize different parts of the semantic embedding space representing different meanings. These grid point vectors are further clustered into macro concepts shown on the map.

SOMs smear the entire data into a 2D grid. Sometimes, however, we do not want to put the projected data on a grid. Additionally, we are not given a natural representation of data in a Euclidean space. In such cases, we use another class of projection method called multidimensional scaling.

The reader can refer to Chap. 14 (Sect. 14.4) in Friedman et al. (2001) to learn more about self-organizing maps.

2.3 Multidimensional Scaling

PCA and SOM are two different kinds of projection/visualization methods. In PCA, we project the data linearly to minimize loss of variance, while in SOM we quantize each data point into a grid point via competitive learning. Another way to map a high dimensional data into a low dimensional data is to find each data point's representatives in the lower dimensions such that the distance between every pair of points in the high dimension matches the distance between their representatives in the projected space. This is known as multidimensional scaling (MDS).

Intuition: The idea of "structure" in data manifests in many ways—correlation, variance, or pairwise distances between data points. In MDS, we find a representative (not a quantization as in SOM) for each data point in the original space such that the distance between two points in the original space is preserved in the MDS projected space. Figure 15.10 shows the basic idea behind MDS.

Formulation: Let $\mathbf{D} = [\delta_{ij}]$ be the $N \times N$ distance matrix between all pairs of N points in the original space. Note that techniques like MDS do not require

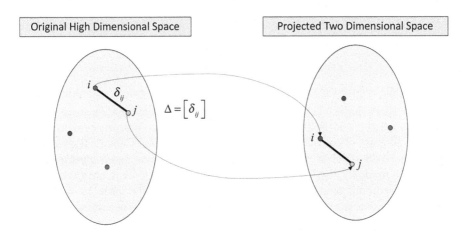

Fig. 15.10 Multidimensional scaling preserves pairwise distances between all pairs of points

the original data to be a multivariate data. As long as we can compute distance between pairs of points (e.g., Euclidian distance between multivariate real-valued vectors, cosine similarity between word or paragraph or document embeddings, TFIDF cosine similarity between two documents, Jaccard coefficient between two market baskets, even a subjective score by "smell expert" on how similar two smells are, or any engineered distance or similarity function) MDS can be applied. In MDS, each data point is associated with a low dimensional representative vector, that is, let $\mathbf{X} = \{x_n\}_{n=1}^{N}$ be the N representative vectors (parameters to be learnt), one for each data point. The goal is to find $\mathbf{X}$ such that the distance between every pair of points $\Delta(x_i, x_j)$ in the MDS space matches as much as possible the actual distance δ_{ij} between the corresponding points, that is,

$$J(\mathbf{X}) = \min_{\mathbf{X}} \sum_{1 \leq i < j \leq N} \left(\Delta\left(x_i, x_j\right) - \delta_{ij}\right)^2$$

Modifications: Different variants of proximity preserving embeddings have been developed over time.

- **Multidimensional Scaling**: The original objective is modified by dividing with the sum squared of all the distances. This is done to make sure that the overall distances between points do not grow.

$$J_{MDS}(\mathbf{X}) = \frac{\sum_{1 \leq i < j \leq N} \left(\Delta\left(x_i, x_j\right) - \delta_{ij}\right)^2}{\sum_{1 \leq i < j \leq N} \Delta(x_i, x_j)^2}$$

- **Sammon Map**: The intuition behind this is that when two points are very far from each other in the original space, then the error between their distances in the projected space and the original space matters less. Only when points are close to each other in the original space that the error matters.

$$J_{SPE}(\mathbf{X}) = \sum_{1 \leq i < j \leq N} \frac{\left(\Delta\left(x_i, x_j\right) - \delta_{ij}\right)^2}{\delta_{ij}}$$

Figure 15.11 shows a 2D map of the various product categories of a grocery store. This map was created by first learning the strength of co-occurrence consistency between all pairs of categories from the point-of-sale data. Consistency measures the degree with which a pair of product categories is purchased *together more often than random*. Two products that are consistently closer to each other (e.g., meat and seafood) land up in close proximity in the 2D space as well. This visualization reveals not only the structure in the purchase co-occurrence grammar of the customers but can also be used to change the store layout to match customer buying patterns or create rules for recommending products, etc.

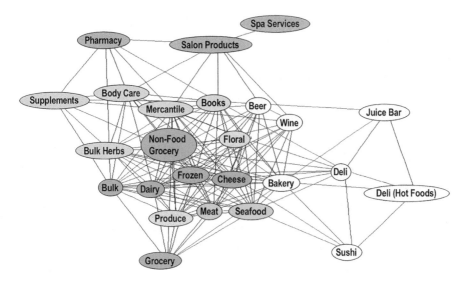

Fig. 15.11 Store layout based on co-occurrence of products from various categories

In this section, we have studied a variety of algorithms that help understand the data better by projecting it to a two- or three-dimensional space and creating different kinds of visualizations around them—histograms, scatter plots, self-organizing-maps, and multidimensional scaling. Next, we explore one of the most popular paradigms called clustering in the unsupervised learning suite of algorithms.

The reader can refer to Chap. 14 (Sects. 14.8 and 14.9) from Friedman et al. (2001) to learn more about multidimensional scaling.

3 Clustering

The fundamental hypothesis in finding structure in data is that while a dataset can be very large, the underlying processes that generated the data has only finite degrees of freedom. There are only a small number of actual latent sources of variations from which the data actually emerged. For example,

- In retail point-of-sale data, we might see a lot of variation from customer to customer but inherently there are only a finite types of customer behaviors based on their lifestyle (e.g., brand savvy, frugal), life-stage (e.g., bachelor, married, has kids, old age), purchase behavior (e.g., when, where, how much, which channel) and purchase intents (grocery, birthday, vacation related, home improvement, etc.). We might not know all these variations or combinations in advance, but we know we are not dealing with an infinite number of such variations and we can discover such quantization if we let the data speak for itself.

- Similarly, while it seems that there are billions of videos on YouTube or billions of pages on the web, or millions of people on LinkedIn and Facebook, the different types of videos (music albums, home videos, vacation videos, talent videos, cat videos, etc.), pages (news, spam, blogs, entertainment, etc.), or people (software engineers, managers, data scientists, artists, musicians, politicians, etc.) is a reasonably finite set. Whether we know all the types or not is another question, but what we definitely know is that the number of such types is not as many as the number of entities.
- Similarly, consider all words in a language. It appears that there are many words in the dictionary but they can again be grouped by, say, parts of speech, root, tense, and meaning, into only a small number of types.
- Finally, consider telematics data while someone is driving the car. Again, the data variation might be very large across all cars but the number of things people do while driving (soft or hard brake, soft or hard acceleration, sharp or comfortable left or right turns, etc.) combined with the number of driving scenarios (pot-holes uphill, downhill, highway, inner-roads, etc.) is still finite.

When these finite "sources" of variation in the data are already known in advance and/or when we have to map the data variations into a specific set of *known* types, this becomes a **classification problem**. We deal with classification in the supervised learning chapter at length. On the other hand, when these variations are *not known* in advance and need to be *discovered*, by grouping similar data points together (whatever "similar" means for that type of data), then it becomes a **clustering problem**.

In many systems of intelligence including our own, we transition from clustering to classification. For example, in early childhood, babies do not know all the variations of what they see or hear so they internally do clustering to quantize these variations. If they see more data of a certain type, the resolution of quantization on those parts of the data becomes fine-grained. At this stage, we do know that this is similar to what I have seen before (quantized symbol number 48), but we do not know yet what to call it. As we grow and language develops, we learn that those quantization have been given names (vertical line, sleeping line or "nose," "eyes," "square," "triangle," etc.). Now we have some *known* quantization that we call **classes**, and when a new experience comes, we first try to map it to a known class (e.g., if a child has never seen a goat before, she might "classify" it as a "dog"), but if this quantization is not "close enough" to any of the known classes, then she might ask the mother—it looks like a dog, is it a dog? And when she gets a new label (no it is called a "goat"), she creates another class in the brain. In this stage, we rely not only on the known set of named classes but are also open to discovering beyond the known. This is where we are in the *hybrid* stage of learning—exploit the known and explore the unknown simultaneously. As we grow older and we have "seen enough," the number of new quantization reduces as we have a sufficiently large number of classes to represent all inputs and nothing seems to surprise us anymore.

Another example of this process of transition from clustering to classification happens in customer feedback when we move from an early stage product to a maturity stage product, say, when we build our first product (an app, a service, a physical product, etc.) and we start to get customer feedback. As we start to go through this feedback, we realize that there are only so many variations that we are seeing. For example, in an online retail business, customers might complain about delivery time, delivery charges, wrong product delivered, login problems, etc. In restaurant business, customers might complain about quality of service, ambience, quality of food, price, etc. In fleet management (Ola, Uber), customers might complain about quality of car, delay in pickup, cancellation by driver, driver rudeness, driving safety, etc. Now initially when we do not know what these categories might be, we just cluster similar text feedback together based on keywords and assign name to these clusters. Once we know these clusters, we can then create a menu system based on the most common types of complaints and can transition to a more structured feedback than unstructured text feedback in the early days of the product. Thus, while clustering and classification are two very different paradigms, they are related to each other. This, in fact, is one of the most important use cases of clustering—to *discover* the quantized states of the system, that is, the sources of variations in the data. In this section, we will explore three broad clustering approaches: partitional, agglomerative, and spectral.

3.1 Partitional Clustering

If we assume a certain number of clusters and try to partition the data into those many clusters, then it is called partitional clustering. Different algorithms, most notably *K-means clustering*—that partitions the data into K clusters—are examples of partitional clustering. Consider a multivariate dataset where we can define Euclidean distance between two points meaningfully, that is, we have already transformed all the features and z-scored them.

Let $\mathbf{X} = \{x_1, x_2, \ldots, x_N\}$ be the N data points that need to be clustered into K clusters ($1 \leq K \leq N$). If $K = 1$, that means the entire data is clustered into one cluster. In that case, the mean of the entire dataset is the cluster center we are looking for. In case $K = N$, then each data point is by itself a cluster center. Both these are valid but not useful extreme cases. Typically, the value of K is somewhere in between. We will first formulate this as an optimization problem using the same process as above—intuition, formulation, modification, and optimization.

Intuition: Clustering is about "grouping similar things (feature vectors representing them) together." There are two equivalent ways to represent a "clustering": *Enumeration* vs *Representation*. In *enumeration*, we can explicitly label each data point with the cluster id it belongs to. Let $\delta_{n,k} \in \{0, 1\}$ be a set of binary labels such that it is ones if n^{th} data point is associated with the kth cluster and zero otherwise. In *representation*, each cluster is represented by a cluster mean of all data points it represents.

Formulation: Let $\mathbf{M} = \{m_1, m_2, \ldots, m_K\}$ be the K cluster means—the representatives of the K clusters—that we are looking for. In a way we are quantizing the raw data by these cluster centers that act as representatives of the data. The objective is to find such representatives that approximate the data the best, that is, the error of approximation is minimum. Now if data point x_n is represented by the cluster center m_k, that is, $\delta_{n,k}$ is 1 and $\delta_{n,\ell} = 0$ for all other $\ell \neq k$, then the objective we are trying to minimize is the sum (squared) of the distance between each data point and *its* representative:

$$J(\mathbf{M}, \mathbf{\Delta}) = \sum_{n=1}^{N} \sum_{k=1}^{K} \delta_{n,k} \|x_n - m_k\|^2$$

Optimization: In the above equation, there are two kinds of parameters—the enumeration parameters, $\mathbf{\Delta} = [\delta_{n,k}]$, that associate a data point with a cluster, and representation parameters, $\mathbf{M}$, the mean of each cluster. Note that, both classes of parameters are interdependent on each other, that is, if $\mathbf{M}$ is known, then $\mathbf{\Delta}$ can be computed, and if $\mathbf{\Delta}$ is known, then $\mathbf{M}$ can be computed. This is an example of a class of optimization problems that can be solved using **Expectation–Maximization (EM)** algorithms that alternate between two steps in each iteration: expectation step and maximization step.

(a) **Expectation step** (the E-step) updates $\mathbf{\Delta}_t$ given the current value of $\mathbf{M}_t$ by *associating* a data point with its *nearest* representative:

$$\delta_{n,k}^{(t)} \leftarrow 1 \text{ if } \left(k = \arg\min_{j=1\ldots K} \left\|x_n - m_j^{(t)}\right\|^2\right), 0 \text{ otherwise.}$$

(b) **Maximization step** (the M-step) updates $\mathbf{M}_{t+1}$ given the current value of $\mathbf{\Delta}_t$ by *maximizing* the above objective function resulting in:

$$\frac{\partial J(\mathbf{M}, \mathbf{\Delta})}{\partial m_k} = 2 \sum_{n=1}^{N} \delta_{n,k}(x_n - m_k) = 0 \Rightarrow m_k^{(t+1)} \leftarrow \frac{\sum_{n=1}^{N} \delta_{n,k}^t x_n}{\sum_{n=1}^{N} \delta_{n,k}^t}$$

Figure 15.12 shows these two steps pictorially on how a complete EM iteration works. Figure 15.12a shows two randomly initialized cluster centers. Figure 15.12b shows how given those two initial cluster centers, the data points are enumerated by the cluster they are closest to (orange vs. green). Figure 15.12c shows how with the new associations the cluster centers are updated using the M-step. Figure 15.12d shows the final association update leading to the desirable clustering of the data.

There are several properties of K-means clustering that are likeable and some that are not:

Sensitivity to Initialization—In case of PCA, the solution to the objective function was what we call a "closed-form solution" because there is only one

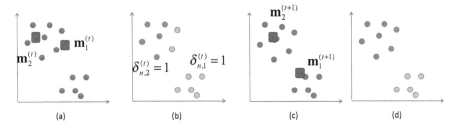

Fig. 15.12 (**a**) Initial cluster centers, (**b**) E-step associating data points with one cluster or the other, (**c**) M-step updating the cluster centers, (**d**) next E-step shows convergence

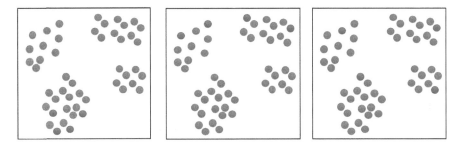

Fig. 15.13 Clustering is sensitive to initialization. Three different possible random initializations (blue) that might either result in different final clusters or more iterations to convergence

optimal answer there. But clustering does not have such an objective function that gives us one final answer. Here, the final clusters learnt depend on the way we have initialized the clustering. Figure 15.13 shows three different initializations of the same data. Depending on the initialization, the final cluster might either be suboptimal or take longer to converge to the optimal even if it is possible. But random initialization could give any of these or other combinations as the initial clusters, and hence K-means is not always guaranteed to give the same clusters. As a general rule, we do not like "non-determinism" in our algorithms—no guarantee that we will get the same results for the same data and the same hyper parameters (number of clusters).

Smart Initialization: There are a number of algorithms that have been proposed to make K-means clustering more "optimal" and "deterministic" from an initialization perspective. One such initialization method is the farthest first point (FFP) initialization where we choose the first cluster center deterministically, that is, pick the data point farthest from the mean of the entire data. Then we choose the second cluster center that is farthest from the first. The third is picked such that it is farthest from the first two, and so on. Figure 15.14 shows one such initialization where first figure shows the first two clusters picked. Middle figure shows how the next cluster is chosen such that it is farthest from the first two, and third is chosen such that it is farthest from the first three. This guarantees good coverage of the space and leads to a decent initialization, resulting in a closer to optimal clustering.

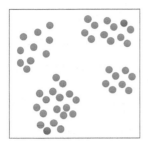 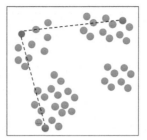 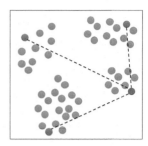

Fig. 15.14 Farthest first point initialization. The first figure shows two cluster centers initialized. Middle figure shows how the third is picked such that it is farthest from both the first two and the fourth is farthest from all three

Scale and Speed—K-means clustering and such partitional algorithms are highly scalable. The overall complexity of K-means is O(NKDT) where N is the number of data points, K is the number of clusters, D is the dimensionality of the data (quantifying the time it takes to compute the distance between a data point and a cluster center), and T is the number of iterations it takes for a K-means to converge. Their linearity in all these dimensions makes such partitional algorithms so popular. Not only that, K-means clustering is also highly parallelizable at two levels. In the map-reduce sense, if (a) we make previous iteration cluster centers available to all mappers, (b) if a map job—while processing one record at a time—generates its nearest cluster index as the key and the record itself as value, and (c) reducer takes the average of all the data points of the same key, then we can achieve K-fold parallelism for K-means clustering. Alternately, the distance computation using Euclidean distances between a data point and all the cluster centers can be parallelized using GPUs.

The Distance Function—The above formulation works well for Euclidean distances, but when the data is not represented as a point in space (only distances between pairs of points are given), we cannot use K-means or its variants. Then, we have to rely on other algorithms. When the data point is not a point in a Euclidean space but an L1 normalized (e.g., clustering probability distributions) or L2 normalized (e.g., TFIDF representation of documents) space, then we need to tweak K-means algorithm slightly. In an L2 normed TFIDF vector space, all documents are in a hyper-sphere with their L2 norms being 1. In **spherical K-means clustering**, the cluster centers are also forced to be in the same "spherical" L2 normed space as the original data. Here instead of computing distance we assign a document to that cluster whose cosine similarity is maximum among all clusters. Second, after computing the mean, we renormalize the mean vectors into L2 normed vectors. These two modifications to K-means clustering make it amenable to L2 normed data.

Number of Clusters: Another problem with K-means clustering is that it requires the number of clusters K as a hyper-parameter. Now without any prior knowledge or understanding of the data, we cannot say what is the right number of clusters. Often this number is decided through a rigorous statistical analysis by trying different values of K and measuring a quantity such as the "gap" between random clustering and actual clustering with a certain value of K (this is also known as the "gap statistic"). There are other such mechanisms that can be deployed to find the right number of clusters for the data. Another option is to let the business limitations decide the number K. For example, if we can only create five unique campaigns for all our customers, then we may want to segment our customers only into five clusters and create one campaign for each.

Heterogeneous Clusters: K-means clustering only discovers homogeneous and spherical clusters. For example, when clusters are of different sizes from each other, of different densities, or of different shapes, then K-means clustering does not do a good job of discovering them. This is because of the Euclidean distance used in K-means which makes clustering look for hyper-spherical clouds in the data space. Other methods such as agglomerative clustering are typically used to discover elongated clusters, and mixture of Gaussians is able to model arbitrary shaped clusters. These will be covered later.

Hard vs. Soft Clustering: Most machine learning algorithms have two variations—the "hard" or brittle version and the "soft" or robust version. In K-means clustering, when it is decided that a certain data point is closest to one of the clusters, it is "hard-assigned" to that cluster only and to none other, that is, $\delta_{n,k} \in \{0, 1\}$. This has two problems: First, it ignores the actual distance between the data point and the cluster center. If the distance between the data point and the cluster center is small, then the "degree-of-belongingness" of this point to this cluster should be higher. Second, if a data point is just at the boundary—only barely closer to one cluster mean than the other—then this assigns that point to the cluster it is closer to—so it does not take the second nearest into account. To alleviate these problems, we do a *softer* version of K-means clustering known as the ***soft K-means clustering*** where instead of doing a hard assignment, we can define a soft or probabilistic assignment $\delta_{n,k} \in [0, 1]$ between each data point and each cluster center. In soft clustering, we define the degree of association of a data point with a cluster as inversely proportional to its distance from the cluster center going from soft to hard as iterations progress.

$$\delta_{n,k}^{(t)} \leftarrow \frac{\exp\left(-\frac{\|x_n - m_k\|^2}{\sigma^2(t)}\right)}{\sum_{j=1}^{K} \exp\left(-\frac{\|x_n - m_j\|^2}{\sigma^2(t)}\right)}$$

To learn more about K-means clustering and expectation maximization, one can read Chap. 10 (Sect. 10.2.1) and Chap. 11 (Sect. 11.1.3) in Han et al. (2011), Chap. 11 (Sect. 11.4) in Murphy (2012), and Chap. 6.12 (Sect. 6.12) in Michalski et al. (2013).

3.2 Hierarchical Clustering

Partitional clustering assumes that there is only one "level" in clustering. But in general, the world is made up of a "hierarchy of objects." For example, the biological classification of species has several levels—domain, kingdom, phylum, class, order, family, genus, and species. All the documents on the web can be clustered into coarse (sports, news, entertainment, science, academic, etc.) to fine grained (hockey, football, ..., or political news, financial news, etc.). To discover such a "hierarchical organization" from data, we do hierarchical clustering in two ways: top-down and bottom-up.

Top-down hierarchical clustering also known as ***divisive clustering*** where we apply partitional clustering recursively first to, say, find K1 clusters at the first level of the hierarchy, then within each find K2 clusters and so on. With the right number of levels and number of clusters at each level (which may be different in different parts of the hierarchy), we can now discover the overall structure in the data in a top-down fashion. This, however, still suffers—at each level in the hierarchy—the problems that a partitional clustering algorithm suffers from—initialization issues, number of clusters at each level, etc. So it can still give a variety of different answers and the problem of non-determinism remains.

Bottom-up hierarchical clustering also known as ***agglomerative clustering*** is the other approach to building the hierarchy of clusters. Here we start with the raw data points themselves at the bottom of the hierarchy, and we find the distances between all pairs of points and merge the two points that are nearest to each other since they make the most sense to "merge." Now the merged point replaces the two points that were merged and we are left with $N - 1$ data points when we started with N data points. The process continues as we keep merging two data points or clusters together until the entire data is merged into a single root node. Figure 15.15 shows the result of a clustering of ten digits (images) in a bottom-up fashion. The structure

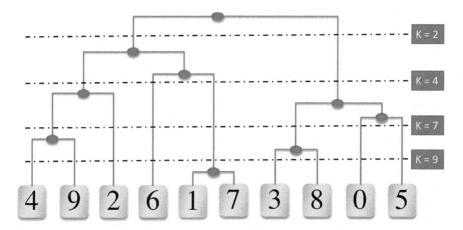

Fig. 15.15 Bottom-up agglomerative clustering of digits—tree structure

is called a **dendrogram** that shows how at each stage two points or clusters are merged together. First, digits 1 and 7 got merged. Then 3 and 8 got merged, then 4 and 9 got merged, then 0 and 5 got merged, then the cluster {3,8} and {0,5} got merged, and so on. The process leads eventually to a binary tree that we can cut at any stage to get any number of clusters we want.

The key to agglomerative clustering is the definition of distance between two "clusters" in general (e.g., clusters {3,8} and {0,5}). Different ways of doing this define different kinds of agglomerative clustering, resulting in different forms of clustering shapes. In the following, let $\mathbf{X} = \{x_1, x_2, \ldots, x_P\}$ be the set of P points in cluster $\mathbf{X}$ and let $\mathbf{Y} = \{y_1, y_2, \ldots, y_Q\}$ be the set of Q points in cluster $\mathbf{Y}$. Note that either P or Q or both can be 1. The distance between the set $\mathbf{X}$ and set $\mathbf{Y}$ can be defined in many ways:

- **Single linkage**—distance between two nearest points across $\mathbf{X}$ and $\mathbf{Y}$ is used as distance between the two clusters. This gives elongated clusters as two clusters with even one point close to one point of another cluster will be merged.

$$\Delta(\mathbf{X}, \mathbf{Y}) = \min_{p=1\ldots P} \left\{ \min_{q=1\ldots Q} \left\{ \Delta(x_p, y_q) \right\} \right\}$$

- **Complete linkage**—the other extreme of single linkage is where distance between two farthest points across $\mathbf{X}$ and $\mathbf{Y}$ is used as distance between clusters. Here the clusters discovered are more rounded as every point of one cluster must be close to every other point of the other cluster.

$$\Delta(\mathbf{X}, \mathbf{Y}) = \max_{p=1\ldots P} \left\{ \max_{q=1\ldots Q} \left\{ \Delta(x_p, y_q) \right\} \right\}$$

- **Average linkage**—is between the single and complete linkage clustering where distance between the two clusters is computed as the average distance between all pair of points among them. This makes clustering robust to noise.

$$\Delta(\mathbf{X}, \mathbf{Y}) = \frac{1}{PQ} \sum_{p=1}^{P} \sum_{q=1}^{Q} \Delta(x_p - y_q)$$

There are several pros and cons of hierarchical agglomerative clustering.

- **Deterministic clusters**—Unlike partitional clustering (e.g., K-means) where the final cluster depends on initialization, agglomerative clustering always gives the same clustering for the same dataset and definition of distance. It does not depend on any initialization since there is no initialization.
- **Feature representation vs. distance function**—Partitional clustering works only on multivariate data where each data point must be a point in a Euclidean

space. Agglomerative clustering can work on datasets where only pairwise distances are given and data has no feature representation.
- **Scale**: One of the drawbacks of agglomerative clustering is that it is quadratic in the number of data points because, to begin with, we have to compute pairwise distances between all pairs of points. This makes it highly impractical as the number of data points increases. For very large datasets, it is possible to first do a large number (e.g., $K = \sqrt{N}$) of clusters to remove fine grained noise and then do hierarchical clustering on these K cluster centers as data points. Thus mixing partitional and hierarchical merges the best of both worlds.
- **Number of clusters**: Finally, agglomerative clustering gives us all the number of cluster we need. We can cut the dendrogram at any level to get that many clusters. This does not require us to start with prior knowledge about the number of clusters as a parameter.

The reader can refer to Chap. 25 (Sect. 25.5) from Murphy (2012), Chap. 14 (Sect. 14.3.12) from Friedman et al. (2001), and Chap. 3 (Sect. 3.4.3) from Han et al. (2011) for additional material on hierarchical clustering.

3.3 Spectral Clustering

Partitional clustering works on data with Euclidean feature spaces. Hierarchical clustering works on pairwise distance functions in a bottom-up fashion and recursive partitional clustering in a top-down fashion. There is another class of clustering algorithm that works on similarity graphs where each node represents an entity and weight on the edge; connecting the nodes quantifies similarity between the two edges. Spectral clustering is a very useful clustering algorithm in domains where it is easier to quantify such similarity measures between entities rather than representing them as a feature vector, for example, two LinkedIn profiles, two songs, two movies, and two stock market returns time series. Again, we will follow the four stages to develop a proper objective function for spectral clustering.

Intuition: Consider a graph with six nodes {a,...,f} shown in Fig. 15.16. Edge weights indicate similarity between pairs of entities. In order to partition this graph into two parts, we must remove a subset of edges. The edges removed constitute "loss of information" which we want to minimize. Clearly removing the smallest weight edges makes sense as shown in Figure 15.16. Removing the three edges between nodes {a,d}, {c,d}, and {c,f} will result in two partitions {a,b,c} and {d,e,f} that by themselves are highly connected to each other.

Formulation: We translate the above intuition into an objective function. Let $\mathbf{W} = [w_{ij}]$ be the symmetric similarity matrix of size $N \times N$ where N is the number or nodes in the graph (i.e., number of LinkedIn profiles or number of movies among which we know similarity). One way to formulate this would be to introduce variables $\mathbf{X} = \{x_1, x_2, \ldots, x_N\}$ where $x_n \in \{1, -1\}$ depending on whether after partitioning this graph into two parts, the node n belongs to the first partition

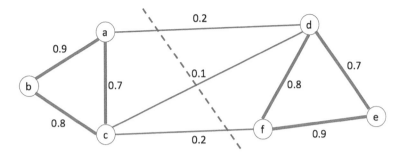

Fig. 15.16 A similarity graph with six nodes. Edge weights are similarity between corresponding pairs of entities. The graph is partitioned into two parts such that total weight of removed edges is minimum

(1) or the second partition (0). Now the intuition suggests that two nodes (i, j) should be in the same partition (i.e., $(x_i - x_j)^2$ is 0) if they are very similar (i.e., w_{ij} is high) and in different partitions (i.e., $(x_i - x_j)^2$ is 1) if they are dissimilar, that is, $(w_{ij}$ is low). We can therefore capture the intuition by maximizing the following objective function.

$$J(\mathbf{X}|\mathbf{W}) = \frac{1}{2} \sum_{1 \leq i,j \leq N}^{N} w_{ij} (x_i - x_j)^2$$

Modification: As we solve the above objective, we get the following:

$$J(\mathbf{X}|\mathbf{W}) = \frac{1}{2} \sum_{1 \leq i,j \leq N}^{N} w_{ij} \left(x_i^2 + x_j^2 - 2x_i x_j\right) = \sum_{i=1}^{N} x_i^2 \left(\sum_{j=1}^{N} w_{ij}\right)$$

$$- \sum_{1 \leq i,j \leq N}^{N} x_i w_{ij} x_j = \mathbf{x}^T (\mathbf{D} - \mathbf{W}) \mathbf{x}$$

where $\mathbf{D}$ is the diagonal matrix whose diagonal elements are sum of the rows of $\mathbf{W}$

$$\mathbf{D} = \begin{bmatrix} d_1 & 0 & 0 \\ 0 & \ddots & 0 \\ 0 & 0 & d_N \end{bmatrix}$$

where $d_n = \left(\sum_{j=1}^{N} w_{nj}\right)$. The matrix $\mathbf{L} = (\mathbf{D} - \mathbf{W})$ is called the unnormalized graph Laplacian of the similarity matrix $\mathbf{W}$. It is a positive semi-definite matrix with smallest eigenvalue 0 and the corresponding eigenvector as all 1's. If the graph has k connected components, that is, each connected component has no link across, then there will be k smallest eigenvalues equal to 0. Assuming the graph has only one connected component, the second smallest eigenvector is used to partition the graph into two parts. We can take the median value of the second smallest eigenvector and partition the graph such that the nodes whose second eigenvector components are above the median are in one partition and the remaining nodes in the other partition. This partitioning can be applied recursively now to break the two components further into two partitions in a top-down fashion.

In this section, we have studied a number of clustering algorithms depending on the nature of the data. One of the open problems in clustering is how to systematically define distance functions when data is not a straightforward multivariate real-valued vector. This is where the critical domain knowledge is required. The next paradigm—density estimation—extends the idea of clustering by allowing us to describe each cluster with a "shape" called its density function.

For further reading, the reader can refer to Chap. 25 (Sect. 25.4) from Murphy (2012) and Chap. 14 (Sect. 14.5.3) from Friedman et al. (2001).

4 Density Estimation

The fundamental hypothesis that data has structure implies that it is not uniformly distributed across the entire space. If it were, it would not have any structure. In other words, all parts of the feature space are not equally *probable*. Consider a space with two features "age" and "education." Let us say age takes a value from 0 to 100 years and "education" from, say, 1 to 20. Now probability P(age = 3, education = PhD) is zero and P(age = 26, education = PhD) is high. Similarly, P(age = 20, education = grade-1) is low and P(age = 5, education = grade-1) is high. Estimating this joint probability, given the data, gives us a sense of which combination of feature values are more likely than others. This is the essence of *structure* in the data, and density estimation captures such joint probability distributions in the data. Density estimation has many applications, for example:

- **Imputation**: If one or more of the feature values is missing, given the others we can estimate the missing value as the value that gives the highest joint probability after substituting it.
- **Bayesian classifiers**: Another application of density estimation is to build a "descriptive" classifier for each class where the descriptor is essentially a class conditional density function $P(\mathbf{x}|c)$.
- **Outlier detection**: Another important application of density estimation is outlier detection used in many domains such as fraud, cyber security, and when dealing with noisy data. A data point with low probability after we have learnt the density function is considered an outlier point.

There are two broad density estimation frameworks. First, is the ***nonparametric density estimation*** where we do not learn a model but use the "memory" of all the known data points to determine the density of the next data point. Parzen Window is an example of a nonparametric density estimation algorithm. Second is the *parametric density estimation* where we first make an assumption about the distribution of the data itself and then fit the parameters of this function using maximum log likelihood optimization. If individual parametric density functions are not enough to represent the complexity of the data (e.g., data is multimodal), then we apply mixture of parametric density models (e.g., mixture of Gaussians).

4.1 Nonparametric Density Estimation

Let us first develop an intuition behind density functions from an example. Imagine that in a room floor we scatter a large number of magnets at specific locations. Each of these magnets has the same "magnetic field of influence" that diminishes as we go away from the magnet. Now imagine if there is a piece of iron at a certain location in the room, it will experience a total magnetic field that is the sum of all the magnets. The magnets that are closer to this piece of iron will have a higher influence than the farther ones.

Let $\mathbf{X} = \{x_1, x_2, \ldots, x_N\}$ be the set of N magnets (data points) scattered in some high dimensional space. Let x be a new data point (iron) whose density (influence by all the magnets), $P(x)$, has to be estimated. In a nonparametric kernel density estimation, we represent this total field of influence as follows:

$$P(x) = \frac{1}{n} \sum_{n=1}^{N} K_\sigma(x, x_n) = \frac{1}{n\sigma\sqrt{2\pi}} \sum_{n=1}^{N} \exp\left(\frac{\|x - x_n\|^2}{2\sigma^2}\right)$$

Here $K_\sigma(x, x_n)$ is the kernel function (chosen to be Gaussian in the above example) that measures the influence of the training data point (magnet) x_n on the test data point (iron) x and σ is the decay with which the field of influence drops (and therefore how wide the field spreads). If σ is too small, then each training point has a very sharp and narrow range field of influence. If σ is too large, then each training point has a very broad field of influence. Like in K-means clustering, K controls complexity, and σ controls the complexity of the density function here. Nonparametric density estimation has the following pros and cons.

- **No prior knowledge**: Nonparametric density estimation does not require that we know the functional form of the density function. This is very handy when there is no domain knowledge about the phenomenon that generates the data. However, we still have to play with σ, the spread of each density function around each training data point. Choosing a small value of σ will model noise in the data and choosing a large value will not capture the signal. There is, like in all hyper-parameter spaces, a sweet spot that we must find through experimentation.
- **Scoring time**: Nonparametric methods are also known as "lazy-learners" since they spend no time "training" a model but at the scoring time their complexity is O(N)—linear in the number of training data points (magnets). This makes them unsuitable for real-time tasks (e.g., if we were to make a real-time decision

about whether a credit card transaction is fraud or not and we are using outlier detection based on density estimations, we cannot use such nonparametric density estimators).
- **Robustness to noise**: Since each training data point has an influence on density estimation of each point, even noisy points get to have their say. It is therefore important to identify and remove the noisy points from the training set or use parametric techniques for highly noisy datasets.

4.2 Parametric Density Estimation

In nonparametric density functions, the data is stored "as is" and is used to compute density using kernel functions. The parametric density estimation functions, on the other hand, first define a parametric form and then find the parameters by optimizing an objective function. We will follow the same four-stage process of intuition, formulation, modification, and optimization to learn parameters.

Intuition: Let $P(x|\theta)$ be a parametric density function where θ is the set of parameters to be learnt. For N data points $\mathbf{X} = \{x_1, x_2, \ldots, x_N\}$, we have to find the set of parameters that "fits" the data best. In other words, we need to find the parameters θ such that the probability of seeing the entire data is maximum.

Formulation: Parametric density function problems are all modeled as optimization problems where we try to find the set of parameters that maximizes the likelihood of seeing the data. Since each data point is identical and independently distributed, the likelihood of seeing the entire data is the product of the likelihood of seeing each data point independently.

$$\theta^* = \arg\max_{\theta} J(\theta|\mathbf{X}) = \arg\max_{\theta} \prod_{n=1}^{N} P(x_n|\theta)$$

Modification: Typically, when any density functional form (e.g., Gaussian or Poisson or exponential) is substituted for $P(x_n|\theta)$, the product term becomes too complex to solve. We therefore modify this to the **log likelihood function** which is monotonic and equivalent to maximizing likelihood function, $\theta^* = \arg\max_{\theta} \ln J(\theta|\mathbf{X}) = \arg\max_{\theta} \sum_{n=1}^{N} \ln P(x_n|\theta)$.

Optimization: Finally, we will optimize this for a few density functions in one-dimensional spaces.

- **Exponential distribution**: where $P(x|\theta) = \theta e^{-\theta x}$, for $x > 0$ and 0 otherwise. So

$$J(\theta|\mathbf{X}) = \sum_{n=1}^{N} \ln P(x_n|\theta) = \sum_{n=1}^{N} [\ln \theta - \theta x_n] = N \ln \theta - \theta \sum_{n=1}^{N} x_n$$

$$\frac{\partial J(\theta|X)}{\partial \theta} = \frac{N}{\theta} - \sum_{n=1}^{N} x_n = 1 \therefore \widehat{\theta} = \frac{1}{\frac{1}{N}\sum_{n=1}^{N} x_n}$$

- **Bernoulli distribution**: where $P(x|\theta) = \theta^x(1-\theta)^{1-x}$, for $x \in \{0,1\}, 0 < \theta < 1$. So

$$J(\theta|\mathbf{X}) = \sum_{n=1}^{N} \ln P(x_n|\theta) = \sum_{n=1}^{N} [x_n \ln \theta + (1-x_n)\ln(1-\theta)]$$

$$\frac{\partial J(\theta|X)}{\partial \theta} = \frac{1}{\theta}\sum_{n=1}^{N} x_n - \frac{1}{1-\theta}\left[N - \sum_{n=1}^{N} x_n\right] = 0 \therefore \widehat{\theta} = \frac{1}{N}\sum_{n=1}^{N} x_n$$

- **Poisson distribution**: where $P(x|\theta) = \frac{\theta^x}{x!}e^{-\theta}$, $x = 0,1,2,\ldots$ and $\theta > 0$. So

$$J(\theta|\mathbf{X}) = \sum_{n=1}^{N} \ln P(x_n|\theta) = \sum_{n=1}^{N} [x_n \ln \theta - \theta - \ln x_n!]$$

$$\frac{\partial J(\theta|X)}{\partial \theta} = \frac{1}{\theta}\sum_{n=1}^{N} x_n - N = 0 \therefore \widehat{\theta} = \frac{1}{N}\sum_{n=1}^{N} x_n$$

- **Normal distribution**: where $P(x|\mu,\sigma^2) = \frac{1}{\sqrt{2\pi}\sigma}\exp\left(-\frac{(x-\mu)^2}{2\sigma^2}\right)$, $\theta = \{\mu,\sigma^2\}$

$$J(\mu,\sigma^2|\mathbf{X}) = \sum_{n=1}^{N} \ln P(x_n|\theta) = -\frac{1}{2}\sum_{n}^{N}\left[\ln \sigma^2 + \frac{x-\mu}{2\sigma^2} + \ln 2\pi\right]$$

$$\frac{\partial J(\mu,\sigma^2|X)}{\partial \mu} = \frac{1}{2\sigma^2}\sum_{n=1}^{N}(x-\mu) = 0 \therefore \mu = \frac{1}{N}\sum_{n=1}^{N} x_n$$

$$\frac{\partial J(\mu,\sigma^2|X)}{\partial \sigma^2} = \frac{1}{2}\sum_{n=1}^{N}\left[\frac{1}{\sigma^2} - \frac{(x-\mu)^2}{\sigma^4}\right] = 0 \therefore \sigma^2 = \frac{1}{N}\sum_{n=1}^{N}(x-\mu)^2$$

The reader can refer to Criminisi et al. (2012) and Robert (2014) to learn more about density estimation.

4.3 Mixture of Gaussians

Often, a single Gaussian is not enough to model the complexity of multimodal data. For example, in case of OCR, the same digit might be written in two or three different ways (e.g., a 7 with a cut in the middle or not, a 9 with a curve at the bottom or not), there could be font or other variations. In speech, there could be multiple accent variations within a language (e.g., the English language has different accents, e.g., American, British, Indian, and Australian). In such cases, it is better to learn a multimodal density function using a mixture of unimodal density functions—one for each variant. If each density function is a Gaussian, then this multimodal density function is called a mixture of Gaussians (MoG).

Insight: In MoG we assume that there are $K > 1$ Gaussians that might generate the data. Each of the Gaussians has its own mean, covariance, and prior. So we first pick one of the Gaussians from the mixture with a certain "prior" and then use that Gaussian to generate a data point with a certain probability that diminishes as we go away from the mean of that Gaussian. Another way to think about MoG is that it is a Bayesian extension of K-means clustering. In K-means clustering, one of the problems was that since we were using only Euclidean distances between a data point and its cluster center, all clusters were spherical and of different shapes and densities were not easy to handle. In MoG, we enable this extra degree of freedom that each cluster, that is, Gaussian, can now have an arbitrary covariance matrix to adjust to the "shape" of the cluster. Second, instead of forcing a data point to be in one cluster only, MoG lets a data point to be influenced by more than one Gaussian depending on its distance from the mean of the Gaussian and the shape of the Gaussian depending on the covariance matrix. This is also connected to Parzen Windows as follows. On one extreme if we model the entire data's density with a single Gaussian, we might get a very simple model that might not capture the essence of the data. If on the other extreme, we treat each data point as its own Gaussian like in Parzen window, then we might be overlearning. But a mixture of Gaussians is giving the right number of Gaussians needed to model the data between these two extremes.

Formulation: Let us say there are K Gaussians that we have to model to explain the data $\mathbf{X} = \{x_1, \ldots, x_N\}$. Each Gaussian has its own prior $\pi_k = P(k)$, mean μ_k, and covariance matrix Σ_k. Let $\Theta = \{\theta_1, \theta_2, \ldots, \theta_K\}$ be the set of parameters where $\theta_k = \{\pi_k, \mu_k, \Sigma_k\}$. Like in K-means clustering, we will use a set of latent parameters: $\Delta = [\delta_{n,k}]$ that quantifies the association of the nth data point with the kth Gaussian. This will morph into the softer posterior probability $P(k|x_n)$. The overall maximum likelihood objective is

$$J(\Theta|\mathbf{X}) = \prod_{n=1}^{N}\prod_{k=1}^{K}[P(x_n,k)]^{\delta_{n,k}} = \ln\prod_{n=1}^{N}\prod_{k=1}^{K}[P(x_n,k)]^{\delta_{n,k}}$$

Modification: We apply two modifications to the above data likelihood objective. First, we convert the joint probability of data and mixture into two parts:

$P(x_n, k) = P(x_n | k) P(k)$ and take the log of the likelihood to make the calculus easy for optimization. Also note that there are constraints on $\delta_{n,k}$ that for each n they must add up to 1. We put all these into the modified objective function:

$$J(\Theta | \mathbf{X}) = \prod_{n=1}^{N} \left(\prod_{k=1}^{K} \delta_{n,k} [\ln P(x_n | k)] + \ln P(k) + \lambda \left(\prod_{k=1}^{K} \delta_{n,k} - 1 \right) \right)$$

Optimization: Similar to the K-means clustering, mixture of Gaussians also uses an EM approach to solve the two sets of parameters alternately resulting in the following iterative solution:

The expectation step becomes the Bayes theorem:

$$\delta_{n,k}^{(t)} = P^{(t)}(k|n) \leftarrow \frac{\pi_k^{(t)} P^{(t)}(x_n|k)}{\sum_{j=1}^{K} \pi_j^{(t)} P_t(x_n|j)} = \frac{\pi_k^{(t)} P^{(t)}\left(x_n | \mu_k^{(t)}, \Sigma_k^{(t)}\right)}{\sum_{j=1}^{K} \pi_j^{(t)} P_t\left(x_n | \mu_j^{(t)}, \Sigma_j^{(t)}\right)}$$

The maximization step when optimized for mean and covariance results in the following updates:

$$\pi_k^{(t+1)} \leftarrow \frac{1}{N} \sum_{n=1}^{N} \delta_{n,k}^{(t)}$$

$$\mu_k^{(t+1)} \leftarrow \frac{\sum_{n=1}^{N} \delta_{n,k}^{(t)} x_n}{\sum_{n=1}^{N} \delta_{n,k}^{(t)}}$$

$$\Sigma_k^{(t+1)} \leftarrow \frac{\sum_{n=1}^{N} \delta_{n,k}^{(t)} \left(x_n - \mu_k^{(t+1)}\right) \left(x_n - \mu_k^{(t+1)}\right)^T}{\sum_{n=1}^{N} \delta_{n,k}^{(t)}}$$

Gaussian mixture models are used extensively in many domains including speech, outlier detection, and building Bayesian classifiers especially when a class has multiple latent subclasses that need to be discovered automatically. MoG still depends on initialization, and one way to do it is to first do farthest first point sampling to initialize K cluster centers. Then do K-means clustering to converge on clusters and use those cluster centers as the initial means and the covariance matrices of those clusters as the initial covariance. Using this as the seed, we learn MoG and further refine those K-means clusters.

Figure 15.17 shows increasing degrees of complexity of parametric densities for the same dataset. In (a) we use a single spherical density (i.e., variance along all dimensions is assumed to be same) -equal diagonal elements (i.e., we ignore correlation among dimensions). In (b) we still use a single Gaussian but now each dimension can have a different variance while still ignoring correlation among

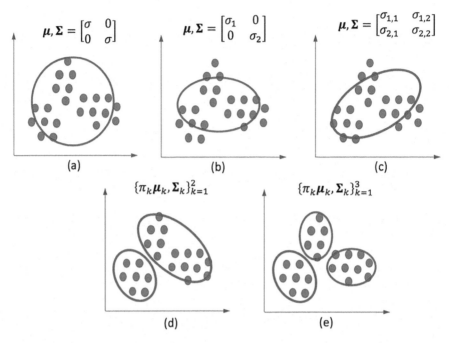

Fig. 15.17 Different complexities of a density function: (**a**) single Gaussian, spherical covariance. (**b**) Single Gaussian, diagonal covariance, (**c**) single Gaussian, full covariance, (**d**) mixture of two full covariance Gaussians, (**e**) mixture of three full covariance Gaussians

dimensions. In (c) we continue to use a single Gaussian to model the density, but we allow a full covariance. In (d) we increase the number of Gaussians to be two as one Gaussian does not seem to be sufficient to model the density of this data. In (e) we finally use three Gaussians to model the density—which seems to be sufficient. Adding more will try to memorize the data and not generalize. See Rasmussen (2004) to read more about mixture of Gaussian.

In this section, we have studied a variety of density estimation paradigms. There are other density estimation frameworks, for example, hidden Markov models for sequence of symbols. Overall, density estimations can give deep insights about "where to look" in the data, which parts of the data matter, and which parts are "surprising" or "anomalous."

5 Frequent Itemset Mining (FISM)

Insofar we have explored a variety of unsupervised learning frameworks that discover different types of structures in the data. We now explore another very common kind of data type—the **itemset data** and see what kind of structure can be found in such data. The itemset data is best described as a dataset where each record or data point is a "set of items from a large vocabulary of possible items." Let us first consider various domains where such data occurs:

- **Market basket data**: One of the most common examples of itemset data is the market basket data or the ***point of sale (POS)*** data where each basket is comprised of a "set" of products purchased by a customer in either a single visit or multiple visits put together (e.g., all products purchased in a week or one quarter, or the whole lifetime). Here the list of all products sold by the retailer is the vocabulary from which a product in the item could come from. We are losing information by just considering the set of products and not include their quantity or the price which would make it a "weighted set" instead. The problem is that in a typical heterogeneous retail environment, the products are not comparable. For example, 1 l of milk, 1 dozen bananas, and 1 fridge are not comparable to each other either in physical or monetary units. Hence, we stick with just the unweighted sets or ***baskets*** rather than weighted sets or ***bags***.
- **Keyword sets**: Another common example of itemset data is a keyword set. Often entities such as images on Flickr, videos on YouTube, papers in conferences, or even movies in IMDB are associated with a set of keywords. These keywords are used to tag or describe the entity they are referring to. Here the vocabulary from which the keywords can come from is predefined (e.g., keyword lists for conference papers) or is taken to be the union of all the keywords associated with all the entities.
- **Token sets**: Itemset data is also present in many other contexts not as keywords or products but arbitrary tokens. For example, hashtags in tweets whether per tweet or per account is an itemset. Another example is the set of skills in each LinkedIn profile is an itemset data. In a user session on YouTube, all videos watched by a user in one session constitute an itemset as well. All WhatsApp groups are itemsets of phone numbers. In a payment app or a credit card account, the set of merchants where a customer shopped in the last n days is also an itemset. It is up to us how we convert any transaction data into an itemset data as it makes sense.
- **Derived itemsets**: Itemset data can also be derived from other datasets. In neuroscience experiments, for example, we might want to discover which neurons fire together in response to different experiences or memories. In such cases, we can consider a moving (overlapping) window of a certain size and all *neurons* that fire in the same window could be considered an itemset. Similarly, in gene expression experiments, all genes that express themselves from the same stimuli could also be considered an itemset.

In all these itemset datasets we are interested in finding patterns of "co-occurrence"—that is, which subsets of items co-occur in the same itemsets. There are many ways to define co-occurrence. In frequent itemset mining (FISM), we are interested in finding "large and frequent" itemsets. While the dataset is simple and the definition of what is a pattern is also very straightforward, what makes this a complex problem is the combinatorial explosion when the vocabulary of possible items is very large. One of the key algorithms that we will develop here called the *apriori algorithm* solves this problem using a very basic insight from set theory.

Intuition: Consider an itemset data shown in Fig. 15.18i, it has a total of 10 data points over a vocabulary of 6 items {a,b,c,d,e,f}. We will first consider itemsets of size 1. There are six itemsets of size 1 (Fig. 15.18ii). For each we can compute the frequency which is the number of itemsets (out of 10) in which that item was present. Now that we have itemsets of size 1 and their frequencies (also known as *support*), we can compute itemsets of size 2 and so on. Now since we only care about the "frequent" itemsets and not all itemsets, we can define a frequency threshold θ_f (also known as support threshold) such that only those itemsets of size k (= 1 for now) will be kept whose frequency is above this threshold and others will be deemed not "supported" (i.e., noisy). This goes with the underlying philosophy of pattern recognition that anything that is high frequency is a pattern worth remembering. Now from itemsets of size 1 we can find itemsets of size 2 and their support, again pruning off those whose support is less, etc.

Formulation: The only problem with this brute-force counting is the following. In order for us to count the frequency of an itemset of size k, we need to maintain a counter with the itemset as the key and its count as value. As we go through a dataset, we check whether this itemset is a subset of the data itemset or not. If so, we increment its counter. Now as the vocabulary size grows and the value of

Item-set Data	Item-sets of Size 1		Candidates of size 2	Item-sets of size 2		Candidates of size 3	Item-sets of size 3	
Data set X	F(1)	Freq	C(2)	F(2)	Freq	C(3)	F(3)	Freq
{a,b}	{a}	6	{a,b}	{a,b}	3	{a,b,e}	{a,b,e}	1
{b,c,e}	{b}	5	{a,c}	{a,c}	3	{a,c,e}	{a,c,e}	1
{a,c,e,f}	{c}	6	{a,e}	{a,e}	3	{a,c,f}	{a,c,f}	2
{c,d,f}	{d}	2	{a,f}	{a,f}	4	{a,e,f}	{a,e,f}	3
{a,e,f}	{e}	6	{b,c}	{b,c}	2	{c,e,f}	{c,e,f}	2
{b,e,f}	{f}	7	{b,e}	{b,e}	3			
{a,b,c,d}			{b,f}	{b,f}	2			
{c,e,f}			{c,e}	{c,e}	3			
{a,c,f}			{c,f}	{c,f}	4			
{a,b,e,f}			{e,f}	{e,f}	5			
(i)	*(ii)*		*(iii)*	*(iv)*		*(v)*	*(vi)*	

Fig. 15.18 The apriori algorithm at work. (i) The dataset where each data point is a set of items from a dictionary of six possible items {a,b,c,d,e,f}. (ii) Frequent itemsets of size 1. If support threshold is 3, then all itemsets of size less than 3 are ignored (i.e., {d}). (iii) Using the apriori trick, all candidate itemsets of size 2 created from frequent itemsets of size 1. (iv) A pass over the data gives frequency of each of the candidates. Note that we did not have to worry about any pair of itemsets involving item d because its frequency count is less than threshold (3). (v) Again applying apriori trick to create candidates of size 3. (vi) Final frequent itemset of size 3 or more is {a,e,f} of size 3 and others of size 2

k grows, the potential number of combinations that we might have to keep in the counter memory grows to $O\binom{N}{k}$. So we apply the famous "apriori trick" here which tames the combinatorial explosion in an intelligent fashion.

Modification: The *Apriori Trick* is based on a simple observation that if $f(\mathbf{s}|\mathbf{X})$ is the frequency of the itemset $\mathbf{s}$ of size k in a dataset $\mathbf{X}$, then its frequency cannot be greater than the frequency of the *least* frequent subset of size $k - 1$ of $\mathbf{s}$. In other words, let us say if $\mathbf{s} = \{a,b,c\}$ and let us say its frequency is *3*, then it must be true that the frequency of all of its subsets, that is, {a,b}, {b,c}, and {a,c}, is at least 3. Otherwise, it will not be possible for {a,b,c} to have a frequency of 3. More formally:

$$f(\mathbf{s}|\mathbf{X}) \leq \min_{i \in \mathbf{s}} \{f(\mathbf{s} \sim i|\mathbf{X})\}$$

Where $\mathbf{s} \sim i$ is the set obtained by removing item i from set $\mathbf{s}$. Using this "apriori trick," the frequent itemset is able to ignore many *itemsets from counting as it knows that they will not be frequent anyway.*

Optimization: *The frequent itemset mining algorithm essentially grows itemsets from size k to size k + 1 as follows using a three-step process.*

- **Candidate Generation Step**: The input to this step is the frequent itemsets (whose support is above a threshold) of size k, $\mathbf{F}_k$. From this frequent itemset we first generate a candidate set of size $k + 1$, $\mathbf{C}_{k+1}$ that satisfy the apriori property, that is, we add all itemsets of size $k + 1$ to $\mathbf{C}_{k+1}$ whose subsets of size k are present in $\mathbf{F}_k$ (Fig. 15.18iii, v).
- **Frequency Counting Step**: The $k + 1$ size itemsets in the candidate set $\mathbf{C}_{k+1}$ are the only itemsets that have a chance to have a frequency above the support threshold, θ_f. All other combination of itemsets of size $k + 1$ are not counted at all. This really reduces the combination of itemsets on which the counter has to run in the next iteration:

$$f(\mathbf{s}|\mathbf{X}) = \sum_{n=1}^{N} \delta(\mathbf{s} \subseteq \mathbf{x}_n), \forall \mathbf{s} \in \mathbf{C}_{k+1}$$

- **Frequency Pruning Step**: Finally, when a pass through the data has been made and all frequencies of candidate itemsets are counted, the itemsets whose frequency is below the support threshold are removed to obtain $\mathbf{F}_{k+1}$, the final frequent itemsets of size $k + 1$.

Figure 15.18 shows the entire process of generating frequent itemsets of size up to 3 from an itemset data with support threshold of 3. Each iteration alternates between the above three steps.

The purpose of creating frequent itemsets is to find rules of the sort: (If *condition* then *trigger*) with some confidence. For example, once we have discovered through the above process that {a,e,f} is a frequent itemset, we can now create rules of the form: {a,e} → {f}, {a,f} → {e}, {e,f} → {a}, {a} → {e,f}, {e} → {a,f}, {f} → {a,e}. Each rule comes with a confidence score computed based on the frequency of the entire set {a,e,f} and the frequency of the condition set, that is,

$$Confidence(\{a,e\} \to \{f\}) = Support(\{a,e,f\})/Support(\{a,e\})$$

In other words, this says that if someone bought both a and e, then the probability that they will also buy f is 1 and can therefore be recommended with a very high confidence. In frequent itemset mining, all such rules are created and a confidence threshold θ_c is used to prune out rules with lower confidence. The output of the frequent itemset algorithm is the set of such rules with high support and confidence.

Frequent itemset mining has been one of the early algorithms that almost gave birth to the field of "data mining." It was the first breakthrough of its kind in mining such itemset data and since then, there have been a number of improvements in smart data structures to store the candidate and frequent itemsets to make it faster and more scalable. It has also been applied to areas beyond retail data mining for which it was originally invented. It has been used to discover "higher order features" of type "sets of items" in various domains including computer vision where each image region could be thought of as a collection of symbols from a vocabulary (HoG or SIFT). If many regions across many images show the same set of items (e.g., face images all show eye, nose, mouth, etc.), then a new object (face) can be created from a set of lower order features. Wherever we have a "set of items" dataset, we can use FISM.

See Chap. 6 (Sect. 6.2) in Han et al. (2011) for additional material on frequent itemset mining.

6 Network Analysis

Now that we understand how to find patterns in sets and multivariate data, we turn our attention to an even more complex yet commonly available data type—a *graph* or *network* data. These graphs may be weighted (i.e., edges have weights) or unweighted (i.e., edges are binary—either present or not), directed (i.e., edges either go from a node to another) or undirected (i.e., there is no direction on edges), or homogeneous (i.e., all nodes and edges are of same types) or heterogeneous (i.e., nodes or edges are of different types). Analyzing graphs for patterns presents very interesting challenges and a lot of opportunities in a wide variety of applications. There are a number of different kinds of patterns that can be discovered in graphs. In this section, we will focus on two kinds of network analyses problems: (1) PageRank, one of the most important algorithms in graph theory that led to the birth of companies like Google, (2) Detecting Cliques in graphs—another commonly used algorithm with many applications.

Graphs or networks are present in many domains. Internet, for example, is a collection of a very large number of web pages (generated at the rate of more than

1000 pages per minute) with links going from one page to another (directed graph). This is perhaps one of the largest graph out there. Social networks are another class of large graphs—LinkedIn, Facebook, telecom networks (e.g., people calling each other above a threshold), financial networks (e.g., based on money transfers), etc.

Weighted graphs can also be created from transaction or co-occurrence data. For example, consider a market basket data where we can quantify the consistency with which two products (a, b) are co-purchased together $P(a, b)$ more often than random $P(a)P(b)$ using, for example, pointwise mutual information.

$$\phi(a, b) = \log \frac{P(a, b)}{P(a)P(b)}$$

Any co-occurrence data can be converted to such weighted graphs where the edges can be removed if these weights are below a threshold. Many measures such as Jaccard coefficient, normalized pointwise mutual information, and cosine similarity can be used to create these weighted graphs. Next we develop two algorithms for network analysis.

6.1 Random Walks (PageRank)

Given a directed graph like the Internet, we are interested in finding out which is the most important node in the graph. The key motivation behind this problem came from Google where they wanted to sort all pages that contained a keyword in an order that "made sense." They posed this as a "random surfer" problem—if a surfer randomly picks a page on the Internet and starts following the links, what would be the probability that he will be at a certain page, and if we average over all such random surfers, which page on the Internet would have the most number of people, the second most number of people, and so on. This distribution over pages in the steady state gives the PageRank of each page on the Internet.

Intuition: Every page on the Internet has a set of incoming edges (shown for node j in Fig. 15.19) and a set of outgoing edges (shown for node i in

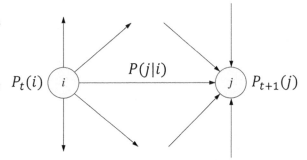

Fig. 15.19 Outgoing edges of node i and **Incoming** edges of node j. The probability of being on node j at time t + 1 depends on the probability of being on node i at time t and making a transition from i to j

Fig. 15.19). When on any page (node i), a random surfer might have some (e.g., equal) probability of going to one of the outgoing edges from this page. Thus the probability of the random surfer to "reach" a page (node j) would be to first "be" at one of the incoming pages (e.g., node i) of this page with a certain probability and then reach this page (node j) with a certain transition probability from that page (node i) in the next iteration as shown in Fig. 15.19.

Formulation: We now formulate this PageRank problem. Let us assume that there are N pages on the Internet $\mathbf{X} = \{x_1, x_2, \ldots, x_N\}$. Let $\mathbf{I}(x_n)$ be the set of in-neighbors of x_n and let $\mathbf{O}(x_n)$ be the set of out-neighbors of x_n. The Link structure is characterized by the transition probabilities: $\mathbf{P} = [P(x_j|x_i)]$, $\forall x_j \in \mathbf{O}(x_i)$. This could be either an equal probability or a weighted probability depending on the nature of the links going from x_i to x_j. For example, this transition probability will depend on whether there are lots of prominent links or a few footnote links going from x_i to x_j vs. x_i to x_k, some other page going out of x_i. Let us assume that there is a prior probability that a user might "start" or "randomly go to" a particular page. This prior depends on, for example, how many people have this page as their home page or how often is this page typed directly in the browser compared to the other pages. Let $Q(x_i)$ be this initial probability of going to this page. Let us say this random jumping to this page happens with a probability $(1 - \lambda)$ and with a probability λ the surfer actually systematically follows the links (browser behavior). Now at any given iteration t, we can compute the probability that a random surfer will be at a certain page:

$$P_{t+1}(x_j) \leftarrow (1-\lambda) Q(x_j) + \lambda \sum_{i=1}^{N} P_t(x_i) P(x_j|x_i) = \frac{1-\lambda}{N} + \lambda \sum_{x_i \in \mathbf{I}(xj)} \frac{P_t(x_i)}{|\mathbf{O}(x_i)|}$$

In the above we made an assumption that $Q(x_j)$ are all equal to $1/N$ and outgoing probabilities $P(x_j|x_i)$ are all equal to $1/|\mathbf{O}(x_i)|$. Once converged, this gives the most "central" or important pages based on the link structure of the graph. Such an analysis can be done not just on the Internet graph but any directed graph. For example, if we have a gene expression graph that suggests which gene affects which other genes, we can find the most important genes in the network. Similarly, if we have an influencer–follower graph on a social network, we can find the most influential people in the social network and so on.

6.2 Maximal Cliques

PageRank was an example of a global network analysis algorithm. Cliques are an important class of patterns that are sought in graphs in many domains. Consider for example, in retail, a product graph of which products "go with" which other products. A clique in such a graph would indicate a "product bundle" that characterizes the latent intent of a user. Similarly a set of keywords that are all connected to each other might indicate a coherent concept as shown in Fig. 15.20.

15 Machine Learning (Unsupervised)

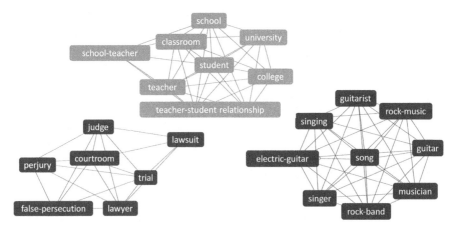

Fig. 15.20 A set of cliques found in keyword–keyword co-occurrence graph created from IMDB dataset

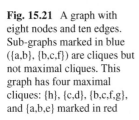

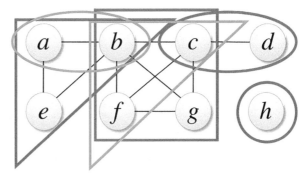

Fig. 15.21 A graph with eight nodes and ten edges. Sub-graphs marked in blue ({a,b}, {b,c,f}) are cliques but not maximal cliques. This graph has four maximal cliques: {h}, {c,d}, {b,c,f,g}, and {a,b,e} marked in red

Here we first create a graph between all pairs of keywords based on how often they co-occur more than random. This graph is then binarized by applying a threshold and then cliques are sought in this graph.

A "**Clique**" is a fully connected subgraph of a binary graph. A "**Maximal Clique**" is a clique that is not a sub-graph of any other clique. Figure 15.21 shows a graph with eight nodes and ten edges. It has four maximal cliques. Finding all maximal cliques in a graph is an NP-hard problem with a known complexity of $O\left(3^{\frac{n}{3}}\right)$ for a graph with n nodes. In this section, we will present a MapReduce algorithm for finding all maximal cliques of a binary graph. Finding such maximal cliques in graphs could help improve our understanding of the graph, find actionable insights in the graph, and even discover higher order structures beyond nodes and edges (e.g., product bundles or communities).

In order to develop the MapReduce algorithm for finding all maximal cliques, we will first introduce a few concepts:

- **Neighborhood of a clique**: For any known clique in the graph (e.g., {b,c}) we define its neighbor as the set of nodes that are connected to *all* nodes in the clique. Here since node f and node g are connected to **both** b and c, they form the neighborhood of the clique {b,c}. In other words:

$$N(\{b, c\}) = \{f, g\}$$

$$N(\{b, c, f\}) = \{g\}$$

$$N(\{a\}) = \{b, e\}$$

$$N(\{a, b\}) = \{e\}$$

- **Neighborhood of a maximal clique**: Note that neighborhood of a maximal clique by this definition is an empty set:

$$N(\{a, b, e\}) = \varnothing$$

$$N(\{b, c, f, g\}) = \varnothing$$

$$N(\{c, d\}) = \varnothing$$

$$N(\{h\}) = \varnothing$$

- **Clique map**: We define a map between a clique (key) and its neighborhood (value) as a Clique Map. This is the main data structure that will be used by the MapReduce algorithm to find all maximal cliques iteratively.

Iterative MapReduce for Finding Maximal Cliques

The key to a MapReduce algorithm is the way we represent each record, what we do with it during the Map step, and how we define the Reduce step. Finding maximal cliques in a graph is accomplished by running a MapReduce algorithm that

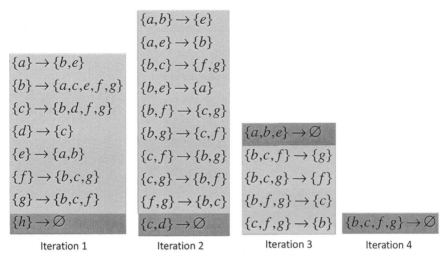

Fig. 15.22 Four MapReduce iterations needed to find all maximal cliques of different sizes for the graph shown in Fig. 15.20. Iteration 1 is the input to the algorithm—it comprises of all cliques of size 1 and their clique neighbors. Iteration 2 is the set of all cliques of size 2 (edges) and their neighbors, iteration 3 is the set of all cliques of size 3 and their neighbors, and so on. In each iteration, a clique whose clique neighbor is empty is deemed a maximal clique and stored

starts with cliques of size 1, that is, each node is a clique. This is stored along with its adjacency list or clique neighbor (forming a clique map shown in Fig. 15.21, iteration 1). Figure 15.22 shows the four iterations of the algorithm where each iteration is the same MapReduce step where we go from clique maps of size k cliques to clique maps of size $k + 1$ cliques. The crux of this algorithm is now the Map and the Reduce steps that will take us from one iteration to the next.

The Map Step

In each iteration of the algorithm, we are given a clique map with a clique and its neighborhood. We want to grow the clique by adding one neighbor at a time to the original clique. We make the following observation about a clique map (e.g., $N(\{b,c\}) = \{f,g\}$): If one element (say f) is removed from the clique neighbor and added to the clique itself ($\{b,c\}$), the resulting set ($\{b,c,f\}$) will also be a clique. This is true because we know that by definition f is connected to both b and c and $\{b,c\}$ is already a clique so $\{b,c,f\}$ will also be a clique. However, also note that we cannot guarantee that what remains on the neighborhood side (i.e., $\{g\}$) is still a neighbor of the new clique ($\{b,c,f\}$) because for that we need to guarantee that g is connected to f, information that is not availale to this mapper.

The Reduce Steps

The output of the mapper is an intermediate key value obtained from clique maps. While the keys of these maps are guaranteed to be cliques of size $K + 1$, the values are not guaranteed to be the neighbors of the corresponding cliques. In order to obtain the clique map of size $K + 1$, the reducer must take an intersection of all the

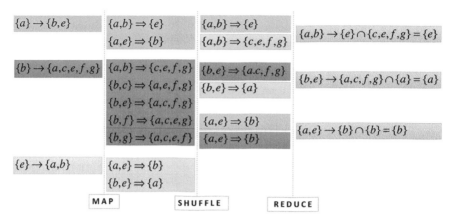

Fig. 15.23 Each MapReduce iteration for finding all maximal cliques in an unweighted graph. The Map step takes clique maps of size K, generates all possible cliques of size $K + 1$ by moving one element at a time to the clique side from the neighborhood side. The Reduce step then takes the intersection of the remaining neighbors for the same clique of size $K + 1$ resulting in clique maps of size $K + 1$

sets of the same clique. Figure 15.23 shows the entire process from Map to Shuffle to Reduce that takes us from clique maps of size 1 to clique maps of size 2. Repeating this process in each iteration results in cliques of various sizes.

While we explored only two broad ideas—one macros and one micro—in network analysis, there are a large number of algorithms especially around community detection where softer variants of cliques—communities are discovered within the networks. Analysis of networks can find interesting structures like fraud syndicates in telecommunication or service networks and financial networks. Link Prediction, another important area in network analysis, is used by LinkedIn and other social networks to suggest more connections to any individual based on their neighborhood structure and so on. Handcock et al. (2007) can be a helpful resource to learn further.

7 Conclusion

In this chapter, we explored a variety of unsupervised learning paradigms—projection, clustering, density estimation, frequent itemset mining, and network analysis. These paradigms are typically used to understand different types of data (multivariate, sets, similarity matrices, graphs). This is an essential first step before we start to build supervised learning models from such data. These algorithms help us visualize the data, remove redundancy in the form of feature correlations, find groups of similar items to quantize the data into "representative" clusters, find objects at higher order of abstractions, and in general help reverse engineer the process that might have generated the data in the first place. In general, unsupervised learning is like "reading the book of data" to get a general lay of the land, a broad understanding of the data, without a particular question being asked of this data.

Supervised learning, on the other hand, starts with a question and forces us to read the book but only with respect to the question. In general, it is always better to explore the data using these unsupervised learning approaches before building supervised learning models on it. The insights derived from these algorithms can be used as is to draw conclusions about the data, make decisions, or serve as features for the next stages of modeling.

Electronic Supplementary Material

More examples, corresponding code, and exercises for the chapter are given in the online appendices to the chapter. All the datasets, code, and other material referred in this section are available in www.allaboutanalytics.net.

References

Criminisi, A., Shotton, J., & Konukoglu, E. (2012). Decision forests: A unified framework for classification, regression, density estimation, manifold learning and semi-supervised learning. *Foundations and Trends® in Computer Graphics and Vision, 7*(2–3), 81–227.

Friedman, J., Hastie, T., & Tibshirani, R. (2001). *The elements of statistical learning (Vol. 1, No. 10). Springer series in statistics*. New York, NY: Springer.

Han, J., Pei, J., & Kamber, M. (2011). *Data mining: Concepts and techniques*. Amsterdam: Elsevier.

Handcock, M. S., Raftery, A. E., & Tantrum, J. M. (2007). Model-based clustering for social networks. *Journal of the Royal Statistical Society: Series A (Statistics in Society), 170*(2), 301–354.

Murphy, K. (2012). *Machine learning – A probabilistic perspective*. Cambridge, MA: The MIT Press.

Michalski, R. S., Carbonell, J. G., & Mitchell, T. M. (Eds.). (2013). *Machine learning: An artificial intelligence approach*. Berlin: Springer Science & Business Media.

Rasmussen, C. E. (2004). Gaussian processes in machine learning. In O. Bousquet, U. von Luxburg, & G. Rätsch (Eds.), *Advanced lectures on machine learning. ML 2003. Lecture notes in computer science* (Vol. 3176). Berlin: Springer.

Robert, C. (2014). Machine learning, a probabilistic perspective. *Chance, 27*(2), 62–63.

Chapter 16
Machine Learning (Supervised)

Shailesh Kumar

Every time we search the Web, buy a product online, swipe a credit card, or even check our e-mail, we are using a sophisticated *machine learning* system, built on a massive cloud platform, driving billions of decisions every day. Machine learning has many paradigms. In this chapter, we explore the philosophical, theoretical, and practical aspects of one of the most common machine learning paradigms—*supervised learning*—that essentially learns a mapping from an *observation* (e.g., symptoms and test results of a patient) to a *prediction* (e.g., disease or medical condition), which in turn is used to make *decisions* (e.g., prescription). This chapter explores the process, science, and art of building supervised learning models. The examples, corresponding code, and exercises for the chapter are given in the online appendices to the chapter.

1 Introduction

We are drowning in data yet starving for knowledge

The last few decades have seen an unprecedented growth in our ability to collect and process large volumes of data in a variety of domains—from science to social media, e-commerce to enterprises, Internet to Internet-of-things, and healthcare

---

Electronic supplementary material The online version of this chapter (https://doi.org/10.1007/978-3-319-68837-4_16) contains supplementary material, which is available to authorized users.

S. Kumar (✉)
Reliance Jio, Navi Mumbai, Maharashtra, India
e-mail: skumar.0127@gmail.com

© Springer Nature Switzerland AG 2019
B. Pochiraju, S. Seshadri (eds.), *Essentials of Business Analytics*, International Series in Operations Research & Management Science 264,
https://doi.org/10.1007/978-3-319-68837-4_16

to human resource management. Today's data-driven decision systems enable us to make intelligent, accurate, and real-time decisions using this data. They have the potential of making research, manufacturing, businesses, processes, enterprises, education, transportation, agriculture, and governance increasingly automated, efficient, and effective.

Today's data-driven decision systems are a result of a serendipitous convergence of three key technologies that matured over the last few decades: First, the *Internet* that made it possible for everyone to contribute to, and connect with the collective human knowledge and services globally. Second, *cloud computing* that made it possible for individuals and enterprises to store and process enormous amounts of data, and third, *machine learning*—the process, science, and art of converting data into insights, insights into predictions, and predictions into decisions. At a high level, there are three broad paradigms in machine learning:

- ***Unsupervised learning*** is typically used to *describe* the structure in the data (e.g., projection, density estimation, clustering) or *discover* latent patterns in it (e.g., communities in networks, topics in a corpus, or frequent item-sets in market basket data). The goal of unsupervised learning is to improve our understanding of the data and derive actionable insights from it.
- ***Supervised learning*** is typically used to learn a mapping from an *observation* (e.g., activities of a user in a bank) to a *prediction* (e.g., is the customer about to churn), leading to a *decision* (e.g., take action to prevent customer churn). Most decision systems today in a variety of domains are based on Supervised Learning models.
- ***Reinforcement learning*** is typically used in sequential decision tasks to predict the best *action* (e.g., next Chess move) from the current *state* (e.g., board position) to maximize immediate (e.g., strengthening the board position) and eventual (e.g., winning the game) *reward*.

In this chapter, we will focus on the philosophical, theoretical, and practical aspects of machine learning in general and supervised learning in particular.

- The ***philosophical*** goal of machine learning is to understand the nature of intelligence and learning itself. Here we will explore the fundamentals of *understanding* and *generalization* in the context of supervised learning.
- The ***theoretical*** goal of machine learning is to build and improve formal learning frameworks and algorithms. Here, we will explore various supervised learning algorithms, relationships among them, and their pros and cons.
- The ***practical*** goal of supervised learning is to blend data, domain knowledge, and learning algorithms to build accurate and, if needed, interpretable prediction models. Here we will explore some of the real-world challenges and practical aspects of building models.

2 The Philosophy: Nature of Intelligence

Our technology, our machines, is part of our humanity. We created them to extend ourselves, and that is what is unique about human beings!—Ray Kurzweil

Ever since the dawn of mankind, we have been trying to extend ourselves in all our faculties: If we could not lift more, we created levers and cranes; if we could not move fast and far, we created horse carts and cars; if we could not see far, we created telescopes; if we could not speak loud enough, we created microphones; if we could not compute fast enough, we created calculators and computers; if we could not talk far enough, we created telephones and mobiles; etc. In this journey, we are also extending one of our most important faculties that make us unique—our intelligence. Using machine learning and Artificial Intelligence, we are now at the early stages of building intelligent machines that can see, listen, speak, read, learn, understand, think, create, plan, and converse like humans.

Before we can build intelligent machines, however, it is essential to understand the nature of intelligence itself. Intelligence has many facets; for example, it is the ability to:

- *Learn* causality or correlation from past data (e.g., Should I approve this loan?)
- *Recognize* structures in the data (e.g., words in speech, objects in images)
- *Understand* semantics using context (e.g., *apple* is healthy, I like *apple* products)
- *Adapt* to novel situation (e.g., network routers react to change in traffic patterns)
- *Reason* about alternate ways of solving a problem (e.g., playing chess)
- *Synthesize* data (e.g., next word in a ??, next utterance in a conversation)

In the context of supervised learning, let us explore two of these notions of intelligence in a little more depth: understanding and generalization.

2.1 Understanding: From Syntax to Semantics

Does the Google Search Engine actually *understand* the Web? Do YouTube or NetFlix understand the videos they store? Do Amazon and Zomato understand the reviews written by their customers? Do our "smart" phones actually understand what we are speaking into them? It is one thing to collect, store, transfer, or index a large amount of data, but it is completely a different thing to actually *understand* it. One of the first fundamental qualities of an intelligent system is its ability to interpret the raw data it is receiving at the right level of abstraction (e.g., pixels, lines, blobs, eyes, face, body). But what does *understanding* mean?

Our language and sensory systems evolved not only to capture and transmit the raw data to the brain, but to actually *understand* it in real time, that is, to *identify* structures, objects, and attributes in them. Our visual system—perhaps the most sophisticated intelligent system so far—*looks* at pixels in the retina but *sees* a fresh red rose or a flying eagle in the brain. Similarly, when we process a sequence of words (e.g., "*Apple filed a suit against Orange*") we *interpret* or assign meaning to each part (word)—for example, "Apple" the company, not fruit—so the whole (sentence) makes sense.

Understanding is a hierarchical process of using **context** to interpret each part so the whole—as a juxtaposition of its parts—makes sense.

A large class of Unsupervised and Supervised Learning algorithms today are understanding algorithms as they try to interpret the raw data, for example, this word is a noun (part of speech tagger), this document is about hockey (document classifier), this article mentions Mahatma Gandhi (information extraction), this video segment shows bungee jumping (activity recognition), this image shows a cat under a table (object recognition in images), this person is Mr. X (speaker recognition from voice, face recognition from image, or fingerprint recognition).

2.2 Generalization: From This to Such

A database that stores even billions of records is neither considered *knowledgeable* (since it does not understand the data it contains) nor *intelligent*. Now consider the letter in Fig. 16.1. We can tell what this letter is immediately in spite of the fact that we have never seen such renditions of this letter before. If we were a database system, we should have seen all possible renditions of this letter to recognize it. But that is not how our brain works: we do not **memorize**; we **generalize**. The second important property of intelligence—the basis of Supervised Learning—is this ability to *generalize* what we have learnt from the past mappings of input to output to new inputs we have never seen before.

In Supervised Learning, we try to mimic this aspect of intelligence. So to build a "classifier" that can recognize say a handwritten letter or an action in a video (e.g., bowling, batting, catching) we first provide it with enough "training examples" of what is the input and what should this input be called (the class label). We learn a

Fig. 16.1 We can recognize this letter immediately without having seen any of these renditions before

"classifier" with this training data that can now recognize new examples that it has never seen before. The basic principle of generalization is that:

*Inputs that belong to the **same** class must be **similar** to each other in some way.*

This brings up another important notion in machine learning called *similarity*. What makes two documents or two gene sequences or two pieces of music or two customers similar? One might say that learning to generalize an input is implicitly the art of defining what makes two inputs similar. For example, if we show a lot of images of cats to a computer vision system where they all differ by the background, the size of the cats, the color of the cat and we keep telling the learning system they are all cats then the model will figure out what is really similar among all these images (furry, big eyes, whiskers, etc.). Now if we distinguish all cats from all dogs the system will further figure out what makes cats more similar to each other, dogs more similar to each other, and cats different from dogs. So, in a way *supervised learning* is the art of *similarity learning* and the ability to determine what aspects to focus on (signal) and what aspects to ignore (noise) given a set of training examples.

Supervised learning learns to make "similar" objects "nearby" and "different" objects "distant" either by *compactly describing similar objects* (**descriptive**) or by *robustly discriminating between different* (**discriminative**) objects.

3 The Supervised Learning Paradigms

At a high-level supervised learning is a mapping between an input (e.g., cause) and an output (e.g., effect). Different paradigms in supervised learning—classification, regression, retrieval, recommendation—differ by the nature of their input and output:

3.1 The Classification Paradigm

A classification model maps a set of ***input features*** to a ***discrete class label*** (e.g., Digit (0–9) or character (A–Z) (class labels) from images; emotions (sad, happy, confused, frustrated, ...) from face images; land-cover (water, marsh, sand, etc.) from remote sensing data; e-mail type (spam, promotion, finance, update) from e-mails (e-mail classifier); words (e.g., words in any language) from speech data (speech to text); objects (cat, dog, car, tree, etc.) in images; computer vision activity (stealing, holding, throwing, etc.) from videos (activity recognition), part-of-speech of a word in a sentence (POS taggers); sentiment in a tweet; or review about an entity (movie, product, etc.)

In all these cases the *input* could take any form—multivariate, text, image, speech, video, sequence of transactions, etc. This raw data is further converted to some meaningful *features*. The output is a discrete class label from a predefined set of labels. In two-class classification problems (e.g., spam vs. not-spam, churn vs. not-churn, conversion probability estimation), the output is typically interpreted as probability of target class (e.g., spam, churn, click). Appropriate thresholds on this probability can be used to make a binary decision. In general, the classes themselves might have hierarchies (e.g., news articles might be labelled as sports, entertainment, politics, science, etc. at the first level of hierarchy while within *sports* class, there might be subclasses for various sports or within science there might be subclasses such as space, medicine, and technology).

3.2 The Regression Paradigm

A regression model maps ***input features*** to a ***real or ordinal value*** (e.g., click-through-rate prediction in search and advertising, lifetime-value prediction of a customer, efficiency prediction from device sensors, capacity of a customer to take loan, and value of a house/property in a local neighborhood): Regression is used in many ways: either to predict a value, to predict *score* in a certain range (e.g., in ordinal regression we might want to predict a score from say 1 to 5 for ratings corresponding to poor, fair, bad, good, and excellent), or for forecasting a value into the future (e.g., demand prediction for products in retail).

3.3 The Recommendation Paradigm

A recommendation model maps a ***past behavior*** into ***future potential activities*** (e.g., which product a customer might buy given what he/she purchased, browsed, etc. in the past, which movie a user might like on NetFlix or YouTube given his/her past content consumption, who will a user like to connect to on Facebook or LinkedIn given his/her current connections and interactions, which news or tweet a user might like given his/her past consumptions, which topic the student should study next given how he/she has fared in past topics). A typical recommendation engine uses a two-stage process:

Creating user profile: In this first stage, user's past behavior is used to build his/her profile (a set of features and their weights). For example, in retail the profile might be built based on the products the user searched, added to wish list, purchased, read a review about, wrote a review about, etc. In education, the profile of a student may be created using the time spent on learning, number of problems solved, and test scores on problems associated with each topic. In YouTube, a user profile may be built based on videos previously watched, liked, and commented by the user. Note that a user may have different types of interactions with the same entity. Each interaction type could be given a different weightage (e.g., purchase is more important than browse, writing a review might be more important than reading one) while creating the user profile. Once the user profile is built it is used to make the actual recommendation.

Making Recommendations: In this second stage, the user's profile is now "matched" with the (properties of the) entities to determine whether a user would like to engage with that entity (product, topic in education, or YouTube video). The matching and profiling can be done at the *id level* (e.g., which movie, which video) or at the *property level* (which genera of movie, which director, etc.). The score may be further refined using a *utility* function based on the business goals; for example, for certain set of customers we might use recommendation for immediate cross-sell, while for others we might use it to maximize their lifetime value.

3.4 The Retrieval Paradigm

A retrieval model maps a ***query*** into a ***sorted list of entities*** (e.g., relevant Web pages on search engines for a given text query, relevant images, videos, news stories on search engine given a text query, relevant images/videos for an image query (content-based image retrieval), relevant song for a given humming or audio snippet, relevant property/car/products on various entity search portals, relevant flights/hotels on various travel portals (structured queries), and relevant gene sequences for a gene snippet query.

In both retrieval and recommendation paradigms, the output is a ***list of entities*** sorted by a score. The key difference is that in recommendation, the (recommendation) score is based on the ***behavior summarized into a profile*** of the user, while in retrieval the (relevance) score is based on the match to a ***query***. For example, in Web search, one might use URL match, title match, anchor text match (text associated with all incoming links to this page), header match, body match, etc. Click feedback data is used to learn the relative importance of various types of matches between query and entity fields to synthesize the final relevance score.

One of the key skills of a data scientist is to ***formulate*** a business problem as one (or more) of these paradigms, pick the right kind of modelling approach within the paradigm, and using data and domain knowledge to learn these models. In the rest of this chapter, we will focus primarily on the classification paradigm. You can refer Chaps. 7 and 8 (Linear and Advanced Regression) for the regression paradigm and Chap. 21 (Social Media and Web Analytics) for examples on the retrieval paradigm.

4 The Process: From Data to Decisions

Data Science is a continuous dialogue between data and business.

One of the primary goals of Data Science is to drive business and operational decisions from data to maximize profitability or efficiency metrics, respectively. Figure 16.2 shows the overall ***process*** typically used to drive decisions from data. We will explore each of the stages of this process in this section.

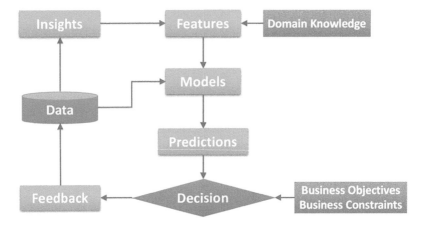

Fig. 16.2 The overall data science process for building and improving models

4.1 The Insights Stage: Dating the Data

> *Often the most effective way to **describe, explore**, and **summarize** a set of **numbers**—even a very large set—is to **look at pictures** of those numbers.*—Edward R. Tufte

The first stage in the data science process is to understand the nuances in the data itself before we start building models. The insights hidden in the data either confirm some of our own hypotheses about the underlying process that generated the data or reveal new aspects of the process that we did not know before. Some of the basic practices for revealing insights in the data include:

Feature Distributions: One of the most basic set of insights comes from individual feature distributions. Most modelling techniques assume normal or well-behaved distributions, while most real-world features are either exponentially or log-normally distributed. Looking at feature histograms reveals such nuances and helps correct for them by, for example, taking the log of those features that are exponentially distributed. Further, looking at feature distributions of different classes reveals whether or not a certain feature would be useful for discriminating various classes. Feature distributions reveal structure in each feature independent of other features.

Scatter Plots: A powerful yet simple technique in understanding feature interactions is scatter plots between all pairs of features. This visually shows correlation among features, if any. Further, color coding each data point with class label reveals combination of features that might help discriminate classes. Figure 16.3 shows scatter plots of IRIS dataset[1] with respect to a few pairs of features. Scatter plots limit us to only visualize the data two or three features at a time.

---

[1]https://archive.ics.uci.edu/ml/datasets/iris (Retrieved August 2, 2018).

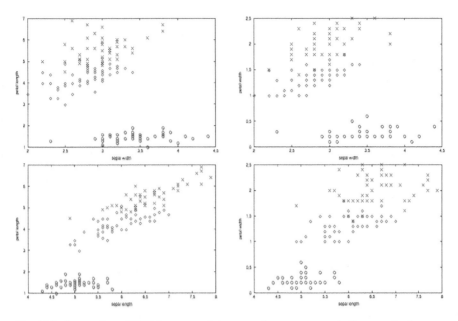

Fig. 16.3 Scatter plots of IRIS data—reveals how two classes are more similar to each other

Principal components analysis (PCA): Principal components analysis projects the data into lower-dimensional spaces by preserving maximal spread or variance in the projected space. Scatter plot of the top two or three principal components projection of the data reveals the "joint" structure in the data across all features.

Fisher discriminant analysis: PCA is an unsupervised projection that only preserves spread of all data points irrespective of their class labels. For classification problems it is far more useful to do a Fisher Discriminant Analysis where projection is done to exaggerate the differences between classes. Figure 16.4 shows the PCA vs. FDA projections of three classes in MNIST data.[2] It shows how Fisher projections try to separate the three classes while PCA projections do not care about the class labels.

Other visualization techniques: A large number of visualization techniques including self-organizing maps, multidimensional scaling, and t-SNE can be used to gain deeper understanding in the data.

---

[2]http://yann.lecun.com/exdb/mnist/ (Retrieved August 2, 2018).

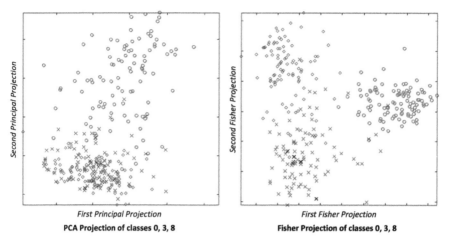

Fig. 16.4 PCA vs. Fisher projection of same data (classes 0, 3, and 8 from MNIST)

4.2 The Feature Engineering Stage: Exaggerate Signals that Matter

One of the most creative parts of the data science process—literally the art of data science—is the feature engineering stage. There are two types of data scientists, viz., feature engineering:

The **feature-centric** data scientists believe in systematically and painstakingly creating meaningful features to make the modelling stage simple. For them "real" data science happens here. They marry their deep understanding of the data (acquired from insights stage) with substantial appreciation of the domain knowledge (acquired from domain experts) to build features. These features are highly interpretable, semantically deeper than the original data, and cover all potential aspects of input-output mapping. This traditional approach to data science is more useful when labelled data is less compared to domain knowledge and interpretability of model output is as important as its accuracy.

The **model-centric** data scientists, on the other hand, believe that throwing a large amount of labelled data and computational resources (e.g., GPUs) will automatically learn the right features (in the lower layers) as well as the mapping between those features and the output (in the higher layers) of a *deep learning* model. Here, the creative process takes the form of designing the right architecture—nature and type of layers in the deep learning models as opposed to designing individual features. This model-centric deep learning approach works well in domains such as text, vision, speech, and time series data where (1) the space of possible features is very large, which makes it impractical to explore it through traditional feature engineering; (2) the amount of data is substantial enough to learn the large number of parameters in deep learning models; and (3) the semantic gap between the raw input (e.g., pixels in images or words in text) to the final output (e.g., activity in video or meaning of a document) is so large that we need a hierarchy of features and not just a single layer of features.

In the rest of this section, we will explore a number of transformations on raw data that constitute traditional feature engineering:

Feature transformation: In a typical model, the different input features might have very different distributions and ranges. Combining them into a model such as logistic regression without first making their distributions "compatible" makes the life of the model miserable. Taking log of certain features (that are exponentially or log-normally distributed), binning the values, or applying any domain-specific transformation (e.g., Fourier transformations or wavelet transformations on time-series data) might help build better models than just shoving the raw inputs into the model. For example, in many models using income as a feature, it might be better to either bin the income or take the log of the income since income is typically exponentially distributed (lots of people have low income, very few have very high income). Using percentile scores or cumulative density binning is also an example of taming the distribution variability in the data.

Feature normalization: Even after proper transformations, the raw inputs might be in different ranges and their values in different units. For example, to predict the value of a house, one might need features such as number of rooms and bathrooms (count), area of the house (square foot), distance from nearest school or places of interest (kilometers), prices of nearby houses sold recently (money), and age of the house (years). While the distribution can be tamed as described above, the values might still need to be brought into comparable ranges. For this, the features might need to be transformed to some min-max range (so min is always 0 and max is always 1) or z-scored values could be used (so the mean of each feature is zero and standard deviation is 1). Such transformations then let the model do the actual job of learning the relative importance of these features instead of forcing them to *also* compensate for these feature differences. Care must be taken to first remove outliers in each feature before learning parameters for min-max or z-score normalization.

Creating invariant features: Often the raw data contains variances in it that are not related to the problem at hand. For example, speech recognition problems have accent variances; images might have illumination, pose, rotation, and scale variances; and transaction and time series data might have seasonal variances. In essence, the final "data" that we see (e.g., sound of a word spoken by a person) is a "joint" of the actual signal in it (e.g., the actual word spoken) with additional factors (accent, tonal quality, loudness, etc.). Keeping what is essential for the task (signal) and ignoring what is not (noise) is the key to good feature engineering. Understanding and removing these variances is perhaps the most intricate part of feature engineering and requires deep understanding of the domain, possible sources of such variances, and the tools to remove these variances. If not removed, the model will become complex and will try to learn these variances instead of doing actual classification.

Ratio Features: A lot of features contain variances that can be removed simply by dividing them with other features. For example, in information retrieval models, query length bias is removed by dividing the total match between query and document field (e.g., title) with query length. In credit models, instead of using total debt it is better to use debt-to-income ratio, instead of using total-payment a better feature would be the percent of EMI paid, and instead of total-credit-taken, percent of credit limit reached might be better features. Such ratio features cannot be "discovered" by the modelling techniques that are only doing linear combination of features (e.g., logistic regression or linear Support Vector Machines). Infusing domain knowledge through ratio features helps model explore the right "space" in which to discriminate classes.

Output feature ratios: Not only the input features, even the output features might also have biases that must be corrected for before trying to predict them. For example, instead of predicting click-through-rate of a document for a query, we might want to first take into account the expected click-through-rate bias at each position (e.g., people are anyway more likely to click on the first result than second and so on irrespective of the query and document). In forecasting sales, instead of predicting the raw sales count, we might want to predict deviation from the expected sales given the context (city, season, etc.). The ratings data (movie or product rating) has inherent "consumer bias." A critical consumer will typically rate most products say 1–3 out of 5 and hardly give a rating of 5, while a generous customer might rate most products between 3 and 5. Now a rating of 4 on a certain product does not mean the same thing for these two customers. It should be "calibrated" correctly to remove individual customer's rating biases to make them "comparable" across customers.

Creating new features: Additional features beyond basic transformations, normalizations, ratios, and bias corrections are also needed in many domains. Consider, for example, four features in a credit card fraud prevention problem: location and time of the last and the current transaction. These four features by themselves put into a logistic regression model might not be able to predict whether the current transaction is fraud or not. But a common sense domain knowledge that "there should be sufficient time between two distant transactions" can be used to translate these four features into say a *velocity* feature, that is, ratio of distance between current and previous transaction to time between current and previous transaction is a single "semantic" feature that can help predict fraud. In speech and vision domain, biologically motivated semantic features are extracted from raw signals.

Defining output variable: In some of the problems the prediction variable might be very obvious (click through rate in search, spam vs. not-spam in Web page or e-mail classification, land-cover type in remote sensing, etc.). However, in many other domains, we might have to first define the output variable itself. For example, in *churn prediction,* we might have to define churn in terms of *future* user behavior (e.g., did not make any purchase in the last 3 months). In credit modelling, we might define a high-risk customer as someone who missed his last three EMIs in a row. In such problems where *future* is to be predicted based on *current and past* observation, defining the future output to be predicted becomes very critical.

Setting the right defaults: A default value is typically associated with a feature if no meaningful value can be assigned. For numeric features, often such default values are zero. Assuming such defaults or not setting them thoughtfully is one of the most common "bugs" in modelling. Consider a feature called *first-occurrence* of a query word in a document field. The earlier the word occurs in the field, the better—so lower the value of first-occurrence, the better. Now if in a field no query word is present, what should be the default value of this feature? If we pick a default value of 0, then it will confuse the model where both for the *best case* (when the query word is the first word (at position 0) of the field) and the *worst case* (where the query word is not at all present in the field) take the same feature value. A better default might be the length of the field plus a constant or a high number. It is essential to deliberate over the default values of all features to make sure that the default value in conjunction with the regular values are "consistent" with the goals of the modelling.

Imputing missing features: One of the realities of real-world data science is the absence of features in the collected data. This happens either because the data was never collected for a period of time and plugins to collect a feature were added later, or the sensor was down for a while, or there are data corruption issues. In these cases, either we use one of the many feature imputation techniques or use modelling techniques (such as decision trees and their variants) that gracefully handle missing features. Again substituting the wrong defaults or simple average value of a feature may not always work.

Feature selection: Once a large number of features have been engineered, we might decide not to use all of them together in the same model because some of them might be highly correlated with each other. Feature selection methods can be model agnostic (aka filter methods) or model centric (aka wrapper methods). In a model-agnostic approach, features are sorted by some measure of "goodness," which is computed based on their discriminative power (e.g., Fisher discriminant) and nonredundancy with other features. The best features are then chosen to build the models. Filter methods are used when we have a large number of features (say tens of thousands) and it is not clear which modelling technique we want to use. In model-centric feature, selection features are added one at a time (forward feature selection) or removed one at a time (backward feature selection) in a greedy manner to maximally increase the model performance (e.g., accuracy). Being model centric, every time a set of features is evaluated, the model has to be trained and evaluated. This makes model-centric feature selection potentially very time-consuming. Feature selection is a classic NP—hard "subset selection problem" where we know how to compute the "goodness" of a "set" of features but there is no simple (polynomial) algorithm to find an optimal set for a given dataset and modelling technique. Many other techniques such as genetic algorithms and simulated annealing have also been explored for feature selection.

4.3 The Modelling Stage: Matching Data and Model Complexity

Over the last several decades, the field of machine learning has given birth to a very large number of modelling techniques—some of which are described in the next section. Each technique has its own pros and cons and was developed to specifically address a set of weaknesses in other modelling techniques or "reformulate" the classification problem differently. In this section, we will explore the common guiding principles typically used for choosing the right modelling technique and using the output of these models correctly to solve the business problems.

Interpretability vs. accuracy: In a number of business problems, it is more important to interpret the output of the prediction model (i.e., give a reason for why the score is high or low) and not just to be accurate at it. For example, credit models are legally required to give top three reasons why a user has been denied a loan. Similarly, in churn prediction models, it might be useful not just to know that a certain customer is about to churn but also the reason why the customer is about to churn. This "reason code" can help address those reasons specifically for each customer. In such cases, it is better to use modelling techniques that are more interpretable and can generate a *reason code* along with a prediction score for each input. In cases where accuracy is more important than interpretability, another class of modelling techniques is preferred.

Scoring time vs. training time: Most models are deployed in high-throughput environments. For example, a search engine must be able to generate the top ten matches within half a second, a credit card fraud model must approve or disapprove each transaction within a second, in autonomous vehicles, the car must respond to the environment in real-time. In taxi hailing services, a cab must be allocated within a few seconds of a request. In all such cases the *scoring throughput* of the model must be high. While part of this is an engineering problem, part of it is also a data science problem where the right modelling technique makes all the difference. Similarly, the training time of a model might also matter when the model has to be updated frequently to compensate for real-time inputs from the data. ETA prediction models in Google Maps, for example, must update their predictions about expected arrival times in real time as new data is fed into the model continuously. Traffic routing models must respond quickly to the changes in the traffic patterns or network issues in real time. Modelling techniques that have a high training time might not be useful here.

Matching data complexity with model complexity: Once the modelling technique is chosen, one of the fine arts in data science is to pick the right complexity of the model. In other words, we must *match* the complexity of the model with the complexity of the data itself. If a more complex model is chosen, it might *memorize* the training data and may not *generalize* well to the unseen data. If a less complex model is chosen, it might not be able to capture the essential causal structure in the data. This principle of picking the right complexity of model is known by many names: bias–variance trade-off, signal-to-noise ratio, or Occam's razor. In essence,

16 Machine Learning (Supervised)

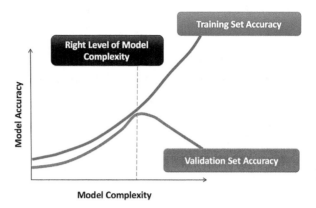

Fig. 16.5 Model complexity is chosen based on the gap between training and validation accuracy

the model needs to be *just complex enough and not any more*. In practice, the right model complexity for a given labelled data and modelling technique is arrived at as follows: We start with a simple model and increase its complexity slowly while measuring the training and test set accuracy—that is, how well it does on the data that was used to build the model and how well it does on the unseen data. As model complexity increases the training and test accuracies will go up. But beyond the point of peak generalization, the test accuracy will start falling as the model will start to learn the noise in the training data. This is a good indication of the right model complexity as shown in Fig. 16.5. Each modelling technique comes with a set of "knobs" to increase their model complexity.

From predictions to decisions: The output of a model is typically a score—for example, the credit score, the fraud score, or the predicted demand in a forecasting model. Machine learning stops where this score is generated. Data science starts where this score is now used to make decisions. Often a number of business constraints and metrics determine how the score should be used. For example, a bank with a higher risk appetite might give loans at a lower score than another. In recommendation engines, for example, we might not just recommend the product with the highest recommendation score but might decide to recommend products that are also highly connected to other products for increasing cross-sell beyond just the current recommendation. Decisions are made, often, with conflicting business metrics in mind and the model prediction outputs serve as key inputs to the overall business logic that tries to solve a complex multiobjective optimization actually make the final decision.

Feedback and continuous learning: Once the model is deployed, feedback is collected on how well it is doing. This feedback is critical for monitoring model performance and continuously updating the models. For example, search engines continuously update their models based on real-time click feedback data by moving the search results up or down based on whether they are getting higher or lower than expected clicks for the result at that position. This feedback data is the real goldmine in any modelling exercise. It is the cheapest and most consistent source of "ground truth" that is very critical for building supervised learning models. This feedback also comes in implicit form. For example, if the model predicted that a customer is about to churn but he/she did not or vice versa then such implicit feedback can be used to continuously improve the models. Modelling is therefore never a one-time exercise. Using this feedback data to automatically and periodically update the model really completes the "continuous learning" loop in real-time, large-throughput systems that evolve as the business processes, customer behavior, and environment evolves.

4.4 The Algorithms: Classification Models

If it looks like a duck, swims like a duck, and quacks like a duck, then it probably is a duck.

In the rest of the chapter, we will focus primarily on the classification paradigm. We will assume that the raw data has already been transformed into a meaningful *feature space* as discussed above.

Definition of a classifier: Essentially, a classifier partitions the feature space into *pure* regions. A region is considered pure if most of the points in that region belong to the same class. There are two ways of characterizing pure regions: Either we learn to "describe" each class (the descriptive classifiers) or we learn to "discriminate" between the classes (the Discriminative Classifiers). Figure 16.6 shows how a descriptive vs. a discriminative classifier approaches the same two-class problem.

We seem to be using both classifiers: as we discover new objects in the world and see one or more examples of it, we build a descriptive classifier that learns the essence of the class. But when we are confused between two classes (e.g., "dog" vs "goat," letter "o" vs. "c"), that is, their descriptions "overlap" quite a bit, then

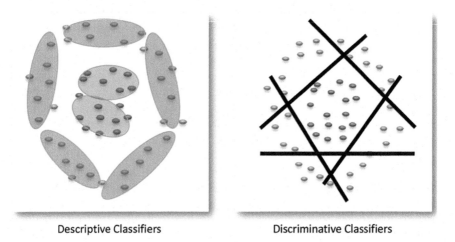

Fig. 16.6 A descriptive classifier learns the shape of each class. A discriminative classifier tries to find the decision boundaries between the classes

we fine-tune these descriptive models to discriminate between them. A number of both descriptive classifiers and discriminative classifiers are discussed below. *The website contains corresponding R code, data, examples, and exercises.*

Rule-Based Classifiers

Rule-based classifiers are the simplest, handcrafted, interpretable classifiers that codify existing knowledge into a set of rules of the form: *If (Condition) then Class*. Such rule-based *descriptive* classifiers occur in many domains including science and medicine (e.g., blood group classifiers (A, B, O, AB), obesity classifiers (underweight, normal, overweight, obese based on simple BMI thresholds), diabetes classifiers (type I vs. type II), symptoms-based disease classifiers (e.g., if fever >103 and throat infection and shivering, then viral infection), periodic table (valence-based element classifier into inert gases, heavy metals, etc.), organic vs. inorganic, hydrophilic vs. hydrophobic, acidic vs. alkaline, etc. Classification of species in a hierarchical fashion is an enormous rule-based classifier. Even businesses and financial institutions have been running for a long time on rule-based systems.

Rule-based classifiers are great at encoding human knowledge (Fig. 16.7). These rules can be simple or complex, depending on one or many features, and can be nested hierarchically such that the class prediction of one rule can become a feature to another rule at the next level in the hierarchy. Such a rule-based system, also known as an ***Expert System*** is the best way to bootstrap a data-starved, knowledge-rich process until it becomes data rich itself and rules can actually be learnt from data. One of the biggest advantages of rule-based systems is that they are highly interpretable and every decision they make for each input can be explained. Rule-based classifiers, however, have a few limitations: The knowledge that these rule-based systems contain may not be complete, adding new knowledge, updating obsolete knowledge, keeping all knowledge consistent in a large knowledge-base is an error-prone cumbersome process, human generated rule-bases might contain "subjective-bias" of the experts, and finally, not all knowledge is deterministic or binary—adding uncertainty or degrees to which a rule is true requires data/evidence.

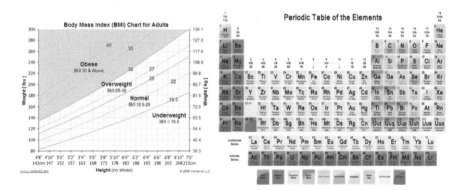

Fig. 16.7 Examples of simple rule-based classifiers in healthcare and nature

As data started to become more and more abundant and the rule-based systems started to become harder and harder to manage and use, a new opportunity of *learning rules from data* emerged. This led to the first algorithm—decision trees—that marked the beginning of machine *learning*. Decision trees combined the interpretability of rule-based classifiers with learnability of data-driven systems that do not need humans to handcraft the rules enabling the discovery of interactions among features that are far more complex for a human to encode.

Decision Trees Classifier

One of the earliest use cases of machine learning was to learn rules directly from data, adapt the rules as data changes, and enable us to even quantify the goodness of the rules given a dataset. Decision trees are an early attempt to learn rules from data. Decision trees follow a simple recursive process of greedily partitioning the feature space, one level at a time, discovering *pure* regions. A region is a part of the feature space represented by a node in the decision tree. The root-node represents the entire feature space.

Purity of a region: In a classification problem, a region is considered "pure" if it contains points only from one class and "impure" if it contains almost equal number of examples from each class. There are several measures of purity that have been used in various decision tree algorithms. Consider a region in the feature space that contains n_c points from class $c \in \{1, 2, \ldots, C\}$ for a C class classification problem. The class distribution $\mathbf{p} = \{p_c\}_{c=1}^{C}$ is given by:

$$p_c = \frac{n_c + \lambda}{\sum_{c'=1}^{C} n_{c'} + C\lambda}$$

where λ is the "Laplacian smoothing" parameter that makes the distribution estimate more robust to small counts. We describe different measures of purity as a function of $\mathbf{p}$.

- **Accuracy**: The first measure of purity is accuracy itself. If the class label assigned to a region is its majority class then the accuracy with which data points in the region are correctly labelled is:

$$Purity_{ACC}(p_1, p_2, \ldots, p_C) = \max_c \{p_c\}$$

- **Ginni index**: In the accuracy measure of purity we *hard*-assign a region to its majority class. This can be brittle to noise in data. If a region is *soft*-assigned to class c with probability p_c then the *expected accuracy* of the region is called the Ginni index of purity:

$$Purity_{GINNI}(p_1, p_2, \cdots, p_C) = \sum_c p_c^2$$

- **Entropy**: An information theoretic measure of *impurity* of a distribution is its Shannon Entropy, which is highest (say 1) when the distribution is uniform and 0 when the entire probability mass is centered on one class. *Purity is an inverse of this entropy*.

$$Purity_{INFO}(p_1, p_2, \cdots, p_C) = 1 - Entropy(p_1, p_2, \cdots, p_C)$$

$$= 1 + \sum_c p_c^2 \log_C p_c$$

Gain in purity: A decision tree recursively partitions the entire feature space into pure subregion using a greedy approach. At each node (starting from the root node), it finds the best feature with which to partition the region into subregions. The "best" feature is the one that maximizes the gain in purity of the subregions resulting from that feature.

Let us say node m is partitioned using feature f (e.g., COLOR) into $K_{m,f}$ children nodes (e.g., RED, GREEN, BLUE): $\left\{R_{f,1}^m, R_{f,2}^m, \ldots, R_{f,K_m,f}^m\right\}$. Let $Purity\left(R_{f,k}^m\right)$ be the purity of subregion $R_{f,k}^m$ and $p\left(R_{f,k}^m\right)$ be the fraction of data at m that goes to the subregion $R_{f,k}^m$. Then purity gain due to feature f at node m is:

$$PurityGain_m(f) = \sum_{k=1}^{K_{m,f}} p\left(R_{f,k}^m\right) \times Purity\left(R_{f,k}^m\right) - Purity(m)$$

Decision tree algorithm: Decision tree recursively partitions each region into subregions by picking that feature at each node that yields the maximum purity gain. Figure 16.8 shows a decision tree over a dataset over five variables {A, B, C, D, E}. Let us say variable A takes two possible values {A1, A2}, variable B takes two values {B1, B2 }, C takes two values {C1, C2}, D takes two values {D1, D2}, and E takes two values {E1, E2}. At the root node, the decision tree algorithm tries all the five variables and picks the one (in this case variable B) that gives the highest purity gain. The entire region is now partitioned into two parts: B = B1, and B = B2. Now that variable B has already been used, the remaining four variables are considered at each of these nodes. In this example, it turns out that under B = B2, variable A is the best choice; under node A = A2, variable C is the best choice; and under C = C1, variable E is the best choice among all the other choices within those regions. Variable D does not increase purity at any node.

The sample data "Decision_Tree_Ex.csv" and R code "Decision_Tree_Ex.R" are available on website.

A leaf node at any time in the growing process is considered for growing further: (1) Its depth (distance from the root node) is less than a depth-threshold, (2) its purity is less than a purity-threshold, and (3) its size (number of data points) is more than a size-threshold. These thresholds (Depth, Purity, and Size) control the *complexity* or

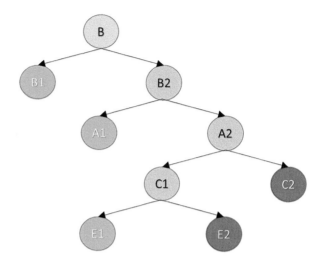

Fig. 16.8 A decision tree over a dataset with five features \{A, B, C, D, E\}

size of the decision tree. Different values of these thresholds might yield a different tree for the same dataset, but it will look the same from the root node onward. Sometimes, a tree is overgrown and pruned to a smaller tree as needed.

Decision trees were created to learn rules from data. A Decision Tree model can be easily written as a collection of highly interpretable rules. For example, the tree in Fig. 16.8 learns the five rules, one for each leaf node. Each rule is essentially an AND of the path from the root node to the leaf node.

- B = B1 ➔ Class = Green
- B = B2 and A = A1 ➔ Class = Green
- B = B2 and A = A2 and C = C2 ➔ Class = Red
- B = B2 and A = A2 and C = C1 and E = E1 ➔ Class = Green
- B = B2 and A = A2 and C = C1 and E = E2 ➔ Class = Red

Apart from interpretability, decision trees are also very deterministic—they generate the same tree given the same data—thanks to their greedy nature. This is essential for robustness, stability, and repeatability. The scoring throughput of decision trees is high. They just have to apply at most D conjunctions, where D is the depth of the tree. Apart from this, decision trees are also known to handle a combination of numeric and categorical features together. Numeric features at any node are partitioned into two by rules like Age <25. Finally decision trees handle missing data gracefully. They either ignore the missing features (so when a feature is missing the training data is ignored for that feature's purity computation) or assume the most likely value of that feature at that node (fine-grained imputation).

One of the key criticisms of decision trees is that they are not guaranteed to yield an optimal partition of the feature space due to their greedy nature. It is possible that a bad feature chosen early in the tree can lead to a pretty suboptimal subtree below that as there is no mechanism of "backtracking" and correcting for a bad-greedy choice made earlier. This is a classic example of the fundamental trade-off in AI between optimality and speed. Decision trees were originally designed for categorical features only. They handle numeric features using thresholding type rules—for example, if (temperature <100 degrees). This often limits them to partitioning the numeric subspace only along the numeric axes. If the required decision boundary is oblique, then decision trees land up learning staircase functions that could lead to very large trees. This can be addressed by learning logistic regression models at each internal node using all the numeric variables available at that node and using a threshold on that logistic model to partition it into two parts. This is a classic example of overcoming model limitations by combining them with other techniques.

Decision tree classifiers have evolved over the last few decades. Ensemble version of decision trees including Random Forest and XGBoost are commonly used for complex supervised learning problems. You can read more about decision trees in Chap. 3 of "Machine Learning" by Carbonell et al. (1983) or Chap. 8 of "Data Mining: Concepts and Techniques" by Han et al. (2011).

4.4.1 k-Nearest Neighbor Classifier (k-NN)

Often when we have to make important decisions, we take the advice of our *near* ones. We are more influenced by the opinions (about products, movies, restaurant, politics, etc.) of our friends, family, social circle, etc. than those of strangers. This principle that *nearby things have a higher influence than far-off things* is the essence of a whole family of algorithms starting with *k*-nearest neighbor (*k*-NN) classifier. In k-NN, the training data is stored as is and there is no modelling. Hence, this is an example of a **nonparametric** classifier. During the scoring phase, first the k-nearest neighbors in the training set (previously seen and labelled examples) are sought. Then the new example is assigned the majority class among these k-nearest neighbors as shown in Fig. 16.9. Let the two classes—blue triangle and red star— training data be stored as is. The new example—the green square—is classified as a blue triangle if k = 5 is chosen because in the top-5 neighbors of the new example, blue triangle is the majority class. While, for the case of k = 10, it is classified as red star class.

Fig. 16.9 k-NN classifier: the new example (green square) is classified based on the majority class in its top neighborhood (i.e., for k = 5, it is classified as blue triangle)

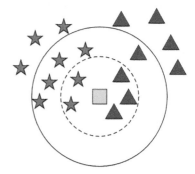

In k-NN, the hyperparameter k determines the "complexity" of the k-NN classifier. For a small k (e.g., 1), a new data point is assigned a class label with only very little "evidence" and will hence be very brittle to noise. For a large k (e.g., 17), we might get a smooth boundary but dependence on too many data points might "average out" the right structure in decision boundaries. Since there is no training, the training time for k-NN classifiers is zero; however, the space complexity of k-NN is high as it just stores the entire training data as "model."

Scoring a new data point is also expensive as it involves computing its distance from all the training data points before finding the k-nearest neighbors. While advancements have been made to store the training data in ways that reduce the time complexity of scoring, still it renders k-NN not practical for high-throughput, real-time scoring. k-NN classifier is brittle to noise as its decision changes abruptly with majority classes. Choosing a high value of k makes it more noise-robust.

One of the key criticisms of k-NN classifier is that it loses information about the actual distance between the training and the test point once the training point is deemed to be within its k-Nearest. A true k-NN classifier should take distances into account wherever it matters. The distance function is the key to the k-NN classifier. For simple multivariate data, Euclidian distances (after proper transformations and normalizations) suffices but for nontrivial data types (e.g., LinkedIn profiles, Gene Sequences, Images, Videos), defining distances is yet another creative process. Read more about k-NN in Chap. 8 in "Machine Learning" by Carbonell et al. (1983) or Chap. 9 in "Data Mining: Concepts and Techniques" by Han et al. (2011).

Parzen Window Classifier (PWC)

In nature, each point mass has a gravitational field of influence. Similarly, each magnet has a magnetic field of influence both of which decrease as one moves away from the source of that influence. Similarly, each data point in a space has a "density field of influence" that decreases with distance. Parzen Window Classifier (PWC) uses this basic intuition to overcome some of the shortcomings of the k-NN classifier and makes a much more robust *softer variant* of the k-NN classifier. Unlike in k-NN where the distance between a training data point and a test data point is used only to *pick* the top k-NN points, PWC uses the actual distances for estimating the influence of *all* training data points on the test example. This makes PWC far more robust to noise and a "Soft variant" of the k-NN classifiers.

Let $X = \{(x_n, c_n)\}_{n=1}^{N}$ denote the training set containing N labelled examples. Let us assume that there is a Kernel field (K) around each of these training points. Now for a test point x we first compute the total influence it is receiving from all the points in each class, c. Influence of a data point x_n on x is given by a kernel function $K_\sigma(\|x - x_n\|)$ that decreases as x goes away from x_n and integrates to one.

$$P(x|c) \alpha \frac{1}{N_c} \sum_{n=1}^{N} \delta(c = c_n) K_\sigma(\|x - x_n\|)$$

The point is then assigned to the class whose cumulative influence on it is maximum.

The complexity of the PWC is controlled by the hyperparameter σ that essentially captures the "spread" of the kernel field. If the field associated with each data point is too wide, then we learn "coarse" decision boundaries that might be very robust to noise but not capture the actual shape of the decision boundaries. On the other hand, if σ is too low, we might land up getting very jagged boundaries that are easily influenced by only a handful of nearby training data points. A low k in k-NN is equivalent to a low σ in PW classifier. As in k-NN, since here also we are not really training a model but just storing the training data as is, there is no training complexity in PW classifier. The scoring complexity, similar to k-NN classifiers is high since to score a new data point, its distance from all the training examples must be computed.

While PWCs address some of the robustness to noise issues of k-NN classifiers, they still do not *learn* anything (nonparametric) and have to depend on the entire training data all the time. Refer to Chap. 14 in "Machine Learning—A Probabilistic Perspective" by Murphy (2012) or Chap. 6 in "The Elements of Statistical Learning" by Friedman et al. (2001) to learn more about PWC.

4.4.2 Bayesian Classifier

The key to PW classifier is the *density function $P(x|c)$* that essentially quantifies whether the point x "looks like" previously seen points that belong to class c. PW *aggregates the influence* of all training points in class c on x to estimate $P(x|c)$. But as humans, we do not classify by first remembering all previous examples of each class and comparing a new example with them. We, on the other hand, build a "representation" of each class by *summarizing* or *describing the essence* of all the data per class into a class "model."

In Bayesian Classifiers, each class c is modelled by (a) its *class prior $P(c)$* that quantifies the probability that an unseen data point would belong to class c and (b) the *class conditional density function $P(x|c)$* that quantifies the probability of having seen "such" a data point from class c in the training data. Unlike in PW, in Bayesian Classifiers, $P(x|c)$ is *modelled* (and not just *computed*) using a parametric density function that takes into account the nature of the data (e.g., multivariate, text, speech) as well as the parametric form used to model it (e.g., Normal distribution). The class prior and class conditional density functions are learnt from the training data. They are then used to compute the *class posteriori probability $P(c|x)$* over all the classes c for a new data point x. This is done by using one of the most celebrated relationships in statistics and probability theory—a relationship between cause and effect, between learning and scoring, between past observations and future predictions, and between data and knowledge—the **Bayes Theorem**:

$$P(c|x) = \frac{P(c)P(x|c)}{P(x)} = \frac{P(c)P(x|c)}{\sum_{c'} P(c')P(x|c')}$$

While computing class priors $P(c)$ is straightforward—all we have to do is normalize the counts of each class in the data, it is in modelling the class conditional density function $P(x|c)$ where a Bayesian Data Scientist spends most of his time. We give a flavor of a few common density functions below.

Unimodal Bayesian classifier (UBC): The simplest Bayesian classifier on multivariate numeric data models each class c as a unimodal (assuming normally distributed) cloud, centered around a mean μ_c and with a certain covariance Σ_c that captures the shape of the cloud, that is, $P(x|c) = N(x|\mu_c, \Sigma_c)$. The mean and covariance are computed by maximizing the log-likelihood of the class data:

$$\mu_c = \tfrac{1}{N}\sum_{n=1}^{N} \delta(c_n = c)\, x_n \text{ and } \Sigma_c = \tfrac{1}{N}\sum_{n=1}^{N} \delta(c_n = c)\, (x_n - \mu_c)(x_n - \mu_c)^T$$

Unimodal Bayesian classifier and PW are two extreme ways of estimating the same statistic: $P(x|c)$. In nonparametric PW each training data point is associated with a Gaussian kernel of a certain width around it. In the parametric unimodal Bayesian classifier, all the data points associated with a class are "described" using a single Gaussian—in terms of its mean and covariance parameters.

Multimodal Bayesian classifier (MBC): Clearly there is a continuum of complexity from PW—that uses one Gaussian per data point—a potential overkill to UBC—that uses one Gaussian per class—which might not be sufficient to describe the class. In many domains, a class might be multimodal, that is, it might have subclasses. For example, the same word might have two very different pronunciations in different accents. An object in image might look very different in different pose, illumination, and scale. A letter in OCR might have different fonts and emphases (bold, italics, etc.). In handwritten digits, for example, people write a digit "7" or "1" or "9" in different ways. In all such cases, the entire class cannot be modelled as a single Gaussian but as a mixture-of-Gaussians (MoG), that is, two or more Gaussians—one representing each subclass. Figure 16.10 shows a two-class problem data where using one Gaussian per class (left) might yield a low accuracy classifier (under trained) but using three Gaussians per class might be the right level of complexity for this dataset. In general, MoG is a generic and powerful way of modelling arbitrarily complex density functions to match complexity of data (number of subclasses per class). An MoG is written as:

$$P(x|c) = \sum_{k=1}^{M_c} \pi_k^{(c)} N\left(x|\mu_k^{(c)}, \Sigma_k^{(c)}\right)$$

where:

- $\pi_k^{(c)}$ is the prior proportion of subclass k in class c.
- $\mu_k^{(c)}$ is the mean of subclass k of class c.
- $\Sigma_k^{(c)}$ is the covariance of subclass k of class c.

These parameters are learnt using the EM algorithm using the data from each class independently. The number of mixture components for each class can vary depending on the number of subclasses it might have. The EM algorithm for learning MoG is described in the unsupervised learning chapter. The unimodal

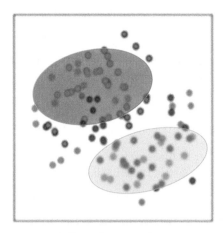 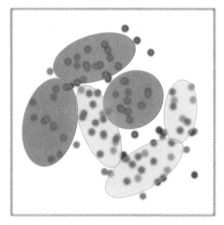

1-Gaussian per class   3-Gaussians per class

Fig. 16.10 Using 3 instead of 1 Gaussian per class: matching data complexity with model complexity

Gaussians, mixture-of-Gaussians, and Parzen window form a continuum of models for modelling each class with a single, multiple, and maximum possible Gaussians.

For more details, the interested reader can refer to Chap. 8 in "The Elements of Statistical Learning" by Friedman et al. (2001) or Chap. 8 in "Data Mining: Concepts and Techniques" by Han et al. (2011).

Naïve Bayes classifier: One of the fundamental problems in estimating density functions in particular and learning robust ML models in general is the *curse of dimensionality*—as the dimensionality of the feature space increases, the volume of data needed to get the same level of robustness increases exponentially. In cases where the number of features is large and the amount of data we have is not sufficient to populate each region in the "joint" space of all features, a *naïve* yet practical assumption is made that *all features are conditionally independent, given the* class. This results in the following simplification of the class conditional density function:

$$P(x|c) = \prod_{d=1}^{D} P(x_d|c)$$

where D is the number of dimensions. This significantly reduces the number of parameters. In case of discrete data with C classes, D features each of which taking M values, the total number of parameters for computing the joint probability density function is $O(CM^D)$. The amount of data needed to have sufficient statistics in each cell would also be enormous. A naïve Bayes classifier on this data only needs to estimate $O(CMD)$ parameters.

A common application for Naïve Bayes classifiers is in text classification where the number of dimensions, that is, words, phrases, bigrams, and trigrams can be very large compared to the size of the labelled corpus. A document is represented as a *bag-of-words* where each document x is represented by the number of times each word w_d occurs in the document—the term frequency: $tf(w_d|x)$. In the training phase, class conditional probabilities of each word are computed from labelled data as follows:

$$P(w_d|c) = \frac{n(w_d, c) + \lambda}{\sum_w n(w, c) + D\lambda}$$

Here $n(w, c) = \sum_{x \in c} tf(w|x)$ is the number of times word w occurs in class c, D is the total number of words in the dictionary, and λ is the Laplacian smoothing constant used to make sure that none of the $P(w_d|c)$ becomes 0.

These estimates can be used to compute $P(x|c)$. A new document x is classified by first computing its class conditional probability density. But since the document lies in a high-dimensional (number of unique words after preprocessing) sparse space (each document only contains a very small fraction of the total words in the dictionary), it is not possible to model the density in the joint space. We therefore make a naïve assumption that all words are independent given the document belongs to a certain class:

$$P(x|c) = \prod_{d=1}^{D} P(w_d|c)^{tf(w_d|x)}.$$

Each occurrence of each word w_d in the document is multiplied to itself $tf(w_d|x)$ times. The Bayes rule is then used to compute the posterior probability of a class given a new document. To prevent numerical underflow issues, instead of computing $P(c|x)$ we compute:

$$\ln P(x|c) \propto \ln P(c) + \sum_d tf(w_d|x) \ln P(w_d|c)$$

where the denominator is left out as it not involved in the classification.

As k-NN classifiers are a good baseline for numeric multivariate classification problems, Naïve Bayes classifiers are a good baseline for text classification problems. Feature engineering in bag-of-words representation of text data involves (1) removing stop words (e.g., articles), (2) doing stemming (so all variants of a word, e.g., "run," "running," "runs," "ran," are mapped to the root word "run") and before stemming, adding higher-order words (e.g., bigrams, trigrams, or better yet adding phrases discovered through other means). One might also move from bag-of-*words* to bag-of-*topics* representation as discussed in Chap. 15 on unsupervised machine learning).

Bayesian classifiers are robust to data and feature noise, they can adapt to feature covariance within each subclass, incorporate class priors systematically where needed, and are well grounded in theory. The creativity in Bayesian classifiers is in learning class conditional probability density function $P(x|c)$.

Discriminant analysis: Earlier we defined classification as the art of partitioning the feature space into pure regions. We can achieve this in two ways. Either by describing each pure region as is done by Bayesian classifiers via class conditional probability density functions $P(x|c)$ or equivalently by characterizing the boundaries between two pure regions that *discriminate* the two classes. Here we will explore the descriptive to discriminative transition through Discriminant Analysis.

The decision whether x belongs to one class or the other depends on which of the two posterior probabilities is higher. In this sense $g_c(x) = P(c|x)$ is called a *discriminant function* and any monotonic variant of this is also a discriminant function. Simplifying this we get:

$$g_c(x) = P(c|x) \propto \ln P(c|x) \propto \ln P(x|c) + \ln P(c)$$

The *decision boundary* between two classes (assuming a two-class problem for simplicity) is the locus of all points x where the two posterior probabilities are same, that is, where the two regions intersect or where the points cannot be classified in one class or the other: $P(c_1|x) = P(c_2|x)$. The discriminant classifiers label a data point into the maximum discriminant value class: $c^*(x) = \arg\max_c \{g_c(x)\}$. The decision boundary can be derived by solving: $g_1(x) = g_2(x)$ or $g_1(x) - g_2(x) = 0$.

Linear discriminant analysis (LDA): In a two-class problem, if we make the assumption that the covariance (shapes of the Gaussians) of the two classes are the same, that is, $\Sigma_1 = \Sigma_2 = \Sigma$ then the decision boundary is given by: $\ln P(x|c_1) + \ln P(c_1) - \ln P(x|c_2) - \ln P(c_2) = 0$, which when simplified leads to a *linear decision boundary*: $w^T x + w_0 = 0$, where:

$w = \Sigma^{-1}(\mu_1 - \mu_2)$ and $w_0 = \frac{1}{2}(\mu_2 - \mu_1) \Sigma^{-1} (\mu_2 - \mu_1) + \ln \frac{P(c_1)}{P(c_2)}$

Quadratic discriminant analysis (QDA): If we allow the two covariance matrices to be arbitrary, that is, the classes take any shape possible within the unimodal constraints, then the decision boundaries become more quadratic: $x^T W_2 x + w_1^T x + w_0 = 0$. Figure 16.11 shows the linear and quadratic discriminant decision boundaries:

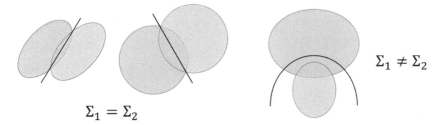

Linear Discriminant Analysis   Quadratic Discriminant Analysis

Fig. 16.11 Linear vs. quadratic discriminant analysis (LDA vs. QDA)

LDA and QDA are a bridge between descriptive and discriminative classifiers. The decision boundary in LDA and QDA is simply *computed*—in terms of class prior, mean, and covariance properties—but not *learnt*. Hence, they are still fundamentally descriptive classifiers yet a bridge between the shape and the boundary of the class.

You may read more about LDA and QDA in Chap. 4 in "Machine Learning— A Probabilistic Perspective" by Murphy (2012) or Chap. 4 in "The Elements of Statistical Learning" by Friedman et al. (2001).

Perceptron: One of the earliest discriminative classifiers is a **perceptron**—a simple, biologically inspired, functional model of what we believe the neuron in the brain does. Our brain contains billions of neurons, each connected to thousands of other neurons both laterally (within the same layer) and hierarchically (across layers). Each neuron does, more or less, functionally the same thing—it aggregates the inputs received from incoming neurons (connected to its dendrites), attenuates the aggregated activation, and makes it available at its axons to pass on to its "children" neurons. While a neuron sitting at the lower layer (e.g., on the retina of the eye) might take raw pixel level input and combine them to detect lines, the *face detecting neuron* at much higher up in the visual cortex hierarchy might be taking inputs from *eye detecting* neurons, *nose detecting* neurons, *mouth detecting* neurons, etc. as inputs and predict whether it is "seeing" a face. The simplicity of each neuron combined with the complexity with which they are arranged and work together makes the brain one of the most mysterious and powerful masterpieces of evolution. This also forms the basis of the modern deep learning paradigms that use a variety of neurons and deep layered architecture to replicate some of the most complex human brain capabilities of vision, speech, and text understanding. All of this complexity starts with the "transistors of the brain"—the neurons (Fig. 16.12).

The basic perceptron algorithm that captures the early essence of a neuron for a two-class problem is very simple. Here, let P and N be the set of positive and negative examples, respectively. Let w_t be the weights of the perceptron in iteration t, initialized randomly (imagine the neurons of a newborn baby that has never seen any data yet but has this powerful infrastructure to learn a hierarchical representation

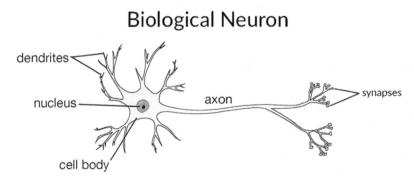

Fig. 16.12 A neuron—the building block of the brain

of the world he/she is about to interact with). Then the perceptron algorithm updates these weights iteratively by (1) sampling a data point, (2) classifying it into one of the two classes based on its current weights, (3) determining whether it has classified it correctly or not given the class label associated with the data point, and (4) update its weights if it made a mistake in the classification:

- Sample a data point $x \in P \cup N$
- If $x \in P$ and $w_t . x \leq 0$ then: $w_{t+1} \leftarrow w_t + x; t \leftarrow t+1$
- If $x \in N$ and $w_t . x \geq 0$ then: $w_{t+1} \leftarrow w_t - x; t \leftarrow t+1$

The perceptron "converges" if we either reach a maximum number of iterations or better yet when no more examples are wrongly classified by the perceptron. Perceptron-based classifiers are mostly useful for two-class problems, they are not very robust to noise, they learn in an online fashion and therefore very sensitive to the order in which the data is presented, and finally they make a *hard* decision—if a point is on the correct side—no matter how far, they will try to self-correct—rendering them brittle. Perceptron is equivalent to k-NN classifier, which is also a hard classifier sharing some of the similar problems that perceptron-based classifiers have.

Logistic Regression: One of the oldest, time-tested, discriminative classifiers in machine learning is Logistic Regression. On the one hand, it is the softer version of the perceptron (pretty much like the Parzen Windows is a softer version of the k-NN and mixture-of-Gaussians is the softer version of k-Means clustering); on the other hand, it is the nonlinear version of linear regression. It models the log-odds ratio of the target class vs. the background class as a linear combination of the inputs.

$$\ln\left(\frac{P(Y=1|x)}{P(Y=0|x)}\right) = w^T x \implies P(Y=1|x) = \frac{1}{1+\exp\left(-w^T x\right)}$$

where $w \in R^{D+1}$ is the set of $D + 1$ parameters including the constant bias term.

Most machine learning is optimization. Every parametric modelling technique optimizes an objective function written in terms of the data (or some statistics on the data) and some parameters. Clustering, for example, minimizes the distance between a data point and *its* cluster center, Fisher discriminant maximizes separation between classes, decision trees try to split a leaf node into the purest possible subregions, and perceptron tries to minimize misclassification error, etc. Modelling is essentially the art of formulating and solving an objective function. Sometimes, the solution is closed form (PCA, Fisher, LDA, QDA, etc.), sometimes it is greedy (e.g., decision trees), and sometimes it is iterative (e.g., perceptron). The objective function too can take multiple forms. Sometimes it is a variant of the sum-squared-error (e.g., K-means clustering), sometimes it is maximizing (log) likelihood of seeing the data.

Logistic regression is the solution of a maximum log-likelihood objective function:

$$J(w) = \ln \prod_{n=1}^{N} P(Y=1|x_n)^{y_n} P(Y=0|x_n)^{1-y_n}$$

Substituting logistic function for $P(Y = 1|x_n)$ and optimizing for θ yields the following update rule:

$$w_t \leftarrow w_{t-1} + \eta \sum_{n=1}^{N} (y_n - P(Y=1|x_n)) x_n$$

where w_t is the weight parameters at iteration t initialized at $t = 0$ to low random values and η is the learning rate that needs to be carefully set for fast convergence with stability. A logistic regression converges when these weights stop changing substantially. Note that when learning rate η is 1 and $P(Y = 1|x_n)$ is either 1 or 0 (hard), the above update rule reduces to that of perceptron algorithm.

Logistic regression is a highly interpretable classifier. The weights (both sign and magnitude) learnt by logistic regression reveal a lot about the nature of relationship between the corresponding feature and the binary output label. The complexity of a logistic regression is controlled by penalizing the magnitude of each weight, that is, no single weight is allowed to dominate the overall decision. This is done by adding a penalty term: $\lambda \|w\|^2$ to the objective function.

One of the limitations of Logistic Regression is that it can only learn linear relationship between input and output. To learn nonlinear decision boundaries using logistic regression, we can add higher-order terms ($x_i x_j$), ratios (x_i/x_j), log transforms ($\ln x_i$), and other complex functions (e.g., ($x_i^2/(x_j + x_k)$)) of the input features. In fact, adding all variants of a variety of such transform (all second-order or third-order terms, log transforms of all features, etc.) and its ability to pick the right combination from this large pool is what makes them simple, powerful, flexible, and highly versatile modelling paradigm. This ability to add new variables is also known as generalized linear models where instead of just considering the raw features, any transformations of raw features can be used as inputs.

$$\ln \left(\frac{P(Y=1|x)}{P(Y=0|x)} \right) = w^T \Phi(x) \implies P(Y=1|x) = \frac{1}{1 + \exp\left(-w^T \Phi(x)\right)}$$

where $\Phi(x)$ is a potentially very large number of features obtained from x.

$$\Phi(x) = \left(\ldots x_i \ldots, \ldots x_i x_j \ldots, \ldots x_i x_j x_k \ldots, \ldots \ln x_i \ldots \right)$$

The other limitation of Logistic Regression is that it can only be used with numeric features. When the dataset contains both numeric and non-numeric features, we can use a 1-hot-encoding (e.g., if R, G, B are three colors then R can be represented by a vector (1 0 0), B with (0 1 0) and G with (0 0 1)) to create a multivariate representation out of these symbolic data. In spite of all the flexibility and simplicity, the onus of "engineering" features to capture nonlinearity in the input-output mapping is still on the modeller. In other words, logistic regression just learns the mapping given the features. It does not learn the features themselves. In that sense, it is only *partially* intelligent. A truly intelligent system should be able to figure out arbitrary relationships between the input and output without us having to even partially engineer that knowledge via feature engineering or guess work it through hit and trial. In other words, we need a *universal function approximator* that can *learn* any arbitrary mapping—both features and decision boundaries, no just latter given former.

Neural Networks: One of the most important breakthroughs in machine learning was precisely such a universal function approximator—an artificial neural network (ANN)—that can learn arbitrary mappings given enough training data, computational resources, and model complexity. Consider the two datasets in Fig. 16.13, each being a two class classification problem: red squares vs blue circles. Clearly, a single logistic regression classifier cannot solve this problem in this feature space. As discussed earlier, there are two mindsets to address this mismatch between data complexity and model complexity:

In the **feature-centric mindset**, the raw input features are first transformed into new features (e.g., $x_3 = \max\{x_1(1 - x_2), (1 - x_1)x_2\}$ (left dataset) or $x_3 = (x_1 - a_1)^2 + (x_2 - a_2)^2$ (right dataset), where (a_1, a_2) is the center of the entire data). While transforming features obviates the need for a complex model, in most real-world problems, it is not clear which and how many such transformed features are needed to make the class separation well behaved or linearly separable.

In the **model-centric mindset**, the problem of learning both the features and the decision boundaries is solved *simultaneously*. Neural networks are a canonical example of this mindset where the overall model is composed of multiple layers such that the layers closer to the input features try to learn better features (e.g., the lines shown in Fig. 16.13 are logistic regression lines representing binary features—indicating which side of the line a data point is on) while the layers closer to the output try to learn the decision boundaries (e.g., in the left case, the final output is red class for a data point that is above the lower line and below the upper line). The beauty of neural networks is that all neurons seem to be doing the same thing—trying to partition the feature space using hyperplanes—but the role they are playing depends on the layer in which they occur and the input they see from the previous layer. ANNs are biologically inspired ML models. Our sensory-brain-mortar system is organized in layers of neurons with both feed forward connections from lower layers (e.g., neurons in our retina) to higher layers (e.g., neurons responsible for recognizing faces) and feedback connections going from higher layers (whole) to lower layers (part).

Neural Network Architecture: In the simplest neural network architecture—a fully connected, feed forward artificial neural network—neurons are organized in $L+1$ layers. Layer 0 is the *input* layer and layer L is the *output* layer with $L - 1$ *hidden* layers in between. Let N_ℓ be the number of neurons in layer ℓ. Figure 16.14 shows a neural network with one input layer (grey) with $N_0 = 2$ input units, $N_1 = 5$ hidden units, and $N_2 = 2$ output units.

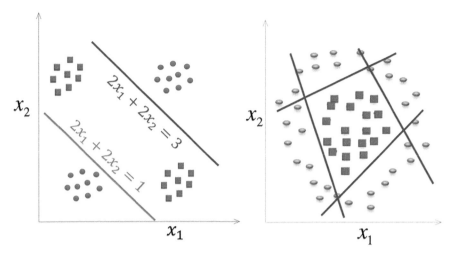

Fig. 16.13 A neural network classifier on two datasets

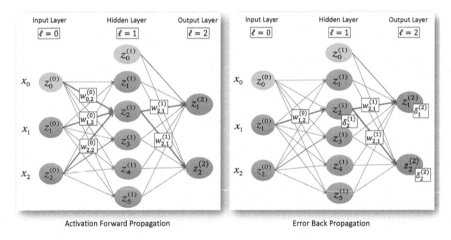

Fig. 16.14 A neural network with one hidden layer

In the **Activation Forward Propagation** (Fig. 16.14, left), the input layers are activated by the input data $z^{(0)} = x$; these activations travel to the subsequent layers to eventually activate the output layer.

Each hidden unit first aggregates the activations of all previous layer neurons and applies the bias term:

$a_i^{(\ell+1)} = w_0^{(\ell)} + \sum_{i=1}^{N_\ell} w_i^{(\ell)} z_i^{(\ell)} = \sum_{i=0}^{N_\ell} w_i^{(\ell)} z_i^{(\ell)}$ (where $z_0^{(\ell)}$, the bias term is always set to 1).

It then transforms these aggregates nonlinearly to generate activations of these hidden units:

$$Z_k^{(\ell+1)} = g\left(a_i^{(\ell+1)}\right) = g\left(w_0^{(\ell)} + \sum_{i=1}^{N_\ell} w_i^{(\ell)} z_i^{(\ell)}\right) = g\left(\sum_{i=0}^{N_\ell} w_i^{(\ell)} z_i^{(\ell)}\right)$$

These activation functions provide nonlinearity and limit the aggregate in a certain range. Examples of some of the activation functions are:

- Sigmoid/Logistic: $g(a) = \frac{1}{1+\exp(-\lambda a)}$ (between [0,1]).
- Hyperbolic: $g(a) = \tanh(\lambda a) = \frac{\exp(\lambda a) - \exp(-\lambda a)}{\exp(\lambda a) + \exp(-\lambda a)}$ (between [−1,1])
- Soft Max: $g(a_i) = \frac{\exp(\lambda a_i)}{\sum_{j=1}^{N_L} \exp(\lambda a_j)}$ (used in output layers for learning posterior probabilities)

The final activations in layer L, $z^{(L)} = \widehat{y}$ is the output that the neural network predicts for the input x for the current set of weights. During the training phase, enough of these input-output pairs are given.

It is in the **Error Back Propagation** (Fig. 16.14, right) that the real magic of neural network learning happens, where the weights are updated using an **Error Backpropagation Algorithm**, perhaps one of the most important algorithms in ML. For a given labelled data (x,y), it minimizes the squared error $E(W) = ||y - \widehat{y}(W)||^2$ between the expected y and the actual $\widehat{y}$ output. This error is *back propagated* from layer ℓ to layer $\ell - 1$ by updating the weights $w_{i,j}^{(\ell)}$ as follows:

$$\Delta w_{i,j}^{(\ell)} \propto -\frac{\partial E(W)}{\partial w_{i,j}^{(\ell)}} = -\left(\frac{\partial E(W)}{\partial a_j^{(\ell+1)}}\right)\left(\frac{\partial a_j^{(\ell+1)}}{\partial w_{i,j}^{(\ell)}}\right) = -\delta_j^{(\ell+1)} z_i^{(\ell)}$$

In other words, the update in the weight $w_{i,j}^{(\ell)}$ is proportional to the activation on its input neuron $z_i^{(\ell)}$ at the error at its output neuron $\delta_j^{(\ell+1)}$. The negative sign indicates that we are trying to minimize the error. The error $\delta_j^{(L)}$ for the output layer is simply:

$$\delta_j^{(L)} = \frac{\partial E(W)}{\partial a_j^{(L)}} = \frac{\partial E(W)}{\partial z_j^{(L)}} \frac{\partial z_j^{(L)}}{\partial a_j^{(L)}} = \left(z_j^{(L)} - y_j\right) g'\left(a_j^{(L)}\right)$$

The error $\delta_j^{(\ell)}$ for all the other layers is given by:

$$\delta_i^{(\ell)} = \frac{\partial E(W)}{\partial a_i^{(\ell)}} = g'\left(a_i^{(\ell)}\right) \sum_{j=1}^{N_{\ell+1}} w_{ij}^{(\ell)} \delta_j^{(\ell+1)}$$

In other words, the total error in each of the nodes in the next layer $\delta_j^{(\ell+1)}$ propagates backward in proportion to the weight $w_{ij}^{(\ell)}$ to form the total the error $\delta_i^{(\ell)}$ at this node.

A wide variety of applications, architectures, and heuristics have been proposed in the last couple of decades that have made neural networks one of the most common and powerful machine learning algorithms especially in two cases: (1) where there is sufficient data to train large networks and (2) where accuracy is more important than interpretability. Recent advances in deep learning have carved a special place for neural networks and their variants—recurrent neural networks (CNN), auto-encoders, convolution neural networks (CNN), and generative adversarial networks—in machine learning. In Chap. 17 on deep learning, a few examples, sample code, and further details are presented. Other interesting books to learn more about ANN are "Machine Learning—A Probabilistic Perspective" by Murphy (2012), "The Elements of Statistical Learning" by Friedman et al. (2001), and "Machine Learning" by Carbonell et al. (1983).

Support Vector Machines: Machine learning is really an art of *formulating an intuition into an objective function*. In classification, the fundamental problem is to find pure regions. So far we have explored a number of classification paradigms that greedily, iteratively, or hierarchically try to find such pure regions in the feature space. A good classifier should be both deterministic and robust to data and label noise. Perceptron, logistic regression, and neural networks are nondeterministic as they depend on model initialization and choice of hyperparameters governing model training. Decision tree, k-NN, Parzen windows, on the other hand, are more deterministic as they yield the same model for a given dataset and hyperparameters but k-NN and Parzen windows could be sensitive to data noise.

Consider the two-class classification problem shown in Fig. 16.15 (left). A perceptron trained on this data, could give an infinite number of solutions depending on its initialization, each of which will have a 100% accuracy. The fundamental question that forms the basis of support vector machines classifier is *which hyperplane is "optimal" among the infinite possibilities.* From a robustness point of view, the "best" hyperplane maximizes the width or margin of the linear decision boundary. This gives a unique solution for this problem (Fig. 16.15, right).

It is easier to understand this using an analogy. Assume that we want to build a straight road between two villages. Let the labelled data points $\{(x_n, y_n)\}_{n=1}^N$ where $y_n \in \{-1, +1\}$ denote the houses of the two villages (classes). The goal is to build the *widest possible straight road* (i.e., maximum margin) that can be built without destroying any house in either of the two villages. This can be done by choosing the center and direction of the road in such a way that as we increase its width equally on either side and stop as soon as it touches the first house on either side. More formally let us say $x^T w + b$ (the solid green line in the right Fig. 16.15) denote the center of the road. The dotted lines parallel to it, on either side denote the boundaries of the roads obtained by extending the road on either side and stopping as soon as it hits a house (data point) on either side.

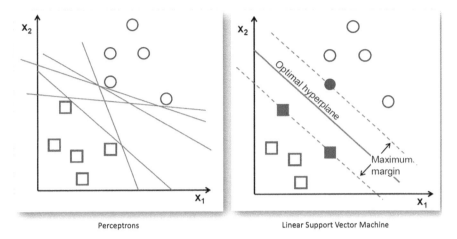

Fig. 16.15 The intuition behind maximum margin classifiers

This can be formulated as the following optimization problem. There are N constraints, one for each data point (house) so that it lies on the correct side of the road, that is,

$$\begin{cases} \boldsymbol{w}^T \boldsymbol{x}_n + b \geq +1 \ \forall n \text{ where } y_n = +1 \\ \boldsymbol{w}^T \boldsymbol{x}_n + b \leq -1 \ \forall n \text{ where } y_n = -1 \end{cases} \implies y_n \left(\boldsymbol{w}^T \boldsymbol{x}_n + b \right) \geq 1, \forall n = 1 \ldots N$$

Note that we can use 1 as a threshold on both sides because any scaling factor can be subsumed in the linear coefficients $\boldsymbol{w}$ and constant term b. Using Geometry, (or examining the distance when there is equality), the width of the road is:

$$J(\boldsymbol{w}) = \left| \frac{(-1-b)}{\|\boldsymbol{w}\|} - \frac{(1-b)}{\|\boldsymbol{w}\|} \right| = \frac{2}{\|\boldsymbol{w}\|}$$

This needs to be maximized. To make the overall function well behaved we write it as:

$$\text{minimize} : \frac{1}{2} \|\boldsymbol{w}\|^2, s.t. y_n \left(\boldsymbol{w}^T \boldsymbol{x}_n + b \right) - 1 \geq 0, \forall n = 1 \ldots N$$

This can then be written using Lagrange Multiplier as the *primal* objective function. Every time the nth constraint is violated (house is broken), a positive penalty α_n is paid. Since we want to **minimize** the objective function (which is obtained by inverting and squaring the margin), whenever $y_n(\boldsymbol{w}^T \boldsymbol{x}_n + b) - 1$ is negative the value of the overall objective should go up. This effect is obtained by the following Lagrange Multiplier terms combining the objective with constraints:

$$L_P(\boldsymbol{w}, b) = \frac{1}{2}\|\boldsymbol{w}\|^2 - \sum_{n=1}^{N} \alpha_n \left[y_n \left(\boldsymbol{w}^T \boldsymbol{x}_n + b \right) - \right]$$

$$= \frac{1}{2}\|\boldsymbol{w}\|^2 - \sum_{n=1}^{N} \alpha_n y_n \left(\boldsymbol{w}^T \boldsymbol{x}_n + b \right) + \sum_{n=1}^{N} \alpha_n$$

The points on either side of the road on which the margin "hinges" are called the Support Vectors—these are highlighted in Fig. 16.15 (right). Further SVM formulation is built on three "SVM Tricks."

SVM Trick 1—Primal to Dual: The primal objective function above contains two types of parameters—the original parameters of the hyperplane ($\boldsymbol{w}$ and b) as well as the Lagrange multipliers α_n. Note that hyperplane parameters can be used to determine the support vectors and similarly knowing the support vectors can determine the hyperplane parameters. Hence, the two sets of parameters are complementary to each other and both need not be present in the same objective function. To clean this up, let us optimize w.r.t. the hyperplane parameters first:

$$\frac{\partial L_P(\boldsymbol{w}, b)}{\partial \boldsymbol{w}} = \boldsymbol{w} - \sum_{n=1}^{N} \alpha_n y_n \boldsymbol{x}_n = 0 \Rightarrow \boldsymbol{w}^* = \sum_{n=1}^{N} \alpha_n y_n \boldsymbol{x}_n$$

$$\frac{\partial L_P(\boldsymbol{w}, b)}{\partial b} = \sum_{n=1}^{N} \alpha_n y_n = 0$$

The solution for $\boldsymbol{w}$ shows how knowing the support vectors and the data can be used to find the hyperplane. The second equation implies that the total penalty associated with the positive class is the same as the total penalty associated with the negative class. Substituting both these back into the primal, and simplifying, we get:

$$L_D(\boldsymbol{\alpha}) = \sum_{n=1}^{N} \alpha_n - \frac{1}{2} \sum_{m<n} \alpha_m \alpha_n y_m y_n \boldsymbol{x}_m^T \boldsymbol{x}_n, \; s.t. \sum_{n=1}^{N} \alpha_n y_n = 0 \; and \; \alpha_n \leq 0, \forall n$$

The dual objective function is *pure*—it is only in terms of the Lagrange multipliers. The solution to this convex optimization problem gives the right support vectors that form the boundaries of the widest possible road we can build without breaking any house. The first term minimizes the total penalty, while the second term uses pairwise dot-products or similarities between all pairs of points.

SVM Trick 2—Slack Variables: More often than not, either because of the nature of the decision boundary or noise in the data, the two classes may not be linearly separable. In such cases we will have to break some houses (violate some constraints) to build a road. Not only that, even if the data is linearly separable, it is possible that we might be able to build a wider road (find a better margin) if we were allowed to break some houses as shown in Fig. 16.16 below. Here for the

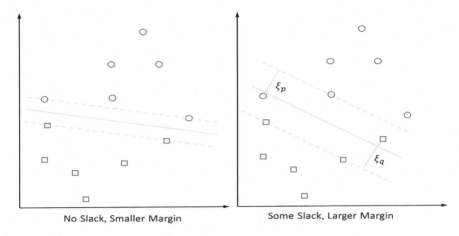

Fig. 16.16 The trade-off between bigger margin and violating the constraints on same dataset

same data if no constraints were allowed to be violated we can only build a smaller margin classifier (left), but if two constraints are allowed to be violated (two houses were allowed to be broken) then we can build a wider margin classifier. This ability to trade-off between maximizing the margin and violating some constraints is the second kernel trick. It is realized by introducing "slack variables" $\{\xi_n \geq 0\}_n^N$ that allow a certain slack on each constraint:

$$\left\{\begin{array}{l} \boldsymbol{w}^T\boldsymbol{x}_n + b \geq +1 - \xi_n \ \forall n \text{ where } y_n = +1 \\ \boldsymbol{w}^T\boldsymbol{x}_n + b \leq -1 + \xi_n \ \forall n \text{ where } y_n = -1 \end{array}\right\} \Longrightarrow y_n\left(\boldsymbol{w}^T\boldsymbol{x}_n + b\right) \geq 1 - \xi_n, \forall n=1\ldots N$$

The Primal Objective function with slack variables has two additional terms. First a cost C associated with the total slack given and a set of terms to ensure that all the slack variables are positive.

$$L_P(\boldsymbol{w}, b) = \frac{1}{2}\|\boldsymbol{w}\|^2 + C\sum_n \xi_n - \sum_{n=1}^N \alpha_n \left[y_n\left(\boldsymbol{w}^T\boldsymbol{x}_n + b\right) - 1 + \xi_n\right] - \sum_n \mu_n \xi_n$$

Here, ξ_n are slack variables and μ_n are Lagrange multipliers on these slack variables. Converting this to the dual, however, gives a very elegant variation of the original dual.

$$L_D(\boldsymbol{\alpha}) = \sum_{n=1}^Z \alpha_n - \frac{1}{2}\sum_{m<n} \alpha_m \alpha_n y_m y_n \boldsymbol{x}_m^T \boldsymbol{x}, \ s.t. \ \sum_{n=1}^N \alpha_n y_n = 1 \text{ and } \alpha_n \leq C, \forall n$$

Note that the **only difference** now is that earlier the penalty of violating a constraint had no upper bound (i.e., $\alpha_n \geq 0$), which means that violating even a single constraint could result in an infinite cost. But with the introduction of the slack variables and a cost C on these slack variables changes the dual in only one way: It just upper-limits the amount of penalty that any single violation can cause, that is, $0 \leq \alpha_n \leq C$. This implies that even if a few constraints are violated, the maximum penalty could at most be C for each such violation and if that leads to a wider margin, so be it. The cost parameter C controls the complexity of the SVM classifiers. A low value of C will allow more constraints to be violated and larger margin, simpler classifier be learnt while a high value of C will allow smaller number of constraints to be violated and smaller margin, complex classifier to be learnt.

SVM Trick 3—Kernel Functions: Machine learning is the art of matching data complexity with model complexity. This is accomplished in two ways: Either we use linear (simple) models with nonlinear (complex) features or nonlinear (complex) models with linear (simple) features. For example, in logistic regression if the raw features are used as-is, we are not able to learn the complex decision boundaries and so we add nonlinear features (via generalized linear models). The third kernel trick is on the same lines. The original SVM formulation is only for two-class problems and learns a linear large margin classifier. To build more complex models than linear, we can introduce nonlinear features and "warp" the space and learn a linear classifier in the warped space. Note, however, that in SVM the only way data points are used in a space is to take their dot-products $x_m^T x_n$. Let us call this the kernel or similarity between these two data points: $K(x_m, x_n)$. SVM classifier really needs (only) this pairwise dot product (the Gram Matrix) as input. Now if there were a class of kernels where it was possible to actually compute this pairwise dot product in the transformed space directly without actually having to first transform the data into that space, then we could use this generalized kernels directly. In other words, let us say

$$K(x_m, x_n) = \phi(x_m)^T \phi(x_n)$$

where $\phi(x)$ is the nonlinear high-dimensional space to which the raw input x is mapped. Some of the common kernels used in SVM are:

- Polynomial Kernels: $K_{c,d}^{poly}(x, x') = (x^T x' + c)^d$ with hyperparameters c and d.
- Radial Basis Function Kernels: $K_\sigma^{rbf}(x, x') = \exp\left(-\frac{\|x-x'\|^2}{2\sigma^2}\right)$

Using such nonlinear kernels to first warp the space into a hypothetical high-dimensional space, building a linear large margin classifier, in that space, and therefore realizing a nonlinear large margin classifier in the original space is the third kernel trick. Together, these three tricks make SVMs one of the most elegant formulations of an intuition into a powerful machine learning algorithm.

Scoring Using SVM
Given a dataset, the cost parameter C, the kernel, the kernel, and its hyperparameters, SVM learns a large margin classifier by finding the support vectors and generate as output the set of support vector weights $\{\alpha_n\}$. A new data point x is scored as:

$$S(x) = \sum_n \alpha_n y_n K(x, x_n)$$

The class label is the sign of the score. Note that this scoring function is similar to Parzen window scoring except that in Parzen windows all training data points are used while in SVM, the weighted sum is taken only w.r.t. the support vectors, hence the time complexity is much lower.

One of the key drawbacks of SVM methods is that their training is quadratic in the number of training data (as they need pairwise cosine similarity) and hence with larger dataset learning an SVM can take much longer and can become quite infeasible. Sampling the data can address this.

In many domains, it is easier and more natural to quantify similarity between two data points than to represent a data point in a multidimensional space. For example, similarity between two LinkedIn or Facebook profiles, two gene sequences, two images or words, or two documents is much more natural than to represent them in a multidimensional feature space. In such cases, kernel-based approaches including SVM, k-NN, and Parzen windows might be more natural to use than traditional models such as decision trees or logistic regressions.

SVM in particular and Kernel methods in general have been applied in a variety of applications. Text classification using TFIDF representation was one of the areas in which they have shown remarkable success. A lot of research has gone into discovering new kernel functions for specific datasets and extending the SVM thinking (large margin) to other domains such as regression and outlier detection as well.

To learn more about SVM, you can refer to Chap. 14 in "Machine Learning—A Probabilistic Perspective" by Murphy (2012), Chap. 12 in "The Elements of Statistical Learning" by Friedman et al. (2001), or Chap. 9 of "Data Mining: Concepts and Techniques" by Han et al. (2011).

Ensemble learning: We have explored two broad approaches of building models: First, extracting better features (i.e., semantic, hierarchical, domain knowledge driven, statistics driven) and building complex models (deeper decision trees, deeper neural networks, nonlinear SVM vs. Linear SVM, neural networks vs. logistic regression, mixture-of-Gaussians vs. single Gaussian per class, etc.). Instead of building a single increasingly complex model (both feature and model complexity), the third approach is to *divide-and-conquer,* that is, break the problem into simpler subproblems, solve each subproblem independently, and combine their solutions. This is called *ensemble* learning, where the models must be different from each other in some ways while being similar to each other with respect to the nature of the modelling technique and complexity. In other words, we neither want to create an ensemble of, say, a neural network and a decision tree—they should all be

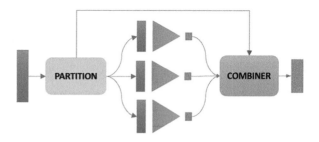

Fig. 16.17 The general architecture for ensemble learning

the same modelling technique—nor do we want to create an ensemble where each model by itself is too powerful. A good ensemble is a collection of diverse, shallow, similar models. The goal of ensemble learning could be accuracy, robustness, and even interpretability in some cases. Figure 16.17 shows the generic architecture of ensemble learning where the Partition layer divides the problem into multiple subproblems, the learners learn the model for each partition and produce an output for any new input. These outputs are then combined by the combiner—which is aware of the nature of the partition.

A number of different ensemble learning frameworks, each using a different way of partitioning and combining the models have been proposed in the past. Here we summarize a few.

Sample-Based Ensemble (Bagging)

Bagging or Bootstrap Aggregation is one of the oldest forms of ensemble learning methods especially used to build robust "average" models when the amount of labelled data is small. In Bagging, multiple samples (say 80%) of the original training data are used to build different models. These models are then "averaged" by taking the average of the output of these models for a given input. This averaging reduces the variance. The Partition is based on random sampling of the data and combiner is just a simple average.

Feature-Based Ensemble (Random Forest)

Instead of sampling a subset of rows (i.e., data points) as in bagging, we can also sample a subset of columns (i.e., features). Feature sampling is especially useful with modelling techniques such as decision trees where models are very sensitive to the set features they are allowed to use. Since decision trees are greedy, given all the features, they will generate the same tree. Lately, Random Forest has become a popular and powerful modelling tool. For a dataset with D features, we sample $\sqrt{D}$ features at a time, build a tree with these features, and then take an average of all these trees to generate the final output. Both sample-based and feature-based bagging are highly parallelizable as each of the models can be built independent of the other.

Accuracy-Based Ensemble (Boosting)

In both sample-based and feature-based bagging, all models are created equal— there is no bias or preference or order among the models. They are as random and independent as possible. The goal in bagging is "model averaging" not "accuracy improvement." Boosting is another class of ensemble learning approach that tries to improve accuracy of the model by building a sequence of models such that the next model focuses on the cumulative weakness of the models built so far. In boosting, the first model gives equal importance to all data points. The second model tries to focus more on (increase weight) those data points for which the first model does not do as well. The third model increases the weights of those data points whose error according to the cumulative first and second model so far is high. Boosting is not amenable to parallelism as the next model depends on the previous $k-1$ models. Nevertheless, it is one of the most powerful techniques for building ensemble models. The key to boosting is again a large number of shallow/weak models. One of the most famous boosting algorithms is XGBoost that applies boosting to decision trees.

Region-Based Ensemble (Mixture-of-Experts)

The bagging and boosting algorithms focused on sampling the data or features randomly. Another class of ensemble learning algorithms is where each model focuses on a different part of the input space instead of building a single model for the entire space. For example, if we were to build credit models for all businesses, one approach would be to build a single complex model for all types (size × vertical) of businesses. Another approach might be to build a separate model for small, medium, and large businesses and also businesses in different verticals. The business size and vertical now become the "partitioning variables" instead of "input features" to the model. And each of the models becomes an "expert" in that cohort of businesses. Such a framework is called a mixture-of-experts. Local linear embedding is an example of a mixture-of-experts where to model a complex regression function, instead of using a high-order polynomial we might use local linear planes where each plane is valid only over a small region in the input space and near the boundaries the outputs of two planes are interpolated to give the final output.

Output-Based Ensemble (Binary Hierarchical Classifiers)

Most machine learning algorithms such as logistic regression or support vector machines are natural at solving two-class problems. But more often than not we are faced with classification problems with more than two classes (digits recognition, remote sensing, etc.). In such cases, we can apply these two-class classification algorithms in creative ways:

1-vs-rest classifier: One approach is to take a C-class problem and break it into C 2-class problems, where each problem takes one of the C classes as a positive class and all the other C-1 classes as negative class. This approach has a few drawbacks as the negative class can become large and create an imbalanced two-class problem each time. If we were to sample the negative class (C-1), choosing the right negative samples becomes critical to build a good 1-vs-rest classifier. Finally, the decision boundary where one class has to be discriminated from all the others might be too complex to learn.

Pairwise classifier: Here, the C-class problem is divided into (C choose 2) two-class problems where a 2-class classifier is built for each pair of classes. The advantage here is that each of the pairwise classifiers can select or engineer its own set of features (e.g., features needed to distinguish digits 1 vs. 7 are very different from the features needed to distinguish digits 3 vs. 8). With such specific features that focus on discriminating just the two classes at a time, the accuracy of these pairwise classifiers can be very high even with simple models. The domain knowledge that is discovered in terms of which features are needed to discriminate which pair of classes is an additional outcome. At the time of scoring, a new data point is first sent through all the pairwise classifiers where each one gives a label from among its class pair. Note that here each of the C classes has equal votes. The majority voting is used to then combine the output. The only drawback here is that the number of classifiers needed to be built is quadratic in number of classes. This can, however, be parallelized. Similarly, at scoring time each new data point has to be sent through all the pairwise classifiers. This again can be parallelized. Pairwise classifiers do not suffer from some of the problems of 1-vs-rest classifiers.

Binary hierarchical classifiers: In hierarchical clustering, data is clustered either in top-down (divisive) or bottom-up (agglomerative) fashion. In the same way, if we have a large number of classes, then the classes themselves can be clustered hierarchically. The distance between two classes can be measured by the accuracy of the pairwise classifier itself. Figure 16.18 shows an example of such a binary tree discovered from classifying letters of the English alphabet using OCR features. Here classes G and Q are merged first, then classes (M, W) and (F, P), etc. are merged. This is based on the training accuracy between those pairwise classes. These two classes are merged together (bottom up) and a new meta-class {G,O} is created. Now we are left with C-1 classes. The process is repeated again and a whole binary tree of classes is created.

Each internal node in this tree is a two-class classifier with its own set of features that best discriminate the two child (meta)classes. When a new data point has to be classified, it first goes to the root node where the root node classifier decides whether it looks more like "left meta-class" or "right meta-class." The data point is

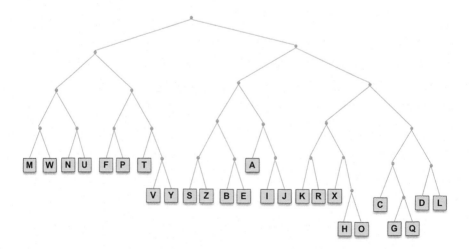

Fig. 16.18 A binary hierarchical classifier for a 26-class classification problem

then passed along that path recursively. This can be done both in a "hard" way, that is, send it either to left or right. Or this can be done in a "soft" way, that is, send it to both left and right with the posterior probability weight. These posterior weights are then multiplied across each path leading from the root node to the leaf node to get the overall posterior probability of each class. Such a classifier eventually needs only C-1 pairwise classifiers (as opposed to C choose 2 for pairwise classifier), it will still use features that discriminate only the two meta-classes at each node, and most importantly, this will automatically discover the class hierarchy as additional domain knowledge that we might not be aware of before.

Ensemble learning is used where individual complex models are not enough and we need to build robust and accurate models. The mixture-of-experts and binary hierarchical classifiers not only improve model accuracy but also improve model interpretation as they focus us on the right features and therefore a simpler yet more accurate decision boundary. In general, ensembles are more reliable than individual models as they explore the possible space of input/output mapping more thoroughly.

To learn more about various ensembling techniques, you can refer to Chap. 16 in "Machine Learning—A Probabilistic Perspective" by Murphy (2012), Chaps. 8, 10, and 15 in "The Elements of Statistical Learning" by Friedman et al. (2001), Chap. 11 in "Machine Learning" by Carbonell et al. (1983), or Chap. 8 of "Data Mining: Concepts and Techniques" by Han et al. (2011).

In this chapter, so far, we have explored the classification paradigm in depth. In this section, we will discuss various aspects of the recommendation engine paradigm.

4.5 The Algorithms: Recommendation Engine

Birds of a feather flock together.

Use-cases for recommendation engines: Today recommendation engines are used across many domains; for example, in the e-commerce domain, offers are recommended; in the media domain, movies, songs, news, videos, TV shows are recommended; in the food app domains, restaurants and dishes are recommended; in the travel domain, hotel and vacation packages are recommended; in the social media domain, people to connect are recommended, and in personalized education, the next concept, the right content, and the next problem the student should attempt are recommended. These recommendation engines serve multiple user needs including *discovery*, *personalization*, *serendipity*, and *optimization*.

Discovery—typically in domains such as e-commerce, media, social networking, or education there are a large number of choices (products, songs, connections, topics to study) for the user to choose from. There are different (and often a combination of) modes in which the applications make it convenient for the user to discover what they are looking for:

- In *Search mode*, the user knows the complete or partial name of the item he/she is looking for (e.g., product, song, movie, or topic name).

- In *browse mode*, the items are *organized* in categories (e.g., electronics vs. sports) and hierarchies (e.g., electronics → cameras → SLR cameras) so the user can navigate through this organization structure to reach the item he/she is looking for.
- In *filter and sort* **mode**, any list of items (obtained from search or browse) is further refined by either filtering (including or excluding) or sorting (in ascending or descending order) the items by various properties (brand, rating, price, etc.). In all three modes, the onus is on the user to discover what they are looking for using these modes.
- In *recommendation mode*, the discovery process becomes proactive where the system itself suggests or pushes the items to the user that he/she is most likely to engage with next. This is one use-case of recommendation engines—build products that enable "intelligent proactive discovery" of items from a large collection of items for the user.

Personalization—most Apps or home pages of services today have an entry point that can be personalized for each user. For example, each user sees a different set of videos when they log into YouTube. They see a different set of suggestions for potential connections when they log into their LinkedIn, Facebook, or Twitter accounts. They also see different Netflix, Amazon Prime, Gaana, Saavn home pages depending on their previous activities on these Apps. The home pages differ not just from other users but also from their last visit. This degree of personalization of home pages of websites or apps is also powered by recommendation engines in the backend.

Serendipity—The epitome of intelligence is not just to do what makes sense but to do what might surprise us *while* it makes sense, to *exceed* our expectations. An essential element of advanced recommendation systems is this serendipity. Search is a zeroth order intelligence where the user already knows what he/she is looking for and the system is just trying to "match the query" with content meta-data; personalization is the first-order discovery where the user is suggested proactively what he/she might be looking for next, but serendipity is really a second-order intelligence where the user is suggested what he/she might not even be looking for but is pleasantly surprised by it. Serendipity opens up gates to a new dimension of exploration for the user. Serendipity in recommendations is like mutation in evolution. It allows for random yet connected exploration.

Optimization—Finally, recommendation engines can also be used to optimize different utility functions depending on the life stage of a customer. For example, when a new customer boards on a business (e.g., a bank, a retailer), the business tries to optimize the relationship of the customer over a period of time. In the beginning, the goal is just to transition a new customer to a loyal customer. Then from a loyal to a valuable customer, and then from a valuable to a retained customer. Each of these stages of a customer journey vis-à-vis the business can apply a different utility function when recommending the next set of products. For example, to make a new customer loyal, a retailer might offer daily use products at a lower price to the customer (milk, soap, groceries, etc.). Once the customer is loyal, the business might want to cross-sell the customer into other categories that are more profitable to the business and relevant to the customer (clothing, shoes, etc.). After that, the business might want to up-sell the customer even more profitable items such as high-end electronics or jewelry, etc. This delicate hand-holding of a customer leads to a long-term lifetime value of a customer. Some of the most advanced recommendation engines fine-tune their recommendations to each customer based not just on the customer's historical and demographic indicators but also the inferred stage of the customer's life stage in this journey.

4.5.1 Recommendation Engine Problem Formulation

Problem statement: Given the past engagement of a user with items in the domain, predict whether or not a particular user will exhibit high or low engagement with a particular item that he/she may or may not have been exposed to yet.

Figure 16.19 shows an example of an Engagement Matrix with six users engaging with eight items. If a user likes an item, the corresponding cell is marked green and if the user does not like an item, that cell is marked red. So, for example, user 2 likes items 2, 4, and 7 but does not like the items 3 and 6. These items could be movies or songs or products, etc. The gray boxes indicate that the corresponding user (row) has not yet interacted with the item (column). Let us say we have this data collected over a large number of users and items and we want to predict whether user-5 will like item-4 and whether user-6 will like item-1. In other words, should we recommend item-4 to user-5 and item-1 to user-6? How will we compute these "recommendation scores"?

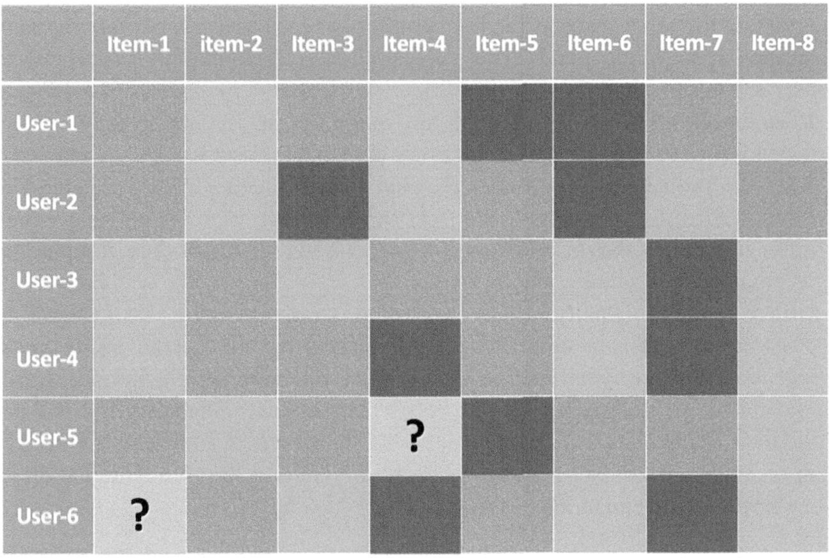

Fig. 16.19 An engagement matrix with six users and eight items. Red indicates that the user did not like the item and green indicates that the user liked the item. Gray indicates the user has not yet interacted with the item

The key components of a recommendation engine are as follows:

- **Interaction data**—captures the transactions/interaction between user and the items. Given the application, there could be many types of interactions such as search, browse, write a comment, like, and share. Capturing all possible interactions makes our applications, not just a tool to deliver value, but also a sensor to capture data from user's likes and dislikes.
- **Domain knowledge**—is the set of properties associated with items (e.g., actors, director, genre of a movie; service, ambience, food quality of a restaurant; or category and attributes of a product; teacher, speed, teaching style, other attributes of an educational video) and users (e.g., demographics).
- **Engagement**—quantifies the degree of affiliation between a user and an item that the user has interacted with. Capturing and quantifying a holistic engagement metric by itself is perhaps the most important art in building a great recommendation engine. We will discuss this in more detail in the next section, "Engagement between a user and an item."
- **Item profile**—there are two broad classes of recommendation engine algorithms. In *memory-based recommendation engines*, each item is treated as a unique ID. Here we only care about "which" items the user liked or not. However, the underlying idea is that every item has attributes and it is because of these attributes the user really liked or disliked the item (actor, music, direction, plot, genera, etc. of a movie). *Model-based* recommendation engines try to understand what "type" of items the user likes. In this case, the more detailed item features we can create, the more accurate model-based recommendation engines will become.
- **User profile**—in the model-based recommendation engines, we create a user profile in the same space as the item profile—so each user is a point in the same attribute space as the item is. These user profiles are a result of the cross between user engagement and the item profiles.
- **Recommendation score**—finally, the output of a recommendation engine depends on a score that predicts for every item that the user has not yet interacted with, the probability that the user would like to interact with the item. This score is then used to suggest the right items to each user. Various algorithms differ in whether they are memory or model based and how they compute the recommendation score.
- **Utility function**—finally, what is important is not necessarily the fact that the customer bought something that was suggested but the overall utility to the business. When a search Engine shows an ad, for example, it is not necessarily the most likely ad the user is going to click but if clicked, the advertiser will pay the most to the publisher. Hence, the utility function here is different. Similarly, in recommendation engines, we often put a layer of utility function that drives the final recommendations.

We will describe how these components come together to create a recommendation engine. We will define some notation that we will use going forward.

- There are N users: $\mathbf{U} = \{u_1, u_2, \ldots, u_N\}$
- There are M items: $\mathbf{I} = \{i_1, i_2 \ldots, i_M\}$
- Let $\mathbf{I}(u_n)$ be the set of items user u_n has interacted with.
  - So in the above example, $\mathbf{I}(u_1) = \{i_2, i_4, i_5, i_6, i_8\}$.
- Let $\mathbf{U}(i_m)$ be the set of users who have interacted with item i_m.
  - So, in the above example, $\mathbf{U}(i_3) = \{u_2, u_3, u_6\}$.

4.5.2 Engagement Between a User and an Item

One of the most creative parts of a recommendation engine is the definition of "engagement" between a user and an item that he/she has interacted with. Engagement can be measured in several ways: explicit vs. implicit feedback, one overall vs. multiaspect feedback.

Explicit Feedback—All intelligent systems improve with feedback—both explicit and implicit. In explicit feedback, user is explicitly asked for either a binary (thumbs up/thumbs down) or multilevel star rating at the end of the activity (e.g., after watching a movie, using a product, visiting a restaurant, or taking a cab ride). This could be even at a fine-grained level, that is, for each aspect of the product or service (e.g., acting, script, direction, music of a movie). Explicit feedback could be reliable if collected unambiguously in the right context (e.g., immediately after the activity). However, most of the time, explicit feedback is insufficient (not everyone gives it), fraught with problems of subjective user bias and contains both deliberate and natural noises. For example, a conservative user might always give ratings 1–3 out of 5 while a liberal user might always give ratings 3–5 for the same level of experience. This implies that a rating of 3 does not mean the same thing for the two types of users. Correction of such subjective bias is essential in explicit feedback-based recommendation systems.

Implicit Feedback—The best feedback is implicit—that is, gleaned from the natural activity of the user on the system (e.g., clicks in search engines, clicks on recommended YouTube videos, purchases on e-commerce sites, likes on Facebook and Twitter, shares on news stories, listening to a song or watching a movie, and acceptance of a connection suggested on LinkedIn or Facebook). The fact that user engaged with an item (a Web page, a news story, a YouTube video, a song, a movie, a product, a person) is itself an indication of his/her affiliation for that item. Implicit feedback is integrated into the user experience, it is abundant—available with every user activity, has no subjective bias, and is highly reliable. The real silent innovation in AI in the last decade was the art of collecting such implicit feedback and using it to continuously improve these services.

Refining the Implicit Feedback—Converting user actions captured in the logs to a dependable measure of engagement is an art that requires deep domain understanding and understanding of the nuances in the logs data. One example of this that we saw earlier was the subjective user bias in the way users rate an item.

We can remove that bias by a simple z-scoring of each user's past rating, that is, for each user, we can create a normalized rating based on his/her mean and standard deviation of all the ratings given in the past and use this normalized rating instead of the raw rating. Another example is in media consumption—songs, movies, videos. If a user clicks on a recommended item but does not finish it, only consumes a few seconds of it and then returns back—this also indicates lack of engagement. So, clicking is not engagement, finishing the experience up to a certain percentage is true engagement. Furthermore, repeated interaction of a user with an item indicates deeper engagement. Temporally, recent interactions should get a higher engagement score.

Combining Multiple Feedbacks—Finally, for the same item, the user might be giving more than one feedback. For example, in e-commerce, the user might be searching for product, spending time browsing the product, reading reviews on the product, adding it to wish list, removing from wish list, purchasing the product, returning the product, writing a review on the product, or responding to a review of the product. In media, the user might be again searching for a content, consuming the content partially or fully, downloading it, liking it, sharing it, etc. Combining all these various interactions both implicit and explicit—normalizing their scales, weighting them appropriately, etc., to come up with the final engagement score is again a fine art.

More formally, let there be K different types of engagements between a user and an item:

$$e(u_n, i_m) = \{e_1(u_n, i_m), e_2(u_n, i_m), \ldots, e_K(u_n, i_m)\}$$

These different engagements are combined systematically using a set of weights to get the final engagement score between a user and an item:

$$e(u_n, i_m) = \sum_{k=1}^{K} w_k e_k(u_n, i_m)$$

Further, these weights combining various engagements could either be chosen or learnt, they could either be global or depend on a user segment, etc. The simplest example of an engagement is that a user rated an item on a scale of say 1–5. In the rest of the chapter, we will assume that each user–item combination has an engagement score obtained by combining the various engagement scores as discussed above.

One of the most basic principles of generalization in machine learning is that *"similar inputs lead to similar outputs."* This principle is used in different ways in classification algorithms discussed above. In the recommendation engine, the same principle will be used slightly differently in the different recommendation engine paradigms presented below.

Collaborative Filtering (CF) Paradigm

The earliest recommendation engines are nonparametric **memory-based recommendation engines** that are heuristics based purely on the user–item engagement matrix. Properties of users or properties of items are not used in these. These are collaborative filtering recommendation engines that are of two types, **user–user** and **item–item** collaborative filtering. Examples of commonly used collaborative filtering methods such as user-based and item-based recommendation systems are given in the online appendix.

User–User Collaborative Filtering: The basic intuition behind a user–user CF is as follows. In the above matrix if we could compute user–user similarity, we will find that user-1 and user-2 are both similar to each other in the way they like and not-like the items that both of them have interacted with. User-5 is also similar to user-1 and user-2. On the same lines, user-3 and user-4 are similar to each other and user-6 is also similar to users 3 and 4. Now since user-5 is more similar to users-1 and 2, and users-1 and 2 both liked item-4, there is a good chance that user-5 will also like item-4. Similarly, since user-6 is more similar to users 3 and 4, and they both liked item-1, there is a high chance that user-6 will also like item-1. This intuition led to the birth of user–user collaborative filtering.

There are two parts to building such a collaborative filtering-based recommendation engine: (a) quantifying user–user similarity and (b) estimating recommendation score using this similarity.

Quantifying User–User Similarity: The only data we have about two users is how they have engaged with (rated) all the items that they have interacted with. If there are items that both users have interacted with and their interactions were "correlated," then the two users will be considered similar to each other. The two common measures of similarity are:

- **Cosine similarity**: L2-normalized dot product between *rows* in the engagement matrix:

$$Sim_{COSINE}(u, v) = \frac{\sum_{i \in \mathbf{I}(u) \cap \mathbf{I}(v)} e(u, i) \times e(v, i)}{\sqrt{\sum_{i \in \mathbf{I}(u)} e(u, i)^2} \sqrt{\sum_{i \in \mathbf{I}(v)} e(v, i)^2}}$$

- **Pearson's correlation**: User-bias removed, L2-normalized over common elements:

$$Sim_{PEARSON}(u, v) = \frac{\sum_{i \in I(u) \cap I(v)} (e(u, i) - e(u)) \times (e(v, i) - e(v))}{\sqrt{\sum_{i \in \mathbf{I}(u) \cap \mathbf{I}(v)} (e(u, i) - e(u))^2} \sqrt{\sum_{i \in \mathbf{I}(u) \cap \mathbf{I}(v)} (e(v, i) - e(v))^2}}$$

where $e(u)$ is the average engagement of user u, that is, $e(u) = \frac{1}{|\mathbf{I}(u)|} \sum_{i \in I(u)} e(u, i)$

Estimating Recommending Score: Once the user–user similarity is computed, we can use these to compute the rating of an item (e.g., item-4 above) for a user (e.g., user-5 above) given his/her similarity with all the other users (e.g., user-3 and user-4) and whether they liked or did not like the item. This can be written as a simple weighted sum as follows:

$$\widehat{e}(u, i) = e(u) + \frac{\sum_{v \in \mathbf{UNeb}(u,i)} |Sim(u, v)| \times \left(e\left(v, i\right) - e(v)\right)}{\sum_{v \in \mathbf{UNeb}(u,i)} |Sim(u, v)|}$$

where **UNeb**(u, i) is the set of *users* that are:

- Similar to the user u (i.e., have a nonzero similarity score)
- These users have an engagement score for item i

In the above example, Neb(user-5, item-4) = {item-1, item-2, item-4, item-6}. The neighbor list can be further pruned by choosing say just the top K most similar users. This measure also normalizes for user's bias by subtracting the average engagement score of the neighboring user $e(v)$ from the actual engagement $e(v, i)$. The estimated recommendation score, $\widehat{e}(u, i)$ is computed for all cells that are empty in the engagement matrix.

Item–Item Collaborative Filtering: Engagement matrix is a user–item matrix. We used it on a user-centric manner to learn user–user similarity, we can also use it in an item-centric manner to learn item–item similarity and then compute recommendation score. There is another reason why item–item similarity might be more practical in some cases. Typically, in a large retailer (e.g., Amazon, Flipkart) or media outlet (e.g., NetFlix, JioCinema), the number of users is O(100M) and computing user–user similarity becomes prohibitively expensive. The number of items, however, is typically O(100K) and it is easier to compute item–item similarity among them. Again, the same two-stage process is applied here:

Quantifying item–item similarity: The user–user similarity matrix was computed from rows of the engagement matrix. Similarly, the item–item similarity matrix can be computed from columns of the engagement matrix.

- **Cosine Similarity**: L2-normalized dot product between columns in the engagement matrix:

$$Sim_{COSINE}(i, j) = \frac{\sum_{u \in \mathbf{U}(i) \cap \mathbf{U}(j)} e(u, i) \times e(u, j)}{\sqrt{\sum_{u \in \mathbf{U}(i)} e(u, i)^2} \sqrt{\sum_{u \in \mathbf{U}(j)} e(u, j)^2}}$$

- **Pearson's Correlation**: Item-bias removed, L2-normalized over common elements:

$$Sim_{PEARSON}(i, j) = \frac{\sum_{u \in \mathbf{U}(i) \cap \mathbf{U}(j)} (e(u, i) - e(i)) \times (e(u, j) - e(j))}{\sqrt{\sum_{u \in \mathbf{U}(i) \cap \mathbf{U}(j)} (e(u, i) - e(i))^2} \sqrt{\sum_{u \in \mathbf{U}(i) \cap \mathbf{U}(j)} (e(v, j) - e(j))^2}}$$

where $e(i)$ is the average engagement of item i, that is, $e(i) = \frac{1}{|U(i)|}\sum_{u \in U(i)} e(u, i)$

Estimating Recommendation Score: We can use the item–item similarity to compute the rating of a user (e.g., user-5 above) for an item (e.g., item-4 above) given the item's similarity with all the other items and whether they were liked or not liked by this user. This can be written as a simple weighted sum as follows:

$$\widehat{e}(u, i) = e(i) + \frac{\sum_{j \in \mathbf{INeb}(u,i)} |Sim(i, j)| \times \left(e(u, j) - e(j)\right)}{\sum_{j \in \mathbf{INeb}(u,i)} |Sim(i, j)|}$$

where $\mathbf{INeb}(u, i)$ is the set of *items* that are:

- Similar to the item i (i.e., have a nonzero similarity score with i)
- These items have an engagement score with user u

The Cold Start Problem: Memory-based recommendation engines are *brute force heuristics* that gives high confidence recommendation scores for users who have enough interactions or items that have enough interactions. But whenever a new user (a more frequent scenario) or a new item (a less frequent scenario) is introduced in the system, the corresponding user's row or item's column in the engagement matrix is very sparse as he/she has had no interactions yet. This problem is typically addressed by recommending the most popular items to the new user in his/her context (e.g., city, country, demographic cohort). As the data on the user grows, we shift from this "default" model slowly to the collaborative filtering model. This transition from some default model for a data poor entity (e.g., new user or new item) to the target model (e.g., CF) as the data increases is another common theme across many practical implementations of ML algorithms.

Clustering Versions of CF: The CF approaches are computationally expensive and are not robust to "engagement noise." Another class of recommendation engines uses standard clustering algorithms to either cluster the users or items or both to first create a smoother, more robust representation of "similar customers" or "similar items" and use these as "representatives" to compute recommendation scores.

Matrix Factorization Approaches

In a way a CF recommendation engine score over the empty cells in the user–item matrix (where there is no interaction yet between a user–item pair) can be interpreted as a "smearing" or "smoothing" or "interpolation" of the cell given the corresponding row (user) and the column (item). The user–user CF was doing a row-centric interpolation and the item–item CF was doing the column-centric interpolation of the cell. What if we want to do both at the same time? A variety of matrix factorization approaches have been proposed in the literature. The most basic matrix factorization approach explored was to take the *singular value decomposition (SVD)* of the user–item engagement matrix. SVD is mathematically the best way to approximate the matrix, but the singular vectors (left and right) could have negative components that made them less interpretable.

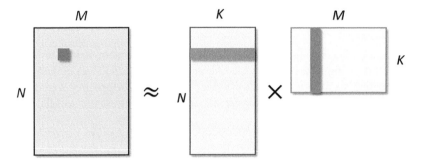

Fig. 16.20 Non-negative matrix factorization for approximating engagement matrix

This was further refined into a *non-negative matrix factorization* as shown in Fig. 16.20 where the original $N \times M$ engagement matrix (with N users and M items) is approximated as a product of two matrices: $\mathbf{E} \approx \mathbf{E}_K = \mathbf{A}_K \times \mathbf{B}_K$ where:

- $\mathbf{A}_K$ is an $N \times K$ matrix with one row corresponding to each user.
- $\mathbf{B}_K$ is a $K \times M$ matrix with one column corresponding to each item.
- Where $K \ll \min\{M, N\}$.

The goal is to find $\mathbf{A}_K$ and $\mathbf{B}_K$ iteratively starting with $\mathbf{A}_1$ and $\mathbf{B}_1$ (i.e., finding the first column of $\mathbf{A}$ and first row of $\mathbf{B}$ that, when multiplied, best approximate $\mathbf{E} = \mathbf{E}_0$. In general, by the time we reach iteration $k - 1$, $\mathbf{E}$ has already been approximated to $\mathbf{E}_{k-1} = \mathbf{A}_{k-1} \times \mathbf{B}_{k-1}$. The remaining error in the engagement matrix that is yet to be modelled, $\Delta \mathbf{E}_{k-1} = \mathbf{E} - \mathbf{E}_{k-1}$, is minimized by learning column $\mathbf{a}_k$ and row $\mathbf{b}_k$ as follows:

$$J(\mathbf{a}_k, \mathbf{b}_k | \Delta \mathbf{E}_{k-1}) = \sum_{(u,i) \in \mathbf{E}} (\Delta e_{k-1}(u, i) - a_k(u) b_k(i))^2 + \lambda_A \sum_u a_k(u)^2 + \lambda_B \sum_i b_k(i)^2$$

Here:

- $(u, i) \in \mathbf{E}$ implies that the summation is over cells where user u has engaged with item i.
- The first term minimizes error of approximation between residual error and parameters.
- The second and third are regularization terms that penalize for high values of parameters.
- Also note that in iteration 1, $\Delta \mathbf{E}_0 = \mathbf{E}$ itself.

Solving for the parameters we get:

$$\frac{\partial J(\mathbf{a}_k, \mathbf{b}_k | \Delta \mathbf{E}_{k-1})}{\partial a_k(v)} = 2 \sum_{i \in \mathbf{I}(v)} (a_k(v) b_k(i)) - \Delta e_{k-1}(v, i) b_k(i) + 2\lambda_A a_k(v) = 0$$

$$a_k^{(t+1)}(v) \leftarrow (1-\eta) a_k^{(t)}(v) + \eta \frac{\sum_{i \in I(u)} \Delta e_{k-1}(v,i) b_k^{(t)}(i)}{\lambda_A + \sum_{i \in I(v)} b_k^{(t)}(i)^2}$$

$$\frac{\partial J(a_k, b_k | \Delta E_{k-1})}{\partial b_k(j)} = 2 \sum_{u \in U(j)} (a_k(u) b_k(j)) - \Delta e_{k-1}(u,j) a_k(u) + 2\lambda_B b_k(j) = 0e$$

$$b_k^{(t+1)}(j) \leftarrow (1-\eta) b_k^{(t)}(j) + \eta \frac{\sum_{u \in U(j)} \Delta e_{k-1}(u,j) a_k^{(t)}(u)}{\lambda_B + \sum_{u \in U(j)} a_k^{(t)}(u)^2}$$

Here, η is the learning rate. Vectors $\mathbf{a}_k$ and $\mathbf{b}_k$ are initialized to small values and learnt via these alternate updates until convergence is achieved. Once we have reached the maximum dimensionality K, we can compute the "interpolated" or "smeared" score for the cells with no engagement scores:

$$\widehat{e}(u,i) = \sum_{k=1}^{K} a_k(u) \times b_k(i)$$

Again, this can be used in two ways:

- Find the most likely users who will highly engage with item i (and have not done so yet)
- Find the most likely items that user i will highly engage with (and has not done so yet)

The cold start problem—when a user (row in the engagement matrix) or a new item (column in the engagement matrix)—is added, the recommendations have to be done using averages.

In both, the memory-based CF (user–user or item–item) and the model-based CF (matrix factorization) we have treated each user and item as an *ID* in a dictionary without considering their properties themselves. This approach models only the "what" and does not model the "why" behind a user engaging or not engaging with an item. Machine learning is really the art of finding the *why* behind the *what* and the CF type approaches do not provide such insights. We next discuss another class of recommendation engines that address this problem.

4.5.3 Profile-Based Recommendation Engines

CF-based recommendation engines answer the point question: "while ***this*** user engaged with ***this*** item." This is fine when we do not know anything about a user and an item and they are just IDs in the two dictionaries. But if we know or infer enough about a user (e.g., demographics, behavior patterns) and if we know enough about

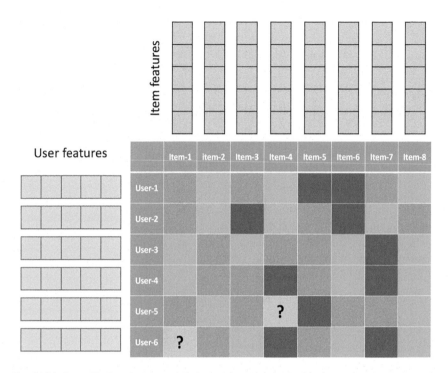

Fig. 16.21 In profile-based recommendation engines, we do not just know the past engagement between users and items, but we also know or infer additional user and item features that can be used to determine which type of users like which type of items

the items (e.g., meta-data), then we can answer the space question: "will *such* user engage with *such* an item." These are also known as profile-based recommendation engines and work with not only the engagement matrix but also the user and item properties beyond just the interaction among them as shown in Fig. 16.21.

There are four stages in building a profile-based recommendation engine.

1. **Characterize the features of users and items**: The meta-data (given or inferred) characterizes the "space" in which users and items live. For example:

    - **User features** include the basic demographics of the user—their age group, income group, gender, location, device, behavior patterns, preferences, etc.
    - **Item features** depend on the nature of the item, for example:
        - Movies—features are actor(s), actress(es), director(s), genre, plot, producer(s), etc.
        - Songs—features are singer(s), music director(s), album, genre, melody, length, etc.
        - News—features are entities, events, location, actions, sentiment, etc.
        - Videos—features are keywords, playlists, source, etc.
        - Tweets—features are hash-tags, keywords, source, sentiments, etc.
        - Clothes—features are brand, the fabric, the color, fashion type, fitting style, etc.

2. **Profile users (items) in item (user) space**: The intuition behind profile-based recommendation engines is as follows: We hypothesize that a user is engaging with an item because of *certain properties of that item*. If we consider all items that a user is heavily engaging with and find out what is common among them—then we start to build a *user profile* in terms of item properties. For example:

   - A movie customer likes movies with certain plots and a certain set of directors.
   - A song customer likes classic songs sung by particular artists.
   - A retail customer likes clothes with bright colors, of a certain fabric, from a certain brand.

   In other words, the explicit engagements (e.g., like, buy, consume, add to list, write comment), of a user with an item (i.e., values in the engagement matrix) can be used to build the *"implicit profile"* of the user (item) characterizing **what kind** *of items* (users) a user (an item) likes (is liked by) instead of **what** items (users) a user (item) likes (is liked by). More formally,

   - Let $\pi(u) = \{\pi_1(u), \pi_2(u), \ldots, \pi_L(u)\}$ be the L properties associated with user u
   - Let $v(i) = \{v_1(i), v_2(i), \ldots, v_K(i)\}$ be the K properties associated with item i

   For now, let us assume these properties are binary *indicator functions* (e.g., "is actor X in movie i"). Like in CF, we could take a user-centric approach (e.g., *user–user* similarity), an item-centric approach (e.g., *item–item* similarly), or joint user–item centric approach (e.g., matrix factorization), here too we can either take a user-centric, item-centric, or a joint approach.

   - **User profiling**: Given the engagement matrix **E** between users and items and item properties, we can build a user profile by aggregating profiles of all items that the user engaged with. This answers the question: What kind of items this user is engaging with?

   $$\phi_k(u) = \frac{\sum_{i \in \mathbf{I}(u)} e(u,i) \times v_k(i)}{\sum_{i \in \mathbf{I}(u)} e(u,i)}, \forall 1 \leq k \leq K$$

   - **Item profiling**: Again, given the engagement matrix **E** and user properties, we can build an item profile by aggregating profiles of all users that engaged with this item. This answers the question: What kind of users engage with this item?

   $$\theta_\ell(i) = \frac{\sum_{u \in \mathbf{U}(i)} e(u,i) \times \pi_\ell(u)}{\sum_{u \in \mathbf{U}(i)} e(u,i)}, \forall 1 \leq \ell \leq L$$

   These user profiles are in item-spaces and the item-profiles are in user-spaces. In a way, we can think of a user profile as a point in the item-space and all items are also points in the same space. Similarly, an item-profile is a point in the user-space and all users are also points in the user-space.

3. **Profile matching**: The next step is to define a distance or similarity between two points in the item- or user-spaces. We can do this, for example, as a cosine similarity between the two.

   - *Similarity between an item and a user profile*: Similarity between an item (a K-dimensional point, $v(i)$, in item-space) and a user profile (a K-dimensional point, $\phi(u)$, in item-space)

   $$Sim\left(v(i), \phi(u)\right) = \frac{\sum_k v_k(i) \times \phi_k(u)}{\sqrt{\sum_k v_k(i)^2}\sqrt{\sum_k \phi_k(u)^2}}$$

   - *Similarity between a user and an item-profile*: Similarity between a user (an L-dimensional point, $\pi(u)$, in user-space) and an item-profile (an L-dimensional point, $\theta(i)$, in user-space)

   $$Sim\left(\pi(u), \theta(i)\right) = \frac{\sum_\ell \pi_\ell(u) \times \theta_\ell(i)}{\sqrt{\sum_k \pi_\ell(u)^2}\sqrt{\sum_\ell \theta_\ell(i)^2}}$$

4. **The final recommendation score**: The final recommendation score between a user u and item i can be computed as a combination of the two similarity scores:

   $$\widehat{e}(u, i) = Sim(\pi(u), \theta(i))^\alpha \, Sim(v(i), \phi(u))^{1-\alpha}$$

There are two key advantages of using profile-based recommendation engines:

- **Better generalization**—Being a model-based approach, these profile-based recommendation engines tend to generalize better as data increases. The key is to define the Profiling and Similarity functions better.
- **Cold start problem**—If we want to recommend a new item to a user who has a profile, all we have to do is compare the properties of the item with the profile of the user. So for example, if a movie is an action movie with a certain actor and we know that the user likes "such" movies (this genre and this actor), then we can recommend it to this user even if this movie has never been seen before.

Advanced Topics in Recommendation Engines

Recommending content, product, and services is a fine art done with different degrees of knowledge about a customer and with different sets of end-goals in mind. Below we highlight some of the other aspects that are typically considered beyond the recommendation score in making real-world recommendations.

1. *Deep content-based recommendations*: Profile-based recommendation engines put an item as a point in a feature space. Typically, this feature space is composed of the meta-data associated with the item as discussed above. In addition to this meta-data we can also extract deeper features from the items themselves that can be augmented with their profiles. This is possible only for a certain kind of

items that are rich in content. For example, if we are recommending a product on an e-commerce portal, we might have access to a number of additional product features but beyond that we do not have much to go on to accentuate the product's profile. But in content-rich items such as songs, videos, movies, news articles, teaching content, etc. we can extract deeper features from the content itself, add them to the meta-data features and then build a more holistic profile of the items to improve their representation. For example:

- *Song recommendation* can be enhanced by learning its melody and style by extracting say frequency characteristics of the songs, the instruments playing, etc. Why a user likes a song is not just because of its meta-data but because of its content.
- *Movie recommendation* can be enhanced by extracting activity features, for example, is there a car chase or an action sequence or a court scene or a cultural scene in the movie. What is the storytelling style or background music or nature of language used, etc.
- *News recommendation* can be enhanced by extracting entities, events, issues, topics, and sentiments about them in the news article and not just by representing it as a bag-of-words. The reader is interested in the real-world stories that the news represents, not just the words.
- *Teaching content recommendations* can be enhanced by identifying the different parts of a teaching content—real-world example, definition, detail, humor, motivation for the topic being taught—and how they are ordered. This will help extract the "teaching style" of the teaching content that must be matched with the "learning style" of the student.

Ultimately, algorithms in machine learning can take us only so far. It is the features that we can extract about our items that makes these algorithms come alive.

2. Strategic recommendations: Often recommendations are done on a trigger and for a purpose. The triggers could be entering the store (imagine a face recognition system recognizes customer) or logging into the online portal, reviewing a product (people who bought this also bought that), or making a final purchase of all items in the basket (printing coupons on the back of the receipt) These recommendations could serve very different purposes, which might be both tactical or strategic. The choice of the "utility function" to apply for a particular recommendation instance depends on the stage and context of the customer. For example:

- *Recommendation for Loyalty*—Often if the goal is just to maximize the loyalty of a new customer, the business might just recommend commonly or repeatedly bought products (e.g., groceries or clothes) at a discounted price to a certain customer.

- *Recommendation for Cross-Sell*—Here the recommendations could increase the market basket by suggesting related products that the user might need along with the main product he/she is purchasing, for example, selling additional cartridges with a printer, additional storage with a camera, or a sun screen with beach wear. This is done through deeper understanding of "what products go together well" and not just item–item correlations as done in CF. This is applicable mostly in product domains where product relationships matter, more than in media domains where the only relationship between content is their similarity and not dependence (unless there are sequels).
- *Recommendation for Upsell*—In order to increase the value of a loyal customer we might recommend products in the same category (e.g., TV, fridge, cell phones) that are of a higher value to the seller than the one that the customer is currently looking for. Giving a higher resolution phone or a bigger TV or a product variant with more features is another kind of recommendation that is typically done at the time of purchase itself.
- *Recommendation for Lifetime Value*—Often, businesses strive to foster a long-term, value-oriented strategic relationships with customers (retailers, banks, cab-services, etc.). Here recommendations are chosen carefully not just for the value the next recommendation will add but how that will help open the gates for the next set of recommendations later. For example, a bank might give a lower rate on the car loan to a customer to recommend a home loan later. A bank might give very good interest rates to a customer on his/her salary account to maximize his/her lifetime association.
- *Recommendation for Preventing Churn*—Finally, some recommendations or offers might have a corrective intonation. For example, in a cab hailing service, if a customer is known to have had a bad experience (driver cancelled a ride or customer gave a poor feedback), then an immediate offer that brings the customer back on the positive side of the business might be useful—again this needs to be relevant to the customer as well.

3. Blended recommendations: The traditional notion of recommendation score is based on a user's past engagement but when a final recommendation page is rendered (e.g., personalized home screen of a user or recommendations of YouTube videos), other biases are also introduced in the final recommendation scores:

- *Engagement bias*: This is the traditional recommendation scores based on past (positive or negative) engagements of the user with other items. This is done using some of the algorithms discussed above.
- *Preference bias*: Often when a user is boarded on, his/her preference for coarse categories (e.g., sports, science, entertainment in news portals or artist, genre in the song) are recorded and if a new item falls into a customer's preference bucket, it is shown.

- *Location bias*: Typically, our apps know our location and if there are events (e.g., news or shows or sale) that are relevant to a user's location and are also relevant with respect to past engagements, then it might be shown to the user.
- *Popularity bias*: If something is becoming suddenly popular because it is either important or trending and even if it is only slightly relevant to the user it might still be shown to him/her. For example, a drastic news event (e.g., a terrorist attack or natural disaster or a major business announcement, best-selling book, a hit movie, or a viral video). This is typically seen in verticals where users give explicit feedback—likes, shares, buys, etc. indicating popularity.
- *Social bias*: Finally, in a social network setting, items that are explicitly popular in one's neighborhood in the social graph might also surface in a customer's recommendations as *birds of a feather flock together.* One might like what his/her friends on the network like.

What a user finally sees might be a combination of all these aspects together giving a ranking of what the user might see. The actual feedback by the user might be used to learn the weights of which biases among the above are more important to the user than others.

4. Cross-domain recommendations: Earlier, each service was focused only on one aspect of the user. For example, banks know only about a user's financial view, retailers know only about their purchase behavior, and cab hailing services know only about a user's travel behaviors within the city and its neighborhood, while airline services know only about the user's air travel. They all have a siloed view of a customer and can only suggest recommendations that are best suited accordingly. The next-generation businesses might provide different types of services to the same customer (e.g., Amazon has both a retail business and a media business) or might have different views on the same customer via different channels (e.g., firms such as Paytm or banks understand from the customer's payment behavior what kind of cross-vertical engagements the customer is having). The public profile of the customer—their Facebook, Twitter, and LinkedIn profiles—can also provide additional insights about a customer. Soon, recommendation engines will be able to combine all these views and suggest the right products and services with a more holistic view of the customer. For example,

- Knowing that customer is booking a flight to a beach resort city (e.g., Florida or Goa) during the summer, one might recommend the right clothes and beach products for the customer.
- Knowing that a customer just bought sports shoes might lead to recommendation of "sporty music" to a customer.
- Knowing that a customer just took a home loan, a bank might recommend home furnishing products from a partner retailer to him/her.

5. Workflow-based recommendations: Finally, recommendation to a user could also be based on a well-defined workflow based on a "prerequisite graph" of items. For example:

   - *Next Concept Recommendation in Personalized Education*: When a student has engaged with the past concepts and mastered them to different degrees, a curriculum personalization engine can decide what next concept the student is ready to learn next. As shown in Fig. 16.22 below, the system might recommend some concepts that the student needs to master well before he/she can move forward or some concepts that he/she is ready to learn next because he/she has already mastered all of their prerequisites well.
   - *Next Action Recommendation in Agriculture*: Decades of agricultural experience and knowledge that mankind has accumulated can be used to recommend the right action to farmers at the right time in the right region depending on the climate, soil, and pest infestations prevalent in the region, etc. When to prepare the soil and how, when to plant the seed, when to put fertilizers and pesticides, when to water the plants, when to worry about the rains, and when to harvest the crop. A personalized workflow-based recommendation engine with flexibility to adapt to changing conditions on the ground can be made available to all farmers to alleviate them from guesswork and ignorance.

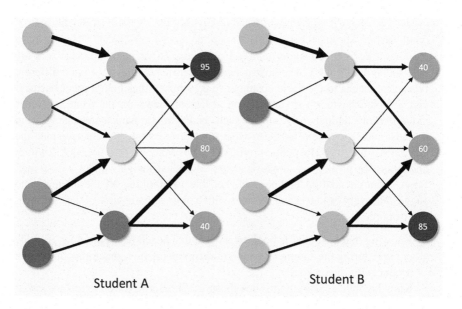

Fig. 16.22 Students A and B have mastered different set of concepts well (green), not so well (orange), and not at all (red). Depending on their mastery levels and the prerequisite or concept dependence graph, the workflow-based recommendation engine might suggest a different set of concepts to learn for student A vs. student B

- *Next Best Career Move*: Career building is a strategic art that requires right choices at the right time. Today, most people make some of the most important career decisions with limited understanding or ad hoc criteria. Each potential move in the career might have different prerequisites (e.g., an MBA school admission might require a minimum of say 3 years of job experience, a job might require one to have a certain set of hard or soft skills, or a profession, e.g., researcher, doctor, professor, might require one to have an advanced degree). A career moves recommendation engine, aware of prerequisite constraints, a user's personality, and aspirations might recommend the best next move—whether it is taking a certain MOOC course, pursuing a degree from a certain college, an internship experience in a certain company, or a volunteer work in a certain organization.

6. Contextual Recommendations: So far, we talked about which item the customer is most likely to engage with, but the success of that engagement might depend not just on the accuracy of the recommendation score but also on the context in which the recommendation is made. For example,

- Recommending one's potential cuisine just before meal times
- Recommending movies just before weekend or holiday starts
- Recommending back-to-school items toward the close of summer vacations
- Recommending a tourist spot when one has just landed in a new city for vacation
- Recommending cartridge exactly 2 months after a customer bought a printer
- Recommending different songs in the morning than evening than weekends

The timing, the triggers, the location, the device, and the channel—are other aspects to be considered when making a relevant recommendation to the user.

Overall, recommendation engines play a very important role for many types of user interactions with the business, for discovering new items, for keeping the user engaged and informed, and helping the users make better choices. These recommendation engines become better with deeper understanding of the user—both where they have been and where they are heading. To read more on recommendation engines, you can refer to Singhal et al. (2017), Li et al. (2011), or Chap. 13 of "Data Mining: Concepts and Techniques" by Han et al. (2011).

5 Conclusion

In this chapter, we have explored the philosophy of generalization, the process of building a classifier, and the theoretical aspects of a wide variety of classifiers. We have explored trade-offs between accuracy and interpretability, hard vs. soft classifiers, descriptive vs. discriminative approaches, and feature-centric vs. model-centric thinking. The real art in building the right classifier comes from understanding the features, the nature of the decision boundary, and picking the right modelling algorithm to match the data complexity to the model complexity. Every decision today has a potential to be driven by data. The real challenge is to find the right insights, engineer the right features, build the right models, and apply the right business optimization to convert model predictions into decisions. Doing all this right will improve our ability to make more accurate, personalized, and real-time decisions, improving our businesses and processes multifold.

Electronic Supplementary Material

All the datasets, code, and other material referred in this section are available in www.allaboutanalytics.net.

- Data 16.1: Decision_Tree_Ex.csv
- Code 16.1: Decision_Tree_Ex.R

More examples, corresponding code, and exercises for the chapter are given in the online appendices to the chapter.

References

Carbonell, J. G., Michalski, R. S., & Mitchell, T. M. (1983). An overview of machine learning. In R. S. Michalski, J. G. Carbonell, & T. M. Mitchell (Eds.), *Machine learning volume 1. Symbolic computation* (pp. 3–23). Berlin: Springer Science & Business Media.

Friedman, J., Hastie, T., & Tibshirani, R. (2001). *The elements of statistical learning (Vol. 1, No. 10) Springer series in statistics*. New York, NY: Springer.

Han, J., Pei, J., & Kamber, M. (2011). *Data mining: Concepts and techniques*. Amsterdam: Elsevier.

Li, L., Chu, W., Langford, J., & Wang, X. (2011). Unbiased offline evaluation of contextual-bandit-based news article recommendation algorithms. In WSDM'11 (Ed.), *Proceedings of the Fourth ACM International Conference on Web Search and Data Mining* (pp. 297–306). New York City, NY: ACM.

Murphy, K. (2012). *Machine learning – A probabilistic perspective*. Cambridge, MA: The MIT Press.

Singhal, A., Sinha, P., & Pant, R. (2017). Use of deep learning in modern recommendation system: A summary of recent works. *International Journal of Computers and Applications, 180*(7), 17–22.

Chapter 17
Deep Learning

Manish Gupta

1 Introduction

Deep learning has caught a great deal of momentum in the last few years. Research in the field of deep learning is progressing very fast. Deep learning is a rapidly growing area of machine learning. Machine learning (ML) has seen numerous successes, but applying traditional ML algorithms today often means spending a long time hand-engineering the domain-specific input feature representation. This is true for many problems in vision, audio, natural language processing (NLP), robotics, and other areas. To address this, researchers have developed deep learning algorithms that automatically learn a good high-level abstract representation for the input. These algorithms are today enabling many groups to achieve groundbreaking results in vision recognition, speech recognition, language processing, robotics, and other areas.

The objective of the chapter is to enable the readers:

- Understand what is deep learning
- Understand various popular deep learning architectures, and know when to use which architecture for solving their business problem
- Know how to perform image analysis using deep learning
- Know how to perform text analysis using deep learning

Electronic supplementary material The online version of this chapter (https://doi.org/10.1007/978-3-319-68837-4_17) contains supplementary material, which is available to authorized users.

M. Gupta (✉)
Microsoft Corporation, Hyderabad, India
e-mail: manishg.iitb@gmail.com

© Springer Nature Switzerland AG 2019
B. Pochiraju, S. Seshadri (eds.), *Essentials of Business Analytics*, International Series in Operations Research & Management Science 264,
https://doi.org/10.1007/978-3-319-68837-4_17

Introduction to Deep Learning

Wikipedia defines deep learning as follows. "Deep learning (deep machine learning, or deep structured learning, or hierarchical learning, or sometimes DL) is a branch of machine learning based on a set of algorithms that attempt to model high-level abstractions in data by using model architectures, with complex structures or otherwise, composed of multiple non-linear transformations." The concept of deep learning started becoming very popular around 2012. This was mainly due to at least two "wins" credited to deep learning architectures. In 2012, Microsoft's top scientist *Rick Rashid* demonstrated a voice recognition program that translated Rick's English voice into *Mandarin Chinese* in Tianjin, China.[1] The high accuracy of the program was supported by deep learning techniques. Similarly, in 2012, a deep learning architecture won the ImageNet challenge for the image captioning task.[2]

Now deep learning has been embraced by companies in a large number of domains. After the 2012 success in speech recognition and translation, there has been across the board deployment of deep neural networks (DNNs) in the speech industry. All the top companies in machine learning including Microsoft, Google, and Facebook have been making huge investments in this area in the past few years. Popular systems like IBM *Watson* have also been given a deep learning upgrade. Deep learning is practically everywhere now. It is being used for image classification, speech recognition, language translation, language processing, sentiment analysis, recommendation systems, etc. In medicine and biology, it is being used for cancer cell detection, diabetic grading, drug discovery, etc. In the media and entertainment domain, it is being used for video captioning, video search, real-time translation, etc. In the security and defense domain, it is being used for face detection, video surveillance, satellite imagery, etc. For autonomous machines, deep learning is being used for pedestrian detection, lane tracking, recognizing traffic signs, etc. This is just to name a few use cases. The field is growing very rapidly—not just in terms of new applications for existing deep learning architectures but also in terms of new architectures.

In this chapter, we primarily focus on three deep supervised learning architectures: multilayered perceptrons (MLPs), convolutional neural networks (CNNs), and recurrent neural networks (RNNs). This chapter is organized as follows. In Sect. 2, we discuss the biological inspiration for the artificial neural networks (ANN), the artificial neuron model, the perceptron algorithm to learn the artificial neuron, the MLP architecture and the backpropagation algorithm to learn the MLPs. MLPs are generic ANN models. In Sect. 3, we discuss convolutional neural networks which are an architecture specially designed to

---

[1] http://deeplearning.net/2012/12/13/microsofts-richard-rashid-demos-deep-learning-for-speech-recognition-in-china/ (accessed on Jan 16, 2018).

[2] https://papers.nips.cc/paper/4824-imagenet-classification-with-deep-convolutional-neural-networks.pdf (accessed on Jan 16, 2018).

learn from image data. Finally, in Sect. 4, we discuss the recurrent neural networks architecture which is meant for sequence learning tasks (mainly text and speech).

2 Artificial Neural Network (ANN) and Multilayered Perceptron (MLP)

2.1 Biological Inspiration and the Artificial Neuron Model

Deep learning is an extension of research in the area of artificial neural networks (ANNs) as discussed in Chap. 16 on supervised learning. In this section, we elaborate on training a simple neuron using the perceptron algorithm.

Training an artificial neuron involves using a set of labeled examples to estimate the values of the weights w_i (a vector of the same size as the number of features). Rosenblatt (1962) proposed the perceptron algorithm to train the weights of an artificial neuron. It is an iterative algorithm to learn the weight vector. The basic idea is to start with a random weight vector and to update the weights in proportion to the error contributed by the inputs. Algorithm 17.1 presents the pseudo-code for the perceptron algorithm.

Algorithm 17.1: The Perceptron Algorithm
1. Randomly initialize weight vector w_0
2. Repeat until error is less than a threshold γ or max_iterations M:

   (a) For each training example (x_i, t_i):
   - Predict output y_i using current network weights w_n
   - Update weight vector as follows: $w_{n+1} = w_n + \eta \times (t_i - y_i) \times x_i$

Note that here η is called as the learning rate, t_i is the true label for the instance x_i, and y_i is the predicted class label for the instance x_i. Thus, $t_i - y_i$ is the error made by the neuron with the current weight vector on the instance x_i. Note that a neuron also takes a bias term b as part of the weights to be learned. The bias term is often folded in into the weight vector w by assuming a dummy input and setting it to 1. In that case, the size of the weight vector is number of features + 1. The correction (or the update) of the weights using the perceptron algorithm is equivalent to translation and rotation of the separating hyper-plane for a binary classification problem.

Minsky and Papert (1969) proved that a single artificial neuron is no better than a linear classifier. To be able to learn nonlinear patterns, one can progress in two ways: change the integration function or consider MLP. One can change the integration function from a simple linear weighted summation to a quadratic function $\left(f = \sum_{j=1}^{m} w_j x_j^2 - b\right)$ or a spherical function $\left(f = \sum_{j=1}^{m} (x_j - w_j)^2 - b\right)$. We will discuss MLP next.

2.2 Multi-layered Perceptrons (MLPs)

Figure 17.1 shows a typical multilayered perceptron architecture. Interestingly such a multilayered perceptron can learn very complex boundaries much more beyond linear boundaries. In fact, it can also learn nonlinear boundaries. It has an input layer, an output layer, and one or more hidden layers. The number of units (neurons) in the input layer corresponds to the dimensionality of the input data. The number of units in the output layer corresponds to the number of unique classes. If there are a large number of hidden layers, the architecture is called a deep learning architecture. The "deep" in "deep learning" refers to the depth of the network due to multiple hidden layers.

In an MLP, each edge corresponds to a weight parameter to be learned. Note that each neuron in a layer k produces an input for every other neuron in the next layer $k + 1$. Thus, this is a case of dense connectivity. Learning the MLP means learning each of these weights. Note that a perceptron cannot be directly used to learn weights for an MLP because there is no supervision available for the output of the internal neurons (neurons in the hidden layers). Thus, we need a new algorithm for training an MLP.

Given a particular fixed weight vector for each edge in the MLP, one can compute the predicted value y_i for any data point x_i. Thus, given a training dataset, one can plot an error surface where each point on the error surface corresponds to a weight configuration.

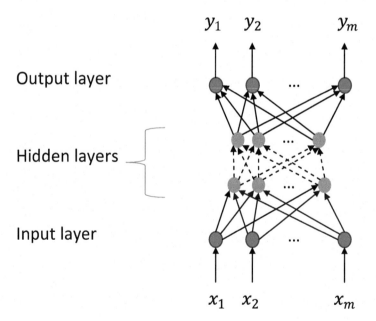

Fig. 17.1 Multilayered perceptron (MLP)

To learn a good weight vector, we present a gradient descent-based algorithm which starts with a random point on this error surface and over multiple iterations moves down the error surface in the hope of finding the deepest valley on this surface. This means that we start with a randomly initialized weight vector and update the weight vector so as to always move in the direction of the negative gradient. Gradient descent algorithms suffer from local minima issues. This means that the valley that we end up at after the gradient descent converges may not be the deepest valley globally. However, just like other algorithms with local optima issues, the problem can be solved by doing multiple runs of gradient descent each with differently initialized weight vectors and then choosing the one with the smallest error. Formally the algorithm is called as back propagation algorithm which works as follows.

Algorithm 17.2: Back Propagation Algorithm
1. Initialize network weights (often small random values).
2. Repeat until error is less than a threshold γ or max_iterations M:

    (a) For each training example (x_i, t_i):

    - Predict output y_i using current network weights w_n (forward pass).
    - Compute error at the output unit: $error = t_i - y_i$.
    - Propagate back error from output units to all the hidden units right until the input layer (backward error propagation step).
    - Update network weights using the gradient descent update equation:
        new weight = old weight $- \eta \times$ gradient of the error with respect to the weight.

The error is backpropagated from a neuron n_2 in layer k to a neuron n_1 in layer $k - 1$ in the ratio of the weight w_{12} on the edge between n_1 and n_2 to the weight on all the inputs to the neuron n_2.

There are multiple variants of the backpropagation algorithm. If the weight update is done after every instance, it is called *stochastic gradient descent*. Often times, batch-wise updates lead to quick convergence of the algorithm, where weights are updated after looking at a batch of instances. In such a case, the algorithm is called as *batch-wise gradient descent*.

The weights can be updated using a constant learning rate. However, if the learning rate is too small, it leads to slow convergence. If the learning rate is too large, it can lead to divergence rather than convergence. Hence, setting the learning rate is tricky. This has led to the development of various update methods (e.g., momentum, averaging, AdaGrad (Duchi et al. 2011), RMSProp (Hinton et al. 2012), Adam (Kingma and Ba 2014), and AdaDelta (Zeiler 2012)). Interested readers can read more about some of these update schedules in the paper by Rumelhart et al. (1986).

2.3 Practical Advice When Using ANNs and an Overview of Deep Learning Architectures

The fundamental difference between ANNs and other traditional classifiers is the following. For building traditional classifiers, a data scientist first needs to perform domain-specific feature engineering and then build models on top of featurized data. This needs domain knowledge, and a large amount of time is spent in coming up with innovative features that could help predict the class variable. In case of ANNs, the data scientist simply supplies the raw data to the ANN classifier. The hope is that the ANN can itself learn both the representation (features) and the weights too. This is very useful in hard-to-featurize domains like vision and speech. Multiple layers of a deep ANN capture different levels of data abstraction.

There are multiple hyper-parameters one has to tune for various deep learning architectures. The best way to tune them is by using validation data. But here are a few tips in using MLPs. The initial values for the weights of a hidden layer i could be uniformly sampled from a symmetric interval that depends on the activation function. For the *tanh activation* function, the interval could be $\left[-\sqrt{\left(\frac{6}{fan_{in}+fan_{out}}\right)}, \sqrt{\left(\frac{6}{fan_{in}+fan_{out}}\right)} \right]$ where fan_{in} is the number of units in the $(i-1)$-th layer and fan_{out} is the number of units in the i-th layer. For the *Sigmoid* function, the suggested interval is $\left[-4\sqrt{\left(\frac{6}{fan_{in}+fan_{out}}\right)}, 4\sqrt{\left(\frac{6}{fan_{in}+fan_{out}}\right)} \right]$. This initialization ensures that, early in training, each neuron operates in a regime of its activation function where information can easily be propagated both upward (activations flowing from inputs to outputs) and backward (gradients flowing from outputs to inputs).

How many hidden layers should one have? How many hidden units per layer? There is no right answer to this. One should start with one input, one hidden, and one output layers. Theoretically this can represent any function. Add additional layers only if the above does not work well. If we train for too long, possible overfitting can happen—the test/validation error increases. Hence, while training, use validation error to check for overfitting. Simpler models are better—try them first (Occam's razor).

Overview of Deep Learning Architectures

A large number of deep learning architectures have been proposed in the past few years. We will discuss just a few of these in this chapter. We mention a partial list of them below for the sake of completeness.

1. Deep supervised learning architectures: classification—multilayered perceptron (MLP); similarity/distance measure—DSSM, convolutional NN; sequence-to-sequence—recurrent neural net (RNN)/long short-term memory (LSTM); question answering and recommendation dialog—memory network (MemNN); reasoning in vector space—tensor product representation (TPR).

Table 17.1 Few popular libraries to build deep learning models

	Caffe	Torch	Theano	Tensorflow	CNTK
Language	C++	Lua	Python	Python	C++
Multi-GPU	Yes	Yes	Ok	Yes	Yes
Readability	Yes	Yes	Very poor	Very poor	Yes
Complex models	No	Yes	Yes	Yes	Yes
Visualization	No	Ok	No	Yes	Yes
Training	Windows/Linux	Linux only	Windows/Linux	Windows/Linux	Windows/Linux

2. Deep unsupervised learning: pre-training—denoising auto-encoder (DA) and stacked DA; energy-based models—restricted Boltzmann machines (RBM) and deep belief networks (DBN).
3. Deep reinforcement learning: an agent to play games, Deep Q-Network (DQN).

Training deep learning models is usually a compute-intensive task. Deep learning models work well when you have large amounts of data to train them. Hence, most people use graphics processing units (GPUs) to train good models. There are a few popular libraries to easily build deep learning models. Table 17.1 presents a comparison of these libraries.

2.4 Summary

ANN is a computational model inspired from the workings of the human brain. Although a perceptron can simply represent linear functions, multiple layers of perceptrons can represent arbitrary complex functions. The backpropagation algorithm can be used to learn the parameters in a multilayered feed-forward neural network. The various parameters of a feed-forward ANN such as learning rate, number of hidden layers, and initial weight vectors need to be carefully chosen. An ANN allows for learning of deep feature representations from raw training data.

2.5 An Example: MNIST Data

The following section explains how to build a simple MLP using the "mxnet" package in R for the MNIST handwritten digit recognition task. The MNIST data comprises of handwritten digits (60,000 in training dataset and 10,000 in test dataset) produced by different writers. The sample is represented by a 28 × 28 pixel map with each pixel having value between 0 and 255, both inclusive. You may refer to the MNIST data website[3] for more details. Here, we provide a sample of only 5000 digits (500 per digit) in the training sample and 1000 digits (100 per digit) in the test dataset. The task is to recognize the digit.

The main stages of the code below are as follows:

1. Download and perform data cleaning.
2. Visualize the few sample digits.
3. Specify the model.

    (a) Fully connected
    (b) Number of hidden layers (neurons)
    (c) Activation function type

4. Define the parameters and run the model.

    (a) "softmax" to normalize the output
    (b) X: Pixel data (X values)
    (c) Y: Dependent variable (Y values)
    (d) ctx: Processing device to be used

5. Predict the model output on test data.
6. Produce the classification (confusion) matrix and calculate accuracy.

Sample code "MLP on MNIST.R" and datasets "MNIST_train_sample.csv" and "MNIST_test_sample.csv" are available on the website.

3 Convolutional Neural Networks (CNNs)

In this section, we discuss a deep learning architecture called as convolutional neural networks. This architecture is mainly applied to image data. However, there have also been some use-cases where CNNs have been applied to embedding matrices for text data. In such cases, a text sequence is mapped onto a matrix where each word in the sequence is represented as a row using the word embedding for the word. Further, such an embedding matrix is treated very similar to an image matrix. We will first talk about ImageNet and various visual recognition problems. After that, we will discuss the technical details of a CNN.

3.1 *ImageNet and Visual Recognition Problems*

ImageNet[4] is an image dataset organized according to the WordNet (Miller 1995) hierarchy. Each meaningful concept in WordNet, possibly described by multiple

---

[3]http://yann.lecun.com/exdb/mnist/ (accessed on Jan 16, 2018).
[4]Imagenet dataset is hosted on http://image-net.org/ (accessed on Aug 1, 2018).

Fig. 17.2 A sample image

words or word phrases, is called a "synonym set" or "synset." There are more than 100,000 synsets in WordNet, majority of them are nouns (80,000+). The ImageNet project is inspired by a growing sentiment in the image and vision research field—the need for more data. There are around 14,197,122 images labeled with 21,841 categories.

This dataset is used for the ImageNet Large Scale Visual Recognition Challenge held every year since 2010. The challenge runs for a variety of tasks including image classification/captioning, object localization, object detection, object detection from videos, scene classification, and scene parsing. The most popular task is image captioning.

The image classification task is as follows. For each image, competing algorithms produce a list of at most five object categories in the descending order of confidence. The quality of a labeling is evaluated based on the label that best matches the ground truth label for the image. The idea is to allow an algorithm to identify multiple objects in an image and not be penalized if one of the objects identified was in fact present but not included in the ground truth (labeled values). For example, for the image in Fig. 17.2, "red pillow" is a good label, but "flying kite" is a bad label. Also, "sofa" is a reasonable label, although it may not be present in the hand-curated ground truth label set.

Table 17.2 shows the winners for the past few years for this task. Notice that in 2010, the architecture was a typical feature engineering-based model. But since 2012 all the winning models have been deep learning-based models. The depth of these models has been increasing significantly as the error has been decreasing over time.

CNNs have been used to solve various kinds of vision-related problems including the image classification challenge. Such tasks include object detection, action classification, image captioning, pose estimation, image retrieval, image segmentation for self-driving cars, traffic sign detection, face recognition, video classification, whale recognition from ocean satellite images, and building maps automatically from satellite images.

Table 17.2 ImageNet challenge winning architectures (compiled by author)

Year	2010	2012	2014	2015	2016	2017
Model, Institution	Linear SVM, NEC-UIUC	AlexNet, SuperVision	Visual Geometry Group (VGG) Oxford, Googlenet	Resnet, MSRA	Trimps-Soushen, The Third Research Institute of the Ministry of Public Security, P.R. China	Squeeze-and-Excitation Networks, NUS-Qihoo_DPNs (CLS-LOC)
#layers	Not a neural network	7 layers	27 layers	152 layers	Ensemble of Inception-v3 (48 layers), Inception-v4 (~114 layers), Residual Network (152 layers), Inception-ResNet-v2 (200+ layers), Wide Residual Network (~16 layers)	Integrated SE blocks to stacked ResNet-152
Accuracy	28%	16%	7%	3.6%	2.99%	2.25%

3.2 Biological Inspiration for CNNs

Hubel and Wiesel (1962) made the following observations about the visual cortex system. Nearby cells in the cortex represented nearby regions in the visual field. Visual cortex contains a complex arrangement of cells. These cells are sensitive to small subregions of the visual field, called a receptive field. The subregions are tiled to cover the entire visual field and may overlap. These cells act as local filters over the input space and are well suited to exploit the strong spatially local correlation present in natural images. Additionally, two basic cell types have been identified. Simple cells respond maximally to specific edge-like patterns within their receptive field. Complex cells have larger receptive fields and are locally invariant to the exact position of the pattern.

The question is how to encode these biological observations into typical MLPs. Fukushima and Miyake (1982) proposed the neocognitron, which is a hierarchical, multilayered artificial neural network, and can be considered as the first CNN in some sense.

Besides the visual cortex system, in general, we tend to think in terms of hierarchy, for example, the vision hierarchy (pixels, edges, textons, motifs, parts, objects), the speech hierarchy (samples, spectral bands, formants, motifs, phones, words), and the text hierarchy (character, word, phrases, clauses, sentences, paragraphs, story). To encode this hierarchical behavior into a neural framework, we will study CNNs in this section.

Why cannot we rely on MLPs for image classification? Consider a simple task where you want to learn a classifier to detect images with dogs versus those without. In the popular CIFAR-10 image dataset, images are of size $32 \times 32 \times 3$ (32 wide, 32 high, 3 color channels) only, so a single fully connected neuron in a first hidden layer of a regular neural network would have $32 \times 32 \times 3 = 3072$ weights. A 200×200 image, however, would lead to neurons that have $200 \times 200 \times 3 = 120,000$ weights. Such network architecture does not take into account the spatial structure of data, treating input pixels which are far apart and close together on exactly the same footing. Clearly, the full connectivity of neurons is wasteful in the framework of image recognition, and the huge number of parameters quickly leads to overfitting. This motivates us to build specific architecture to deal with images, as discussed below.

3.3 Technical Details of a CNN

Figure 17.3 shows four kinds of layers that a typical CNN has: the convolution (CONV) layer, the rectified linear units (RELU) layer, the pooling (POOL) layers, and the fully connected (FC) layers. FC layers are the ones that we have seen so far in MLPs. In this section, we will discuss the other three layers (CONV, RELU, and POOL) in detail one by one.

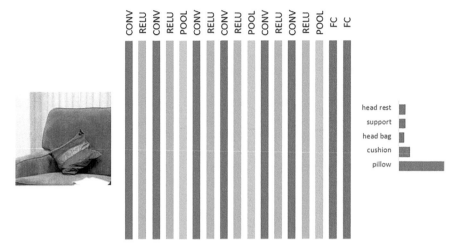

Fig. 17.3 ConvNet: CONV, RELU, POOL, and FC layers

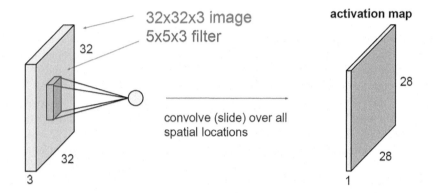

Fig. 17.4 Convolution layer (Source: CS231N Stanford course slides)

CONV Layer

Let us start by understanding the convolution layer. Given an original image, the convolution layer applies multiple filters on the image to obtain *feature maps*. Filters are rectangular in nature and always extend the full depth of the input volume. For example, in Fig. 17.4, the input image has a size of $32 \times 32 \times 3$, and a filter of size $5 \times 5 \times 3$ is being applied. To get the entire *feature map*, the filter is convolved with the image by sliding over the image spatially and computing the dot products. The sliding can be done one-step or multiple steps at a time; this is controlled using a parameter called the stride. Filters are like features defined over the input volume. Rather than just using one filter, we could use multiple filters. The final output volume depth depends on the number of filters used. For example, if we had six

$5 \times 5 \times 3$ filters, we will get six different activation maps each of size $28 \times 28 \times 1$[5] leading to an output volume size of $28 \times 28 \times 6$. Note that an activation map can also be seen as a 28×28 sheet of neuron outputs where each neuron is connected to a small region in the input, and all of them share parameters.

The elements of the filters are the weights that are learned using backpropagation during training. The convolution layer helps us implement two important concepts in a CNN:

1. **Sparse Connectivity**: Convolution layer enforces a local connectivity pattern between neurons of adjacent layers. The inputs of hidden units in layer m are from a subset of units in layer $m - 1$, units that have spatially contiguous receptive fields.
2. **Shared Weights**: In CNNs, each filter is replicated across the entire visual field. These replicated units share the same parameterization (weight vector and bias) and form a feature map. Gradient descent can still be used to learn such shared parameters, with only a small change to the original algorithm. The gradient of a shared weight is simply the sum of the gradients of the parameters being shared. Replicating units in this way allows for features to be detected regardless of their position in the visual field. Weight sharing increases learning efficiency by greatly reducing the number of free parameters being learnt.

Convolution can be done by sliding the filter across the entire space of the input volume with a stride of 1 or larger stride values. Larger stride values lead to small output volumes. Also, sometimes, the original input volume is padded with zeroes at the border to prevent the loss of information at the border. In general, it is common to see CONV layers with stride 1, filters of size $F \times F$, and zero padding with $(F - 1)/2$. For example, if a $32 \times 32 \times 3$ image is padded by two zeros all around, then the activation map size will be $((36 - 5)/1) + 1 = 32$. So now there is no loss of the information (at the borders) because the whole image is covered.

Due to weight sharing, the number of weights to be learned in a CONV layer is much lesser compared to the weight in a layer in an MLP.

RELU Layer

Next, we discuss about the RELU (rectified linear units) layer. This is a layer of neurons that applies the activation function $f(x) = max(0,x)$. It increases the nonlinear properties of the decision function and of the overall network without affecting the receptive fields of the convolution layer. Other functions are also used to increase nonlinearity, for example, the hyperbolic tangent $f(x) = tanh(x)$ and the sigmoid function. This layer clearly does not involve any weights to be learned.

POOL Layer

There are several nonlinear functions to implement pooling among which max pooling is the most common. It partitions the input image into a set of nonoverlapping

---

[5] Activation Map Size = ((image size − filter size)/stride) + 1. Here, Image size is 32. Filter Size is 5. Stride = 1. Activation Map size = $((32 − 5)/1) + 1$ which is equal to 28.

Fig. 17.5 Pooling example

rectangles and, for each such subregion, outputs the maximum. The intuition is that the exact location of a feature is less important than its rough location relative to other features. The pooling layer serves to progressively reduce the spatial size of the representation, to reduce the number of parameters and amount of computation in the network, and hence to also control overfitting. Figure 17.5 shows an example of max pooling with a pool size of 2×2.

Finally, after several convolutional and max pooling layers, the high-level reasoning in the neural network is done via fully connected layers. Neurons in a fully connected layer have connections to all activations in the previous layer, as seen in regular MLPs.

3.4 Summary

In summary, we have discussed an interesting deep learning architecture, CNNs, for images in this section. CNNs are very popular these days across a large variety of image processing tasks. Convolution networks are inspired by the hierarchical structure of the visual cortex. Things that differentiate CNNs from DNNs are sparse connectivity, shared weights, feature maps, and pooling.

3.5 An Example: MNIST Data (Similar to MLP Approach)

The main stages of the code are as follows:

1. Download and perform data cleaning.
2. Visualize few sample digits.
3. Specify the model:

    (a) First convolution layer and specifying kernel
    (b) Activation function type
    (c) Pooling layer and specifying the type of pooling (max or average)
    (d) Second convolution layer, activation function, and pooling layer
    (e) First fully connected and specifying the number of hidden layers (neurons)
    (f) Second fully connected
    (g) Applying softmax to normalize the output

4. Define the parameters and run the model:

   (a) lenet: pointer to the last computation node in the network definition
   (b) X: pixel data (X values)
   (c) Y: dependent variable (Y values)
   (d) ctx: processing device to be used
   (e) num.round: maximum number of iterations over the dataset
   (f) array.batch.size: batch size for batch-wise gradient descent
   (g) Learning rate
   (h) Momentum: for momentum based gradient descent updates
   (i) WD: weight decay

5. Predict the model output on test data.
6. Produce confusion matrix and calculate accuracy.

The sample code helps understand how to build a CNN using the "mxnet" R package. The code "Mxnet-MNIST_CNN.R" and the datasets "MNIST_train_sample.csv" and "MNIST_test_sample.csv" are available on the website.

4 Recurrent Neural Networks (RNNs)

In this section, we will discuss a deep learning architecture to handle sequence data, RNNs. We will first motivate why sequence learning models are needed. Then we will talk about technical details of RNNs (recurrent neural networks) and finally discuss about their application to image captioning and machine translation.

4.1 Motivation for Sequence Learning Models

Sequences are everywhere. Text is a sequence of characters. Speech is a sequence of phonemes. Videos are sequences of images. There are many important applications powered by analytics on top of sequence data. For example, machine translation is all about transforming a sequence written in one language to another. We need a way to model such sequence data using neural networks. Humans do not start their thinking from scratch every second. As you read this section, you understand each word based on your understanding of previous words. You do not throw everything away and start thinking from scratch again. Your thoughts have persistence. Thus, we need neural networks with some persistence while learning. In this chapter, we will discuss about RNNs as an architecture to support sequence learning tasks. RNNs have loops in them which allow for information to persist.

Language models are the earliest example of sequence learning for text sequences. A language model computes a probability for a sequence of words:

$P(w_1, \ldots, w_m)$. Language models are very useful for many tasks like the following: (1) *next word prediction*: for example, predicting the next word after the user has typed this part of the sentence. "Stocks plunged this morning, despite a cut in interest rates by the Federal Reserve, as Wall ..."; (2) *spell checkers*: for example, automatically detecting that minutes has been spelled incorrectly in the following sentence. "They are leaving in about fifteen minuets to go to her house"; (3) *mobile auto-correct*: for example, automatically suggesting that the user should use "find" instead of "fine" in the following sentence. "He is trying to fine out."; (4) *speech recognition*: for example, automatically figuring out that "popcorn" makes more sense than "unicorn" in the following sentence. "Theatre owners say unicorn sales have doubled..."; (5) *automated essay grading*; and (6) *machine translation*: for example, identifying the right word order as in *p(the cat is small) > p(small the is cat)*, or identifying the right word choice as in *p(walking home after school) > p(walking house after school)*.

Traditional language models are learned by computing expressing probability of an entire sequence using the chain rule. For longer sequences, it helps to compute probability by conditioning on a window of n previous words. Thus, $P(w_1, \ldots, w_m) = \Pi_{i=1}^{m} P(w_i|w_1, \ldots w_{i-1}) \approx \Pi_{i=1}^{m} P(w_i|w_{i-(n-1)}, \ldots, w_{i-1})$. Here, we condition on the previous n values instead of previous all values. This approximation is called the Markov assumption. To estimate probabilities, one may compute unigrams, bigrams, trigrams, etc., as follows, using a large text corpus with T tokens.

Unigram model: $p(w_1) = \frac{count(w_1)}{T}$
Bigram models: $p(w_2|w_1) = \frac{count(w_1, w_2)}{count(w_1)}$
Trigram models: $p(w_3|w_1, w_2) = \frac{count(w_1, w_2, w_3)}{count(w_1, w_2)}$

Performance of n-gram language models improves as n for n-grams increases. Smoothing, backoff, and interpolation are popular techniques to handle low frequency n-grams. But the problem is that there are a lot of n-grams, especially as n increases. This leads to gigantic RAM requirements. In some cases, the window of past consecutive n words may not be sufficient to capture the context. For instance, consider a case where an article discusses the history of Spain and France and somewhere later in the text, it reads, "The two countries went on a battle"; clearly the information presented in this sentence alone is not sufficient to identify the name of the two countries.

Can we use MLPs to model the next word prediction problem? Figure 17.6 shows a typical MLP for the next word prediction task as proposed by Bengio et al. (2003). The MLP is trained to predict the t-th word based on a fixed size context of previous $n - 1$ words. This network assumes that we have a mapping C from any word i in the vocabulary to a distributed feature vector like word2vec.[6] Thus, if m is the

---

[6]Word2vec is an algorithm for learning a word embedding from a text corpus. For further details, read Mikolov et al. (2013).

17 Deep Learning

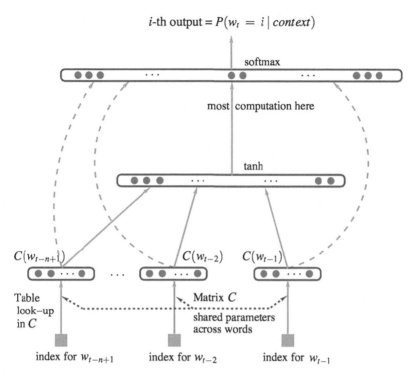

Fig. 17.6 MLP for next word prediction task (Source: Bengio et al. 2003)

dimension for the feature vector representation, and $|V|$ is vocabulary size, C is a $|V| \times m$ sized matrix. $C(w_{t-i})$ is the vector representation of the word that came i words ago. C could also be learned along with the other weights in the network. Further, the model contains a hidden layer with a nonlinearity. Finally, at the output layer, a softmax is performed to return the probability distribution of size $|V|$ which is expected to be as close as possible to the one-hot encoded representation of the actual next word.

In all conventional language models, the memory requirements of the system grow exponentially with the window size n making it nearly impossible to model large word windows without running out of memory. But in this model, the RAM requirements grow linearly with n. Thus, this model supports a fixed window of context (i.e., n). There are two drawbacks of this model: (1) the number of parameters increase linearly with the context size, and (2) it cannot handle contexts of different lengths. RNNs help address these drawbacks.

4.2 Technical Details of RNNs

RNNs is a deep learning neural architecture that can support next word prediction with variable n. RNNs tie the weights at each time step. This helps in conditioning

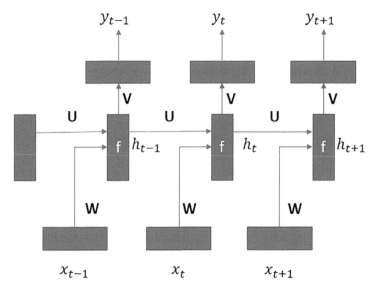

Fig. 17.7 Basic RNN architecture

the neural network on all previous words. Thus, the RAM requirement only scales with the number of words in the vocabulary. Figure 17.7 shows the architecture of a basic RNN model with three units. **U**, **V**, and **W** are the shared weight matrices that repeat across multiple time units. Overall the parameters to be learned are **U**, **V**, and **W**.

RNNs are called recurrent because they perform the same task for every element of a sequence. The only thing that differs is the input at each time step. Output is dependent on previous computations. RNNs can be seen as neural networks having "memory" about what has been calculated so far. The information (or the state) h_t at any time instance t is this memory. In some sense, h_t captures a thought that summarizes the words seen so far. RNNs process a sequence of vectors x by applying a recurrence formula at every time step: $h_t = f_{U,W}(h_{t-1}, x_t)$, where h_t is the new state, $f_{U,W}$ is some function with parameters U and W, h_{t-1} is the old state, and x_t is the input vector at current time step. Notice that the same function and the same set of parameters are used at every time step.

The weights for an RNN are learned using the same backpropagation algorithm, also called as backpropagation through time (BPTT) in the context of RNNs. The training data for BPTT should be an ordered sequence of input-output pairs $\langle x_0, y_0 \rangle, \langle x_1, y_1 \rangle, \ldots, \langle x_{n-1}, y_{n-1} \rangle$. An initial value must be specified for the hidden layer output h_0 at time t_0. Typically, a vector of all zeros is used for this purpose. BPTT begins by unfolding a recurrent neural network through time. When the network is unfolded through time, the unfolded network contains k instances of *a unit, each containing an input, a hidden layer, and an output.* Training then proceeds in a manner similar to training a feed-forward neural network with backpropagation, except that each epoch must run through the observations, y_t, in sequential order. Each training pattern consists of $\langle h_t, x_t, x_{t+1}, x_{t+2}, \ldots, x_{t+k-1}, y_{t+k}\rangle$. Typically, backpropagation is applied in an online manner to update the weights as each training pattern is presented. After each pattern is presented, and the weights have been updated, the weights in each instance of *U*, *V*, *and* *W* are averaged together so that they all have the same weights, respectively. Also, h_{t+1} is calculated as $h_{t+1} = f_{U,W}(h_t, x_{t+1})$, which provides the information necessary so that the algorithm can move on to the next time step, $t + 1$. The output y_t is computed as follows: $y_t = \text{softmax}(V\ h_t)$. Usually the cross entropy loss function is used for the optimization: Given an actual output distribution y_t and a predicted output distribution $\widehat{y}_t$, cross entropy loss is defined as $-\sum_{j=1}^{|V|} y_{t,j} \log \widehat{y}_{(t,j)}$. Note that y_t is the true vector; it could be a one-hot encoding of the expected word or a word2vec representation of the expected word at the *t*-th time instant.

4.3 Example: Next Word Prediction

The following pseudo-code shows how to build an RNN using the "mxnet" R package for the next word prediction task. Below are the main code stages:

1. Download the data and perform cleaning.
2. Create Word 2 Vector, dictionary, and lookup dictionary.
3. Create multiple buckets for training data.
4. Create iterators for multiple buckets data.
5. Train the model for multiple bucket data with the following parameters:

    (a) Cell_type = "lstm" #Using lstm cell which can hold the results
    (b) num_rnn_layer = 1
    (c) num_embed = 2
    (d) num_hidden = 4 #Number of hidden layers
    (e) loss_output = "softmax"
    (f) num.round = 6

6. Predict the output of the model on "Test" data.
7. Calculate the accuracy of the model.

The sample code helps understand how to build an RNN using the "mxnet" R package. The code "Next_word_RNN.R" and the datasets "corpus_bucketed_train.rds" and "corpus_bucketed_test.rds" are available on the website.

The basic RNN architecture can be extended in many ways. Bidirectional RNNs are RNNs with two hidden vectors per unit. The first hidden vector maintains the state of the information seen so far in the sequence in the forward direction, while the other hidden vector maintains the state representing information seen so far in the sequence in the backward direction. The number of parameters in bidirectional RNNs is thus twice the number of parameters in the basic RNN.

RNNs could also be deep. Thus, a deep RNN has stacked hidden units, and the output neurons are connected to the most abstract layer.

4.4 Applications of RNNs: Image Captioning and Machine Translation

Recurrent networks offer a lot of flexibility. Thus, they can be used for a large variety of sequence learning tasks. Such tasks could be classified as one-to-many, many-to-one, or many-to-many depending on the number of inputs and the number of outputs. An example of a one-to-many application is image captioning (image → sequence of words). An example of many-to-one application is sentiment classification (sequence of words → sentiment). An example of "delayed" many-to-many application is machine translation (sequence of words → sequence of words). Finally, an example of the "synchronized" many-to-many case is video classification on frame level.

In the following, we will discuss two applications of RNNs: image captioning and machine translation. Figure 17.8 shows the neural CNN-RNN architecture for the image captioning task. First a CNN is used to obtain a deep representation for the image. The representation is then passed on to the RNN to learn captions. Note that the captions start with a special word START and end with a special word END. Unlike image classification task where the number of captions is limited, in image captioning, the number of captions that can be generated are many more since rather than selecting one of say 1,000 captions, here the task is to generate captions.

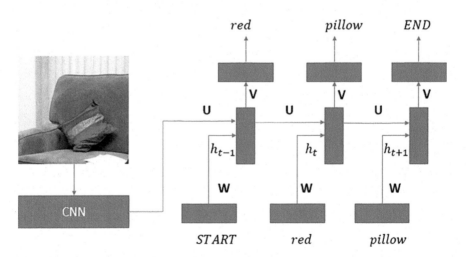

Fig. 17.8 CNN-RNNs for image captioning task

17 Deep Learning

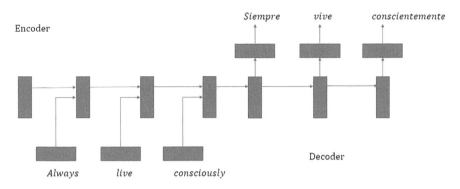

Fig. 17.9 RNNs for machine translation

As shown in Figure 17.8, a CNN trained on the ImageNet data is first used. Such a CNN was discussed in Sect. 3. The last fully connected layer of the CNN is thrown away, and the result from the CNN's penultimate layer is fed to the first unit of the RNN. One-hot encoding of the special word START is fed as input to the first unit of the RNN. At the training time, since the actual image captions are known, the corresponding word representations are actually fed as input for every recurrent unit. However, at test time, the true caption is unknown. Hence, at test time, the output of the k-th unit is fed as input to the $(k + 1)$-th unit. This is done for better learning of the order of words in the caption. Cross entropy loss is used to compute error at each of the output neurons. Microsoft COCO[7] is a popular dataset which can be used for training such a model for image captioning. The dataset has about 120K images each with five sentences of captions (Lin et al. 2014).

Lastly let us discuss about application of RNNs to machine translation. Figure 17.9 shows a basic encoder–decoder architecture for the machine translation task using RNNs. The encoder RNN tries to encode all the information from the source language into a single hidden vector at the end. Let us call this last hidden vector of the encoder as the "thought" vector. The decoder RNN uses information from this thought vector to generate words in the target language. The architecture tries to minimize the *cross entropy error* for all target words conditioned on the source words.

There are many variants of this architecture as follows. (1) The encoder and the decoder could use shared weights or different weights. (2) Hidden state in the decoder always depends on the hidden state of the previous unit, but it could also optionally depend on the thought vector and predicted output from the previous unit. (3) Deep bidirectional RNNs could be used for both encoder and decoder.

Beyond these applications, RNNs have been used for many sequence learning tasks. However, RNNs suffer from vanishing gradients problem. In theory, RNN

---

[7]Microsoft COCO dataset http://www.mscoco.org/ (accessed on Aug 1, 2018) or http://cocodataset.org/ (accessed on Aug 1, 2018).

can memorize in hidden state, that is, h_t, all the information about past inputs. But, in practice, standard RNN cannot capture very long-distance dependency. Vanishing/exploding gradient problem in backpropagation: gradient signal can end up being multiplied a large number of times (as many as the number of time steps) by the weight matrix associated with the connections between the neurons of the recurrent hidden layer. If the weights in transition weight matrix are small (or, more formally, if the leading eigenvalue of the weight matrix is smaller than 1.0), it can lead to vanishing gradients where the gradient signal gets so small that learning either becomes very slow or stops working altogether. It can also make more difficult the task of learning long-term dependencies in the data. Conversely, if the weights in this matrix are large (or, again, more formally, if the leading eigenvalue of the weight matrix is larger than 1.0), it can lead to a situation where the gradient signal is so large that it can cause learning to diverge. This is often referred to as exploding gradients. A solution to this problem is long short-term memory (LSTM) which are deep learning architectures similar to RNNs but with explicit memory cells. The main idea is to keep around memories to capture long-range dependencies and to allow error messages to flow at different strengths depending on the inputs. The intuition is that memory cells can keep information intact, unless inputs make them forget it or overwrite it with new input. The memory cell can decide to output this information or just store it. The reader may refer to Hochreiter and Schmidhuber (1997) for further details about LSTMs.

4.5 Summary

In summary, recurrent neural networks are powerful in modeling sequence data. But they have the vanishing/exploding gradient problem. LSTMs are better since they avoid the vanishing/exploding gradient problem by introducing memory cells. Overall, RNNs and LSTMs are really useful in many real-world applications like image captioning, opinion mining, and machine translation.

5 Further Reading

The advances being made in this field are continuous in nature due to the practice of sharing information as well as cooperating with researchers working in labs and in the field. Therefore, the most recent information is available on the Web and through conferences and workshops. The book[8] by Goodfellow et al. (2016), the

---

[8]https://www.deeplearningbook.org/ (accessed on Aug 1, 2018).

deep learning tutorials,[9] and specialization in deep learning[10] offered by Andrew Ng are good starting points for learning more.

Additional reference material (accessed on Aug 1, 2018):

- Good Introductory Tutorial: http://web.iitd.ac.in/~ sumeet/Jain.pdf
- A Brief Introduction to Neural Networks: http://www.dkriesel.com/en/science/neural_networks

CNN feature visualization:

- http://people.csail.mit.edu/torralba/research/drawCNN/drawNet.html?path=imagenetCNN
- http://www.wildml.com/2015/09/recurrent-neural-networks-tutorial-part-1-introduction-to-rnns/
- http://karpathy.github.io/2015/05/21/rnn-effectiveness/
- http://colah.github.io/posts/2015-08-Understanding-LSTMs/
- Hochreiter, S., & Schmidhuber, J. (1997). Long short-term memory. *Neural Computation*, 9(8), 1735–1780. http://www.bioinf.jku.at/publications/older/2604.pdf
- http://jeffdonahue.com/lrcn/
- https://github.com/kjw0612/awesome-rnn#theory

Electronic Supplementary Material

All the datasets, code, and other material referred in this section are available in www.allaboutanalytics.net.

- Data 17.1: MNIST_train_sample.csv
- Data 17.2: MNIST_test_sample.csv
- Data 17.3: corpus_bucketed_test.rds
- Data 17.4: corpus_bucketed_train.rds
- Code 17.1: MLP_MNIST.R
- Code 17.2: MXNET_MNIST_CNN.R
- Code 17.3: Next_word_RNN.R

Exercises

Ex. 17.1 Which of these is false?

(a) Deep learning needs large amounts of data for learning.

---

[9]http://deeplearning.net/reading-list/tutorials/ (accessed on Aug 1, 2018).
[10]https://www.coursera.org/specializations/deep-learning (accessed on Aug 1, 2018).

(b) Deep learning classifiers are all linear in nature.
(c) Deep learning needs a lot of compute power.
(d) Deep learning consists of multiple model architectures.

Ex. 17.2 What is the algorithm used to train a single neuron?

(a) Backpropagation
(b) Forward propagation
(c) Perceptron
(d) None of the above

Ex. 17.3 How can you make an artificial neuron learn nonlinear patterns?

(a) Change integration function to be nonlinear
(b) Use multilayered perceptrons
(c) Both of the above
(d) None of the above

Ex. 17.4 What is the weight update equation in perceptron?

(a) New w = old w + (learning rate) × (error) × (instance vector)
(b) New w = old w - (learning rate) × (error) × (instance vector)
(c) New w = (learning rate) × (error) × (instance vector)
(d) New w = (learning rate) × (error) × (instance vector)

Ex. 17.5 If an MLP has an input layer with 10 features, hidden layer with 20 neurons, and output layer with 1 output, how many parameters are there?

(a) $10 \times 20 + 20 \times 1$
(b) $(10 + 1) \times 20 + (20 + 1) \times 1$
(c) $(10 - 1) \times 20 + (20 - 1) \times 1$
(d) 10×20

Ex. 17.6 Why cannot the perceptron algorithm work for MLPs?

(a) We never discussed this in the class!
(b) MLPs have too many parameters, and perceptron is not very efficient when there are too many parameters.
(c) Supervision is not available for neurons in the hidden layers of an MLP.
(d) Perceptrons are meant to learn only linear classifiers, while MLPs can learn more complex boundaries.

Ex. 17.7 We discussed three different activation functions. Which of the following is not an activation function?

(a) Step function
(b) Spherical function
(c) Ramp function
(d) Sigmoid function

Ex. 17.8 What is false among the following?

(a) MLPs have fully connected layers, while CNNs have sparse connectivity.
(b) MLPs are supervised, while CNNs are usually used for unsupervised algorithms.
(c) MLPs have more weights, while CNNs have fewer number of weights to be learned.
(d) MLP is a general modeling architecture, while CNNs specialize for images.

Ex. 17.9 Given an image of $32 \times 32 \times 3$, a single fully connected neuron will have how many weights to be learned?

(a) $32 \times 32 \times 3 + 1$
(b) 32
(c) 3
(d) 32×32

Ex. 17.10 What is the convolution operation closest to?

(a) Jaccard similarity
(b) Cosine similarity
(c) Dot product
(d) Earth mover's distance

Ex. 17.11 How many weights are needed if the input layer has 32×32 inputs and the hidden layer has 20×20 neurons?

(a) $(32 \times 32 + 1) \times 20 \times 20$
(b) $(20 + 1) \times 20$
(c) $(32 + 1) \times 20$
(d) $(32 + 1) \times 32$

Ex. 17.12 Consider a volume of size $32 \times 32 \times 3$. If max pooling is applied to it with pool size of 4 and stride of 4, what are the number of weights in the pooling layer?

(a) $(32 \times 32 \times 3 + 1) \times (4 \times 4)$
(b) $4 \times 4 + 1$
(c) 0
(d) $32 \times 32 \times 3$

Ex. 17.13 Which among the following is false about the differences between MLPs and RNNs?

(a) MLPs can be used with fixed-sized sequences, while RNNs can handle variable-sized sequences.
(b) MLPs have more weights, while RNNs have fewer number of weights to be learned.
(c) MLP is a general modeling architecture, while RNNs specialize for sequences.
(d) MLPs are supervised, while RNNs are usually used for unsupervised algorithms.

Ex. 17.14 We looked at two neural models for next word prediction: an MLP and an RNN. Given a vocabulary of 1000 words, and a hidden layer of size 100, a context of size 6 words, what are the number of weights in an MLP?

(a) $(6 \times 1000 + 1) \times 100 + (100 + 1) \times 1000$
(b) $(1000 + 1) \times 100 + (100 + 1) \times 100 + (100 + 1) \times 1000$
(c) $(6 \times 6 + 1) \times 100 + (6 \times 6 + 1) \times 1000$
(d) $(1000 + 1) \times (100 + 1) \times 6$

Ex. 17.15 How does backpropagation through time differ from typical backpropagation in MLPs?

(a) Weights on edges supposed to have shared weights must be averaged out and set to the average after every iteration.
(b) Backpropagation in MLPs uses gradient descent, while backpropagation through time uses time series modeling.
(c) Backpropagation in MLPs has two iterations for every corresponding iteration in backpropagation through time.
(d) None of the above.

Answer in Length

Ex. 17.16 Define deep learning bringing out its five important aspects.

Ex. 17.17 Describe the backpropagation algorithm.

Ex. 17.18 RNNs need input at each time step. For image captioning, we looked at a CNN-RNN architecture.

(a) What is the input to the first hidden layer of the RNN?
(b) Where do the other inputs come from?
(c) How is the length of the caption decided?
(d) Does it generate new captions by itself or only select from those that it had seen in training data?
(e) If vocab size is V, hidden layer size is h, and average sequence size is "s," how many weights are involved in an RNN?

Hands-On Exercises

Ex. 17.19 Create a simple logistic regression-based classifier for the popular iris dataset in mxnet.

Ex. 17.20 Create an MLP classifier using three hidden layers of sizes 5, 10, 5 for the MNIST digit recognition task using mxnet. (Hint: Modify the code from Sect. 2.5 appropriately).

Ex. 17.21 Create a CNN classifier using two CONV layers each with twenty 5 × 5 filters with padding as 2 and stride as 1. Also use pooling layers with 2 × 2 filters with stride as 2. Do this for the MNIST digit recognition task using mxnet. (Hint: Modify the code from Sect. 3.5 appropriately).

Ex. 17.22 Train an RNN model in mxnet for the next word prediction task. Use a suitable text corpus from https://en.wikipedia.org/wiki/List_of_text_corpora. (Hint: Modify the code from Sect. 4.2 appropriately).

References

Bengio, Y., Ducharme, R., Vincent, P., & Jauvin, C. (2003). A neural probabilistic language model. *Journal of Machine Learning Research, 3*, 1137–1155.
Duchi, J., Hazan, E., & Singer, Y. (2011). Adaptive subgradient methods for online learning and stochastic optimization. *Journal of Machine Learning Research, 12*, 2121–2159.
Fukushima, K., & Miyake, S. (1982). Neocognitron: A self-organizing neural network model for a mechanism of visual pattern recognition. In S. Amari & A. Michael (Eds.), *Competition and cooperation in neural nets* (pp. 267–285). Berlin: Springer.
Goodfellow, I., Bengio, Y., & Courville, A. (2016). *Deep learning* (Vol. 1). Cambridge: MIT Press.
Hinton, G., Srivastava, N., & Swersky, K. (2012). Lecture 6d—A separate, adaptive learning rate for each connection. *Slides of lecture neural networks for machine learning*. Retrieved Mar 6, 2019, from https://www.cs.toronto.edu/~tijmen/csc321/slides/lecture_slides_lec6.pdf
Hochreiter, S., & Schmidhuber, J. (1997). Long short-term memory. *Neural Computation, 9*(8), 1735–1780.
Hubel, D. H., & Wiesel, T. N. (1962). Receptive fields, binocular interaction and functional architecture in the cat's visual cortex. *The Journal of Physiology, 160*(1), 106–154.
Kingma, D. P., & Ba, J. (2014). Adam: A method for stochastic optimization. arXiv:1412.6980.
Lin, T. Y., Maire, M., Belongie, S., Hays, J., Perona, P., Ramanan, D., et al. (2014). Microsoft coco: Common objects in context. In D. Fleet, T. Pajdla, B. Schiele, & T. Tuytelaars (Eds.), *European conference on computer vision* (pp. 740–755). Cham: Springer.
Mikolov, T., Chen, K., Corrado, G., & Dean, J. (2013). Efficient estimation of word representations in vector space. arXiv:1301.3781
Miller, G. A. (1995). WordNet: A lexical database for English. *Communications of the ACM, 38*(11), 39–41.
Minsky, M. L., & Papert, S. (1969). *Perceptrons*. MIT Press, Cambridge, MA.
Rosenblatt, F. (1962). *Principles of neurodynamics*. Wuhan: Scientific Research Publishing.
Rumelhart, D. E., Hinton, G. E., & Williams, R. J. (1986). Learning internal representations by error propagation. In D. Rumelhart & J. McClelland (Eds.), *Parallel distributed processing. Explorations in the microstructure of cognition* (Vol. 1, pp. 318–362). Cambridge, MA: Bradford Books.
Zeiler, M. D. (2012). ADADELTA: An adaptive learning rate method. arXiv:1212.5701.

Part III
Applications

Chapter 18
Retail Analytics

Ramandeep S. Randhawa

1 Introduction

1.1 Background

Retail is one of the largest sectors in today's economy. The global retail sector is estimated to have revenues of USD 28 trillion in 2019 (with approximately USD 5.5 trillion sales in the USA alone). This sector represents 31% of the world's GDP and employs billions of people throughout the globe.[1] A large and growing component of this is e-commerce or e-tail, which includes products and services ordered via the Internet, with sales estimated to be about USD 840 billion in 2014, and expected to grow at a rate of about 20% over the subsequent years.[2] Analytics is gaining increasing prominence in this sector with the retail analytics market size being estimated at over USD 3.52 billion in 2017 and is expected to grow at a CAGR of over 19.7% over the next few years.[3]

Electronic supplementary material The online version of this chapter (https://doi.org/10.1007/978-3-319-68837-4_18) contains supplementary material, which is available to authorized users.

---

[1] https://www.businesswire.com/news/home/20160630005551/en/Global-Retail-Industry-Worth-USD-28-Trillion (accessed on Jul 31, 2018).

[2] https://www.atkearney.com/consumer-goods/article?/a/global-retail-e-commerce-keeps-on-clicking (accessed on Mar 1, 2019).

[3] https://www.marketsandmarkets.com/Market-Reports/retail-analytics-market-123460609.html (accessed on Jul 31, 2018).

R. S. Randhawa (✉)
Marshall School of Business, University of Southern California, Los Angeles, CA, USA
e-mail: Ramandeep.Randhawa@marshall.usc.edu

Retail acts as the last stop in the supply chain by selling products directly to customers. Given that retailers are focused on this aspect, collecting data on customer behavior and preferences and incorporating these into business decisions are quite natural. And so, retail has indeed been an early adopter of analytics methodologies and focuses heavily on advancing knowledge in this domain.

1.2 What Is Retail Analytics?

Retail analytics is an umbrella term that comprises various elements which assist with decision-making in the retail business. Typically, this includes data collection and storage (data warehousing), data analysis that involves some statistical or predictive modeling, and decision-making. Traditionally, the analysis of data was limited to monitoring and visualizing some key performance indicators (KPIs) retrospectively.

One may use the term *business intelligence* to refer to the gamut of activities that underlie intelligent business decision-making. However, typically this term is used to refer to the collection and presentation of historical information in an easy-to-understand manner, via reports, dashboards, scorecards, etc. The term *advanced analytics* is typically reserved for when predictive modeling is applied to data via statistical methods or machine learning. Our focus in this chapter will be on the later, *advanced analytics*, methodologies that can significantly assist in the decision-making process in retail.

To understand the role analytics plays in retail, it is useful to break down the business decisions taken in retail into the following categories: consumer, product, workforce, and advertising.

1. *Consumer*: Personalization is a key consumer-level decision that retail firms make. Personalized pricing by offering discounts via coupons to select customers is one such decision. This approach uses data collection via loyalty cards to better understand a customer's purchase patterns and willingness to pay and uses that to offer personalized pricing. Such personalization can also be used as a customer retention strategy. Another example is to offer customers a personalized sales experience: in e-tail settings, this entails offering customers a unique browsing experience by modifying the products displayed and suggestions made based on the customer's historical information.
2. *Product*: Retail product decisions can be broken down into single product and group of product decisions. Single or individual product decisions are mostly inventory decisions: how much stock of the product to order, and when to place the order. At the group level, the decisions are typically related to pricing and assortment planning. That is, what price to set for each product in the group and how to place the products on the store-shelves, keeping in mind the variety of products, the number of each type of product, and location. To make these decisions, predictive modeling is called for to forecast the product

demand and the price-response function, and essentially the decision-maker needs to understand how customer reacts to price changes. A fine understanding of consumer choice is also needed to understand how a customer chooses to buy a certain product from a group of products.
3. *Human resources*: The key decisions here are related to the number of employees needed in the store at various times of the day and how to schedule them. To make these decisions, the overall work to be completed by the employees needs to be estimated. Part of this is a function of other decisions, such as the effort involved in stocking shelves, taking deliveries, changing prices, etc. There is additional work that comes in as a function of the customer volume in the store. This includes answering customer questions and manning checkout counters.
4. *Advertising*: In the advertising sphere, companies deal with the typical decisions of finding the best medium to advertise on (online mediums such as *Google Adwords*, *Facebook*, *Twitter*, and/or traditional mediums such as print and newspaper inserts) and the best products to advertise. This may entail cultivating some "loss-leaders" that are priced low to entice customers into the store, so they may also purchase other items which have a greater margin.

We refer the reader to the survey article by Bradlow et al. (2017). It reviews big data and predictive analytics practices in retailing. They discuss several statistical issues and methods, including Bayesian analysis, which are important in collecting, processing, modeling, and analysis of data. In addition, they emphasize ethical and privacy issues.

1.3 *Examples of Retail Analytics in Action*

- Analytics has revealed that a great number of customer visits to online stores fail to convert at the last minute, when the customer has the item in their shopping basket but does not go on to confirm the purchase. Theorizing that this was because customers often cannot find their credit or debit cards to confirm the details, Swedish e-commerce platform Klarna moved its clients (such as Vistaprint, Spotify, and 45,000 online stores) onto an invoicing model, where customers can pay after the product is delivered. Sophisticated fraud prevention analytics are used to make sure that the system cannot be manipulated by those with devious intent.
- Trend forecasting algorithms comb social media posts and Web browsing habits to elicit what products may be causing a buzz, and ad-buying data is analyzed to see what marketing departments will be pushing. Brands and marketers engage in "sentiment analysis," using sophisticated machine learning-based algorithms to determine the context when a product is discussed. This data can be used to accurately predict what the top selling products in a category are likely to be.

- Russian retailers have found that the demand for books increases exponentially as the weather gets colder. So retailers such as Ozon.ru increase the number of book recommendations which appear in their customers' feeds as the temperature drops in their local areas.[4]
- The US department store giant, Macy's, recently realized that attracting the right type of customers to its brick-and-mortar stores was essential. Due to its analytics showing up a dearth of the vital millennials demographic group, it recently opened its "One Below" basement[5] at its flagship New York store, offering "selfie walls" and while-you-wait customized 3D-printed smartphone cases. The idea is to attract young customers to the store who will hopefully go on to have enduring lifetime value to the business.
- Amazon has proposed using predictive shipping analytics[6] to ship products to customers before they even click "add to cart." According to a recent trend report by DHL, over the next 5 years, this so-called psychic supply chain will have far reaching effects in nearly all industries, from automotive to consumer goods. It uses big data and advanced predictive algorithms to enhance planning and decision-making.

1.4 Complications in Retail Analytics

There are various complications that arise in retail scenarios that need to be overcome for the successful use of retail analytics. These complications can be classified into (a) those that affect predictive modeling and (b) those that affect decision-making.

Some of the most common issues that affect predictive modeling are demand censoring and inventory inaccuracies (DeHoratius and Raman 2008). Typically, retail firms only have access to sales information, not demand information, and therefore need to account for the fact that when inventory runs out, actual demand is not observed. Ignoring this censoring of information can result in underestimating demand. There is also a nontrivial issue of inventory record inaccuracies that exists in retail stores—the actual number of products in an inventory differs from the number expected as per the firm's IT systems (DeHoratius 2011). Such inaccuracy may be caused by theft, software glitches, etc. This inaccuracy needs to be incorporated into demand estimation because it confounds whether demand is low or appears low due to product shortage. Inaccuracy also affects decision-

---

[4]https://www.forbes.com/sites/bernardmarr/2015/11/10/big-data-a-game-changer-in-the-retail-sector/#651838599f37 (accessed on Jul 31, 2018).

[5]https://www.bloomberg.com/news/articles/2015-09-25/this-is-macy-s-idea-of-a-millennial-wonderland (accessed on Jul 31, 2018).

[6]https://www.forbes.com/sites/onmarketing/2014/01/28/why-amazons-anticipatory-shipping-is-pure-genius/#178cd4114605 (accessed on Jul 31, 2018).

making by impacting the timing of order placement. Some of the other factors that affect decision-making are constraints on changing prices, physical constraints on assortments, supplier lead times, supplier contracts, and constraints on workforce scheduling. In particular, retail firms deal with many constraints on changing prices. Some of these are manpower constraints: changing assortments requires a reconfiguration of store shelves, and changing prices may involve physically tagging products with the new price (a labor-intensive process). To make it easier to change prices, many stores such as Kohl's are turning to electronic shelf labels as a means of making the price-changing process efficient.[7] There are additional nonphysical constraints that a firm may need to deal with. For instance, in fashion, prices are typically only marked down, and once a price is lowered, it is not increased. There are also limits to how often prices may be changed, for instance, twice a week. There are many supplier-based constraints that need to be considered as well, for instance, lead times on any new orders placed and any terms agreed to in supplier contracts.

In this chapter, our focus will be on use of retail analytics for product-based decision-making. We continue in Sect. 2, by exploring the various means of data collection that are in use by retailers and those that are gaining prominence in recent times. In Sect. 3, we will discuss some key methodologies that are used for such decision-making. In particular, we will discuss various statistics and machine learning methodologies for demand estimation, and how these may be used for pricing, and techniques for modeling consumer choice for assortment optimization. In Sect. 4, we will focus on the many business challenges and opportunities in retail, focusing on both e-tail settings and the growth in retail analytic startups.

2 Data Collection

Retail data can be considered as both structured (spreadsheet with rows and columns) and unstructured (images, videos, and other location-based data). Traditional retail data has been structured and derived mostly from point-of-sale (POS) devices and data supplied by third parties. POS data typically captures sales information, number of items sold, prices, and timestamps of transactions. Combined with inventory record keeping, this data provides a rich trove of information about products sold and, in particular, product baskets (collection of items in the cart) sold. Retailers tend to use loyalty programs to attach customer information to this information, so that customer level sales data can be analyzed. Third-party data typically consists of competitor information, such as prices and product assortments. It also consists of some broad information about the firm's customers, such as their demographics and location.

---

[7]https://www.wsj.com/articles/now-prices-can-change-from-minute-to-minute-1450057990 (accessed on Jul 31, 2018).

The recent trend is to capture more and more unstructured data. There now exists technology that can help retailers collect information not only about direct customer sales but also about product comparisons, that is, what products were compared by the customer in making decisions. Video cameras coupled with image detection technology can help collect data on customer routes through a store. This video data can also be used to collect employee data (e.g., what tasks are employees doing, how are customers being engaged, and how much time does a customer needing assistance have to wait for the assistance to be provided). Recently, many firms have also employed eye-tracking technology in controlled environments to collect data on how the store appears from a customer's perspective; a major downside of this technology is that it requires the customer to wear specialized eyeglasses.

With the advent of *Internet of Things* (IoT), the potential to collect in-store data has increased. Walmart began using radio-frequency identification (RFID) technology about a decade ago. Initially, the main goal of using this technology was to track inventory in the supply chain. However, increasingly, retailers are finding it beneficial to track in-store inventory. RFID tags are far easier to read than barcodes because they do not require direct line-of-sight scanning. This ease of tracking allows the tags to be used to collect data on the movement of products through the store. For instance, in fashion retail, the retailer can track the items that make their way to the fitting rooms; the combination of items tried can also be tracked, and finally it can easily be detected whether the items were chosen or not. All of this provides a rich set of data to feed into the system for analytics.

Near-field communication (NFC) chips are also being used by retailers to simplify the shopping experience. Most of the current NFC usage is targeted at payments. However, several retailers are also using NFC scanning as a means to provide customers with additional information about the product. This helps collect information about the products a customer is considering. Because NFC readers are not present in all smartphones, some retailers also use Quick Response (QR) codes for their products that customers can typically scan using an app for similar functionality.

Another new method of collecting customer data is via *Bluetooth beacons*. Beacons use Bluetooth Low Energy, a technology built into recent smartphones. The beacons are placed throughout the store and can detect the Bluetooth signal from a customer's smartphone that is in the vicinity. These devices can send information to the smartphone via specialized apps. In this sense, the beacons provide a lot of flexibility for the retailer to engage with and interact with the customer (assuming that the customer has the specialized app). This can be used to push notifications about products, coupons, etc. in real time to the customer. Furthermore, because the customer interacts with the app to utilize this information, the effect of sending the information to the customer can also be tracked immediately. This technology seems to have a lot of potential for personalizing the retail experience for customers, as well as for collecting information from the customer. As per Kline,[8] nearly a quarter of US retailers have implemented such beacons. Macy's and Rite Aid are some of the prominent retailers to complete a rollout of beacons into most of its stores in 2015.

Some of the most exciting potentials for data collection can be seen in the recently launched *Amazon Go* retail store. The store allows customers to simply grab items and go, without needing to formally check out at a counter. The customer only needs to scan an app while entering the store. The use of a large number of video cameras coupled with *deep learning*-based algorithms make this quite plausible. Deep learning is an area of machine learning that has gained considerable attention recently because of its state-of-the-art ability to decipher unstructured data, especially for image recognition; see Chap. 17 on deep learning. In the retail context, the video cameras capture customers and their actions, and the deep learning algorithms decipher what the actions mean: what items are they grabbing from the shelves and if they are putting back any items from their bag. Such an approach would revolutionize customers' retail experience. However, at the same time, it provides the firm with large amounts of data beyond customer routes. It allows the firm to pick up on moments of indecision, products that were compared, especially when one product is replaced by a similar product.

3 Methodologies

We will focus on product-based analytics to support inventory decisions, assortment, and pricing decisions. The key elements of such analytics are to estimate consumer demand for products, include the case of groups of products, and then take decisions by optimizing over the relevant variables.

Some of the fundamental decisions a retailer makes is to decide on the inventory level and set the price for each SKU. Typically this involves forecasting the demand distribution and then optimizing the decision variable based on the retailer's objective. Forecasting demand is a topic that has received a lot of attention and has a long history of methodologies.

3.1 Product-Based Demand Modeling

Typical forecasting methods consider the univariate time series of sales data and use time-series-based methods such as exponential smoothing and ARIMA models; see Chap. 12 on forecasting analytics. These methods typically focus on forecasting

---

[8]https://www.huffingtonpost.com/kenny-kline/how-bluetooth-beacons-wil_b_8982720.html (accessed on May 10, 2018).

sales and may require uncensoring to be used for decision-making. Recently, there have been advances that utilize statistical and machine learning approaches to deal with greater amounts of data.

As the number of predictors grow, estimating demand becomes statistically complicated because the potential for overfitting increases. Typically, one deals with such a situation by introducing some "regularization." Penalized L_1 regularization is a common, extremely successful methodology developed to deal with high dimensionality as it performs variable selection. Penalized L_1 regression called LASSO (least absolute shrinkage and selection operator) was introduced in Tibshirani (1996) and in the context of the typical least squares linear regression can be understood as follows: suppose the goal is to predict a response variable $y \in R^n$ using covariates $X \in R^{n \times p}$, then the LASSO objective is to solve

$$\min_{\substack{p \\ \beta \in R}} ||y - X\beta||_2 + \lambda ||\beta||_1 \tag{18.1}$$

where $||\cdot||_x$ represents the L_x norm of the expression in parentheses. Such a formulation makes the typical least squares estimator biased because of the regularization term; however, by selecting the regularizer appropriately, the variance can be reduced, so that on the whole, the estimator performs better for prediction. The use of the L_1-norm facilitates sparsity and leads to "better" variable selection. This is especially useful in the case of high-dimensional settings in which the number of parameters p may even exceed the data points n. Prior to the introduction of LASSO, L_2-based regularization, also called ridge regression, was a common way to alleviate overfitting. More recently the elastic net has been proposed that uses both the L_1- and L_2-norms as a regularizer. We direct the reader to the open source book[9] by James et al. (2013) for a detailed discussion of these methodologies. A description is also contained in Chap. 7 on regression analysis.

Recently, Ma et al. (2016) used a LASSO-based approach (along with additional feature selection) to estimate SKU-based sales in a high-dimensional setting, in which the covariates included cross-category promotional information, involving an extremely large number of parameters. They found that including cross-category information improves forecast accuracy by 12.6%.

Ferreira et al. (2015) recently used an alternate machine learning-based approach for forecasting demand in an online fashion retail setting using regression trees. Regression trees are a nonparametric method that involves prediction in a hierarchical manner by creating a number of "splits" in the data. For instance, Fig. 18.1 (reproduced from Ferreira et al. 2015) displays a regression tree with two such splits.

Demand is then predicted by answering the questions pertaining to each of the splits: first, whether the price of the product is less than $100. If not, then the demand is predicted as 30. Otherwise, the following question is asked: whether the relative price of competing styles is less than 0.8 (i.e., is the price of this style

---

[9] http://www-bcf.usc.edu/~gareth/ISL/ (accessed on Jul 31, 2018).

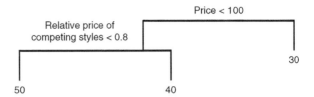

Fig. 18.1 Regression tree example (Reproduced from Ferreira et al. 2015)

less than 80% of the average price of competing styles); if the answer is no, then the demand is predicted as 40. Otherwise, it is predicted as 50. The paper further uses the variance reduction method of bootstrap aggregation or bagging, in which an ensemble of trees is "grown," with each tree trained on a random sampling of the dataset, that is, if the data has N records, then each tree is trained on $m < N$ records randomly sampled from the dataset with replacement. This reduces the interpretability of the model but improves the performance. We refer the reader to the previously cited book on statistical learning and Chap. 16, Machine Learning (Supervised), for details on this methodology. A closely related method is that of random forests, which is similar to bagged trees, except each tree is only allowed to split on a subset of the parameters; this reduces the correlation between the trees and lowers the variance further. Random forests are extremely good out-of-the-box predictors; however, because each tree only uses a subset of parameters for training, its overall interpretability is quite limited.

Recently, (artificial) neural networks (NN) have also been employed for demand forecasting (Au et al. 2008). A neural network is a large group of nodes that are arranged in a layered manner, and the arcs in the network are associated with weights (Sect. 2.1 in Chap. 17 on deep learning). The input to the network is transformed sequentially layer by layer—the input to a layer is used to compute the output of each node in the layer based on the inputs to the node, and this serves as an input to the next layer. In this manner the neural network produces an output for any given input. The weights of the neural network are "trained" typically by gradient descent methods to minimize a loss function that relates to the error between the output and the input. Neural networks can model highly nonlinear dependencies and as such work extremely well in detecting patterns and trends in complex scenarios. Neural networks have been around for a long time; see Chaps. 16 and 17. The well-known logistic regression function can be represented by a single-layer neural network. However, more interesting networks are obtained by creating a large number of layers; hundred-layer neural nets are not uncommon. Deep neural networks are notoriously difficult to train, as they require a lot of data and computational power. With recent advances in data collection and computing, it has become possible to harness the potential of these networks. Initial research demonstrates that NN can be used effectively in the context of predicting fashion demand, and these illustrate the potential for using such methods for demand forecasting in the future. The performance of NNs (and their sophistication) increases as the amount of training data increases. With the spurt in data collection, especially that of unstructured data, NNs provide an exciting potential for demand forecasting.

Turning to decision-making, the application of these statistical and machine learning methodologies generates new challenges and the potential for generating better decisions. For instance, the work of Ferreira et al. (2015) studies the problem of optimizing the prices of the products (at fixed inventory levels). In this case, the split-based demand prediction approach implies that the optimization program becomes an integer program, where, for a fixed set of prices, the decision variable is a binary variable relating to whether a particular <product, price> pair is offered or not. The paper proposes an efficient approximation solution by using a linear programming relaxation. The solution is tested in a field experiment and found to generate a revenue lift of about 10%.

Example—Price and Plan Inventory for a Reseller of Hand Tools:

The sample dataset given in "handtools_reseller.csv" contains information about the demand for products offered by a reseller of refurbished hand tools. The data contains information about the department, the category, the average price of competing products in the category, the MSRP (manufacturer's suggested retail price), the number of competing styles, the total sales events (for all products), and the number of sales events for the product. The data provides the "demand" for all products unlike the article which estimates demand for products that run out of stock. The first five lines of data are shown in Table 18.1.

The regression tree used to train and predict demand included all these variables. It based the fit on a 70–30 split between train and test samples. It grew a full tree and pruned it to match the lowest error on the test set. The details of the approach used are given in Chap. 16 on supervised learning. The final tree looks similar to the one given below (your tree might be slightly different depending on the test and train samples).

In this tree, moving to the left means answering "Yes." A naïve method for setting price and ordering inventory can be based upon this tree. For example, one can focus on introducing a new product into a category and ask what price to charge and how much inventory to order. Suppose the store is considering introducing a new product within a category that already has five competing styles in it. The average price (average) is 125, and the number of sales events (total sale) is 4. The MSRP for the product is also 125. Notice that the tree does not need other information to

Table 18.1 Demand for refurbished hand tools (sample from handtools_reseller.csv)

Product	Department	Category	MSRP	Price	Average price competing	Number of competing styles	Total sales events	Past 12 months sales events	Demand
9728	3	3	417	261	215.4	4	4	3	9
9131	3	2	290	124	133	7	5	1	18
2102	3	1	122	21	50.2	3	1	1	40
1879	1	2	258	84	135.6	3	6	1	38
1515	3	1	133	128	98.6	8	3	2	6

18 Retail Analytics   609

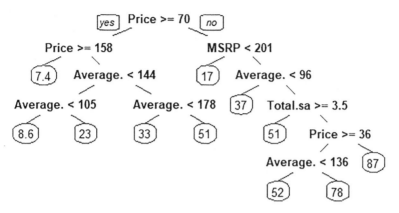

Fig. 18.2 Regression tree output

Table 18.2 Demand prediction model (Tree_Regression_Example.xlsx)

Price	MSRP	Average	Total sale	Price <70	Price >70	Demand	Number of competing styles	Revenue
$157.97	125	125	4	0	23	23	5	$3634.00

Given a price and other variables calculates the revenue
Does not adjust the average if the price changes nor computes impact on other products

predict the demand except the price of the product. We can then create an Excel model shown in Table 18.2.

In this simple model, one can set a price for the product and drop it down the tree to predict the demand using a series of "IF" statements. The demand if price <70 and price >70 are computed separately to minimize the number of "IF" blocks. The demand is the total of the two. Using *Solver*, an add-in to Microsoft Excel, one can maximize the revenue by varying the price. The "optimal" price is $158 and the optimal revenue is $3634. The reader is asked to make several enhancements to the model shown above in the chapter exercises.

The work by Ferreira et al. (2015) takes the typical approach of first estimating demand and then optimizing decisions. From the perspective of decision-making based on available data, the sequential nature of this approach is unnecessary, and one can conceive directly optimizing the decision of interest based on all available data. Indeed, Liyanage and Shanthikumar (2005) prove that formulating the decision as a function of the data and then directly optimizing it can lead to improved performance compared with the sequential approach.

The work by Ban and Rudin (2018) studies such an approach in the context of optimal inventory choice for a single product (at fixed prices). This is the setting of the classical newsvendor problem in which the goal is to select the stock level to minimize shortfall and holding costs (this dates back to Edgeworth 1888 and is a building block for more sophisticated models of stochastic inventory optimization; see Porteus 2002, for more background). The work by Ban and Rudin (2018) considers a feature-based approach in which the demand is assumed to be a function of many features which are observable before the order is placed, such as season and weather. Thus, the newsvendor's goal is to optimize the order quantity based on the observed features, when the demand is uncertain and dependent on these observable features. The paper sets up the optimization problem as a machine learning problem, including regularization as discussed before when dealing with a large number of features, and proves that the problem can be solved using LP, MIP, or QCQP programs. The methodology is applied to the case of nurse staffing in a hospital emergency room and shown to reduce costs by about 24% relative to existing benchmarks.

3.2 Incorporating Consumer Choice in Demand Modeling

The previous discussion focuses on exogenous demand models without getting into specifics on how demand arises. In retail situations, product substitution is abundant wherein when a customer's preferred item is out of stock, they substitute by selecting a different product. (The work reported in Gruen et al. 2002 suggests that such substitution rates can be significant.)

This directly motivates modeling customer preferences over all the products carried by the retailer. One of the workhorse models for such consumer choice modeling is the multinomial logit (MNL); also see Chap. 8 on advanced regression analysis and Chap. 23 on pricing analytics for further applications. The MNL model considers each consumer as selecting a product that maximizes their utility. In particular, the model describes each product-j from a set of numbered products $1,...,N$ via an average utility v_j so that the utility of a consumer i, from choosing product-j, is then given by $u_{i,j} = v_j + \varepsilon_{ij}$, where ε_{ij} is a zero-mean idiosyncratic noise term. In the MNL, the noise is modeled as a Gumbel distribution,[10] and for this distribution we can define the probability that customer i chooses product-j when given a subset of products A as follows:

$$P(Customer\ chooses\ product\ j) = \frac{\exp(v_j)}{1 + \sum_{k \in A} \exp(v_k)} \quad (18.2)$$

Mixed MNL is a natural extension of this model, in which customer heterogeneity is incorporated. Moreover, utility can be a function of the attributes of the product. In particular, the coefficient β applied to the attributes is not fixed but assumed to differ across customers. The coefficients are typically modeled as arising from a certain distribution with an unknown parameter that is used as part of the estimation process.

---

[10] An alternative has been proposed in Alptekinoglu and Semple (2016) that considers exponentially distributed noise, and the model is called the expononial choice (EC) model.

This MNL model has inbuilt an "independence of irrelevant alternatives," and the model becomes invalid in situations where this effect does not hold (for further details, refer to the book on choice models: Train 2009). To alleviate this, the Nested Logit model has been proposed in which the products are organized into categories in a tree structure, and the customer selects a product by iteratively selecting one of the many categories that the customer needs at each step (i.e., the customer selects the branch of tree). This continues until the customer selects a product (i.e., arrives at a leaf node of the tree).

The MNL model has proved to be extremely valuable in demand estimation when dealing with situations with product substitutability. It has been used extensively in the marketing literature, which uses panel data that includes both household- and store-level data aggregated over time. Such an approach traces back to Guadagni and Little (1983), Chandukala et al. (2008), and Wierenga et al. (2008), who present a detailed overview of choice modeling using panel data.

A small example of inventory planning under static substitution, assortment planning at AbD (inspired by the lecture notes of Dorothee Honhon at UT Dallas):

AbD is a wholesaler of granite blocks that offers a huge assortment of slabs. However, luckily, the slabs are grouped into types of stones, and even within stones, classified based on color and quality. For example, within granite slabs, they offer four grades that are priced differently. These grades come in light to dark shades. There are two modes of supply and two possible assumptions about demand. In the first mode of supply, AbD orders from Brazil after the customer places an order. This method is used only for the highest quality of slabs because the delivery commands a premium price. This method is labeled Make to Order or MTO. For MTO, the demand depends on the assortment offered and is static. There is no carrying cost. However, some customers might not opt to wait that long!

For the rest of the market, AbD uses a make to stock (MTS) approach. In this model, stone is ordered in anticipation of demand. For online sales, AbD has seen that the demand is somewhat predictable within each group. However, because the slabs are ordered in advance, AbD has to carry inventory. Moreover, the customer rarely sees the "stock in hand" and simply has to go with what is offered on the website. AbD has seen that there is little overlap between the variety preferred on the Internet channel and the colors preferred by walk-in customers. Sales also take place to walk in customers. Such in-store sales are highly random. Typically, AbD has seen the variance of the sales in a category to be close to the mean demand and therefore assumes a Poisson distribution for the demand. AbD faces the risk of carrying unsold inventory as well as that of running out of stock. In this case, the customer also sees the inventory. Therefore, the customer (who is typically impatient) might switch to another product within the category or walk out of the store.

For illustration purposes, we consider a sample (Table 18.3) within the dark category of slabs. The name of the slab, the selling price, and the carrying cost are listed. A market survey has identified the customer "gross value" attached to each of the products. The "value" of no purchase is also provided. All figures are in dollars.

Table 18.3 Sample from the dark category of slabs

Name of slab	Price per slab	Value attached by customer	Carrying cost per period
Roman Blue	1440	1750	14.4
Niagara	2323	2500	20
Forever Black	1717	1987	19.87
Violet Black	1744	1800	18

Table 18.4 AbD sells MTO

	Demand, D	5			Stocking cost	175		
	Scale factor, s	100						
	Price (p)	Value (v)	Hold cost	v − p	Exp (v − p)	Prob. purchase	Profit	Less stock cost
Roman Blue	1440	1750	14.4	310	22.19795128	0.4857	$ 664.49	$ 489.49
Niagara	2323	2500	20	177	5.870853361	0.1285	$ 285.58	$ 110.58
Forever Black	1717	1987	19.87	270	14.87973172	0.3256	$ 526.71	$ 351.71
Violet Black	1744	1800	18	56	1.7506725	0.0383	$ 63.36	$ (111.64)
						0.9781		
	The scale factor is chosen based on the no purchase probability (here = 0.0219)							
					Expected profit			$ 840.15

Profit = D * Prob. purchase (Price − Cost − hold cost)
Note that cost = 80% price

For simplicity, assume that the value of no purchase is normalized to zero. The net value to a customer from a purchase of a slab is equal to (Value − Price). Also, assume that each customer will purchase exactly one slab. The cost of a slab is 80% of the price. There is a per unit holding cost on account of interest, storage, breakage, etc. There is a fixed cost of 175 per slab stocked. The average number of customers per period in all examples is 5.

Consider the case when the product is sold as MTO. We use the MNL choice model in this example: In the simplest version of the model the probability that a customer chooses a product i is given by $\pi_i = e^{((v_i - p_i)/s)}/(1 + \sum_j e^{((v_l - p_j)/s)})$; where p_i stands for the price and v_i for the value attached to product-i, s is a scale factor, and the sum is taken over all products which are compared with product-i. Chapter 23, Pricing Analytics, has further details about this model of consumer choice. Let D be the average demand. Assume the scale factor is 100. The expected profit is then given by

$$D \times \Sigma_j(0.2 \, \pi_j p_j - \text{Hold cost}) - \text{number of products stocked} \times \text{stocking cost}$$

The expected profitability from offering all four products in the assortment is computed as shown (the profit is computed after subtracting the stocking cost) in Table 18.4.

The reader is asked to verify these calculations, as well as evaluate whether this is the optimal MTO assortment to offer in the exercises.

Continuing the discussion of consumer choice-based demand modeling, the work in Vulcano et al. (2012) analyzes a model of demand that combines an MNL choice model with nonhomogeneous Poisson model of arrivals. The paper jointly estimates the customer preferences and parameters of the (demand) arrival process using sales data, which includes product availability information. The paper uses an expectation–maximization (EM) algorithm and demonstrates the efficacy of its method using industry datasets, including that from an actual retail chain.

While the MNL is a great parametric class of models, it does have additional shortcoming in modeling product substitution. Some examples of papers that consider alternative models of product substitution are Anupindi et al. (1998), Chong et al. (2001), Bell et al. (2005), Kök and Fisher (2007), and Fisher and Vaidyanathan (2014).

Recently, a Markov chain-based choice model that generalizes the MNL has been proposed in Blanchet et al. (2016). The paper uses state transitions of the Markov chain to model product substitutions and to approximate general models, including a mixture of MNLs and nested logit.

An alternative to these models is the class of nonparametric choice models that have recently been studied in Haensel and Koole (2011), Farias et al. (2013), and Van Ryzin and Vulcano (2017). In these papers, the customer types are defined via a direct ranking of the product types, and when faced with choosing from a set of offered products, the customers go down their rank list and pick the highest-ranked available product (the list includes the no-purchase option, so customers may leave without a purchase). Farias et al. (2013) take a robust optimization approach by solving for customer type distribution that leads to worst-case revenue, while imposing the observed choice data as a constraint. A key result therein is that the demand model obtained is approximately the sparsest choice model. They show a 20% improvement in accuracy over benchmark models for an automobile sales dataset. Haensel and Koole (2011) and Van Ryzin and Vulcano (2017) use EM methods to estimate the unknown parameters. A recent paper (Jagabathula and Vulcano 2017) also proposes a nonparametric approach that focuses on repeated consumer interactions wherein customer preferences may be altered by price or display promotions, and customers exhibit bounded rationality by only considering an unobserved subset of the offered products, referred to as consideration sets. Using a grocery dataset, they show 40% improvement in prediction accuracy over state-of-the-art benchmarks based on variants of the MNL commonly used in current industry practice. Recently, Jagabathula and Rusmevichientong (2016) add to this literature by incorporating pricing by associating customers with a price threshold in addition to the consideration sets.

Turning to decision-making, there is a large collection of papers that have looked at assortment optimization using such choice models. Overall, there are three decisions here: assortment to offer; the inventory levels of each product in assortment; and the price of each item in the assortment. Assortment optimization introduces a complexity because it represents a combinatorial problem that may require enumeration of the different products. One of the pioneering papers by Ryzin and Mahajan (1999) optimizes the inventory and assortment using a MNL-based demand framework. In the paper, a nesting feature is observed, so that the products can be ranked by their value and the optimal assortment only requires considering nested subsets of the ranked products. Empirically, such optimization has been seen to improve a firm's financials significantly, for instance, Kök and Fisher (2007) estimate a 50% increase in profit to a retailer, Fisher and Vaidyanathan (2014) report a sales lift of 3.6% and 5.8% for two different product categories, and Farias et al. (2013) and Jagabathula and Rusmevichientong (2016) estimate about 10% increases in revenue. We direct the reader to Rusmevichientong et al. (2006), Smith et al. (2009), Honhon et al. (2010), and Honhon et al. (2012) for optimized decision-making under variants of rank-based choice models. More recently, there has been a growing interest in optimizing assortments using nonparametric methods. For instance, Bertsimas and Mišic (2015) use a nonparametric choice model related to Farias et al. (2013) but forego the robust approach to directly estimate the choice model by efficiently solving a large-scale linear optimization problem using column generation and then solve the assortment optimization piece based on the solution to a practically tractable mixed integer optimization problem. The previously referenced Jagabathula and Rusmevichientong (2016) solves a joint assortment and pricing problem (which is known to be NP-hard) using an approximation algorithm with a provable performance guarantee based on a DP formulation.

Interestingly, as we move to dynamic assortments, which become more relevant in the context of e-tail, Bernstein et al. (2015) and Golrezaei et al. (2014) solve this problem in a limited inventory setting. The latter, in fact, consider a very general consumer choice model and propose an algorithm that does not require knowledge of customer arrival information.

Example: Assortment over Internet

Consider the previous example and the case when the assortment is offered over the Internet. In this case the assortment is changed every selling season. We model this as MTS problem with *no* substitution. Once the product runs out, the customer who asks for it gets a message that it is out of stock, and s/he walks away with no purchase. We are also told that disposing of unsold product at the end of a period (or selling season) recovers 85% of the cost of the product. This cost is estimated as the cost of the item less the cost of shipping to a discount outlet and holding cost for selling the product at the end of the season. The demand is assumed to be distributed Poisson with mean equal to 5. Though we can calculate the expected profit analytically, to illustrate an alternative method, we will use simulations! For doing this, we generate demand 1000 times. In each simulation, we draw the number of customers according to the Poisson distribution. Then determine which slab is preferred by each of the customers. Given a stocking quantity, it is straightforward but a little tedious to calculate the expected profit. The "Assortment_Examples.xlsx" sheet contains the simulation. The sample summary results when stocking one slab of each type is shown below (data and average profit above and first few rows of simulation below in Tables 18.5 and 18.6).

For example, in the first simulation, three customers arrived. All wanted Roman Blue. The actual sale was for one slab of Roman Blue and one slab each of Niagara,

18 Retail Analytics

Table 18.5 AbD sells MTS

	Demand, D		5		Stocking cost	175		
	Scale factor, s		100					
	Price (p)	Value (v)	Stock cost	v-p	Exp (v-p)	Prob. Purchase	Profit	Less stock cost
Roman Blue	1440	1750	14.4	310	22.19795128	0.4857	$ 685.07	$ 510.07
Niagara	2323	2500	20	177	5.870853361	0.1285	$ 284.03	$ 109.03
Forever Black	1717	1987	19.87	270	14.87973172	0.3256	$ 544.66	$ 369.66
Violet Black	1744	1800	18	56	1.7506725	0.0383	$ 52.41	$ (122.59)
						0.9781		
	The scale factor is chosen based on the no purchase probability (here = 0.0219)							
					Expected profit			$ 866.16

Table 18.6 MTS profit using simulation

Simulation	Random	Profit	Inventory	1 Roman Blue	1 Niagara	1 Forever Black	1 Violet Black	Total No. of customers
1	0.17144837	$ (1106.08)		3	0	0	0	3
2	0.73233508	$ 195.36		1	3	0	2	6
3	0.98212926	$ 744.80		6	1	2	1	10
4	0.48677441	$ 1.44		3	0	1	1	5
5	0.13994017	$ (556.64)		2	0	1	0	3

Forever Black, and Violet Black had to be salvaged at a loss. The reader is asked to verify the simulation setup in the exercise and then make optimal choices.

AbD sells to walk-in customers or sells MTS with substitution. We can already see how this problem becomes vastly more complicated when there is substitution! In addition to creating arrivals, we have to keep track of the sequence in which the customers arrive and then see if there is stock. If there is stockout, we need to model whether there is substitution from the remaining products or the customer leaves the store empty-handed. Exercise 18.4 gives a simple example to illustrate these ideas.

In many settings, customer preferences are not known, and one may need to learn these while simultaneously optimizing the assortment. Caro and Gallien (2007) and Rusmevichientong et al. (2010) were among the first to study this problem. Caro and Gallien (2007) undertook a Bayesian learning approach in which the underlying primitives have a certain distribution, and the Bayesian approach is used to learn these parameters. On the other hand, Rusmevichientong et al. (2010) used an adaptive learning approach in which such priors are not assumed, and an explore–exploit paradigm is used: in this approach, the decision-maker balances "exploration" to collect relevant data points, with "exploitation" to generate revenue based on the data observed thus far. More recently, the notion of *personalized assortments* is becoming prevalent, especially in e-tail settings, wherein a customer could be shown an assortment of items based on customer-specific information, such as historical data on preferences, demographics, etc. In such a situation, estimating customer choice becomes a high-dimensional estimation problem (dimensionality equals the number of customers times the number of products). A recent paper (Kallus and Udell 2016) considers such a problem and proposes a low-rank mixed multinomial logit choice model in which the customer choice matrix is assumed to have a low-rank latent representation. The paper proposes a nuclear-norm regularized maximum likelihood estimator for learning this model and shows that it can learn the model with few customer interactions. Broadly, with the growing data collection capabilities, we expect further proliferation in such models that estimate individual customer-level choices and use it to make personalized decisions.

4 Business Challenges and Opportunities

4.1 Omni-Channel Retail

The tremendous success of e-commerce has led many retailers to augment their brick-and-mortar stores with an online presence, leading to the advent of multichannel retail. In this approach the retailer has access to multiple channels to engage with and sell to customer. Typically, each of these channels is managed separately. This multichannel approach has been overshadowed by what is commonly referred to as *omni-channel* retail, in which the firm integrates all the channels to provide a seamless experience to customers. A good example of such an approach is the "buy online, pick up in store" (BOPS) approach that has become quite commonplace. This seamless approach inarguably improves the customer experience and overall sales; however, it can lead to unintended outcomes. For instance, Bell et al. (2014) show that such a strategy can reduce online sales and instead lead to an increase in store sales and traffic. In that context, the authors find that additional sales are generated by cross-selling in which the customers who use the BOPS functionality buy additional products in the stores, and further there is a channel effect as well in which online customers may switch to becoming brick-and-mortar customers.

The benefits of the omni-channel approach are even spurring online retailers to foray into physical stores. For instance, recent studies show how, by introducing an offline channel via display showrooms, WarbyParker.com was able to increase both overall and the online channel's demand (see Bell et al. 2014).

Thus, there is significant value for a retailer to foray into omni-channel. However, while doing so, it is crucial for the firm's retail analytics to transcend to omni-channel analytics for correct estimation and optimal decision-making.

4.2 Retail Startups

There has been a spurt in retail analytics startups recently. A majority of these companies can be classified as those using technology to aid in data collection and those that are using sophisticated means of analyzing the data itself.

In terms of data collection, there are many startups that cater to the range of retailers both small and large. Some illustrative examples here are Euclid Analytics, which uses in-store Wi-Fi to collect information on customers via their smartphones. The company is able to collect in-store behavior of customers and also data on repeat visits. Collection of Wi-Fi-based information also allows the retailer to track what customers do online while in store. This lets the retailer better understand their customer base, including what products they are researching on their smartphones (showrooming). Another recent startup is Dor, which is targeted at smaller retailers and sells a device that counts the foot traffic in the store. It then provides a dashboard view of the traffic and provides insights to optimize staffing. Startups like Brickstream and Footmarks produce sensors that monitor foot traffic, so associates can react to shoppers in real time. Swirl works with brands like Urban Outfitters, Lord & Taylor, and Timberland to monitor shopper behavior with beacons. Point Inside provides beacons to Target. Startups like Estimote, Shelfbucks, and Bfonics leverage beacons for in-store proximity marketing, such as sending mobile notifications to shoppers about the products they are currently browsing.

Video analytics is another exciting area that startups are getting into. For instance, Brickstream uses video analytics to capture in-depth behavior intelligence. RetailNext is one of the larger startups in this domain and covers a wide gamut of data collection, starting with Wi-Fi-based data collection to more sophisticated methods using video camera feeds. RetailNext also delves into an array of analytics solutions including staffing schedules and some A/B testing. Video analytics helps in heat mapping customer paths and can also be used for loss prevention.

Oak Labs builds interactive touchscreen mirrors that aim to revolutionize the fitting room. The mirror allows customers to explore product recommendations and digitally seek assistance from store associates. The use of technology here enhances the customer experience and collects valuable data at the same time.

PERCH produces interactive digital displays used by brands like Kate Spade and Cole Haan. Blippar focuses on integrating digital and physical domains by using an app that unlocks content upon scanning products. Aila manages interactive in-aisle tablets for stores that provide customers with detailed product information upon scanning a product barcode. These startups focus on improving customer experience by providing more information while also collecting information on how customers choose products that can help the retailer make smarter decisions.

Turning to analytics, some startups such Sku IQ and Orchestro focus on providing a unified view to multichannel firms. Sku IQ provides a unified view of inventory, sales, and customers from all channels. Orchestro focuses on demand estimation by combining data from POS systems, internal ERP, and third-party data into a common view of demand. Orchestro was recently acquired by E2open, a cloud-based supply chain network management systems provider.

Turning to advanced analytics-focused startups, Blue Yonder focuses on analytics to optimize price and inventory replenishment, while Celect focus on using machine learning to optimize product assortments. Finally, Stitch Fix provides an interesting perspective on how personalization for customers may have enormous business potential. Stitch Fix curates fashion apparel and gives recommendations to its members. To do so, it uses machine learning algorithms that make suggestions to human stylists who then use their experience and knowledge to make recommendations to the end customers.

5 Conclusion and Further Reading

This chapter has set out various models and approaches used in the retail industry. The survey is not exhaustive, simply because of changes that take place every day in design, manufacture, and delivery of products and services. Retailing and supply chains are joined together, and any progress in one will lead to changes in the other. The changes do not occur synchronously—due to constant experimentation, opening of new markets, new channels, and proliferation of supply sources—the approach has been opportunistic. The references and the journals that published the papers cited in this chapter are a good starting point for learning more on the subject and staying on top of the developments.

Electronic Supplementary Material

All the datasets, code, and other material referred in this section are available in https://www.allaboutanalytics.net.

- Data 18.1: handtools_reseller.csv
- Data 18.2: Assortment_Examples.xlsx
- Data 18.3: Tree_Regression_Example.xlsx

Exercises

Ex. 18.1 Reseller of Hand Tools:

(a) Replicate the regression tree shown in Fig. 18.2. The details of the procedure can be found in Chap. 16 on supervised learning.
(b) Enhance the model to consider the impact of product price on the average price of competing products.

(c) Enhance the model to not only consider the new product revenue but also the revenue of competing products. To do so, assume the other products have prices 75, 100, 125, 150, and 175, and their MSRPs are equal to their prices.
(d) The store management would like to ensure that the total sales of the new product is at least 25 units. How does this change affect your solution?
(e) The store wishes to order inventory based on the forecast demand. The store manager argues that the demand prediction is just one number! He says, "We need an interval forecast." How would you modify the model to predict an interval (such as (20,26))? What inventory would you order? (Hint: Look at the error associated with the prediction at a node of the regression tree. Can you use this?)

Ex. 18.2 AbD Sells MTO:

(a) Compute the expected profit (shown in the Table 18.4) using the formulae provided in the chapter. You can consult AssortmentExamples.xlsx MTO sheet.
(b) Is this the optimal assortment to offer?
(c) What would change if the stocking cost were to increase to $230 per product stocked?
(d) How would AbD evaluate whether to add a new product to this category? Create your own example and show the analysis.

Ex. 18.3 AbD Sells MTS:

(a) Compute the expected profit (shown in Table 18.5) using the formulae provided in the chapter. Verify the simulation.
(b) Is this the optimal assortment to offer? Can you optimize the expected profit?
(c) What would change if the stocking cost were to increase to $230 per product stocked?
(d) How would AbD evaluate the impact of a 5% increase in price of all products?

Ex. 18.4 AbD Sells to Walk-Ins:

AbD decides to offer only two types of slabs to walk-in customers, Roman Blue and Forever Black. Assume that the choice probabilities are 0.4857 and 0.3256, respectively. If a customer does not find the desired product, s/he will switch to the other product with half the original probability (0.2428 and 0.1628). AbD keeps just one slab of each as inventory. Use the same costs and prices as in the previous exercise. Calculate the expected profit when exactly one customer arrives and also when exactly two customers arrive. (Hint: Enumerate all possible sequences in which a product is demanded.)

References

Alptekinoğlu, A., & Semple, J. H. (2016). The exponomial choice model: A new alternative for assortment and price optimization. *Operations Research, 64*(1), 79–93.
Anupindi, R., Dada, M., & Gupta, S. (1998). Estimation of consumer demand with stock-out based substitution: An application to vending machine products. *Marketing Science, 17*(4), 406–423.

Au, K. F., Choi, T. M., & Yu, Y. (2008). Fashion retail forecasting by evolutionary neural networks. *International Journal of Production Economics, 114*(2), 615–630.

Ban, G.-Y., & Rudin, C. (2018). *The big data newsvendor: Practical insights from machine learning*. Operations Research. Available at SSRN: https://ssrn.com/abstract=2559116. doi.org/10.2139/ssrn.2559116

Bell, D. R., Bonfrer, A., & Chintagunta, P. K. (2005). Recovering stockkeeping-unit-level preferences and response sensitivities from market share models estimated on item aggregates. *Journal of Marketing Research, 42*(2), 169–182.

Bell, D. R., Gallino, S., & Moreno, A. (2014). *Offline showrooms and customer migration in omni-channel retail*. Retrieved August 1, 2018, from https://courses.helsinki.fi/sites/default/files/course-material/4482621/17.3_MIT2014%20Bell.pdf

Bernstein, F., Kök, A. G., & Xie, L. (2015). Dynamic assortment customization with limited inventories. *Manufacturing & Service Operations Management, 17*(4), 538–553.

Bertsimas, D., & Mišic, V. V. (2015). *Data-driven assortment optimization*. Operations Research Center, MIT.

Blanchet, J., Gallego, G., & Goyal, V. (2016). A Markov chain approximation to choice modeling. *Operations Research, 64*(4), 886–905.

Bradlow, E. T., Gangwar, M., Kopalle, P., & Voleti, S. (2017). The role of big data and predictive analytics in retailing. *Journal of Retailing, 93*(1), 79–95.

Caro, F., & Gallien, J. (2007). Dynamic assortment with demand learning for seasonal consumer goods. *Management Science, 53*(2), 276–292.

Chandukala, S. R., Kim, J., Otter, T., Rossi, P. E., & Allenby, G. M. (2008). Choice models in marketing: Economic assumptions, challenges and trends. *Foundations and Trends® in Marketing, 2*(2), 97–184.

Chong, J. K., Ho, T. H., & Tang, C. S. (2001). A modeling framework for category assortment planning. *Manufacturing & Service Operations Management, 3*(3), 191–210.

DeHoratius, N. (2011). Inventory record inaccuracy in retail supply chains. *Wiley encyclopedia of operations research and management science* (pp. 1–14).

DeHoratius, N., & Raman, A. (2008). Inventory record inaccuracy: An empirical analysis. *Management Science, 54*(4), 627–641.

Edgeworth, F. Y. (1888). The mathematical theory of banking. *Journal of the Royal Statistical Society, 51*(1), 113–127.

Farias, V. F., Jagabathula, S., & Shah, D. (2013). A nonparametric approach to modeling choice with limited data. *Management Science, 59*(2), 305–322.

Ferreira, K. J., Lee, B. H. A., & Simchi-Levi, D. (2015). Analytics for an online retailer: Demand forecasting and price optimization. *Manufacturing & Service Operations Management, 18*(1), 69–88.

Fisher, M., & Vaidyanathan, R. (2014). A demand estimation procedure for retail assortment optimization with results from implementations. *Management Science, 60*(10), 2401–2415.

Golrezaei, N., Nazerzadeh, H., & Rusmevichientong, P. (2014). Real-time optimization of personalized assortments. *Management Science, 60*(6), 1532–1551.

Gruen, T. W., Corsten, D. S., & Bharadwaj, S. (2002). *Retail out-of-stocks: A worldwide examination of extent, causes and consumer responses*. Washington, DC: Grocery Manufacturers of America.

Guadagni, P. M., & Little, J. D. (1983). A logit model of brand choice calibrated on scanner data. *Marketing Science, 2*(3), 203–238.

Haensel, A., & Koole, G. (2011). Estimating unconstrained demand rate functions using customer choice sets. *Journal of Revenue and Pricing Management, 10*(5), 438–454.

Honhon, D., Gaur, V., & Seshadri, S. (2010). Assortment planning and inventory decisions under stockout-based substitution. *Operations Research, 58*(5), 1364–1379.

Honhon, D., Jonnalagedda, S., & Pan, X. A. (2012). Optimal algorithms for assortment selection under ranking-based consumer choice models. *Manufacturing & Service Operations Management, 14*(2), 279–289.

Jagabathula, S., & Rusmevichientong, P. (2016). A nonparametric joint assortment and price choice model. *Management Science, 63*(9), 3128–3145.

Jagabathula, S., & Vulcano, G. (2017). A partial-order-based model to estimate individual preferences using panel data. *Management Science, 64*(4), 1609–1628.

James, G., Witten, D., Hastie, T., & Tibshirani, R. (2013). *An introduction to statistical learning* (Vol. 112). New York: Springer.

Kallus, N., & Udell, M. (2016). Dynamic assortment personalization in high dimensions. *arXiv preprint arXiv, 1610*, 05604.

Kök, A. G., & Fisher, M. L. (2007). Demand estimation and assortment optimization under substitution: Methodology and application. *Operations Research, 55*(6), 1001–1021.

Liyanage, L. H., & Shanthikumar, J. G. (2005). A practical inventory control policy using operational statistics. *Operations Research Letters, 33*(4), 341–348.

Ma, S., Fildes, R., & Huang, T. (2016). Demand forecasting with high dimensional data: The case of SKU retail sales forecasting with intra-and inter-category promotional information. *European Journal of Operational Research, 249*(1), 245–257.

Porteus, E. L. (2002). *Foundations of stochastic inventory theory*. Stanford, CA: Stanford University Press.

Rusmevichientong, P., Shen, Z. J. M., & Shmoys, D. B. (2010). Dynamic assortment optimization with a multinomial logit choice model and capacity constraint. *Operations Research, 58*(6), 1666–1680.

Rusmevichientong, P., Van Roy, B., & Glynn, P. W. (2006). A nonparametric approach to multiproduct pricing. *Operations Research, 54*(1), 82–98.

Ryzin, G. V., & Mahajan, S. (1999). On the relationship between inventory costs and variety benefits in retail assortments. *Management Science, 45*(11), 1496–1509.

Smith, J. C., Lim, C., & Alptekinoglu, A. (2009). Optimal mixed-integer programming and heuristic methods for a Bilevel Stackelberg product introduction game. *Naval Research Logistics, 56*(8), 714–729.

Tibshirani, R. (1996). Regression shrinkage and selection via the lasso. *Journal of the Royal Statistical Society. Series B (Methodological)*, 267–288.

Train, K. E. (2009). *Discrete choice methods with simulation*. Cambridge: Cambridge University Press. https://eml.berkeley.edu/~train/distant.html.

van Ryzin, G., & Vulcano, G. (2017). An expectation-maximization method to estimate a rank-based choice model of demand. *Operations Research, 65*(2), 396–407.

Vulcano, G., Van Ryzin, G., & Ratliff, R. (2012). Estimating primary demand for substitutable products from sales transaction data. *Operations Research, 60*(2), 313–334.

Wierenga, B., van Bruggen, G. H., & Althuizen, N. A. (2008). Advances in marketing management support systems. In B. Wierenga (Ed.), *Handbook of marketing decision models* (pp. 561–592). Boston, MA: Springer.

Chapter 19
Marketing Analytics

S. Arunachalam and Amalesh Sharma

1 Introduction

It is very hard to ignore the potential of analytics in bringing robust insights to the boardroom in order to make effective firm, customer, and product/brand level decisions. Advance analytics tools, available data, and allied concepts have enormous potential to help design effective business and marketing strategies. In such a context, understanding the tools and their various implications in various different contexts is essential for any manager. Indeed, the robust use of the analytics tools has helped firms increase performance in terms of sales, revenues, profits, customer satisfaction, and competition. For details of how marketing analytics can help firms increase its performance, please refer to Kumar and Sharma (2017).

Marketing analytics can help firms realize the true potential of data and explore meaningful insights. Marketing analytics can be defined as a "high technology-enabled and marketing science model-supported approach to harness the true values of the customer, market, and firm level data to enhance the effect of marketing strategies" (Kumar and Sharma 2017; Lilien 2011).

Basically, marketing analytics is the creation and use of data to measure and optimize marketing decisions. Marketing analytics comprises tools and processes

---

Electronic supplementary material The online version of this chapter (https://doi.org/10.1007/978-3-319-68837-4_19) contains supplementary material, which is available to authorized users.

S. Arunachalam (✉)
Indian School of Business, Hyderabad, Telangana, India
e-mail: S_Arunachalam@isb.edu

A. Sharma
Texas A&M University, College Station, TX, USA

© Springer Nature Switzerland AG 2019
B. Pochiraju, S. Seshadri (eds.), *Essentials of Business Analytics*, International Series in Operations Research & Management Science 264, https://doi.org/10.1007/978-3-319-68837-4_19

that can inform and evaluate marketing decisions right from product/service development to sales (Farris et al. 2010; Venkatesan et al. 2014). According to CMO survey, spending on analytics will increase from 4.6% to 22% in the next 3 years.[1] This shows the increasing importance of analytics in the field of marketing. Top marketers no longer rely on just intuition or past experience to make decisions. They want to make decisions based on data. But in the same survey, "Lack of processes or tools to measure success through analytics" and "Lack of people who can link to marketing practice" have been cited as the top two factors that prevent marketers from using advanced marketing analytic tools in the real world.

This chapter is a step toward closing these two gaps. We present tools that both inform and measure the success of marketing activities and strategies. We also hope that this will help current and potential marketers get a good grasp of marketing analytics and how it can be used in practice (Lilien et al. 2013; Kumar et al. 2015). Since analytics tools are vast in numbers and since it is not feasible to explain the A–Z of every tool here, we select a few commonly used tools and attempt to give a comprehensive understanding of these tools. Interested readers can look at the references in this chapter to get more advanced and in-depth understanding of the discussed tools.

The processes and tools discussed in this chapter will help in various aspects of marketing such as target marketing and segmentation, price and promotion, customer valuation, resource allocation, response analysis, demand assessment, and new product development. These can be applied at the following levels:

- *Firm*: At this level, tools are applied to the firm as a whole. Instead of focusing on a particular product or brand, these can be used to decide and evaluate firm strategies. For example, data envelopment analysis (DEA) can be used for all the units (i.e., finance, marketing, HR, operation, etc.) within a firm to find the most efficient units and allocate resources accordingly.
- *Brand/product*: At the brand/product level, tools are applied to decide and evaluate strategies for a particular brand/product. For example, conjoint analysis can be conducted to find the product features preferred by customers or response analysis can be conducted to find how a particular brand advertisement will be received by the market.
- *Customer*: Tools applied at customer level provide insights that help in segmenting and targeting customers. For example, customer lifetime value is a forward-looking customer metric that helps assess the value provided by customers to the firm (Fig. 19.1).

Before we move further, let us look at what constitutes marketing analytics. Though it is an ever-expanding field, for our purpose, we can segment marketing analytics into the following processes and tools:

---

[1] https://cmosurvey.org/2017/02/cmo-survey-marketers-to-spend-on-analytics-use-remains-elusive/ (accessed on Jul 6, 2018).

Fig. 19.1 Levels of marketing analytics application

1. Multivariate statistical analysis: It deals with the analysis of more than one outcome variable. Cluster analysis, factor analysis, perceptual maps, conjoint analysis, discriminant analysis, and MANOVA are a part of multivariate statistical analysis. These can help in target marketing and segmentation, optimizing product features, etc., among other applications.
2. Choice analytics: Choice modeling provides insights on how customers make decisions. Understanding customer decision-making process is critical as it can help to design and optimize various marketing mix strategies such as pricing and advertising. Largely, Logistic, Probit, and Tobit models are covered in this section.
3. Regression models: Regression modeling establishes relationships between dependent variables and independent variables. It can be used to understand outcomes such as sales and profitability, on the basis of different factors such as price and promotion. Univariate analysis, multivariate analysis, nonlinear analysis, and moderation and mediation analysis are covered in this section.
4. Time-series analytics: Models stated till now mainly deal with cross-sectional data (however, choice and regression models can be used for panel data as well). This section consists of auto-regressive models and vector auto-regressive models for time-series analysis. These can be used for forecasting sales, market share, etc.
5. Nonparametric tools: Non parametric tools are used when the data belongs to no particular distribution. Data envelopment analysis (DEA) and stochastic frontier analysis (SFA) are discussed in this section and can be used for benchmarking, resource allocation, and assessing efficiency.
6. Survival analysis: Survival analysis is used to determine the duration of time until an event such as purchase, attrition, and conversion happens. Baseline hazard model, proportional hazard model, and analysis with time varying covariates are covered in this section.

7. Salesforce/sales analytics: This section covers analytics for sales, which includes forecasting potential sales, forecasting market share, and causal analysis. It comprises various methods such as chain ratio method, Delphi method, and product life cycle analysis.
8. Innovation analytics: Innovation analytics deals specifically with new products. New product analysis differs from existing product analysis as you may have little or no historical data either for product design or sales forecasting. Bass model, ASSESSOR model, conjoint analysis can be used for innovation analytics.
9. Conjoint analysis: This section covers one of the most widely used quantitative methods in marketing research. Conjoint (trade-off) analysis is a statistical technique to measure customer preferences for various attributes of a product or service. This is used in various stages of a new product design, segmenting customers and pricing.
10. Customer analytics: In this section, we probe customer metrics such as customer lifetime value, customer referral value, and RFM (recency, frequency, and monetary value) analysis. These can be used for segmenting customers and determining value provided by customers.

This chapter provides a comprehensive understanding of the marketing analytics field. We explain a part of it through basic and advanced regression, data envelopment analysis, stochastic frontier analysis, innovation analytics, and customer analytics.

Each section in this chapter has marketing related problems and solutions to reinforce concepts and enhance understanding. We describe how various tools can be implemented for a particular objective. We also discuss limitations of the tools under consideration. You should familiarize yourself with all the processes and tools so that you can choose appropriate ones for your analysis.

2 Methods of Marketing Analytics

2.1 Regression

Interaction Effect

This section covers the steps for conducting interactions between continuous variables in multiple regression (MR). We provide an intuitive understanding of this concept and share a step-by-step method for statistically conducting and probing an interaction effect. Chapter 7 on linear regression analysis contains detailed explanation of the theory underlying MR.

Interaction effects, also called moderator effects, are the combined effects of multiple independent variables on the dependent variable. They represent situations where the effect of a variable on a dependent variable is contingent on another variable (Baron and Kenny 1986). Let us consider a hypothetical example where a manager is trying to understand the effect of advertisement and price on sales of a product. It can be argued that price and advertisement may have interaction effect on sales, that is, when both advertisement and price change it may have a different effect on product sales as compared to when either advertisement or price changes independently. To capture such effects, we require the following MR equation (error is omitted for ease of presentation in all equations):

$$Y = i_0 + a\,X + b\,Z + c\,XZ + \text{Error} \qquad (19.1)$$

where

$Y =$ continuous dependent variable (sales in the above example)

$X, Z =$ continuous independent variables (advertisement and price in the above example)

$XZ =$ new variable computed as the product of X and Z (product of advertisement and price in the above example)

$i_0 =$ intercept

a, b, and c $=$ slopes

An important aspect to remember in specifying an equation with interaction terms is that the lower order terms should always be present in that equation. That is, it is incorrect to test for interaction by omitting X and Z and just having the XZ term alone. Here we assume that the independent variables, X and Z, are mean-centered. To mean-center a variable the value of a variable is subtracted by the mean of the variable in the sample (mean-centered variable = variable − mean [variable]). The product term XZ is computed after mean-centering both X and Z. The dependent variable, Y, is generally not mean-centered. Mean-centering helps in ease of interpretation of effects and also removes nonessential multicollinearity. More details about benefits of mean-centering (it is very useful to know) can be found from the book Aiken and West (1991) and from other sources mentioned in the references section (Cohen et al. 2003).

How does the presence of the interaction term, XZ, help in addressing the managerial question above? Why is it needed and why cannot interaction effects be uncovered using the simple MR equation: $Y = i_0 + a\,X + b\,Z$ that does not contain the interaction term, XZ? To understand this let us start with the simple MR equation that does not have the XZ term. Here the slopes "a" and "b" are the effect of X and Z on Y, respectively. The regression of Y on X (i.e., slope "a") is constant across legit (explained below) values of Z and vice versa for regression of Y on Z (i.e., slope "b"). This means that the regression of Y on X is independent of Z because for any value of Z, value of the slope would be "a." Hence, this does not answer the managerial question asked above as it does not capture the variation in the effect of X according to Z and vice versa. However, this is not the case if we consider Eq. (19.1) that has the interaction term XZ. To understand why and how let us rewrite Eq. (19.1) in this way:

$$Y = (\boldsymbol{a} + \boldsymbol{c}\,\mathbf{Z})\,X + (\boldsymbol{i_0} + \boldsymbol{b}\,\mathbf{Z}) \qquad (19.2)$$

This expanded view of Eq. (19.1) helps answer the two questions we raised above. It is seen from Eq. (19.2) that the slope of the regression of Y on X is now the term ($a + c\ Z$), unlike just "a" from the simple equation. This term ($a + c\ Z$) is called simple slope in the moderation literature as the effect of Y on X is now conditional on the value of Z! Therefore, the effect is now dependent on Z rather than being independent as noted above. To go back to our managerial example this would mean that effect of advertisement (X) on sales (Y) is dependent on price (Z). This is precisely what we would like to achieve or test. This also means that for every value of Z, regression of Y on X would have a different line, which is called a simple regression line. Equation (19.2) could also be rewritten by pulling out Z and having a simple slope that is (b + c X), which would help understand how the effect of price (Z) on sales (Y) is dependent on advertisement (X).

Estimation: Estimating Eq. (19.2) using any statistical software like *STATA (or R)*, is straightforward. The command *regress (lm)* can be used to estimate the equation after mean-centering the independent variables and creating the XZ term. Interpretation and probing of interaction effects need more understanding (Preacher et al. 2006). Let us assume that the slope "c" of XZ term is statistically significant (at $p < 0.05$). What does this mean? How to interpret this significant effect?

There are two very important takeaways when the slope "c" is significant. First, it means that simple slopes for any pair of simple regression lines obtained using two different legit values of Z are statistically different. Second, for any specific value of Z the researcher has to test whether the simple regression line is significantly (i.e., statistical significance) different from zero. These two points are better understood and observed by plotting graphs of the interaction. We recommend that the researcher should always plot graphs to thoroughly understand the interaction effects. Before we probe these effects through graphs let us understand the meaning of "legit values" of Z. Let us recall that we have mean-centered the variables X and Z and hence the mean of these two variables would be zero and the range would be from negative to positive values around zero. It is always a good practice to choose two values of Z—one low value and another high value as one standard deviation below and above the mean (which is zero now) respectively, to plot the simple regression lines. The reason is that these two values would be within the range of values of Z in our sample. This is important because we should not choose a value that is not within the sample range. The researcher is free to choose any two values that are within the range of Z. If we choose an arbitrary value and if that value is outside of the range of Z, then we are testing the effects for out-of-sample data, which is incorrect. Therefore, researchers should pay attention in choosing legit values of Z.

Let us assume we choose two legit values of Z, namely, Z-low (Z_L) and Z-high (Z_H). So the simple slopes from Eq. (19.2) that determine the simple regression lines are ($a + c\ Z_L$) and ($a + c\ Z_H$). Therefore, we can plot two regression lines by substituting the two values of Z in Eq. (19.2). These two lines would then represent the effect of advertisement (X) on sales (Y) when the price is low (Z_L) and when the price is high (Z_H). Deriving from Eq. (19.2) the simple regression equations are:

19 Marketing Analytics

$$\text{Sales for lower price}: Y_L = (a + c\mathbf{Z_L})X + (i_0 + b\mathbf{Z_L})$$
$$\text{Sales for higher price}: Y_H = (a + c\mathbf{Z_H})X + (i_0 + b\mathbf{Z_H})$$
(19.3)

Let us assume we have estimated Eq. (19.2) from a sample dataset that contained values for Y, X, and Z and that we have this prediction equation (intercept i_0 is omitted for pedagogical reason; X and Z are mean-centered):

$$\widehat{Y} = 0.12\,X + 0.09Z + 0.46XZ \tag{19.4}$$

The parameters "a," "b," and "c" are statistically significant ($p < 0.05$). Please note it is not necessary for "a" and "b" to be significant to probe the interaction effect. It is however important to have "c," the estimate for the interaction term XZ, to be statistically significant to probe the interaction effect (refer above for the two important takeaways). Let us also assume that one standard deviations of X and Z are 0.5 and 0.5. The following steps help in plotting the graphs.

We calculate the predicted value of Y (sales) for the two legit values of X and Z. So we take low and high values of X and Z and plug it in Eq. (19.4). As one standard deviation of mean-centered X and Z are 0.5 and 0.5, legit low and high values would be −0.5 and 0.5, respectively. When these values are used to compute the predicted sales (Y) for low advertisement (X_L) and low price (Z_L), Eq. (19.4) becomes:

$$\widehat{Y} = 0.12 * (-0.5) + 0.09 * (-0.5) + 0.46 * (-0.5 * -0.5)$$

This leads to predicted sales of 0.01 at low advertisement and low price. Similarly, predicted values for all other combinations of X and Z can be obtained. We recommend using an excel sheet to compute this 2 * 2 table as shown below (Table 19.1).

Once we have the above table, it becomes easy to plot the two regression lines corresponding to low and high values of price (Z). The two rows in the table are the two regression lines. The graph can be plotted as a line graph in excel (Fig. 19.2).

(Excel approach: To plot, click the "Insert" option and select line graph and choose the two data points for low and high to get the two regression lines as shown below.)

Interpretation of the graph: The line titled "High Price" shows the effect of advertisement on sales when the price is also increased to a high value. The line shows that the slope is positive; that is, when price is increased, as advertisement increases, the effect on sales is positive. The line titled Low Price shows the effect

Table 19.1 Predicted sales values

Predicted sales value in each cell			
		\multicolumn{2}{c}{X (Advertisement)}
		Low	High
Z (Price)	Low	0.01	−0.1
	High	−0.13	0.22

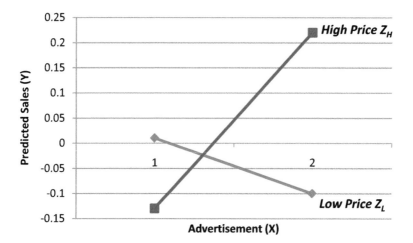

Fig. 19.2 Regression lines corresponding to price (Z)

of advertisement on sales when price is decreased to a low value. Interestingly, the slope of the line is negative; that is, when price is decreased to a low value, as advertisement increases, the effect on sales is negative. These two plots are important takeaways for a manager as it shows that price and advertisement are complements to each other (needless to say, this effect is for the sample dataset we have). Therefore, a manager should not increase advertisement and decrease price as this causes lower sales (the regression line for Z_L is negative).

This way we can test and understand interaction effects thoroughly. Effects in marketing studies are rarely just main effects. Most often, variables interact to show complex effects on the outcome variable of interest. We have shown one way to interpret this effect for interaction of two continuous variables. Similar steps can be undertaken for interaction between a categorical variable and a continuous variable as well. This strategy can also be extended to three-way interaction effects, that is, effects where three variables interact to produce differential effect on the outcome. These are advanced concepts but can be directly extended using the steps narrated above. We recommend interested readers to learn more about these advanced concepts by following the materials in the references.

Curvilinear Relationships

Many a time, in marketing studies, relationships between independent variables (X) and dependent variables (Y) are complex instead of linear. Most often, they represent a curvilinear relationship and studies tend to hypothesize a U-shaped or an inverted U-shaped effect. Rather than thinking about curvilinear relationships as just U-shaped (or inverted U), it would help to appreciate curvilinear relationships according to the effect of X on Y. The effect of X on Y could be (1) increasing at an increasing rate, (2) increasing at a decreasing rate, or (3) decreasing at a decreasing rate, thus altering the slope away from a simple linear trend.

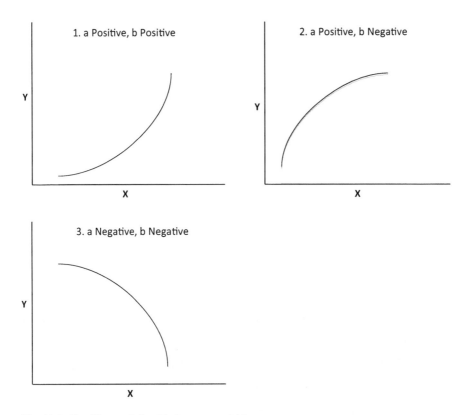

Fig. 19.3 Curvilinear relationship between variables

The three relationships mentioned above can be better understood as a regression equation. For testing any hypothesized curvilinear relationship between X and Y, the following equation helps:

$$Y = i_0 + a\,X + b\,X^2 \tag{19.5}$$

(Note: X is mean-centered)

As noted in the section on interaction effect, lower order term (i.e., X) should be present when a higher order term, formed as product of X*X, is present. The magnitude and the sign of slopes "a" and "b" of Eq. (19.5) contain a wealth of information that helps the researcher in understanding what could be the shape of this specification. Refer Fig. 19.1 to understand the expected shapes depending on the sign (Fig. 19.3).

Going back to the example of the effect of advertisement and price on sales (Y); let us now consider (for pedagogical reason) only the effect of advertisement (X) on sales (Y). A manager could argue that as advertisement increases, product sales tend to increase. Also, beyond a point the effect of advertisement on sales might not increase at the same rate. The manager is actually talking about a curvilinear relationship between X and Y, wherein the effect of X on Y increases, but beyond a point it increases at a decreasing rate. This leads to a prediction line as depicted in Graph (19.3, 2) above, wherein slope "a" is positive and "b" is negative.

As we learnt in the interaction section, we could try to compute the simple slope of Eq. (19.5) by restructuring it. However, this method is incorrect and cannot be used for equations that have curvilinear effects. We have to use simple calculus to derive the simple slope. As we all know, the first partial derivative with respect to (w.r.t) X is the simple slope of regression of Y on X. This technique can be applied to the interaction equation we dealt with earlier as well. So, going back to Eq. (19.5), the first partial derivative is:

$$\partial Y/\partial X = a + 2bX \tag{19.6}$$

As seen in Eq. (19.6), the simple slope depends or is conditional on the value of X itself. This along with the sign of "a" and "b" is precisely the reason for the effect of X on Y being curvilinear and not just linear. Again, going to back to our steps in choosing legit values of X, we can take one standard deviation below and above the mean (which is zero) and compute the values of the simple slope.

If a manager would like to investigate how price interacts with the curvilinear relationship of advertisement on sales, we can follow the same steps we did in interaction. However, to derive the simple slope, we have to use calculus and not just simple restructuring of the equation for reasons stated above. Interaction effects on curvilinear relationship are complex and advanced concepts. So we urge the interested readers to peruse the resources in the reference section for a complete understanding. We provide some ideas to help understand some basics of interaction effects in curvilinear relationships below.

The full equation after including price (Z) would be:

$$Y = i_0 + aX + bX^2 + cZ \tag{19.7}$$

Now unlike simple interaction effects, researchers have to think deeply on whether price (Z) interacts with just advertisement (X) or Z interacts with the higher order term (X^2) as well. Conceptually, these two effects are different and so the researcher has to specify the correct equation based on the hypothesis being tested. We should also remember to always include the lower order terms while trying to introduce higher order terms that have interaction effects. Let us consider one example, where we introduce XZ term. Then Eq. (19.7) will change to the following:

$$Y = i_0 + aX + bX^2 + cZ + dXZ \tag{19.8a}$$

To derive the simple slope, which shows the effect of X on Y, we take the first partial derivative w.r.t X:

$$\partial Y/\partial X = a + 2bX + dZ \qquad (19.8b)$$

It can be seen from the above equation that not only does the effect depend on X (due to the presence of the term $2bX$) but it also depends on Z (due to the term dZ). This means that the curvilinear effect is conditional on Z as well.

Furthermore, if a manager hypothesizes that Z alters both the level and the shape of the curvilinear effect, it can be tested by introducing the X^2Z term. This term is the product of X^2 and Z. Then Eqs. (19.8a and 19.8b) will change to the following:

$$Y = i_0 + aX + bX^2 + cZ + d\,XZ + e\,X^2Z \qquad (19.9a)$$

Please note the presence of all lower order terms (which are X, X^2, Z, and XZ) of the highest order term (X^2Z). Neglecting any the lower order terms leads to incorrect interpretation of equation (19.9a and 19.9b). Again, to calculate the simple slope we do the following:

$$\partial Y/\partial X = a + 2bX + dZ + 2e\,XZ \qquad (19.9b)$$

This shows that Z affects not just the level but the shape of the curve as well.

Mediation

Mediation is used to study relationships among a set of variables by estimating a system of equations. The objective of a mediation analysis is to extract the mechanism behind the effect of X on Y. Mediation analysis is useful in understanding the intervening mechanism that actually causes the effect of X on Y (MacKinnon 2008; Preacher and Hayes 2004). To find this intervening mechanism, an intervening variable called the mediator (M) is introduced. We can imagine this mediation as a pathway: X→M→Y. Therefore, the total effect of X on Y is now partitioned into an indirect effect via M and a direct effect of X on Y.

In mediation analysis, there is a series of regressions or a set of equations that has to be estimated. First is the effect of X on the mediator M (X→M, slope a), next is the effect of mediator on the dependent variable Y (M→Y, slope b), and finally the direct effect of X on Y (X→Y. slope c'). The researcher should note that now there are two dependent variables, namely, Y and the newly introduced intervening mediator M. It can be easily understood through the diagram given below (Fig. 19.4).

The diagram (Fig. 19.4) can be represented by the following set of regression equations:

$$Y = i_0 + bM + c'X \qquad (19.10)$$

$$M = j_0 + aX \qquad (19.11)$$

where i_0 and j_0 are intercepts and are ignored hereafter for pedagogical purpose.

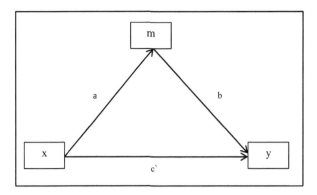

Fig. 19.4 Mediation effect through "M"

Now the mediating effect, also called the indirect effect of X on Y via M, can be obtained by inserting Eq. (19.11) in Eq. (19.10):

$$Y = ab\,X + c'\,X \qquad (19.12)$$

This shows that the indirect effect of X on Y is the product of two slopes captured as "a * b" and the direct effect of X on Y is "c'."

The slopes "a," "b," and "c'" can be estimated using any statistical software and the product of the slopes "a * b" can be computed to find the strength of the mediation. However, we recommend using software that supports path analysis or structural equation modeling (SEM) techniques (STATA or Mplus or R) to estimate both the equations simultaneously without having to run multiple regressions. This practice can help the researcher easily progress from the simple model narrated here to more complex mediational models that are often the case in research. One important question still remains—how to test for the statistical significance of the product term "a * b"? The statistical test of the simple slopes "a," "b," and "c'" using standard errors is straightforward. However, for the product of the slopes, recent research in mediational analysis suggests that deriving the standard error (using a technique called delta method) for a higher order product term is inaccurate and that statistical significance has to be tested using nonparametric techniques like bootstrapping. The intuition behind this is that though the individual slopes "a" and "b" are assumed to be normally distributed, the product of these slopes cannot be. Hence, as the derivation of standard errors through parametric methods (like OLS or maximum likelihood) assumes normality, researchers are advised to use nonparametric resampling procedures like bootstrapping to derive the standard errors. The logic behind bootstrapping procedure is straightforward. The estimates of the indirect effect "a * b" are repeatedly obtained from samples drawn from the original sample with replacement. The standard deviation of those estimates is the standard error, which is then used to build a nonsymmetric 95% confidence interval to test the significance of the indirect effect. If the 95% confidence interval of the indirect effect does not contain zero the effect is statistically significant. In this case, we can conclude that there is a significant mediation effect.

Let us go back to the example of advertisement (X) having an effect on product sales (Y). Now a manager might be interested in knowing the reason behind the "increase in advertisement having a positive effect on sales." One could argue that, as advertisement increases (X), customers' awareness (M) about the product also increases, which encourages them to buy more, leading to an increase in product sales (Y). Therefore, here the mediator is consumer awareness (M). If somehow the manager can measure or capture this variable, it can be used, along with the X and Y variables, to test this hypothesized mediation effect. But what is the managerial implication? What really is the "extra" insight derived by doing this mediation analysis? An important insight is that the manager is now able to identify one relevant mechanism that actually causes the effect of X on Y. The manager can then increase resource allocation to marketing strategies that help improve consumer awareness levels. Furthermore, this manager could be intrigued to investigate whether there are other mediators or intervening mechanisms like awareness. Understanding and finding the reasons behind effects are critical to decision making. Armed with the knowledge of the mechanism behind the cause-and-effect relationship, executives can make informed decisions which lead to greater success.

In this section, we have covered simple mediation analysis. Interested readers are strongly recommended to consult the resources provided in the reference section for advanced mediation topics like mediation analysis with multiple mediators and multiple dependent variables. One interesting and important advance in mediation analysis in recent times has been the ability to integrate moderation with mediation analysis. This is called moderated mediation analysis (Preacher et al. 2007). This is an important advancement because it resonates well with the purpose of theory testing in marketing or in general any social science domain. Theory is important primarily for two reasons: to provide arguments to explain the reason behind why and how a variable affects another and to uncover the boundary condition under which this effect could change. This is essentially the spirit of any moderated mediation test as it helps to simultaneously understand the mechanism (mediation) and the conditional effect (moderation) of that mediation pathway. That is, to investigate if the indirect effect of '$X \to M \to Y$' as a unit is conditional on the level of a moderator variable, say Z. This would imply how the product term "*ab*" is conditional on legit values of Z (say low and high levels of Z). For example, considering the example of how "*consumer awareness*" *(M)* channelizes the effect of advertisement (X) on product sales (Y), a manager can also understand how the strength of this pathway varies depending on whether the product price (Z) is low or high. As the mediation pathway is conditional on another variable, or mediation effect is moderated by another variable, this test is termed moderated mediation test. The theory and empirics behind moderated mediation has exploded in recent times. Readers are requested to consult the references for a deeper understanding of this technique.

2.2 Data Envelopment Analysis (DEA)

Suppose you are a manager of a big MNC and you want to identify the efficient units (finance, marketing, operations, HR, international business, etc.) within your organization. You may want to decide on allocating resources, hiring, increasing the benefits to the employees of a specific unit, based on the efficiency of the unit. How are you going to decide on the efficiency? You may use observational data for such decisions. However, such an approach may not be great while designing critical strategies such as resource allocation or hiring.

A very prominent analytic tool that can be used in such decision making is "data envelopment analysis" (Cook and Seiford 2009; Cooper et al. 2004). DEA is designed to help managers measure and improve the performance of their organizations. As the quest for efficiency is a never-ending goal, DEA can help capture the efficiency of each unit and suggest the potential factors that may be responsible for making units efficient.

DEA allows managers to take into account all the important factors that affect a unit's performance to provide a complete and comprehensive assessment of efficiency (Charnes et al. 1978). DEA does this by converting multiple inputs and outputs into a single measure of efficiency. By doing so, it identifies those units that are operating efficiently and those that are not. The efficient units, that is, units making best use of input resources to produce outputs, are rated as being 100% efficient, while the inefficient ones obtain lower scores.

Let us understand the technical aspects of DEA in a simple way. Suppose, a manager of ICICI bank wants to measure the efficiency of HR, finance, marketing, operation, public relation, and accounting departments to allocate resources for the following year (Sherman and Franklin 1985). Managers generally have information regarding the number of employees, total issues handled, raises given, and the contribution of these units toward the operational profits of the firm. Managers can use DEA with number of employees, total issues handled, and raises given as the inputs and contribution of these units toward the operational profits as the output. DEA analysis will provide a score of either 1 or less than 1 to each unit. A score of 1 means the unit is perfectly efficient and anything less than 1 means that the unit has the room to grow and be more efficient. Based on the relative importance of the inputs, units may be asked to work on specific inputs such that the units become efficient. Further, managers can now allocate the resources they want to the efficient units.

DEA also provides guidance to managers on the reduction/increment required in the inputs of the units to become efficient, helping managers answer questions, such as "How well are the units doing?" and "How much could they improve?" It suggests performance targets, such as marketing unit should be able to produce 15% more output with their current investment level or HR unit should be able to reduce churn by 25% and still produce the same level of outputs. It also identifies the best performing units. One of the most interesting insights from DEA is that one can test the operating practices of such units and establish a guide to "best practices" for others to emulate.

Let us understand using a practical example. Suppose a bank such as ICICI wants to measure the efficiencies of their branches across India. The goal of the management is to ensure that each of these branches achieves the best possible performance; the problem though is deciding what that means and how best to go about measuring it. The outputs of the branches such as sales, sales growth, accounts, and market share can be studied and compared. Similarly, a branch's inputs, such as staff, office space, and materials costs can be measured. Managers can then develop performance ratios such as sales per member of staff or profit per unit of office space utilized.

However, all these attempts to measure performance may not produce a clear picture as branches may exhibit considerable variation depending on the performance indicator chosen. This is where DEA helps as it provides a more comprehensive measure of efficiency by taking into account all the important factors that affect a branch's performance.

Although the choice of inputs and outputs largely depends on the availability of the data, the beauty of DEA lies in its ability to handle multiple inputs and multiple outputs to give a comprehensive score about the efficiency of the units. DEA can be used within a specific organization (e.g., measuring the efficiency of various divisions of ICICI bank); across organizations in an industry (e.g., comparing the efficiency of all the private sector banks in India); and across industries (e.g., measuring efficiency of banks, FMCG companies). However, such analysis will require availability of similar inputs and outputs for all organizations.

Technical Details of DEA

DEA can be analyzed either as constant return to scale (CRS) or variable return to scale (VRS) (Banker and Thrall 1992; Banker et al. 2004; Seiford and Zhu 1999). Let us first consider the CRS model. Let us assume that there are N decision-making units (DMUs) (DMUs are various banks in our example) with K inputs and 2 outputs (profits and customer satisfaction). These are represented by x_i and y_i for ith DMU. The purpose of DEA is to construct a nonparametric envelopment frontier over the data points such that all observed points lie on or below the production frontier (Charnes et al. 1985; Boussofiane et al. 1991; Chen and Soo 2010). We can express the problem as a linear program as follows:

$$\max_{\theta, \lambda} \theta$$
$$\text{such that } -\theta y_i + Y\lambda \geq 0, \lambda \geq 0$$
$$x_i - X\lambda \geq 0, \lambda \geq 0$$

X is the K x N input matrix and Y is 2 x N output matrix; θ is a scalar and λ is an N x 1 vector of constants. The efficiency score for ith DMU will be the value of $(1/\theta)$. The DMU with efficiency score $= 1$ will be called efficient DMU. To get the efficiency score for each DMU (i.e., various divisions of ICICI bank) in the sample, we will need to solve the linear programing for the specific DMU under consideration. Note that we can use CRS DEA when the underlying assumption is that the DMUs are operating at an optimal scale. However, this assumption might not hold in reality, most DMUs in a sample may not work at the optimal scale. In such a context, CRS may not be an ideal tool. To avoid any potential scale effects, one may use VRS. One can compute the technical efficiency with VRS that excludes the scale effects. The CRS linear programming problem can be modified to account for VRS by adding the convexity constraint $e_\lambda = 1$ (where e_λ is N X 1 vector of ones). The additional constraints give the frontier piecewise linear and concave characteristics.

$$\max_{\theta,\lambda} \theta$$
$$\text{such that } -\theta y_i + Y\lambda \geq 0, e_\lambda = 1, \lambda \geq 0$$
$$x_i - X\lambda \geq 0, e_\lambda = 1, \lambda \geq 0$$

The CRS and VRS provide different values for the efficiency score of a DMU; it indicates the presence of scale inefficiency in the DMU. One can compute the scale inefficiency by taking the ratio of efficiency score obtained from CRS to the efficiency score obtained from VRS (Caudill et al. 1995; Gagnepain and Ivaldi 2002; Greene 2010).

Example

Suppose that we are interested in evaluating the efficiency of the hospital units (Ouellette and Valérie 2004) of a chain based on a number of characteristics: the total number of employees, the size of units in square meters, the number of patients each unit serves, total number of specialists, total revenue, and satisfaction. It becomes obvious that finding the most efficient units requires us to compare records with multiple features.

To apply DEA, we must define our inputs (X) and outputs (Y). In the case of a hospital chain, X can be the total number of employees, the size of units in square meters, the number of patients each unit serves, total number of specialists; and Y can be total revenue and satisfaction. If we run DEA, we will estimate the output to input ratio for every hospital under the ideal weights (ideal weights are weights that consider the values that each unit puts on inputs and outputs). Once we have their ratios, we will rank them according to their efficiency (Banker and Morey 1986).

STATA/R Code

DEA can be analyzed using multiple statistical programing software. Here, we are providing the syntax required to conduct the analysis in STATA[2] or R[3]. Although STATA does not have a built-in function, one can use user-written command (dea) to do the analysis. In case you find it difficult, please type "help dea" in STATA command window and you will get a step-by-step explanation for the analysis.

---

[2]https://www.cgdev.org/sites/default/files/archive/doc/stata/MO/DEA/dea_in_stata.pdf (accessed on Jan 30, 2019).

[3]https://www.rdocumentation.org/packages/TFDEA/versions/0.9.8.3/topics/DEA (accessed on Jan 30, 2019).

19 Marketing Analytics

To do the analysis in STATA, you need to download user-written command (type net install st0193).

"dea ivars=ovars [if] [in] [, options]"

Options:

- rts(crs|vrs|drs|nirs) specifies the returns to scale. The default is rts(crs)
- ort(in|out) specifies the orientation. The default is ort(in)
- stage(1|2) specifies the way to identify all efficiency slacks. The default is stage (2)

To do the analysis in R, you need to use *"Benchmarking"* package.

```
> install.packages("Benchmarking")   # to install package for the
  first time
> library(Benchmarking)   # to load package
# DEA code
> eff <- dea(x,y, RTS="crs")   # where x is input vector and y
  is output vector
> eff   # where "eff" will give the efficiency of each unit
# RTS options allow us to specify which return of scale we want
    • "crs" - constant return to scale
    • "vrs" - varying return to scale
    • "drs" - decreasing return to scale
    • "irs" - increasing return to scale
```

DEA in Practice

There are eight units of a restaurant chain (largely in Northern India) and the manager wants to measure the efficient unit for the best restaurant award. The manager has insights only on the total number employees and the revenue (in 100,000) of each unit as follows:

Units	1	2	3	4	5	6	7	8
Employee	5	6	11	15	20	9	12	10
Revenue	4	4.8	8	14	18	5	11	9

In order to find the efficiency of each unit, we first calculate the efficiency by dividing the revenue by employee as follows:

Units	1	2	3	4	5	6	7	8
Employee	5	6	11	15	20	9	12	10
Revenue	4	4.8	8	14	18	5	11	9
Revenue/employee	0.8	0.8	0.727	0.93	0.9	0.55	0.91	0.9

As per the efficiency measurement, we find that Unit 4 is the efficient one and Unit 6 is the least efficient one.

Now, to compute the relative efficiency, we need to divide the efficiency of the units by the efficiency of the most efficient units, that is, the relative efficiency of units is measured by taking the ratio of efficiency of each unit and the efficiency of most efficient unit as shown below.

$0 \le$ revenue *per employee for each* unit/revenue *per employee for the most efficient* unit ≤ 1.

Units	1	2	3	4	5	6	7	8
Employee	5	6	11	15	20	9	12	10
Revenue	4	4.8	8	14	18	5	11	9
Revenue/employee	0.8	0.8	0.727	0.93	0.9	0.55	0.91	0.9
Relative efficiency	0.860215	0.860215	0.78172	1	0.967742	0.591398	0.978495	0.967742

Data envelopment analysis shows that the Unit 4 is the most efficient relative to all other units; and the Unit 6 is the least efficient. That means Unit 4 will be on the frontier and rest will be within the frontier. One can also show the frontier pictorially.

2.3 Stochastic Frontier Analysis (SFA)

Stochastic frontier analysis is a parametric approach, largely used to estimate the production or costs in economics (Baccouche and Kouki 2003; Fenn et al. 2008). Data envelopment analysis and SFA are competing approaches (Jacobs 2001; Cullinane et al. 2006); however, there is no single approach that unifies both. If a manager wants to find what causes the inefficiency in a firm in its operation, the manager may want to adopt SFA. SFA relies on the assumption that decision-making units (such as banks and hospitals) behave suboptimally and they can maximize or minimize their respective objective functions (costs, profits, operational efficiency, etc.) in order to improve (Parsons 2002).

Let us now discuss the components of an SFA (Kumbhakar and Lovell 2003; Bera and Sharma 1999). The stochastic production frontier was first proposed independently by Aigner et al. (1977) and Meeusen and Van den Broeck (1977). In their specification (which is largely different from the standard production function), there are two distinct error terms in the model and can be shown as follows:

$$y_i = \beta x_i - u_i + v_i$$

where x_i are inputs, y_i is the output; u_i captures inefficiency, shortfall from maximal output govern by production function βx_i; and v_i is the error term or the outside influences beyond the control of the producer. In sum, the SFA has two components: a stochastic production frontier serving as a benchmark against which firm efficiency is measured, and a one-sided error term with independent and identical distribution across observations and captures technical inefficiency across production units.

If a manager allows the inefficiencies to depend on the firm level factors (employees, experts, size, alliances, acquisitions, etc.), the manager can examine the determinants of inefficiencies. Such understanding helps to implement different policy interventions to improve efficiency. Managers can also modify the inputs and incorporate in the production function to reduce the inefficiencies.

Now the question is whether one should use SFA over DEA? (Wadud and White 2000; Koetter and Poghosyan 2009). Imagine that there are random variations in the inputs. These variations can make the DEA analysis unstable. A potential advantage of the SFA over DEA is that random variations in inputs can be accommodated.[4] Although there are some benefits, SFA also suffers from multiple disadvantages including its complications in handling multiple outputs. Further, it also requires stochastic multiple output distance functions, and it raises problems for outputs that take zero values.

Technical Details of SFA

The technological relationship between a few input variables and corresponding output variables is given by a "*production function*" (Aigner et al. 1997). Econometrically, if we use data on observed outputs and inputs, the production function will indicate the average level of outputs that can be produced from a given level of inputs (Schmidt 1985). One can estimate production functions at either an individual or an aggregate level.

The literature on production function suggests that the implicit assumption of production functions is that all firms are producing in a technically efficient manner, and the representative (average) firm therefore defines the frontier (Førsund et al. 1980; Farrell and Fieldhouse 1962). The estimation of the production frontier assumes that the boundary of the production function is defined by "best practices" units. It therefore indicates the maximum potential output for a given set of inputs along a straight line from the origin point. The error term in the SFA model represents any other reason firms would be away from (within) the boundary. Observations within the frontier are deemed "inefficient (Hjalmarsson et al. 1996; Reinhard et al. 2000)."

The SFA Model and STATA/R Code

Restating the stochastic frontier model:

$$y_i = \beta x_i - u_i + v_i, u_i = |U|$$

where $|U|$ is the aggregated (in)efficiency.

---

[4]http://www.fao.org/docrep/006/Y5027E/y5027e0d.htm (accessed on Jul 6, 2018).

In this area of study, estimation of the model parameters is usually not the primary objective. Estimation and analysis of the inefficiency of individuals in the sample and of the aggregated sample are usually of greater interest.

STATA has a built-in command to estimate SFA.[5] This is essentially a regression analysis where the error term consists of a random error and an inefficiency term, and again can be estimated for both production and cost functions. The syntax of the frontier command is:

```
frontier depvar [indepvars] [if] [in] [weight] [, options]
```

The most important options are:

Distribution (`distname`) specifies the distribution for the inefficiency term as half-normal (`hnormal`), `exponential`, or truncatednormal (`tnormal`). The default is `hnormal`.

`cm (varlist)` fits conditional mean model; may be used only with distribution (`tnormal`). `uhet(varlist)` explanatory variables for technical inefficiency variance function

`vhet(varlist)` explanatory variables for idiosyncratic error variance function

`cost` fit cost frontier model; default is production frontier model

`vce(vcetype)` vcetype may be `oim`, `opg`, `bootstrap`, or `jackknife`.

To do the analysis in R, you can use "Benchmarking" package and sfa function as below.

```
> sfa(x,y)         # where x is a n x k matrix of k inputs
for n units, and y is a n x 1 vector of outputs for n units
> summary(sfa(x,y))
> e <- sfa(x,y)              # Estimate efficiency for each unit
> eff(e)
```

SFA in Practice

A manager can find the effect of inefficiency and the drivers of inefficiency by adopting an SFA approach. For example, given the availability of the data on the improper implementation of the marketing-mix elements, and the within-unit corruption, a manager can model the inefficiency as a function of improper implementation and corruption and see their effect on the overall productivity (e.g., revenue contribution) of each unit. In marketing, there has been several applications of SFA; see, for example, Feng and Fay (2016) who apply it to evaluate salesperson capability, Parsons (2002) who comments upon its application to sales people and retails outlets and observes that unlike explaining just the mean performance SFA provides explanation for the gap between the unit and the nest performer, and Vyt (2008) who uses SFA for comparing retailers' geo-marketing.

---

[5]https://www.stata.com/manuals13/rfrontier.pdf (accessed on Jan 30, 2019).

2.4 Conjoint Analysis

Conjoint analysis is a marketing research technique to determine consumer preferences and potential customers. Knowing customers' preference provides invaluable information about how customers think and make their decisions before purchasing products. Thus, it helps firms devise their marketing strategies including advertising, promotion, and sales activities (Ofek and Toubia 2014).

Conjoint analysis can be defined as a technique to analyze and determine how consumers value various attributes of a product and the trade-off they are willing to make among the features that comprise a product. These Attributes are inherent characteristics/features of a product. For example, price, color, RAM, camera, screen size, and battery power can be the attributes of a smartphone. Different values that a product attribute can take represent the various levels of that attribute. Black, gold, and white may be levels of the attribute color. Similarly, attribute RAM may have the levels 4GB, 8GB, and 16GB. The importance of each attribute can be inferred by experimentally manipulating the features of a product in various product bundles and observing the consumer's ratings for that product or choices among competing products in the market. The insights provided by conjoint analysis help in developing clearly defined products by identifying those features that are more appealing to the consumers. Conjoint analysis can be productively employed in the following areas:

- Designing products that maximize the measured utilities for customers in a target segment
- Modifying existing products and developing new products
- Selecting market segments for which a given product delivers high utility
- Planning competitive strategy
- Analyzing pricing policies

This technique is extremely useful in new product development process. When firms clearly define products at early stages they are 3.3 times more likely to be successful. It is often used for several types of products such as consumer goods, electrical and electronic products, retirement housing and air travel.

Types of Conjoint Analysis

The basic principle behind any type of conjoint analysis is the same—products are broken down into product attributes, and customers are faced with trade-offs when deciding which combination of attributes and their corresponding levels to purchase. There are different types of conjoint analyses depending on the response type, questioning approach, type of design and whether all the attributes are included in every question of the survey questionnaire. The four types of conjoint analysis based on response type are—(1) rating-based conjoint, (2) best–worst conjoint, (3) rank-based conjoint, and (4) choice-based conjoint. Standard conjoint and adaptive choice-based conjoint are based on the questioning approach. Generic conjoint and brand-specific conjoint are based on the type of design. Full-profile conjoint and partial-profile conjoint are classified depending on the inclusion of all attributes in every question of the survey.

Choice-based conjoint analysis is the most common type of conjoint analyses. In choice-based conjoint, respondents are presented with several questions with each question comprising 2–5 products and are asked to choose their preferred option. The results are then used to calculate a numerical value (known as a "utility score," "utility," or "partworth") that measures how much each attribute and level influenced the customer's decision to make that choice.

Conducting a Conjoint Analysis

A conjoint analysis comprises three stages:

Experimental Design: This stage is the most crucial part of the study as it ensures that the study includes all attributes and the values of attributes (levels) that will be tested. Three steps in the design are:

- Selection of attributes relevant to the product or service—Attributes are selected on the basis of required information. Only attributes that can be controlled by the firm should be included. Selecting too many attributes should be avoided as it leads to longer questionnaires, which are time consuming and difficult to answer. Ambiguous attributes must be avoided. For example, selecting vibe as an attribute for smartphone may lead to misinterpretation on the part of respondents.
- Selection of levels for the attributes—Levels for the attributes must be understandable and clear to the respondents. Ambiguity must be avoided while specifying levels. For example, specifying 4.5″–5.5″ as a smartphone screen size level should be avoided because it leaves room for interpretation. For quantitative variables the distance between two levels should not be so large that the evaluation of different levels becomes too easy.
- Development of product bundles to be evaluated—In this step, the number of product bundles that the respondent would evaluate are decided based on fractional factorial designs which are used to reduce the number of products.

Data Collection: The type of conjoint analysis used dictates the nature of data to be collected. The goal is to obtain respondents' preferences for a carefully selected set of product bundles (Profiles). Generally, this data is collected by using a paper based or online survey. The data size depends upon the number of attribute levels to be tested. The more the number of levels of attributes more is the data to be collected for better accuracy. The first step in this process is to design a data collection procedure using one of the following methodologies:

- Pairwise evaluations of product bundles: The respondent considers two or more products at the same time and allocates 100 points among the options.
- Rank-order product bundles: The respondent sorts the products presented and assigns rank 1 to the most preferred option and a rank equal to the number of products presented to the least preferred option.

- Product evaluation on a rating scale: The respondent evaluates each product on a scale (e.g., 1–100), giving a rating of 100 to the most preferred product.
- Choice method: The respondent considers multiple products that are defined on all attributes in the study and chooses the best in the best scaling method and chooses the best and worst product in best–worst scaling method.

Decision Exploration: The last stage in conducting conjoint analysis is decision exploration, that is, evaluating product design options by segmenting customers based on their partworth functions, testing the likely success of a new product by simulating the market conditions and transforming the partworths into product choices that consumers are most likely to purchase.

Interpreting Conjoint Results

The conjoint output gives estimated utilities or partworths corresponding to average consumer preferences for the level of any given attribute.

Let us illustrate the partworths estimation to determine a consumer's preference for pizza. We define four attributes—cheese type, toppings, drink, and price with 3, 3, 2, and 3 levels, respectively, for each attribute.

Cheese	Cheddar
	Mozzarella
	Parmesan
Toppings	Onion
	Mushroom
	Pepperoni
Drink	Coke
	Pepsi
Price	$5
	$7
	$10

Consumer's preferences are quantified using linear regression for various attributes of pizza. In the survey, the respondent indicates her preferences for various pizzas defined by different combinations of the given attributes and their levels.

Sample responses of a single consumer for multiple pizza profiles

Cheddar	Mozzarella	Parmesan	Onion	Mushrooms	Pepperoni	Coke	Pepsi	Five$	Seven$	Ten$	Rating
0	0	1	1	0	0	0	1	0	0	1	4
0	1	0	0	0	1	0	1	0	0	1	1
0	0	1	0	1	0	0	1	1	0	0	6
1	0	0	0	1	0	0	1	0	1	0	1
0	1	0	1	0	0	1	0	0	1	0	3
0	0	1	0	0	1	1	0	0	1	0	6
1	0	0	0	0	1	1	0	0	0	1	6
0	0	1	1	1	0	1	0	1	0	0	7
1	1	0	0	0	0	0	1	1	0	0	2
0	1	0	0	1	1	1	0	0	0	0	1
0	0	0	0	1	0	1	0	0	0	1	3
1	0	0	1	0	0	0	1	0	1	0	2

Now, the approach is to regress the consumer's preferences for pizza based on her ratings of the various attribute levels.

We run the regression on categorical dummy variables. We set one attribute level as the baseline and remove the corresponding level from regression. Our baseline in this example is a pizza with Cheddar cheese, onion toppings, coke, and a price of 5$.

The utility of this baseline pizza is captured in the intercept from the regression output. The other coefficients give the partworths for various attribute levels. Depending on the type of information needed further insights can be derived from the partworths calculated above.

Conjoint Analysis Applications

There are many possible applications of conjoint analysis. However, the four common applications are trade-off analysis, market share forecasting, determining relative attribute importance, and comparing product alternatives. All other applications are generally variants of these four applications.

Trade-off analysis: Utilities from conjoint analysis are used to analyze whether average consumers would be willing to give up on one particular attribute to gain improvements in another.

For the given example, the partworths are estimated as follows:

Partworths			
Intercept		3.28	Best Bundle
Cheese	Cheddar	0	Parmesan
	Mozzarella	−1.11	
	Parmesan	2.78	
Toppings	Onion	0	Pepperoni
	Mushroom	0.78	
	Pepperoni	0.89	
Drink	Coke	0	Coke
	Pepsi	−2	
Price	Dollar5	0	10$
	Dollar7	−0.11	
	Dollar10	0.44	

When analyzing a group of consumers, the partworths may be interpreted as the average preferences in the group. For example, the utilities can be used to justify changes in one particular attribute at the expense of another.

Market share forecasting: This relies on the use of a multinomial logit model. To use conjoint output for market share prediction, the following conditions must be satisfied: The firm must know that we do not need other competing products in the market besides its own offering that a customer is likely to consider before selecting a product in the category. Each of these competitive products' important features must be included in the experimental design. The utilities of the competing product should also be evaluated. Based on product utilities, the market share for product i can be calculated as:

$$Share_i = \frac{e^{U_i}}{\sum_{j=1}^{n} e^{U_j}}$$

U_i is the estimated utility of product i,
U_j is the estimated utility of product j,
n is the total number of products in the competitive set, including product i.

Determining attribute importance: The range of estimated partworths within a given attribute tells how significant the attribute is in the decision process of a consumer. To calculate the importance of any attribute, the difference between the highest and lowest utility level of that attribute is divided by the sum of the differences between the highest and lowest utility level of all the attributes including the one for which the importance is being calculated. The resulting number is generally interpreted as the % decision weight of an attribute in the overall choice process. Attribute importance is measured using the following metric:

$$I_i = \frac{U_i - \overline{U}_i}{\sum_{i=1}^{n}(U_i - \overline{U}_i)}$$

I_i = importance of any given attribute i
U_i = highest utility within a given attribute
$\overline{U}_i$ = lowest utility level within a given attribute
n = total number of attributes

In the example discussed above, attribute importance for cheese can be calculated as follows:

Range = Max (levels) − Min (levels) = 2.78 − (−1.11) = 3.89

Attribute importances		
Attribute	Range	Importance
Cheese	3.89	0.53
Toppings	0.89	0.12
Drink	2	0.27
Price	0.55	0.08

Importance of Cheese = Range/ (Sum of ranges of all the attributes)
= 3.89/ (3.89 + 0.89 + 2 + .55) = 0.53

Similarly, the relative importance of other attributes can be calculated as shown in the table above.

Comparing product alternatives: Conjoint analysis can also be used to determine consumer choice between two alternative products. Based on the utility derived from each product offering, we can predict how a consumer would choose between the hypothetical profiles.

Let us consider two product profile ratings of a single respondent and predict her choice among these two profiles.

Product	Cheddar	Mozzarella	Parmesan	Onion	Mushrooms	Pepperoni	Coke	Pepsi	Five$	Seven$	Ten$	Rating
A	0	0	1	1	0	0	0	1	0	0	1	4
B	0	1	0	0	0	1	0	1	0	0	1	1

Utility of a product profile is calculated using the estimated partworths. The profile elements are each multiplied by the corresponding partworths and summed up to get the utility. The calculated utilities are then compared and the one with the higher value would be the consumer's choice.

Coefficients of the attribute levels:

Cheddar	Mozzarella	Parmesan	Onion	Mushrooms	Pepperoni	Coke	Pepsi	Five$	Seven$	Ten$
0	−1.11	2.78	0	0.78	0.89	0	−2	0	−0.11	0.44

Utility of Product A = $3.28 + 0.00^*0 + (-1.11)^*0 + (2.78)$
$$^*1 + (0)^*1 + (0.78)^*0 + (0.89)^*0 + (0)^*0 + (-2.00)$$
$$^*1 + 0^*0 + (-0.11)^*0 + 0.44^*1 = 4.5$$

Utility of Product B = $3.28 + 0.00^*0 + (-1.11)^*1 + (2.78)^*0 + (0)$
$$^*0 + (0.78)^*0 + (0.89)^*1 + (0)^*0 + (-2)$$
$$^*1 + 0^*0 + (-0.11)^*0 + (0.44)^*1 = 1.5$$

Utility of Product A > Utility of Product B

Therefore, the predicted choice of this particular consumer is Product A.

Product	Cheddar	Mozzarella	Parmesan	Onion	Mushrooms	Pepperoni	Coke	Pepsi	Five$	Seven$	Ten$	Utility	Choice
A	0	0	1	1	0	0	0	1	0	0	1	4.5	A
B	0	1	0	0	0	1	0	1	0	0	1	1.5	

Conjoint analysis has several advantages, such as uncovering hidden drivers that may not be apparent to the respondents themselves and evaluating the choice at an individual level. It can be used to obtain brand equity by determining the popularity of a brand. However, it has certain disadvantages such as added complexity in the experimental design because of the inclusion of a large number of attributes. This increases respondent's fatigue in taking the survey and thus compromises the accuracy of the result. Also, the validity of conjoint analysis depends on the completeness of attributes. Conjoint analysis equates a customer's overall utility for a product with the sum of her utilities for the component parts. Hence, a highly valued option on one attribute can compensate for unattractive options on another attribute and thus give misleading results.

2.5 Customer Analytics

Customer Lifetime Value (CLV)

Customers are the basis of a firm's existence. A firm creates and provides value (product or service) to customers and in return customers provide value (revenue/profits) to firms. This section will focus on one such measure for analyzing the value provided by customers to firms—customer lifetime value (CLV).

Before we understand CLV, let us look at some other customer metrics that were (are still being) used extensively before paving the way for the much superior one of CLV. One way to determine customer value is to look at metrics such as average customer revenue or average customer profit. Unless used for a specific segment, these metrics put all customers at equal level, which is incorrect and certainly not of interest to marketers who celebrate customer heterogeneity. Some customers do provide more profits than others and some customers actually lead to losses. Another interesting way to analyze the value provided by a customer is to look at their purchase behavior through RFM analysis (recency, frequency, monetary value). In RFM analysis, one looks at the following parameters:

- Recency—How recently was the last purchase made?
- Frequency—How frequently do they purchase?
- Monetary value—How much do they spend?

Customers can be rated on each parameter and classified from most valuable (highest recency, frequency, and monetary value) to least valuable (lowest recency, frequency, and monetary value). This is a simple model that can help segment customers and predict their future behavior. But all these are backward looking metrics—that is, these metrics do not take into account future potential of a customer. These methods were good enough when we did not have access to great data and advanced statistical software to analyze data. Now we have the capability to collect and analyze data at the lowest level, which allows more sophisticated analysis. CLV is one such metric that directly accounts for future value. CLV can be defined as the total financial contribution from the current period into the future—that is, revenues minus costs—of a customer over his/her future lifetime with the company and therefore reflects the future profitability of the customer (Kumar 2010). It is a forward looking customer metric that not only takes into account the current value but also the future value provided by customers. It provides a dollar value for customer relationship. It helps distinguish customers according to the value provided by them over the life of their business with the firm. Future marketing strategies can then be planned accordingly for both current and future customers.

Now that we have established that CLV is an important metric we will try to calculate it. CLV is calculated by discounting all the current and future profits expected from the customer's side. It deals with profit margins from each customer instead of revenue. It also takes into account the percentage of customers retained by the firm. CLV can be calculated as:

$$\text{CLV} = \sum_t \frac{m_t r_t}{(1+i)^t} \quad (19.13)$$

where m_t is the profit margin during year/duration t
r_t is the retention rate during year/duration t
i is the constant discount rate
t is year/duration

When the onetime acquisition cost of a customer is subtracted from Eq. (19.13) we get the CLV of that particular customer. If we assume an infinite horizon, a constant profit margin, a constant retention rate, and a constant discount rate Eq. (19.13) can be simplified to:

$$CLV = m\left(\frac{r}{1+i-r}\right) \quad (19.14)$$

Here $\left(\frac{r}{1+i-r}\right)$ is the margin multiple and depends on the retention rate and discount rate. The higher the retention rate and lower the discount rate, the more valuable is a customer. Table 19.2 provides the margin multiple based on typical retention and discount rates. Table 19.2 is a very useful "back of the envelope" calculation of CLV!

CLV is the maximum value provided by customers to the firm. Hence, it can be used as the upper limit of any customer-centric activity. For example, if a potential customer's CLV is $5 then a firm should not spend more than $5 on acquiring this customer. This holds for customer retention and development activities as well. As mentioned before, the dollar value provided by CLV can help distinguish cohorts of customers. So it also helps distinguish future prospects that are similar to currently profitable customers.

As with any metric, CLV also has some limitations. First, it is very difficult to calculate profit margins at individual level. Sometimes, revenue and/or cost cannot be attributed to a single customer, making it difficult to calculate profit margins. Similarly, it is difficult to accurately calculate retention rates because the rate calculation requires sophisticated analysis. A small increase in retention rate can

Table 19.2 Margin multiple

Retention rate (r)	Discount rate (i)			
	10%	12%	14%	16%
60%	1.20	1.15	1.11	1.07
70%	1.75	1.67	1.59	1.52
80%	2.67	2.50	2.35	2.22
90%	4.50	4.09	3.75	3.46

substantially increase the margin multiple, which in turn increases the CLV. Hence, accuracy of retention rate is extremely important for calculating CLV.

Despite the limitations mentioned above, CLV helps in making important marketing decisions. It is a customer metric that takes into account both the current and future profitability of a customer. But CLV should not be the only decision-making criteria. Firms should take into account factors such as reference/influence of the customer, and brand reputation, along with the CLV to make marketing decisions.

Customer Referral Value (CRV)

There are many resource-intensive ways to go about acquiring new businesses. While most firms are engaged in the traditional ways of attracting customers and hence gaining profits, there are effective ways to bring the business at no or little cost. One such way is customer referral. Think about your friends who are loyal customers to Tata Motors. Tata Motors can use your friends to refer you for a new model of car, or simply to the firm itself. In such situations, will Tata Motors incur cost to acquire? Probably not... or may be a little. Conditional on the tangible and intangible aspects of reference, you may end up buying a new car from Tata Motors. And that is where the power of metrics such as customer referral value (CRV) lies.

Let us think about a few firms that engage in the referral programs: *Dropbox referral program* secured 4 million users in just 15 months. Inspired by PayPal, who literally gave free money for referrals, *Dropbox* added double-sided referral programs, where both referrer and referee get rewarded. Amazon prime, PayPal, Airbnb, and Uber have recently seen huge success of referral programs in their business strategies.[6] Whether it is a B2B or a B2C business, customer referral has seen significant success in recent times. According to LinkedIn, 84% of B2B buying decisions start with a referral. Further, customer referral has a good conversion rate. The probability of a referred customer getting converted is 30% higher as compared to a lead generated through traditional channels.[7]

Now the question is how to capitalize on the referrals and design business strategies. Literature on customer management suggests that firms should engage in measuring the values of each referral and then decide the follow-up strategies (e.g., Kumar 2010; Kumar et al. 2007). Accordingly, customer referral value is defined as an estimate of lifetime values of any type-one referrals—people who would have not purchased or become customers without referrals (e.g., Kumar et al. 2007). Firms should also include the value of type-two referrals—people who would have become customers anyway. This has implications for managing marketing efforts to acquire new customers.

CRV is more complicated than computing customer lifetime value (CLV). Computation of CRV requires the estimation of the average number of successful referrals a customer makes after providing some incentive from the firm's side. For

---

[6]https://www.referralcandy.com/blog/47-referral-programs/ (accessed on May 19, 2018).
[7]https://influitive.com/blog/9-stellar-referral-program-examples/ (accessed on May 19, 2018).

that, we need to look at the past behavior, which must include enough variance in the number of referrals for proper empirical modeling and accuracy. Computation of the CRV requires understanding of the time that can go by and still be sure that a customer's referrals are actually prompted by a firm's referral incentives. Further, it is critical to understand the conversion rate of referrals to actual customers. Finally, a customer's referral value is the present value of her type-one referrals plus present value of her type-two referrals (Kumar et al. 2007). Customer referral value of a customer is the monetary value associated with the future profits given by each referred prospect, discounted to present value.

CRV can be calculated by summing up the value of the customers who joined because of the referral and the value of the customers that would have joined anyway discounted to present value. We can compute the CRV of customer[8] i as

$$CRV_i = \sum_{t=1}^{T}\sum_{y=1}^{n1} \frac{(A_{ty} - a_{ty} + M_{ty} + ACQ1_{ty})}{(1+r)^t} + \sum_{t=1}^{T}\sum_{y=1}^{n2} \frac{(ACQ_{ty})}{(1+r)^t}$$

where

A_{ty} = contribution margin by customer y who otherwise would not buy the product

a_{ty} = cost of the referral for customer y

$ACQ1_{ty}$ = savings in acquisition cost from customers who would not join w/o the referral

$ACQ2_{ty}$ = savings in acquisition cost from customers who would have joined anyway

T = number of periods that will be predicted into the future (e.g., years)

n1 = number of customers who would not join w/o the referral

n2 = number of customers who would have joined anyway

M_{ty} = marketing costs required to retain customers

A firm can first compute the CRV of their customers and then categorize them based on the value of CRV. A firm may find a particular group of customers to have significantly higher CRV than that of others. The firm can market to or provide incentives to this set of customers to increase the referrals and hence new customer acquisitions. Moreover, a firm can look at profiles, similar to that of the high CRV group, who have not referred yet and induce them to refer by providing some incentives. Given the available data, CRV can be computed with any standard statistical software.

Customer Influence Value (CIV)

Imagine yourself to be looking at the best car that is affordable within your budget. Also, imagine that you do not have much knowledge on the technical aspects of a car. Probably you will ask your friends and colleagues or search online for advice.

---

[8]Formula for computing CRV is adopted from Kumar et al. (2007).

What else can you do to reach to your decision? With growing power of social platforms (online or offline), you may want to post a question on Facebook or ask a question to an expert auto-blogger, or follow someone on Instagram whose ideas, comments, and feedback about the auto industry influence you. Just about everything from big firms to kids have some sort of strategies, tips, experiences, and attribution that drive others' decision, sales, impression, etc. That said, most firms cannot fully grasp the value of societal connections strategies and tactics, which goes beyond traditional marketing ploys and tactics.[9] It is critical to compute this influence while determining the value of your customers. Social influence can play a significantly larger role in your decision to buy a new car or putting your kids in a particular school, or deciding on which dating app to go for. Hence, value of a customer can go beyond her purchase value, or referral values. In addition to CLV and CRV, value of a customer can stem from her influence on other customers. Value of a customer's influence refers to the monetary value of the profits associated with the purchases generated by a customer's social media influence on other acquired customers and prospects, discounted to present value (Kumar 2013).

Understanding CIV can be of great value for most firms. For an ice cream retailer, Kumar and Mirchandani (2012) show that a firm can harness the true value of customer influence. They design a seven-step process to identify the influencers in online social network, observe their influences over time, and substantially improve the firm performance. Indeed customers' influence has significant value for firms and firms should measure and implement CIV in their business strategies. For a detailed understanding of CIV computation, refer Kumar et al. (2013), Kumar and Mirchandani (2012).

3 Applications

Marketing Analytics has evolved over the last century of applications, research and data collection. Some might even say that marketing was the first consumer of large "business" data! Wedel and Kannan (2016) provide a readable summary of the evolution of marketing analytics as well as pose several questions for researchers. They classify the applications into customer relationship management (CRM), marketing mix analytics, personalization, and privacy & data security. In this book, too, there are many applications of the tools: the chapter on social media analytics has applications to online advertising, A/B experiments, and digital attribution; the chapters on forecasting analytics, retail analytics, pricing analytics, and supply chain analytics contain applications to their specialized settings, such as demand forecasting, assortment planning, and distribution planning; and the case study "InfoMedia Solutions" contains an application to media-mix planning. Other

---

[9]CLV: http://www.customerlifetimevalue.co/ and CIV: https://www.mavrck.co/resources/ (accessed on Sep 15, 2018).

applications covered in this book are recommendation engines, geo-fencing, market segmentation, and search targeting. The rapid changes in data availability and tools will continue to spur the development of new applications in marketing. To keep abreast of the new developments, researchers may follow topics in "Marketing Science Institute[10]" and their research priorities.

Electronic Supplementary Material

All the datasets, code, and other material referred in this section are available in www.allaboutanalytics.net.

- Data 19.1: exercise_inter.csv
- Data 19.2: exercise_curvilinear.csv
- Data 19.3: exercise_mediation.csv
- Data 19.4: ABC_hospital_group.csv
- Data 19.5: restaurant_chain_data.csv ("DEA in practice" section)
- Data 19.6: pizza.csv ("Conjoint Analysis Interpretation" section)
- Data 19.7: product_profile_ratings.csv ("Comparing product alternatives" section in conjoint analysis)

Exercises

Ex. 19.1 Use the data file titled "*exercise_inter.csv*" and answer the following questions:

a) Why are independent variables mean-centered? Mean-center advertising, discount, and promotion variables.
b) Is the effect of advertising on sales contingent on the level of discount? Plot a graph to interpret the interaction effect.
c) Is the effect of advertising on sales contingent on the level of promotion? Plot a graph to interpret the interaction effect.

Ex. 19.2 Use the data file titled "*exercise_curvilinear.csv*" and answer the following questions:

a) Is the effect of advertising on profit curvilinear or linear? Plot a graph if the relationship is curvilinear.
b) Is the effect of sales promotion on profit curvilinear or linear? Plot a graph if the relationship is curvilinear.
c) Is the effect of rebates and discount on profit curvilinear or linear? Plot a graph if the relationship is curvilinear.

---

[10]www.msi.org (accessed on Jul 6, 2018).

Ex. 19.3 Use *"exercise_mediation.csv"* data and answer the following questions:

a) Does recall mediate the effect of advertising on market share?
b) Why is bootstrapping used in mediation analysis?

Ex. 19.4 Find the most efficient hospital unit of ABC hospital group given the following information:

Unit	1	2	3	4	5	6	7	8
Profit (crore)	120	160	430	856	200	320	189	253
Number of Specialists	150	100	120	180	220	90	140	160
Area (sq feet)	21,000	32,650	40,000	18,780	19,870	50,000	33,000	19,878

References

Aigner, D. J., Lovell, C. A. K., & Schmidt, P. (1977). Formulation and estimation of stochastic frontier production functions. *Journal of Econometrics, 6*(1), 21–37.
Aiken, L.S., and West, S.G, 1991. *Multiple regression: Testing and interpreting interactions.*
Baccouche, R., & Kouki, M. (2003). Stochastic production frontier and technical inefficiency: A sensitivity analysis. *Econometric Reviews, 22*(1), 79–91.
Banker, R. D., Cooper, W. W., Seiford, L. M., Thrall, R. M., & Zhu, J. (2004). Returns to scale in different DEA models. *European Journal of Operational Research, 154*, 345–362.
Banker, R. D., & Morey, R. (1986). Efficiency analysis for exogenously fixed inputs and outputs. *Operation Research, 34*, 513–521.
Banker, R. D., & Thrall, R. M. (1992). Estimation of returns to scale using data envelopment analysis. *European Journal of Operational Research, 62*(1), 74–84.
Baron, R. M., & Kenny, D. A. (1986). The moderator–mediator variable distinction in social psychological research: Conceptual, strategic, and statistical considerations. *Journal of Personality and Social Psychology, 51*(6), 1173.
Bera, A. K., & Sharma, S. C. (1999). Estimating production uncertainty in stochastic frontier production function models. *Journal of Productivity Analysis, 12*(2), 187–210.
Boussofiane, A., Dyson, R. G., & Thanassoulis, E. (1991). Applied data envelopment analysis. *European Journal of Operational Research, 52*(1), 1–15.
Caudill, S. B., Ford, J. M., & Gropper, D. M. (1995). Frontier estimation and firm-specific 1076 inefficiency measure in the presence of heteroskedasticity. *Journal of Business & Economic Statistics, 13*(1), 105–111.
Charnes, A., Cooper, W. W., & Rhodes, E. (1978). Measuring the efficiency of decision making units. *European Journal of Operational Research, 2*, 429–444.
Charnes, A., Clark, T., Cooper, W. W., & Golany, B. (1985). A developmental study of data envelopment analysis in measuring the efficiency of maintenance units in the U.S. air forces, in: R. Thompson and R.M. Thrall (eds.). *Annals of Operational Research, 2*, 95–112.
Chen, C.-F., & Soo, K. T. (2010). Some university students are more equal than others: efficiency evidence from England. *Economics Bulletin, 30*(4), 2697–2708.
Cohen, J., Cohen, P., West, S. G., & Aiken, L. S. (2003). *Applied multiple regression/correlation analysis for the behavioral sciences.* Hillsdale, NJ: Lawrence Erlbaum Associates.

Cook, W. D., & Seiford, L. M. (2009). Data envelopment analysis (DEA)–Thirty years on. *European Journal of Operational Research, 192*(1), 1–17.

Cooper, W. W., Seiford, L. M., & Zhu, J. (2004). *Data envelopment analysis. Handbook on data envelopment analysis* (pp. 1–39). Boston, MA: Springer.

Cullinane, K., Wang, T. F., Song, D. W., & Ji, P. (2006). The technical efficiency of container ports: comparing data envelopment analysis and stochastic frontier analysis. *Transportation Research Part A: Policy and Practice, 40*(4), 354–374.

Farrell, M. J., & Fieldhouse, M. (1962). Estimating efficient production functions under increasing returns to scale. *Journal of the Royal Statistical Society. Series A (General), 125*, 252–267.

Farris, P. W., Bendle, N. T., Pfeifer, P. E., & Reibstein, D. J. (2010). *Marketing metrics: The definitive guide to measuring marketing performance, Introduction* (pp. 1–25). London: Pearson.

Feng, C., & Fay, S. A. (2016). Inferring salesperson capability using stochastic frontier analysis. *Journal of Personal Selling and Sales Management, 36*, 294–306.

Fenn, P., Vencappa, D., Diacon, S., Klumpes, P., & O'Brien, C. (2008). Market structure and the efficiency of European insurance companies: A stochastic frontier analysis. *Journal of Banking & Finance, 32*(1), 86–100.

Førsund, F. R., Lovell, C. K., & Schmidt, P. (1980). A survey of frontier production functions and of their relationship to efficiency measurement. *Journal of Econometrics, 13*(1), 5–25.

Gagnepain, P., & Ivaldi, M. (2002). Stochastic frontiers and asymmetric information models. *Journal of Productivity Analysis, 18*(2), 145–159.

Greene, W. H. (2010). A stochastic frontier model with correction for sample selection. *Journal of Productivity Analysis, 34*(1), 15–24.

Hjalmarsson, L., Kumbhakar, S. C., & Heshmati, A. (1996). DEA, DFA and SFA: a comparison. *Journal of Productivity Analysis, 7*(2-3), 303–327.

Jacobs, R. (2001). Alternative methods to examine hospital efficiency: Data envelopment analysis and stochastic frontier analysis. *Health Care Management Science, 4*(2), 103–115.

Koetter, M., & Poghosyan, T. (2009). The identification of technology regimes in banking: Implications for the market power-fragility nexus. *Journal of Banking & Finance, 33*, 1413–1422.

Kumar, V. (2010). *Customer relationship management*. Hoboken, NJ: Wiley Online Library.

Kumar, V. (2013). *Profitable customer engagement: Concept, metrics and strategies*. Thousand Oaks, CA: SAGE Publications India.

Kumar, V., Andrew Petersen, J., & Leone, R. P. (2007). How valuable is word of mouth? *Harvard Business Review, 85*(10), 139.

Kumar, V., Bhaskaran, V., Mirchandani, R., & Shah, M. (2013). Practice prize winner-creating a measurable social media marketing strategy: Increasing the value and ROI of intangibles and tangibles for Hokey Pokey. *Marketing Science, 32*(2), 194–212.

Kumar, V., & Mirchandani, R. (2012). Increasing the ROI of social media marketing. *MIT Sloan Management Review, 54*(1), 55.

Kumar, V., & Sharma, A. (2017). Leveraging marketing analytics to improve firm performance: insights from implementation. *Applied Marketing Analytics, 3*(1), 58–69.

Kumar, V., Sharma, A., Donthu, N., & Rountree, C. (2015). Practice prize paper-implementing integrated marketing science modeling at a non-profit organization: Balancing multiple business objectives at Georgia Aquarium. *Marketing Science, 34*(6), 804–814.

Kumbhakar, S. C., & Lovell, C. K. (2003). *Stochastic frontier analysis*. Cambridge: Cambridge university press.

Lilien, G. L., Rangaswamy, A., & De Bruyn, A. (2013). *Principles of marketing engineering*. State College, PA: DecisionPro.

Lilien, G. L. (2011). Bridging the academic–practitioner divide in marketing decision models. *Journal of Marketing, 75*(4), 196–210.

MacKinnon, D. P. (2008). *Introduction to statistical mediation analysis*. Abingdon: Routledge.

Meeusen, W., & van den Broeck, J. (1977). Efficiency estimation from Cobb-Douglas production functions with composed error. *International Economic Review, 18*(2), 435–444.

Ofek, E., & Toubia, O. (2014). *Conjoint analysis: A do it yourself guide*. Harvard Business School, note, 515024.

Ouellette, P., & Vierstraete, V. (2004). Technological change and efficiency in the presence of quasi-fixed inputs: A DEA application to the hospital sector. *European Journal of Operational Research, 154*(3), 755–763.

Parsons, J. L. (2002). Using stochastic frontier analysis for performance measurement and benchmarking. *Advances in Econometrics, 16*, 317–350.

Preacher, K. J., Curran, P. J., & Bauer, D. J. (2006). Computational tools for probing interactions in multiple linear regression, multilevel modeling, and latent curve analysis. *Journal of Educational and Behavioral Statistics, 31*(4), 437–448.

Preacher, K. J., & Hayes, A. F. (2004). SPSS and SAS procedures for estimating indirect effects in simple mediation models. *Behavior Research Methods, 36*(4), 717–731.

Preacher, K. J., Rucker, D. D., & Hayes, A. F. (2007). Addressing moderated mediation hypotheses: Theory, methods, and prescriptions. *Multivariate Behavioral Research, 42*(1), 185–227.

Reinhard, S., Lovell, C., & Thijssen, G. (2000). Environmental efficiency with multiple environmentally detrimental variables; estimated with SFA and DEA. *European Journal of Operational Research, 121*(3), 287–303.

Schmidt, P. (1985). Frontier production functions. *Econometric Reviews, 4*(2), 289–328.

Seiford, L. M., & Zhu, J. (1999). An investigation of returns to scale under data envelopment analysis. *Omega, 27*, 1–11.

Sherman, H. D., & Gold, F. (1985). Bank branch operating efficiency: Evaluation with data envelopment analysis. *Journal of Banking & Finance, 9*(2), 297–315.

Venkatesan, R., Farris, P., & Wilcox, R. T. (2014). *Cutting-edge marketing analytics: Real world cases and data sets for hands on learning*. London: Pearson Education.

Vyt, D. (2008). Retail network performance evaluation: A DEA approach considering retailers' geomarketing. *The International Review of Retail, Distribution and Consumer Research*, 235–253.

Wadud, A., & White, B. (2000). Farm household efficiency in Bangladesh: a comparison of stochastic frontier and DEA methods. *Applied Economics, 32*(13), 1665–1673.

Wedel, M., & Kannan, P. K. (2016). Marketing analytics for data-rich environments. *Journal of Marketing, 80*, 97–121.

Other Resources

Kristopher, J. Preacher (2018). *Preacher's website*. Retrieved May 19, 2018, from http://quantpsy.org/medn.htm.

Retrieved May 19, 2018, from www.conjoint.online.

Software: STATA, Mplus. Retrieved May 19, 2018, from www.statmodel.com.

Chapter 20
Financial Analytics

Krishnamurthy Vaidyanathan

1 Part A: Methodology

1.1 Introduction

Data analytics in finance is a part of quantitative finance. Quantitative finance primarily consists of three sectors in finance—asset management, banking, and insurance. Across these three sectors, there are four tightly connected functions in which quantitative finance is used—valuation, risk management, portfolio management, and performance analysis. Data analytics in finance supports these four sequential building blocks of quantitative finance, especially the first three—valuation, risk management, and portfolio management.

Quantitative finance can be dichotomized into two branches or subsets having mild overlaps. The first is the risk-neutral world or the Q-world, and the second, wherein data analytics is used extensively, is the risk-averse world or the P-world. Quant professionals in the Q-world are called the Q-quants, and those of the P-world are called P-quants. Before we delve into the methodology of data analysis in finance which we structure as a three-stage process in this chapter, we briefly highlight the processes, methodologies, challenges, and goals of these two quant-worlds and also look at the history and origins of these two dichotomized worlds of quantitative finance.

Electronic supplementary material The online version of this chapter (https://doi.org/10.1007/978-3-319-68837-4_20) contains supplementary material, which is available to authorized users.

K. Vaidyanathan (✉)
Indian School of Business, Hyderabad, Telangana, India
e-mail: vaidya_nathan@isb.edu

1.2 Dichotomized World of Quant Finance

One can paraphrase Rudyard Kipling's poem *The Ballad of East and West* and say, "Oh, the Q-world is the Q-world, the P-world is the P-world, and never the twain shall meet." Truth be told, Kipling's lofty truism is not quite true in the Quant world. The Q-world and P-world do meet, but they barely talk to each other. In this section, we introduce the P- and Q-worlds and their respective theoretical edifices.

1.2.1 Q-Quants

In the Q-world, the objective is primarily to determine a fair price for a financial instrument, especially a derivative security, in terms of its underlying securities. The price of these underlying securities is determined by the market forces of demand and supply. The demand and supply forces come from a variety of sources in the financial markets, but they primarily originate from buy-side and sell-side financial institutions. The buy-side institutions are asset management companies—large mutual funds, pensions funds, and investment managers such as PIMCO who manage other people's money—both retail and corporate entities' money. The sell-side institutions are market makers who make money on the margin they earn by undertaking market making. That is, they are available to buy (bid) a financial instrument for another market participant who wants to sell and make available to sell (offer) a financial instrument for somebody wanting to buy. They provide this service for a commission called as bid–offer spread, and that is how they primarily make their money from market making. The trading desks of large investment banks such as Goldman Sachs, JPMorgan, Citibank, and Morgan Stanley comprise the sell-side. The Q-quants primarily work in the sell-side and are price-makers as opposed to P-quants who work in the buy-side and are typically price-takers.

The Q-quants borrow much of their models from physics starting with the legendary Brownian motion. The Brownian motion is one of the most iconic and influential ingress from physics into math finance and is used extensively for pricing and risk management. The origins of the Q-world can be traced to the influential work published by Merton in 1969 (Merton 1969)—who used the Brownian motion process as a starting point to model asset prices. Later in 1973, Black and Scholes used the geometric Brownian motion (GBM) to price options in another significant work for which they eventually won the Nobel Prize in 1997 (Black and Scholes 1973). This work by Black and Scholes gave a fillip to pricing of derivatives as these financial instruments could be modeled irrespective of the return expectation of the underlying asset. In simple terms, what that meant was that even if I think that the price of a security would fall while you may think that the price of that security would increase, that is, the return expectations are different, yet we can agree on the price of a derivative instrument on that security. Another important edifice in the Q-space is the fundamental theorem of asset pricing by Harrison and Pliska (1981). This theorem posits that the current price of a security is fair only if there exists a stochastic process such as a GBM with constant expected value for all future points in time. Any process that satisfies this property is called a martingale. Because the expected return is the same for all financial instruments, it implies that there is no extra reward for risk taking. It is as if, all the pricing is done in a make-believe world called the risk-neutral world, where irrespective of the risk of a security, there is no extra compensation for risk. All financial instruments in this make-believe world earn the same return regardless of their risk. The risk-free instruments earn the risk-free rate as do all risky instruments. In contrast, in the P-world, the economic agents or investors are, for all intents and purposes, risk-averse, as most people are in the real world.

The Q-quants typically have deep knowledge about a specific product. So a Q-quant who, for instance, trades credit derivatives for a living would have abundant knowledge about credit derivative products, but her know-how may not be very useful in, say, a domain like foreign exchange. Similarly, a Q-quant who does modeling of foreign exchange instruments may not find her skillset very useful if she were to try modeling interest rates for fixed income instruments. Most of the finance that is done in the Q-world is in continuous time because as discussed earlier, the expectation of the price at any future point of time is equal to its current price. Given that this holds for all times, the processes used in the Q-world are naturally set in continuous time. In contrast, in the P-world, the probabilities are for a risk-averse investor. This is because in the real world, people like you and me need extra return for risk-taking. Moreover we measure returns over discrete time intervals such as a day, a week, a month, or a year. So most processes are modeled in discrete time. The dimensionality of the problem in the P-world is evidently large because the P-quant is not looking at a specific instrument or just one particular asset class but multiple asset classes simultaneously. The tools that are used to model in the P-world are primarily multivariate statistics which is what concerns data analysts.

1.2.2 P-Quants

We now discuss the P-world and their origins, tools, and techniques and contrast them with the Q-world. The P-world started with the mean–variance framework by Markowitz in 1952 (Markowitz 1952). Harry Markowitz showed that the conventional investment evaluation criteria of net present value (NPV) needs to be explicitly segregated in terms of risk and return. He defined risk as standard deviation of return distribution. He argued that imperfect correlation of return distribution of stocks can be used to reduce the risk of a portfolio of stocks. He introduced the concept of diversification which is the finance equivalent of the watchword—do not put all your eggs in one basket. Building on the Markowitz model, the next significant edifice of the P-world was the capital asset pricing model (CAPM) by William Sharpe in 1964 (Sharpe 1964). William Sharpe converted the Markowitz "business school" framework to an "economics department" model. Sharpe started with a make-believe world where all investors operate in the Markowitz framework. Moreover, all investors in the CAPM world have the same expectation of returns and variance–covariance. Since all the risky assets in the financial market must be held by somebody, it turns out that in Sharpe's "economics department" model, all investors end up holding the market portfolio—a portfolio where each risky security has the weight proportional to how many of them are available. At the time when William Sharpe postulated the model, the notion of market portfolio was new. Shortly thereafter, the financial industry created mutual funds which hold a diversified portfolio of stocks, mostly in the proportion of stocks that are available. It was as if nature had imitated art, though we all know it is almost always the other way around. In 1990, Markowitz and Sharpe won the Noble Prize in Economics for developing the theoretical basis for diversification and CAPM.

The next significant edifice came in 1970 by Eugene Fama called the efficient market hypothesis (EMH). The EMH hypothesizes that no trading strategy based on already available information can generate super-normal returns (Fama 1970). The EMH offered powerful theoretical insights into the nature of financial markets. More importantly, it lent itself to empirical investigation which was imperative and essential for finance—then a relatively nascent field. As a result, the efficient market hypothesis is probably the most widely and expansively tested hypothesis in all social sciences. Eugene Fama won the Nobel Prize in 2013 for his powerful insights on financial markets.

Another important contribution in the mid-1970s was the arbitrage pricing theory (APT) model by Stephen Ross (1976). The APT is a multifactor model used to calculate the expected return of a financial asset. Though both CAPM and APT provided a foundational framework for asset pricing, they did not use data analytics because their framework assumed that the probability distribution in the P-world is known. For instance, if financial asset returns follow an elliptical distribution, every investor would choose to hold a portfolio with the lowest possible variance for her chosen level of expected returns. Such a portfolio is called a minimum variance portfolio, and the framework of portfolio analysis is called the mean–variance framework. From a finance theory viewpoint, this is a convenient assumption to make, especially the assumption that asset returns are jointly normal, which is a special case of an elliptical distribution. Once the assumption that the probability distribution is already known is made, theoretical implications can be derived on how the asset markets should function. Tobin proposed the separation theorem which postulates that the optimal choice of investment for an investor is independent of her wealth (Tobin 1958). Tobin's separation theorem holds good if the returns of the financial assets are multinormal. Extending Tobin's work, Stephen Ross postulated the two-fund theorem which states that if investors can borrow and lend at the risk-free rate, they will possess either the risk-free portfolio or the market portfolio (Ross 1978). Ross later generalized it to a more comprehensive k-fund separation theorem. The Ross separation theorem holds good if the financial asset returns follow any elliptical distribution. The class of elliptical distribution includes the multivariate normal, the multivariate exponential, the multivariate Student t-distribution, and multivariate Cauchy distribution, among others (Owen and Rabinovitch 1983).

However, in reality, the probability distribution needs to be estimated from the available financial information. So a very large component of this so-called information set, that is, the prices and other financial variables, is observed at discrete time intervals, forming a time series. Analyzing this information set requires manifestly sophisticated multivariate statistics of a certain spin used in economics called as econometrics, wherein most of the data analytics tools come into play. In contrast to this, in the Q-world, quants mostly look at the pricing of a specific derivative instrument and get the arbitrage-free price of the derivative instrument based on the underlying fundamental instrument and other sets of instruments. However, in the P-world, we try to estimate the joint distribution of all the securities that are there in a portfolio unlike in the Q-world wherein we are typically concerned with just one security. The dimensionality in the Q world is small, but in the P-world it is usually a lot larger. So a number of dimension reduction techniques, mostly linear factor models like principal component analysis (PCA) and factor analysis, have a central role to play in the P-world. Such techniques achieve parsimony by reducing the dimensionality of the data, which is a recurring objective in most data analytic applications in finance. Since some of these data analytic techniques can be as quantitatively intense or perhaps more intense than the financial engineering techniques used in the Q-world, there is now a new breed of quants called the P-quants who are trained in the data analytic methodologies. Prior to the financial crisis of 2008, the Q-world attracted a lot of quants in finance. Many having PhDs in physics and math worked on derivative pricing. Till the first decade of the twenty-first century culminating in the financial crisis, all the way back from the 1980s when the derivatives market started to explode, quantitative finance was identified with the Q-quants. But in recent years, especially post-crisis, the financial industry has witnessed a surge of interest and attention in the P-world, and there is a decrease of interest in the Q-world. This is primarily because the derivatives markets have shrunk. The second-generation and third-generation types of exotic derivatives that were sold pre-crisis have all but disappeared. In the last decade, there has been a severe reduction in both the volume and complexity of derivatives traded. Another reason why the P-quants have a dominant role in finance is that their skills are extremely valuable in risk management, portfolio management, and actuarial valuations, while the Q-quants work mostly on valuation. Additionally, a newfound interest in data analytics, in general and mining of big data specifically, has been a major driver for the surge of interest in the P-world.

1.3 Methodology of Data Analysis in Finance: Three-Stage Process

The methodology of data analysis in the P-space in this chapter is structured as a three-part process—asset price estimation, risk management, and portfolio analysis. The first part, asset price estimation, is split into five sequential steps; the second part, risk management, into three steps; and the third part, portfolio analysis, into two steps. The methodology of the three-stage framework for data analysis in finance is shown in Fig. 20.1. The first among the estimation steps is the process of

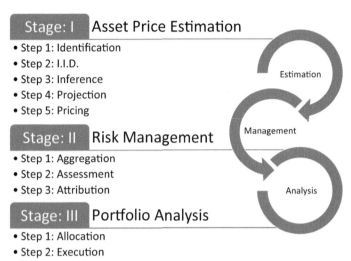

Fig. 20.1 Methodology of the three-stage framework for data analysis in finance

variable identification. Variable identification is an important first step because the variable to be modeled is different for different asset classes as shown in Fig. 20.1. The second step is to transform the identified variable into an independent and identically distributed (i.i.d.) process. The third step is to infer the joint distribution of these i.i.d. processes of multiple financial variables. The fourth step is to forecast the variable using past information, and the final fifth step in asset price estimation is to derive the forecasted price from the i.i.d. variables' joint distribution.

The next two stages in the modeling process constitute analysis for risk and portfolio management. The first three steps within the second part of analysis pertain to risk management, and the remaining two steps apply to portfolio management. Within risk management, we aggregate the entire portfolio (Stage II Step 1, Sect. 1.3.8) and then evaluate the portfolio (Stage II Step 2, Sect. 1.3.9) in terms of various risk metrics such as Value at Risk (VaR) and threshold persistence. Then, the return and risk of the portfolio is attributed to several risk factors using various decomposition techniques (Stage II Step 3, Sect. 1.3.10).

The final two steps constitute portfolio management where we will look at the design of the optimal portfolio or, from a firm-wide context, design of an optimal policy to allocate the portfolio of assets of a firm (Stage III Step 1, Sect. 1.3.12). In the next step, we execute and allocate assets according to the optimal portfolio benchmark determined in the previous step (Stage III Step 1, Sect. 1.3.12). This is done these days mostly programmatically using algorithms. Each of the steps are explained using simple stand-alone examples. Suitable references are provided at the appropriate steps. A comprehensive application of the framework that combines multiple steps is provided in Sect. 20.2. We now discuss the three-stage methodology starting with variable identification for different asset classes such as equities, foreign exchange, fixed income, credit, and commodities.

1.3.1 Stage I: Asset Price Estimation

The objective of the first stage is to estimate the price behavior of an asset. It starts with identification of the financial variable to model.

Step 1: Identification

The first step of modeling in the P-world is to identify the appropriate variable which is different for distinct asset classes. The basic idea is to find a process for the financial variable where the residuals are essentially i.i.d. The most common process used for modeling a financial variable x is the random walk:

$$x_t = x_{t-1} + \varepsilon_t$$

where ε_t is the error term and is random. The postulation where a financial variable follows a random walk is called as random walk hypothesis and is consistent with efficient market hypothesis (Fama 1965). What it means in simple terms is that financial variables are fundamentally unpredictable. However, if one looks at any typical stock price, the price changes in such a way that the order of magnitude of the change is proportional to the value of the stock price. This kind of behavior conflicts with homogeneity across time that characterizes a financial variable when it follows a random walk. As a way out, the variable that is actually modeled is the logarithm (log) of the stock price, and it has been observed that the log of the stock price behaves as a random walk. A simple random walk has a constant mean and standard deviation, and its probability distribution does not change with time. Such a process is called as a stationary process.

Similar to stock prices, the log of foreign exchange rates or the log of commodity prices behaves approximately as a random walk. The underlying variable itself, that is, the stock price, currency rate, or commodity price, is not exactly stationary, but the log of stock price, currency rate, or commodity price conducts itself just about as a haphazard random walk. A stationary process is one whose probability distribution does not change with time. As a result, its moments such as variance or mean are not time-varying.

However, choosing the right financial variable is as important as the modification made to it. For example, in a fixed income instrument such as a zero-coupon bond, the price converges to the face value as the bond approaches maturity. Clearly, neither the price itself nor its log can be modeled as a random walk. Instead, what is modeled as a random walk is the yield on bonds, called yield to maturity. Simply put, yield is the internal rate of return on a bond that is calculated on the cash flows

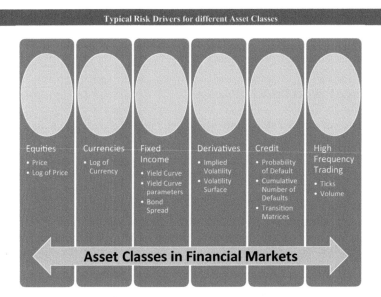

Fig. 20.2 Typical risk drivers for different asset classes

the bond pays till maturity. And this variable fits a random walk model adequately. The financial variables that are typically modeled in the different asset classes are shown in Fig. 20.2.

Step 2: I.I.D.

Once the financial variable that is of interest is identified, the next step in data preparation is to obtain a time series of the variables that are of interest. These variables should display a homogenous behavior across time, and as shown in Fig. 20.2, the variables are different for different asset classes. For instance, in equities, currencies, or commodities, it is the log of the stock/currency/commodity price. For fixed income instruments, the variable of interest may not be the price or the log of price. The variable of interest would be the yield to maturity of the fixed income security.

Once we have the raw data of the financial variable, we test if the financial variable follows a random walk using statistical tests such as the Dickey–Fuller test or the multiple variance ratio test (Dickey and Fuller 1979; Campbell et al. 1997). A lot of times they may follow less random processes and therefore may be predictable to some degree. This is known as non-random walk hypothesis. Andrew Lo and Craig Mackinlay, at MIT Sloan and Wharton, respectively, in their book *A Non-Random Walk Down Wall Street* present a number of tests and studies that validate that there are trends in financial markets and that the financial variables identified in Step 1 are somewhat predictable. They are predictable both in cross-sectional and time series terms. As an example of the ability to predict using cross-sectional data, the Fama–French three-factor model postulated by Eugene Fama and Kenneth French uses three factors to describe stock returns (Fama and French 1993). An example of predictability in time series is that the financial variable may display some kind of mean reversion tendency. What that means is that if the value of the financial variable is quite high, it will have a propensity to decrease in value and vice versa. For example, if yields become very high, there may be propensity for them to come back to long-run historical average levels. In such cases, the features that cause deviation from the random walk are extracted out so that the residuals display i.i.d. behavior. The models used would depend on the features displayed by the financial variable. For example, volatility clustering like mean reversion is a commonly observed feature in financial variables. When markets are extremely volatile, the financial variable fluctuates a lot, and there may be a higher probability of a large variability than otherwise. Techniques like autoregressive conditional heteroscedasticity model (ARCH) or generalized autoregressive conditional heteroscedasticity model (GARCH) are used to factor out volatility clustering (Engle 1982). If the variable displays some kind of mean reversion, one might want to use autoregressive moving average (ARMA) models if it is a univariate case or use vector autoregression (VAR) models in multivariate scenarios (Box et al. 1994; Sims 1980). These are, in essence, econometric models which can capture linear interdependencies across multiple time series and are fundamentally a general form of autoregressive models (AR) (Yule 1927). We could also use stochastic volatility models, and those are comparatively commonly used in volatility clustering. Long memory processes primarily warrant fractional integration models (Granger and Joyeux 1980). Fractional integration displays a long memory which principally means that the increments of the financial variable display autocorrelation. The increments therefore are not i.i.d., and these autocorrelations persist across multiple lags. For instance, the value of the random variable at time $t + 1$ is a function of time $t, t-1, t-2$, and so on. The lags decrease very gradually and are therefore called long memory processes. Such trends, be they long memory, volatility clustering, or mean reversion, are modeled using techniques such as fractional integration, GARCH, or AR processes, respectively. After such patterns are accounted for, we are left with i.i.d. shocks with no discernible pattern.

Step 3: Inference

The third step in estimation after the financial variable is identified and after we have gotten to the point of i.i.d. shocks is to infer the joint behavior of i.i.d. shocks. In the estimation process, we typically determine those parameters in the model which gets us to an i.i.d. distribution. We explain the first three steps using data on S&P 500 for the period from October 25, 2012, to October 25, 2017. The first step is to identify the financial variable of interest. We work with returns rather than with

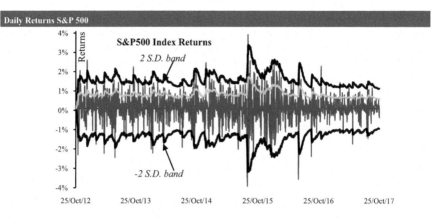

Fig. 20.3 Daily returns of the S&P 500

absolute index levels of S&P 500 for reasons mentioned in Step 1. From the daily index levels, the 1-day returns are calculated as follows:

$$r_t = \log(p_t) - \log(p_{t-1})$$

This return r_t itself is not distributed in an i.i.d. sense. Neither are the daily returns identical nor are they independent. We can infer from a quick look at the graph of the returns data in Fig. 20.3 that the returns are not i.i.d.

One may refer to Chap. 5, on data visualization, for a better understanding on how to interpret the graph. One of the things to observe from the graph is that if the return in a given day was either extremely high or low, it normally followed that return on the subsequent day was also quite high/low. That is, if the return was volatile at time t, the probability of it being more volatile is higher than it being stable at time $t + 1$. So in this case, the data seems to suggest that the financial variable is conditionally heteroscedastic, which means that the standard deviation is neither independent nor identical across time periods. To accommodate for conditional heteroscedasticity, we can use the GARCH(1,1) model (Bollerslev 1986). This model accounts for autocorrelation and heteroscedasticity, that is, for correlations among errors at different time periods t and different variances of errors at different times. The way we model variance σ_t^2 is:

$$\sigma_t^2 = \omega + \alpha\, r_{t-1}^2 + \beta \sigma_{t-1}^2$$

We have to estimate the parameters ω, α, β. The estimation technique we use is maximum likelihood estimation (MLE). As the name implies, we maximize the likelihood of estimating the true values of the parameters ω, α, β. If GARCH(1,1) model is the right specification and if the parameters ω, α, β are estimated correctly, then $\frac{r_t}{\sigma_t}$ will be a sequence of random i.i.d. variables. In this case we assume the average daily returns to be zero. Using the Gaussian distribution, the likelihood or the probability of $\frac{r_t}{\sigma_t}$ being normally distributed is given by:

$$\frac{1}{\sqrt{2\pi\sigma_t^2}}e^{-\frac{1}{2}\left(\frac{r_t}{\sigma_t}\right)^2}$$

This completes Step 2 of reducing the variable to an i.i.d. process. The next step is to compute the joint distribution. Since the variables $\frac{r_t}{\sigma_t}$ across time are independent, the joint likelihood L of the sample is calculated as the product of the above likelihood function using the property of independence across the time series of n data points. Therefore:

$$L = \prod_{t=1}^{n} \frac{1}{\sqrt{2\pi\sigma_t^2}}e^{-\frac{1}{2}\left(\frac{r_t}{\sigma_t}\right)^2}$$

Since the above product would be a very small number in magnitude, the natural log of the above likelihood is maximized in MLE. This log-likelihood is given by:

$$\ln(L) = -\frac{1}{2}\sum_{t=1}^{n}\left\{\ln\left(2\pi\sigma_t^2\right) + \left(\frac{r_t}{\sigma_t}\right)^2\right\}$$

The joint log-likelihood is a function of the parameters ω, α, β. The value of the parameters that maximizes this joint likelihood is the correct estimate. The above steps are explained in the spreadsheet "Financial Analytics Steps 1, 2 and 3.xlsx" (available on the book's website).

Step 4: Projection

The fourth step is projection. We explain this step using a simple example from foreign exchange markets. Let us say that the financial variable is estimated using a technique such as MLE, GMM, or Bayesian estimation (Hansen 1982). The next step is to project the variable using the model. Say the horizon is 1 year, and we want to calculate the expected profit or loss of a certain portfolio. A commonly used technique for this is the Monte Carlo simulation, which is another ingress from physics (Fermi and Richtmyer 1948). We project the financial variable in the Q-space using risk-neutral parameters and processes. This also helps us to understand how the P- and Q-worlds converge.

Let us say we want to project the value of the exchange rate of Indian Rupee against the US Dollar (USD/INR). USD/INR as of end October 2017 is 65. We assume that the returns follow a normal distribution characterized by its mean and standard deviation. The projection could be done either in the P-space or in the Q-space. In the P-space, the projection would be based on the historical average annual return and the historical annualized standard deviation, and we would use these first and second moments to project the USD/INR. The equivalent method in Q-world would be to calculate using the arbitrage-free drift.

To estimate the arbitrage-free drift, let us assume that the annualized USD interest rate for 1 year is 1.7% and that the 1-year INR rate is 6.2%. A dollar parked in a savings account in the USA should yield the same return as that dollar being converted to rupee, parked in India, and then reconverted to USD. This is needed to ensure no arbitrage in an arbitrage-free foreign exchange market. This implies that the exchange rate of USD/INR should depreciate at 4.5%. This can also be understood from the uncovered interest rate parity criterion for a frictionless global economy (Frenkel and Levich 1981). The criterion specifies that real interest rates should be the same all over the world. Let us assume that real interest rate globally is 0.5% and that the US inflation is 1.2%, implying nominal interest rate is 1.7%. Likewise, inflation in India is 5.7% implying a nominal interest rate of 6.2%. Inflation in India is 5.7% and that in the USA is 1.2%. Therefore, the currency in India (Rupee) should get depreciated against the US currency (Dollar) by the differential of their respective inflations 4.5% (=5.7%—1.2%). Let us assume that the standard deviation of USD/INR returns is 10%. Once we have the mean and standard deviation, we can run a Monte Carlo simulation to project the financial variable of interest. Such an exercise could be of interest if revenues are in dollars and substantial portion of expenditure is in dollars. For a more detailed reading of applicability of this exercise to the various components of earnings such as revenues and cost of goods sold in foreign currency, please refer to "Appendix C: Components of earning of the CreditMetrics™" document by JPMorgan (CreditMetrics 1999).

Monte Carlo Simulation

We first pick a random number from the standard normal distribution say x. We then scale (multiply) x by standard deviation and add average return to get a random variable mapped to the exact normal distribution of returns.

$$R = x^* \, (10\%) \, /\mathrm{sqrt}(365) + (4.5\%) \, /365$$

Note that the average return and standard deviation are adjusted for daily horizon by dividing with 365 and square root of 365, respectively. After scaling the variable, we multiply price of USD/INR at t with $(1 + R)$ to project the value of USD/INR to the next day. The above steps are explained in the spreadsheet "Financial Analytics Steps 4 and 5.xlsx" (available on the book's website). An example of the above simulation with USD/INR at 65 levels at day 1 is run for seven simulations and 10 days in Table 20.1 and Fig. 20.4.

Once we have the projections of the currency rates in forward points in time, it is an easy task to then evaluate the different components of earnings that are affected by the exchange rate.

Table 20.1 Simulation of the dollar to rupee exchange rate

Date	1	2	3	4	5	6	7
27-10-2017	65.00	65.00	65.00	65.00	65.00	65.00	65.00
28-10-2017	65.08	65.25	64.89	65.59	64.72	65.13	64.86
29-10-2017	65.57	64.74	65.56	65.37	64.41	65.16	64.58
30-10-2017	64.78	65.09	65.52	65.58	64.31	65.58	64.41
31-10-2017	65.27	64.65	65.28	65.51	64.92	65.72	63.98
01-11-2017	65.04	64.15	65.08	65.80	64.30	65.71	63.47
02-11-2017	64.97	64.31	65.34	65.82	64.64	65.47	63.62
03-11-2017	65.76	64.49	65.39	65.98	64.43	65.26	63.56
04-11-2017	65.63	64.24	65.22	66.35	63.97	65.73	63.19
05-11-2017	65.64	64.46	65.41	66.32	64.12	65.04	63.25
06-11-2017	65.00	64.39	65.29	66.34	64.29	64.92	64.17
07-11-2017	64.59	64.26	65.97	65.87	64.46	64.96	64.82

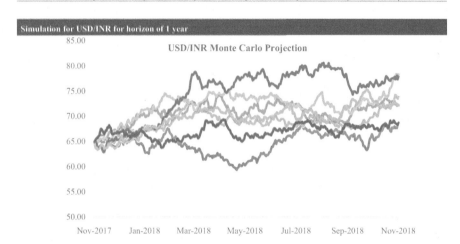

Fig. 20.4 Simulation of the dollar to rupee exchange rate

Step 5: Pricing

The fifth step is pricing which logically follows from projection. The example that we used in Step 4 was projection of USD/INR for a horizon of 1 year. What pricing allows us to do is arrive at the ex-ante expected profit or loss of a specific instrument based on the projections done in Step 4. In a typical projection technique like Monte Carlo, each of the steps is equally likely. So the probability of each of these steps is given by $1/n$, where n is the number of simulations done. The ex-ante profit or loss of the instrument is given by $1/n$ times the probability of profit or loss of the instrument. For instance, in the case of a forward contract on USD/INR that pays off 1 year from now, the payoff would be calculated at the end of 1 year as the projected value of USD/INR minus the forward rate, if it is a long forward contract and vice versa for a short forward contract. A forward contract is a contract between two parties to buy or sell an asset at a specified future point in time. In this case, the asset is USD/INR, and the specified future point in time is 1 year. The party that buys USD is supposed to be a "Long" forward contract, while the other party selling USD 1 year from now is a "Short" forward contract. The expected ex-ante payoff is the summation of the payoff in all the scenarios divided by the number of simulations. After pricing, we move on to the next stage of risk management.

1.3.2 Stage II: Risk Management

The second stage of data analytics in finance concerns risk management. It involves analysis for risk aggregation, risk assessment, and risk attribution. The framework can be used for risk analysis of a portfolio or even for an entire financial institution.

Step 1: Aggregation

The first of the three steps in risk management is risk aggregation. The aggregation step is crucial because all financial institutions need to know the value of the portfolio of their assets and also the aggregated risk exposures in their balance sheet. After we have priced the assets at the instrument level, to calculate the value of the portfolio, we need to aggregate them keeping in view the fact that the risk drivers are correlated. The correlation of the various risk drivers and the financial instruments' risk exposure is thereby aggregated. We exposit aggregation using one of the commonly used tools for risk aggregation called copula functions (Ali et al. 1978). Copula functions, especially Gaussian copulas, are used extensively by the financial industry and the regulators due to their analytical tractability. The Basel Committee on Banking Supervision relies exclusively on Gaussian copula to measure risk capital of banks globally. A copula function is a multivariate probability distribution for which the marginal distributions are known. Copula function illustrates the dependence between these correlated random variables. Copula in Latin means to link or to tie. They are widely used in both the P-world and the Q-world for risk aggregation and optimization. The underlying edifice for copula function is the Sklar's theorem (Sklar 1959). This theorem posits that any multivariate joint distribution of risk drivers can be described in terms of the univariate marginal distributions of the individual risk drivers. A copula function describes the dependence structure between these correlated random variables for the univariate marginal distributions. As usual, we discuss this step using an example. Let us say that there is a loan portfolio comprising N number of exposures. To keep the example computationally simple, we keep $N = 5$. So we have a bank which has lent money to five different corporates which we index $i = 1, 2, \ldots, 5$. We assume for simplicity that Rs. 100 is lent to each of the five firms. So the loan portfolio is worth Rs. 500. The way we will go about aggregating the risk of this Rs. 500 loan portfolio is that we will first describe the marginal distribution of credit for each of the five corporates. We will then use the Gaussian copula function to get the joint distribution of the portfolio of these five loans.

Let us assume for simplicity that each corporate has a probability of default of 2%. Therefore, there is a 98% chance of survival of the corporate in a year. The horizon for the loan is 1 year. Assume that in the event of a default, the bank can recover 50% of the loan amount. The marginal distributions are identical in our example for ease of exposition, but the copula models allow for varying distributions as well. What we want to do is that based on the correlation structure, we want to calculate the joint distribution of credit of each of these corporates. We model this using a one-factor model. The single factor is assumed to be the state of the economy M, which is assumed to have a Gaussian distribution.

To generate a one-factor model, we define random variables x_i ($1 \leq i \leq N$):

$$x_i = \rho_i M + \sqrt{1 - \rho_i^2} Z_i$$

In the above equation, the single factor M and the idiosyncratic factor Z_i are independent of each other and are standard normal variables with mean zero and unit standard deviation. The correlation coefficient ρ_i satisfies $1 \leq \rho_i < 1$. The above equation defines how the assets of the firm are correlated with the economy M. The correlation between the assets x_i of firm i and assets x_j of firm j is $\rho_i \rho_j$.

Let H be the cumulative normal distribution function of the idiosyncratic factor Z_i. Therefore:

$$\text{Probability}\,(x_i < x | M) = H\left(\frac{x - \rho_i M}{\sqrt{1 - \rho_i^2}}\right)$$

The assets of each of these corporates are assumed to have a Gaussian distribution. Note that the probability of default is 2%, corresponding to a standard normal value of -2.05. If the value of the asset standardized with the mean and its standard deviation is more than -2.05, the entity survives, else it defaults. The conditional probability that the ith entity will survive is therefore:

$$S_i\,(x_i < x | M) = 1 - H\left(\frac{x - \rho_i M}{\sqrt{1 - \rho_i^2}}\right)$$

The marginal distribution of each of the assets is known, but we do not know the joint distribution of the loan portfolio. So, we model the portfolio distribution using copulas based on the correlations that each of these corporates has. The performance of the corporate depends on the state of the economy. There is a correlation between these two variables. This can be explained by noting that certain industries such as steel and cement are more correlated with the economy than others like fast-

Table 20.2 State probabilities for one-factor Gaussian copula

States of the economy	Midpoint of range	Probability of state
1	−3.975	7.40435E-06
2	−3.925	9.02075E-06
3	−3.875	1.09626E-05
4	−3.825	1.32891E-05
5	−3.775	1.60692E-05
6	−3.725	1.93824E-05
7	−3.675	2.33204E-05
8	−3.625	2.79884E-05
9	−3.575	3.3507E-05
10	−3.525	4.00135E-05

moving consumer goods. Assume that the correlation of the first corporate with the economy is 0.2, the second is 0.4, the third is 0.5, the fourth is 0.6, and the fifth is 0.8. So the pairwise correlation can be calculated as the product of the two correlations to the single factor, which in our example is the economy. We model the state of the economy as a standard normal random variable in the range from −3.975 to 3.975 in intervals of 0.05. We take the mid-point of these intervals. Table 20.2 shows these values for the first ten states of the economy. The probability of the economy being in those intervals is calculated in column 3 of Table 20.2 using the Gaussian distribution. This is given by:

$$Prob\left\{m - \frac{\Delta}{2} \leq M \leq m + \frac{\Delta}{2}\right\}$$

where M follows the standard normal distribution, m is the mid-point of the interval, and Δ is the step size. The way to interpret the state of the economy is that when it is highly negative such as −2, then the economy is in recession. And if it is high such as greater than 2, the economy is booming, and if it is close to zero, then the health of the economy is average. Once we have the probabilities for the state of the economy (Table 20.2), we calculate the conditional probability of a corporate defaulting, and this again depends on the correlation between its asset values and the states of the economy.

Let $\pi(k)$ be the probability that exactly k firms default in the N-firm loan portfolio. Depending on the state of the economy, the conditional probabilities of M are independent. Therefore, the conditional probability that all the N firms will survive is:

$$\pi(0|M) = \prod_{i=1}^{N} S_i(x_i < x|M)$$

Similarly,

$$\pi(1|M) = \pi(0|M) \sum_{i=1}^{N} \frac{1 - S_i(x_i < x|M)}{S_i(x_i < x|M)}$$

Define

$$w_i = \sum_{i=1}^{N} \frac{1 - S_i(x_i < x|M)}{S_i(x_i < x|M)}$$

Conditioned on the state of the economy, the chance of exactly k firms defaulting is given by the combinatorial probability

$$\pi(k|M) = \pi(0|M) \sum_{i=1}^{N} w_{q(1)} w_{q(2)} \ldots w_{q(k)}$$

where $\{q(1), q(2), \ldots, q(k)\}$ is the combinatorial interpretation of the number of ways of k default among N firms $\{1, 2, \ldots, N\}$ and the summation is taken over the

$$q(k) = \frac{N!}{k!(N-k)!}$$

different ways in which k firms can default among N firms.

$\pi(k|M)$ is the combinatorial probability of k defaults, and $\sum_{i=1}^{N} w_{q(1)} w_{q(2)} \ldots w_{q(k)}$ represents summation over all possible combinations of k defaults among N firms. This is tabulated in Table 20.3 for the first ten states.

Table 20.3 Conditional survival probabilities for one-factor Gaussian copula

States of economy	Corporate_1	Corporate_2	Corporate_3	Corporate_4	Corporate_5
1	0.623	0.278	0.143	0.053	0.000
2	0.632	0.292	0.155	0.060	0.001
3	0.642	0.306	0.167	0.068	0.001
4	0.651	0.320	0.180	0.076	0.001
5	0.660	0.335	0.194	0.085	0.002
6	0.669	0.350	0.208	0.095	0.002
7	0.678	0.365	0.222	0.106	0.003
8	0.687	0.381	0.237	0.118	0.004
9	0.696	0.397	0.253	0.130	0.005
10	0.705	0.412	0.269	0.144	0.007

Table 20.4 Conditional joint survival probabilities for one-factor Gaussian copula

States of economy	$k=0$	$k=1$	$k=2$	$k=3$	$k=4$	$k=5$
1	5.37E-07	0.001336	0.035736	0.24323	0.134507	0.220673
2	9.99E-07	0.001746	0.041595	0.258263	0.136521	0.206578
3	1.83E-06	0.002267	0.048212	0.273353	0.138099	0.192794
4	3.3E-06	0.002926	0.055647	0.288374	0.139211	0.179355
5	5.84E-06	0.003755	0.063955	0.30319	0.139833	0.166295
6	1.02E-05	0.004791	0.073188	0.317651	0.139941	0.153647
7	1.75E-05	0.006079	0.08339	0.331596	0.139519	0.141441
8	2.96E-05	0.007672	0.094599	0.344857	0.138554	0.129705
9	4.93E-05	0.00963	0.106837	0.357256	0.13704	0.118463
10	8.08E-05	0.012025	0.120116	0.368611	0.134975	0.107738

Then we calculate the discrete joint distribution of survival of all the firms together. There are five possibilities—all firms survive, one firm fails, two firms fail, three firms fail, four firms fail, and all five firms fail. This is tabulated in Table 20.4 for the first ten states.

The above steps are explained in the spreadsheet "Financial Analytics Step 6.xlsx" (available on the book's website). For each outcome, we have the losses corresponding to that precise outcome. So using the copula functions we have effectively used the information on marginal distribution of the assets of each firm and their correlation with the economy, to arrive at the joint distribution of the survival outcomes of the firms. We thus are able to aggregate the risk of the portfolio even though as our starting point we only had the marginal probability distribution of only individual loans.

Step 2: Assessment

We now move on to the second step of risk management which is assessment of the portfolio. Assessment of the portfolio is done by summarizing it according to a suitable statistical feature. More precisely, assessment is done by calculating the ex-ante risk of the portfolio using metrics such as threshold persistence (TP) or value at risk (VaR) and sometimes sensitizing it using methods like stress-testing. Threshold persistence is defined as follows: Given the time frame for which a portfolio would remain constant and unchanged (T), the threshold level of cumulative portfolio return (β) and the horizon over which the cumulative return remains below the threshold β. VaR, on the other hand, is a measure of the risk of a portfolio under normal market conditions over a certain time horizon, typically a year, for most asset classes. VaR is used by regulators and firms to assess out how much loss can possibly happen in a certain portfolio and how much asset value is required to cover for this loss. Since VaR is intuitive and is comparable on an apple to apple basis across asset classes, it is widely popular both with the regulators and the market participants. VaR is defined for a given confidence level—usually 99%. This means that the risk manager can be confident that 99 times out of 100, the loss from the portfolio will not exceed the VaR metric. This metric is also used for financial reporting and for calculating the regulatory capital of financial institutions. VaR is an ex-ante assessment in the Bayesian sense—the VaR number is a value that is ex-ante assessed as the loss that can possibly result for the portfolio. It only incorporates information available at the time of computation. VaR is used for governance in pension plans, endowments, trusts, and other such risk-averse financial institutions where the investment mandate often defines the maximum acceptable loss with given probabilities. A detailed description of how Value at Risk has been used to calculate capital can be found in Chapter 3, "VAR-Based Regulatory Capital," of the book *Value at Risk: The New Benchmark for Managing Financial Risk* by Philippe Jorion. This particular measure incorporates the previous steps of portfolio aggregation. We will understand the step using an example. We will examine the VaR computation with a simple portfolio comprising 1 USD, 1 EUR, 1 GBP, and 100 JPY. The value of the portfolio in INR terms is Rs. 280 (1 USD = Rs. 64, 1 Euro (EUR) = Rs. 75, 1 Sterling (GBP) = Rs. 82, 100 Yen (JPY) = Rs. 59). We want to calculate at the end of 1 year what is the possible loss or gain from this particular portfolio. To aggregate the risk, we make use of the correlation matrix between the currencies as described in Table 20.5.

We will use Cholesky decomposition—which fundamentally decomposes the correlation matrix into a lower triangular matrix and an upper triangular matrix (Press et al. 1992). The only condition is that the correlation matrix should be positive definite Hermitian matrix. This decomposition is almost akin to computing the square root of a real number.

$$A = LL^*$$

A is a positive definite Hermitian matrix, L is a lower triangular matrix, and L^* is the transpose conjugate of L. The Cholesky decomposed matrix for the correlation matrix of Table 20.5 is shown in Table 20.6.

Table 20.5 Currency correlation matrix

	USD/INR	EUR/INR	GBP/INR	JPY/INR
USD/INR	1	0.9	0.5	0.5
EUR/INR	0.9	1	0.5	0.5
GBP/INR	0.5	0.5	1	0.2
JPY/INR	0.5	0.5	0.2	1

Table 20.6 Cholesky decomposed lower triangular matrix

	USD/INR	EUR/INR	GBP/INR	JPY/INR
USD/INR	1	0	0	0
EUR/INR	0.9	0.43589	0	0
GBP/INR	0.5	0.114708	0.858395	0
JPY/INR	0.5	0.114708	−0.07358	0.855236

Table 20.7 Simulation of portfolio gain/loss

Simulation number	Log prices				Prices in INR				Gain/loss
	USD	EUR	GBP	JPY	1 USD	1 EUR	1 GBP	1 JPY	
0	4.2	4.3	4.4	4.1	64.0	75.0	82.0	59.0	280.0
1	4.2	4.4	4.6	4.2	67.2	81.6	96.1	64.2	29.0
2	4.2	4.4	4.5	4.2	68.6	79.8	88.5	63.5	20.4
3	4.2	4.3	4.5	4.0	63.6	70.9	88.2	53.8	−3.6
4	4.3	4.5	4.6	4.2	74.3	89.6	95.8	64.3	44.0
5	4.2	4.4	4.4	4.2	69.1	77.6	83.7	65.6	16.0
6	4.2	4.4	4.4	4.3	66.2	78.6	83.4	71.4	19.6
7	4.2	4.3	4.4	4.0	63.7	74.7	81.0	55.3	−5.3
8	4.1	4.2	4.4	4.0	62.6	69.2	82.1	56.3	−9.9
9	4.2	4.3	4.4	4.2	64.6	71.4	83.4	63.8	3.2
10	4.2	4.2	4.4	4.0	65.8	69.1	81.4	56.6	−7.1

For each currency we then simulate a random number drawn from a standard normal distribution. These are independently drawn. This vector of independent draws can be converted to a vector of correlated draws by multiplying with the decomposed matrix.

$$Y = LX$$

where Y is the vector of correlated prices and X is the vector of i.i.d. draws.

This process is repeated multiple times to arrive at a simulation of correlated draws. Using Step 4 we project the log of the prices of USD/INR, EUR/INR, GBP/INR, and JPY/INR. We price the exchange rate and aggregate the portfolio and subtract from the original value to get the portfolio loss or gain. These steps are repeated for a given number of simulations as shown in Table 20.7.

We then calculate the VaR at 99% level from the simulated gains or losses. The above steps are explained in the spreadsheet "Financial Analytics Step 7.xlsx" (available on the book's website). For a simulation run 100 times on the above data, a VaR of −38 INR was obtained at 1% confidence level.

Step 3: Attribution

The third step in risk management analysis is attribution. Once we have assessed the risk of the portfolio in the previous step, we need to now attribute the risk to different risk factors. For instance, the combined risk of Rs. 38 of the portfolio in the previous example can be attributed to each of the individual assets. Like for a portfolio, this can be done at a firm level as well. What financial institutions typically do is to attribute risk along a line of business (LoB). This is because banks and financial institutions are interested in measuring the capital consumed by various activities. Capital is measured using the Value at Risk metric. VaR has become an inalienable tool for risk control and an integral part of methodologies that seek to allocate economic and/or regulatory capital. Its use is being encouraged by the Reserve Bank of India (RBI), the Federal Reserve Bank (Fed), the Bank for International Settlements, the Securities and Exchange Board of India (SEBI), and the Securities and Exchange Commission (SEC). Stakeholders including regulators and supervisory bodies increasingly seek to assess the worst possible loss (typically at 99% confidence levels) of portfolios of financial institutions and funds. A detailed description of how Value at Risk has been used to calculate capital can be found in Chapter 3, "VAR-Based Regulatory Capital," of the book *Value at Risk: The New Benchmark for Managing Financial Risk* by Philippe Jorion. There are three commonly employed measures of VaR-based capital—stand-alone, incremental, and component. It has been found that different banks globally calculate these capital numbers differently, but they follow similar ideas behind the measures.

Stand-Alone Capital

Stand-alone capital is the amount of capital that the business unit would require, if it were viewed in isolation. Consequently, stand-alone capital is determined by the volatility of each LoB's earnings.

Incremental Capital

Incremental capital measures the amount of capital that the business unit adds to the entire firm's capital. Conversely, it measures the amount of capital that would be released if the business unit were sold.

Component Capital

Component capital, sometimes also referred to as allocated capital, measures the firm's total capital that would be associated with a certain line of business. Attributing capital this way has intuitive appeal and is probably the reason why it is particularly widespread.

We use a simplified example to understand how attribution is done using metrics such as stand-alone, incremental, and component capital. Let us assume that there is a bank that has three business units:

- Line of Business 1 (LoB1)—Corporate Banking
- Line of Business 2 (LoB2)—Retail Banking
- Line of Business 3 (LoB3)—Treasury Operations

For ease of calculation, we assume that the total bank asset is $A = $ Rs. 3000 crores. We also assume for the sake of simplicity that each of the LoBs has assets worth $A_i = $ Rs. 1000 crores, $i = 1, 2, 3$. The volatility of the three lines of businesses is:

$$\sigma = \sqrt{\sigma_1^2 + \sigma_2^2 + \sigma_3^2 + 2\rho_{12}\sigma_1\sigma_2 + 2\rho_{23}\sigma_2\sigma_3 + 2\rho_{31}\sigma_3\sigma_1}$$

where σ_i is the volatility of the ith line of business and ρ_{ij} is the correlation between the ith and j^{th} LoB. The volatility of all three LoBs is calculated in Table 20.8, while that of each LoB is calculated in Table 20.9.

Table 20.8 Total capital calculation for the entire business

	Assets	Volatility	Capital
Bank	3000	4.534%	316.9142629

Table 20.9 Capital calculation attribution for LoB1, LoB2, and LoB3

	Assets	Volatility	Standalone capital	Incremental capital	Component capital
LoB1	1000	$\sigma_1 = 5\%$	116.50	51.25	69
LoB2	1000	$\sigma_2 = 7\%$	163.10	67.05	102
LoB3	1000	$\sigma_3 = 9\%$	209.70	89.81	146
Total	3000	4.53%	489	208	317
Unattributed			(172)	109	–

LoB1 is moderately correlated with that of LoB2 ($\rho_{12}=30\%$) and less correlated to LoB3 ($\rho_{31}=10\%$). LoB2 is uncorrelated to LoB3 ($\rho_{23}=0$).

The capital required at 99% ($z = 2.33$) calculated as Value at Risk is given by $2.33 A_i \sigma_i$. The stand-alone capital required for the first line of business is $2.33 \times \text{Rs.}1000$ crores $\times 5\% = 116.50$ crores. The stand-alone capital required for the second line of business is $2.33 \times \text{Rs.}1000$ crores $\times 7\% = 163.10$ crores. The stand-alone capital required for the third line of business is $2.33 \times \text{Rs.}1000$ crores $\times 9\% = 209.70$ crores.

The total capital is given by:

$$C = 2.33 A\sigma$$

$$= 2.33\sqrt{A_1^2\sigma_1^2 + A_2^2\sigma_2^2 + A_3^2\sigma_3^2 + 2\rho_{12}A_1A_2\sigma_1\sigma_2 + 2\rho_{23}A_2A_3\sigma_2\sigma_3 + 2\rho_{31}A_3A_1\sigma_3\sigma_1}$$

The incremental capital for LoB1 is calculated as the total capital less the capital of LoB2 and LoB3. It measures the incremental increase in capital from adding LoB1 to the firm. The incremental capital for LoB1 is therefore:

$$= 2.33 \left\{ \sqrt{A_1^2\sigma_1^2 + A_2^2\sigma_2^2 + A_3^2\sigma_3^2 + 2\rho_{12}A_1A_2\sigma_1\sigma_2 + 2\rho_{23}A_2A_3\sigma_2\sigma_3 + 2\rho_{31}A_3A_1\sigma_3\sigma_1} \right.$$

$$\left. - \sqrt{A_2^2\sigma_2^2 + A_3^2\sigma_3^2 + 2\rho_{23}A_2A_3\sigma_2\sigma_3} \right\}$$

The incremental capital for LoB2 is therefore:

$$= 2.33 \left\{ \sqrt{A_1^2\sigma_1^2 + A_2^2\sigma_2^2 + A_3^2\sigma_3^2 + 2\rho_{12}A_1A_2\sigma_1\sigma_2 + 2\rho_{23}A_2A_3\sigma_2\sigma_3 + 2\rho_{31}A_3A_1\sigma_3\sigma_1} \right.$$

$$\left. - \sqrt{A_1^2\sigma_1^2 + A_3^2\sigma_3^2 + 2\rho_{31}A_3A_1\sigma_3\sigma_1} \right\}$$

The incremental capital for LoB3 is calculated as:

$$= 2.33\left\{\sqrt{A_1^2\sigma_1^2+A_2^2\sigma_2^2+A_3^2\sigma_3^2+2\rho_{12}A_1A_2\sigma_1\sigma_2+2\rho_{23}A_2A_3\sigma_2\sigma_3+2\rho_{31}A_3A_1\sigma_3\sigma_1} \right.$$

$$\left. -\sqrt{A_1^2\sigma_1^2+A_2^2\sigma_2^2+2\rho_{12}A_1A_2\sigma_1\sigma_2}\right\}$$

The component capital for LoB1 is calculated as:

$$A_1\sigma_1\frac{\partial C}{\partial (A_1\sigma_1)} = A_1\sigma_1\frac{(A_1\sigma_1 + \rho_{12}A_2\sigma_2 + \rho_{31}A_3\sigma_3)}{A\sigma}$$

This is because $\frac{\partial \sigma}{\partial \sigma_1} = \frac{(\sigma_1+\rho_{12}\sigma_2+\rho_{31}\sigma_3)}{\sigma}$.

Similarly, the component capital for LoB2 is calculated as:

$$A_2\sigma_2\frac{\partial C}{\partial (A_2\sigma_2)} = A_2\sigma_2\frac{(A_2\sigma_2 + \rho_{12}A_1\sigma_1 + \rho_{23}A_3\sigma_3)}{A\sigma}$$

Likewise, the component capital for LoB3 is:

$$A_3\sigma_3\frac{\partial C}{\partial (A_3\sigma_3)} = A_3\sigma_3\frac{(A_3\sigma_3 + \rho_{13}A_1\sigma_1 + \rho_{23}A_2\sigma_2)}{A\sigma}$$

The component capital of each LoB always sums to the total capital. Please refer to the spreadsheet "Financial Analytics Step 8.xlsx" (available on the book's website) for the specificities of the calculation. Readers interested in total capital calculation for the entire business may refer to the RiskMetrics™ framework developed by JPMorgan (RiskMetrics 1996).

1.3.3 Stage III: Portfolio Analysis

The third stage of data analytics in finance concerns portfolio risk management. It involves optimal allocation of risk and return as well as the execution required to move the portfolio from a suboptimal to an optimal level.

Step 1: Allocation

After having aggregated the portfolio, assessed the risk, and then attributed the risk to different lines of businesses, we move on to changing the portfolio for the entire firm, for a division or an LoB for optimal allocations. So if we continue with the previous example where we have three lines of business, the amount is essentially kept the same—Rs. 1000 crores. If we analyze the results from Step 3 of risk management, we find that risk attribution from all three metrics—standalone, incremental, and component capital—indicates that the lowest attribution of risk happens along the first line of business. If the Sharpe ratio (excess return as a proportion of risk) for LoB1 is the highest (followed by that of LoB2 and LoB3 respectively), then it is optimal for the firm to allocate more capital to the first line of business and then to the second line of business. LoB3 is perhaps the most expensive in terms of risk-adjusted return. Step 1 of portfolio analysis involves optimally allocating the assets such that the overall risk of the firm is optimal. Readers interested in optimal allocation of assets may refer to the RiskMetrics framework developed by JPMorgan (RiskMetrics 1996).

Step 2: Execution

The last step is execution. Having decided to change the portfolio from its current level to a more optimal level, we have to execute the respective trades for us to be able to get to the desired portfolio risk levels. Execution happens in two steps. The first step is order scheduling which is basically a planning stage of the execution process. Order scheduling involves deciding how to break down a large trade into smaller trades and timing each trade for optimal execution. Let us say a financial institution wants to move a large chunk of its portfolio from one block to the other. This is called as a parent order which is further broken down into child orders. The timescale of the parent order is in the order of a day known as volume time. In execution, the way time is measured is not so much in calendar time (called wall-clock time) but in what is called as activity time. Activity time behaves as a random walk. In this last step, we are coming back to Step 1 where we said that we need to identify the risk drivers. For execution, the variable to be modeled is the activity time. This behaves approximately as a random walk with drift and activity time as a risk driver in the execution world, especially in high-frequency trading.

There are two kinds of activity time—tick time and volume time. Tick time is the most natural specification for activity time on very short timescales which advance by 1 unit whenever a trade happens. The second type—volume time—can be intuitively understood by noting that volume time lapses faster when more trading activity happens, that is, the trading volume is larger. After the first step of order scheduling, the second step in order execution is order placement which looks at execution of child orders, and this is again addressed using data analytics. The expected execution time of child orders is of the order of a minute. The child orders—both limit orders and market orders—are based on real-time feedback using opportunistic signals generated from data analytic techniques. So, in order placement, the timescale of limit and market orders is of the order of milliseconds, and the time is measured by tick-time which is discrete. These two steps are repeated in execution algorithms after concluding the first child order called scheduling. It is executed by placing limit and market orders. Once the child order is fully executed, we update the parent order with the residual amount to be filled. We again compute the next child order and execute. This procedure ends when the parent order is exhausted. Execution is almost always done programmatically using algorithms and is known as high-frequency trading (HFT). The last step thus feeds back into the first step of our framework.

1.4 Conclusion

To conclude, the framework consists of three stages to model, assess, and improve the performance of a financial institution and/or a portfolio. The first five steps pertain to econometrical estimation. The next three steps concern risk management and help measure the risk profile of the firm and/or the portfolio. The last two steps are about portfolio management and help in optimizing the risk profile of the financial institution and/or the portfolio. Following these sequential steps across three stages helps us avoid common pitfalls and ensure that we are not missing important features in our use of data analytics in finance. That being said, not every data analysis in the finance world involves all the steps across three stages. If we are only interested in estimation, we may just follow the first five steps. Or if we are only interested in risk attribution, it may only involve Step 3 of risk management. The framework is all encompassing so as to cover most possible data analysis cases in finance. Other important aspects outside the purview of the framework like data cleaning are discussed in Sect. 20.2.

2 Part B: Applications

2.1 Introduction

This chapter intends to demonstrate the kind of data science techniques used for analysis of financial data. The study presents a real-world application of data analytic methodologies used to analyze and estimate the risk of a large portfolio over different horizons for which the portfolio may be held. The portfolio that we use for this study consists of nearly 250 securities comprising international equities and convertible bonds. The primary data science methods demonstrated in the case study are principal component analysis (PCA) and Orthogonal GARCH. We use this approach to achieve parsimony by reducing the dimensionality of the data, which is a recurring objective in most data analytic applications in finance. This is because the dimensionality of the data is usually quite large given the size, diversity, and complexity of financial markets. We simultaneously demonstrate common ways of taking into account the time-varying component of the volatility and correlations in the portfolio, another common goal in portfolio analysis. The larger objective is to demonstrate how the steps described in the methodology framework in the chapter are actually implemented in financial data analysis in the real world.

The chapter is organized as follows. The next section describes the finance aspects of the case study and its application in the financial world. Section 2.3 also discusses the metrics used in the industry for assessing risk of the portfolio. In Sect. 2.4, the data used and the steps followed to make the data amenable for financial analysis are described. Section 2.5 explains the principles of principal component analysis and its application to the dataset. Section 2.6 explains the Orthogonal GARCH approach. Section 2.7 describes three different types of GARCH modeling specific to financial data analysis. The results of the analysis are presented in Sect. 2.8.

2.2 Application of Data Science in the World of Investing

For most non-finance professionals, investments especially in hedge funds are shrouded in secrecy. The sensational stories of Ponzi hedge funds like that of Bernard Madoff make for great headlines and even greater storytelling. In fact, the chronicle of Bernie Madoff's Ponzi scheme is now a Hollywood movie called "The Wizard of Lies" starring Robert De Niro which got released in May 2017. But Hollywood movies do little to advance data analytics education or explain how data science can be used to investigate the portfolio risk of a hedge fund. Not all asset managers have the resources of a Harvard or Yale endowment fund to apply sophisticated models to detect market risk in hedge fund portfolios. This does not mean that we cannot use econometric models to estimate, measure, and assess market risk in portfolios, as we will demonstrate in this case study.

Before the advent of data science, measurement of market risk in relation to hedge funds was considered difficult, if not unmanageable. Large endowment funds like that of Harvard and Yale had the resource to engage econometricians to do the quantitative risk assessment and measurement, but it was mostly the preserve of a select few. Additionally, for a long time, hedge funds were engaged by "golden aged" investment managers who had no understanding of data science. These investment managers were mostly statistically challenged and therefore had more than their fair share of skepticism with regard to data science. They had the good old perspective that hedge fund risks and returns are based on fund managers' talent and that quantitative risk measures are not capable of measuring such complex risks. As a result, the most common risk assessment technique was extensive due diligence carried out by a dedicated set of risk professionals.

However, since the explosion of data science techniques and methodologies in the last decade, there has been a tectonic shift in how data science is viewed in the investment management world. If baseball matches and election outcomes can be predicted using data science, surely hedge fund risks too can be assessed using econometric tools. Another practical challenge facing the investment management industry has been the increase in the size and number of the hedge funds. As per Bloomberg estimates, there are more than 10,000 hedge funds available for investment. It is humanly impossible to carry out due diligence of more than 10,000 hedge funds by any one asset management company (AMC).

Apart from developments in data science and the vastness of hedge fund universe, another important driver in the use of data analytics in asset management has been the advancements in robust risk quantification methodologies. The traditional measures for risk were volatility-based Value at Risk and threshold persistence which quantified downside deviation. These risk metrics are described in the next section. The problem with a simple volatility-based Value at Risk is that it assumes normality. So the assumption made is that financial market returns distribution is symmetrical and that the volatility is constant and does not change with time. It implicitly assumes that extreme returns, either positive or negative, are highly unlikely. However, history suggests that extreme returns, especially extreme negative returns, are not as unlikely as implied by the normal distribution. The problem with downside measures such as threshold persistence is that, although they consider asymmetry of returns, they do not account for fat tails of distributions. These criticisms have resulted in the development of robust risk measures that account for fat tails and leverage such as GJR and EGARCH (see Sect. 2.7.5). So, nowadays all major institutional investors who have significant exposure to hedge funds employ P-quants and use data analytic techniques to measure risk. The exceptions of the likes of Harvard and Yale endowment funds have now become the new norm. Consolidation of market risk at the portfolio level has become a standard practice in asset management. In this chapter, we present one such analysis of a large portfolio comprising more than 250 stocks (sample data in file: tsr.txt) having different portfolio weights (sample data in file: ptsr.txt) and go through the steps to convert portfolio returns into risk metrics. We use Steps 1–6 of the data analysis methodology framework. We first identify the financial variable to model as stock returns. We reduce the dimensionality of the data using principal component analysis from 250 stock returns to about ten principal components. We then use GARCH, GJR, and EGARCH (described in Step 3 of "Part A—Methodology") to make suitable inference on portfolio returns. We estimate the GARCH, GJR, and EGARCH parameters using maximum likelihood estimation. We then project the portfolio returns (Step 4 of the methodology) to forecast performance of the hedge fund. We finally aggregate the risks using Step 6 of the framework and arrive at the key risk metrics for the portfolio. We now describe the risk metrics used in the investment management industry.

2.3 *Metrics for Measuring Risk*

As described in the "Financial Analytics: Part A—Methodology," two metrics are used for measuring the risk of the portfolio: value at risk and threshold persistence.

2.3.1 Value at Risk (VaR)

Value at Risk has become one of the most important measures of risk in modern-day finance. As a risk-management technique, Value at Risk describes the loss in a portfolio that can occur over a given period, at a given confidence level, due to exposure to market risk. The market risk of a portfolio refers to the possibility of financial loss due to joint movement of market parameters such as equity indices, exchange rates, and interest rates. Value at Risk has become an inalienable tool for risk control and an integral part of methodologies that seek to allocate economic and/or regulatory capital. Its use is being encouraged by the Reserve Bank of India (RBI), the Federal Reserve Bank (Fed), the Bank for International Settlements, the Securities and Exchange Board of India (SEBI), and the Securities and Exchange Commission (SEC). Stakeholders including regulators and supervisory bodies increasingly seek to assess the worst possible loss (typically at 99% confidence levels) of portfolios of financial institutions and funds. Quantifying risk is important to regulators in assessing solvency and to risk managers in allocating scarce economic capital in financial institutions.

2.3.2 Threshold Persistence (TP)

Given a threshold level of return for a given portfolio, traders and risk managers want to estimate how frequently the cumulative return on the portfolio goes below this threshold and stays below this threshold for a certain number of days. Traders also want to estimate the minimum value of the cumulative portfolio return when the above event happens. In order to estimate both these metrics, two factors specify a threshold, namely, financial market participants define a metric called threshold persistence.

Threshold persistence is defined as follows: Given the time frame for which a portfolio would remain constant and unchanged (T), two factors specify a threshold, namely, cumulative portfolio return (β) and the horizon over which the cumulative return remains below the threshold β. For the purposes of this chapter, we label this threshold horizon as T'. The threshold persistence metrics are defined as:

(a) The fraction of times the net worth of the portfolio declines below the critical value (β) vis-à-vis the initial net worth of the portfolio and remains there for T' days beneath this critical value
(b) The mean decline in the portfolio net worth value compared to the initial critical level conditional on (a) occuring

To clarify the concept, consider the following example. Say $T = 10$ days, $\beta = -5\%$, $T' = 2$ days, and the initial net worth of the portfolio is Rs. 100. We simulate the portfolio net worth (please refer to Step 4 of the methodology framework to understand how simulation is performed), and, say, we obtain the following path (Table 20.10):

Table 20.10 Threshold persistence example

Day 1	Day 2	Day 3	Day 4	Day 5	Day 6	Day 7	Day 8	Day 9	Day 10
102	98	**94**	**90**	**93**	96	98	90	95	97

The pertinent progression here for calculating (a) and (b) are the net worth of the portfolio in days 3, 4, and 5 since the net worth of the portfolio is lower than Rs. 95 on all these three days. Observe that the decline to Rs. 90 on Day 8 would not be reckoned as an applicable occurrence here since $T' = 2$ and the net worth of the portfolio came back above the critical value on Day 9 (the critical time span is 2 days, and it reverted above the critical level before 2 days). Let us suppose that we simulate ten paths in all and in not one of the remaining paths of the simulation does the portfolio value dip below the critical value and stays below the critical value over the 2-day horizon. Therefore, the proportion of times the value of the portfolio goes below the critical value is 1/10. Given that such a dip happens over the critical time period of over 2 days, the drop would be -10%.

2.4 Data

The data that is normally available from secondary sources are the prices of the various securities in the sample portfolio. The prices would be in local currencies—US securities in US dollars, Japanese equity in Japanese yen, and so on. In the case study, there is data from ten different currencies.

The data that is available from financial information services providers such as Bloomberg or Thomson Reuters (the two largest providers in the global financial markets), more often than not, is not "ready-made" for analysis. The foremost limitation in the data made available by financial information services providers is that they require considerable data cleaning before the data analytic methodologies can be applied. The data cleaning process is usually the most time-consuming and painstaking part of any data analysis, at least with financial data. The portfolio that we use for this study consists of nearly 250 securities. We use the data in the context of the study to describe in general the steps taken to make the data amenable for financial analysis:

- The prices of almost all securities are in their native currencies. This requires conversion of the prices into a common currency. Globally, the currency of choice is US dollars, which is used by most financial institutions as their base currency for reporting purposes. This is a mandatory first step because the prices and returns converted into the base currency are different from those in their native currencies.
- Securities are traded in different countries across the world, and the holidays (when markets are closed) in each of these countries are different. This can lead to missing data in the time series. If the missing data is not filled, then

this could manifest as spurious volatility in the time series. Hence, the missing data is normally filled using interpolation techniques between the two nearest available dates. The most common and simplest interpolation methodology used in financial data is linear interpolation.
- Some securities may have no price quotes at all because even though they are listed in the exchange, there is no trading activity. Even when there is some trading activity, the time periods for which they get traded may be different, and therefore the prices that are available can vary for different securities. For instance, in the portfolio that we use for this study, some securities have ten years of data, while others have less than 50 price data points available. Those securities which do not have at least a time series of prices spanning a minimum threshold number of trading days should be excluded from the analysis. For the purpose of this case study, we use 500 price data points.
- While too few price points is indeed a problem from a data analysis perspective, many a times, a long time series can be judged to be inappropriate. This is because in a longer time series, the more historical observations get the same weights as the recent observations. Since recent observations have more information relevant to the objective of predicting future portfolio risk, a longer time series can be considered inappropriate. In the case study, the time series used for analysis starts in May 2015 and ends in May 2017 thus giving us 2 years of data (in most financial markets, there are approximately 250 trading days in a year) or 500 time series observations.
- Prices are customarily converted into continuously compounded returns using the formula $r_t = \ln(P_t/P_{t-1})$. As explained in Step 1 of the methodology in the "Financial Analytics: Part A—Methodology," we work with returns data rather than price data. Time series analysis of returns dominates that using prices because prices are considerably non-stationary compared to returns.
- Portfolio returns are computed from the security returns as discussed in Step 6 of the methodology framework. In the case study, two portfolios—an equally weighted portfolio and a value-weighted portfolio (calculated by keeping the number of shares in each of the security in the portfolio constant)—are used for the analysis.

2.5 Principal Component Analysis

2.5.1 Eigenvectors and Eigenvalues

For the purposes of the case study, readers need to understand PCA, the way it is computed, and also the intuition behind the computation process. We explain the intermediate steps and the concepts therein to make Sect. 2 of the chapter self-contained. Further discussion on PCA is found in Chap. 15 on unsupervised learning.

20 Financial Analytics

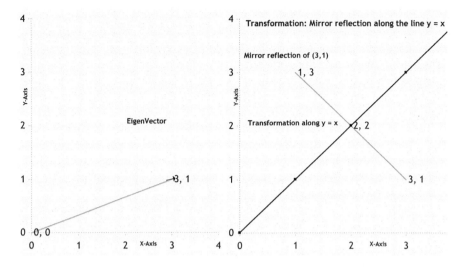

Fig. 20.5 Pictorial description of eigenvectors

From basic matrix algebra we know that we can multiply two matrices together, provided that they are of compatible sizes. Eigenvectors are a special case of this. Consider the two multiplications between a matrix and a vector below.

$$\begin{pmatrix} 0 & 1 \\ 1 & 0 \end{pmatrix} * \begin{pmatrix} 1 \\ 3 \end{pmatrix} = \begin{pmatrix} 3 \\ 1 \end{pmatrix}$$

$$\begin{pmatrix} 0 & 1 \\ 1 & 0 \end{pmatrix} * \begin{pmatrix} 2 \\ 2 \end{pmatrix} = \begin{pmatrix} 2 \\ 2 \end{pmatrix}$$

In the first multiplication, the resulting matrix $\begin{pmatrix} 3 \\ 1 \end{pmatrix}$ is not an integer multiple of the original matrix $\begin{pmatrix} 1 \\ 3 \end{pmatrix}$. In the second multiplication, the resulting matrix is a multiple (of 1) of the original matrix $\begin{pmatrix} 2 \\ 2 \end{pmatrix}$. The first matrix is not an eigenvector of the matrix $\begin{pmatrix} 0 & 1 \\ 1 & 0 \end{pmatrix}$, while the second one is an eigenvector. Why is it so? The reason is that the eigenvector remains a multiple of itself after the transformation. It does not get transformed after multiplication like the first one.

One can think of the matrix $\begin{pmatrix} 1 \\ 3 \end{pmatrix}$ as a vector in two dimensions originating from (0,0) and ending at (1,3) as shown in Fig. 20.5.

For ease of visual imagination, we have employed the matrix $\begin{pmatrix} 0 & 1 \\ 1 & 0 \end{pmatrix}$ in the discussion above. This matrix can be thought of as the following transformation: reflection of any vector along the line $y = x$. For instance, a vector (1,3) after multiplication by this matrix becomes (3,1), that is, a reflection of the vector itself along the $y = x$ line. However, the reflection of the vector (2,2) would be the vector itself. It would be a scalar multiple of the vector (in this case, the scalar multiple is 1). Thus, an eigenvector even after transformation remains a scalar multiple of itself. The scalar multiple is called the eigenvalue "λ." In other words, an eigenvector remains itself when subject to some transformation and hence can capture a basic source of variation. When more than one eigenvector is put together, they can constitute a basis to explain complex variations.

In general, an $n \times n$ dimension matrix can have a maximum of n eigenvectors. All the eigenvectors of a matrix are orthogonal to each other, no matter how many dimensions they have. This is important because it means that we can represent the data in terms of these perpendicular eigenvectors, instead of expressing them in terms of the original assets. This helps to reduce dimensionality of the problem at hand considerably, which characteristically for financial data is large.

2.5.2 PCA Versus Factor Analysis

Having understood the mathematical intuition behind PCA, we are in a position to appreciate why PCA is a dominant choice compared to factor analysis in financial data analysis. The objective of PCA is to be able to explain the variation in the original data with as few (and important!) dimensions (or components) as possible. Readers could argue that the dimensions can be reduced through factor analysis as well. Why use PCA instead?

To illustrate graphically, in Fig. 20.6 the red dots are the data points for a hypothetical time series of two dimensions. Factors 1 and 2 together explain a major portion of the variation in the data, and also those two factors put together fit the data better than PCA1 and PCA2, but Factors 1 and 2 are not orthogonal as can be seen in the graph above. Hence, their covariance matrix would have non-zero diagonal elements. In other words, Factors 1 and 2 would covary. Therefore, we would not only have to estimate the factors but also the covariance between them, which in a time series of 200 odd securities (>200 dimensions) can be onerous (>20,000 covariances!). When these covariances have to be dynamically modeled, the procedure becomes considerably inefficient in most financial data analysis.

In contrast, PCA 1 and 2 explain equally the variation in the data, and yet they have zero covariance. In the analysis of time series of yield curves, for example, Factors 1 and 2 are thought of as duration and convexity. These two factors help explain a lot of variation in yield curves across time, but they are not orthogonal

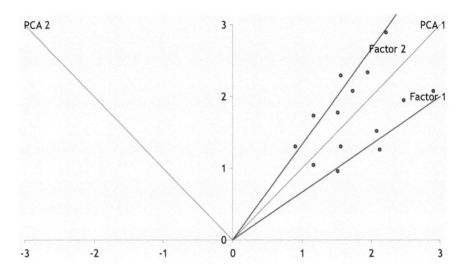

Fig. 20.6 Pictorial description of PCA and factor analysis

because convexity changes with duration. However, level of the yield curve and its slope are orthogonal components which can explain variation in bond prices equally well.

2.5.3 PCA Results

The results of the PCA for the portfolio of securities is detailed below. The first 21 principal components explain 60% nearly half of the variation as seen from Table 20.11.

Figure 20.7 shows that the first ten principal components capture the variation in the portfolio returns quite accurately. These ten principal components explain close to 50% of the variation in the security returns. As can be seen from Table 20.11, the first ten principal components explain 47.5% of the variation, while the next thirteen components explain less than 15% of the variation. Adding more principal components presents a trade-off between additional accuracy and the added dimensionality of the problem in most financial data analyses.

In the data in the portfolio that we study, the principal components from 11 onward each help explain less than 2% of the additional variation. However, adding one more principal component adds to the dimensionality by 10% and results in a commensurate increase in the computational complexity. Hence, we can limit to ten principal components for the subsequent analysis. This reduces the dimensionality of the data from 250 to 10.

As the histogram in Fig. 20.8 shows, the difference between the actual portfolio returns and the returns replicated using the ten principal components are, for the

Table 20.11 Contribution of various principal components

Principal components	Variance contribution	Cumulative variance
X1	16.03%	16.03%
X2	7.81%	23.85%
X3	5.11%	28.96%
X4	4.19%	33.15%
X5	3.50%	36.64%
X6	2.62%	39.27%
X7	2.38%	41.64%
X8	2.20%	43.85%
X9	1.96%	45.81%
X10	1.71%	47.53%
X11	1.64%	49.16%
X12	1.56%	50.72%
X13	1.36%	52.08%
X14	1.32%	53.40%
X15	1.23%	54.63%
X16	1.14%	55.76%
X17	1.07%	56.83%
X18	1.01%	57.84%
X19	0.99%	58.82%
X20	0.94%	59.76%
X21	0.93%	60.69%
X22	0.88%	61.57%
X23	0.86%	62.43%

most part, small. Hence, we can limit our subsequent analysis to ten principal components.

2.5.4 Stationarity of the Principal Components

To rule out spurious predictability in a time series, stationarity of the predicting variables is extremely important in most financial data analysis. For example, if a time series has a time trend to it, then it is rare that the time trend would repeat itself in the future. When a time series with a time trend is estimated, the time series would produce brilliant results (extremely high R-squared for the regression and high t-statistics for the coefficients of the predicting variables) leading a less careful data scientist to infer a well-fitted model capable of high levels of predictability. However, the predictability underlying such a time series is spurious and misleading. Thus, it is important to rule out such time trends in the predicting variables which are the principal components in our case. Figure 20.9 shows visually the absence of time trends in the ten principal components.

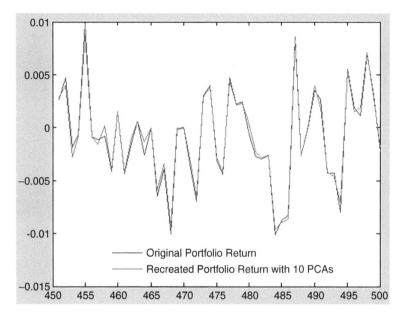

Fig. 20.7 Replicated time series using ten principal components vs original portfolio returns

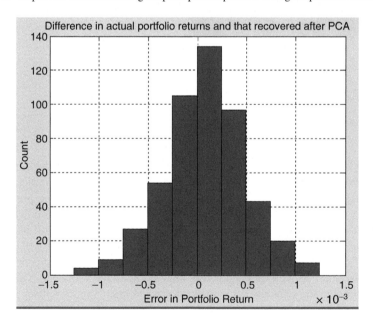

Fig. 20.8 Difference in actual portfolio returns vs PCA

Table 20.12 shows the results of the augmented Dickey–Fuller test (ADF) of stationarity for all the ten principal components. The ADF is used to test for the

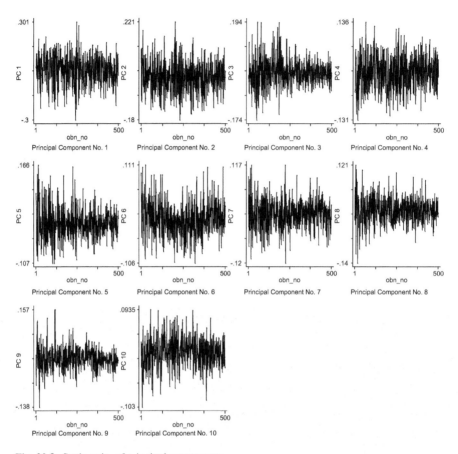

Fig. 20.9 Stationarity of principal components

presence of unit roots in the time series. If $y_t = y_{t-1} + e_t$, then the time series will blow up as the number of observations increases. Further, the variance of the time series will be unbounded in this case. In order to rule out the presence of unit roots, the augmented Dickey–Fuller test runs regressions of the following kind: $y_t - y_{t-1} = \rho\, y_t - 1 + e_t$. If the time series has a unit root, then ρ will be equal to zero. The ADF essentially tests the null hypothesis that $\rho = 0$ versus $\rho \neq 0$. As the results of the tests in Table 20.12 indicate, none of the principal components have a unit root. Also, as we examined earlier, they do not have a time trend either. So, predictability in the principal components is not spurious. This completes Step 2 of our framework in the chapter.

Let D.PCi indicate the first difference of the respective principal component. The absence of a unit root (which is the test of stationarity) is indicated by the coefficient of the lag of PC(i) being different from zero. Data scientists in the financial domain at times use the MacKinnon probability value to indicate the probability that the test statistic is different from the augmented Dickey–Fuller critical values.

Table 20.12 Augmented Dickey–Fuller test for stationarity of principal components

	D.PC1	D.PC2	D.PC3	D.PC4	D.PC5	D.PC6	D.PC7	D.PC8	D.PC9	D.PC10
Lag of PC(i)	−0.847*	−1.061*	−1.034*	−1.000*	−1.016*	−0.915*	−1.043*	−0.951*	−0.866*	−0.837*
	(19.11)	(23.76)	(23.06)	(22.31)	(22.73)	(20.62)	(23.28)	(21.24)	(19.49)	(18.91)
Constant	0.007	−0.003	0.004	−0.002	0.003	−0.008*	−0.001	0	0.008*	0.006*
	(1.80)	(0.89)	(1.85)	(0.74)	(1.44)	(4.57)	(0.74)	(0.18)	(5.51)	(4.38)
Obns	499	499	499	499	499	499	499	499	499	499
R-sqd	0.42	0.53	0.52	0.5	0.51	0.46	0.52	0.48	0.43	0.42
P-value	0.000	0.000	0.000	0.000	0.000	0.000	0.000	0.000	0.000	0.000

Absolute value of t statistics in parentheses; * indicates significance of coefficient at 1% level

2.6 Orthogonal GARCH

Given the large number of factors that typically affect the position of a large portfolio, estimating the risk of the portfolio becomes very complex indeed. At the heart of most data analytics models for estimating risk is the covariance matrix which captures the volatilities and the correlations between all the risk factors. Typically hundreds of risk factors encompassing equity indices, foreign exchange rates, and yield curves need to be modeled through the dynamics of the large covariance matrix. In fact, without making assumptions about the dynamics of these risk factors, implementation of models for estimating risk becomes quite cumbersome.

Orthogonal GARCH is an approach for estimating risk which is computationally efficient but captures the richness embedded in the dynamics of the covariance matrix. Orthogonal GARCH applies the computations to a few key factors which capture the orthogonal sources of variation in the original data. The approach is computationally efficient since it allows for an enormous reduction in the dimensionality of the estimation while retaining a high degree of accuracy. The method used to identify the orthogonal sources of variation is principal component analysis (PCA). The principal components identified through PCA are uncorrelated with each other (by definition they are orthogonal). Hence, univariate GARCH models can be used to model the time-varying volatility of the principal components themselves. The principal components along with their corresponding GARCH processes then capture the time-varying covariance matrix of the original portfolio. Having described principal component analysis and Orthogonal GARCH, we now illustrate the different variants of GARCH modeling.

2.7 GARCH Modeling

After the dimensionality of the time series is reduced using PCA, we now proceed to Step 3 of our framework in the chapter with modeling the covariance using GARCH on the principal components.

We first motivate the use of GARCH for measuring risk of a portfolio. Most common methodologies for estimating risk, through a Value at Risk calculation, assume that portfolio returns follow a normal distribution as shown in Fig. 20.10. This methodology of calculating VaR using normal distribution implicitly assumes that the mean and standard deviation of the portfolio returns remain constant.

However, ample empirical evidence in finance shows that security returns exhibit significant deviations from normal distributions, particularly volatility clustering and fat tail behavior. There are certain other characteristics of equity markets which are not adequately accounted for in a normal distribution. Data scientists

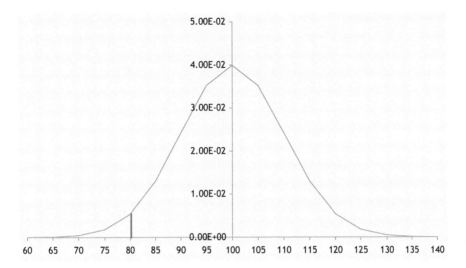

Fig. 20.10 Normal distribution

in finance therefore use GARCH models as they are devised to encapsulate these characteristics that are commonly observed in equity markets.

2.7.1 Volatility Clustering

Equity returns series usually exhibit this characteristic in which large changes tend to follow large changes and small changes tend to follow small changes. For instance, if markets were more volatile than usual today, there is a bias toward they being more volatile tomorrow than they typically are. Similarly, if markets were "quiet" today, there is a higher probability that they may be "quiet" tomorrow compared to they being unusually volatile. In both cases, it is difficult to predict the change in market activity from a "quiet" to a "volatile" scenario and vice versa. In GARCH, significant perturbations, either for good or for worse, are intrinsic part of the time series we use to predict the volatility for the next time period. These large perturbations and shocks, both positive and negative, persist in the GARCH model and are factored in the future forecasts of variance for future time periods. They are sometimes also called persistence and model a process in which successive disturbances, although uncorrelated, are nonetheless serially dependent.

An examination of the time series of principal components reveals that periods of high volatility are often clustered together. This has to be taken into account using a GARCH model.

Table 20.13 Shapiro–Wilk test

Variable	Shapiro–Wilk test statistic W	Prob. (normal)
pc1	0.99	0.01
pc2	1.00	0.63
pc3	0.97	0.00
pc4	1.00	0.39
pc5	0.98	0.00
pc6	0.99	0.00
pc7	1.00	0.12
pc8	0.99	0.00
pc9	0.97	0.00
pc10	1.00	0.44

2.7.2 Leverage Effects

Asset returns are often observed to be negatively correlated with changes in volatility. Meaning, markets tend to be more volatile when there is a sell-off vis-à-vis when markets rally. This is called leverage—volatility tends to rise in response to lower than expected returns and to fall in response to higher than expected returns. Asymmetric GARCH models are capable of capturing the leverage effect.

2.7.3 Fat Tails or Excess Kurtosis

The tail of distributions of equity returns are typically fatter compared to a normal distribution. In simple terms, the possibility of extreme fluctuations in returns is understated in a normal distribution, and these can be captured with GARCH models.

This lack of normality in our portfolio is tested by analyzing the distribution of the principal components using quantile plots as shown in Fig. 20.11. Fat tails are evident in the distribution of principal components as seen from the quantile plots since the quantiles at both the extremes deviate from the quantiles of a normal distribution. To further test whether the distributions for the principal components are normal, the Shapiro–Wilk test of normality is usually performed on all the principal components. The results of the Shapiro–Wilk test are provided in Table 20.13.

As is evident from Table 20.13, pc1, pc3, pc5, pc6, pc8, and pc9 exhibit substantial deviations from normality, while the remaining principal components are closer to being normally distributed. Since six of the ten principal components exhibit deviations from normality, it is important to model fat tails in the distribution of principal components. Figure 20.11 depicts quantiles of principal components plotted against quantiles of normal distribution (45% line). A look at the plot of the time series of the principal components reveals that periods of volatility are often clustered together. Hence, we need to take into account this volatility clustering using GARCH analysis.

20 Financial Analytics

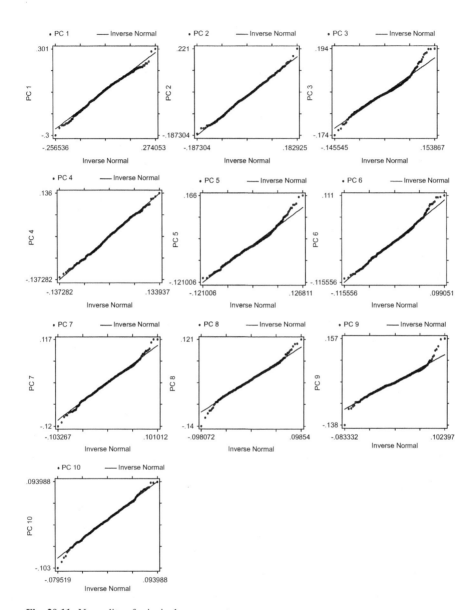

Fig. 20.11 Normality of principal components

Now that we have discussed why we use GARCH in financial data analysis, let us try to understand it conceptually. GARCH stands for generalized autoregressive conditional heteroscedasticity. Loosely speaking, one can think of heteroscedasticity as variance that varies with time. Conditional implies that future variances depend

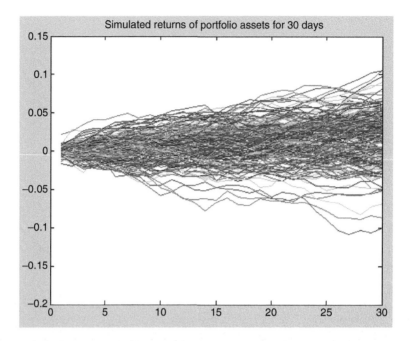

Fig. 20.12 Simulated returns of portfolio assets with GJR model, Gaussian distribution

on past variances. It allows for modeling of serial dependence of volatility. For the benefit of those readers who are well versed with econometric models and for the sake of completeness, we provide the various models used in the case study. Readers may skip this portion without losing much if they find it to be too mathematically involved.

2.7.4 Conditional Mean Model

This general ARMAX(*R*,*M*,*Nx*) model for the conditional mean applies to all variance models.

$$y_t = C + \sum_{i=1}^{R} \varphi_i y_{t-i} + \varepsilon_t + \sum_{j=1}^{M} \theta_j \varepsilon_{t-j} + \sum_{k=1}^{Nx} \beta_k X(t,k)$$

with autoregressive coefficients φ_i, moving average coefficients θ_j, innovations ε_t, and returns y_t. X is an explanatory regression matrix in which each column is a time series and $X(t,k)$ denotes the tth row and k^{th} column.

In the case study, we implement the ARMA(1,1) model for all the principal components since a parsimonious model which captures both the autoregressive and moving average components of the conditional mean is desirable. Please see Chap. 12 for a brief description of the ARMA model and determining its parameters.

2.7.5 Conditional Variance Models

The conditional variance of the innovations, σ_t^2, is

$$\text{Var}_{t-1}(y_t) = \text{E}_{t-1}\left(\varepsilon_t^2\right) = \sigma_t^2$$

The conditional variance quantifies the amount of variation that is left if we use the conditional expectation to predict y_t. The key insight of GARCH lies in the distinction between conditional and unconditional variances of the innovations process ε_t. The term *conditional* implies unambiguous dependence on a past time series of observations. The term *unconditional* is more concerned with long-term behavior of a time series and assumes no explicit knowledge of the past.

The various GARCH models characterize the conditional distribution of ε_t by imposing alternative parameterizations to capture serial dependence on the conditional variance of the innovations.

GARCH(P,Q) Conditional Variance

The general GARCH(P,Q) model for the conditional variance of innovations is

$$\sigma_t^2 = K + \sum_{i=1}^{P} G_i \sigma_{t-i}^2 + \sum_{j=1}^{Q} A_j \varepsilon_{t-j}^2$$

GJR(P,Q) Conditional Variance

The general GJR(P,Q) model for the conditional variance of the innovations with leverage terms is

$$\sigma_t^2 = K + \sum_{i=1}^{P} G_i \sigma_{t-i}^2 + \sum_{j=1}^{Q} A_j \varepsilon_{t-j}^2 + \sum_{j=1}^{Q} L_j S_{t-j}^- \varepsilon_{t-j}^2$$

where

$$S_{t-j}^- = \begin{cases} 1 & \varepsilon_{t-j} < 0 \\ 0 & \text{otherwise} \end{cases}$$

EGARCH(P,Q) Conditional Variance

The general EGARCH(P,Q) model for the conditional variance of the innovations with leverage terms and an explicit probability distribution assumption is

$$\log\left(\sigma_t^2\right) = K + \sum_{i=1}^{P} G_i \log\left(\sigma_{t-i}^2\right) + \sum_{j=1}^{Q} A_j \left[\frac{|\varepsilon_{t-j}|}{\sigma_{t-j}} - E\left\{\frac{|\varepsilon_{t-j}|}{\sigma_{t-j}}\right\}\right]$$
$$+ \sum_{j=1}^{Q} L_j \left(\frac{\varepsilon_{t-j}}{\sigma_{t-j}}\right)$$

Models Used in the Case Study

We use the ARMA(1,1) model ($R = 1$, $M = 1$ in the equation for conditional mean) for conditional mean along with GARCH(1,1) and GJR(1,1) for our case study analysis. We employ the normal distribution and the Student's t-distribution to model the fat tails in the portfolio returns.

Although the above models are simple, they have several benefits. These represent parsimonious models that require estimation of at most eight parameters. The fewer the parameters to estimate, the greater the accuracy of these parameters. Complicated models in financial data analysis, more often than not, do not offer tangible benefits when it comes to predicting financial variables.

GARCH Limitations

While it is easy to be impressed with the mathematical expositions of the models, and the fact that GARCH models provide insights into a wide range of financial market applications, they do have limitations:

- The GARCH model at the end of the day is a parametric specification. The parameters remain stable only if the underlying market conditions are stable. GARCH models are good at capturing heterscedastic variances. That being said, they cannot capture tempestuous fluctuations in the market. Till now there does not exist a well-accepted model which can model market crashes as they are extremely unpredictable and unique.
- Asset returns have fat tails, i.e., large deviations from average are quite likely. GARCH models are not equipped to capture all of these fat tail returns that are observed in financial time series of returns. Time-varying volatility does explain a limited portion of this fat tail but, given the limitations of the normal distribution, cannot explain all of it. To offset for this constraint, data analysts more often than not implement Student's t-distribution in GARCH modeling.

2.8 Results

The calculation of Value at Risk for large portfolios presents a trade-off between speed and accuracy, with the fastest methods relying on rough approximations and the most realistic approach often too slow to be practical. Financial data scientists try to use the best features of both approaches, as we try to do in this case study.

Tables 20.14, 20.15, 20.16, 20.17, 20.18, 20.19, 20.20, and 20.21 show the calculation of Value at Risk and threshold persistence using four different models—GARCH with Gaussian distribution, GARCH with Student's t-distribution, GJR with Student's t-distribution, and EGARCH with Student's t-distribution. This is done for both the value-weighted portfolio and equi-weighted portfolio as below:

- Table 20.14 shows the calculation of Value at Risk and threshold persistence for market value-weighted GARCH model with a Gaussian distribution.
- Table 20.15 shows the calculation of Value at Risk and threshold persistence for market value-weighted GARCH model with a Student's t-distribution.
- Table 20.16 shows the calculation of Value at Risk and threshold persistence for market value-weighted GJR model with a Student's t-distribution.
- Table 20.17 shows the calculation of Value at Risk and threshold persistence for market value-weighted EGARCH model with a Student's t-distribution.
- Table 20.18 shows the calculation of Value at Risk and threshold persistence for equi-weighted GARCH model with a Gaussian distribution.
- Table 20.19 shows the calculation of Value at Risk and threshold persistence for equi-weighted GARCH model with a Student's t-distribution.

Table 20.14 Market value-weighted, GARCH, Gaussian

Horizon	Value 1	Value 2	Value 3	Value 4	Value 5
	Value at Risk				
5	−2.42723%	−2.40828%	−2.51901%	−2.43652%	−2.36555%
10	−3.48908%	−3.43362%	−3.36722%	−3.22902%	−3.27812%
20	−4.58238%	−4.71451%	−4.45732%	−4.47809%	−4.48455%
60	−7.28079%	−7.27208%	−7.30652%	−7.13126%	−7.28960%
125	−9.84899%	−9.21783%	−9.61347%	−9.95102%	−9.26511%
250	−11.90397%	−11.27439%	−12.28956%	−11.69521%	−11.58802%
	Percentage of times portfolio is below beta				
5	0.01000%	0.00000%	0.00000%	0.01000%	0.00000%
10	0.10000%	0.08000%	0.07000%	0.05000%	0.02000%
20	0.70000%	0.71000%	0.62000%	0.70000%	0.51000%
60	7.07000%	6.69000%	6.72000%	7.23000%	6.76000%
125	15.11000%	14.76000%	14.75000%	14.98000%	15.06000%
250	22.15000%	21.72000%	22.16000%	22.53000%	22.38000%
	Average drop in portfolio when level drops below beta				
5	−6.08684%	0.00000%	0.00000%	−5.99193%	0.00000%
10	−6.42005%	−5.82228%	−6.09030%	−5.79600%	−5.36812%
20	−6.19519%	−6.21074%	−6.33539%	−6.04260%	−5.97896%
60	−6.74850%	−6.73923%	−6.79577%	−6.65510%	−6.71048%
125	−7.48904%	−7.39024%	−7.52920%	−7.50248%	−7.34893%
250	−8.34098%	−8.25300%	−8.37886%	−8.26895%	−8.25156%

Table 20.15 Market value-weighted, GARCH, Student's *t*

Horizon	Value 1	Value 2	Value 3	Value 4	Value 5
	Value at Risk				
5	−2.37854%	−2.39676%	−2.40350%	−2.38341%	−2.43340%
10	−3.34525%	−3.32570%	−3.40838%	−3.33190%	−3.32166%
20	−4.53003%	−4.56522%	−4.67216%	−4.49673%	−4.43257%
60	−7.28625%	−7.36338%	−7.19435%	−6.95804%	−7.09243%
125	−9.07297%	−9.50501%	−9.43058%	−9.32454%	−9.33755%
250	−11.33888%	−11.74615%	−11.71408%	−11.35302%	−11.60062%
	Percentage of times portfolio is below beta				
5	0.01000%	0.00000%	0.00000%	0.00000%	0.00000%
10	0.08000%	0.08000%	0.05000%	0.05000%	0.07000%
20	0.70000%	0.64000%	0.60000%	0.57000%	0.60000%
60	6.71000%	6.75000%	7.16000%	6.48000%	6.94000%
125	14.85000%	14.65000%	14.92000%	14.31000%	15.24000%
250	21.63000%	21.58000%	22.16000%	21.74000%	22.49000%
	Average drop in portfolio when level drops below beta				
5	−5.60456%	0.00000%	0.00000%	0.00000%	0.00000%
10	−6.17866%	−5.98915%	−5.65985%	−6.35440%	−6.02021%
20	−5.96225%	−6.06412%	−6.32215%	−6.26961%	−6.05488%
60	−6.71141%	−6.72747%	−6.74299%	−6.64512%	−6.64694%
125	−7.33353%	−7.37906%	−7.47784%	−7.33874%	−7.37931%
250	−8.15825%	−8.24734%	−8.27726%	−8.16879%	−8.22191%

Table 20.16 Market value-weighted, GJR, Student's *t*

Horizon	Value 1	Value 2	Value 3	Value 4	Value 5
	Value at Risk				
5	−2.64120%	−2.67966%	−2.56970%	−2.72138%	−2.71597%
10	−3.82148%	−3.80847%	−3.73202%	−3.86924%	−3.82415%
20	−5.33825%	−5.41227%	−5.26804%	−5.26190%	−5.39392%
60	−8.50145%	−8.79663%	−8.54043%	−8.80623%	−8.73182%
125	−11.53589%	−11.68212%	−11.20231%	−11.61595%	−11.51297%
250	−15.96105%	−15.13785%	−14.87603%	−15.85603%	−14.83327%
	Percentage of times portfolio is below beta				
5	0.02000%	0.00000%	0.02000%	0.02000%	0.00000%
10	0.17000%	0.26000%	0.19000%	0.18000%	0.19000%
20	1.59000%	1.48000%	1.29000%	1.39000%	1.51000%
60	10.12000%	10.74000%	10.29000%	9.78000%	9.76000%
125	21.56000%	21.63000%	21.51000%	21.27000%	20.89000%
250	32.62000%	32.44000%	32.70000%	32.65000%	31.98000%
	Average drop in portfolio when level drops below beta				
5	−6.29212%	0.00000%	−6.46780%	−5.83254%	0.00000%
10	−7.07093%	−6.76246%	−6.49871%	−6.47768%	−6.24227%
20	−6.57822%	−6.59474%	−6.74424%	−6.55687%	−6.45223%
60	−7.25348%	−7.16626%	−7.14942%	−7.19248%	−7.19982%
125	−8.02682%	−7.96397%	−7.93543%	−7.98327%	−7.99186%
250	−9.18003%	−9.17600%	−9.04975%	−9.15641%	−9.10158%

Table 20.17 Market value-weighted, EGARCH, Student's t

Horizon	Value 1	Value 2	Value 3	Value 4	Value 5
	Value at Risk				
5	−2.79077%	−2.85296%	−2.77644%	−2.69456%	−2.72527%
10	−3.93194%	−3.94460%	−3.83798%	−3.83507%	−3.87852%
20	−5.40504%	−5.28543%	−5.31962%	−5.33853%	−5.39825%
60	−8.58980%	−8.38608%	−8.37522%	−8.48768%	−8.71635%
125	−11.54664%	−11.21148%	−11.22397%	−11.52668%	−11.24505%
250	−14.78795%	−14.88799%	−14.44216%	−14.46602%	−14.44222%
	Percentage of times portfolio is below beta				
5	0.01000%	0.01000%	0.03000%	0.00000%	0.00000%
10	0.23000%	0.20000%	0.13000%	0.16000%	0.18000%
20	1.85000%	1.54000%	1.48000%	1.55000%	1.43000%
60	10.73000%	9.78000%	10.17000%	10.70000%	10.75000%
125	21.42000%	21.09000%	20.72000%	21.30000%	21.10000%
250	32.14000%	31.55000%	31.17000%	31.80000%	31.33000%
	Average drop in portfolio when level drops below beta				
5	−6.64370%	−6.70843%	−6.19737%	0.00000%	0.00000%
10	−6.16354%	−6.24120%	−6.27418%	−6.43948%	−6.07836%
20	−6.28625%	−6.43126%	−6.39376%	−6.50772%	−6.33898%
60	−7.08317%	−7.08325%	−7.08800%	−7.10921%	−7.19341%
125	−7.97817%	−7.88085%	−7.99144%	−8.00227%	−8.00791%
250	−9.05973%	−9.00021%	−9.05599%	−9.06243%	−9.00664%

Table 20.18 Equi-weighted, GARCH, Gaussian

Horizon	Value 1	Value 2	Value 3	Value 4	Value 5
	Value at Risk				
5	−3.02645%	−3.12244%	−3.10246%	−3.15969%	−3.08626%
10	−4.22080%	−4.34003%	−4.23744%	−4.39152%	−4.36571%
20	−5.77279%	−5.99089%	−5.89724%	−5.84402%	−6.00114%
60	−9.49140%	−9.72849%	−9.46143%	−9.57403%	−9.76131%
125	−13.00547%	−12.91606%	−13.16552%	−12.80902%	−13.03503%
250	−16.19272%	−17.10488%	−17.15668%	−16.31266%	−16.92225%
	Percentage of times portfolio is below beta				
5	0.01000%	0.01000%	0.01000%	0.01000%	0.04000%
10	0.27000%	0.45000%	0.29000%	0.38000%	0.36000%
20	2.24000%	2.77000%	2.45000%	2.50000%	2.63000%
60	14.98000%	14.71000%	14.69000%	14.50000%	15.09000%
125	26.76000%	26.31000%	26.23000%	26.49000%	27.08000%
250	35.85000%	35.38000%	36.05000%	35.87000%	35.99000%
	Average drop in portfolio when level drops below beta				
5	−6.30389%	−6.15856%	−5.35099%	−7.11069%	−5.63802%
10	−6.01000%	−6.16643%	−6.19892%	−6.46273%	−6.09211%
20	−6.35667%	−6.38641%	−6.30022%	−6.38837%	−6.44409%
60	−7.36676%	−7.40805%	−7.34800%	−7.37507%	−7.43454%
125	−8.36959%	−8.50622%	−8.45060%	−8.38819%	−8.41793%
250	−9.69286%	−9.85419%	−9.74382%	−9.63297%	−9.78937%

Table 20.19 Equi-weighted, GARCH, Student's t

Horizon	Value 1	Value 2	Value 3	Value 4	Value 5
	Value at Risk				
5	−3.09375%	−3.14080%	−3.10339%	−3.07041%	−3.11094%
10	−4.36277%	−4.21170%	−4.38308%	−4.27252%	−4.39058%
20	−5.79307%	−5.92853%	−5.95257%	−5.78613%	−6.01026%
60	−9.49324%	−9.41315%	−9.67444%	−9.57806%	−9.44238%
125	−12.96079%	−13.42418%	−12.86986%	−13.34371%	−12.92972%
250	−16.55830%	−16.94585%	−16.56531%	−16.59782%	−16.18167%
	Percentage of times portfolio is below beta				
5	0.01000%	0.02000%	0.03000%	0.01000%	0.01000%
10	0.40000%	0.27000%	0.40000%	0.24000%	0.36000%
20	2.56000%	2.34000%	2.62000%	2.48000%	2.57000%
60	14.93000%	14.54000%	15.28000%	14.27000%	14.20000%
125	26.55000%	26.00000%	26.80000%	25.64000%	25.82000%
250	36.04000%	34.73000%	35.91000%	34.59000%	35.32000%
	Average drop in portfolio when level drops below beta				
5	−5.48361%	−6.01500%	−5.60246%	−5.88618%	−6.43174%
10	−6.00148%	−6.38804%	−6.26810%	−6.37318%	−6.13003%
20	−6.38101%	−6.47291%	−6.52129%	−6.41859%	−6.44267%
60	−7.31137%	−7.31075%	−7.43181%	−7.44125%	−7.33935%
125	−8.38976%	−8.49580%	−8.42735%	−8.48043%	−8.35742%
250	−9.67619%	−9.72951%	−9.68775%	−9.79860%	−9.54100%

Table 20.20 Equi-weighted, GJR, Student's t

Horizon	Value 1	Value 2	Value 3	Value 4	Value 5
	Value at Risk				
5	−3.41649%	−3.37054%	−3.43304%	−3.35435%	−3.48986%
10	−4.89835%	−4.84086%	−4.86367%	−4.79133%	−5.00438%
20	−6.97760%	−6.83118%	−6.85524%	−6.72537%	−6.99981%
60	−11.39598%	−11.26982%	−11.43001%	−11.77783%	−11.78486%
125	−15.86655%	−16.52494%	−15.94209%	−15.45319%	−16.09934%
250	−20.67736%	−21.36881%	−22.31839%	−20.82328%	−21.46509%
	Percentage of times portfolio is below beta				
5	0.03000%	0.05000%	0.09000%	0.06000%	0.10000%
10	0.68000%	0.65000%	0.76000%	0.62000%	0.86000%
20	4.16000%	4.32000%	3.88000%	3.98000%	3.72000%
60	20.00000%	20.72000%	19.83000%	19.46000%	19.59000%
125	34.93000%	36.08000%	34.81000%	34.70000%	34.74000%
250	48.44000%	49.80000%	48.28000%	48.03000%	48.37000%
	Average drop in portfolio when level drops below beta				
5	−7.36444%	−7.72945%	−6.35243%	−5.81271%	−6.14997%
10	−6.56317%	−6.69730%	−6.58803%	−6.36042%	−6.67660%
20	−6.85652%	−6.77818%	−6.97909%	−6.77189%	−7.14428%
60	−7.87520%	−7.83506%	−7.91032%	−7.91935%	−7.88042%
125	−9.18330%	−9.18761%	−9.21253%	−9.17098%	−9.21382%
250	−10.85028%	−10.95301%	−11.02630%	−10.94284%	−10.95912%

Table 20.21 Equi-weighted, EGARCH, Student's t

Horizon	Value 1	Value 2	Value 3	Value 4	Value 5
	Value at Risk				
5	−3.57247%	−3.56529%	−3.50317%	−3.52430%	−3.41424%
10	−5.03825%	−4.94431%	−4.94004%	−4.95409%	−4.90723%
20	−6.80683%	−6.90082%	−6.80978%	−7.15053%	−7.04633%
60	−11.54241%	−11.56384%	−11.17854%	−11.59858%	−11.47896%
125	−16.02569%	−15.78560%	−15.09081%	−16.25870%	−15.86065%
250	−21.40973%	−21.42843%	−19.87466%	−20.98633%	−20.91555%
	Percentage of times portfolio is below beta				
5	0.05000%	0.05000%	0.03000%	0.04000%	0.05000%
10	0.87000%	0.74000%	0.67000%	0.83000%	0.76000%
20	4.18000%	4.10000%	4.18000%	4.42000%	3.96000%
60	19.67000%	20.03000%	20.09000%	19.69000%	20.00000%
125	33.91000%	34.78000%	34.46000%	34.42000%	34.52000%
250	47.81000%	48.03000%	46.65000%	47.48000%	47.87000%
	Average drop in portfolio when level drops below beta				
5	−6.45992%	−6.25018%	−7.06294%	−6.51616%	−6.25605%
10	−6.47289%	−6.46955%	−6.53686%	−6.39210%	−6.61915%
20	−6.77121%	−6.81445%	−6.71482%	−6.87959%	−6.89216%
60	−7.93299%	−7.95354%	−7.88962%	−7.96126%	−7.92641%
125	−9.23481%	−9.20317%	−9.10547%	−9.27140%	−9.19768%
250	−10.87354%	−10.92524%	−10.85239%	−10.95101%	−10.92249%

- Table 20.20 shows the calculation of Value at Risk and threshold persistence for equi-weighted GJR model with a Student's t-distribution.
- Table 20.21 shows the calculation of Value at Risk and threshold persistence for equi-weighted EGARCH model with a Student's t-distribution.

2.8.1 Value-Weighted vis-à-vis Equi-weighted

In general, the value-weighted portfolio has less dispersion than equi-weighted portfolio. This is to be expected because in general traders have a higher weightage for assets which have less volatility given similar expected returns. This is consistent with the results in the tables obtained for percentage of times the portfolio value hits the threshold level (−5.00%) and the average drop in the portfolio given that the portfolio hits this threshold. Both the values are lower for value-weighted portfolio vis-à-vis equi-weighted portfolio.

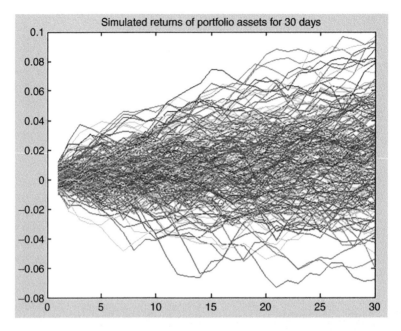

Fig. 20.13 Simulated returns of portfolio assets with GJR model, Student's t-distribution

2.8.2 Gaussian vis-à-vis Student's t-Distribution

For GARCH models, the difference in dispersion for the two distributions is small. However, the tables report the results for GJR and EGARCH using the two different-distributions, and, in general, as is to be expected, the Student's t-distribution tends to have a higher dispersion than in the case of Gaussian distribution.

Figures 20.11 and 20.12 show the simulated paths for the equal-weighted portfolio over the 30-day horizon with GJR model with a Gaussian distribution and Student's t-distribution. As is clearly visible, the fat-tailed Student's t-distribution generates greater variation at the extremes than the normal distribution.

2.8.3 GARCH, GJR, and EGARCH

GARCH tends to underestimate the VaR and persistence measure vis-à-vis GJR and EGARCH. Again this is to be expected given that GJR and EGARCH factor in the leverage effect which GARCH fails to do. GJR and EGARCH return similar results which again is to be expected.

Five values are exhibited for each parameter to show the measure of dispersion. Standard errors for the estimates are computed as also the t-statistics, and both are found to be statistically acceptable.

Each value itself is an average of 10,000 paths. The horizon is mentioned in the first column. The threshold horizon is taken as 2 days for consistency across horizon.

VaR and persistence measures are also computed for horizons of 2 years, 3 years, 4 years, and 5 years. The range of Value at Risk is between 20 and 22% for these horizons. The probability of the portfolio remaining below the threshold level β for 2 or more days is about 42–47%, whereas the average drop in the portfolio given that this happens is about 11–13%.

2.8.4 Principal Component Analysis Results

- Table 20.22 shows the PCA analysis for GARCH model with a Gaussian distribution.
- Table 20.23 shows the PCA analysis for GARCH model with a Student's t-distribution.
- Table 20.24 shows PCA analysis for GJR model with a Student's t-distribution.
- Table 20.25 shows the PCA analysis for EGARCH model with a Student's t-distribution.

2.9 Conclusion

The case study provides an overview of statistical portfolio risk analysis as is practiced in the investment management world. The approach employs econometric modeling of stock price returns and explains the econometric theory behind the application so as to make the chapter self-contained. The risk measurement is formulated using industry standard risk metrics such as Value at Risk and threshold persistence. We use robust methods that account for fat tails and leverage such as GJR and EGARCH to measure risk. One fundamental implication of data analysis for financial markets is that risk regimes change. So a GJR or EGARCH may be apt for this data set but may not be universally appropriate for risk measurement of other kinds of financial data. Since risk regimes change in abrupt and unforeseen ways, a P-quant needs to understand and communicate the assumptions and limitations of data analytics to consumers of risk reports. For instance, it may not be out of place to keep reminding the consumers of risk reports that worst outcomes like Value at Risk and threshold persistence look singularly at the extreme left tail of the portfolio loss distribution. They are therefore less tractable and stable than a simpler metric like variance that is computed over a long-time horizon.

That said, risk measurement and reporting in financial institutions, in general, has moved away from long descriptive type discussions to providing more quantitative information so that risk professionals can make their own assessment. Additionally, the frequency of reporting has changed significantly. Chief risk officers (CROs) in AMCs typically receive reports that contain VaR and threshold persistence estimates

Table 20.22 GARCH, Gaussian

	Coeff C	Coeff AR	Coeff MA	Coeff K	Coeff GARCH	Coeff Arch	Coeff leverage	Log likelihood estimate
PCA 1	0.006191	−0.037002	0.252324	0.002363	0.581073	0.150042	0.000000	482.661472
PCA 2	0.003177	0.019312	−0.090477	0.000727	0.673067	0.159143	0.000000	664.383809
PCA 3	0.000307	1.000000	−0.832486	0.000062	0.533216	0.327323	0.000000	1275.084082
PCA 4	0.001868	−0.099412	−0.037937	0.000177	0.880774	0.042033	0.000000	812.619731
PCA 5	0.000127	−0.781957	0.813060	0.000005	0.965923	0.028736	0.000000	941.624211
PCA 6	−0.000491	0.790801	−0.808419	0.000618	0.416907	0.092563	0.000000	961.156845
PCA 7	−0.000602	−0.254338	0.235336	0.000008	0.978600	0.013823	0.000000	967.958283
PCA 8	−0.007667	−0.942045	0.994696	0.000029	0.906187	0.063275	0.000000	1026.846244
PCA 9	0.003197	−0.802868	0.771076	0.000042	0.854067	0.103361	0.000000	1035.681025
PCA 10	−0.000173	0.935406	−0.996544	0.000607	0.240479	0.096681	0.000000	1042.554168

20 Financial Analytics

Table 20.23 GARCH, Student's *t*

	Coeff C	Coeff AR	Coeff MA	Coeff K	Coeff GARCH	Coeff ARCH	Coeff leverage	Log likelihood estimate
PCA 1	0.006368	−0.044536	0.258300	0.002355	0.582574	0.149677	0.000000	482.593216
PCA 2	0.003167	0.024280	−0.095032	0.000733	0.671451	0.159680	0.000000	664.281732
PCA 3	0.000307	1.000000	−0.831038	0.000062	0.532482	0.328004	0.000000	1274.948566
PCA 4	0.001870	−0.097163	−0.040524	0.000177	0.880851	0.042276	0.000000	812.438789
PCA 5	0.001090	−0.735297	0.769962	0.000010	0.955841	0.036009	0.000000	946.274938
PCA 6	−0.000612	0.774787	−0.794976	0.000619	0.418905	0.096472	0.000000	964.726132
PCA 7	0.000830	−0.987641	1.000000	0.000009	0.974729	0.017016	0.000000	971.144066
PCA 8	−0.007482	−0.946004	0.994906	0.000034	0.888511	0.078636	0.000000	1028.985630
PCA 9	0.003049	−0.810527	0.778840	0.000042	0.853885	0.103836	0.000000	1035.893574
PCA 10	−0.004129	−0.569986	0.622669	0.000544	0.337492	0.085532	0.000000	1042.300629

Table 20.24 GJR, Student's *t*

	Coeff C	Coeff AR	Coeff MA	Coeff K	Coeff GARCH	Coeff ARCH	Coeff leverage	Log likelihood estimate
PCA 1	0.003418	−0.025396	0.239486	0.001970	0.659505	0.000000	0.221961	488.076790
PCA 2	0.003857	0.067171	−0.149177	0.000968	0.614463	0.235522	−0.156928	665.908440
PCA 3	0.000358	1.000000	−0.827733	0.000061	0.558873	0.367855	−0.141102	1276.062104
PCA 4	0.001946	−0.101657	−0.035439	0.000184	0.875252	0.051476	−0.013321	812.486859
PCA 5	0.000893	−0.737373	0.771080	0.000009	0.958213	0.028510	0.010643	946.341024
PCA 6	−0.003602	−0.445124	0.440511	0.000620	0.420793	0.127517	−0.069612	964.670930
PCA 7	−0.000854	−0.320778	0.305520	0.000011	0.970576	0.010650	0.018427	969.812064
PCA 8	−0.006819	−0.946771	0.995566	0.000037	0.883832	0.108470	−0.054267	1029.653034
PCA 9	0.000362	0.783815	−0.822747	0.000043	0.851227	0.101817	0.006893	1036.241450
PCA 10	−0.004473	−0.588531	0.639748	0.000432	0.455036	0.050880	0.075837	1042.590481

20 Financial Analytics

Table 20.25 EGARCH, Student's t

	Coeff C	Coeff AR	Coeff MA	Coeff K	Coeff GARCH	Coeff ARCH	Coeff leverage	Log likelihood estimate
PCA 1	0.003997	−0.064723	0.281886	−1.036871	0.782946	0.185203	−0.163616	484.578627
PCA 2	0.003834	0.092554	−0.174779	−0.994841	0.820174	0.252133	0.084304	667.115909
PCA 3	0.000333	1.000000	−0.830915	−0.415517	0.947283	0.386928	0.067537	1272.729986
PCA 4	0.002047	−0.094801	−0.046898	−0.443169	0.927416	0.105747	0.032806	813.616259
PCA 5	0.000785	−0.758420	0.793959	−0.001776	1.000000	0.041367	−0.003358	945.009720
PCA 6	−0.000627	0.773375	−0.792651	−3.274587	0.510104	0.231433	0.035319	965.878475
PCA 7	−0.000004	0.982168	−1.000000	−0.060656	0.991111	0.036622	0.001679	971.161642
PCA 8	−0.006313	−0.944539	0.995196	−0.282069	0.959395	0.154341	0.021652	1028.909113
PCA 9	0.000353	0.759004	−0.806882	−0.374390	0.946651	0.220938	−0.018483	1034.961690
PCA 10	−0.004564	−0.590409	0.646866	−2.425124	0.652703	0.153982	−0.082664	1043.202155

for 1-Week, 1-Month, and 1-Year horizon. For instance, from Table 20.14 the CRO can infer that if stock returns follow approximately Gaussian distribution, there is a 1% chance (Value at Risk at 99% confidence level) that the portfolio might lose around 12% of its value over a 1-year horizon. Using the metric of threshold persistence, the CRO can infer that over a 1-year horizon, there is a 22% chance of the portfolio dipping below 5%. And given that such a dip happens over the critical time period of over 2 days, the drop in the portfolio value would be approximately 8%. The other tables quantify risk of portfolio when asset returns have excess kurtosis or when there are causal mechanisms at play between returns and volatility such as leverage effects.

Some CROs and investment management boards prefer to receive only summary risk reports. The summary report is typically short so as to make it less likely that the risk numbers will be missed by the board members. Most P-quant CROs choose to receive both the summary and detailed risk reports. It is not usual for the modern-day CROs to receive daily MIS (management information system) reports that contain analysis from Table 20.14 to Table 20.21 on a daily basis. In the last few years, most CROs come from the P-world and are quantitatively well equipped to understand and infer risks from the detailed risk reports.

Apart from the senior management and the board, the other principal audience of risk reports are regulators. Regulators like the Fed and the RBI mandate all financial institutions that they regulate to upload their risk reports in a prescribed templete at the end of each business day. Regulators normally prescribe templates for risk reporting so that they can do an apples-to-apples comparison of risk across financial institutions. Regulators themselves use systems to monitor the change in risk of a given financial insitution over time. More importantly, it helps them aggregate risk of all financial institutions that they regulate so as to assess the systemic risk in the financial industry. With rapid advances in data sciences, it is envisaged that application of analytics in finance would get increasingly more sophisticated in times to come.

Electronic Supplementary Material

All the datasets, code, and other material referred in this section are available in www.allaboutanalytics.net.

- Data 20.1: Financial Analytics Steps 1, 2 and 3.xlsx
- Data 20.2: Financial Analytics Steps 4 and 5.xlsx
- Data 20.3: Financial Analytics Step 6.xlsx
- Data 20.4: Financial Analytics Step 7.xlsx
- Data 20.5: Financial Analytics Step 8.xlsx
- Data 20.6: nifty50.txt
- Data 20.7: ptsr.txt
- Data 20.8: randomgaussian.txt
- Data 20.9: randomgaussiancurrency.txt
- Data 20.10: tsr.txt

Exercises

Ex. 20.1 Stage I Step 3 Inference: We consider the National Stock Exchange index NIFTY-50 recorded daily for the period November 1, 2016–October 31, 2017. Let p_t be the NIFTY-50 index and r_t be the log return $\{r_t = \log(p_t) - \log(p_{t-1})\}$. Load the data from the file nifty50.txt into Matlab.

- Draw graphs of the stock index, the log returns, and the squared log returns.
- Do the graphs indicate GARCH effects?
- Estimate the GARCH parameters (ω, α, β).

Ex. 20.2 Stage I Step 4 Projection: Load the data on random numbers from randomgaussian.txt into Matlab. The standard deviation is 20%, while the average return is 5%. Note that the average return and standard deviation should be adjusted for daily horizon by dividing with 365 and square root of 365, respectively. Project the value of USD/INR for a horizon of 1 year.

Ex. 20.3 Stage II Step 2 Aggregation: Assume a loan portfolio of Rs. 500 lent to five different corporates for Rs. 100 each. Aggregate the risk of this Rs. 500 loan portfolio using one-factor Gaussian copulas. Assume each corporate has a probability of default of 4%. The horizon for the loan is 1 year. Assume that in the event of a default the bank can recover 75% of the loan amount. Assume the single factor to be the economy. The correlation of each firm's asset to the economy is given in the table below. Calculate the joint distribution of credit of each of these corporates using a one-factor model.

	Corporate_1	Corporate_2	Corporate_3	Corporate_4	Corporate_5
C_i	1.00%	1.00%	1.00%	1.00%	1.00%
D_i	75.00%	75.00%	75.00%	75.00%	75.00%
a_i	0.200000	0.400000	0.500000	0.600000	0.800000
X_i	4.00%	4.00%	4.00%	4.00%	4.00%

Ex. 20.4 Stage II Step 2 Assessment: Load the data on random numbers from randomgaussiancurrency.txt into Matlab. Compute VaR for a portfolio of 1 USD, 1 EUR, 1 GBP, and 100 JPY. The value of the portfolio in INR terms is Rs. 280 (1 USD = Rs. 64, 1 EUR = Rs. 75, 1 GBP = Rs. 82, 100 JPY = Rs. 59). Calculate the possible loss or gain from this portfolio for a 1-year horizon. To aggregate the risk, use the correlation matrix below between the currencies:

	USD/INR	EUR/INR	GBP/INR	JPY/INR
USD/INR	1	0.9	0.6	0.4
EUR/INR	0.9	1	0.5	0.5
GBP/INR	0.6	0.5	1	0.2
JPY/INR	0.4	0.5	0.2	1

Ex. 20.5 Stage II Step 3 Attribution: A bank comprises three lines of businesses:

- Line of Business 1 (LoB1)—Corporate Banking
- Line of Business 2 (LoB2) —Retail Banking
- Line of Business 3 (LoB3) —Treasury Operations

LoB1 has a correlation of 0.5 with LoB2 and has a correlation of 0.2 with LoB3. LoB2 is uncorrelated to LoB3. The total bank assets are Rs. 6000 crores. Each of the LoBs has assets worth Rs. 2000 crores.

	Assets	Volatility
LoB1	2000	$\sigma_1 = 5\%$
LoB2	2000	$\sigma_2 = 7\%$
LoB3	2000	$\sigma_3 = 9\%$

(a) Determine the total economic capital for the bank.
(b) Attribute the capital consumed by LoB1, LoB2, and LoB3 on a stand-alone basis.
(c) Attribute the capital consumed by LoB1, LoB2, and LoB3 on incremental basis.
(d) Attribute the capital consumed by LoB1, LoB2, and LoB3 using a component approach.

Ex. 20.6 For multiple horizons of 5, 10, 20, 60, 125, and 250 trading days, for a market value-weighted portfolio, calculate the following:

(a) The Value at Risk of the portfolio using GJR model with Gaussian distribution
(b) The Value at Risk of the portfolio using EGARCH model with Gaussian distribution
(c) The percentage of times portfolio is below beta using GJR model with Gaussian distribution
(d) The percentage of times portfolio is below beta using EGARCH model with Gaussian distribution
(e) The average drop in portfolio when level drops below beta using GJR model with Gaussian distribution
(f) The average drop in portfolio when level drops below beta using EGARCH model with Gaussian distribution

Ex. 20.7 Repeat Exercise 20.6 for an equi-value weighted portfolio.

References

Ali, M. M., Mikhail, N. N., & Haq, M. S. (1978). A class of bivariate distributions including the bivariate logistic. *Journal of Multivariate Analysis, 8*, 405–412.

Black, F., & Scholes, M. (1973). The pricing of options and corporate liabilities. *The Journal of Political Economy, 81*(3), 637–654.

Bollerslev, T. (1986). Generalized autoregressive conditional heteroskedasticity. *Journal of Econometrics, 31*(3), 307–327.

Box, G., Jenkins, G. M., & Reinsel, G. C. (1994). *Time series analysis: Forecasting and control.* Upper Saddle River, NJ: Prentice-Hall.

Campbell, J. Y., Lo, A. W., & MacKinlay, A. C. (1997). *The econometrics of financial markets.* Princeton: Princeton University Press.

CreditMetrics. (1999). *Technical document* (1st ed.). New York: J.P. Morgan.

Dickey, D. A., & Fuller, W. A. (1979). Distribution of the estimators for autoregressive time series with a unit root. *Journal of the American Statistical Association, 74*(366), 427–431.

Engle, R. F. (1982). Autoregressive conditional heteroscedasticity with estimates of the variance of United Kingdom inflation. *Econometrica, 50*(4), 987–1007.

Fama, E. F., & French, K. R. (1993). Common risk factors in the returns on stocks and bonds. *Journal of Financial Economics, 33*, 3.

Fama, E. F. (1965). Random walks in stock market prices. *Financial Analysts Journal, 21*(5), 55–59.

Fama, E. F. (1970). Efficient capital markets: A review of theory and empirical work. *Journal of Finance, 25*(2), 383–417.

Fermi, E., & Richtmyer, R. D. (1948). Note on census-taking in Monte Carlo calculations. *LAM, 805*(A).

Frenkel, J. A., & Levich, R. M. (1981). Covered interest arbitrage in the 1970s. *Economics Letters, 8*(3).

Granger, C. W. J., & Joyeux, R. (1980). An introduction to long-memory time series models and fractional differencing. *Journal of Time Series Analysis, 1*, 15–30.

Hansen, L. P. (1982). Large sample properties of generalized method of moments estimators. *Econometrica, 50*(4), 1029–1054.

Harrison, J. M., & Pliska, S. R. (1981). Martingales and stochastic integrals in the theory of continuous trading. *Stochastic Processes and their Applications, 11*(3), 215–260.

Markowitz, H. (1952). Portfolio selection. *The Journal of Finance, 7*(1), 77–91.

Merton, R. C. (1969). Lifetime portfolio selection under uncertainty: The continuous-time case. *The Review of Economics and Statistics, 51*(3), 247–257.

Owen, J., & Rabinovitch, R. (1983). On the class of elliptical distributions and their applications to the theory of portfolio choice. *Journal of Finance, 38*, 745–752.

Press, W. H., Teukolsky, S. A., Vetterling, W. T., & Flannery, B. P. (1992). *Numerical recipes in C: The art of scientific computing* (p. 994). Cambridge: Cambridge University.

RiskMetrics. (1996). *Technical document* (4th ed.). New York: J.P. Morgan.

Ross, S. A. (1978). Mutual fund separation in financial theory—The separating distributions. *Journal of Economic Theory, 17*, 254–286.

Ross, S. (1976). The arbitrage theory of capital asset. Pricing. *Journal of Economic Theory, 13*, 341–360.

Sharpe, W. (1964). Capital asset prices: A theory of market. equilibrium under conditions of risk. *Journal of Finance, 19*, 425–442.

Sims, C. (1980). Macroeconomics and reality. *Econometrica, 48*(1), 1–48.

Sklar, A. (1959). Fonctions de répartition à n dimensions et leurs marges. *Publications de l'Institut de Statistique de L'Université de Paris, 8*, 229–231.

Tobin, J. (1958). Liquidity preference as behaviour towards risk. *Review of Economic Studies, 25*, 65–86.

Yule, G. U. (1927). On a method of investigating periodicities in disturbed series, with special reference to Wolfer's sunspot numbers. *Philosophical Transactions of the Royal Society of London, Series A, 226*, 267–298.

Chapter 21
Social Media and Web Analytics

Vishnuprasad Nagadevara

1 Introduction

Social media has created new opportunities to both consumers and companies. It has become one of the major drivers of consumer revolution. Companies can analyze data available from the web and social media to get valuable insights into what consumers want. Social media and web analytics can help companies measure the impact of their advertising and the effect of mode of message delivery on the consumers. Companies can also turn to social media analytics to learn more about their consumers. This chapter looks into various aspects of social media and web analytics.

2 What Is Social Media and Web Analytics?

2.1 Why Is It Different? What All Does It Cover?

Social media analytics involves gathering information from social networking sites such as Facebook, LinkedIn and Twitter in order to provide businesses with better understanding of customers. It helps in understanding customer sentiment, creating

---

Electronic supplementary material The online version of this chapter (https://doi.org/10.1007/978-3-319-68837-4_21) contains supplementary material, which is available to authorized users.

V. Nagadevara (✉)
IIM-Bangalore, Bengaluru, Karnataka, India
e-mail: nagadevara_v@isb.edu

© Springer Nature Switzerland AG 2019
B. Pochiraju, S. Seshadri (eds.), *Essentials of Business Analytics*, International Series in Operations Research & Management Science 264,
https://doi.org/10.1007/978-3-319-68837-4_21

customer profiles and evolving appropriate strategies for reaching the right customer at the right time. It involves four basic activities, namely, listening (aggregating what is being said on social media), analyzing (such as identifying trends, shifts in customer preferences and customer sentiments), understanding (gathering insights into customers, their interests and preferences and sentiments) and strategizing (creating appropriate strategies and campaigns to connect with customers with a view to encourage sharing and commenting as well as improving referrals). One of the major advantages of social media analytics is that it enables businesses to identify and encourage different activities that drive revenues and profits and make real-time adjustments in the strategies and campaigns. Social media analytics can help businesses in targeting advertisements more effectively and thereby reduce the advertising cost while improving ROI.

On the other hand, web analytics encompasses the process of measuring, collecting and analyzing web traffic data to understand customer behaviour on a particular website. Web analytics can help in improving user experience leading to higher conversion rates. Companies can tweak the design and functionality of their websites by understanding how users interact with their websites. They can track user behaviour within the website and how users interact with individual elements on each page of the website. Web analytics can also help in identifying the most profitable source of traffic to the website and determining which referrals are important in terms of investing in marketing efforts. Google Analytics is probably the best tool available free of cost to any website owner for tracking and analyzing the traffic to their website. These analytics include sources of traffic, bounce rates, conversions, landing pages and paid search statistics. It is easy to integrate this with Google AdWords.

It is obvious that social media analytics is very different from web analytics. The data sources are different. These two complement each other and when used together in tandem can provide deep insights into the traffic patterns, users and their behaviour. For example, one can measure the volume of visitors to the website based on referrals by different social networks. By ranking these social networks based on the traffic generated, one can determine how to focus on the right networks. We can even determine the influencers in the networks and their behaviour on our website.

2.2 What Additional Information/Details Can It Provide?

Many companies have started leveraging the power of social media. A particular airline keeps the customers informed through social media about the delays and the causes for such delays. In the process, the airline is able to proactively communicate the causes for the delays and thereby minimize the negative sentiment arising out of the delays. In addition, the airline company is also able to save much of the time of its call centre employees, because many customers already knew about the delays as well as the reasons associated with the delays and hence do not make calls to the call centre.

Social media analytics can be effectively used to gauge customers' reaction to a new product or service introduced into the market. Social media gives an opportunity to listen to the person on the street. The simplest way to do this is to scan various social media sites such as Twitter, Facebook or LinkedIn or various discussion forums and blogs. By analyzing these messages and blogs, the company can understand the customer perceptions about the new product or service. It is also possible to analyze the trend of customer perceptions overtime as well as the response to various advertising and marketing campaigns. This response can be measured almost on a real-time basis.

Search engine optimization (SEO) is another technique to acquire customers when they are looking for a specific product or service or even an organization. For example, when a customer initiates a search for a product, say a smartphone, there is a possibility of getting overloaded and overwhelmed with the search results. These results contain both "paid listings" and "organic listings". Paid search listings are advertisements that businesses pay to have their ads displayed when users do a search containing specific keywords. On the other hand, "Organic search listings" are the listings of web pages returned by the search engine's algorithms that closely match the search string. Companies can make use of SEO in order to get their websites listed higher in the search results, especially higher among the organic listings. They can also initiate online advertisements and achieve higher visibility.

It is worthwhile for the companies to monitor the postings on various social media. One can use the posts as well as various blogs and reviews to constantly monitor customer opinions through opinion mining and sentiment analysis. Constant analysis of the opinions and sentiment analysis can help companies not only to understand the customer sentiments but also to develop strategies to further promote positive sentiment or carry out damage control in case of negative sentiment. Companies can also learn from negative comments to improve their products and services. In addition, they can monitor customers' response to the corrective actions taken by them.

Websites can use cookies as well as IP addresses to identify past history of customers and their preferences. Based on past history and browsing habits, customers can be served appropriate ads. For example, if we can predict that the customer likes outdoor sports, he or she can be targeted with an offer for a trekking package. The predictions can be made even with respect to a new arrival on the website. For example, if the customer has a Facebook account and arrives through viewing an ad on Facebook, we can identify his or her hobbies and interests and present appropriate ads.

It has become very common for customers to look at product (or service) reviews and recommendations that are posted on various websites, before making a decision. Many websites (such as TripAdvisor) present an analysis of the reviews as well as a summary of the reviews to help customers make the appropriate decision. While presenting the summary of the reviews, companies can make appropriate offerings to facilitate quick acceptance.

Analysis of frequently used keywords on search engines can help in identifying the current issues and concerns. It can reveal public sentiment towards emerging issues as well as public attitude towards political parties. It can help in identifying pockets of political support. It can also help in creating appropriate campaigns and policy formulation to take full advantage of the prevailing sentiments.

It is very common to have leaders and followers in social media. By using network analytics (such as network and influence diagrams), organizations can identify the most influential persons in the network. Organizations can take the help of such influential persons to promote a new product or service. They can be requested to contribute an impartial review or article about the product or service. Such a review (or a recommendation) can help in promoting the new product or service by resolving uncertainty or hesitancy in the minds of the followers.

Network and influence diagrams are used very effectively in identifying key players in various financial frauds or drug laundering cartels. One such study identified the perpetrators of 9/11 attacks by using a network diagram based on the information available in the public domain. The network diagram could clearly identify the key players involved in 9/11 starting from the two Cole bombing suspects who took up their residences in California as early as 1999[1].

Social media analytics can also be used for public good. Data from mobile telephones are used to identify and predict traffic conditions. For example, Google Traffic analyzes data sourced from a large number of mobile users. Cellular telephone service providers constantly monitor the location of the users by a method called "trilateration" where the distance of each device to three or more cell phone towers is measured. They can also get the exact location of the device using GPS. By calculating the speed of users along a particular length of road, Google generates a traffic map highlighting the real-time traffic conditions. Using the existing traffic conditions across different routes, better alternative routes can be suggested.

With the advent of smartphones, mobile devices have become the most popular source for consuming information. This phenomenon has led to the development of new approaches to reach consumers. One such approach is geo-fencing. It involves creating a virtual perimeter for a geographical area and letting companies know exactly when a particular customer (or potential customer) is likely to pass by a store or location. These virtual perimeters can be dynamically generated or can be predefined boundaries. This approach enables companies to deliver relevant information or even pass on online coupons to the potential customer. The concept of geo-fencing can be combined with other information based on earlier search history, previous transactions, demographics, etc. in order to better target the customer with the right message.

In the rest of the chapter, we describe some applications in greater detail: display advertising in real time, A/B experiments for measuring value of digital media and handling e-retailing challenges, data-driven search engine advertising, analytics of digital attribution and strategies and analytics for social media and social enterprises and mobile analytics and open innovation.

---

[1] Valdis Krebs, "Connecting the Dots; Tracking Two Identified Terrorists" available at http://www.orgnet.com/tnet.html (last accessed on Jan 18, 2018).

3 Display Advertising in Real Time

The Internet provides new scope for creative approaches to advertising. Advertising on the Internet is also called online advertising, and it encompasses display advertisements found on various websites and results pages of search queries and those placed in emails as well as social networks. These display advertisements can be found wherever you access the web. As in the case of any other mode of advertising, the objective of display advertisement is to increase sales and enhance brand visibility. The main advantage of display advertisements is that all the actions of the user are trackable and quantifiable. We can track various metrics such as the number of times it was shown, how many clicks it received, post-click-and-view data and how many unique users were reached. These display advertisements can be all pervasive and be placed anywhere and on any type of web pages. They can be in the form of text, images, videos and interactive elements. Another advantage of display advertisements is that the conversion and sales are instantaneous and achieved with a single click.

There are different types of display advertisements. The most popular one is the banner advertisement. This is usually a graphic image, with or without animation, displayed on a web page. These advertisements are usually in the GIF or JPEG images if they are static, but use Flash, JavaScript or video if there are animations involved. These banner advertisements allow the user to interact and transact. These can be in different sizes and can be placed anywhere on the web page (usually on the side bar). You need to carry out appropriate tests or experimentation (refer to the section on Experimental Designs in this chapter) to know what works best for you. You can design your banner as a single- or multiple-destination URL(s).

There are banners that appear between pages. As you move from one page to the next through clicks, these advertisements are shown before the loading of the next page. These types of advertisements are referred to as interstitial banners.

The display advertisements can be opened in a new, small window over the web page being viewed. These are usually referred to as pop-ups. Once upon a time, these pop-up advertisements were very popular, but the advent of "pop-up blockers" in the web browsers had diminished the effectiveness of the pop-up advertisements. Users can be very selective in allowing the pop-ups from preselected websites. There are also similar ones called pop-under where an ad opens a new browser window under the original window.

Some of the local businesses can display an online advertisement over a map (say Google Maps). The placement of the advertisement can be based on the search string used to retrieve the map. These are generally referred to as map advertisements.

Occasionally, an advertisement appears as a translucent film over the web page. These are called floating advertisements. Generally, these advertisements have a close button, and the user can close the advertisement by clicking on the close button. Sometimes, these advertisements float over the web page for a few seconds before disappearing or dissolving. In such a case, there is no click-through involved, and hence it is difficult to measure the effectiveness of such ads.

Websites are generally designed to display at a fixed width and in the centre of the browser window. Normally, this leaves considerable amount of space around the page. Some display advertisements take advantage of these empty spaces. Such advertisements are called wallpaper advertisements.

3.1 How to Get the Advertisements Displayed?

There are many options for getting the advertisements displayed online. Some of these are discussed below.

One of the most popular options is placing the advertisements on social media. You can get your ads displayed on social media such as Facebook, Twitter and LinkedIn. In general, Facebook offers standard advertisement space on the right-hand side bar. These advertisements can be placed based on demographic information as well as hobbies and interests which can make it easy to target the right audience. In addition to these ads, Facebook also offers engagement ads which will facilitate an additional action point such as a Like button or Share button or a button to participate in a poll or even to add a video. Sponsored stories or posts can also be used to promote a specific aspect of a brand. These stories or posts can appear as news feeds. You can even publicize an existing post on Facebook.

Twitter also allows advertisements. Some promotional tweets appear at the top of the user's time line. The section "Who to Follow" can also be used to have your account recommended at a price. Usually, the payment is made when a user follows a promoted account. The Trends section of Twitter is also available for advertisements. While this section is meant for the most popular topics at any particular time, the space is also available for full-service Twitter ads customers.

LinkedIn allows targeted advertisements with respect to job title, job function, industry, geography, age, gender, company name and company size, etc. These advertisements can be placed on the user's home page, search results page or any other prominent page.

Online advertisements can be booked through a premium media provider, just like one would book from a traditional advertising agency. A premium media provider usually has access to high-profile online space and also can advise on various options available.

Another option is to work with an advertising network. Here, a single sales entity can provide access to a number of websites. This option works better if the collection of websites are owned or managed by a common entity, such as HBO or Times Inc. The Google Display Network is another such entity. Usually, the advertising network can offer targeting, tracking and preparing analytic reports in addition to providing a centralized server which is capable of serving ads to a number of websites. This advertising network can also advise you based on various factors that influence the response, such as demographics or various topics of interest.

If you are looking for advertising inventory (unsold advertising space), advertising exchanges can help. The publishers place their unsold space on the exchange, and it is sold to the highest bidder. The exchange tries to match the supply and demand through bidding.

One of the fastest-growing forms of display advertising is mobile advertising. Mobile advertising includes SMS and MMS ads. Considering that mobile is an intensely interactive mass media, advertisers can use this media for viral marketing. It is easy for a recipient of an advertisement to forward the same to a friend. Through this process, users also become part of the advertising experience.

There are blind networks such as www.buzzcity.com which will help you to target a large number of mobile publishers. A special category of blind networks (can be called as premium blind networks) can be used to target high-traffic sites. The sites www.millennialmedia.com and www.widespace.com are good examples of premium blind networks.

Ad servers play a major role in display advertisement. These servers can be owned by publishers themselves or by a third party and are used to store and serve advertisements. The advantage of the ad servers is that a simple line of code can call up the advertisement from the server and display it on the designated web page. Since the ad is stored at one single place, any modifications are to be carried out at one place only. In addition, the ad servers can supply all the data with respect to the number of impressions, clicks, downloads, leads, etc. These statistics can be obtained from multiple websites. One example of a third-party ad server is *Google DoubleClick*.

3.2 Programmatic Display Advertising

Programmatic advertising is "the automation of the buying and selling of desktop display, video, FBX, and mobile ads using real-time bidding. Programmatic describes how online campaigns are booked, flighted, analyzed and optimized via demand-side software (DSP) interfaces and algorithms"[2].

Traditionally, it is assumed that online visitors are exposed to a display ad when they somehow arrive on a website. The targeting of the ad is based on limited knowledge with respect to the relevance of the content, geography, type of device, etc. The advertiser enters into an agreement with the publisher with respect to the number of insertions at a certain price. In other words, the display ad is bought in the old-fashioned way, negotiating the number of exposures/insertions and the price, and served to a large number of potential consumers with a fond hope that somehow it will reach the right people.

Today, programmatic display delivers display ads in real time with specific messages based on each individual consumer's profile and behaviour. With the

---

[2]Gurbaksh Chahal, Chairman & CEO of RadiumOne.

evolution of recent technologies, the focus is shifting to understanding each individual customer and exposing him or her to the right display. This process is driven by real-time technologies using specially designed algorithms. The result is programmatic display advertising.

There are opportunities to display billions of ads every day, and advertisers can use them effectively to improve conversion rates. The technology enables the advertiser to obtain immediate information about where the ad was displayed, to whom it was displayed and for how long. This helps the advertiser to analyze and review the display ad very quickly and take any action that is required, including stopping of the ad, if it is not performing as expected.

The main components of programmatic display advertising are as follows:

(a) Supply-Side Platform (SSP)

The SSP helps the publishers to better manage and optimize their online advertising space and advertising inventory. SSP constantly interacts with the ad exchange and the demand-side platform (DSP). Admeld (www.admeld.com) and Rubicon (https://rubiconproject.com/) are two examples of SSPs.

(b) Demand-Side Platform (DSP)

The DSP enables the advertisers to set and apply various parameters and automate the buying of the displays. It also enables them to monitor the performance of their campaigns. Turn (now Amobee, https://www.amobee.com/), AppNexus (https://www.appnexus.com/) and Rocket Fuel (https://rocketfuel.com/) are some of the DSPs.

(c) Ad Exchange

Ad exchanges such as Facebook Ad Exchange or DoubleClick Ad Exchange facilitate purchase of available display inventory through auctions. These auctions are automated and take place within milliseconds, before a web page loads on the consumer's screen. These enable the publishers to optimize the price of their available inventory.

(d) Publisher

Publishers are those who provide the display ad inventory.

(e) Advertiser

The advertiser bids for the inventory in real time depending on the relevance of the inventory.

Real-time bidding (RTB) is an integral process of programmatic display advertising and is described in Fig. 21.1.

Whenever a consumer makes a request for a web page and if there is available space for a display ad, the information regarding the consumer/visitor, context of the web page requested and earlier web behaviour is sent to an ad exchange through the publisher. The ad exchange auctions the available space to various DSPs, and the winning bid/display ad is passed on to the publisher. The ad is displayed on the consumer's screen while the requested page is being loaded. This process is completed within a few milliseconds so that there is no delay in the page loading adversely affecting the user experience.

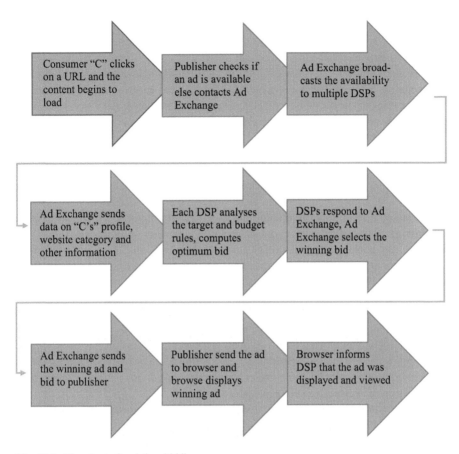

Fig. 21.1 Flowchart of real-time bidding

The entire process described above takes less than half a second. In other words, the entire process is completed and the display ad is shown while the browser is loading the requested page on the consumer's screen.

This process of matching the right display to the right consumer is completely data driven. Data with respect to the context, who is to see the display, the profile of the consumer, who is a good target, etc. is part of the process. In order to complete the process within a very short time span, it is necessary to build all the rules in advance into the system. The advertiser would have analyzed all the data and identified the appropriate profile, the possible number of exposures of the ad and the rules for bidding beforehand, based on his own analytics. The rules try to match the information about the context, profile of the consumer and space available received from the publisher with the requirement and trigger automatic bidding.

The programmatic display advertisement benefits both the publishers and advertisers. The advertisers benefit by effectively targeting only those who match the existing profiles. These profiles can be obtained by analyzing their own data or from third parties. With the context built into the process, the advertiser can reach consumers who are browsing content that is most relevant to the product or service that the advertiser is offering. They can be very selective in terms of the website and the right context. They do not have to tie themselves to a pre-negotiated price and quantity. They will be paying only for the relevant exposure that was made to the most relevant target audience. The advertiser can analyze data with respect to visits to the website, bounce rates, earlier marketing efforts, etc. in order to improve the conversions. It also enables the advertiser to quickly review and revise the advertisement strategies instead of waiting for the entire campaign to be completed.

The publishers gain by maximizing the revenue through auctioning the available space. Each ad is auctioned to the highest bidder based on the context and consumer profile. It also allows them to optimize the available advertising space.

Programmatic display advertising opens up yet another opportunity in display advertising. The advertisers can use dynamic creative optimization (DCO) to improve the conversion rates. DCO involves breaking an ad into a number of components and creating different versions of each component. These components can be with respect to content, visuals, colours, etc. These components are then dynamically assembled to suit a particular consumer based on the context, profile, demographics as well as earlier browsing history. These details along with any other information available (such as time of the day or weather at that particular time) are fed into the DCO platform. The display ad is assembled based on this information, using predetermined rules before it is sent to the publisher's server. Thus, DCO can take advantage of the targeting parameters received from the ad exchange to optimally create (assemble) the appropriate ad.

It is expected that programmatic display advertising in the USA alone will reach more than $45 billion by 2019 (Fig. 21.2).

4 A/B Experiments for Measuring Value of Digital Media and Handling E-Retailing Challenges

Web technology companies such as Amazon, Facebook, eBay, Google and LinkedIn are known to use A/B testing in developing new product strategies and approaches. A/B testing (also called A/B splits or controlled experimentation) is one of the widely used data-driven techniques to identify and quantify customer preferences and priorities. This dates back to Sir Ronald A. Fisher's experiments at the *Rothamsted Agricultural Experimental Station* in England in the 1920s. It was called A/B splits because the approach was to change only one variable at a time. The publication of Ronald Fisher's book *The Design of Experiments* changed the approach where the values of many variables are changed simultaneously and the impact of each of the variables is estimated. These experimental designs use the concept of ANOVA (discussed in the earlier chapter) extensively, with appropriate modifications.

Fig. 21.2 US programmatic digital display ad spending

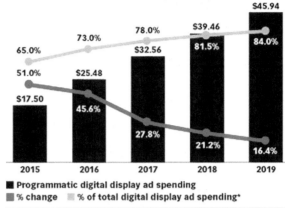

4.1 Completely Randomized Design

The simplest of the experimental designs is the completely randomized design (CRD). Internet experiments that present one of several advertising messages to users of search engines are typical examples of CRD. While there are a predetermined set of k advertisement messages offered to the users, each user is exposed to one and only one message. The response is measured by the hit ratio (proportion of those who click on the message to access the particular website in response to the message). Consider a scenario where an experiment is conducted to evaluate the effect of a display advertisement. Three different types of advertisements (namely, A, B and C) were designed and displayed randomly on five different search engines, and the experiment is carried for 3 weeks. The users are taken to a specific web page when the users click on the display advertisement. The number of visitors to the landing page is counted and used as a metric for the effectiveness of the display. Here, the three types of displays are called the treatments. A particular type of display is assigned randomly for a particular search engine in a given week for displaying to the users. The summary of the visitors based on the three different displays is presented in Table 21.1.

The above observations come from three different treatment groups, each group having five observations (replications). In general, each observation can be denoted by Y_{ij} where i represents the treatments and j represents the replication number. Even though the above example has an equal number of replications for each

Table 21.1 Number of visitors to the web page based on type of display

	Type of display		
	A	B	C
Number of visitors	2565	2295	2079
	864	2430	3051
	1269	2133	2619
	2025	1350	3105
	2241	864	3348
n	5	5	5
Average	1793	1814	2840
Variance	497,032	457,593	250,193
Number of clicks	574	592	501
	175	575	630
	240	449	507
	444	302	587
	405	160	547

treatment, it is not necessary to have an equal number of replications. The above experiment is called balanced because each treatment has an equal number of replications.

The above data can be analyzed for any significant difference between population means using ANOVA.

H_0: $\mu 1 = \mu 2 = \mu 3$.
H1: at least one μ is different.

The F test shown in Table 21.3 indicates that there is significant difference in the response to different displays.

The CRD can be represented as a regression model. Since everyone is familiar with regression analysis and many statistical packages have regression modules, representing and analyzing CRD as a regression model makes it easy to analyze and interpret the results.

The regression model for CRD is

$$Y_{ij} = \mu + \beta_i + \varepsilon_{ij}$$

where Y_{ij} is the response corresponding to ith treatment and jth replication
  μ is the overall mean
  β_i is the treatment effect and
  ε_{ij} is the random error

Since the treatments are nominal variables, these are represented as dummy variables. As there are three treatments, these will be represented by two dummy variables. Type of Display "A" is represented by $D1 = 1$ and $D2 = 0$, "B" is represented by $D1 = 0$ and $D2 = 1$, and "C" is represented by $D1 = 0$ and $D2 = 0$. The data reformatted with dummy variables as required for the regression and the results of regression analysis are presented in Table 21.2 (a) and (b).

Table 21.2 (a) Reformatted data with dummy variables for the CRD experiment. (b) Regression results of the CRD experiment

(a) Reformatted data with dummy variables for the CRD experiment		
Yij	D1	D2
2565	1	0
864	1	0
1269	1	0
2025	1	0
2241	1	0
2295	0	1
2430	0	1
2133	0	1
1350	0	1
864	0	1
2079	0	0
3051	0	0
2619	0	0
3105	0	0
3348	0	0
(b) Regression results of the CRD experiment		
Regression statistics		
Multiple R	0.6531	
R square	0.4265	
Adjusted R square	0.3309	
Standard error	633.7240	
Observations	15	

	df	SS	MS	F
Regression	2	3,584,347	1,792,174	4.4625
Residual	12	4,819,273	401,606.1	
Total	14	8,403,620		
	Coefficients	Standard error	t stat	P-value
Intercept	2840.4	283.41	10.0222	0.0000
D1	−1047.6	400.8022	−2.6138	0.0226
D2	−1026.0	400.8022	−2.5599	0.0250

It can be seen that the ANOVA table calculated in Table 21.3 is identical to the ANOVA table obtained from the regression analysis. In addition, the regression coefficients corresponding to the dummy variables D1 and D2 are negative and statistically significant. This implies that Display C (which was left out in creating the dummy variables) has resulted in significantly higher visitors than Displays A and B. The intercept and other regression coefficients can be interpreted as the differences between mean responses of the three displays. The pairwise differences in the treatment effects can be obtained by post hoc tests.

Table 21.3 ANOVA table

Source of variation	SS	df	MS	F	P-value	F crit
Between groups	3,584,347	2	1,792,174	4.4625	0.0356	3.8853
Within groups	4,819,273	12	401,606			
Total	8,403,620	14				

Table 21.4 Click-through rates

Treatment	Block (search engine)					Average
	SE1	SE2	SE3	SE4	SE5	
A	0.2236	0.2021	0.1892	0.2193	0.1806	0.2030
B	0.2580	0.2365	0.2107	0.2236	0.1849	0.2227
C	0.2408	0.2064	0.1935	0.1892	0.1634	0.1987
Average	0.2408	0.2150	0.1978	0.2107	0.1763	0.2081

4.2 Randomized Complete Block Design

In the above experiment, it is possible that there is an effect of the search engine, in addition to the effect of the type of display. In other words, there are two sources of variation, the type of display and the search engine. Since the displays are randomly assigned to the search engines and weeks, it is possible that all the 3 weeks selected for the first search engine could have been assigned Display A, while no week is assigned for Display A for the second search engine. It is necessary to run the experiment in "blocks" if we need to isolate the effect of display as well as the effect of the search engine. Such a design is called the randomized complete block design. In other words, the experiment should be run in blocks such that the three types of display are tested on each search engine on each of the weeks. When the visitors come to the landing page, they can "click" on the page to obtain additional information. The number of clicks is recorded, and the click-through rate is calculated as Click-through rate (CTR) = Number of clicks/Number of visitors.

The design and the results are shown in Table 21.4.

In a general randomized block design, there are k treatments and b blocks. The observations are represented by Y_{ij} where i represents the treatments and j represents the blocks.

The above data can be analyzed for differences between means by ANOVA, except that the sums of squares corresponding to treatments and blocks have to be estimated separately. The formulae for calculating these sums of squares are presented as follows:

$$SS(Treatment) = b \sum_{i=1}^{k} \left(\overline{Y}_{i.} - \overline{Y}_{..} \right)^2$$

$$SS(Block) = k \sum_{i=1}^{b} \left(\overline{Y}_{.j} - \overline{Y}_{..}\right)^2$$

$$SS(Total) = \sum_{i=1}^{k} \sum_{j=1}^{b} \left(Y_{ij} - \overline{Y}_{..}\right)^2$$

$$SS(Error) = SS(Total) - SS(Treatment) - SS(Block)$$

$$SS(Error) \text{ can also be calculated as } \sum_{i=1}^{k} \sum_{j=1}^{b} \left(Y_{ij} - \overline{Y}_{i.} - \overline{Y}_{.j} + \overline{Y}_{..}\right)^2$$

where

Y_{ij} is the measurement corresponding to ith treatment and jth block,

$\overline{Y}_{i.}$ is the average of ith treatment,

$\overline{Y}_{.j}$ is the average of jth block and

$\overline{Y}_{..}$ is the overall mean.

The degree of freedom for treatment is $k - 1$ and that of block is $b - 1$. While the total degree of freedom is $kb - 1$, the error degree of freedom is $(k - 1)(b - 1)$. The ANOVA table for the above data is presented in Table 21.5.

It can be concluded from the ANOVA table that there are significant differences in responses between types of display as well as the search engines. The statistical significance is much higher (p-value = 0.0008) in the case of block (search engine) than that of treatment (type of display).

Table 21.5 ANOVA table for the experiment data

ANOVA						
Source of variation	SS	df	MS	F	P-value	F crit
Treatment	0.0016	2	0.0008	7.6239	0.0140	4.4590
Block	0.0067	4	0.0017	15.5385	0.0008	3.8379
Error	0.0009	8	0.0001			
Total	0.0092	14				

The regression model for the randomized complete block design can be represented by

$$Y_{ij} = \mu + \beta_i + \pi_j + \varepsilon_{ij}$$

where Y_{ij} is the response for ith treatment and jth block
μ is the mean
β_i is the effect of Treatment i
π_j effect of Block j and
ε_{ij} is the random error

The coding of the above data for regression analysis is shown in Table 21.6. The dummy variables corresponding to the treatments (types of display) are represented by T_i, and those corresponding to blocks (search engines) are represented by SE_j.

As usual, the dummy variables corresponding to T3 and SE5 are omitted. The regression output is presented in Table 21.7.

It can be seen that the regression sum of squares is equal to the sum of the "treatment sum of squares" and the "block sum of squares". Consequently, the "F" value is different. It can also be seen that all the treatment effects and block effects are significant indicating that the effects of T1 and T2 are significantly better than that of T3. Similarly, the effects of the first four search engines (SE1 to SE4) are significantly better than that of SE5. The pairwise comparisons can be obtained by running post hoc tests.

Table 21.6 Coding for regression analysis

		Y	T1	T2	SE1	SE2	SE3	SE4
A	SE1	0.2236	1	0	1	0	0	0
B	SE1	0.2580	0	1	1	0	0	0
C	SE1	0.2408	0	0	1	0	0	0
A	SE2	0.2021	1	0	0	1	0	0
B	SE2	0.2365	0	1	0	1	0	0
C	SE2	0.2064	0	0	0	1	0	0
A	SE3	0.1892	1	0	0	0	1	0
B	SE3	0.2107	0	1	0	0	1	0
C	SE3	0.1935	0	0	0	0	1	0
A	SE4	0.2193	1	0	0	0	0	1
B	SE4	0.2236	0	1	0	0	0	1
C	SE4	0.1892	0	0	0	0	0	1
A	SE5	0.1806	1	0	0	0	0	0
B	SE5	0.1849	0	1	0	0	0	0
C	SE5	0.1634	0	0	0	0	0	0

Table 21.7 Results of regression analysis

Regression statistics					
Multiple R	0.9520				
R square	0.9063				
Adjusted R square	0.8361				
Standard error	0.0104				
Observations	15				
ANOVA					
	df	SS	MS	F	Significance F
Regression	6	0.0084	0.0014	12.9003	0.0010
Residual	8	0.0009	0.0001		
Total	14	0.0092			
	Coefficients	Standard error	t stat	P-value	
Intercept	0.1668	0.0071	23.4828	0.0000	
T1	0.0043	0.0066	0.6537	0.5316	
T2	0.0241	0.0066	3.6608	0.0064	
SE1	0.0645	0.0085	7.5955	0.0001	
SE2	0.0387	0.0085	4.5573	0.0019	
SE3	0.0215	0.0085	2.5318	0.0352	
SE4	0.0344	0.0085	4.0510	0.0037	

4.3 Analytics of Multivariate Experiments

The real power of experimental designs is felt when we have to estimate the effects of a number of variables simultaneously. For example, consider a scenario where a company is contemplating a particular display advertisement. They have identified three different variables (factors), each having two levels, to test. These are font (traditional font vs. modern), background colour (white vs. blue) and click button design (simple "Okay" vs. large "Click Now to Join"). The ad copies are randomly displayed to each viewer, and the conversion rate (defined as those who click to reach the website and join as members (free of cost)) is calculated. In the traditional experimental design, such as CRD, we will first decide which one is likely to be most important. Let us say the click button is the most important. Then, we would combine each of the two levels of click button with one of the other two factors (say, traditional font and white background) and run four replications. This actually involves eight runs (four each of simple button + traditional font + white background and large button + traditional font + white background). The resulting conversion rate can be used to decide which type of click button is more effective (say, large button). Now, we will select the background colour for experimentation. Since we already have the combination with white background, we will now select blue background and combine it with "large button" (since it was more effective) and traditional font and run four replications. Suppose the results show that blue background is more effective. Now, we select the combination of large button and

Table 21.8 Coding factorial design

Factor	Level 1 (code)	Level 2 (code)
Font	Traditional (−1)	Modern (+1)
Background colour	White (−1)	Blue (+1)
Click button design	Simple "Okay" (−1)	Large "Click Now to Join" (+1)

blue background and combine it with traditional font and modern font. We already have four runs of traditional font, blue colour and large button. Now we have to carry out four runs of the combination of modern font, blue colour and large button. Thus, we have a total of 16 runs. These 16 runs will help us to estimate the effects of font, background colour and type of click button. But it is also possible that there can be interaction effects between these factors. For example, a combination of small button with blue background and modern font could be much more effective than any other combination. Notice that we did not experiment with this particular combination at all, and hence, we have no way of estimating this effect. Same is true with many other interactions.

The factorial design developed by Ronald Fisher is a much better approach. With three factors and two levels for each factor, there are eight possible combinations. These eight combinations can be displayed randomly to the viewers and the conversion rate calculated. It is important that each combination is displayed with equal probability. This process involves only eight runs instead of 16 runs required in the earlier approach. Table 21.8 shows the factorial design of the above experiment with two replications. The levels of each factor are represented by +1 and −1. The coding is as follows:

There are only two levels for each of the factors in our experiment. Hence, these designs are called two-level factorial designs (since there are three factors, this design is referred to as 2^3 factorial design). In general, there can be many more levels for each factor. The change in the response (conversion rate) when the level of the factor is changed from −1 to +1 is called the "main effect". For example, the main effect for the factor "font" is the change in the conversion rate when the font is changed from "traditional (−1)" to "modern (+1)". When the effect of one factor is influenced by another factor (a typical example is water and fertilizer in agricultural experiments), it implies that there is a synergy between these two factors. Such effects are called interaction effects. In Table 21.9, the coding of interaction variables is the multiplication of the corresponding columns.

In order to isolate the effects of each factor and the interactions, we need to calculate the sum of squares corresponding to each main effect and interaction effect. The ANOVA table for the data of the above experiment is presented in Table 21.10. These results are obtained by running the model in R using the following code:

```
> Twoway_anova <- aov(Conversion_Rate ~ Font + Background
    + Click + FB + FC + BC + FBC, data=factorial_experiment)
> summary.aov(Twoway_anova)
```

21 Social Media and Web Analytics

Table 21.9 Factorial design with two replications

	Main effects			Interaction effects				
Conversion rate (%)	Font (F)	Background colour (B)	Click button (C)	FB	FC	BC	FBC	
1.36	−1	−1	−1	+1	+1	+1	−1	REPLICATION 1
2.04	+1	−1	−1	−1	−1	+1	+1	
1.53	−1	+1	−1	−1	+1	−1	+1	
1.87	+1	+1	−1	+1	−1	−1	−1	
3.23	−1	−1	+1	+1	−1	−1	+1	
5.78	+1	−1	+1	−1	+1	−1	−1	
2.04	−1	+1	+1	−1	−1	+1	−1	
5.61	+1	+1	+1	+1	+1	+1	+1	
0.68	−1	−1	−1	+1	+1	+1	−1	REPLICATION 2
2.38	+1	−1	−1	−1	−1	+1	+1	
0.51	−1	+1	−1	−1	+1	−1	+1	
2.38	+1	+1	−1	+1	−1	−1	−1	
2.04	−1	−1	+1	+1	−1	−1	+1	
4.59	+1	−1	+1	−1	+1	−1	−1	
2.55	−1	+1	+1	−1	−1	+1	−1	
6.12	+1	+1	+1	+1	+1	+1	+1	

Table 21.10 ANOVA table for 2^3 factorial experiment

Dependent variable: conversion rate						
	Df	Sum sq.	Mean sq.	F value	Pr(>F)	
Font (F)	1	17.703	17.703	54.149	7.93E−05	***
Background colour (B)	1	0.016	0.016	0.05	0.8291	
Click button (C)	1	23.064	23.064	70.547	3.07E−05	***
FB	1	0.219	0.219	0.669	0.4372	
FC	1	3.658	3.658	11.188	0.0102	*
BC	1	0.045	0.045	0.138	0.7198	
FBC	1	0.305	0.305	0.934	0.3622	
Residuals	8	2.615	0.327			

Signif. codes: ***, 0.001; **, 0.01; *, 0.05
R squared = 0.945 (adjusted R squared = 0.897)

It can be seen from the above table that the main effects of "font" and "click button" are significant and the effect of "background colour" is not significant. In addition, only the interaction between the "font" and "click button" is significant. All other interactions are not significant.

The mean conversion rates for each level of the factors and the interactions are presented in Table 21.11.

The mean effects of different factors and the interactions are also presented in Table 21.11 (overall column). The way these effects need to be interpreted is that the conversion rate will go up by 2.1038 when we change the font from traditional

Table 21.11 Mean conversion rates

	Mean conversion rates		
	1	−1	Overall (level 2 − level 1)
Font (F)	3.8463	1.7425	2.1038
Background colour (B)	2.8263	2.7625	0.0637
Click button (C)	3.9950	1.5938	2.4013
FB	2.9113	2.6775	0.2338
FC	3.2725	2.3163	0.9563
BC	2.8475	2.7413	0.1063
FBC	2.9325	2.6563	0.2763

Table 21.12 Results of regression analysis

Regression statistics				
Multiple R	0.9722			
R square	0.9451			
Adjusted R square	0.8970			
Standard error	0.5718			
Observations	16.0000			
ANOVA				
	df	SS	MS	F
Regression	7	45.0099	6.4300	19.6677
Residual	8	2.6155	0.3269	
Total	15	47.6254		
	Coefficients	Standard error	t stat	P-value
Intercept	2.7944	0.1429	19.5486	0.0000
Font (F)	1.0519	0.1429	7.3586	0.0001
Background colour (B)	0.0319	0.1429	0.2230	0.8291
Click button (C)	1.2006	0.1429	8.3992	0.0000
FB	0.1169	0.1429	0.8176	0.4372
FC	0.4781	0.1429	3.3448	0.0102
BC	0.0531	0.1429	0.3716	0.7198
FBC	0.1381	0.1429	0.9663	0.3622

to modern. The increase in the conversion rate is only 0.0637 when we change the background colour from white to blue.

The conclusion is that the company should use modern font with large click button with "Click Now to Join". The background colour does not matter.

More or less similar information could be obtained by carrying out a regression on the conversion rate with the columns of main effects and interaction effects in Table 21.9. The results of the regression analysis are presented in Table 21.12.

You can notice that the p-values of each effect in the regression analysis and the ANOVA in Table 21.10 match exactly. The intercept is nothing but the overall mean, and the regression coefficients corresponding to each factor or interaction are the shifts (positive or negative) from the overall mean.

The above example deals with two-factorial experimental design. The same model can be expanded to scenarios where there are more than two levels for the factors. The real problem will be the number of possible runs needed for the experiment. If there are six factors with two levels each, then the experiment will require 64 runs, not counting the replications. In such situations, one can use "fractional factorial designs". The discussion on fractional factorial designs is beyond the scope of this book. Interested students can read any textbook on experimental designs.

The interaction effects can be gauged better by drawing the interaction graphs. Figure 21.3 shows the interaction graphs for the three two-factor interactions (FB, FC and BC). When the two lines in the graph run parallel to each other, it indicates that there is no interaction between the two factors. A comparison between the two graphs, FB and FC, indicates that the conversion rate increases significantly when both font and click button are set at +1 level.

4.4 Orthogonality

An experimental design is said to be orthogonal if for any two design factors, each factor level combination has the same number of runs. The design specified above is an orthogonal design. Consider any two factors in the experiment and the effect on the response variable is studied for four possible combinations. There are exactly two runs (not counting the replications) for each combination. In addition, if you take any two columns (other than the response column) in Table 21.9 and multiply the corresponding elements and total them, the total is always zero. This also implies that the correlation between any two columns in Table 21.9 (other than the response column) is zero. This is a characteristic of the two-level factorial design. Because of this orthogonal nature of the design, all the effects can be estimated independently. Thus, the main effect of "font" does not depend on the main effect of "click button".

Experimental designs are extensively used in many social networks such as Facebook, Twitter and LinkedIn to make data-driven decisions. LinkedIn actually created a separate platform called XLNT (pronounced as Excellent) to carry out experiments on a routine basis. The platform can support more than 400 experiments per day with more than 1000 metrics. They have been using this platform for deploying experiments and analyzing them to facilitate product innovation. Their experiments range from visual changes in the home pages to personalizing the subject lines in emails[3].

---

[3]Ya Xu et. al., "From Infrastructure to Culture: A/B Testing Challenges in Large Scale Social Networks", KDD'15, 11–14 August 2015, Sydney, NSW, Australia.

Fig. 21.3 Interaction graphs

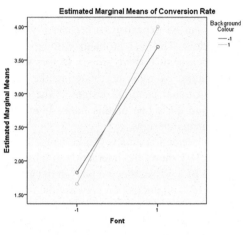
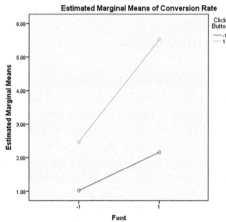
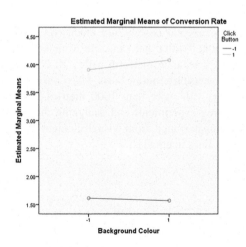

4.5 Data-Driven Search Engine Advertising

Today, data-driven marketing has become ubiquitous. Data-driven marketing strategies are based on insights obtained from data collected through customer touch points, customer interactions and customer demographics. This data is analyzed to understand customer behaviour and make predictions. These become inputs to better marketing and advertising efforts. The advertising efforts are aimed at enhancing customer experience through personalization. The approach helps companies to convey the right message at the right time to the right target. Companies can create highly customized campaigns delivering a personalized and focussed message to the consumer. With the amount of data at their disposal, companies can identify the groups of consumers that can be easily targeted, based on the consumer profiles and by identifying actionable strategies.

Consumers keep moving from one device to another (mobile to laptop to digital TV to tablet). They move from e-commerce sites to social websites. Data-driven advertising enables companies to follow the customer through various devices as well as different platforms. Companies can target the consumer with consistent messages across different channels. They can make relevant offers, specially designed to a particular consumer. For example, if a customer has purchased a mobile, he or she can be offered various accessories that can go with the specific model of mobile purchased.

Using techniques such as big data analytics, companies can stay up to date with the changing trends and preferences of consumers. One can even predict future changes in tastes and preferences. This data, coupled with appropriate marketing and advertising strategies, can actually help make the best changes.

Through data-driven advertising, marketers are able to reach consumers online irrespective of their physical locations. Individual consumers can be identified and selected to receive specific, highly focussed messages based on their behavioural patterns that will ultimately facilitate conversion. Implementing dynamic advertisement online which enables the consumer to interact can be a great source of obtaining first-hand information from the customer.

Today's consumers extensively rely on search engines to obtain information that they need before purchasing any product or service. By understanding and identifying what these consumers are searching for, marketers can create focussed messages. Once they identify what consumers are searching for, they can optimize the paid searches as well as create better content on their own websites that are in line with the frequent consumer searches and ultimately drive the traffic to their own websites.

Example: How Is Paid Search Done?

Let us take an example with Google AdWords to understand how paid search works. When a customer starts a search, Google starts the process by identifying the keywords that are relevant for the search string entered by the customer. These keywords need not necessarily be part of the search string entered by the customer. For example, if the search string entered is "social media marketing", keywords such as "digital marketing" and "search engine marketing" may be found to be relevant. Then, it starts looking at the available AdWords from the pool of advertisers and determines if an auction for the available advertisement is possible.

On the other side, advertisers identify various keywords that they want to bid on, along with the amount that they are willing bid. They also pair each of the keywords with specific advertisements. These advertisements can either be predesigned or put together by combining different objects based on various parameters received from the search engine (Google, in this case).

Then, Google enters the keywords it deems relevant into the auction. Google determines the ranks of different possible keywords and the associated advertisements based on two factors, namely, the maximum bid and the quality score. The maximum bid is what is specified by the advertiser for a particular keyword. The quality score is usually determined by the click-through rate (CTR), relevance of the ad and the landing page. The rank of the advertisement is determined by multiplying the quality score with the maximum bid amount. This rank is used to decide on the position of the advertisement in the search results page.

One of the interesting aspects of the above model is that the bid amount of an advertiser need not be the highest among the competing advertisers in order to get the highest rank. If a particular advertisement (paired with a specific keyword) has a high-quality score, even a much lower bid amount can result in a top rank. Google understands this and correspondingly calculates the cost per click (CPC).

Let us consider an example where a customer entered a search string "detergent for baby bottles". Based on this search string, Google determines that the keywords "dish washing", "dishwasher soap" and "baby soap" are relevant and identifies four possible advertisers. The details of the bid amounts and quality scores of each of the four advertisers are given in Table 21.13.

Advertiser 1 is ranked as number 1 even though bid amount is the lowest. This is because of the high-quality score. Similarly, the highest bidder, Advertiser 4, is ranked lowest because of the low-quality score. Based on the above, the advertisement of Advertiser 1 is displayed.

When the winning ad is displayed and the customer clicks on it, the cost per click of the winning ad is calculated using the rank of the advertiser who is ranked just next to the winning ad and the quality score of the winning ad. In the above example, the winning score of the next best rank is 18 (Advertiser 2), and the quality score of the winner is 10. The CPC is calculated as $18/10 + 0.01 = 1.81$. The presumption here is that even if Advertiser 1 bids an amount of \$1.81, its rank will be 18.1 (given

Table 21.13 Bid amounts and quality scores of advertisers

Advertiser	Maximum bid (\$)	Quality score	Rank = bid*quality score	Actual CPC
Advertiser 1	2	10	20	1.81
Advertiser 2	3	6	18	2.01
Advertiser 3	4	3	12	3.51
Advertiser 4	5	2.1	10.5	5

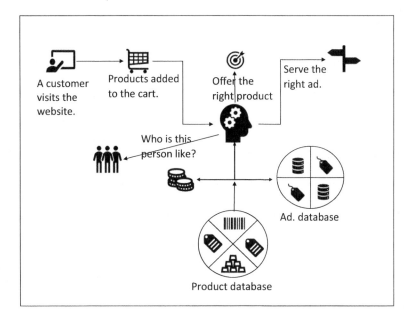

Fig. 21.4 Flowchart of recommendations

that the quality score is 10) which is marginally higher than the next best rank. The addition of one cent ($0.01) in the calculation formula is to ensure that the winning ad is marginally higher and not tied with the next best ranked ad.

Prediction Model for Ad and Product

Similar methods can be employed to display appropriate ads and/or make appropriate recommendations when customers access e-commerce sites for purchasing a product or service. The process is briefly presented in Fig. 21.4. Let us consider a customer who visits a particular e-commerce website looking for a stroller. She logs in and goes through different models and selects an infant baby stroller and adds the same to her shopping cart. Her past search history indicates that she had also been searching for various products for toddlers. In addition, her demographic profile is obtained from her log-in data as well as her previous purchasing decisions. This information can be fed into a product database which identifies a "duo stroller (for an infant and a toddler)" which is marginally higher in cost as compared to the one she had already selected. An advertisement corresponding to the duo stroller is picked up from an ad database and displayed to the customer along with a discount offer. The customer clicks on the ad, gets into the company website of the duo stroller, checks the details and goes through some reviews. She goes back to the e-commerce website and buys the duo stroller with the discount offered.

A similar process can be applied to make recommendations such as "those who purchased this item also bought these" to various customers based on simple market basket analysis. Other product recommendations can be made based on customer profiles, past browsing history, or similar items purchased by other customers.

4.6 Analytics of Digital Attribution

Traditionally, TV and print media advertising had been considered as the most effective marketing medium. But, in the recent years, digital ads have been outperforming all other media. With technology enabling advertisers to track consumers' digital footprints and purchasing activities in the digital world, advertisers are able to gain more insights into the behaviour of the consumers. Nevertheless, the consumers are simultaneously exposed to many other types of advertising in the process of making online purchases. Digital media constantly interacts with other media through multichannel exposures, and they complement each other in making the final sale. In such a scenario, attribution of the share of various components of digital media as well as other media is becoming more and more important.

The consumer today is exposed to multiple channels, each providing a different touch point. Imagine a consumer, while watching a TV show, comes across an ad for Samsung S8 smartphone and searches for "Samsung smartphones" on Google and comes across a pop-up ad for Samsung C9. He clicks on the ad and browses the resulting website for various comments and reviews on C9. He watches a couple of YouTube videos on C9 by clicking on a link given in one of the reviews. Then he goes to a couple of e-commerce sites (say, Amazon and Flipkart) and does a price comparison. A couple of days later, he receives email promotions on C9 from both the e-commerce sites. He comes across a large POS advertisement for C9 at a neighbourhood mobile store and stops and visits the store to physically see a C9 that is on display. A couple of days later, he receives a promotional mailer from Axis Bank which offers a 10% cashback on C9 at a particular e-commerce site (say Amazon) and finally buys it from the site.

The question here is: How much did each of these ads influence the consumer's decision and how should the impact of each of these channels be valued? These questions are important because the answers guide us in optimizing our advertisement spend on different channels.

Today, the explosion of data from various sources provides unmatched access to consumer behaviour. Every online action on each and every website is recorded, including the amount of time spent on a particular web page. Data from transactions from retail stores, credit card transactions, call centre logs, set-top boxes, etc. are available. All this information can be analyzed to not only understand consumer behaviour but also evaluate the contribution of each marketing channel.

An attribution model is the rule, or set of rules, that determines how credit for sales and conversions is assigned to touch points in conversion paths. Attribution is giving credit to different channels that the company employs to promote and broadcast the advertising message. Digital attribution pertains to attributing credit to various components for providing the marketing message in the online world using various forms of media and online platforms. It is true that the digital world offers phenomenal opportunities for measurability and accountability. Nevertheless, it is more challenging in the digital world to disentangle the impacts of various forms of advertising and those of different platforms employed. Some of the actions are not

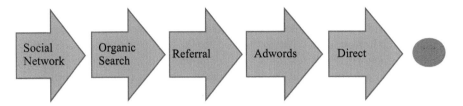

Fig. 21.5 Funnel showing different steps (channels) till conversion

controllable by the advertiser. For example, a search-based ad can be displayed only when the consumer initiates a search for a product or service on a particular search engine. It is always challenging to understand what triggers the consumer to initiate such search, leave alone influencing the consumer to initiate such an action!

There are different models for digital attribution, each having their own advantages and disadvantages. Before discussing the merits and demerits of each of these models, let us consider the following funnel which resulted in a conversion (Fig. 21.5).

Attribution Models

(i) Last Interaction/Last Click Attribution Model

This is one of the very early models of attribution and is used as a standard model in many web analytics because it is the simplest among all models. Here, the entire credit is given to the last interaction or click just before the conversion. In the above funnel, "Direct" is given the entire 100% credit. This model is used mainly because there is no ambiguity regarding the final action that resulted in the conversion. Since only the last touch point is considered, the cookie expiration period becomes irrelevant. This model is easy to implement and becomes appropriate when the funnels take a long time from initiation to conversion. The disadvantage is that there are many other interactions before the last interaction, and it does not recognize any of those. On the other hand, it is more appropriate when no information is available for visitor-centric analysis.

(ii) Last Non-direct Click Attribution Model

This model underplays the direct traffic. It gives credit to the campaign that was initiated just before conversion. These campaigns can be referring sites, emails, displays on search engines, social websites or even organic searches. "Direct channel" is often defined as when the visitor manually enters the URL in his browser. Most often, any visitor who does not have any referral tag is treated as "Direct". In other words, direct channel is where there is no referral tag and hence, at the same time, attributing the conversion to "Direct" can be misleading. This model avoids these issues involved with direct channel. There are many instances where a consumer actually types the URL because of the brand name and image. Since this model ignores the direct traffic, it is tantamount to undervaluing the brand recognition and brand value. All the efforts that generally go into creating such brand recognition are ignored by this model.

In the above funnel, "AdWords" is given the entire 100% credit.

(iii) Last AdWords Click Attribution Model

This model gives 100% credit to the most recent AdWords ad before converting. This model ignores the contribution of all other channels in the funnel. It looks at only the paid search. Since it ignores all other AdWords, it becomes difficult to optimize the AdWords campaign by identifying and valuing other AdWords that have contributed to the conversion. Needless to say, "AdWords" is given the entire 100% credit in the above funnel.

(iv) First Interaction/First Click Attribution Model

This is reverse of the last interaction/last click attribution model. The entire credit is given to the so-called first click. While there is no ambiguity about the last click, there is no certainty about the first click. Nevertheless, this model gives 100% credit to that particular market channel that brought the customer to the website and, consequently, results in overemphasizing one single part of the funnel, especially the top of the funnel channel. In our example, "Social Network" is given the complete credit. It may take a long time between the first touch point and the conversion. It is possible that the tracking cookie expires during this period, and consequently, the credit is given to a touch point which happens to be first in the cookie expiration window.

(v) Linear Attribution Model

In this model, each step involved, starting from the first interaction to the last interaction or click, is given equal weightage. Thus, this is the simplest of the multi-touch attribution models. The positive aspect of this model is that it considers all the channels involved in the funnel. While this recognizes the importance of each of the intermediate steps, it is very unlikely that the contribution of each step is the same. Each of the channels in the above funnel gets 20% credit. On the other hand, the major role played in the entire process of conversion could be that of social network influence, which actually started the funnel. Nevertheless, this model is better than attributing the entire credit to one single channel.

(vi) Time Decay Attribution Model

The model assigns different credits to different steps involved. It assigns the maximum credit to the last touch point before the conversion, and the touch point just before that gets less credit and so on. It assumes that the touch points that are closest to the conversion are the most influential. There are instances where the first touch point that initiated the entire process is not likely to get appropriate credit, especially if it is far away from conversion. One possible distribution of the credit for conversion in the above funnel could be Social Network 10%, Organic Search 15%, Referral 20%, AdWords 25% and Direct 30%.

(vii) Position-Based Attribution Model

This model, by default, attributes 40% of the credit to the first and last touch points each, and the remaining 20% is distributed among the remaining touch points. It is obviously not necessary to use the default values, and these weightages can be adjusted based on the objective (data-centric) or subjective analysis. By combining this with the time decay model, one can create a customized model. This model is best suited when the importance is on lead generation which got the visitor in and the touch point which clinched the conversion. At the same time, it does not ignore other touch points.

The "Social Network" and the "Direct" touch points get 40% each, whereas the remaining three will get 6.67% each, in the above funnel.

(viii) Custom or Algorithmic Attribution Model

Based on the merits and demerits of the models described earlier, the analyst can build a custom attribution model. These custom models are built based on data on customer behaviour obtained at different stages. It is being acknowledged that these "evidence-based" models are better and more realistic. Unfortunately, these models are not easy to build. The main principle is to estimate which of the touch points contribute to what extent based on the available customer data so that it represents a more accurate picture of the customer's journey from the initiation to conversion. This is called a custom or algorithmic attribution model. These models not only make use of customer data but also use statistical techniques which can lead to continuous optimization of the advertisement budget.

A simple approach for building an algorithmic attribution model starts with identifying the key metrics that are to be used to measure the relative effectiveness of each channel. It is important to associate an appropriate time period with this. This is to be followed with the cost per acquisition for each channel.

With the present technology, the data with respect to various touch points is obtained without much difficulty. This data can be used to build models to predict conversion or otherwise. Generally, the target variable is binary, whether the conversion is successful or not. The predictor variables are the various touch points. To start with, the metric for effectiveness of the model can be the prediction accuracy. Based on the contribution of each of the predictor variables, the attribution to various touch points can be calculated. Consider building a logistic regression model for predicting the success of conversion. Once we achieve the prediction accuracy levels for both the training and testing datasets, the coefficients of the logistic regression can be used to calculate the percentage attribution to various touch points or channels[4].

While it is true that any of the predictive models can be employed for this purpose, models such as logistic regression or discriminant analysis are more amicable since the coefficients corresponding to different predictor variables are available with these models. We can also use black box methods such as artificial neural networks or support vector machines. In such a case, we can assign the attribution values to different predictor variables based on the "sensitivity index" obtained after building the model.

---

[4]For more details, see Shao, Xuhui and Lexin Li, "Data-driven Multi-touch Attribution Models", KDD'11, 21–24 August 2011, San Diego, California, USA.

Other algorithmic models that can be used for calculating the attribution are as follows:

- Survival analysis
- Shapley value
- Markov models
- Bayesian models

We can optimize the budget allocation for each of the channels by using the attribution values (obtained from the predictive models) and the costs estimated earlier (CPA, CPM, etc.). Even a simple optimization technique such as budget-constrained maximization can yield significant results.

Let us consider an example of calculating cost of acquisition (CoA) under different attribution models. Let us consider a customer who purchased an "Amazon Home" through the following process:

1. Yajvin, the customer, first clicks on AdWords and visits the website.
2. Then he visits his Facebook page and clicks on the ad displayed on Facebook, visits the website again and checks out the functionality of the device.
3. Afterwards, he visits the website again through his Twitter account and looks at the technical details.
4. Then, he directly comes to the website and checks on various reviews.
5. Finally, he clicks on an offer with a discount that he received on email and purchases the device.

Let us assume that the advertisement expenditure is as follows: AdWords, $12; Facebook, $15; Twitter, $20; direct, $0; email, $8.

The cost of acquisition under each of the attribution models can be calculated based on the above information. Table 21.14 provides the details about the ad spend on each channel, the weightages for each channel under each model and the calculated CoA.

Similarly, data-based attribution models can be used to estimate the contribution of each channel. This is important to understand which channels are actually driving the sales. Based on this, advertisers can spend money more effectively and maximize ROI. The example below demonstrates the attribution across three different channels using the Shapley value approach.

Table 21.14 Ad spend, weightages and CoA for various channels

Channel	Ad spend	First touch (%)	Last touch (%)	Linear (%)	Position based (%)	Time decay (%)
AdWords	$12.00	100	0	20	40	10
Facebook	$15.00	0	0	20	7	15
Twitter	$20.00	0	0	20	7	20
Direct	$0.00	0	0	20	7	25
Email	$8.00	0	100	20	40	30
CoA		$12.00	$8.00	$11.00	$10.33	$9.85

Table 21.15 Watches sold based on channels

Channel	Number sold
AdWords + Facebook + email	256
AdWords + Facebook	192
AdWords + email	190
Facebook + email	128
AdWords	180
Facebook	120
Email	64

Table 21.16 Permutations of channels and their contributions

Permutation	Contribution of channel by order in the permutation			Contribution of specific channel		
	First channel	Second channel	Third channel	A (AdWords)	B (Facebook)	C (email)
ABC	180	12	64	180	12	64
ACB	180	10	66	180	66	10
BAC	120	72	64	72	120	64
BCA	120	8	128	128	120	8
CAB	64	126	66	126	66	64
CBA	64	64	128	128	64	64
Total contribution across all permutations				814	448	274
Average contribution across all permutations				135.67	74.67	45.67
Percentage contribution				52.99%	29.17%	17.84%

Let us consider an example where the company uses three channels for promoting its product, a smartwatch. These channels are AdWords (Channel A), Facebook (Channel B) and email (Channel C). Based on the data, the number of watches sold through each channel (and each possible combination of channels) is obtained and summarized in Table 21.15.

The company managed to sell 256 smartwatches when the customers exposed (used) all the three channels, while it could sell only 64 smartwatches when the customers used only email and nothing else. These numbers are obtained based on the analysis of purchase data through different channels. Considering that there are three channels (A, B and C) in this example, there are six possible permutations for combining these three channels. These permutations are A→B→C, A→C→B, B→A→C, B→C→A, C→A→B and C→B→A.

In the first permutation, Channel A contributes 180 (contribution of AdWords alone), Channel B contributes 12 (channels A and B together contribute 192, and hence the contribution of B is 192 − 180 = 12), and Channel C contributes 64 (all the three channels together contribute 256, while A and B together contribute 192, and hence the contribution of C is 256 − 192 = 64). Similarly, the contribution of each of the channels corresponding to each permutation is calculated and presented in Table 21.16.

Across all the six permutations, the contribution of AdWords is 180, 180, 72, 128, 126 and 128. The total of these values is 814, and the average is 135.67. Similarly, the averages for Facebook and email are 74.67 and 45.67, respectively. These values are converted into percentages which are presented in the table. These values are referred to as "Shapley values" (named after the Nobel Laureate Lloyd Shapley).

Based on the above analysis, the company should invest more in AdWords and least in email. As a matter of fact, the advertisement budget can be distributed across the three channels in the same ratio as the percentage contributions.

The above approach requires large amounts of data. The company needs to obtain data with respect to each and every combination (all possible subsets as well as individual channels) of the channels employed. If there are n channels, the data has to be obtained for $2^n - 1$ subsets. Implementing experimental designs could be a possible approach to obtain the required data. Once an optimization strategy for budget allocation across different channels is evolved and implemented, constant monitoring of the channels is necessary for further fine-tuning.

5 Strategies for Mobile Devices

With the popularity of smart mobiles in the recent days, more than half of search traffic started to emanate from mobiles. These devices are also the popular medium for interacting within social networks. In addition, Google's mobile ranking algorithm includes mobile-friendly and mobile usability factors as well as availability of mobile apps in its indexing. Consequently, those with mobile-friendly websites and/or mobile apps get much higher ranks and appear at the top of the search results. Consequently, it is becoming more and more important for businesses to evolve a mobile-oriented strategy in order to improve effectiveness of their marketing campaigns. It is becoming necessary to create mobile marketing strategies which improve customer experience while using mobiles at every stage of the customer purchase funnel.

Two important strategies that businesses need to adopt are to create mobile-friendly websites and mobile apps. Users in the initial stages of the purchase funnel are most likely to be using the website rather than downloading and installing the app. On the other hand, mobile apps allow for better interaction and facilitate more creativity in engaging the customer. In other words, it is necessary for businesses to create their own mobile-friendly websites as well as create specific apps.

A mobile website is a website which is designed and created for specifically viewing on a smartphone or a tablet. It needs to be optimized so that it responds or resizes itself to suit the display based on the type of device. We need to understand that customers use these devices at different stages in the purchase funnel. Businesses can accelerate the purchase process through sales alerts, display advertisements, providing QR codes, extending special discounts and issuing discount coupons. It is easy to integrate the mobile-based campaigns with different social media sites so that the customers can interact with others regarding the products and services within their social networks. The websites need to be optimized so that they load faster and they are easy to navigate, and click buttons need to be large enough and have short menu structures. The websites and apps should also ensure that there is minimum amount of typing required. It is also a good idea to allow for maps showing the location since many customers tend to use mobiles when they are on the go.

The website or app should allow users to connect to various social media platforms. This should also include a feature which will make it easy for customers to share the information with others in the network. The apps have an additional advantage. The app stays on the mobile screen, whether the customer is using the app or not. Every time the customer looks at the screen, they see the name of the app or name of the brand which acts as a constant reminder.

Geolocation is an important aspect of the mobile strategy. It is easy to integrate this into mobile apps. Businesses will be able to identify the location of the customer at any particular moment. Data can be collected on places that the customer visits on a regular basis (such as where the customer generally takes a walk). With this kind of information, the app can automatically provide various promotions or exclusive offers that are currently available at a store that is located nearest to the customer. Many of the mobile devices today come equipped with "near-field communication (NFC)". NFC can be useful in locating the customer within a particular store or facility, and the app can draw the user's attention to any items nearby or special discounts based on the past browsing/search/purchase behaviour of the customer through SMS or notifications. This is especially useful when the customer is physically close to the product and at a stage where he or she is ready to make a decision.

It is also important for the app to be able to operate offline. For example, the user could download the catalogue and browse through the offerings without relying on Wi-Fi or the mobile signal.

Ultimately, full benefit of a mobile strategy can be extracted only when the mobile channel is integrated with other channels. The customer should be able to seamlessly move from his or her mobile to any other channel and reach the brand or product or service.

Thus, the mobile strategy should be such that it provides enough customization to leverage the advantages of a mobile or tablet device while integrating with other channels so that the customer has a coherent experience across all channels.

6 The Future of Social Media Analytics

The past decade has seen a phenomenal growth of social media which has changed personal and professional lives of people all over the world. As networking through social media grew, businesses started leveraging social media platforms to reach out to customers directly to attract and retain them. Business organizations found innovative ways to listen to customers' voices through social media and better understand their needs. At the same time, development of technologies provided opportunities to analyze large amounts of unstructured data generated by social media so that the businesses can become more responsive. Today, social media analytics is being used to obtain deeper insights into customer preferences. It also opened new avenues to innovate and experiment. The same social media analytics can be used to communicate with customers at exactly the right moment with the right message to influence the decision-making process. In addition, it opened a new approach to reach the customer through multiple channels. Customer engagement through multiple channels has not only become the need of the hour but imperative to drive home the message and influence decision-making. The available technology is making it easier to understand the behaviour of the customer through his/her use of social media and also to understand the effectiveness of each channel in contributing to the final decision of the customer. Today, social media analytics is still evolving and yet to mature. New applications are emerging on a daily basis while throwing up additional challenges in customer engagement. At the same time, privacy concerns are becoming serious issues, both from ethical and legal positions. The next few years will be challenging to find a proper balance between privacy concerns and the needs of businesses to engage customers more effectively.

Electronic Supplementary Material

All the datasets, code, and other material referred in this section are available in www.allaboutanalytics.net.

- Data 21.1: adspends.csv
- Data 21.2: antiques_devices.csv
- Data 21.3: bid_qs_advt.csv
- Data 21.4: bid_qs_gmat_orgs.csv
- Data 21.5: factorial_experiment.csv
- Data 21.6: furnimart.csv
- Data 21.7: global_time.csv
- Data 21.8: indo_american.csv
- Data 21.9: membership_drive_isha.csv
- Data 21.10: modern_arts.csv
- Data 21.11: watches_sales.csv
- Data 21.12: webpage_visitors.csv
- Data 21.13: SMWA_Solutions.csv
- Code 21.1: factorial_experiment.R

Exercises

Ex. 21.1 Modern Arts (India) has initiated a special email campaign with three different subject lines. These are as follows:

(a) Subject line 1: "Welcome to Modern Arts"
(b) Subject line 2: "Special Invitation to Our Modern Art Exhibition"
(c) Subject line 3: "Are You a Fan of Modern Art?"

A total of 30,000 unique email IDs were selected and divided randomly into four groups of 7500 each. Each of these groups was further randomly divided into three groups of 2500 each. Emails with exactly the same font and content (body) were sent to each of these groups with one of the three subject lines, and the response was recorded. In other words, the experiment consisted of three treatments and four replications. The responses are presented below:

Subject line	Sends	Opens	Clicks
Replication 1			
Subject line 1	2500	278	96
Subject line 2	2500	405	136
Subject line 3	2500	222	62
Replication 2			
Subject line 1	2500	314	87
Subject line 2	2500	461	155
Subject line 3	2500	187	59
Replication 3			
Subject line 1	2500	261	82
Subject line 2	2500	421	147
Subject line 3	2500	216	58
Replication 4			
Subject line 1	2500	289	79
Subject line 2	2500	436	128
Subject line 3	2500	192	41

There are two response rates, namely, "open rate" and "click-through rate". Test to find out which subject line is the best with respect to each of the response rates.

Ex. 21.2 It was revealed that each replication was sent to a different mail account. All the 7500 emails of Replication 1 were actually addressed to Gmail accounts. Similarly, all mails of Replication 2 were sent to Outlook mail accounts. All mails of Replication 3 were sent to Yahoo mail accounts. All mails of Replication 4 were sent to AOL mail accounts. Given this information, Modern Arts (India) decided to consider this as a completely randomized block design in order to look at the effect of treatments and blocks.

Test if there is a significant block effect. Which is the best subject line? Carry out the analysis for both the response rates.

Ex. 21.3 FurniMart is a hub for furniture enthusiasts to both sell and buy specially designed furniture. FurniMart operates through its own website with an online catalogue. The traffic to the website comes mainly from three sources, those who type the URL and reach the website (direct), those who come through AdWords and those who respond to display advertisements from social networks. It had been their experience that many customers added products to their carts directly from the catalogue pages. The existing design of the website displays a graphic 🛒⁺ to facilitate adding products to the cart. Customers select a particular product and click on the graphic button to add the item to their shopping carts. It was felt that a bigger "call-to-action" (CtA) button is likely to lead to better conversion rates.

FurniMart decided to experiment with three different types of buttons. These are displayed below:

	Treatment 1	Treatment 2	Treatment 3
Design of the call-to-action button	Add to Cart	+ CART	🛒⁺

FurniMart created three different websites, namely, HTTP://FurniMart.COM/T1, HTTP://FurniMart.COM/T2 and HTTP://FurniMart.COM/T3, each displaying a different CtA button. The traffic coming from each of the three sources is randomly diverted to each of the three sites such that the total traffic from each source is equally distributed to the three websites. The conversion rates are summarized in the table below:

Source of traffic	Treatment 1	Treatment 2	Treatment 3
Direct	12.20	14.14	11.65
AdWords	13.59	14.42	11.93
Social networks	12.48	12.20	10.54

Analyze the above data to identify which CtA button is the best for conversion.

Ex. 21.4 Consider the above data (Question 3) as a completely randomized block design in order to look at the effect of treatments and blocks. Test if there is a significant block effect. Which is the best treatment?

Ex. 21.5 Global Time is a local dealer for the "Ultimate" brand of smartwatches. Whenever any potential customer searches for smartwatches, Ultimate bids along with other smartwatch sellers. When any customer who is located within the geographical area of Global Time clicks on Ultimate's ad, the visitor is taken to Global Time's website using the geolocation feature. Global Time is trying to revamp its website in order to improve its conversion rates. They have identified three different aspects (treatments) of the website that they want to tweak. These aspects are as follows:

(a) Currently, there is no video on the home page. The proposal is to add a 90 s video showing the features of "Ultimate" smartwatch.

(b) At present, the "Buy" button is at the right side of the web page, vertically centred. The proposal is to shift it to the bottom right so that there is more space for more visuals on the page.
(c) At present, the page displays testimonial in text form. The proposal is to include a small photo of the customer who had given the testimonial.

Global Time decided to carry out A/B testing on these three aspects. The details of the treatments and the corresponding conversion rates are given in the table below:

Treatments			Number of visitors	
Video	Location of "Buy" button	Testimonials	Reaching Global's website	Placing an order for Ultimate
Replication 1				
No video	Right centre	Text only	2590	115
Add video	Right centre	Text only	2458	205
No video	Bottom right	Text only	2406	165
Add video	Bottom right	Text only	2557	144
No video	Right centre	Text with photo	2409	101
Add video	Right centre	Text with photo	2458	103
No video	Bottom right	Text with photo	2561	161
Add video	Bottom right	Text with photo	2587	181
Replication 2				
No video	Right centre	Text only	2519	170
Add video	Right centre	Text only	2574	193
No video	Bottom right	Text only	2476	112
Add video	Bottom right	Text only	2546	125
No video	Right centre	Text with photo	2595	100
Add video	Right centre	Text with photo	2459	110
No video	Bottom right	Text with photo	2562	99
Add video	Bottom right	Text with photo	2445	190

What should "Global Time" do with respect to the three aspects? Are there any interaction effects between these three aspects?

Ex. 21.6 Akshita started a search on Google for organizations which provide GMAT training. A quick analysis of the relevant AdWords by Google found that there are five advertisements that are available for display on the search results page. The bid amounts as well as the quality scores are presented in the table below.

Calculate the ranks and identify the first two advertisers whose ads will be displayed to Akshita. Also, calculate the CPC for each of the advertisers.

Ex. 21.7 Indo-American Consultancy Services (IACS) specializes in placing Indian graduates with considerable work experience with clients in the USA. They advertise their services with display ads in Twitter, LinkedIn and AdWords. When the potential customers click on the display ad, they are taken to the company's website,

Advertiser	Maximum bid ($)	Quality score
Elite Management Training Institute	2	10
Gem Management Training	7	2
International Coaching Institute	4	3
Management Trainings Unlimited	5	2.1
National Management Center	3	6

and the customers are encouraged to register on the website and upload their CVs. Once the potential customer uploads the CV, it is considered as conversion.

Channel	Number of conversions
Twitter + LinkedIn + AdWords	2944
Twitter + LinkedIn	2185
Twitter + AdWords	2208
LinkedIn + AdWords	1472
Twitter	736
LinkedIn	1380
AdWords	2070

Based on the above data, carry out appropriate attribution to each of the three channels.

Caselet 21.1: Membership Drive at ISHA[5]

Vishnuprasad Nagadevara

Initiative for Special Healthcare Access (ISHA) is an organization started by Karthik to provide easy access to private corporate health services at reasonable cost to its members. Karthik, after completing his MBA with health services management specialization, joined a major corporate hospital chain headquartered in Bangalore. He has worked with the chain for 15 years and finally reached the level of chief administrative officer for the entire chain. During his tenure at the hospital chain, he had toyed with an idea of making healthcare accessible to as many persons as possible at affordable rates. In the process, he has set up an organization called "Initiative for Special Healthcare Access (ISHA)" which invites membership of individuals. ISHA made contractual arrangements with a number of private healthcare providers to offer various services to its members. He had put together contractual agreements with different corporate hospitals, diagnostic

---

[5]For classroom discussion only.

service centres and pharmacists. The members of ISHA benefit from steeply discounted services from these private healthcare providers. The service providers benefit from economies of scale through increased demand for their services from the members of ISHA.

As a part of the agreement, members of ISHA receive a certain number of free consultations from any doctor at major hospitals including dental consultations. In addition, they also get two free dental cleaning and scaling at certain dental clinics. The members are also eligible for two free "complete health check-ups" per year. The participating pharmacists give a minimum discount of 20% on medicines subject to minimum billed amount. The members also get discounts on various diagnostic tests including radiology tests.

These benefits are available to the members of ISHA. The membership is available on an individual as well as family basis. The family membership covers a maximum of four members. Additional members can be added into the family membership by paying an additional amount per person.

The economics of the entire model depends on acquiring a critical mass of members. ISHA decided to take advantage of the increasing web access to push its membership drive. They have initiated an email campaign to enrol members with very limited success. They have also realized that campaigns in print media cannot be targeted in a focussed manner leading to high campaign costs and low conversion rates. Karthik finally decided to resort to web-based campaign using display advertising with organic search as well as with AdWords. ISHA hired "Web Analytics Services India Limited (WASIL)" that has expertise in creating, testing and running web-based advertising campaigns.

WASIL is willing to work with ISHA on a result-based payment model. The fees that are payable to WASIL will depend on the success of the campaign in terms of acquiring the members. WASIL has put together a team of three members to create and run the campaign. Rajeev is heading the team with Subbu and Krithika as the other two members. The team decided to first design a set of objects that can be put together into a display advertisement based on keywords in search strings.

"Blue is always associated with health, like it is with BlueCross", said Subbu. "We should have blue colour in the border. That will make it noticeable and will definitely lead to better click-through rates". Subbu and other members of his team are discussing the changes that need to be made in the display advertisement for ISHA. Currently, ISHA does not use any colour in its advertisement. It is very simple with plain white background and bold font in black. It does not have a click button either. The potential visitor can click anywhere on the ad, and it will take the visitor to ISHA's home page.

Rajeev agreed with Subbu that there should be a border with a colour. Rajeev is passionate about green and feels that it gives a soothing feeling which could be easily associated with health services. Krithika suggested that they can use two different versions, one with blue and the other with green. Since all of them agreed that a colour in the border is absolute necessary, the A/B testing can be done with the two colours.

"We need to have a CTA (Call-to-Action) button. Since we are already experimenting with blue and green on the border, the CTA has to be red", Krithika said. "Here again, we have ample scope for experimentation. Should we try two different colours?" asked Rajeev. The rest of the team members did not agree with different colours. They felt that there are already two colours, one in the border and the other on the CTA. The text will have a different colour, at least black. They felt that having more than three colours in a display ad can make it look gaudy and can also be jarring to the visitor. Rajeev said, "We can't leave it as a button. We need to put some 'call-to-action text' on the button, like 'Click Here to Save'. It will draw more attention. There are enough studies to show that an associated text will always reinforce and yield better results". Subbu felt that they should add some more information into the click-to-action text. "Putting a number such as 'Save up to 70%' will get better response". Rajeev was not convinced. He felt that highlighting such a large percentage saving might possibly make people suspicious of the benefits. Many people think that such large discounts are not possible in healthcare, even though the labs do offer 70% discount on specified lab tests. After a prolonged discussion, the team decided to try out both the versions as another treatment in A/B testing.

The present design of ISHA's ad is that the visitors, when they click on the ad, are taken to the home page. It is expected that the visitor will first look at the home page and navigate from there to other relevant pages. The team felt that it is not enough to make the visitor click on the ad and reach the website. The real conversion is when the visitor becomes a paid member, at least on trial basis. Any member can withdraw his or her membership within 15 days from the registration and get full refund. The team felt that the landing page will have a major impact on conversion. The team members also believed that the landing page should be dynamic. It should be based on the search string or the AdWords leading to the display ad. If the visitors are looking for lab tests, the landing page should accordingly be the one with lab tests, with a listing of laboratories located in a particular town and the corresponding discounts. On the other hand, if the visitor was searching for physician consultation, the landing page should correspond to physicians or respective clinics. Finally, the team agreed that the landing page will also be treated as a treatment with home page being the control.

The team decided to use one more treatment in their experimentation. It was understood that ISHA is a new player with a relatively new concept. The organization is not yet an established one. The team members as well as Karthik felt that putting ISHA's logo on the display ad will increase the visibility. The top line of the ad will show either ISHA without logo or ISHA with logo on the right side. The general layout of the ad is shown in Fig. 21.6.

The team concluded that they should go ahead with the four treatments. The team summarized the final proposal as shown in the table below.

WASIL and Karthik approved the team's proposal, and WASIL ran the campaign with the above four treatments over a period of 1 month. Data was collected in terms of number of potential visitors exposed to each type of advertisement (sample

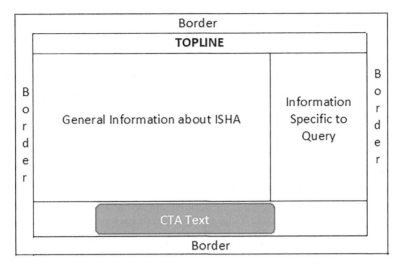

Fig. 21.6 General layout of the ad

Treatment details		
Treatment	Level 1 (−1)	Level 2 (+1)
Border colour	Blue	Green
Top line	Without logo	With logo
CTA	"Click Here to Save"	"Save up to 70%"
Landing page	Home page	Relevant page

size), number of visitors who clicked on the ad (clicks) and, finally, the number of conversions (members). The dataset "membership_drive_isha.csv" is available on the book's website.

Questions

(a) Which of the four treatments is more effective? What is the right level for each of the treatments?
(b) Are there any interaction effects between the treatments?
(c) What would be the final design for the ad in order to maximize the conversions?

Caselet 21.2: Antiques in Modernity[6]

Vishnuprasad Nagadevara

The Meeting

"I think it is Wanamaker who said '*Half the money I spend on advertising is wasted; the trouble is I don't know which half*'. I can't accept that today. Our advertising cost is almost 12% of our revenue, and I need to know where it is going", said Yajvin. "I mean we know exactly where we are spending the money, but I need to know what we are getting back. What is the effect of each of the channels that we are investing in?"

Yajvin is briefing Swetha on her new assignment. Yajvin is the CEO of the online company "Antiques in Modernity (AiM)". Antiques in Modernity specializes in modern electronic devices which have the look of antiques. Swetha heads the analytics consulting company "NS Analytics Unlimited", which provides consultancy services to various companies. In addition to providing consultancy services, NS Analytics also provides training to the client companies so that they can become self-sufficient as much as possible in terms of analytics. Their motto is "We make ourselves redundant by building expertise". NS Analytics has been hired by Antiques in Modernity to analyze their advertising spend and also advise them on how best to improve the ROI on their advertising investment.

"Our products are very special. Look at this turntable. It looks like an antique, and you can actually play a Beatles' gramophone record on this. But it can be used as a Bluetooth speaker; it can connect to you home assistant such as Amazon Echo or Google Home; it can even connect to your mobile. You can stream music from this turntable to any other device even in your backyard!" said Yajvin (Fig. 21.7).

"Similar is the case with our antique wall telephone. It can be used as a wired telephone or as a cordless phone. We are in the process of redesigning it so that you

Fig. 21.7 AiM turntable

---

[6]For classroom discussion only.

can carry the receiver outside your house and use it as a telephone. But let us come back to our problem. As I said, we invest a lot of money in advertising in different channels. We need to find out the effect of each of these channels. I do understand that many of these channels do complement each other in today's markets. Can we somehow isolate the effects of each, so that our ad spend can be optimized?" asked Yajvin.

Swetha responded saying that there are many models that can be used in order to address the problem, but such models require large amounts of reliable data. She also said that each of these models can give different results, and one needs to understand the assumptions involved in each of these models so that the one which is most applicable to a particular scenario can be picked. Yajvin put her in touch with his chief information officer, Skanda, so that Swetha can get a feel for the type of data that is available with the company and also explain her data requirements.

Skanda explained the data available with them. "We mainly depend on online advertisement. We do invest a small amount in the print media, but most of our advertising is on the social networking websites, AdWords and the like. We also get good amount of direct traffic into our website. Since all our sales are through online only, it makes sense for us to work this way", said Skanda. "We do have a system of tracking our potential customers through different channels. We try to collect as much data, reliably, as possible."

Antiques in Modernity

Antiques in Modernity was a start-up, started in Sunnyvale, California, by Yajvin and Skanda 3 years ago. Yajvin graduated from one of the prestigious technical institutes in India with a major in electronics and went on to do his MBA from a well-known management school in Europe. After his MBA, he worked as a product development manager in a European company. After 5 years with the company, he moved to San Jose as a marketing manager for a software product company. While he moved around from one company to another in the next 12 years, he remained in the marketing function.

Yajvin's hobby was to collect antiques, especially very old electronic products which were in full working condition. He used to collect old electronic products from whatever source possible. If they were not in a working condition, he would work on them and get them into working condition. He also used to sell some of them and also gift some of them away. In the process, he got bitten by the entrepreneur bug and decided to float a company with an aim to manufacture specially designed electronic products with latest technology, but look like antiques. The company designs the products, sources the components from China and Far East and assembles them in the USA. Skanda was his classmate from India and joined him in the venture. They decided that since these are niche products, it will be safer to market them through their own website. They felt that such a strategy will give them complete control on the way the products can be advertised and marketed.

The fact that both of them are comfortable with web-based marketing had a major role to play in making the decision. They had also decided to use as much of online advertising as possible.

AiM uses display ads on social networks, especially LinkedIn, Twitter and Facebook. They also use Google AdWords in order to display their ads based on the search strings used by potential customers. They also keep track of customers who reach their site through organic search. Since their products are sold only through online from their own website, the final conversion is when the customer places the order. They use various methods to trace the channels from which customers reach their website.

The Data

During their meeting, Skanda promised Swetha that he can provide details of each and every funnel starting from the first visit to their website, as well as the referrer by a customer (or potential customer). Swetha felt that there is no reason to look at the funnels that are incomplete which may or may not lead to conversion at some later date. She requested for data only on funnels which resulted in final conversion. There are also many repeat customers who type the URL directly and reach the website and make purchases. Similarly, there are some who purchase items on their very first visit to the website. Skanda told her that he can provide details of each funnel corresponding to each and every conversion. He felt that such detailed data could be useful because AiM sells different products and the profit margins are different for different products. On the other hand, Swetha felt that such detail was not necessary because the advertisements are not specific to any particular product. Even the display ads which are put together on the fly, by AiM based on AdWords or search strings, are not product specific. The main theme of these ads is that their products are latest in technology, but packaged as antiques. They are not really antiques either. Hence, she suggested that Skanda summarize the data "funnel-wise". She also suggested that all the social network channels can be clubbed into one channel for the purpose of initial analysis. "We can drill down into different social networks separately at a later date. As a matter of fact, you will be able to do it yourself after we train your people", she said.

Finally, they have agreed to concentrate on four channels: social networks (Channel A), AdWords (Channel B), organic search (Channel C) and direct (Channel D). It was also decided to maintain the actual order of channels within each funnel. Each funnel is to be read as the sequence of the channels. For example, ABCD implies $A \rightarrow B \rightarrow C \rightarrow D$. Swetha explained that the order becomes important for estimating the contribution of each channel under different models. Then there was a question of the final metric. Should the final metric for conversion be revenue, profit margin or just the number of items sold? AiM is currently going through a major costing exercise, especially in terms of assigning the fixed/non-variable costs to different products. It was felt that the option of profit margin is not appropriate until the costing exercise is completed. Skanda and Yajvin felt that the initial exercise can be made based on the sales quantity (number of items sold) and the method can easily be extended to revenue at a later date. Swetha assured them that they will just have to change the values in a simple spreadsheet and everything else will get recalculated automatically!

Swetha received the summarized data as required by her 2 days after her meeting with Yajvin and Skanda. The data "antiques_devices.csv" is available on the book's website.

Further Readings

Abhishek, V., Despotakis, S., & Ravi, R. (2017). *Multi-channel attribution: The blind spot of online advertising*. Retrieved March 16, 2018, from https://papers.ssrn.com/sol3/papers.cfm?abstract_id=2959778.

Fisher, T. (2018). ROI in social media: A look at the arguments. *Database Marketing & Customer Strategy Management, 16*(3), 189–195. Tracy L. Tuten, Michael R. Solomon, Social Media Marketing, Sage Publishing.

Ganis, M., & Kohirkar, A. (2016). *Social media analytics*. New York, NY: IBM Press.

Gardner, J., & Lehnert, K. (2016). What's new about new media? How multi-channel networks work with content creators. *Business Horizons, 59*, 293–302.

Hawn, C. (2017). Take two aspirin and tweet me in the morning: How twitter, facebook, and other social media are reshaping health care. *Health Affairs, 28*(2), 361.

Kannan, P. K., Reinartz, W., & Verhoef, P. C. (2016). The path to purchase and attribution modeling: Introduction to special section. *International Journal of Research in Marketing, 33*, 449–456.

Ledolter, J., & Swersey, A. J. (2007). *Testing 1 - 2 - 3: Experimental design with applications in marketing and service operations*. Palo Alto, CA: Stanford University Press.

Oh, C., Roumani, Y., Nwankpa, J. K., & Hu, H.-F. (2017). Beyond likes and tweets: Consumer engagement behavior and movie box office in social media. *Information & Management, 54*(1), 25–37.

WilliamRibarsky, D. X. W., & Dou, W. (February 2014). Social media analytics for competitive advantage. *Computers & Graphics, 38*, 328–331.

Zafarani, R., Abbasi, M. A., & Liu, H. (2014). *Social media mining*. Cambridge: Cambridge University Press.

Chapter 22
Healthcare Analytics

Maqbool (Mac) Dada and Chester Chambers

1 Introduction to Healthcare Analytics: Simulation Models of Clinics in Academic Medical Centers

Ancient understanding of biology, physiology, and medicine was built upon observations of how the body reacted to external stimuli. This indirect approach of documenting and studying the body's reactions was available long before the body's internal mechanisms were understood. While medical advances since that time have been truly astounding, nothing has changed the central fact that the study of medicine and the related study of healthcare must begin with careful observation, followed by the collection, consideration, and analysis of the data drawn from those observations. This age-old approach remains the key to current scientific method and practice.

1.1 Overview of Healthcare Analytics

The development of technologies related to information capture and analysis over the past 20 years has begun to revolutionize the use of data in all branches of medicine. Along with better and easier methods for the collection, storage,

---

Electronic supplementary material The online version of this chapter (https://doi.org/10.1007/978-3-319-68837-4_22) contains supplementary material, which is available to authorized users.

M. (Mac) Dada (✉) · C. Chambers
Carey Business School, Johns Hopkins University, Baltimore, MD, USA
e-mail: mdada1@jhu.edu

© Springer Nature Switzerland AG 2019
B. Pochiraju, S. Seshadri (eds.), *Essentials of Business Analytics*, International Series in Operations Research & Management Science 264,
https://doi.org/10.1007/978-3-319-68837-4_22

and interpretation of data, the new technologies have spawned a number of new applications.[1] For example, data analysis allows earlier detection of epidemics,[2] identification of molecules (which will play an unprecedented role in the fight against cancer[3]), and new methods to evaluate the efficacy of vaccination programs.[4]

While the capacity of these tools to increase efficiency and effectiveness seems limitless, their applications must account for their limitations as well as their power. Using modern tools of analytics to improve medicine and care delivery requires a sound, comprehensive understanding of the tools' strengths and their constraints. To highlight the power and issues related to the use of these tools, the authors of this book describe several applications, including telemedicine, modeling the physiology of the human body, healthcare operations, epidemiology, and analyzing patterns to help insurance providers.

One problem area that big data techniques are expected to revolutionize in the near future involves the geographical separation between the patient and the caregiver. Historically, diagnosing illness has required medical professionals to assess the condition of their patients face-to-face. Understanding various aspects about the body that help doctors diagnose and prescribe a treatment often requires the transmission of information that is subtle and variable. Hearing the rhythm of a heart, assessing the degradation in a patient's sense of balance, or seeing nuances in a change in the appearance of a wound are thought to require direct human contact. Whether enough of the pertinent data can be transmitted in other ways is a key question that many researchers are working to answer.

The situation is rapidly changing due to the practice of telemedicine. Market research firm Mordor Intelligence expects telemedicine, already a burgeoning market, to grow to 66.6 billion USD by 2021, growing at a compound annual growth rate of 18.8% between 2017 and 2022.[5] New wearable technologies can assist caregivers by collecting data over spans of time much greater than an office visit or hospital stay in a wide variety of settings. Algorithms can use this data to suggest alternate courses of action while ensuring that new or unreported symptoms are not missed. Wearable technologies such as a Fitbit or Apple Watch are able to continuously track various health-related factors like heart rate, body temperature, and blood pressure with ease. This information can be transmitted to medical

---

[1]The article in *Forbes* of October 2016 provided many of the data in this introduction—https://www.forbes.com/sites/mikemontgomery/2016/10/26/the-future-of-health-care-is-in-data-analytics/#61208ab33ee2 (accessed on Aug 19, 2017).

[2]https://malariajournal.biomedcentral.com/articles/10.1186/s12936-017-1728-9 (accessed on Aug 20, 2017).

[3]http://cancerres.aacrjournals.org/content/75/15_Supplement/3688.short (accessed on Aug 20, 2017).

[4]https://www.ncbi.nlm.nih.gov/pmc/articles/PMC4287086 (accessed on Aug 20, 2017).

[5]https://www.mordorintelligence.com/industry-reports/global-telemedicine-market-industry (accessed on Aug 23, 2017).

personnel in real time. For example, military institutions use chest-mounted sensors to determine the points at which soldiers reach fatigue and can suggest tactical options based on this information.

While some wearable technologies are on the cutting edge and are thus often expensive, telemedicine can use cheap, sturdy hardware to make diagnoses easier. Electronic kits such as the Swasthya Slate,[6] which is used in community clinics in New Delhi, can be used by doctors to conduct blood sugar tests and electrocardiograms and monitor a patient's temperature and blood pressure.[7] In Kigali, Rwanda, digital health company Babylon is testing a service that will allow patients to video-call doctors rather than wait in lines at hospitals. In economies which suffer from large burdens on healthcare and a scarcity of trained professionals, interventions such as these can help save time, money, and lives.

The proper application of such technologies can prevent the wearer's lack of expertise from clouding data collection or transmission. Normally, doctors learn about physical symptoms along with the patient's experiences via face-to-face interactions. This adds a gap between the experience and its discussion, as well as the subjective interpretation of the patient. Its effectiveness also depends on the patient's ability to relay the information accurately. Direct recording of data can bridge these gaps ensuring that the doctor receives objective information while also understanding the patient's specific circumstances. This information transmission can be combined with additional elements including web-cameras and online voice calling software. This allows doctors and patients to remain in contact regarding diagnoses without the need for the patient to physically travel to a hospital or office. Thus, new metrics become possible, accuracy is increased, and time is saved, while costs are reduced. Additionally, such solutions may help provide proactive care in case of medical emergency.

The benefits of data analytics are not just limited to diagnosis. Data analytics also facilitates the leverage of technology to ensure that patients receive diagnoses in a timely fashion and schedule treatment and follow-up interactions as needed. Analytics already plays a key role in scheduling appointments, acquiring medicines, and ensuring that patients do not forget to take their medications.

The advantage of using big data techniques is not limited to the *transmission* of data used for diagnosis. Data analysis is key to understanding the fundamental mechanisms of the body's functions. Even in cases where the physical function of the body is well understood, big data can help researchers analyze the myriad ways in which each individual reacts to stimuli and treatment. This can lead to more customized treatment and a decrease in side effects. By analyzing specific interactions between drugs and the body, data analytics can help fine-tune dosages, reduce side effects, and adjust prescriptions on a case-by-case basis. Geno Germano,

---

[6]https://www.thebetterindia.com/49931/swasthya-slate-kanav-kahol-delhi-diagnostic-tests/ (accessed on Jul 16, 2018).

[7]https://www.economist.com/international/2017/08/24/in-poor-countries-it-is-easier-than-ever-to-see-a-medic (accessed on Sep 15, 2018).

former group president of Pfizer's Global Innovative Pharma Business, said in 2015 that doctors might (in the near future) use data about patients' DNA in order to come up with personalized, specific treatment and health advice that could save time and ensure better outcomes.[8]

The ability to analyze multiple streams of related data in real time can be applied to create simulations of the human body. Using such simulations allows researchers to conduct experiments and gather information virtually, painlessly, and at low cost. In constructing artificial models of all or some parts of the body, big data techniques can harness computational power to analyze different treatments. Initiatives such as the *Virtual Physiological Human Institute*[9] aim to bring together diverse modeling techniques and approaches in order to gain a better, more holistic understanding of the body and in turn drive analysis and innovation.

Analytics is being used in the study and improvement of *healthcare operations* to enhance patient welfare, increase access to care, and eliminate wastes. For example, by analyzing patient wait times and behavior, data scientists can suggest policies that reduce the load on doctors, free up valuable resources, and ensure more patients get the care they need when and where they need it. Simulation techniques that predict patient traffic can help emergency rooms prepare for increased number of visitations,[10] while systems that track when and where patients are admitted make it easier for nurses and administrators to allocate beds to new patients.[11] Modern technologies can also help in the provision of follow-up care, that is, after the patient has left the hospital. Software and hardware that track important physical symptoms can notice deviation patterns and alert patients and caregivers. By matching such patterns to patient histories, they can suggest solutions and identify complications. By reminding patients regarding follow-up appointments, they can reduce rehospitalization.

Analytics is also needed to guide the use of information technologies related to updating patient records and coordinating care among providers across time or locations. Technologies like Dictaphones and digital diaries are aimed at collecting and preserving patient data in convenient ways. Careful analysis of this data is key when working to use these technologies to reduce redundant efforts and eliminate misunderstandings when care is handed from one provider to another.

There are many applications of analytics related to the *detection of health hazards and the spread of disease*: Big data methods help insurers isolate trends in illness and behavior, enabling them to better match risk premiums to an individual buyer's risk

---

[8] https://www.forbes.com/sites/matthewherper/2015/02/17/how-pfizer-is-using-big-data-to-power-patient-care/#7881a444ceb4 (accessed on Aug 21, 2017).

[9] http://www.vph-institute.org (accessed on Sep 1, 2017).

[10] http://pubsonline.informs.org/doi/10.1287/msom.2015.0573 (accessed on Aug 21, 2017).

[11] http://www.bbc.com/news/business-25059166 (accessed on Aug 21, 2017).

profile. For instance, a US-based health insurance provider[12] offers Nest Protect,[13] a smoke alarm and carbon monoxide monitor, to its customers and also provides a discount on insurance premiums if they install these devices. Insurers use the data generated from these devices to determine the premium and also in predicting claims.

Information provider *LexisNexis* tracks socioeconomic variables[14] in order to predict how and when populations will fall sick. Ogi Asparouhov, the chief data scientist at LexisNexis, suggests that socioeconomic lifestyle, consumer employment, and social media data can add much value to the healthcare industry.

The use of *Google Trends* [c] data, that is, Internet search history, in healthcare research increased sevenfold between 2009 and 2013. This research involves a wide variety of study designs including causal analysis, new descriptive statistics, and methods of surveillance (Nuti et al. 2014). *Google Brain*,[15] a research project by Google, is using machine learning techniques to predict health outcomes from a patient's medical data.[16] The tracking of weather patterns and their connection to epidemics of flu and cold is well documented. The World Health Organization's program *Atlas of Health and Climate*[17] is such an example regarding the collaboration between metrological and public health communities.

By gathering diverse kinds of data and using powerful analytical tools, insurers can better predict fraud, determine appropriate courses of action, and regulate payment procedures. A comprehensive case study (Ideal Insurance) is included in Chap. 25 that describes how analytics can be used to create rules for classifying claims into those that can be settled immediately, those that need further discussion, and those that need to be investigated by an external agency.

These techniques, however, are not without their challenges. The heterogeneity of data in healthcare and privacy concerns have historically been significant stumbling blocks in the industry. Different doctors and nurses may record identical data in different ways, making analysis more difficult. Extracting data from sensors such as X-ray and ultrasound scans and MRI machines remains a continuing technical challenge, because the quality of these sensors can vary wildly.[18]

Big data techniques in healthcare also often rely on real-time data, which places pressure on information technology systems to deliver data quickly and reliably.

---

[12]https://www.ft.com/content/3273a7d4-00d2-11e6-99cb-83242733f755 (accessed on Sep 2, 2017).

[13]https://nest.com/smoke-co-alarm/overview (accessed on Sep 2, 2017).

[14]http://cdn2.hubspot.net/hubfs/593973/0116_Predictive_Modeling_News.pdf?t=1453831169463 (accessed on Sep 3, 2017).

[15]https://research.google.com/teams/brain (accessed on Sep 3, 2017).

[16]https://www.cnbc.com/2017/05/17/google-brain-medical-records-prediction-illness.html (accessed on Sep 3, 2017).

[17]http://www.who.int/globalchange/publications/atlas/en (accessed on Sep 2, 2017).

[18]https://pdfs.semanticscholar.org/61c8/fe7effa85345ae2f526039a68db7550db468.pdf (accessed on Aug 21, 2017).

Despite these challenges, big data techniques are expected to be a key driver of technological change and innovation in this sector in the decades to come. The rest of this chapter will discuss in detail the use of data and simulation techniques in academic medical centers (AMCs) to improve patient flow.

2 Methods of Healthcare Analytics: Using Analytics to Improve Patient Flow in Outpatient Clinics

Demands for increased capacity and reduced costs in outpatient settings create the need for a coherent strategy on how to collect, analyze, and use data to facilitate process improvements. Specifically, this note focuses on system performance related to patient flows in outpatient clinics in academic medical centers that schedule patients by appointments. We describe ways to map these visits as we map processes, collect data to formally describe the systems, create discrete event simulations (DESs) of these systems, use the simulations as a virtual lab to explore possible system improvements, and identify proposals as candidates for implementation. We close with a discussion of several projects in which we have used our approach to understand and improve these complex systems.

2.1 Introduction

As of 2016, the Affordable Care Act (ACA) extended access to health insurance coverage to roughly 30 million previously uninsured Americans, and that coverage expansion is linked to between 15 and 26 million additional primary care visits annually (Glied and Ma 2015; Beronio et al. 2014). In addition, the number of people 65 and older in the USA is expected to grow from 43.1 million in 2012 to 83.7 million by 2050 (Ortman et al. 2014). This jump in the number of insured Americans coupled with the anticipated growth in the size of the population above the age of 65 will correlate with rising demand for healthcare services.

At the same time, Medicare and other payers are moving away from the older "fee-for-service" model toward "bundled payment" schemes (Cutler and Ghosh 2012). Under these arrangements, providers are paid a lump sum to treat a patient or population of patients. This fixes patient-related revenue and means that these payments can only be applied to fixed costs if variable costs are less than the payment. We expect the continued emergence of bundled payment schemes to accelerate the gradual move away from inpatient treatment to the delivery of care through outpatient settings that has been taking place for over 20 years. Consequently, a disproportionate share of the growth in demand will be processed through outpatient clinics, as opposed to hospital beds. This evolution is also seen as one of the key strategies needed to help get healthcare cost in the USA closer to the costs experienced in other developed countries (Lorenzoni et al. 2014).

An additional complicating factor is that healthcare delivery in the USA is often interspersed with teaching and training of the next generation of care providers. In 2007, roughly 40 million outpatient visits were made to teaching hospitals known as academic medical centers (AMCs) (Hing et al. 2010). Inclusion of the teaching component within the care process dramatically increases the complexity of each patient visit. The classic model of an outpatient visit where a nurse leads the patient to an examination room and the patient is seen by the physician and then leaves the clinic is not a sufficient description of the process in the AMC. Adding a medical resident or fellow (trainee) into the process introduces steps for interactions between the trainee and the patient as well as interactions between the trainee and the attending physician (attending). These added steps increase flow times, the number and levels of resources deployed, and system congestion (Boex et al. 2000; Franzini and Berry 1999; Hosek and Palmer 1983; Hwang et al. 2010). The delays added are easy to understand when one considers the fact that the trainee typically takes longer than the attending to complete the same task and many teaching settings demand that both the trainee and the attending spend time with each patient on the clinic schedule (Williams et al. 2007; Taylor et al. 1999; Sloan et al. 1983).

The addition of the teaching mission is not simply adding steps to a well-managed process. The added complexity is akin to changing from a single-server queueing system to a hybrid system (Williams et al. 2012, 2015). The trainee may function as a parallel (but slower) server, or the trainee and attending may function as serial servers such that a one-step activity becomes a two-step process, or decisions on how the trainee is intertwined in the process may be made dynamically, meaning that the trainee's role may change depending on system status.

In short, we are asking our current healthcare system to improve access to care to a rapidly growing and aging population as demand is shifted from inpatient to outpatient services in teaching hospitals using delivery models that are not well understood. While the extent to which this is even possible is debatable (Moses et al. 2005), it is quite clear that efforts to make this workable require thoughtful data analysis and extremely high-quality operations management (Sainfort et al. 2005).

The primary objective of this chapter is to lay out a strategy toward gaining an understanding of these complex systems, identifying means to improve their performance, and predicting how proposed changes will affect system behavior. We present this in the form of a six-step process and provide some details regarding each step. We close with a discussion of several projects in which our process has been applied.

2.2 *A Representative Clinic*

To make the remainder of our discussion more concrete, let us introduce a representative unit of analysis. Data associated with this unit will be taken from a composite of clinics that we have studied, but is not meant to be a complete representation of

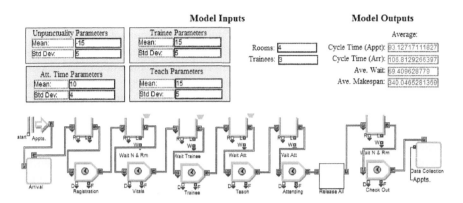

Fig. 22.1 Copied from DES of representative clinic

any particular unit. Consider a patient with an appointment to see the attending at a clinic within an AMC. We will work with a discrete event simulation (DES) of this process. DES is the approach of creating a mathematical model of the flows and activities present in a system and using this model to perform virtual experiments seeking to find ways to improve measurable performance (Benneyan 1997; Clymer 2009; Hamrock et al. 2013; Jun et al. 1999). A screenshot from such a DES is presented in Fig. 22.1 and will double as a simplified process map. By simplified, we mean that several of the blocks shown in the figure actually envelop multiple blocks that handle details of the model. Versions of this and similar models along with exercises focused on their analysis and use are **linked to this chapter**.

Note that the figure also contains a sample of model inputs and outputs from the simulation itself. We will discuss several of these metrics shortly.

In this depiction, a block creates work units (patients) according to an appointment schedule. The block labeled "Arrival" combines these appointment times with a random variable reflecting patient unpunctuality to get actual arrival times. Once created, the patients move to Step 1. Just above Step 1, we show a block serving as a queue just in case the resources at Step 1 are busy. In Step 1, the patient interacts with staff at the front desk. We will label this step "Registration" with the understanding that it may include data collection and perhaps some patient education. In Step 2, a nurse leads the patient into an examination room, collects data on vital signs, and asks a few questions about the patient's condition. We will label this step "Vitals." In Step 3, a trainee reviews the patient record and enters the examination room to interact with the patient. We label this step "Trainee." In Step 4, the trainee leaves the exam room and interacts with the attending. We label this step "Teach." During this time, the trainee may present case information to the attending, and the pair discusses next steps, possible issues, and the need for additional information. In Step 5, the trainee and attending both enter the exam room and interact with the patient. We label this step "Attending." Following this step, the trainee, attending, and room are "released," meaning that they are free to be assigned to the next patient. Finally, the patient returns to the front desk for "Check Out." This step may include collection of payment and making an appointment for a future visit.

In order to manage this system, we need an understanding of its behavior. This behavior will be reflected in quantifiable metrics such as cycle times wait times, and how long it will take to complete the appointment schedule (makespan). Note that cycle times may be calculated based on appointment times or patient arrival times. Both of these values are included among the model outputs shown here. While this model is fairly simple, some important questions may be addressed with its use. For example, we may make different assumptions regarding the attending's processing time and note how this changes the selected output values. This is done by altering the parameters labeled "Att. Time Parameters" among the model inputs. For this illustration, we assume that these times are drawn from a log-normal distribution and the user is free to change the mean and standard deviation of that distribution. However, one benefit of simulation is that we may use a different distribution or sample directly from collected activity time data. We will discuss these issues later. This model can also be used as part of a more holistic approach to address more subtle questions, including how the added educational mission affects output metrics and what is the best appointment schedule for this system. In the next section, we lay out a more complete approach to handling more complex questions such as these.

2.3 How to Fix Healthcare Processes

Much of the early development of research-oriented universities in the USA was driven by the need for research related to healthcare (Chesney 1943). Consequently, when working with physicians and other healthcare professionals in the AMC, a convenient starting point for the discussion is already in place. Research in most parts of healthcare addresses questions using randomized trials or pilot implementations. These typically center on formal experiments which are carefully designed and conducted in clinical or laboratory settings. This experiment-based approach to research has proven to be highly effective and is assumed by many to be the best way to produce evidence-based results on medical questions relating to issues including the efficacy of a new drug or the efficiency of a new technique. One way to get buy-in from practitioners in the AMC is to take a very similar approach to issues related to patient flow.

At the same time, operations research (*OR*) has had a long history of using tools to improve service delivery processes. *OR* employs a predictive modeling investigative paradigm that uses mathematical equations, computer logic, and related tools to forecast the consequences of particular decision choices (Sainfort et al. 2005). Typically, this is done in abstraction without a formal experiment. This approach permits the consideration of alternative choices to quickly be evaluated and compared to see which are most likely to produce preferred outcomes. Many traditional areas of *OR* are prevalent in clinic management. These topics include appointment scheduling (Cayirli et al. 2006), nurse rostering problems (Burke et al. 2004), resource allocation problems (Chao et al. 2003), capacity planning (Bowers and Mould 2005), and routing problems (Mandelbaum et al. 2012).

Given this confluence of approaches and needs, it seems natural for those working to improve healthcare processes to employ *OR* techniques such as DES to conduct controlled, virtual experiments as part of the improvement process. However, when one looks more closely, one finds that the history of implementations of results based on *OR* findings in AMCs is actually quite poor. For example, a review of over 200 papers that used DES in healthcare settings identified only four that even claimed that physician behavior was changed as a result (Wilson 1981). A more recent review found only one instance of a publication which included a documented change in clinic performance resulting from a simulation-motivated intervention (van Lent et al. 2012).

This raises a major question: Since there is clearly an active interest in using DES models to improve patient flow and there is ample talent working to make it happen, what can we do to make use of this technique in a way that results in real change in clinic performance? Virtually any operations management textbook will provide a list of factors needed to succeed in process improvement projects such as getting all stakeholders involved early, identifying a project champion, setting clear goals, dedicating necessary resources, etc. (Trusko et al. 2007). However, we want to focus this discussion on two additional elements that are a bit subtler and, in our experience, often spell the difference between success and failure when working in outpatient clinics in the AMC.

First, finding an important problem is not sufficient. It is critically important to think in terms of finding the right question which also addresses the underlying problem. As outside agents or consultants, we are not in a position to pay faculty and staff extra money to implement changes to improve the system. We need a different form of payment to motivate their participation. One great advantage in the *AMC* model is that we can leverage the fact that physicians are also dedicated researchers. Thus, we can use the promise of publications in lieu of a cash payment to induce participation.

Second, we need to find the right combination of techniques. Experiments and data collection resonate with medical researchers. However, the translation from "lab" to "clinic" is fraught with confounding factors outside of the physician's control. On the other hand, *OR* techniques can isolate a single variable or factor, but modeling by itself does not improve a system, and mathematical presentations that feel completely abstract do not resonate with practitioners. The unique aspect of our approach is to combine *OR* tools with "clinical" experiments. This allows clinicians to project themselves into the model in a way that is more salient than the underlying equations could ever be. The key idea is that value exists in finding a way to merge the tools of *OR* with the methodologies of medical research to generate useful findings that will actually be implemented to improve clinic flow.

2.4 The Process Improvement Process

Given this background, we need a systematic approach to describing, analyzing, and predicting improvements in performance based on changes that can be made to these systems. In order to do this, we need to accomplish at least six things, which form the statement of our method:

1. Describe processes that deliver care and/or service to patients in a relevant way.
2. Collect data on activity times, work flows, and behavior of key agents.
3. Create a DES of the system under study.
4. Experiment with both real and virtual systems to identify and test possible changes.
5. Develop performance metrics of interest to both patients and care providers.
6. Predict changes in metrics which stem from changes in process.

We now turn to providing a bit more detail about each of these steps.

Step 1: Process Description

Much has been written concerning process mapping in healthcare settings (Trusko et al. 2007; Trebble et al. 2010). In many instances, the activity of process mapping itself suggests multiple changes that may improve process flow. However, some insights related to the healthcare-specific complications of this activity warrant discussion.

Perhaps the most obvious way to develop a process map is to first ask the agents in the system to describe the work flow. We have found that this is absolutely necessary and serves as an excellent starting point but is never sufficient. Agents in the system often provide misleading descriptions of process flow. In many cases, physicians are not fully aware of what support staff do to make the process work, and trainees and staff are often quite careful to not appear to contradict more senior physicians. To get high-quality process descriptions, we must gather unbiased insights from multiple levels of the organization. Ideally, this will include support staff, nurses, trainees, and attendings. In some cases, other administrators are valuable as well, especially if there is a department manager or some other person who routinely collects and reports performance data. It is ideal to have all of these agents working on the development of a process map as a group. However, if this cannot be done, it is even more vital to carefully gather information about process flows from as many different angles as possible.

Second, we have found that no matter how much information about the process has been gathered, direct observation by outside agents working on the process improvement process is *always* required. We have yet to find a process description created by internal agents that completely agrees with our observations. Healthcare professionals (understandably) put patient care above all other considerations. Consequently, they make exceptions to normal process flows routinely without giving it a second thought. As a result, their daily behavior will almost always include several subtleties that they do not recall when asked about process flow.

Step 2: Data Collection

In our experience, this is the most time-consuming step in the improvement process. Given a process map, it will be populated with some number of activities undertaken by various agents. The main question that must be asked at this stage is how long each agent spends to complete each step. This approach makes sense for several reasons: First, the dominant patient complaint in outpatient settings is wait times. Thus, time is a crucial metric from the patient's perspective. Second, many systems have been developed which accumulate costs based on hourly or minute-by-minute charges for various resources (Kaplan and Anderson 2003; Kaplan and Porter 2011; King et al. 1994). Consequently, time is a crucial metric from the process manager's perspective as well. Therefore, how long each step takes becomes the central question of interest.

We have utilized four ways to uncover this information. First, agents within the system can be asked how long a process step takes. This is useful as a starting point and can be sufficient in some rare instances. On the other hand, quizzing agents about activity times is problematic because most people think in terms of averages and find it difficult to measure variances. This can only be done after a sufficient number of observations are in hand.

We have also used a second approach in which the caregivers record times during patient visits. For example, in one clinic, we attached a form to each patient record retrieved during each clinic session. In Step 1, staff at the front desk record the patient arrival time and appointment time. The nurse then records the start and end times of Step 2 and so on. This approach can be automated through the use of aids such as phones or iPad apps, where applicable. However, this approach introduces several issues. Recording data interrupts normal flow, and it is not possible to convince the participants that data recording compares in importance to patient care. As a consequence, we repeatedly see instances where the providers forget to record the data and then try to "fill it in" later in the day when things are less hectic. This produces data sets where mean times may be reasonable estimates, but the estimates of variances are simply not reliable.

A third approach to data collection often used in AMCs is to use paid observers to record time stamps. This approach can generate highly reliable information as long as the process is not overly complex and the observer can be physically positioned to have lines of sight that make this method practical. This approach is common in AMCs because they are almost always connected to a larger university and relatively high-quality, low-cost labor is available in the form of students or volunteers. While we have used this technique successfully on multiple occasions, it is not without its problems. First, the observers need to be unobtrusive. This is best done by having them assigned to specific spaces. If personnel travel widely, this becomes problematic. For example, a radiation oncology clinic that we studied had rooms and equipment on multiple floors, so tracking became quite complex. Second, the parties serving patients know they are being observed. Many researchers have reported significant improvements to process flow using this approach, only to find that after the observers left, the system drifted back to its previous way of functioning and the documented improvement was lost.

We have also used a fourth approach to data collection. Many hospitals and clinics are equipped with real-time location systems (RTLS). Large AMCs are often designed to include this capability because tracking devices and equipment across hundreds of thousands of square feet of floor space are simply not practical without some technological assistance. Installations of these systems typically involve placing sensors in the ceilings or floors of the relevant spaces. These sensors pick up signals from transmitters that can be embedded within "tags" or "badges" worn by items or people being tracked. Each sensor records when a tag comes within range and again when it leaves that area. When unique tag numbers are given to each caregiver, detailed reports can be generated at the end of each day showing when a person or piece of equipment moved from one location to another. This approach offers several dramatic advantages. It does not interfere with the care delivery process, the marginal cost of using it is virtually 0, and since these systems are always running, the observation periods can begin and end as needed.

In closing, we should highlight three key factors in the data collection process: (1) data collection needs to be done in a way that does not interfere with care delivery; (2) audits of the data collection system are needed to ensure accuracy; and (3) sufficient time span must be covered to eliminate any effects of the "novelty" of the data collection and its subsequent impact on agent behaviors.

Step 3: Create a DES of the System

We have often found it useful to create DES models of the systems under study as early in the process as possible. This can be a costly process in that a great deal of data collection is required and model construction can be a nontrivial expense. Other tools such as process mapping and queueing theory can be applied with much less effort (Kolker 2010). However, we have repeatedly found that these tools are insufficient for the analysis that is needed. Because the variances involved in activity times can be extremely high in healthcare, distributions of the metrics of interest are important findings. Consequently, basic process analysis is rarely sufficient and often misleading.

Queueing models do a much better job of conveying the significance of variability. However, many common assumptions of these models are routinely violated in clinic settings, including that some processing times are not exponentially distributed, that processing times are often not from the same distribution, and that if arrivals are based on appointments, inter-arrival times are not exponentially distributed.

However, none of these issues pose the largest challenge to applying simple process analysis or queuing models in outpatient clinics. Consider two additional issues. First, the basic results of process analysis or queueing models are only averages which appear in steady state. A clinic does not start the day in steady state—it begins in an empty state. It takes some time to reach steady state. However, if one plots average wait times for a clinic over time, one quickly sees that it may take dozens or even hundreds of cases for the system to reach steady state. Clearly, a clinic with one physician is not going to schedule hundreds of patients for that resource in a single session. Thus, steady-state results are often not informative.

Second, if activity times and/or the logic defining work flow changes in response to job type or system status, then the results of simple process analysis or queueing models become invalid. We have documented such practices in multiple clinics that we have studied (Chambers et al. 2016; Conley et al. 2016). Consequently, what is needed is a tool that can account for all of these factors simultaneously, make predictions about what happens when some element of the system changes, and give us information about the broader distribution of outcomes—not just a means for systems in steady state. DES is a tool with the needed capabilities.

A brief comment on the inclusion of activity times in DES models is warranted here. We have used two distinct approaches. We can select an activity time at random from a collection of observations. Alternatively, we can fit a distribution to collected activity time data. We have found both approaches to work satisfactorily. However, if the data set is sufficiently large, we recommend sampling directly from that set. This generates results that are both easier to defend to statisticians and more credible to practitioners.

Step 4: Field and Virtual Experiments

It is at this point that the use of experiments comes into play, and we merge the *OR* methodology of DES with the experimental methods of medical research. The underlying logic is that we propose an experiment involving some process change that we believe will alter one or more parameters defining system behavior. We can use the DES to predict outcomes if our proposal works. In other cases, if we have evidence that the proposed change works in some settings, we can use the DES to describe how that change will affect system metrics in other settings. The construction of these experiments is the "art" of our approach. It is this creation that leads to publishable results and creates novel insights.

We will provide examples of specific experiments in the next section. However, at this juncture we wish to raise two critical issues: confounding variables and unintended consequences. Confounding variables refer to system or behavioral attributes that are not completely controlled when conducting an experiment but can alter study results. For example, consider looking at a system before an intervention, collecting data on its performance, changing something about the system, and then collecting data on the performance of the modified system. This is the ideal approach, but it implicitly assumes that nothing changed in the system over the span of the study other than what you intended to change. If data collection takes place over a period of months, it is quite possible that the appointment schedule changed over that span of time due to rising or falling demand. In this example, the change in demand would be a confounding variable. It is critically important to eliminate as many confounding variables as you can before concluding that your process change fully explains system improvement. DES offers many advantages in this regard because it allows you to fix some parameter levels in a model even if they may have changed in the field.

It is also critical to account for unintended consequences. For example, adding examination rooms is often touted as a way to cut wait times. However, this also makes the relevant space larger, increasing travel times as well as the complexity of resource flows. This must be accounted for before declaring that the added rooms actually improved performance. It may improve performance along one dimension while degrading it in another.

DES modeling has repeatedly proven invaluable at this stage. Once a DES model is created, it is easy to simulate a large number of clinic sessions and collect data on a broad range of performance metrics. With a little more effort, it can also be set up to collect data on the use of overtimes or wait times within examination rooms. In addition, DES models can be set up to have patients take different paths or have activity times drawn from different distributions depending on system status. Finally, we have found it useful to have DES models collect data on subgroups of patients based on system status because many changes to system parameters affect different groups differently.

Step 5: Metrics of Interest

A famous adage asserts, "If you can't measure it, you can't manage it." Hence, focusing on measurements removes ambiguity and limits misunderstandings. If all parties agree on a metric, then it is easier for them to share ideas on how to improve it. However, this begs an important question—what metrics do we want to focus on? In dealing with this question, Steps 4 and 5 of our method become intertwined and cannot be thought of in a purely sequential fashion. In some settings, we need novel metrics to fit an experiment, while in other settings unanticipated outcomes from experiments suggest metrics that we had not considered earlier.

Both patients and providers are concerned with system performance, but their differing perspectives create complex trade-offs. For example, researchers have often found that increase in face time with providers serves to enhance patient experience (Thomas et al. 1997; Seals et al. 2005; Lin et al. 2001), but an increase in wait time degrades that experience (Meza 1998; McCarthy et al. 2000; Lee et al. 2005). The patient may not fully understand what the care provider is doing, but they can always understand that more attention is preferable and waiting for it is not productive. Given a fixed level of resources, increases in face time result in higher provider utilization, which in turn increases patient wait times. Consequently, the patient's desire for increased face time and reduced wait time creates a natural tension and suggests that the metrics of interest will almost always include both face time and wait time.

Consider one patient that we observed recently. This patient arrived 30 min early for an appointment and waited 20 min before being lead to the exam room. After being led to the room, the patient waited for 5 min before being seen by a nurse for 5 min. The patient then waited 15 min before being seen by the resident. The trainee then spoke with the patient for 20 min before leaving the room to discuss the case with the attending. The patient then waited 15 min before being seen by the resident and the attending together. The attending spoke with the patient for 5 min before being called away to deal with an issue for a different patient. This took 10 min. The attending then returned to the exam room and spoke with the patient for another 5 min. After that, the patient left. By summing these durations, we see that the patient was in the clinic for roughly 100 min. The patient waited for 20 min in the waiting room. However, the patient also spent 45 min in the exam room waiting for service. Time in the examination room was roughly 80 min of which 35 min was spent in the presence of a service provider. Thus, we can say that the overall face time was only 35 min. However, of this time only 10 min was with the attending physician. Consideration of this more complete description suggests a plethora of little-used metrics that may be of interest, such as:

1. Patient punctuality
2. Time spent in the waiting room before the appointment time
3. Time spent in the waiting room after the appointment time
4. Wait time in the examination room
5. Proportion of cycle time spent with a care provider
6. Proportion of cycle time spent with the attending

The key message here is that the metrics of interest may be specific to the problem that one seeks to address and must reflect the nuances of the process in place to deliver the services involved.

Step 6: Predict Impact of Process Changes

Even after conducting an experiment in one setting, we have found that it is extremely difficult to predict how changes will affect a different system simply by looking at the process map. This is another area where DES proves quite valuable. For example, say that our experiment in Clinic A shows that by changing the process in some way, the time for the Attending step is cut by 10%. We can then model this change in a different clinic setting by using a DES of that setting to predict how implementing our suggested change will be reflected in performance metrics of that clinic in the future. This approach has proven vital to get the buy-in needed to facilitate a more formal experiment in the new setting or to motivate implementation in a unit where no formal experiment takes place.

2.5 Experiments, Simulations, and Results

Our work has included a collection of experiments that have led to system improvements for settings such as that depicted in Fig. 22.1. We now turn to a discussion of a few of these efforts to provide context and illustrations of our approach. Figure 22.1 includes an arrival process under an appointment system. This is quickly followed by activities involving the trainee and/or nurse and/or attending. Finally, the system hopes to account for all of these things when searching for an optimized schedule. We discuss a few of these issues in turn.

Arrival Process

We are focusing on clinics which set a definite appointment schedule. One obvious complication is that some patients are no-shows, meaning that they do not show up for the appointment. No-show rates of as much as 40% have been cited in prior works (McCarthy et al. 2000; Huang 1994). However, there is also a subtler issue of patients arriving very early or very late, and this is much harder to account for. Early work in this space referred to this as patient "unpunctuality" (Bandura

1969; White and Pike 1964; Alexopoulos et al. 2008; Fetter and Thompson 1966; Tai and Williams 2012; Perros and Frier 1996). Our approach has been used to address two interrelated questions: Does patient unpunctuality affect clinic performance, and can we affect patient unpunctuality? To address these questions, we conducted a simple experiment. Data on patient unpunctuality was collected over a six-month period. We found that most patients arrived early, but patient unpunctuality ranged from −80 to +20. In other words, some patients arrived as much as 80 min early, while others arrived 20 min late. An intervention was performed that consisted of three elements. In reminders mailed to each patient before their visit, it was stated that late patients would be asked to reschedule. All patients were called in the days before the visit, and the same reminder was repeated over the phone. Finally, a sign explaining the new policy was posted near the registration desk. Unpunctuality was then tracked 1, 6, and 12 months later. Additional metrics of interest were wait times, use of overtime, and the proportion of patients that were forced to wait to be seen (Williams et al. 2014).

This lengthy follow-up was deemed necessary because some patients only visited the clinic once per quarter, and thus the full effect of the intervention could not be measured until after several quarters of implementation. To ensure that changes in clinic performance were related only to changes in unpunctuality, we needed a way to control for changes in the appointment schedule that happened over that time span. Our response to this problem was to create a DES of the clinic, use actual activity times in the DES, and consider old versus new distributions of patient unpunctuality, assuming a fixed schedule. This allowed us to isolate the impact of our intervention.

Before the intervention, 7.7% of patients were tardy and average tardiness of those patients was 16.75 min. After 12 months, these figures dropped to 1.5% and 2 min, respectively. The percentage of patients who arrived before their appointment time rose from 90.4% to 95.4%. The proportion who arrived at least 1 min tardy dropped from 7.69% to 1.5%. The range of unpunctuality decreased from 100 to 58 min. The average time to complete the session dropped from 250.61 to 244.49 min. Thus, about 6 min of overtime operations was eliminated from each session. The likelihood of completing the session on time rose from 21.8% to 31.8%.

Our use of DES allowed us to create metrics of performance that had not yet been explored. For example, we noticed that the benefits from the change were not the same for all patients. Patients that arrived late saw their average wait time drop from 10.7 to 0.9 min. Those that arrived slightly early saw their average wait time increase by about 0.9 min. Finally, for those that arrived very early, their wait time was unaffected. In short, we found that patient unpunctuality can be affected, and it does alter clinic performance, but this has both intended and unintended consequences. The clinic session is more likely to finish on time and overtime costs are reduced. However, much of the benefit in terms of wait times is actually realized by patients that still insist on arriving late.

Physician Processing Times

Historically, almost all research on outpatient clinics assumed that processing times were not related to the schedule or whether the clinic was running on time. Is this indeed the case? To address this question, we analyzed data from three clinic settings. One was a low-volume clinic that housed a single physician, another was a medium-volume clinic in an AMC that had one attending working on each shift along with two or three trainees, and the last one was a high-volume service that had multiple attendings working simultaneously (Chambers et al. 2016).

We categorized patients into three groups: Group A patients were those who arrived early and were placed in the examination room before their scheduled appointment time. Group B patients were those who also arrived early, but were placed in the examination room after their appointment time, indicating that the clinic was congested. Group C patients were those who arrived after their appointment time. The primary question was whether the average processing time for patients in Group A was the same as that for patients in Group B. We also had questions about how this affected clinic performance in terms of wait times and session completion times.

In the low-volume clinic with a single physician, average processing times and standard errors (in parentheses) were 38.31 (3.21) for Group A and 26.23 (2.23) for Group B. In other words, the physician moved faster when the clinic was behind schedule. Similar results have been found in other industries, but this was the first time (to the best of our knowledge) that this had been demonstrated for outpatient clinics.

In the medium-volume clinic, the relevant values were 65.59 (2.24) and 53.53 (1.97). Again, the system worked faster for Group B than it did for Group A. Note the drop in average times is about 12 min in both settings. This suggests that the finding is robust, meaning that it occurs to a similar extent in similar (but not identical) settings. Additionally, remember that the medium-volume clinic included trainees in the process flow. This suggests that the way that the system got this increase in speed might be different. In fact, our data show that the average amount of time the attending spent with the patient was no more than 12 min to begin with. Thus, we know that it was not just the behavior of the attending that made this happen. The AMC must be using the trainees differently when things fall behind schedule.

In the high-volume clinic, the parallel values were 47.15 (0.81) and 17.59 (0.16). Here, we see that the drop in processing times is much more dramatic than we saw before. Again, the message is that processing times change when the system is under stress and the magnitude of the change implies that multiple parties are involved in making this happen. In hindsight, this seems totally reasonable, but the extent of the difference is still quite startling.

As we saw in the previous section, there is an unintended consequence of this system behavior as it relates to patient groups. Patients that show up early should help the clinic stay on schedule. This may not be so because these patients receive longer processing times. Thus, their cycle times are longer. Patients that arrive late have shorter wait times and shorter processing times. Thus, their cycle times are shorter. If shorter cycle times are perceived as a benefit, this seems like an unfair reward for patient tardiness and may explain why it will never completely disappear.

Impact of the Teaching Mission

The result from the previous section suggests that the way that the trainee is used and managed within the clinic makes a difference when considering system performance. To explore this point further, we wanted to compare a clinic without trainees with a similar clinic that included trainees. This is difficult to do as an experiment, but we were lucky when looking at this question. An attending from a clinic with no trainees was hired as the director of a clinic in the AMC that included trainees. Thus, we could consider the same attending seeing the same patients in both settings. One confounding variable was that the two clinics used different appointment schedules (Williams et al. 2012).

We collected data on activity times in both settings. Given these times, we could seed DES models of both clinics and compare results. Within the DES, we could look at both settings as though they had the same appointment schedule. If we consider the two settings using the schedule in place for the AMC, we see that the average cycle time in the AMC was 76.2 min and this included an average wait time of 30.0 min. The average time needed to complete a full schedule was 291.9 min. If the same schedule had been used in the private practice model, the average cycle time would be 129.1 min and the average wait time would be 83.9 min.

The capacity of the AMC was clearly greater than it was in the private practice model. This is interesting because the flow times in the private practice setting using the schedule that was optimized for that setting were much lower. It turns out that the total processing time for each patient was greater in the AMC, but the capacity was higher. This is explained using parallel processing. In the AMC setting, the attending spent time with one patient, while trainees simultaneously worked with other patients. We were able to conduct a virtual experiment by changing the number of trainees in the DES model. We found that having one trainee created a system with cycle times that were much greater than the private practice model. Using two trainees produced cycle times that were about the same. Using three trainees created the reduced cycle times that we noticed in practice. Using more than three trainees produced no additional benefit because both clinics had only three available exam rooms. This enabled us to comment on the optimal number of trainees for a given clinic.

The use of DES also highlighted a less obvious result. It turns out that the wait time in this system was particularly sensitive to the time taken in the step we labeled "Teach." This is the time that the trainee spends interacting with the attending after interacting with the patient. In fact, we found that reducing this time by 1 min served to reduce average wait time by 3 min. To understand this phenomenon, recall that when the trainee and the attending are discussing the case while the patient waits in the examination room for 1 min, the three busiest resources in the system (the trainee, the attending, and the examination room) are simultaneously occupied for that length of time. Thus, it is not surprising that wait times are sensitive to the duration of this activity, although the degree of this sensitivity is still eye-opening.

Preprocessing

Given that wait times are extremely sensitive to teaching times, we created an experiment designed to alter the distribution of these times. Instead of having the trainee review the case after the patient is placed in the examination room and then having the first conversation about the case with the attending after the trainee interacts with the patient, we can notify both the trainee and attending in advance which patient each trainee will see. That way, the trainee can review the file before the session starts and have a conversation with the attending about what should happen upon patient arrival. We also created a template to guide the flow and content of this conversation. We refer to this approach as "preprocessing" (Williams et al. 2015).

We recorded activity times using the original system for 90 days. We then introduced the new approach and ran it for 30 days. During this time, we continued collecting data on activity times.

Before the intervention was made, the average teach time was 12.9 min for new patients and 8.8 min for return patients. The new approach reduced these times by 3.9 min for new patients and 2.9 min for return patients. Holding the schedule as a constant, we find that average wait times drop from 36.1 to 21.4 min and the session completion time drops from 275.6 to 247.4 min.

However, in this instance, it was the unintended consequences that proved to be more important. When the trainees had a more clearly defined plan about how to handle each case, their interactions with the patients became more efficient. The trainees also reported that they felt more confident when treating the patients than they had before. While it is difficult to measure this effect in terms of time, both the trainees and the attending felt that the patients received better care under the new protocol.

Cyclic Scheduling

Considering the works mentioned above, one finding that occurred repeatedly was that the way the trainee was involved in the process had a large impact on system performance and how that was done was often state dependent. Recall that we found that the system finds ways to move faster when the clinic is behind schedule. When a physician is working alone, this can be done simply by providing less face time to patients. When the system includes a trainee, an additional response is available in that either the attending or the trainee can be dropped from the process for one or more patients. Our experience is that doctors strongly believe that the first approach produces huge savings and they strongly oppose the second.

Our direct observation of multiple clinics produced some insights related to these issues. Omitting the attending does not save as much time as most attendings think because the trainee is slower than the attending. In addition, the attending gets involved in more of these cases than they seem to realize. Many attendings feel compelled to "at least say hi" to the patients even when the patients are not really on their schedule, and these visits often turn out to be longer than expected. Regarding the second approach, we have noticed a huge variance in terms of how willing the attending is to omit the trainee from a case. Some almost never do it, while others do it quite often. In one clinic we studied, we found that the trainee was omitted from roughly 30% of the cases on the clinic schedule. If this is done, it might explain why a medium-volume or high-volume clinic within the AMC could reduce cycle times after falling behind schedule to a greater extent than the low-volume clinic can achieve. This can be done by instructing the trainee to handle one case while the attending handles another and having the attending exclude the trainee from one or more cases in an effort to catch up to the clinic schedule.

Accounting for these issues when creating an appointment schedule led us to the notion of cyclic scheduling. The idea is that the appointment schedule can be split into multiple subsets which repeat. We label these subsets "cycles." In each cycle, we include one new patient and one return patient scheduled to arrive at the same time. A third patient is scheduled to arrive about the middle of the cycle. If both patients arrive at the start of the cycle, we let the trainee start work on the new patient, and the attending handles the return patient without the trainee being involved. This was deemed acceptable because it was argued that most of the learning comes from visits with new patients. If only one of the two patients arrives, the standard process is used.

Process analysis tools produce some results about average cycle times in this setting, but since wait times are serially correlated, we want a much clearer depiction of how each patient's wait time is related to that of the following patients. Considering the problem using a queuing model is extremely difficult because the relevant distribution of activity times is state dependent and the number of cycles is small. Consequently, steady-state results are misleading. Studying this approach within a DES revealed that average makespan, wait times, and cycle times are significantly reduced using our cyclic approach and the trainee is involved in a greater proportion of the cases scheduled.

3 Conclusion

While a great deal of time, effort, and money has been spent to improve healthcare processes, the problems involved have proven to be very difficult to solve. In this work, we focused on a small but important sector of the problem space—that of appointment-based clinics in academic medical centers. One source of difficulty is that the medical field favors an experimental design-based approach, while many *OR* tools are more mathematical and abstract. Consequently, one of our core messages is that those working to improve these systems need to find ways to bridge this gap by combining techniques. When this is done, progress can be made and the insights generated can be spread more broadly. Our use of DES builds on tools of process mapping that most managers are familiar with and facilitates virtual experiments that are easier to control and use to generate quantitative metrics amenable to the kinds of statistical tests that research physicians routinely apply.

However, we would be remiss if we failed to emphasize the fact that data-driven approaches are rarely sufficient to bring about the desired change. Hospitals in AMCs are often highly politicized environments with a hierarchical culture. This fact can generate multiple roadblocks that no amount of "number crunching" will ever overcome. One not so subtle aspect of our method is that it typically involves embedding ourselves in the process over some periods of time and interacting repeatedly with the parties involved. We have initiated many projects not mentioned above because they did not result in real action. Every project that has been successful involved many hours of working with faculty, physicians, staff, and technicians of various types to collect information and get new perspectives. We have seen dozens of researchers perform much more impressive data analysis on huge data sets using tools that were more powerful than those employed in these examples, only to end up with wonderful analysis not linked to any implementation. When dealing with healthcare professionals, we are often reminded of the old adage, "No one cares how much you know. They want to know how much you care." While we believe that the methodology outlined in this chapter is useful, our experience strongly suggests that the secret ingredient to making these projects work is the attention paid to the physicians, faculty, and especially staff involved who ultimately make the system work.

Electronic Supplementary Material

All the datasets, code, and other material referred in this section are available in www.allaboutanalytics.net.

- Model 22.1: Model1.mox
- Model 22.2: Model1A.mox
- Model 22.3: Model2.mox
- Model 22.4: Model3.mox

Exercises

In "Using Analytics to Improve Patient Flow in Outpatient Clinics," we laid out a six-step approach to improving appointment-based systems in outpatient academic medical centers. These exercises involve simplified versions of discrete event simulations (DESs) of such settings. Their purpose is to illustrate and conceptualize the process. Completion of these exercises should highlight many issues and subtleties of these systems and help the reader develop ideas that best fit with their setting of interest.

Introduction
Simplified versions of several models referenced in Sect. 22.2.5 of the reading can be considered to explore the issues discussed there. These DES models have been developed in ExtendSim version 9.0.[19] Complete versions of the underlying software are available from the vendor, and a variety of academic pricing models are

---

[19] Download trial version from https://www.extendsim.com/demo (accessed on Jul 16, 2018).

22 Healthcare Analytics

available. A wide variety of texts and tools are also available to assist the potential user with details of software capabilities including Strickland (2010) and Laguna and Marklund (2005). However, the models utilized in this reading are fairly simple to construct and can be easily adapted to other packages as the reader (or instructor) sees fit. For ease of exposition and fit with the main body of the reading, we present exercises corresponding to settings described earlier. Hints are provided in the Hints for Solution word file (refer to book's website) that should help in going through the exercises. The exercises allow the reader to explore the many ideas given in the chapter in a step-by-step manner.

A Basic Model with Patient Unpunctuality

Service providers in many settings utilize an appointment system to manage the arrival of customers/jobs. However, the assumption that the appointment schedule will be strictly followed is rarely justified. The first model (Model 1; refer to book's website) presents a simplified process flow for a hypothetical clinic and embeds an appointment schedule. The model facilitates changes to the random variable that defines patient punctuality. In short, patients arrive at some time offset from their appointment time. By adjusting the parameters which define the distribution of this variable, we can represent arrival behavior. You may alter this model to address the following questions:

Ex. 22.1 Describe clinic performance if all patients arrive on time.

Ex. 22.2 Explain how this performance changes if unpunctuality is included. For this example, this means modeling actual arrival time as the appointment time plus a log-normally distributed variable with a mean of μ and a standard deviation of σ minutes. A reasonable base case may include $\mu = -15$ min, and $\sigma = 10$ min. (Negative values of unpunctuality mean that the patient arrives prior to the appointment time, which is the norm.) Note how changes to μ and σ affect performance differently.

Ex. 22.3 Explain how you would create an experiment (in an actual clinic) to uncover how this behavior changes and how it affects clinic performance.

Ex. 22.4 Explain how you would alter Model 1 to report results for groups of patients such as those with negative unpunctuality (early arrivers), those with positive unpunctuality (late arrivers), and those with appointment times near the end of the clinic session.

Ex. 22.5 The DES assumes that the patient with the earlier appointment time is always seen first, even if they arrived late. How would you modify this model if the system "waits" for late patients up to some limit, "w" minutes rather than seeing the next patient as soon as the server is free?

An Academic Model with Distributions of Teaching Time

The process flow within the academic medical center (AMC) differs from Model 1 in that it includes additional steps and resources made necessary by the hospital's teaching mission. Simple process analysis is useful in these settings to help identify the bottleneck resource and to use management of that resource to improve system performance. However, such simple models are unable to fully account for the impact of system congestion given this more complex flow. For example, idle time is often added because one resource is forced to wait for the availability of another.

Using a DES of such systems may be particularly valuable in that they facilitate various forms of sensitivity analysis which can produce novel insights about these issues. Use Model 2 (refer to book's website) of the AMC to address the following questions:

Ex. 22.6 How do the average values of cycle time, wait time, and makespan respond to changes in teach time?

Ex. 22.7 Describe the linkage between utilization of the trainees in this system and the amount of time they spend with patients. How much of their busy time is not explained by value-adding tasks?

Ex. 22.8 Describe the linkage between the number of trainees and the utilization of other key resources in the system.

Ex. 22.9 Explain how you would create an experiment (in an actual clinic) to uncover how changing the educational process is linked to resident productivity.

Ex. 22.10 How would you alter Model 2 to reflect a new approach to trainee education aimed at increasing the share of their time that adds value to the patient?

State-Dependent Processing Times

Experience with many service systems lends support to the notion that the service provider may be motivated to "speed up" when the system is busy. However, common sense also suggests that this is not sustainable forever. With these facts in mind, it is important to think through how we might measure this behavior and how we may monitor any unintended consequences from such an approach. With this in mind, Model 3 (refer to book's website) includes a reduction to processing times for the attending when the system is busy. Consider this model to address the following questions:

Ex. 22.11 How do average values of cycle time, wait time, and makespan change when the attending gets faster in a busy system?

Ex. 22.12 Instead of reducing face time, consider adding examination rooms to the system instead. Is there any evidence produced by the DES to suggest that one approach is better than the other?

Ex. 22.13 Describe the comparison between decreasing processing times when the system is busy to changing processing times for all settings.

Ex. 22.14 Explain how you would create an experiment (in an actual clinic) to explore how this behavior affects patient flow and service quality. What extra factors do you need to control for?

Ex. 22.15 How would you alter Model 3 to separate the effects of patient behavior (including unpunctuality) from the effects of physician behavior (including changing processing times)?

Cyclic Scheduling

Personnel creating an appointment schedule are likely to favor having a simple template to refer to when patients request appointment times. Consequently, there is administrative value in having a logic that is easy to explain and implement. Again, this is more difficult to do in the AMC since the process flow is more complex. Return to the use of Model 2 and modify it as needed to address the following questions:

Ex. 22.16 Study the existing appointment schedule. Develop the "best" schedule if there is no variability to consider. (You may assume that average activity times are always realized.)

Ex. 22.17 How does your schedule perform when patient unpunctuality is added, and how will you adjust your schedule to account for this?

Ex. 22.18 Assuming that patients are always perfectly punctual and only attending time is variable, look for a schedule that works better than the one developed in Exercise 22.16.

Ex. 22.19 Explain how you would create an experiment (in an actual clinic) to explore ways to reduce this variability. What extra factors do you need to control for?

Ex. 22.20 How would you alter Model 2 to include additional issues such as patient no-shows, emergencies, work interruptions, and open-access scheduling?

Conclusion

It is important to note that DES models are only one tool that can be applied to develop a deeper understanding of the behavior of complex systems. However, adding this approach to the "toolbox" of the clinic manager or consultant should provide ample benefits and support for ideas on how to make these systems better meet the needs of all stakeholders.

References

Alexopoulos, C., Goldman, D., Fontanesi, J., Kopald, D., & Wilson, J. R. (2008). Modeling patient arrivals in community clinics. *Omega, 36*, 33–43.

Bandura, A. (1969). *Principles of behavior modification.* New York, NY: Holt, Rinehart, & Winston.

Benneyan, J. C. (1997). An introduction to using computer simulation in healthcare: Patient wait case study. *Journal of the Society for Health Systems, 5*(3), 1–15.

Beronio, K., Glied, S. & Frank, R. (2014) J Behav Health Serv Res. 41, 410. https://doi.org/10.1007/s11414-014-9412-0

Boex, J. R., Boll, A. A., Franzini, L., Hogan, A., Irby, D., Meservey, P. M., Rubin, R. M., Seifer, S. D., & Veloski, J. J. (2000). Measuring the costs of primary care education in the ambulatory setting. *Academic Medicine, 75*(5), 419–425.

Bowers, J., & Mould, G. (2005). Ambulatory care and orthopaedic capacity planning. *Health Care Management Science, 8*(1), 41–47.

Burke, E. K., De Causmaecker, P., Berghe, G. V., & Van Landeghem, H. (2004). The state of the art of nurse rostering. *Journal of Scheduling, 7*(6), 441–499.

Cayirli, T., Veral, E., & Rosen, H. (2006). Designing appointment scheduling systems for ambulatory care services. *Health Care Management Science, 9*(1), 47–58.

Chambers, C. G., Dada, M., Elnahal, S. M., Terezakis, S. A., DeWeese, T. L., Herman, J. M., & Williams, K. A. (2016). Changes to physician processing times in response to clinic congestion and patient punctuality: A retrospective study. *BMJ Open, 6*(10), e011730.

Chao, X., Liu, L., & Zheng, S. (2003). Resource allocation in multisite service systems with intersite customer flows. *Management Science, 49*(12), 1739–1752.

Chesney, A. M. (1943). *The Johns Hopkins Hospital and John Hopkins University School of Medicine: A chronicle.* Baltimore, MD: Johns Hopkins University Press.

Clymer, J. R. (2009). *Simulation-based engineering of complex systems* (Vol. 65). New York, NY: John Wiley & Sons.

Conley, K., Chambers, C., Elnahal, S., Choflet, A., Williams, K., DeWeese, T., Herman, J., & Dada, M. (2018). *Using a real-time location system to measure patient flow in a radiation oncology outpatient clinic*, Practical radiation oncology.

Cutler, D. M., & Ghosh, K. (2012). The potential for cost savings through bundled episode payments. *New England Journal of Medicine, 366*(12), 1075–1077.

Fetter, R. B., & Thompson, J. D. (1966). Patients' wait time and doctors' idle time in the outpatient setting. *Health Services Research, 1*(1), 66.

Franzini, L., & Berry, J. M. (1999). A cost-construction model to assess the total cost of an anesthesiology residency program. *The Journal of the American Society of Anesthesiologists, 90*(1), 257–268.

Glied, S., & Ma, S. (2015). *How will the Affordable Care Act affect the use of health care services?* New York, NY: Commonwealth Fund.

Hamrock, E., Parks, J., Scheulen, J., & Bradbury, F. J. (2013). Discrete event simulation for healthcare organizations: A tool for decision making. *Journal of Healthcare Management, 58*(2), 110.

Hing, E., Hall, M. J., Ashman, J. J., & Xu, J. (2010). National hospital ambulatory medical care survey: 2007 Outpatient department summary. *National Health Statistics Reports, 28*, 1–32.

Hosek, J. R., & Palmer, A. R. (1983). Teaching and hospital costs: The case of radiology. *Journal of Health Economics, 2*(1), 29–46.

Huang, X. M. (1994). Patient attitude towards waiting in an outpatient clinic and its applications. *Health Services Management Research, 7*(1), 2–8.

Hwang, C. S., Wichterman, K. A., & Alfrey, E. J. (2010). The cost of resident education. *Journal of Surgical Research, 163*(1), 18–23.

Jun, J. B., Jacobson, S. H., & Swisher, J. R. (1999). Application of discrete-event simulation in health care clinics: A survey. *Journal of the Operational Research Society, 50*(2), 109–123.

Kaplan, R. S., & Anderson, S. R. (2003). *Time-driven activity-based costing*. SSRN 485443.

Kaplan, R. S., & Porter, M. E. (2011). How to solve the cost crisis in health care. *Harvard Business Review, 89*(9), 46–52.

King, M., Lapsley, I., Mitchell, F., & Moyes, J. (1994). Costing needs and practices in a changing environment: The potential for ABC in the NHS. *Financial Accountability & Management, 10*(2), 143–160.

Kolker, A. (2010). Queuing theory and discrete event simulation for healthcare: From basic processes to complex systems with interdependencies. In Abu-Taieh, E., & El Sheik, A. (Eds.), Handbook of research on discrete event simulation technologies and applications (pp. 443–483). Hershey, PA: IGI Global.

Laguna, M., & Marklund, J. (2005). *Business process modeling, simulation and design*. Upper Saddle River, NJ: Pearson Prentice Hall.

Lee, V. J., Earnest, A., Chen, M. I., & Krishnan, B. (2005). Predictors of failed attendances in a multi-specialty outpatient centre using electronic databases. *BMC Health Services Research, 5*(1), 1.

van Lent, W. A. M., VanBerkel, P., & van Harten, W. H. (2012). A review on the relation between simulation and improvement in hospitals. *BMC Medical Informatics and Decision Making, 12*(1), 1.

Lin, C. T., Albertson, G. A., Schilling, L. M., Cyran, E. M., Anderson, S. N., Ware, L., & Anderson, R. J. (2001). Is patients' perception of time spent with the physician a determinant of ambulatory patient satisfaction? *Archives of Internal Medicine, 161*(11), 1437–1442.

Lorenzoni, L., Belloni, A., & Sassi, F. (2014). Health-care expenditure and health policy in the USA versus other high-spending OECD countries. *The Lancet, 384*(9937), 83–92.

Mandelbaum, A., Momcilovic, P., & Tseytlin, Y. (2012). On fair routing from emergency departments to hospital wards: QED queues with heterogeneous servers. *Management Science, 58*(7), 1273–1291.

McCarthy, K., McGee, H. M., & O'Boyle, C. A. (2000). Outpatient clinic wait times and non-attendance as indicators of quality. *Psychology, Health & Medicine, 5*(3), 287–293.

Meza, J. P. (1998). Patient wait times in a physician's office. *The American Journal of Managed Care, 4*(5), 703–712.

Moses, H., Thier, S. O., & Matheson, D. H. M. (2005). Why have academic medical centers survived. *Journal of the American Medical Association, 293*(12), 1495–1500.

Nuti, S. V., Wayda, B., Ranasinghe, I., Wang, S., Dreyer, R. P., Chen, S. I., & Murugiah, K. (2014). The use of Google trends in health care research: A systematic review. *PLoS One, 9*(10), e109583.

Ortman, J. M., Velkoff, V. A., & Hogan, H. (2014). *An aging nation: The older population in the United States* (pp. 25–1140). Washington, DC: US Census Bureau.

Perros, P., & Frier, B. M. (1996). An audit of wait times in the diabetic outpatient clinic: Role of patients' punctuality and level of medical staffing. *Diabetic Medicine, 13*(7), 669–673.

Sainfort, F., Blake, J., Gupta, D., & Rardin, R. L. (2005). *Operations research for health care delivery systems. WTEC panel report*. Baltimore, MD: World Technology Evaluation Center, Inc..

Seals, B., Feddock, C. A., Griffith, C. H., Wilson, J. F., Jessup, M. L., & Kesavalu, S. R. (2005). Does more time spent with the physician lessen parent clinic dissatisfaction due to long wait times. *Journal of Investigative Medicine, 53*(1), S324–S324.

Sloan, F. A., Feldman, R. D., & Steinwald, A. B. (1983). Effects of teaching on hospital costs. *Journal of Health Economics, 2*(1), 1–28.

Strickland, J. S. (2010). *Discrete event simulation using ExtendSim 8*. Colorado Springs, CO: Simulation Educators.

Tai, G., & Williams, P. (2012). Optimization of scheduling patient appointments in clinics using a novel modelling technique of patient arrival. *Computer Methods and Programs in Biomedicine, 108*(2), 467–476.

Taylor, D. H., Whellan, D. J., & Sloan, F. A. (1999). Effects of admission to a teaching hospital on the cost and quality of care for Medicare beneficiaries. *New England Journal of Medicine, 340*(4), 293–299.

Thomas, S., Glynne-Jones, R., & Chait, I. (1997). Is it worth the wait? a survey of patients' satisfaction with an oncology outpatient clinic. *European Journal of Cancer Care, 6*(1), 50–58.

Trebble, T. M., Hansi, J., Hides, T., Smith, M. A., & Baker, M. (2010). Process mapping the patient journey through health care: An introduction. *British Medical Journal, 341*(7769), 394–397.

Trusko, B. E., Pexton, C., Harrington, H. J., & Gupta, P. (2007). *Improving healthcare quality and cost with six sigma*. Upper Saddle River, NJ: Financial Times Press.

White, M. J. B., & Pike, M. C. (1964). Appointment systems in out-patients' clinics and the effect of patients' unpunctuality. *Medical Care*, 133–145.

Williams, J. R., Matthews, M. C., & Hassan, M. (2007). Cost differences between academic and nonacademic hospitals: A case study of surgical procedures. *Hospital Topics, 85*(1), 3–10.

Williams, K. A., Chambers, C. G., Dada, M., Hough, D., Aron, R., & Ulatowski, J. A. (2012). Using process analysis to assess the impact of medical education on the delivery of pain services: A natural experiment. *The Journal of the American Society of Anesthesiologists, 116*(4), 931–939.

Williams, K. A., Chambers, C. G., Dada, M., McLeod, J. C., & Ulatowski, J. A. (2014). Patient punctuality and clinic performance: Observations from an academic-based private practice pain centre: A prospective quality improvement study. *BMJ Open, 4*(5), e004679.

Williams, K. A., Chambers, C. G., Dada, M., Christo, P. J., Hough, D., Aron, R., & Ulatowski, J. A. (2015). Applying JIT principles to resident education to reduce patient delays: A pilot study in an academic medical center pain clinic. *Pain Medicine, 16*(2), 312–318.

Wilson, J. C. T. (1981). Implementation of computer simulation projects in health care. *Journal of the Operational Research Society, 32*(9), 825–832.

Chapter 23
Pricing Analytics

Kalyan Talluri and Sridhar Seshadri

1 Introduction

One of the most important decisions a firm has to take is the pricing of its products. At its simplest, this amounts to stating a number (the price) for a single product. But it is often a lot more complicated than that. Various pricing mechanisms such as dynamic pricing, promotions, bundling, volume discounts, segmentation, bidding, and name-your-own-price are usually deployed to increase revenues, and this chapter is devoted to the study of such mechanisms. Pricing and revenue optimization is known by different names in different domains, such as revenue management (RM), yield management, and pricing analytics. One formal definition of revenue management is the study of **how a firm should set and update pricing and product availability decisions across its various selling channels in order to maximize its profitability**. There are several key phrases in this definition: Firms should not only set but also update prices; thus, price setting should be dynamic and depend on many factors such as competition, availability of inventory, and updated demand forecasts. Firms not only set prices but also make product availability decisions; in other words, firms can stop offering certain products at a given price

---

Electronic Supplementary Material The online version of this chapter (https://doi.org/10.1007/978-3-319-68837-4_23) contains supplementary material, which is available to authorized users.

K. Talluri
Imperial College Business School, South Kensington, London, UK
e-mail: kalyan.talluri@imperial.ac.uk

S. Seshadri (✉)
Gies College of Business, University of Illinois at Urbana-Champaign, Champaign, IL, USA
e-mail: sridhar@illinois.edu

(such as the closing of low-fare seats on airlines) or offer only certain assortments in certain channels. Firms might offer different products at different prices across selling channels—the online price for certain products might be lower than the retail price!

The application of pricing and revenue management analytics in business management began in the 1970s. Airline operators like British Airways (then British Overseas Airways Corp.) and American Airlines began to offer differentiated fares for essentially the same tickets. The pioneer of this technique, called yield management, was Bob Crandall. Crandall, who eventually became chief executive of American Airlines, spearheaded a revolution in airline ticket pricing, but its impact would be felt across industries. Hotel chains, such as Marriott International, and parcelling services, like United Parcel Service, have used it to great effect.

These techniques have only become more refined in the decades since. The advent of big data has revolutionized the degree to which analytics can predict patterns of customer demand, helping companies adapt to trends more quickly than ever. Retail chains such as Walmart collect petabytes of data daily, while mobile applications like Uber rely on big data to provide the framework for their business model.

Yet even in its simplest form (a simple posted-price mechanism), pricing is tricky. If you set it too low or too high, you are losing out on revenue. On the other hand, determining the right price, either before or after the sale, may be impossible. Analytics helps; indeed, there are few other areas where data and analytics come together as nicely to help out the manager. That is because pricing is inherently about data and numbers and optimization. There are many unobservable factors such as a customer's willingness to pay and needs, so modeling plays a critical role. Here too, we restrict ourselves by and large to monopoly models, folding in, whenever possible, competitive prices and product features, but do not explicitly model strategic reactions and equilibria. We cover modeling of pricing optimization which by necessity involves modeling customer behavior and constrained optimization.

Moreover, the application of big data techniques to pricing methods raises concerns of privacy. As models become better at understanding customers, companies may find themselves rapidly entering an uncanny valley-like effect, where their clients find themselves disoriented and put off by the amount of precision with which they can be targeted. The European Union's General Data Protection Regulation is explicitly aimed at limiting the use and storage of personal data, necessitating a wide set of reforms by companies across sectors and industries.

The two building blocks of revenue management are developing quantitative models of customer behavior, that is, price-response curves, demand forecasts, market segmentation, etc., and tools of constrained optimization. The first building block is all about capturing details about the consumers at a micro-market level.For example, one might consider which customers shop at what times for which products at a given store of a food retailer. Then, one might model their sensitivity to price, product assortments, and product bundles. This data can be combined with inventory planning system information to set prices. The second building block reflects the fact that price should depend on availability. Therefore, capacity constraints play an important role in price optimization. In addition, there could be other simple constraints, such as inventory availability, route structure of an airline, network constraints that equate inflow and inventory to outflows, and consumption and closing inventory. More esoteric constraints are used to model customer switching behavior when presented with a choice of products or even the strategic behavior of customers in anticipation of a discount or price increase.

What sorts of questions does RM help answer? We have provided a partial list as follows:

- A hotel chain wants guidelines on how to design products for different customer segments. Price is not the only distinguishing feature. For example, hotels sell the same room as different products and at different prices, such as no refund, advance payment required, full refund, breakfast included, access to executive lounge included, etc!
- The owner of a health club wants to know whether the profits will increase if he sets different prices at different times and for different customers.
- A car manufacturer bidding on supply of a fleet of cars would like to know how to bid for a contract based on past bid information, current competition, and other factors to maximize expected profitability.
- A retail chain needs to decide when and how much to discount prices for a fashion good during a selling season to maximize expected revenue.
- In a downtown hotel, business travelers book closer to the date of stay than leisure travelers. Leisure travelers are more price sensitive than business travelers. The hotel manager has to decide how many rooms to save for business travelers.
- A hotel manager has to determine how to price a single-day stay vs. a multiple-day stay.
- A car rental agency has to decide whether it is profitable to transport cars from one location to another in anticipation of demand surge.
- A basketball franchise wants to explore differential pricing. It wants to evaluate whether charging different prices for different days, different teams, and different times of the day will increase revenue.
- How does the freedom to name your own price (invented by Priceline) work?

The analytics professional will recognize the opportunity to employ almost every tool in the analytics toolkit to solve these problems. First, data is necessary at the right granularity and from different sources including points of sales and reservation systems, surveys, and social media chatter. Information is also required on competitive offerings and prices. Data has to be gathered not only about sales but also no-shows and cancellations. Many a times, bookings are done in groups. These bookings have their own characteristics to record. Second, these data have to be organized in a form that reveals patterns and trends, such that revenue managers, product managers, and operations managers can coordinate their actions to change in demand and supply. Third, demand has to be forecast well into the future and at every market level. Some recent systems claim to even predict demand at a granularity of a single customer. The optimal RM solutions of prices and product availability have to be made available in an acceptable format to sales persons, agents, auction houses, etc. Thus, RM requires information, systems, technologies, and training, as well as disciplined action to succeed.

In the rest of the chapter, we provide a glimpse into the more commonly used RM techniques. These include capacity control, overbooking, dynamic pricing, forecasting for RM, processes used in RM, and network RM. We conclude with suggestions for further reading.

2 Theory

The factors that affect pricing are as follows:

1. The nature of the product or service (features, delivery, conditions of sale, channel) and the competing alternatives (both direct and indirect)
2. Customers' valuation of the product, needs, and purchasing behavior

The reader might have noticed that we did not include costs. That is because in this chapter, we do not discuss simple cost-based pricing such as markup rules (e.g., 10% margin), not because it is not practiced—indeed, it perhaps is the most popular methodology due to its simplicity—but because there is not much to say about such simple rules. Rather, we concentrate on market-based pricing that sets the price based on products and competition and what the consumer is willing to pay. Cost does play a role as a lower bound on the price, but the real decision is in setting the margin above the cost as a function of the market, product features, and customer preferences.

So we need a model of a market and the customer purchasing behavior. Both are somewhat abstract and imprecise concepts and only partially observable, but we do our best in modeling them so as to extract insight from observed data.

2.1 Basic Price Optimization Model

Let p represent price (of a single product) and $D(p)$ the demand at that price (assuming all other features are held the same). Revenue optimization is to find the price p that maximizes $R(p) = pD(p)$, and profit optimization is to maximize $(p - c)D(p)$ when c is the cost of producing one unit.

$D(p)$ is called the demand function, and it is natural to assume that it decreases as we increase price. It is also customary to assume it has some functional form, say

$D(p) = a - bp$ or $D(p) = ap^b$ where a and b are the *parameters* of the model that we estimate based on observed data.

> Example: Say, based on data, we estimate that demand for a certain product is $D(p) = 35.12 - 0.02p$ (i.e., demand is assumed to have a linear form $D(p) = a - bp$, where we calibrated $a = 35.12$ and $b = 0.02$). The revenue optimization problem is to maximize $p \times (35.12 - 0.02p)$. From calculus (take the derivative of the revenue function and set it to 0, so $35.12 - 2 \times 0.02p = 0$, and solve it for p), we obtain the optimal price to be $p^* = \frac{35.12}{2 \times 0.02} = 878$.

Capacity restrictions introduce some complications, but, at least for the single product case, are still easy to handle. For instance, in the above example, if price is $878, the demand estimate is $35.12 - 0.02 \times 878 = 17.56$. If, however, we have only ten units, it is natural to raise the price so that demand is exactly equal to 10, which can be found by solving $10 = 35.12 - 0.02p$ or $p = 1256$.

2.2 Capacity Controls

In this section, we look at the control of the sale of inventory when customers belong to different types or, using marketing terminology, segments. The segments are assumed to have different willingness to pay and also different preferences as to when and how they purchase. For example, a business customer for an airline may prefer to purchase close to the departure date, while a leisure customer plans well ahead and would like a guaranteed flight reservation. The original motivation of revenue management was an attempt to make sure that we set aside enough inventory for the late-coming, higher-paying business customer, yet continue selling at a cheaper price to the price-sensitive leisure segment.

We assume that we created products with sale restrictions (such as advance purchase required or no cancellations or weekend stay), and we label each one of these products as *booking classes*, or simply *classes*. All the products share the same physical inventory (such as the rooms of the hotel or seats on a flight). In practice, multiple RM products may be grouped into classes for operational convenience or control system limitations. If such is the case, the price attached to a class is some approximation or average of the products in that class.

From now on, we assume that each booking request is for a single unit of inventory.

2.2.1 Independent Class Models

We begin with the simplest customer behavior assumption, the *independent class* assumption: Each segment is identified with a single product (that has a fixed price), and customers purchase only that product. And if that product is not available for sale, then they do not purchase anything. Since segments are identified one-to-one with classes, we can label them as class 1 customers, class 2 customers, etc.

The goal of the optimization model is to find booking limits—the maximum number of units of the shared inventory we are willing to sell to that product—that maximize revenue.

Let's first consider the two-class model, where class 1 has a higher price than class 2, that is, $f_1 > f_2$, and class 2 bookings come first. The problem would be trivial if the higher-paying customers come first, so the heart of the problem is to decide a "protection level" for the later higher-paying ones and, alternately, a "booking limit" on when to stop sales to the lower-paying class 2 customers.

Say we have an inventory of r_0. We first make forecasts of the demand for each class, say based on historic demand, and represent the demand forecasts by D_j, $j = 1, 2$.

How many units of inventory should the firm protect for the later-arriving, but higher-value, class 1 customers? The firm has only a probabilistic idea of the class 1 demand (the problem would once more be trivial if it knew this demand with certainty).

The firm has to decide if it needs to protect r units for the late-arriving class 1 customers. It will sell the rth unit to a class 1 customer if and only if $D_1 \geq r$, so the *expected marginal revenue* from the rth unit is $f_1 P(D_1 \geq r)$. Intuitively, the firm ought to accept a class 2 request if and only if f_2 exceeds this marginal value or, equivalently, if and only if

$$f_2 \geq f_1 P(D_1 \geq r). \tag{23.1}$$

The right-hand side of (23.1) is decreasing in r. Therefore, there will be an optimal protection level for class 1, denoted r_1^*, such that we accept class 2 if the remaining capacity exceeds r_1^* and reject it if the remaining capacity is r_1^* or less. Formally, r_1^* satisfies

$$f_2 < f_1 P(D_1 \geq r_1^*) \text{ and } f_2 \geq f_1 P(D_1 \geq r_1^* + 1). \tag{23.2}$$

In practice, there are usually many products and segments, so consider $n > 2$ classes. We continue with the independent class assumption and that demand for the n classes arrives in n stages, one for each class in order of revenue with the highest-paying segment, class 1, arriving closest to the inventory usage time.

Let the classes be indexed so that $f_1 > f_2 > \cdots > f_n$. Hence, class n (the lowest price) demand arrives in the first stage (stage n), followed by class $n - 1$ demand in stage $n - 1$, and so on, with the highest-price class (class 1) arriving in the last stage (stage 1). Since, there is a one-to-one correspondence between stages and classes, we index both by j.

We describe now a heuristic method called the expected marginal seat revenue (EMSR) method. This heuristic method is used because solving the n class problem optimally is complicated. The heuristic method works as follows:

Consider stage $j+1$ in which the firm wants to determine protection level r_j for class j. Define the aggregated future demand for classes $j, j-1, \ldots, 1$ by

$$S_j = \sum_{k=1}^{j} D_k,$$

and let the weighted-average revenue (this is the heuristic part) from classes $1, \ldots, j$, denoted $\bar{f}_j$, be defined by

$$\bar{f}_j = \frac{\sum_{k=1}^{j} f_k E[D_k]}{\sum_{k=1}^{j} E[D_k]}, \tag{23.3}$$

where $E[D_j]$ denotes the expected class j demand.

Then, the EMSR protection level for class j and higher, r_j, is chosen by (23.2), to satisfy

$$P(S_j > r_j) = \frac{f_{j+1}}{\bar{f}_j}. \tag{23.4}$$

It is convenient to assume demand for each class j is normally distributed with mean μ_j and variance σ_j^2, in which case

$$r_j = \mu + z\sigma,$$

where $\mu = \sum_{k=1}^{j} \mu_k$ is the mean and $\sigma^2 = \sum_{k=1}^{j} \sigma_k^2$ is the variance of the aggregated demand to come at stage $j+1$ and $z = \Phi^{-1}(1 - f_{j+1}/\bar{f}_j)$ and $\Phi^{-1}(\cdot)$ is the inverse of the standard normal c.d.f. One repeats this calculation for each j.

The EMSR heuristic method is very popular in practice as it is very simple to program and is robust with acceptable performance (Belobaba 1989). One can do the calculation easily enough using Excel, as it has built-in functions for the normal distribution and its inverse.

2.3 Overbooking

There are many industries where customers first reserve the service and then use it later. Some examples are hotels, airlines, restaurants, and rental cars. Now, when a customer reserves something for future use, their plans might change in the meantime. A *cancellation* is when the customer explicitly cancels the reservation, and a *no-show* is when they do not notify the firm but just do not show up at the scheduled time for the service. What the customer does depends on the reservation

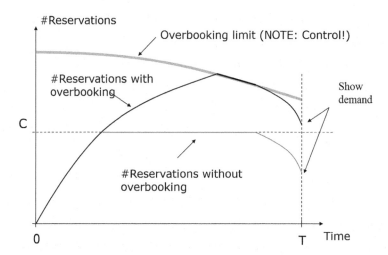

Fig. 23.1 Evolution of the overbooking limit

policies. If there is an incentive like partial or full refund of the amount, customers tend to cancel. If there is no refund, they will just opt for no-show.

Overbooking is a practice of taking more bookings than capacity, anticipating that a certain fraction of the customers would cancel or opt for no-show. This leads to better capacity utilization, especially on high-demand days when the marginal value of each unit of inventory is very high. The firm, however, has to be prepared to handle a certain number of customers who are denied service even though they have paid for the product and have a reservation contract. In many industries, the benefits of better capacity utilization dominate the risk of denying service, and overbooking has become a common practice.

Firms control overbooking by setting an upper limit on how much they overbook (called the overbooking limit). Typically as they come closer to the inventory usage time (say a flight departing), they have a better picture of demand, and they reduce the risk of overbooking if there appears to be high demand with few cancellations. Figure 23.1 shows the dynamics of a typical evolution of the overbooking limit. As the usage date nears and the firm fears that it might end up denying service to some customers, it brings down the overbooking limit toward physical capacity faster (can even be less than the current number of reservations on-hand also, to prevent new bookings).

Overbooking represents a trade-off: If the firm sells too many reservations above its capacity, it risks a scenario where more customers show up than there is inventory and the resulting costs in customer goodwill and compensation. If it does not overbook enough, it risks unsold inventory and an opportunity cost. Overbooking models are used to find the optimal balance between these two factors. We describe one such calculation below that, while not completely taking all factors into consideration, highlights this trade-off mathematically. It is reminiscent of the classical newsvendor model from operations management.

Let C_{DB} denote the *cost of a denied boarding*, that is, the estimated cost of denying service to a customer who has a reservation (which, as we mentioned earlier, includes compensation, loss of goodwill, etc.). Let C_u denote the opportunity cost of underused capacity, typically taken as the expected revenue for a unit of inventory. The overbooking limit we have to decide then is $\theta > C$, where C is the physical capacity.

For simplicity, we assume the worst and that demand will exceed overbooking limit, that is, we will be conservative in setting our limit. Let N be the number of no-shows/cancellations. Since we are not sure of the number of cancellations or no-shows, we model it as a random variable, say as a binomial random variable with parameters θ, p where p is the probability of a cancellation or no-show. Then, the number of customers who actually show up is given by $\theta - N$ (recall demand is conservatively assumed to be always up to θ).

Next, we pose the problem as the following marginal decision: Should we stay at the current limit θ or increase the limit to $\theta + 1$, continuing the assumption that demand is high and will also exceed $\theta + 1$? Two mutually exclusive events can happen: (1) $\theta - N < C$. In this case by moving the limit up by 1, we would increase our profit, or in terms of cost by $-C_u$. (2) $\theta - N \geq C$, and we incur a cost of C_{DB}. So the expected cost per unit increase of θ is

$$-C_u Pr(\theta - N < C) + C_{DB} Pr(\theta - N \geq C).$$

Note that this quantity starts off negative (when $\theta = C$) as $Pr(\theta - N \geq C) = 0$ at that point, but as we keep increasing θ, it decreases, and the C_{DB} risk increases. So that tells us that we can increase profit as long as this is negative but incur a cost if it is positive, and the best decision is to stop when this quantity is 0. This results in a nice equation to determine the optimal θ,

$$-C_u Pr(\theta - N < C) + C_{DB} Pr(\theta - N \geq C) = -C_u(1 - Pr(\theta - N \geq C))$$
$$+ C_{DB} Pr(\theta - N \geq C)0,$$

or set θ, such that

$$Pr(\theta - N \geq C) = \frac{C_u}{C_u + C_{DB}}.$$

If we let $S(\theta)$ be the number of people who show up, an alternate view is that we need to set θ such that $Pr(S(\theta) \leq C) = \frac{C_u}{C_u + C_{DB}}$. If no-shows happen with probability p, shows also follow a binomial distribution with probability $1 - p$. So set θ such that (writing in terms of $\leq$ to suit Excel calculations)

$$Pr(S(\theta) \leq C) = 1 - \frac{C_u}{C_u + C_{DB}}.$$

Fig. 23.2 Computations for the overbooking example

	A	B	C	D	E	F
1		Critical ratio		0.3 p=		0.1
2	theta	X such that PR(Show (Theta-No-shows) >= X) = 0.3				
3	100	92				
4	101	93				
5	102	93	BINOM.INV(A3, 1-$E$1, 1-$C$1)			
6	103	94				
7	104	95	BINOM.INV function			
8	105	96	Returns the smallest value for which the			
9	106	97	cumulative binomial distribution is greater than			
10	107	98	or equal to a criterion value			
11	108	99				
12	109	100 C	So we take Pr(Shows <= C) = 1 - 0.3.			
13	110	101	If no-shows occur with probability, shows occur			
14	111	102	with probability 0.9.			
15	112	103	So we use BINOM.INV(theta, 1-p, 1- Critical			
16	113	103	Ratio). We want to find the theta such that this			
17	114	104	value is equal to our capacity C			
18	115	105				
19	116	106				
20	117	107				
21	118	108				
22	119	109				
23	120	110				
24	121	111				
25	122	112				
26	123	113				
27	124	113				

Example (Fig. 23.2): Say $C_u = \$150$ and $C_{DB} = \$350$. To calculate the overbooking limit, we first calculate the critical ratio:

$$\frac{C_u}{C_u + C_{DB}} = \frac{150}{500} = 0.3.$$

If we assume the distribution of N is approximately normal (quite accurate when $\theta p \geq 5$) with mean= θp and standard deviation = $\sqrt{\theta p}$, we can in fact use the InverseNormal (NORMINV in Excel) to do the calculations. Suppose $C = 100$ and $p = 0.1$ (that is 10% probability that a reservation will eventually cancel). The optimal solution is to overbook nine seats.

2.4 Dynamic Pricing

Dynamic pricing is a version of revenue management, simpler in some sense, but also requiring some close monitoring. There are usually no explicit virtual products aimed at different segments.[1] Rather, the firm changes the price of the product overtime as it observes changes in the many factors that would affect demand: such as time itself (e.g., because higher-valuation customers come after lower-valuation customers), weather, the customer mix, and competition.

---

[1] Although there could be, but let us not complicate unnecessarily.

Over the last few years, dynamic pricing has taken on three distinct flavors: surge pricing, as practiced by Uber, Lyft, and utility companies; repricing, or competition-based pricing, as practiced by sellers on Amazon marketplace; and, finally, markdown or markup pricing, where prices are gradually decreased (as in fashion retail) or increased (as in low-cost airlines) as a deadline approaches.

2.4.1 Surge Pricing: Matching Demand and Supply

This is the newest and perhaps most controversial of dynamic pricing practices. The underlying economic reason is sound and reasonable. When there is more demand than supply, the price has to be increased to clear the market—essentially the good or service is allocated to those who value it the most. When asked to judge the fairness of this, most consumers do not have a problem with this principle, for instance, few consider auctions to be unfair.

However, when applied to common daily items or services, many consumers turn indignant. This is due to many reasons: (1) There is no transparency on how the prices move to balance demand and supply. (2) As the prices rise when a large number of people are in great need of it, they are left with a feeling of being price-gouged when they need the service most. (3) The item or service is essential or life-saving, such as pharmaceutical or ambulance service. Uber was a pioneer in introducing surge pricing into an industry used to a regulated fixed-price system (precisely to bring transparency and prevent price-gouging and also to avoid the hassle of bargaining). While initial reactions[2] have been predictable, it has, in a space of a few years, become a fact of life. This shows the importance of a firm believing in the economic rationale of dynamic pricing and sticking to the practice despite public resistance. Of course, consumers should find value in the service itself—as the prices are lower than alternatives (such as regular taxi service) during off-peak times, eventually consumers realize the importance and necessity of dynamic pricing.

2.4.2 Repricing: Competition-Driven Dynamic Pricing

A second phenomenon that has recently taken hold of is called "repricing" used in e-commerce marketplaces such as Amazon.com. It is essentially dynamic pricing driven by competition.

Many e-commerce platforms sell branded goods that are identical to what other sellers are selling. The seller's role in the entire supply chain is little more than stocking and shipping as warranties are handled by the manufacturer. Service does play a role, but many of the sellers have similar reviews and ratings, and often price is the main motivation of the customer for choosing one seller over the other, as the e-commerce platform removes all search costs.

---

[2] https://nyti.ms/2tybWiV, https://nyti.ms/2uwkLXR. Accessed on May 21, 2018.

Prices however fluctuate, the reasons often being mysterious. Some possible explanations are the firms' beliefs about their own attractiveness (in terms of ratings, reviews, and trust) compared to others and their inventory positions—a firm with low inventories may want to slow down sales by pricing higher. Another possible reason is that firms assess the profiles of customers who shop at different times of day and days of the week. A person shopping late at night is definitely not comparison-shopping from within a physical store, so the set of competitors is reduced.

Note that an e-commerce site would normally have more variables to price on, such as location and the past customer profile, but a seller in a marketplace such as Amazon or eBay has only limited information and has to put together competitor information from scraping the website or from external sources.

Repricing refers to automated tools, often just rules based, that change the price because of competitor moves or inventory. Examples of such rules are given in the exercise at the end of this chapter.

2.4.3 Markdown and Markup Pricing: Changing Valuations

Markdown pricing is a common tactic in grocery stores and fashion retailing. Here, the product value itself deteriorates, either physically because of limited shelf-life or as in the fashion and consumer electronics industry, as fresh collections or products are introduced. Markdowns in fashion have a somewhat different motivation from the markdowns of fresh produce. In fashion retail, the products cannot be replenished within the season because of long sales cycles, while for fresh groceries, the sell-by date reduces the value of the product because fresher items are introduced alongside.

Markdown pricing, as the name indicates, starts off with an initial price and then, gradually at various points during the product life cycle, reduces the price. The price reductions are often in the form of 10% off, 20% off, etc., and sometimes coordinated with advertised sales. At the end of the season or at a prescribed date, the product is taken off the shelf and sold through secondary channels at a steeply discounted price, sometimes even below cost. This final price is called the salvage value.

The operational decisions are how much to discount and when. There are various restrictions and business rules one has to respect in marking down, the most common one being once discounted, we cannot go back up in price (this is what distinguishes markdown pricing from promotions). Others limit the amount and quantities of discounting.

The trade-off involved is similar to what is faced by a newsvendor: Discounting too late would lead to excess inventory that has to be disposed, and discounting too soon will mean we sell off inventory and potentially face a stock-out.

In contrast to markdown pricing where prices go down, there are some industries that practice dynamic pricing with prices going up as a deadline approaches. Here, the value of the product does go down for the firm, but the customer mix may be changing when customers with higher valuations arrive closer to the deadline (either the type and mix of customers might be changing or even for the same customer, their uncertainty about the product may be resolved).

2.5 Forecasting and Estimation

For reasons of practicality, we try to keep models of demand simple. After all, elaborate behavioral models of demand would be useless if we cannot calibrate them from data and to optimize based on them. In any case, more complicated models do not necessarily mean they predict the demand better, and they often are harder to manage and control.

In this section, we concentrate on three simple models of how demand is explained as a function of price. All three are based on the idea of a potential population that is considering purchase of our product. The size of this population is M. Note that M can vary by day or day of week or time of day. Out of this potential market, a certain fraction purchase the product. We model the fraction as a function of price, and possibly other attributes as well.

Let $D(p)$ represent demand as a function of price p.

- In the *additive model of demand*,

$$D(p) = M(a + bp)$$

where a and b are parameters that we estimate from data. If there are multiple products, demand for one can affect the other. We can model demand in the presence of multiple products as

$$D(p_i) = M(a_i + b_i p_i + \sum_{j \neq i} b_{ij} p_j).$$

That is, demand for product i is a function of not just the price of i but also the prices of the other products, p_j, $j \neq i$. The parameters a_i, b_i, and b_{ij} are to be estimated from data.

This model can lead to problems at the extremes as there is no guarantee that the fraction is between 0 and 1.

- In the *multiplicative model of demand*,

$$D(p) = M(ap^b).$$

When there are multiple products, the model is

$$D(p_i) = M(a_i \prod p_j^{b_{ij}}).$$

This is usually estimated by taking the natural logarithms on both sides, so it becomes linear in the parameters. However, this model has to be handled with care in optimization as it can give funny results—essentially when we are optimizing both revenue and profits, the problem can become unbounded.

- *Choice model of demand*: In the choice model, each customer is assumed to make a *choice* among the available products. The following is sometimes called the multinomial-logit (MNL) model,

$$D(p_i) = M \frac{e^{a_i+b_i p_i}}{1 + \sum e^{a_j+b_j p_j}},$$

where e stands for the base of the natural logarithm. Note that this model has far fewer parameters than either the additive or multiplicative model and naturally limits the fraction to always lie between 0 and 1! This is the great advantage of this model.

We show in the exercises how these models can be used for price optimization. The case study on airline choice modeling (see Chap. 26), has a detailed exercise on estimation and use of choice models for price optimization and product design.

2.6 Processes for Capacity Control

Fixing prices for each product aimed at a segment, as outlined in Sect. 2.2.1, and controlling how much is sold at each price requires that we monitor how many bookings have been taken for each product and closing sales at that price whenever we sold enough.

So the sequence is (1) forecasting the demand for each RM product for a specific day and then (2) optimizing the controls given the forecasts and (3) controlling real-time sales for each product so they do not exceed the booking limits for that product.

We list below the main control forms used in RM industries. Because of the limitations of distribution systems that were designed many years ago, the output of our optimization step has to conform to these forms.

- *Nested allocations* or *booking limits*: All the RM products that share inventory are first ranked in some order, usually by their price.[3] Then, the remaining capacity is allocated to these classes, but the allocations are "nested," so the higher class has access to all the inventory allocated to a lower class. For example, if there are 100 seats left for sale and there are two classes, Y and B, with Y considered

---

[3] As we mentioned earlier, it is common to group different products under one "class" and take an average price for the class. In the airline industry, for instance, the products, each with a fare basis code (such as BXP21), are grouped into fare classes (represented by an alphabet, Y, B, etc.).

a higher class, then an example of a nested allocation would be Y100 B54. For example, if 100 Y customers were to arrive, the controls would allow sale to all of them. If 60 B customers were to show up, only 54 would be able to purchase. B is said to have an *allocation* or a *booking limit* of 54. Another terminology that is used is (nested) *protections*: Y is said to have (in this example) a protection of 46 seats.

The allocations are posted on a central reservation system and updated periodically (usually overnight). After each booking, the reservation system updates the limits. In the above example, if a B booking comes in, then (as the firm can sell up to 54 seats to B) it is accepted, so the remaining capacity is 99, and the new booking limits are Y99 B53. Suppose a Y booking comes in and is accepted, there are a couple of ways the firm can update the limits: Y99 B54 or Y99 B53. The former is called *standard* nesting and the latter *theft* nesting.

- *Bid prices*: For historic reasons, most airline and hotel RM systems work with nested allocations, as many global distribution systems (such as Amadeus or Sabre) were structured this way. Many of these systems allow for a small number of limits (10–26), so when the number of RM products exceeds this number, they somehow have to be grouped to conform to the number allowed by the system.

  The booking limit control is perfectly adequate when controlling a single resource (such as a single flight leg) independently (independent of other connecting flights, for instance), but we encounter its limitations when the number of products using that resource increases, say to more than the size of inventory. Consider network RM, where the products are itineraries, and there could be many itineraries that use a resource (a flight leg)—the grouping of the products a priori gets complicated and messy (although it has been tried, sometimes called *virtual nesting*).

  A more natural and appropriate form of control, especially for network RM, is a threshold-price form of control called *bid price* control. Every resource has a non-negative number called a bid price associated with it. A product that uses a combination of resources is sold if the price of the product exceeds the sum of the bid prices of the resources that the product uses. The bid prices are frequently updated as new information comes in or as the inventory is sold off. The next section illustrates the computation of bid prices.

2.7 Network Revenue Management

The Need for Network Revenue Management: In many situations, the firm has to compute the impact of a pricing, product offering, or capacity management decision on an entire network of resources. Consider the case of an airline that offers flights from many origins and to many destinations. In this case, passengers who are flying to different origin-destination (OD) pairs might use the same flight leg.

Say an airline uses Chicago as a hub. It offers itineraries from the East Coast of the USA, such as New York and Boston, to cities in the West, such as LA and San Francisco. Passengers who fly from New York to Chicago include those who travel directly to Chicago and also those traveling via Chicago to LA, San Francisco, etc. The firm cannot treat the flight booking on the New York to Chicago flight independently but has to consider the impact on the rest of the network as it reduces the capacity for future customers wishing to travel to LA and San Francisco.

Similarly, there are inter-temporal interactions when we consider multi-night-stay problems, for example, a car or a hotel room when rented over multiple days. Hence, the Monday car rental problem impacts the Tuesday car rental problem and so forth. Other examples of network revenue management include cargo bookings that consume capacity on more than one OD pair or television advertising campaigns that use advertisement slots over multiple shows and days.

Suboptimality of Managing a Network One Resource at a Time: It is easy to demonstrate that it is suboptimal to manage each resource separately. Consider a situation in which the decision maker knows that the flight from city A to B, with posted fare of 100, will be relatively empty, whereas the connecting flight from B to C, fare of 200, will be rather full. Some passengers want to just travel from A to B, and there are others who want to fly from A to C. Both use the AB leg. In this case, what will be the order of preference for booking a passenger from A to B vis-a-vis one who wants to travel from A to C and pays 275? Intuitively, we would remove the 200 from 275 and value the worth of this passenger to the airline on the AB leg as only 75. Therefore, total revenue might not be a good indicator of value. In this example, allocation of the 275 according to the distance between A–B and B–C might also be incorrect if, for example, the distances are equal. Allocations based on utilization or the price of some median-fare class would also be inappropriate. Therefore, any formula that allocates the value to different legs of the itinerary has to consider both the profitability of each leg and the probability of filling the seat.

An Example to Illustrate an Inductive Approach: Consider a simple example in which, as above, there is a flight from city A to city B and a connecting flight from B to C. The single-leg fares are 200 and 200, whereas the through fare from A to C is 360. There is exactly one seat left on each flight leg. Assume, as is done typically in setting up the optimization problem, time is discrete. It is numbered backward so that time n indicates that there are n time periods left before the first flight takes place. Also, the probability of more than one customer arrival in a time period is assumed to be negligible. Thus, either no customer arrives or one customer arrives. We are given there are three time periods left to go. In each period, the probability of a customer who wants to travel from A to B is 0.2, from B to C is 0.2, and from A to C is 0.45; thus, there is a probability of zero arrivals equal to 0.15. In this example, the arrival probabilities are the same in each period. It is easy to change the percentages overtime. What should be the airline's booking policy with one seat left on each flight?

This problem is best solved through backward induction. Define the state of the system as (n, i, j) where n is the period and i and j the numbers of unsold seats on legs AB and BC.

Consider the state $(1, 1, 1)$. In this state, in the last period, the optimal decision is to sell to whichever customer who arrives. The expected payoff is $0.4 \times 200 + 0.45 \times 360 = 242$. We write the value in this state as $V(1, 1, 1) = 242$. The expected payoff in either state $(1, 0, 1)$ or $(1, 1, 0)$ is $0.2 \times 200 = 40$. We write $V(1, 0, 1) = V(1, 1, 0) = 40$. For completeness, we can write $V(n, 0, 0) = 0$.

When there are two periods to go, the decision is whether to sell to a customer or wait. Consider the state $(2, 1, 1)$ and the value of being in this state, $V(2, 1, 1)$. Obviously, it is optimal to sell to an AC customer. Some calculations are necessary for whether we should sell to an AB or BC customer:

If an AB customer arrives: If we sell, we get $200 + V(1, 0, 1)$ (from selling the seat to a BC customer if they arrive in the last period) $= 240$. Waiting fetches $V(1, 1, 1) = 242$. Therefore, it is best to not sell.

If a BC customer arrives: Similar to the case above, it is better to wait.

If an AC customer arrives: Sell. We get 360.

Thus, $V(2, 1, 1) = 0.4 \times 242 + 0.45 \times 360 + 0.15 \times V(1, 1, 1) = 295.1$.

We can compute $V(2, 1, 0) (= V(2, 0, 1)) = 0.2 \times 200 + 0.8 \times V(1, 1, 0) = 72$.

In period 3, in the state $(3, 1, 1)$, it is optimal to sell if an AC customer arrives. If an AB (or BC) customer arrives, by selling we get $200 + V(2, 0, 1)$ (or $V(2, 1, 0)) = 272$. This is smaller than $V(2, 1, 1)$. Therefore, it is better to wait. This completes the analysis.

The reader can easily generalize to the case when there are different combinations of unsold seats. For example, having solved entirely for the case when a maximum of (k, m) seats are left in the last period, one can use backward induction to solve for the same when there are two periods to go, etc.

The backward induction method is called dynamic programming and can become quite cumbersome when the network is large and the number of periods left is large. It is stochastic in nature because of the probabilities. The astute reader might have noticed that these probabilities can be generated by using an appropriate model of customer choice that yields the probability of choosing an itinerary when presented with a set of options.

Bid Price Approach: Recall the bid price control of Sect. 2.6. The operating rule is to accept an itinerary if its fare exceeds the sum of bid prices on each leg used by the itinerary, if there is sufficient capacity left. The bid prices can be thought of as representing the marginal value of the units of capacity remaining.

But how do we calculate these bid prices? Many different heuristic approaches have been proposed and analyzed, both in the academic and in the practitioner literature (see, e.g., the references at the end of this chapter). These range from solving optimization models such as a deterministic linear program (DLP), a stochastic linear program (SLP), and approximate versions of the dynamic program (DP) illustrated above to a variety of heuristics. (The usual caveat is that the use of bid prices in this manner need not result in the optimal expected revenue. Take, for example, the decision rule that we derived using the dynamic program with three periods to go and one seat that is available on each flight leg. We need two bid prices (one per leg) such that each is greater than the fare on the single leg but their sum is less than the fare on the combined legs. Thus, we need prices b_1 and b_2 such that $b_1 > 200, b_2 > 200, b_1 + b_2 \leq 360$. Such values do not exist.)

In this chapter, we illustrate the DLP approach as it is practical and is used by hotels and airlines to solve the problem. In order to illustrate the approach, we shall first use a general notation and then provide a specific example. We are given a set of products, indexed by $i = 1$ *to* I. The set of resources is labeled $j = 1$ *to* J. If product i uses resource j, let $a_{ij} = 1$ *else* 0. Let the revenue obtained from selling one unit of product i be R_i. We are given that the demand for product i is D_i and the capacity of resource j is C_j. Here, the demand and revenue are deterministic.

In the context of an airline, the products would be the itineraries, the resources are the flight legs, the coefficient $a_{ij} = 1$ if itinerary i uses flight leg j else 0, the capacity would be the unsold number of seats of resource j, and the revenue would be the fare of product i.

In a hotel that is planning its allocations of rooms for the week ahead, the product could be a stay that begins on a day and ends on another day, such as check-in on Monday and checkout on Wednesday. The resource will be a room night. The capacity will be the number of unsold rooms for each day of the week. The coefficient $a_{ij} = 1$ if product i requires stay on day j (e.g., a person who stays on Monday and Tuesday uses one unit of capacity on each room night). The revenue will be the price charged for the complete stay of product i. For a car rental problem, replace room night with car rental for each day of the week. Note that it is possible that two products use the same set of resources but are priced differently. Examples of these include some combination of room sold with/without breakfast, allowing or not allowing cancellation, taking payment ahead or at end of stay, etc.

The problem is to decide how many reservations X_i to accept of each product i. The general optimization problem can be stated as follows (DLP):

$$\max_{X} \sum_{i=1\,to\,I} R_i X_i$$

s.t

$$\sum_{i} a_{ij} X_i \leq C_j, \quad j = 1\,to\,J, \tag{23.5}$$

$$X_i \leq D_i, \quad i = 1\,to\,I, \tag{23.6}$$

$$X_i = 0, 1, 2, \ldots, I$$

Here, constraints (23.5) make sure we don't sell more reservations than the capacity on each flight (on average); constraints (23.6) ensure that the number of reservations for an itinerary is less than the demand for that itinerary (mean of the demand—

remember this is just an approximation). The value of this optimization problem can in fact be shown to be an upper bound on the maximum expected revenue.

The following data is necessary to solve this problem: The demands have to be forecast. The capacity that is available will depend on the number of reservations that have already been accepted and has to be computed. The prices might be somewhat unknown because they fluctuate depending on the market conditions and the deals that are to be negotiated. The DLP will require estimates of the expected price. Moreover, it is assumed that there are no cancellations. It is also assumed there are no no-shows and that overbooking is not allowed. Some variations of the basic DLP do account for these factors.

Despite its limitations and simplicity, DLP is often used in practice because it is fast, convenient, and it uses readily available data. Frequent re-optimization and use of the most recent solution can yield good results. A concrete example will help illustrate the approach.

2.7.1 Case Study

A small hotel is planning its allocation of rooms for the week after the next week. For the purpose of planning, it assumes that the customers who stay on weekends belong to a different segment and do not stay over to Monday or check in before Saturday. It sells several products. Here, we consider the three most popular ones that are priced on the average at $125, $150, and $200. These rates are applicable, respectively, if the customer (1) pays up front, (2) provides a credit card and agrees to a cancellation charge that applies only if the room reservation is cancelled with less than 1 day to go, and (3) is similar to (2) but also provides for free Internet and breakfast (that are virtually costless to the hotel). Customers stay for 1, 2, or 3 nights. The demand forecasts and rooms already booked are shown in Table 23.1. The hotel has a block of 65 rooms to allocate to these products.

In this example, there are 45 products and five resources. Each demand forecast pertains to a product. The available rooms on each of the 5 days constitute the five different resources. The Monday 1-night-stay product uses one unit of Monday capacity. The Monday 2-night-stay product uses one unit each of Monday and Tuesday room capacity, etc. The rooms available are 65 minus the rooms sold.

Table 23.1 Demand forecasts and rooms already booked

	125			150			200			Rooms sold
Monday	7	12	17	17	6	4	9	5	1	12
Tuesday	17	3	10	2	7	4	8	1	1	22
Wednesday	2	19	15	3	3	4	2	4	2	31
Thursday	15	11	0	20	9	0	6	0	0	24
Friday	20	0	0	9	0	0	4	0	0	15

		125	125	125	150	150	150	200	200	200	Rooms Sold		
	Nights Stay	1	2	3	1	2	3	1	2	3			
	Monday	7	12	17	17	6	4	9	5	1	12		
	Tuesday	17	3	10	2	7	4	8	1	1	22		
	Wednesday	2	19	15	3	3	4	2	4	2	31		
	Thursday	15	11	7	20	9	6	6	0	0	24		
	Friday	20	0	7	9	5	7	4	0	0	15		
	Decision Variables										Rooms Sold	Rooms Available	Revenue
	Monday	1	1	1	1	1	1	1	1	1	9	53	2850
	Tuesday	1	1	1	1	1	1	1	1	1	15	43	2850
	Wednesday	1	1	1	1	1	1	1	1	1	18	34	2850
	Thursday	1	1	1	1	1	1	1	1	1	18	41	2850
	Friday	1	1	1	1	1	1	1	1	1	18	50	2850
			Total Revenue	14250									

Fig. 23.3 Decisions variables

Fig. 23.4 Excel Solver setup

	125	125	125	150	150	150	200	200	200	Rooms Sold		
Nights Stay	1	2	3	1	2	3	1	2	3			
Monday	7	12	17	17	6	4	9	5	1	12		
Tuesday	17	3	10	2	7	4	8	1	1	22		
Wednesday	2	19	15	3	3	4	2	4	2	31		
Thursday	15	11	0	20	9	0	6	0	0	24		
Friday	20	0	0	9	0	0	4	0	0	15		
Decision Variables										Rooms Sold	Rooms Available	Revenue
Monday	7	4	0	17	6	4	9	5	1	53	53	12425
Tuesday	0	0	0	2	7	4	8	1	1	43	43	6800
Wednesday	0	0	0	3	1	4	2	4	2	34	34	5750
Thursday	0	2	0	8	9	0	6	0	0	41	41	5600
Friday	20	0	0	9	0	0	4	0	0	50	50	4650
		Total Revenue	35225									

Fig. 23.5 DLP solution

There are 45 decision variables in this problem. The screenshots of the data, decision variables (yellow), and Excel Solver setup are shown in Figs. 23.3 and 23.4.

Solving this problem as a linear program or LP (choose linear and non-negative in Solver), we obtain the solution shown in Fig. 23.5.

Table 23.2 Shadow price of constraints

Cell	Name	Final value	Shadow price	Constraint RH side	Allowable increase	Allowable decrease
$M11$	Monday rooms sold	53	100	53	2	0
$M12$	Tuesday rooms sold	43	150	43	0	2
$M13$	Wednesday rooms sold	34	150	34	2	1
$M14$	Thursday rooms sold	41	150	41	12	7.999999999
$M15$	Friday rooms sold	50	100	50	7.999999999	2.000000001

Note: Shadow prices were rounded to the nearest integer value

Table 23.3 Is rate class open or closed?

	125			150			200		
Night stay	1	2	3	1	2	3	1	2	3
Monday	OPEN	OPEN	CLOSED	OPEN	OPEN	OPEN	OPEN	OPEN	OPEN
Tuesday	CLOSED	CLOSED	CLOSED	OPEN	OPEN	OPEN	OPEN	OPEN	OPEN
Wednesday	CLOSED	CLOSED	CLOSED	OPEN	OPEN	OPEN	OPEN	OPEN	OPEN
Thursday	CLOSED	OPEN	OPEN	OPEN	OPEN	OPEN	OPEN	OPEN	OPEN
Friday	OPEN	OPEN	OPEN	OPEN	OPEN	OPEN	OPEN	OPEN	OPEN

The optimal solution is to not accept many bookings in the $125 rate class, except on Monday and Tuesday. Even some of the demand in the $150 rate class is turned away on Wednesday and Thursday. One might simply use this solution as guideline for the next few days and then re-optimize based on the accepted bookings and the revised demand forecasts. Two potential opportunities for improvement are as follows: (1) The solution does not consider the sequence of arrivals, for example, whether the $125 rate class customer arrives prior to the $150. (2) The solution does not take into account the stochastic aspect of total demand. These can be partially remedied by use of the dual prices provided by the sensitivity analysis of the solution. The sensitivity analysis of the solution to the LP is obtained from any traditional solver including Excel. The sensitivity analysis of the room capacities is given in Table 23.2.

There is one shadow price per resource and day of stay. This can be used as the bid price for a room for that day. For example, if a customer were willing to pay $225 for a 2-night stay beginning Monday, we would reject that offer because the price is less than the sum of the bid prices for Monday and Tuesday (100 + 150), whereas the hotel should accept any customer who is willing to pay for a 1-night stay on Monday or Friday if the rate exceeds $100. One might publish what rate classes are open based on this logic as shown in Table 23.3.

We can also compute the minimum price for accepting a booking (or a group): In order to create the minimum price (see Table 23.4), we have rounded the shadow price manually to integer value. We emphasize that the bid price is an internal control mechanism that helps decisions makers in deciding whether to accept a customer. The bid price need not bear resemblance to the actual price. Also, note

Table 23.4 Minimum price based on shadow prices

Night stay	1	2	3
Monday	100	250	400
Tuesday	150	300	450
Wednesday	150	300	400
Thursday	150	250	250
Friday	100	100	100

Table 23.5 Tuesday night single-resource analysis

Product	Total revenue	Revenue for Tuesday
Monday 125 2 nights	250	150
Monday 125 3 nights	375	125
Monday 150 2 nights	300	200
Monday 150 3 nights	450	200
Monday 200 2 nights	400	300
Monday 200 3 nights	600	350
Tuesday 125 1 night	125	125
Tuesday 125 2 nights	250	100
Tuesday 125 3 nights	375	75
Tuesday 150 1 night	150	150
Tuesday 150 2 nights	300	150
Tuesday 150 3 nights	450	150
Tuesday 200 1 night	200	200
Tuesday 200 2 nights	400	250
Tuesday 200 3 nights	600	300

that even though the $150 rate class for 1-night stay is open on Thursday, the LP solution does not accept all demand. Thus, the bid price is valid only for small change in the available capacity. Moreover, we may need to connect back to the single-resource problem to determine the booking limits for different rate classes. To see this, consider just the resource called Tuesday. Several different products use the Tuesday resource. Subtracting the bid price for the other days from the total revenue, we arrive at the revenue for Tuesday shown in Table 23.5.

Based on this table, we can infer that the DLP can also provide relative value of different products. This can be used in the single-resource problem to obtain the booking limits. We can also group products into different buckets prior to using the booking limit algorithm. Products with Tuesday revenue greater than or equal to 300 can be the highest bucket; the next bucket can be those with revenue between 200 and 250; the rest are into the lowest bucket.

Uses and Limitations of Bid Prices for Network Revenue Management: There are many practical uses of the bid prices. First and foremost, the approach shifts the focus of forecasting to the product level and away from the single-resource level. Thus, the decision maker generates demand forecasts for 1-night and 2-night stays separately instead of forecast for Tuesday night stay. The bid prices can help in route planning, shifting capacity if some flexibility is available, running promotions/shifting demand, identifying bid price trends, etc. For example, the management might decide not to offer some products on certain days, thereby shifting demand to other products. If there is some flexibility, a rental car company might use the bid price as guideline to move cars from one location with a low price to another with a high price. The product values might reveal systematic patterns of under- and over-valuation that can help decide whether to run a promotion for a special weekend rate or to a particular destination. Bid price trends that show a sustained increase over several weeks can indicate slackening of competitive pressure or advance bookings in anticipation of an event.

Several limitations of the approach have been mentioned in the chapter itself. More advanced material explaining the development of the network revenue management can be found in the references given in the chapter.

3 Further Reading

There are several texts devoted to revenue optimization. Robert Cross' book (2011) is one of the earliest ones devoted to the art and science of revenue management in a popular style. Many ideas discussed in this chapter and many more find a place in the book. Robert Phillips' book (2005) and Talluri and Van Ryzin's book (2006) contain a graduate level introduction to the subject. In addition, we have borrowed ideas from the papers listed at the end of the chapter (Bratu 1998; Lapp and Weatherford 2014; Talluri and van Ryzin 1998; Williamson 1992). The INFORMS Revenue Management and Pricing Section website[4] contains several useful references. Finally, there is a *Journal of Revenue and Pricing Management*[5] that is devoted to the topic.

Electronic Supplementary Material

All the datasets, code, and other material referred in this section are available in www.allaboutanalytics.net.

- Data 23.1: Opera.xls

Exercises

Ex. 23.1 (Protection Level) An airline offers two fare classes for economy class seats on its Monday morning flight: one class is sold at $400/ticket and another at

---
[4]http://www.informs.org/Community/revenue-mgt. Accessed on May 22, 2018.
[5]http://www.palgrave-journals.com/rpm/index.html. Accessed on May 22, 2018.

$160/ticket. There are 225 economy seats on the aircraft. The demand for the $400 fare (also called full-fare) seats has a mean of 46, a standard deviation of 16. Assume it follows a normal distribution. The demand for cheaper seats has an exponential distribution with mean of 177. A seat can be sold to either class. Further, the demand for the two fare classes can be assumed to be independent of one another. The main restriction is that the cheaper tickets must be purchased 3 weeks in advance.

(a) How many seats would you protect for the $400 class customers?
(b) The forecast for *cheaper* class passengers has changed. It is now assumed to be less than 190 with probability 1. How many seats would you protect for full-fare customers given this information?
(c) Go back to the original problem. Suppose that unsold seats may sometimes be sold at the last minute at $105. What effect will this have on the protection level (will you protect more or less seats or the same number of seats)? Why?
(d) Will your original answer change if the demands for the two classes are not independent of one another. Explain your answer if possible using an example.

Ex. 23.2 (Bid Price) Please see the data in the Excel sheet *Opera.xls* (available on website). The question is also given in the spreadsheet. It is reproduced below. All data is available in the spreadsheet.

Please carry out the following analysis based on the opera data. You are provided the cumulative booking for 1 year for two ticket classes. Assume that the opera house sells two types of tickets for their floor seats. The first is sold at $145, and the ticket is nonrefundable. The second is for $215 but refundable. The opera house has 245 floor seats. This data is given in two sheets in the spreadsheet.

You may verify (or assume) that the booking pattern is the same for most days. This is because we have normalized the data somewhat and got rid of peaks and valleys. The booking pattern is given 14 days prior to the concert onward. The final entry shows how many persons actually showed up for the concert on each day. Here is a sample of the data for $145 seats:

	−1	0	1	2	3		
11/30/2011	143	143	133	124	116		

For example, today is November 30, 2011. For this date, 116 persons had booked seats with 3 days to go, 124 with 2 days to go, 133 with 1 day to go, and 143 the evening before the concert. Finally, 143 persons showed up on November 30 which was the day of the concert.

We have created a **forecast** for the demand for the two types of seats for the next ten days, December 1 through December 10. We have used the additive method to estimate the pickup (PU).

(In this method, we computed the difference between the average seats sold and seats booked with 1, 2, 3, ... days to go. That is the PU with 1, 2, 3, ... days to go). **See rows 40–44 in the first sheet of the spreadsheet** for the forecast.

23 Pricing Analytics

Answer questions (a)–(d):
(a) Remember the opera has already sold some seats for the next 10 days. Compute the **available capacity** for the next 10 days (December 1 through December 10).
(b) Determine **how many seats to sell** at each price for the next 10 days. You have to set up a linear program for doing this.
(c) Comment on your solution. (How to use the shadow prices? What do the shadow prices reveal? What is necessary for implementing the solution?)
(d) Based on the data, can you provide advice on how to determine the overbooking level? Provide if possible an example using the data and any necessary assumption of the overbooking level and how it will be used by you in the optimization.

Ex. 23.3 (Choice Model) Daisy runs a small store in rural Indiana. Customers who come have to shop in her store or drive miles to go elsewhere. She has heard about revenue optimization! She always wondered at the rate at which customers gobbled her candy bars and always wondered whether she was pricing them right. The three best sellers are Milky Way, Starburst, and Coconut Bar. By gently asking some of her varied but trusted customers, she estimates their willingness to pay is around $2.20, $2.60, and $2.00 for the three types of candy bars. The variance seems to be around 0.10 for each of these willingness-to-pay values. Currently, she charges $2.00 for any of the candy bars. Typically, 100 customers visit her store every day.

(a) Estimate Daisy's current average sales and revenue.
(b) Daisy wants to run a promotion in her store by giving 10% off on one type of bar to customers. Which bar should she discount?
(c) What should be Daisy's optimal uniform price for the three types of candy bars? Would you recommend the price change?

Hint: Use the MNL model of choice. In this model, customers are assumed to be homogenous. They have an underlying utility U_i for product i. Each product is priced at p_i, $i=1, 2, \ldots, n$. The probability they will purchase product i is given by the following calculations:

$$\mu = \frac{\sqrt{(\text{variance} * 6)}}{\pi}$$

U_i = gross utility of product i (assume equal to willingness to pay)

$$v_i = e^{((U_i - p_i)/\mu)} \qquad \text{Prob (Purchase } i) = v_i/(1 + v_1 + v_2 + \ldots + v_n)$$

Ex. 23.4 (Dynamic Pricing) Mike is the revenue management manager at Marriott Hotel on 49th St., Manhattan, New York. He is contemplating how to respond to last-minute "buy it now" requests from customers. In this sales channel, customers can bid a price for a room, and Mark can either take it or wait for the next bid. Customers are nonstrategic (in the sense, they don't play games with waiting to bid). Mark has observed that typically he gets at most one request every hour. Analysis indicates that he gets a request in an hour with probability 0.2. He is looking at the last 3 h of the decision before the booking for the next day closes. For example, if booking closes at midnight, then he is looking at requests between 9 and 10 PM, 10 and 11 PM, and 11 and midnight. Customers either ask for a low rate or a high rate. Typically, half of them ask for a room for $100 and the rest for $235 (which is the posted rate).

Help Mark structure his thoughts and come up with a decision rule for accepting or turning down bids. It may help to think that with 3 h to go he can at most sell three rooms, with 2 h to go he can sell at most two rooms, and with an hour to go he can sell at most one room. (Thus, he can give away excess rooms at any price beyond these numbers, etc.) Use the dynamic programming example.

Ex. 23.5 (Overbooking) Ms. Dorothy Parker is the admissions director at Winchester College that is a small liberal arts college. The college has a capacity of admitting 200 students a year. The classrooms are small and the college wants to maintain a strict limit. Demand is robust with over 800 applications the previous year, out of which 340 students were offered a place on a rolling basis and the target of 200 admissions was met.

However, 17 students who accepted the offer of admission did not show up. Subsequent enquiries revealed that four of them had a last-minute change of heart about their college choice, three decided to take a gap year, and there was no reply from the rest. They paid the deposit and forfeited the amounts by college rules. Admissions contacted those on the waiting list, but it was too late as most already joined other institutions. As a result, the cohort comprised only 183 students stressing the budgets.

Ms. Parker decided that a change of policy was needed, and for the next year, the college will overbook, that is, admit a cohort larger than the capacity of 200. The question is how many. The tuition fee for 1 year of study is $34,500.

(a) What data should Ms. Parker be collecting to make a decision on how many students to admit beyond the limit of 200?
(b) Can we assume that the cost of falling short by a student is the 4 years' worth of tuition revenue? Argue why or why not.
(c) What is the cost of taking on a student over the 200 limit? Explain how you came up with your number.
(d) Ms. Parker decided after some analysis that the lost revenue from a student was $100,000, and the cost of having more students than capacity is as follows:

Students	Cost
201	$10,000
202	$22,000
203	$40,000
204	$70,000
205	$100,000
206	$140,000

23 Pricing Analytics

Beyond that, it is $50,000 per student.

Is this data enough to set a target number of admissions? What other data would be useful? Based only on this data, how many students would you admit?

(e) Analyzing the previous 5 years of data, Ms. Parker observed that with the policy of admitting exactly 200 each year, the final number of students who showed up was as given below:

Admitted	Showed up
200	200
200	195
200	197
200	190
200	192
200	183

If Ms. Parker was to naively admit 217 students based on this year's observation of no-shows, what would be the expected cost? Based on the data, what is the optimal number to overbook?

Ex. 23.6 (Markdown Optimization) Xara is a speciality fashion clothing retail store focusing on the big-and-tall segment of the market. This year, it is selling approximately 12,000 SKUs, with each SKU further classified by sizes. The initial prices for each item are usually set by the headquarters, but once the shipment reaches the stores, the store managers have the freedom to mark down the items depending on sales. Store managers are evaluated based on the total revenue they generate, so the understanding is that they will try to maximize revenue.

The demand for the new line of jeans was estimated based on historical purchases as follows:

$$D(p) = 10,000(1 - 0.0105p)$$

Here, 10,000 stands for the potential market, and the interpretation of $(1 - 0.0105p)$ is the probability of purchase of each member of the market. That is, demand at price p is given by the preceding formula, where p is in the range of 0–$95 (i.e., beyond $95, the demand is estimated to be 0).

The season lasts 3 months, and leftover items have a salvage value of 25% of their initial price. The headquarters sets the following guidelines: Items once marked down cannot have higher prices later. Prices can only be marked down by 10, 20, 30, or 40%. It is assumed demand comes more or less uniformly over the 3 month season.

(a) Based on the demand forecast, what should be the initial price of the jeans, and how many should be produced?

(b) The manager of the store on Portal de l'Angel in Barcelona obtained an initial consignment of 300 jeans, calculated to be the expected demand at that store. After a while, he noticed that the jeans were selling particularly slowly. He had a stock of 200 items still, and it was already 2 months into the season, so it is likely the potential market for the store area was miscalculated. Should he mark down? If so, by how much? (Hint: Based on the expected demand that was initially calculated for the store, you need to derive the demand curve for the store.)

Ex. 23.7 (Repricing) Meanrepricer.com offers a complex rule option where you can set prices according to the following criteria:

- My Item Condition: the condition of your item
- Competitor Item Condition: the condition of your competitors' product
- Action: the action that needs to be taken when applying a rule
- Value: the difference in prices which needs to be applied when using a certain rule

Here are some sample rules. Discuss their rationale (if any) and how effective they are.

(a) If our price for Product A is 100 and our competitors' price for Product A is $100, then the repricer will go ahead and reduce our price by 20% (i.e., from $100 to $80).
(b) In case your competitors' average feedback is lower than 3, chosen condition will instruct the repricer to increase your price by two units.
(c) Sequential rules, where the first applicable rule is implemented:

  (i) Reduce our price by two units if our competitors' product price is within a range of 300–800 units.
  (ii) Increase our price by two units if our competitors' product price is within a range of 500–600 units.

References

Belobaba, P. P. (1989). Application of a probabilistic decision model to airline seat inventory control. *Operations Research, 37*(2), 183–197.
Bratu, S. (1998). *Network value concept in airline revenue management.* Cambridge, MA: Massachusetts Institute of Technology.
Cross, R. G. (2011). *Revenue management: Hard-core tactics for market domination.* New York: Crown Business.
Lapp, M., & Weatherford, L. (2014). Airline network revenue management: Considerations for implementation. *Journal of Revenue and Pricing Management, 13*(2), 83–112.
Phillips, R. L. (2005). *Pricing and revenue optimization.* Palo Alto, CA: Stanford University Press.
Talluri, K., & van Ryzin, G.J. (1998). An analysis of bid-price controls for network revenue management. *Management Science, 44*(11), 1577–1593.

Talluri, K. T., & Van Ryzin, G. J. (2006). *The theory and practice of revenue management* (Vol. 68). New York: Springer.
Williamson, E. L. (1992). *Airline network seat inventory control: Methodology and revenue impacts*. (Doctoral dissertation, Massachusetts Institute of Technology).

Chapter 24
Supply Chain Analytics

Yao Zhao

1 Introduction

Through examples and a case study, we shall learn how to apply data analytics to supply chain management with the intention to diagnose and optimize the value generation processes of goods and services, for significant business value.

A supply chain consists of all activities that create value in the form of goods and services by transforming inputs into outputs. From a firm's perspective, such activities include buying raw materials from suppliers (buy), converting raw materials into finished goods (make), and moving and delivering goods and services to customers (delivery).

The twin goals of supply chain management are to improve cost efficiency and customer satisfaction. Improved cost efficiency can lead to a lower price (increases market share) and/or a better margin (improves profitability). Better customer satisfaction, through improved service levels such as quicker delivery and/or higher stock availability, improves relationships with customers, which in turn may also lead to an increase in market share. However, these twin goals have the potential to affect each other conversely. Improving customer satisfaction often requires a higher cost; likewise, cost reduction may lower customer satisfaction. Thus, it is a challenge to achieve both goals simultaneously. Despite the challenge, however, those companies that were able to achieve them successfully (e.g., Walmart,

---

Electronic supplementary material The online version of this chapter (https://doi.org/10.1007/978-3-319-68837-4_24) contains supplementary material, which is available to authorized users.

Y. Zhao (✉)
Rutgers University, Newark, NJ, USA
e-mail: yaozhao@business.rutgers.edu

Amazon, Apple, and Samsung) enjoyed a sustainable and long-term advantage over their competition (Simchi-Levi et al. 2008; Sanders 2014; Rafique et al. 2014).

The twin goals are hard to achieve because supply chains are highly complex systems. We can attribute some of this complexity to the following:

1. Seasonality and uncertainty in supply and demand and internal processes make the future unpredictable.
2. Complex network of facilities and numerous product offerings make supply chains hard to diagnose and optimize.

Fortunately, supply chains are rich in data, such as point-of-sale (POS) data from sales outlets, inventory and shipping data from logistics and distribution systems, and production and quality data from factories and suppliers. These real-time, high-speed, large-volume data sets, if used effectively through supply chain analytics, can provide abundant opportunities for companies to track material flows, diagnose supply disruptions, predict market trends, and optimize business processes for cost reduction and service improvement. For instance, descriptive and diagnostic analytics can discover problems in current operations and provide insights on the root causes; predictive analytics can provide foresights on potential problems and opportunities not yet realized; and finally, prescriptive analytics can optimize the supply chains to balance the trade-offs between cost efficiency and customer service requirement.

Supply chain analytics is flourishing in all activities of a supply chain, from buy to make to delivery. The Deloitte Consulting survey (2014) shows that the top four supply chain capabilities are all analytics related. They are optimization tools, demand forecasting, integrated business planning and supplier collaboration, and risk analytics. The Accenture Global Operations Megatrends Study (2014) demonstrated the results that companies achieved by using analytics, including an improvement in customer service and demand fulfillment, faster and more effective reaction times to supply chain issues, and an increase in supply chain efficiency. This chapter shall first provide an overview of the applications of analytics in supply chain management and then showcase the methodology and power of supply chain analytics in a case study on delivery (viz., integrated distribution and logistics planning).

2 Methods of Supply Chain Analytics

Supply chain management involves the planning, scheduling, and control of the flow of material, information, and funds in an organization. The focus of this chapter will be on the applications and advances of data-driven decision-making in the supply chain. Several surveys (e.g., Baljko 2013; Columbus 2016) highlight the growing emphasis on the use of supply chain analytics in generating business value for manufacturing, logistics, and retailing companies. Typical gains include more accurate forecasting, improved inventory management, and better sourcing and transportation management.

It is relatively easy to see that better prediction, matching supply and demand at a more granular level, removing waste through assortment planning, and better category management can reduce inventory without affecting service levels. A simple thought exercise will show that if a retailer can plan hourly sales and get deliveries by the hour, then they can minimize their required inventory. One retailer actually managed to do that—"Rakuten" was featured in a television series on the most innovative firms in Japan (Ho 2015). The focus on sellers and exceptional customer service seems to have paid off. In 2017, Forbes listed Rakuten among the most innovative companies with sales in excess of $7 billion and market cap more than $15 billion.[1] Data analytics can achieve similar results, without the need for hourly planning and delivery, and it can do so not only in retail but also in global sourcing by detecting patterns and predicting shifts in commodity markets. Clearly, supply chain managers have to maintain and update a database for hundreds of suppliers around the globe on their available capacity, delivery schedule, quality and operations issues, etc. in order to procure from the best source. On transportation management, one does not have to look further beyond FedEx and UPS for the use of data and analytics to master supply chain logistics at every stage, from pickup to cross docking to last-mile delivery (Szwast 2014). In addition, there are vast movements of commodities to and from countries in Asia, such as China, Japan, and Korea, that involve long-term planning, sourcing, procurement, logistics, storage, etc., many involving regulations and compliance that simply cannot be carried out without the tools provided by supply chain analytics (G20 meeting 2014).

The supply chain is a great place to apply analytics for gaining competitive advantage because of the uncertainty, complexity, and significant role it plays in the overall cost structure and profitability for almost any firm. The following examples highlight some key areas of applications and useful tools.

2.1 Demand Forecasting

Demand forecasting is perhaps the most frequent application of analytics to supply chains. According to the Chief Supply Chain Officer Report (O'Marah et al. 2014), 80% of executives are concerned about the risks posed to their supply chain by excessive customer demand volatility. Demand volatility causes problems and waste in the entire supply chain from supply planning, production, and inventory control to shipping. In simple terms, demand forecasting is the science of predicting the future demand of products and services at every level of an organization, be it a store, a region, a country, or the world. Demand forecasting is essential in planning for sourcing, manufacturing, logistics, distribution, and sales. The sales and operations planning modules of ERP systems help to bring several disciplines together so that forecasts can be created and shared to coordinate different activities

---

[1] https://www.forbes.com/companies/rakuten/ (accessed on Mar 26, 2018).

in the supply chain. These include the obvious ones such as inventory levels, production schedules, and workforce planning (especially for service industries). The less obvious ones are setting sales targets, working capital planning, and supplier capacity planning (Chap. 4, Vollmann et al. 2010). Several techniques used for forecasting are covered in Chap. 12 on "Forecasting Analytics."

One notable example of the use of forecasting is provided by Rue La La, a US-based flash-sales fashion retailer (Ferreira et al. 2015) that has most of its revenues coming from new items through numerous short-term sales events. One key observation made by managers at Rue La La was that some of the new items were sold out before the sales period was over, while others had a surplus of leftover inventory. One of their biggest challenges was to predict demand for items that were never sold and to estimate the lost sales due to stock-outs. Analytics came in handy to overcome these challenges. They developed models using which demand trends and patterns over different styles and price ranges were analyzed and classified, and key factors that had an impact on sales were identified. Based on the demand and lost sales estimated, inventory and pricing are jointly optimized to maximize profit. Chapter 18 on retail analytics has more details about their approach to forecasting and inventory management.

Going forward, firms have started to predict demand at an individual customer level. In fact, personalized prediction is becoming increasingly popular in e-commerce with notable examples of Amazon and Netflix, both of which predict future demand and make recommendations for individual customers based on their purchasing history. Several mobile applications can now help track demand at the user level (Pontius 2016). An example of the development, deployment, and use of such an application can be found in remote India (Gopalakrishnan 2016). As part of the prime minister's Swachha Bharat (Clean India) program, the Indian government sanctioned subsidies toward constructing toilets in villages. A volunteer organization called Samarthan has built a mobile app which helps track the progress of the demand for construction of toilet through various agencies and stages. The app has helped debottleneck the provision of toilets.

2.2 Inventory Optimization

Inventory planning and control in its simplest form involves deciding when and how much to order to balance the trade-off between inventory investment and service levels. Service levels can be defined in many ways, for example, fill rate measures the percentage of demand satisfied within the promised time window. Inventory investment is often measured by inventory turnover, which is the ratio between annual cost of goods sold (COGS) and average inventory investment. Studies have shown that there is a significant correlation between overall manufacturing profitability and inventory turnover (Sangam 2010).

Inventory management often involves the planning and coordination of activities across different parts of the supply chain. The lack of coordination can lead to excessive cost and poor service levels. For example, the bullwhip effect (Lee et al. 1997) is used to describe the upstream amplification and variability in demand of a supply chain due to reactive orders placed by wholesalers, distributors, and factory planners. There are modern tools that can help reduce the effect of such actions by increasing demand visibility and sharing of information (Bisk 2016).

A study by IDC Manufacturing Insights found that many organizations that utilized inventory optimization tools reduced inventory levels significantly in 1 year (Bodenstab 2015). Inventory optimization plays a critical role in the high-tech industry where most products and components become obsolete quickly but demand fluctuates significantly. The ability to predict demand and optimize inventory or safety stock is essential for survival because excess inventory may have to be written off and incur a direct loss. For instance, during the tech bubble burst in 2001, the network giant Cisco wrote off $2.1 billion of inventory (Gilmore 2008).

Inventory management can be improved through acquiring better information and real-time decision-making. For example, an American supermarket chain headquartered in Arkansas had a challenge to improve customer engagement within several of their brick-and-mortar locations. Managers were spending hours in getting the inventory of products in position instead of spending time on customer engagement. The R&D division developed a mobile app that fed real-time information to concerned employees. This mobile app provided a holistic view of sales, replenishment, and other required data that were residing in multiple data sources. They also developed an app for the suppliers to help them gain a better understanding of how their products were moving. Likewise, one of the leading retail chains of entertainment electronics and household appliances in Russia was able to process POS data in real time, which helped in avoiding shortages and excessive stock (Winckelmann 2013). The processing of inventory data of over 9000 items in 370 stores and four distribution centers is complex and time consuming. Their use of the SAP HANA[2] solution with in-memory real-time processing and database compression was a significant asset in improving the results.

2.3 *Supply Chain Disruption*

One of the biggest challenges to supply chain managers is managing disruption. It is important to predict potential disruptions and respond to them quickly to minimize their impact. Supply chain disruption can have a significant impact on corporate performance. At a very high level, firms impacted by supply chain disruptions have 33–40% lower stock returns relative to their benchmark and suffer a 13.5% higher volatility in share prices as compared to a previous year where there was no

---

[2]SAP HANA is an in-memory RDBMS of SAP, Inc.

disruption. Disruptions can have a significant negative impact on profitability—a typical disruption could result in a 107% drop in operating income, 7% lower sales growth, and 11% higher cost (Hendricks and Singhal 2015).

Supply chain disruptions can be caused by either uncontrollable events such as natural disasters or controllable events such as man-made errors. Better information and analytics can help predict and avoid man-made errors. For example, one shipping company that was facing challenges of incomplete network visibility deployed a supply chain technology that gave them a seamless view of the system. The technology enabled managers to get shipping details and take preventive or corrective actions in real time. In this example, prescriptive analytics could have also provided better decision support for the managers to assess and compare various options and actions. The benefits of improved efficiency, more reliable operations, and better customer satisfaction could have aided in the expansion of their customer base and business growth (Hicks 2012).

Connected Cows is a widely reported example of technology being used to aid farmers in better monitoring their livestock (Heikell 2015). The cows are monitored for well-being 24 h a day. This technology not only helps in taking better care of the livestock but also in reducing the disruptions in the production of dairy houses. Connected Cows helps farmers to determine which cows are sick and take timely action to nurse them back to good health with minimal effect on production. A similar concept can be applied to other assets of an organization where creating connected assets can draw valuable insights and provide needed preventive or corrective actions that minimize supply chain disruptions.

All of the aforementioned examples have had considerable historical data that helped in identifying supply chain disruptions and risk assessment. At times, this is not the case, and rare events such as Hurricane Katrina, epidemics, and major outages due to fire accidents may occur. Such events have high impact but low probability without much historical data, and hence the traditional approach cannot be used. The HBR review paper (Simchi-Levi et al. 2014) has addressed this issue by developing a model that assesses the impact of such events rather than their cause. In these extreme cases, the mitigation strategy takes center stage. They visualize the entire supply chain as a network diagram with nodes for the supplier, transportation center, distribution center, etc. where the central feature is the time to recovery (TTR)—"the time it would take for a particular node to be fully restored to functionality after a disruption." Using linear optimization, the model removes one node at a time to determine the optimal response time, and it generates the performance index (PI) for each node. There are many benefits for this approach; most importantly, managers gain a thorough understanding on the risk exposure of each node. The risk can subsequently be categorized as high, medium, and low, and corresponding prescriptive actions can be initiated. This model also depicts some of the dependencies among the nodes and the bottlenecks. There are certain cases where the total spending is low but the overall impact of disruption is significant—a carmaker's (Ford) spending on valves is low; however, the supply disruption of these components would cause production line to be shut down. This methodology was used by the Ford Motor Company to assess its exposure to supply chain disruptions.

2.4 Commodity Procurement

The price and supply of commodity can fluctuate significantly over time. Because of this uncertainty, it becomes difficult for many companies that rely on commodity as raw materials to ensure business continuity and offer a constant price to their customers. The organizations that use analytics to identify macroeconomic and internal indicators can do a more effective job in predicting which way prices might go. Hence, they can insulate themselves through inventory investment and purchases of future and long-term contracts. For example, a sugar manufacturer can hedge itself from supply and demand shocks by multiple actions, such as contracting out production on a long-term basis, buying futures on the commodity markets, and forward buying before prices upswing.

Another example is the procurement of ethanol that is used in medicines or drugs. Ethanol can be produced petrochemically or from sugar or corn. Prices of ethanol are a function of its demand and supply in the market, for which there is good degree of volatility. The price of ethanol is also affected by the supply of similar products in the market. As such, there are numerous variables that can impact the price of ethanol. Data analytics can help uncover these relationships to plan the procurement of ethanol. The same analytics tools and models can be extended to other commodity-based raw materials and components (Chandrasekaran 2014).

The last example is the spike in crop price due to changing climate. Climate change is likely to affect food and hunger in the future due to its impact on temperature, precipitation, and CO_2 levels on crop yields (Willenbockel 2012). Understanding the impact of climate change on food price volatility in the long run would be useful for countries to take necessary preventive and corrective actions. Computable general equilibrium (CGE) is used by researchers to model the impact of climate change, which has the capability to assess the effects external factors such as climate can have on an economy. The baseline estimation of production, consumption, trade, and prices by region and commodity group takes into account the temperature and precipitation (climate changes), population growth, labor force growth, and total factor productivity growth in agricultural and nonagricultural sectors. The advanced stage simulates various extreme weather conditions and estimates crop productivity and prices subsequently.

The examples provided barely touch upon the many different possible applications in supply chain management. The idea of the survey is to provide guidance regarding main areas of applications. The references at the end of the chapter contain more examples and descriptions of methods. In the next section, we describe in detail an example that illustrates inventory optimization and distribution strategies at a major wireless carrier.

3 VASTA: Wireless Service Carrier—A Case Study

Our case study was set in 2010 where VASTA was one of the largest wireless service carriers in the USA and well known for its reliable national network and superior customer service. In the fiscal year of 2009, VASTA suffered a significant inventory write-off due to the obsolescence of handsets (cell phones). At the time, VASTA carried about $2 billion worth of handset inventory in its US distribution network with a majority held at 2000+ retail stores. To address this challenge, the company was thinking to change its current "push" inventory strategy, in which inventory was primarily held at stores, toward a "pull" strategy, where the handset inventory would be pulled back from the stores to three distribution centers (DCs) and stores would alternatively serve as showrooms. Customers visiting stores would be able to experience the latest technology and place orders, while their phones would be delivered to their homes overnight from the DCs free of charge. The pull strategy had been used in consumer electronics before (e.g., Apple), but it had not been attempted by VASTA and other US wireless carriers as of yet (Zhao 2014a, b).

As of 2010, the US wireless service market had 280 million subscribers with a revenue of $194 billion. With a population of about 300 million in the USA, the growth of the market and revenue were slowing down as the market became increasingly saturated. As a result, the industry was transitioning from the "growth" business model that chased revenue growth to an "efficiency" model that maximized operational efficiency and profitability.

The US wireless service industry was dominated by a few large players. They offered similar technology and products (handsets) from the same manufacturers, but competed for new subscribers on the basis of price, quality of services, reliability, network speeds, and geographic coverage. VASTA was a major player with the following strengths:

- Comprehensive national coverage
- Superior service quality and reliable network
- High inventory availability and customer satisfaction

These strengths also led to some weaknesses:

- Lower inventory turnover and higher operating cost when compared to competitors.
- Services and products priced higher than industry averages due to the higher operating costs

The main challenge faced by VASTA was its cost efficiency, especially in inventory costs. VASTA's inventory turnover was 28.5 per year, which was very low compared to what Verizon and Sprint Nextel achieved (around 50–60 turns per year). Handsets have a short life cycle of about 6 months. A $2 billion inventory investment in its distribution system posed a significant liability and cost for VASTA due to the risk of obsolescence. In the following sections, we will analyze VASTA's proposition for change using sample data and metrics.

3.1 Problem Statement

To maintain its status as a market leader, VASTA must improve its cost efficiency without sacrificing customer satisfaction. VASTA had been using the "push" strategy, which fully stocked its 2000+ retail stocked to meet customer demand. The stores carried about 60% of the $2 billion inventory, while distribution centers carried about 40%.

The company was thinking to change the distribution model from "push" to "pull" which pulled inventory back to DCs. Stores would be converted to showrooms, and customers' orders would be transmitted to a DC which then filled the orders via express overnight shipping. Figures 24.1 and 24.2 depict the two strategies. In these charts, a circle represents a store and a triangle represents inventory storage.

The "push" and "pull" strategies represent two extreme solutions to a typical business problem in integrated distribution and logistics planning, that is, the strategic positioning of inventory. The key questions are as follows: Where to place inventory in the distribution system? And how does it affect all aspects of the system, from inventory to transportation and fulfillment to customer satisfaction?

Clearly, the strategies will have a significant impact not only on inventory but also on shipping, warehouse fulfillment, new product introduction, and, most importantly, consumer satisfaction. The trade-off is summarized in Table 24.1.

While the push strategy allowed VASTA to better attract customers, the pull strategy had the significant advantage of reducing inventory and facilitating the fast introduction of new handsets, which in turn reduced the cost and risk of inventory obsolescence. However, the pull strategy did require a higher shipping and warehouse fulfillment cost than the push strategy. In addition, VASTA had to renovate stores to showrooms and retrain its store workforce to adapt to the change.

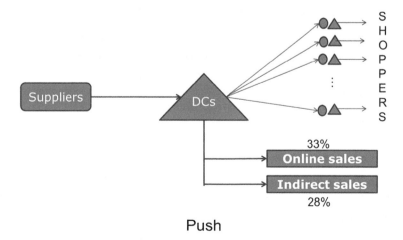

Fig. 24.1 VASTA's old distribution model. Source: Lecture notes, "VASTA Wireless—Push vs. Pull Distribution Strategies," by Zhao (2014b)

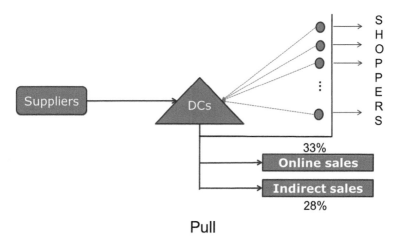

Fig. 24.2 VASTA's proposed new distribution model. Source: Lecture notes, "VASTA Wireless—Push vs. Pull Distribution Strategies," by Zhao (2014b)

Table 24.1 Pros and cons of the two distribution models

	Pros	Cons
Push	• Customer satisfaction • Batch picking at DCs • Batch, 2-day shipping to stores	• Significant inventory investment • Risk of obsolescence
Pull	• Significant inventory reduction • Faster switch to new handsets	• Customers have to wait for delivery • Unit picking at DCs • Unit, express overnight shipping to individual customers

Intuitively, the choice of pull versus push strategies should be product specific. For instance, the pull strategy may be ideal for low-volume (high uncertainty) and expensive products due to its relatively small shipping and fulfillment cost but high inventory cost. Conversely, the push strategy may be ideal for high-volume (low uncertainty) and inexpensive products. However, without a quantitative (supply chain) analysis, we cannot be sure of which strategy to use for the high-volume and expensive products and the low-volume and inexpensive products; nor can we be sure of the resulting financial impact.

3.2 Basic Model and Methodology

We shall evaluate the push and pull strategies for each product at each store to determine which strategy works better for the product–store combination from a cost perspective. For this purpose, we shall consider the total landed cost for product

i at store j, C_{ij}, which is the summation of store inventory cost, IC_{ij}; shipping cost, SC_{ij}; and DC fulfillment cost, FC_{ij}:

$$C_{ij} = IC_{ij} + SC_{ij} + FC_{ij} \tag{24.1}$$

The store inventory cost is represented by

$$IC_{ij} = h_i \times I_{ij} \tag{24.2}$$

where h_i is the inventory holding cost rate for product i (per unit inventory per unit of time) and I_{ij} is the average inventory level of product i at store j.

The shipping cost is represented by

$$SC_{ij} = s_j \times V_{ij} \tag{24.3}$$

where s_j is the shipping cost rate (per unit) incurred for demand generated by store j and V_{ij} is the sales volume per unit of time for product i at store j. Under the push strategy, s_j is the unit shipping cost to replenish inventory at store j by the DCs; under the pull strategy, s_j is the unit shipping cost to deliver the handsets to individual customers from the DCs.

Finally, the DC fulfillment cost is represented by

$$FC_{ij} = f\left(V_{ij}\right) \tag{24.4}$$

where f is an increasing and concave function representing economies of scale in picking and packing.

We shall ignore the difference in DC inventory levels between push and pull because under both strategies, the DCs face the same aggregated demand and must provide the same inventory availability. We summarize the calculation in Table 24.2.

We need to estimate all cost parameters and sales (demand) and inventory level statistics for each product–store combination from the data.

Table 24.2 Basic model of costs

For product i at store j	Costs (per unit of time)
Average inventory level	I_{ij}
Inventory cost	$IC_{ij} = h_i \times I_{ij}$
Weekly sales volume	V_{ij}
Shipping cost	$SC_{ij} = s_j \times V_{ij}$
Fulfillment cost	$FC_{ij} = f(V_{ij})$
Total cost	$IC_{ij} + SC_{ij} + FC_{ij}$

3.3 Cost Parameter Estimates

To calculate the costs, such as store inventory, shipping, and DC fulfillment (e.g., picking and packing) cost for each product–store combination, we need to estimate the inventory holding cost rate, h_i; the shipping cost rate, s_j; and the fulfillment cost function, $f(V_{ij})$. We will use a previously collected data set of sales (or demand, equivalently) and inventory data at all layers of the VASTA's distribution system for 60 weeks. One period will equal 1 week because inventory at both the stores and DCs is reviewed on a weekly basis.

Inventory cost rate:

Inventory holding cost per week = capital cost per week + depreciation cost per week

Capital cost per week = Annual capital cost/Number of weeks in a year

Depreciation cost per week = [Product value − Salvage value]/Product life cycle

VASTA carried two types of handsets: smartphones and less expensive feature phones with parameters and inventory holding cost per week, h_i, as in Table 24.3.

Shipping cost rate: Clearly, the shipping rates are distance and volume dependent. Here, we provide an average estimate for simplicity. The pull strategy requires shipping each unit from DCs to individual customers by express overnight freight. Quotation from multiple carriers returned the lowest flat rate of $12/unit. The push strategy, however, requires weekly batch shipping from DCs to stores by standard 2-day freight. Overnight express rate is typically 2.5 times the 2-day shipping rate; with a volume discount of 40%, we arrive at an average of $2.88/unit. Table 24.4 summarizes the shipping rates.

DC fulfillment cost: Distribution centers incur different costs for batch picking and packing relative to unit picking and packing due the economies of scale. For VASTA's DCs, the pick of the first unit of a product costs on average $1.50. If more than one unit of the product is picked at the same time (batch picking), then the cost of picking any additional unit is $0.1. We shall ignore the packing cost as it is negligible relative to the picking cost.

Table 24.3 Features of phones sold by VASTA

Smartphones (expensive)	Feature (inexpensive) phones
• Average product value: $500	• Average product value: $200
• Salvage value at store: 0%	• Salvage value at store: 0%
• Annual capital cost: 7%	• Annual capital cost: 7%
• Inventory cost/week: $19.90	• Inventory cost/week: $7.96

Table 24.4 Shipping costs of phones

	Pull	Push
Shipping method	Overnight express to customers	2-day batch to stores
Shipping cost rate s_j	$12/unit	$2.88/unit

Under the push strategy, the stores are replenished on a weekly basis. Let V_{ij} be the weekly sales volume. Because of batch picking, the weekly fulfillment cost for product i and store j is

$$f\left(V_{ij}\right) = \$1.50 + \left(V_{ij} - 1\right) \times \$0.1 \quad for \ V_{ij} > 0. \tag{24.5}$$

Under the pull strategy, each demand generated by a store must be fulfilled (picked) individually. Thus, the fulfillment cost for product i and store j is

$$f\left(V_{ij}\right) = V_{ij} \times \$1.50 \quad for \ V_{ij} > 0. \tag{24.6}$$

3.4 Analysis, Solution, and Results

To simplify the analysis, we shall group products with similar features together based on their sales volume and cost. There are essentially two types of phones: smartphones and feature phones. The average cost for a smartphone is $500, and the average cost of a feature phone is $200. Thus, we shall classify products into four categories as follows:

- High-volume and expensive products, that is, hot-selling smartphones
- High-volume and inexpensive products, that is, hot-selling feature phones
- Low-volume and expensive products, that is, cold-selling smartphones
- Low-volume and inexpensive products, that is, cold-selling feature phones

Using the data of a representative store and a representative product from each category (Table 24.5), we shall showcase the solution, analysis, and results.

In the pull model for high-volume products, we assume a per-store inventory level of five phones—these are used for demonstration and enhancing customer experience. Table 24.6 compares the total cost and cost breakdown between the push and pull strategies for the representative high-volume and expensive product.

The calculation shows that we can save 46.51% of the total landed cost for this high-volume and expensive product if we replace the push strategy by the pull strategy. This is true because the savings on inventory cost far exceeds the additional cost incurred for shipping and DC fulfillment.

Table 24.5 Representative sales of types of phone

	Average Weekly sales volume (unit)	Average On-hand inventory (unit)
High volume and expensive (hot-smart)	99	120
High volume and inexpensive (hot-feature)	102	110
Low volume and expensive (cold-smart)	2.5	15
Low volume and inexpensive (cold-feature)	7.3	25

Table 24.6 Savings for "hot-smart" phones between pull and push strategies

High volume and expensive (hot-smart)	Pull	Push
Inventory level	5 (I_{ij})	120 (I_{ij})
Inventory cost	$99.52 $(I_{ij} \times h_{ij} = 5 \times \$19.90)$	$2388.46 $(I_{ij} \times h_{ij} = 120 \times \$19.90)$
Weekly sales volume	99 (V_{ij})	99 (V_{ij})
Shipping cost	$1188 $(V_{ij} \times s_{ij} = 99 \times \$12/unit)$	$285.12 $(V_{ij} \times s_{ij} = 99 \times \$2.88/unit)$
Fulfillment cost	$148.50 $(V_{ij} \times \$1.50 = 99 \times \$1.50)$	$11.30 ($1.50 + (V_{ij} - 1)*0.1 = \$1.50 + 98*0.1)$
Total cost	$1436.02	$2684.88
Savings	–	46.51%

Table 24.7 Savings for "hot-feature" phones between pull and push strategies

High volume and inexpensive (hot-feature)	Pull	Push
Inventory level	5 (I_{ij})	110 (I_{ij})
Inventory cost	$39.81 $(I_{ij} \times h_{ij} = 5 \times \$7.96)$	$875.77 $(I_{ij} \times h_{ij} = 110 \times \$7.96)$
Weekly sales volume	102 (V_{ij})	102 (V_{ij})
Shipping cost	$1224 $(V_{ij} \times s_{ij} = 102 \times \$12/unit)$	$293.76 $(V_{ij} \times s_{ij} = 102 \times \$2.88/unit)$
Fulfillment cost	$153.00 $(V_{ij} \times \$1.50 = 102 \times \$1.50)$	$11.60 ($1.50 + (V_{ij} - 1)*0.1 = \$1.50 + 101*0.1)$
Total cost	$1416.81	$1181.13
Savings	–	−19.95%

Next, we consider the high-volume and inexpensive product. As shown in Table 24.7, the pull strategy does not bring any savings but incurs a loss of about 20% relative to the push strategy. Clearly, the saving on the inventory cost in this case is outweighed by the additional spending on shipping and DC fulfillment.

Now, we consider the low-volume products and assume two store copies for demonstration in the pull method's showroom. Table 24.8 shows the calculation for the low-volume and expensive product.

For the low-volume and inexpensive product, refer to Table 24.9.

Table 24.8 Savings for "cold-smart" phones between pull and push strategies

Low volume and expensive (cold-smart)	Pull	Push
Inventory level	2 (I_{ij})	15 (I_{ij})
Inventory cost	$38.81 ($I_{ij} \times h_{ij} = 2 \times \$19.90$)	$298.56 ($I_{ij} \times h_{ij} = 15 \times \$19.90$)
Weekly sales volume	2.5 (V_{ij})	2.5 (V_{ij})
Shipping cost	$30.00 ($V_{ij} \times s_{ij} = 2.5 \times \$12$/unit)	$7.20 ($V_{ij} \times s_{ij} = 2.5 \times \$2.88$/unit)
Fulfillment cost	$3.75 ($V_{ij} \times \$1.50 = 1.5 \times \$1.50$)	$1.65 ($\$1.50 + (V_{ij} - 1)*0.1$ $= \$1.50 + 1.5*0.1$)
Total cost	$73.56	$307.41
Savings	–	76.07%

Table 24.9 Savings for "cold-feature" phones between pull and push strategies

Low volume and inexpensive (cold-feature)	Pull	Push
Inventory level	2 (I_{ij})	25 (I_{ij})
Inventory cost	$15.92 ($I_{ij} \times h_{ij} = 2 \times \$7.96$)	$199.04 ($I_{ij} \times h_{ij} = 120 \times \$7.96$)
Weekly sales volume	7.3 (V_{ij})	7.3 (V_{ij})
Shipping cost	$87.60 ($V_{ij} \times s_{ij} = 7.3 \times \$12$/unit)	$21.02 ($V_{ij} \times s_{ij} = 7.3 \times \$2.88$/unit)
Fulfillment cost	$10.95 ($V_{ij} \times \$1.50 = 7.3 \times \$1.50$)	$2.13 ($\$1.50 + (V_{ij} - 1)*0.1$ $= \$1.50 + 6.3*0.1$)
Total cost	$114.47	$222.19
Savings	–	48.48%

Table 24.10 Savings for all types of phones between pull and push strategies

	Cold-smart	Cold-feature	Hot-smart	Hot-feature
% Savings	76.07%	48.48%	46.51%	–19.95%

Table 24.10 summarizes the percentage savings as we move from "push" to "pull" for the representative store and representative products of all four categories.

Consistent to our intuition, the pull strategy brings the highest savings for the low-volume and expensive product (cold-smart), and the lowest savings (even a loss) for the high-volume and inexpensive product (hot-feature). In general, the pull strategy tends to bring less savings for products with a higher volume and/or a less cost.

To assess the impact of the pull strategy on store inventory, we quantify the reduction of inventory investment per store. For the representative store, Table 24.11 shows the number of products in each category and their corresponding inventory level reduction. Specifically, there are 22 products in the hot-smart category, 20 in the hot-feature category, 15 in the cold-smart category, and 11 in the cold-feature category. The store inventory investment can be calculated for both the pull and push strategies.

From this table, we can see that inventory investment per store under the pull strategy is only about 5% of that under the push strategy. Thus, the pull strategy can reduce the store-level inventory by about 95%. Given that store inventory accounts for 60% of the $2 billion total inventory investment, the pull strategy will bring a reduction of at least $1 billion in inventory investment as compared to the push strategy.

Despite the significant savings in inventory, the pull strategy can increase the shipping and DC fulfillment costs substantially. To assess the net impact of the pull strategy, we shall aggregate the costs over all products for each cost type (inventory, shipping, and fulfillment) in the representative store and present them in Table 24.12.

The table shows that the inventory cost reduction outweighs the shipping/picking cost inflation and thus the pull strategy results in a net savings per store of about 31% relative to the push strategy.

Table 24.11 Store inventory investment for pull and push strategies

Category	# of products	Pull Inventory level	Pull Inventory investment	Push Inventory level	Push Inventory investment
Hot-smart	22	5	$5 \cdot \$500 \cdot 22$ = \$55,000	120	$120 \cdot \$500 \cdot 22$ = \$1,320,000
Hot-feature	20	5	$5 \cdot \$200 \cdot 20$ = \$20,000	110	$110 \cdot \$200 \cdot 20$ = \$440,000
Cold-smart	15	2	$5 \cdot \$200 \cdot 20$ = \$15,000	15	$110 \cdot \$200 \cdot 20$ = \$112,500
Cold-feature	11	2	$2 \cdot \$200 \cdot 11$ = \$4,400	25	$25 \cdot \$200 \cdot 11$ = \$55,000
Total	68	–	\$94,400	–	\$1,927,500

Table 24.12 Total costs for pull and push strategies

Per store per week	Pull	Push
Total inventory cost	\$3757.85	\$76,729.33
Total shipping cost	\$52,029.60	\$12,487.10
Total picking cost	\$6,503.70	\$528.78
Total cost	\$62,291.15	\$89,745.21

3.5 Advanced Model and Solution

As shown by our prior analysis, the pull strategy does not outperform the push strategy for all products. In fact, for high-volume and inexpensive products (hot-feature phones), it is better to satisfy a portion of demand at stores. Thus, the ideal strategy may be hybrid, that is, the store should carry some inventory so that a fraction of demand will be met in-store, while the rest will be met by overnight express shipping from a DC. The question is how to set the store inventory level to achieve the optimal balance between push and pull.

To answer this question, we shall introduce more advanced inventory models (Zipkin 2000). Consider the representative store and a representative product. Store inventory is reviewed and replenished once a week. The following notation is useful:

- T: the review period
- $D(T)$: the demand during the review period:
- $E[D(t)] = \mu$: the mean of the demand during the review period
- $STDEV[D(t)] = \sigma$: the standard deviation of the demand during the review period

The store uses a base-stock inventory policy that orders enough units to raise the inventory (on-hand plus on order) to a target level S at the beginning of each period. The probability of satisfying all the demand in this period via store inventory alone is α (α is called the Type 1 service level). Assuming that $D(t)$ follows a normal distribution, $Normal(\mu, \sigma)$, and lead time is negligible (as it is true in the VASTA case), then

$$S = \mu + z_\alpha \sigma, \qquad (24.7)$$

where z_α is the normal distribution standard score of α. Clearly, if $\alpha = 0$, we return to the pull strategy where all demand is met by the DCs, only. If $\alpha = 100\%$, then we return to the push strategy where all demand is met by the stores, only. A hybrid strategy will set α such that $0 < \alpha < 1$.

The expected store demand met by the DCs via overnight express shipping can be written as

$$E[D_1] = E[\max\{0, D(T) - S\}] = \sigma[\phi(z_\alpha) - z_\alpha(1-\alpha)], \qquad (24.8)$$

where $\phi(x) = \frac{1}{\sqrt{2\pi}} e^{-\frac{x^2}{2}}$ is the standard normal probability density function.

The expected store demand met by store inventory is

$$E[D_2] = E[\min\{D(T), S\}] = E[D(T)] - E[D_1]. \qquad (24.9)$$

At the end of the period, the expected inventory level is

$$EI = E\left[\max\{0, S - D(T)\}\right] = S - \mu + E[D_1]. \quad (24.10)$$

Because the inventory level at the beginning of the period is S, the average on-hand inventory during the period can be approximated by

$$I = \frac{S + EI}{2}. \quad (24.11)$$

Using α as the decision variable, we can calculate the total landed cost by Eq. (24.1),

$$C_{ij} = IC_{ij} + SC_{ij} + FC_{ij},$$

where $IC_{ij} = h_i \times I_{ij}$ and I_{ij} come from Eq. (24.11).

SC_{ij} is the sum of two parts:

1. Batch and 2-day shipping of the quantity $E[D_2]$ from DCs to the store
2. Express overnight shipping of the quantity $E[D_1]$ from DCs to customers

FC_{ij} is also the sum of two parts:

1. Batch picking at DCs for the quantity of $E[D_2]$
2. Individual picking at DCs for the quantity of $E[D_1]$

To identify the optimal strategy, we shall solve the following nonlinear optimization problem for product i at store j. That is, we shall find the α_{ij} ($0 \leq \alpha_{ij} \leq 1$) such that the total cost C_{ij} is minimized

$$\text{Min}_{0 \leq \alpha_{ij} \leq 1} C_{ij}.$$

To solve this problem, we shall need demand variability (or uncertainty) information in addition to averages, such as the standard deviation of demand per unit of time. Table 24.13 provides the estimates of the representative products at the representative store.

The results for the representative hot-smartphone are plotted in Fig. 24.3. It shows how the total cost varies with α (store type 1 service level). Clearly, a hybrid strategy is best for the representative hot-feature phone (better than both the pull and push strategies). However, the pull strategy is still the best for the representative hot-smartphone.

Similarly, the results for the low-volume products are plotted in Fig. 24.4. The pull strategy works the best for the cold-smartphone, while a hybrid strategy is the best for the cold-feature phone.

24 Supply Chain Analytics

Table 24.13 Estimates of representative phones at a representative store

	Weekly sales average	Weekly sales standard deviation	On-hand inventory
High volume and expensive (hot-smart)	99	40	120
High volume and inexpensive (hot-feature)	102	53	110
Low volume and expensive (cold-smart)	2.5	2.3	15
Low volume and inexpensive (cold-feature)	7.3	9.1	25

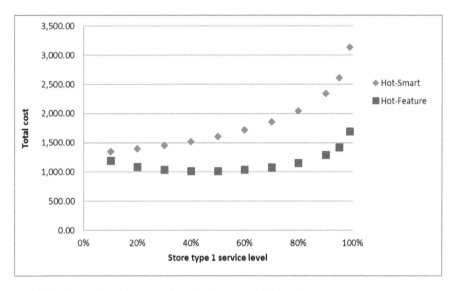

Fig. 24.3 Comparison between push and pull strategies for hot phones

Table 24.14 Savings from moving to pull strategy

	Hot-feature	Cold-feature
The best Type 1 service (α)	50%	50%
% demand met by store	79%	50%
Saving from pull	28%	8.4%

We can draw the following conclusions from these results:

- The pull strategy is best for both the hot- and cold-smartphones.
- For feature phones, it is best to use a hybrid strategy (refer to Table 24.14).

From the store perspective, the savings gained from switching from pull to hybrid is 13.4% (or $8081.90) per store per week.

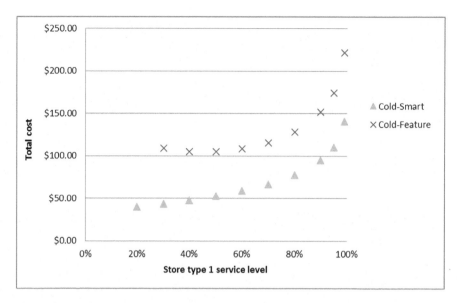

Fig. 24.4 Comparison between push and pull strategies for cold phones

3.6 Customer Satisfaction and Implementation

So far, our analysis focuses on the total landed cost, which is smaller under the pull strategy than the push strategy. Despite this cost efficiency, a fundamental issue remains: Will customers accept the pull strategy? More specifically, will customers be willing to wait for their favorite cell phones to be delivered to their doorstep overnight from a DC?

An analysis of the online sales data shows that in the year 2010, one out of three customers purchased cell phones online. While this fact implies that a large portion of customers may be willing to wait for delivery, it is not clear how the rest two-thirds of customers may respond to the pull strategy. It is also unclear how to structure the delivery to minimize shipping cost while still keeping it acceptable to most customers. The available delivery options include the following:

1. Overnight free of charge
2. Overnight with a fee of $12
3. Free of charge but 2 days
4. Free of charge but store pickup

Different options have significantly different costs and customer satisfaction implications; they must be tested in different market segments and geographic regions. To ensure customer satisfaction, VASTA had decided to start with option 1 for all customers.

Table 24.15 Implementation plan of pull strategy

Phase	
Phase I	• Implement the pull strategy for one DC and some target stores • Negotiate shipping contracts with carriers • Review savings, service levels, and impact on customers
Phase II	• Implement the pull strategy to all stores served by the DC • Experiment the options of store pickup and 2-day free home shipping
Phase III	• Full-scale implementation of the pull strategy to all three DCs • Review savings, service levels, and impact on customers

Implementation of the pull strategy requires three major changes in the distribution system:

1. Converting retail stores to showrooms and retraining sales workforce
2. Negotiating with carriers on the rate and service of the express shipping
3. A massive transformation of the DCs that will transition from handling about 33% individual customer orders to nearly 72% individual customer orders (the indirect sales, through third-party retail stores such as Walmart, can be fulfilled by batch picking and account for 28% of total sales)

Despite the renovation costs and training expenses, showrooms may enjoy multiple advantages over stores from a sales perspective. For instance, removing inventory can save space for product display and thus enhance customers' shopping experiences. Showrooms can increase the breadth of the product assortment and facilitate faster adoption to newer handsets and thus increase sales. Finally, they can also help to reduce store-level inventory damage and theft, thereby minimizing reverse logistics.

Negotiation with carriers needs to balance the shipping rate and the geographic areas covered as a comprehensive national coverage may require a much higher shipping rate than a regional coverage. Important issues such as shipping damages and insurance coverage should also be included in the contract. The hardest part of implementation is the DC transformation, especially given the unknown market response to the pull strategy. Thus, a three-phase implementation plan (see Table 24.15) had been carried out to slowly roll out the pull strategy in order to maximize learning and avoid major mistakes.

3.7 Epilogue

In 2011, VASTA implemented the pull strategy in its US distribution system. FedEx overnight was used. System inventory reduced from $2 billion to $1 billion. Soon after, other US wireless carriers followed suit, and the customer shopping experience of cell phones completely changed in the USA from buying in stores to ordering in stores and receiving delivery at home. In the years after, VASTA continued to fine-tune the pull strategy into the hybrid strategy and explored multiple options of express shipping depending on customers' preferences. VASTA remains as one of the market leaders today.

4 Summary: Business Insights and Impact

In this chapter, we showcase the power of supply chain analytics in integrated distribution and logistics planning via a business case in the US wireless services industry. The company, VASTA, suffered a significant cost inefficiency despite its superior customer service. We provide models, methodology, and decision support for VASTA to transform its distribution strategy from "push" to "pull" and eventually to "hybrid" in order to improve its cost efficiency without sacrificing customer satisfaction. The transformation resulted in $1 billion savings in inventory investment and helped the company to maintain its leadership role in an increasingly saturated marketplace.

Supply chains are complex systems and data rich. We have shown that by creatively combining such data with simple analytics, we can achieve the twin goals of cost efficiency and customer satisfaction while making a significant financial impact. This chapter also reveals three business insights that one should be aware of when employing supply chain analytics:

- *The conflicting goals* of cost efficiency and customer satisfaction are hard to sort out qualitatively. Quantitative supply chain analysis is necessary to strike the balance.
- *System thinking*: Distribution strategies can have a significant impact on all aspects of a system: inventory, shipping, customer satisfaction, as well as in-store and warehouse operations. We must evaluate all aspects of the system and assess the net impact.
- *One size does not fit all*: We should customize the strategies to fit the specific needs of different products and outlets (e.g., stores).

Electronic Supplementary Material

All the datasets, code, and other material referred in this section are available in www.allaboutanalytics.net.

- Data 24.1: Vasta_data.xls

Exercises

Ex. 24.1 Reproduce the basic model and analysis on the representative store for the comparison between push and pull strategies.

Ex. 24.2 Reproduce the advanced model and analysis on the hybrid strategy for the representative store.

Ex. 24.3 For NYC and LA stores, use the basic and advanced models to find out which strategy to use for each type of product, and calculate the cost impact relative to the push strategy.

References

Accenture Global Operations Megatrends Study. (2014). *Big data analytics in supply chain: Hype or here to stay?* Dublin: Accenture.

Baljko, J. (2013, May 3). *Betting on analytics as supply chain's next big thing.* Retrieved December 30, 2016, from http://www.ebnonline.com/author.asp?section_id=1061&doc_id=262988&itc=velocity_ticker.

Bisk. (2016). *How to manage the bullwhip effect on your supply chain.* Retrieved December 30, 2016, from http://www.usanfranonline.com/resources/supply-chain-management/how-to-manage-the-bullwhip-effect-on-your-supply-chain/.

Bodenstab, J. (2015, January 27). Retrieved December 30, 2016, from http://blog.toolsgroup.com/en/multi-echelon-inventory-optimization-fast-time-to-benefit.

Chandrasekaran, P. (2014, March 19). *How big data is relevant to commodity markets.* Retrieved December 30, 2016, from http://www.thehindubusinessline.com/markets/commodities/how-big-data-is-relevant-to-commodity-markets/article5805911.ece.

Columbus, L. (2016, December 18) *McKinsey's 2016 analytics study defines the future of machine learning.* Retrieved December 30, 2016, from http://www.forbes.com/sites/louiscolumbus/2016/12/18/mckinseys-2016-analytics-study-defines-the-future-machine-learning/#614b73d9d0e8.

Deloitte Consulting. (2014). *Supply chain talent of the future findings from the 3rd annual supply chain survey.* Retrieved December 27, 2016, from https://www2.deloitte.com/content/dam/Deloitte/global/Documents/Process-and-Operations/gx-operations-supply-chain-talent-of-the-future-042815.pdf.

Ferreira, K. J., Lee, B. H. A., & Simchi-Levi, D. (2015). Analytics for an online retailer: Demand forecasting and price optimization. *Manufacturing & Service Operations Management, 18*(1), 69–88.

G20 Trade Ministers Meeting. (2014, July 19). *Global value chains: Challenges, opportunities, and implications for policy.* Retrieved December 30, 2016, from https://www.oecd.org/tad/gvc_report_g20_july_2014.pdf.

Gilmore, D. (2008, August 28). *Supply chain news: What is inventory optimization?* Retrieved December 30, 2016, from http://www.scdigest.com/assets/firstthoughts/08-08-28.php.

Gopalakrishnan, S. (2016, July 22). *App way to track toilet demand.* Retrieved December 30, 2016, from http://www.indiawaterportal.org/articles/app-way-track-toilet-demand.

Heikell, L. (2015, August 17). *Connected cows help farms keep up with the herd.* Microsoft News Center. Retrieved December 30, 2016, from https://news.microsoft.com/features/connected-cows-help-farms-keep-up-with-the-herd/#sm.00001iwkvt0awzd5ppu5pahjfsks0.

Hendricks, K., & Singhal, V. R. (June 2015). *The effect of supply chain disruptions on long-term shareholder value, profitability, and share price volatility.* Retrieved January 7, 2017, from http://www.supplychainmagazine.fr/TOUTE-INFO/ETUDES/singhal-scm-report.pdf.

Hicks, H. (2012, March). *Managing supply chain disruptions.* Retrieved December 30, 2016, from http://www.inboundlogistics.com/cms/article/managing-supply-chain-disruptions/.

Ho, J. (2015, August 19). *The ten most innovative companies in Asia 2015.* Retrieved December 30, 2016, from http://www.forbes.com/sites/janeho/2015/08/19/the-ten-most-innovative-companies-in-asia-2015/#3c1077d6465c.

Lee, H. L., Padmanabhan, V., & Whang, S. (1997, April 15). *The bullwhip effect in supply chains.* Retrieved December 30, 2016, from http://sloanreview.mit.edu/article/the-bullwhip-effect-in-supply-chains/.

O'Marah, K., John, G., Blake, B., Manent, P. (2014, September). *SCM World's the chief supply chain officer report.* Retrieved December 30, 2016, from https://www.logility.com/Logility/files/4a/4ae80953-eb43-49f4-97d7-b4bb46f6795e.pdf.

Pontius, N. (2016, September 24). *Top 30 inventory management, control and tracking apps.* Retrieved December 30, 2016, from https://www.camcode.com/asset-tags/inventory-management-apps/.

Rafique, R., Mun, K. G., & Zhao, Y. (2014). *Apple vs. Samsung – Supply chain competition. Case study*. Newark, NJ; New Brunswick, NJ: Rutgers Business School.

Sanders, N. R. (2014). *Big data driven supply chain management: A framework for implementing analytics and turning information into intelligence*. Upper Saddle River, NJ: Pearson Education, Inc..

Sangam, V. (2010, September 2). *Inventory optimization*. Supply Chain World Blog. Retrieved December 30, 2016.

Simchi-Levi, D., Kaminsky, F., & Simchi-Levi, E. (2008). *Designing and managing the supply chain: Concepts, strategies, and case studies*. New York, NY: McGraw-Hill Irwin.

Simchi-Levi, D., Schmidt, W., & Wei, Y. (2014). From superstorms to factory fires - Managing unpredictable supply-chain disruptions. *Harvard Business Review., 92*, 96 Retrieved from https://hbr.org/2014/01/from-superstorms-to-factory-fires-managing-unpredictable-supply-chain-disruptions.

Szwast, S. (2014). *UPS 2014 healthcare white paper series – Supply chain management*. Retrieved December 30, 2016, from https://www.ups.com/media/en/UPS-Supply-Chain-Management-Whitepaper-2014.pdf.

Vollmann, T., Berry, W., Whybark, D. C., & Jacobs, F. R. (2010). *Manufacturing planning and control systems for supply chain management* (6th ed.). Noida: Tata McGraw-Hill Chapters 3 and 4.

Willenbockel, D. (2012, September). *Extreme weather events and crop price spikes in a changing climate*. Retrieved January 7, 2017, from https://www.oxfam.org/sites/www.oxfam.org/files/rr-extreme-weather-events-crop-price-spikes-05092012-en.pdf.

Winckelmann, L. (2013, January 17). *HANA successful in mission to the Eldorado Group in Moscow*. Retrieved January 7, 2017, from https://itelligencegroup.com/in-en/hana-successful-in-mission-to-the-eldorado-group-in-moscow.

Zhao, Y. (2014a). *VASTA wireless – Push vs. pull distribution strategies. Case study*. Newark, NJ; Brunswick, NJ: Rutgers Business School.

Zhao, Y. (2014b). *Lecture notes VASTA wireless – Push vs. pull distribution strategies*. Newark, NJ; Brunswick, NJ: Rutgers Business School.

Zipkin, P. (2000). *Foundations of inventory management*. New York, NY: McGraw-Hill Higher Education.

Chapter 25
Case Study: Ideal Insurance

Deepak Agrawal and Soumithri Mamidipudi

1 Introduction

Sebastian Silver, the Chief Finance Officer of *Ideal Insurance Inc.*, was concerned. The global insurance industry was slowing, and many firms like his were feeling the pressure of generating returns. With low interest rates and increase in financial volatility in world markets, Sebastian's ability to grow the bottom line was being put to test.

Sebastian started going through the past few quarters' financial reports. He was worried about the downward trend in numbers and was trying to identify the root causes of the shortfall in order to propose a strategy to the board members in the upcoming quarterly meeting. To support his reasoning, he started looking through industry reports to examine whether the trend was common across the industry or whether there were areas of improvement for his company. Looking at the reports, he observed that both profit from core operations, that is, profit from insurance service, and customer satisfaction rate were surprisingly lower than industry

---

Electronic supplementary material The online version of this chapter (https://doi.org/10.1007/978-3-319-68837-4_25) contains supplementary material, which is available to authorized users.

D. Agrawal (✉) · S. Mamidipudi
Indian School of Business, Hyderabad, Telangana, India
e-mail: a.deepak@outlook.com

© Springer Nature Switzerland AG 2019
B. Pochiraju, S. Seshadri (eds.), *Essentials of Business Analytics*, International Series in Operations Research & Management Science 264,
https://doi.org/10.1007/978-3-319-68837-4_25

standard. The data was at odds with the company's claim settlement ratio,[1] which Sebastian knew was higher than that of his rivals, and claim repudiation ratio,[2] which was lower than industry average. He also observed that claim settlement was taking longer than expected.

The head of the claims department at *Ideal Insurance*, Rachel Morgan, told Sebastian that there was a tremendous shortage of manpower in the settlement team and added that there was no focus on making innovation and improvements in the claim investigation process. She also reminded Sebastian that she had proposed an in-house department of analysts who could help improve the claim settlement process and support other improvement initiatives. Sebastian promised he would review the proposal, which had been submitted to Adam Berger's HR team in the beginning of the year, and set up a meeting with Rachel in the following week. He also asked her to contact an expert in claim settlement and investigation to provide a detailed review of performance and to suggest a road map for changes.

Following Sebastian's suggestion, Rachel reached out to an independent consultant to verify and analyze Ideal's healthcare policy claims. She knew that fraud prevention, one of the biggest reasons for profit leakage in the sector, would have to be a key priority area. However, she was also aware that improving the probability of detecting fraudulent claims could hurt genuine policyholders. The challenge facing Rachel was how to balance the need to deliver swift responses to customers with the knowledge that too many fraudulent claims would severely hurt the bottom line. It was to solve this challenge that the consultant advised Rachel to consider using advanced analytical techniques. These techniques, such as artificial intelligence and machine learning, could help to make claims processing more efficient by identifying fraud, optimizing resource utilization, and uncovering new patterns of fraud. The consultant added that such applications would improve customer perception of the company because genuine claims would be identified and processed more quickly.

2 The Insurance Industry

The global insurance industry annually writes trillions of dollars of policies. The nature of the insurance industry means that insurers are incredibly sensitive to fluctuations in local and global financial markets, because many policies involve coverage for decades. Premiums that are paid to insurance companies are therefore invested for the long term in various financial assets in order to generate returns and to use capital efficiently. This necessarily means that understanding the global

---

[1]Claims settlement ratio is calculated as the percentage of claims settled in a period out of total claims notified in the same period. The higher the settlement ratio, the higher the customer satisfaction.

[2]Claims repudiation ratio is calculated as the percentage of claims rejected (on account of missing or wrong information) in a period out of total claims notified in the same period.

insurance industry involves not just understanding the nature of the companies that operate in the sector but also its interconnections with financial markets.

Perhaps the most important distinction regarding insurance companies is the nature of the policies written. Large (measured by assets and geographical coverage) firms such as AXA and Prudential usually have an array of policies in every segment. Smaller companies, however, may restrict themselves to writing only in the life or non-life segments. Insurers may thus choose only to be involved in motor, disaster, or injury-linked claims. Companies may also reinsure the policies of other insurance companies—Swiss Re, Munich Re, and Berkshire Hathaway all engage in reinsuring the policies of other insurers. The operating results of any specific insurance company, therefore, will depend not just on the geography in which it operates but also on the type(s) of policy(ies) that it underwrites.

Insurance to a layperson is nothing but the business of sharing risk. The insurance provider agrees to share the risk with the policyholder in return for the payment of an insurance premium. Typically, an underwriter assesses the risk based on age, existing health conditions, lifestyle, occupation, family history, residential location, etc. and recommends the premium to be collected from the potential customer. The policyholder secures his unforeseen risk by paying the premium and expects financial support in case the risk event takes place.

The insurance business depends for survival and profitability on spreading the risk over proper mix and volume and careful planning over a long horizon. The collective premium is either used to honor the claims raised by policyholders or invested in long-term assets in expectation of significant profit. Thus, the insurance provider has two major avenues to earn profit, namely, profit from the core operations of risk sharing and profit from investments. It has been observed that profit from core operations is generally very low or sometimes even negative; however, overall profits are high due to investment strategies followed by the firm—such as value investing, wealth management, and global diversification. Most insurance businesses have an asset management entity that manages the investment of such collected premium. The insurance providers also protect themselves from huge losses from the core business by working with reinsurers such as Swiss Re and Berkshire Hathaway, who share the risks among the insurance providers.

3 The Healthcare Insurance Business

The health and medical insurance industry is a fast-growing segment of the non-life insurance industry. The compound annual growth rate of the global health insurance market is expected to be around 11% during 2016–2020.[3] The revenue from the global private health insurance industry, which was around US$1.45 trillion in 2016,

---

[3] https://www.technavio.com/report/global-miscellaneous-global-health-insurance-market-2016-2020 (accessed on Aug 17, 2018).

is likely to double by 2025 (Singhal et al. 2016). While the USA occupies the top rank in gross written premium revenues, with more than 40% of the market share, an aging population and growing income are expected to lead to major increases in the demand for healthcare services and medical insurance in Latin America and Asia-Pacific regions in the coming years. The major driving forces and disruptive changes in health and medical insurance markets are due to the increase in health risk from the rise in noncommunicable diseases and chronic medical conditions, advances in digital technology and emerging medical sciences, improved underwritings and changes in regulatory environments, and increased consumerism and rise in aging population in developing economies.

Health insurance accounts for billions of dollars of spending by insurance companies. The *Centers for Disease Control and Prevention* estimates that about 15% of spending by insurers is on healthcare plans.[4] Advances in healthcare technology are likely to be balanced by increased demand. As populations continue to age in the USA, Europe, and Japan, spending on this sector will remain a cornerstone of the insurance business for decades to come.

Most healthcare insurance divides the cost of care between the policyholder and the insurer. Depending on the type of policy and the nature of treatment, this division can take place in a number of ways. Some policies pay out once spending crosses a certain threshold—called a "deductible." Others split the cost with the policyholder in a defined ratio—called a "copayment."

4 Claims Processing

Speed is at the heart of designing the processes in a health insurance firm. By its definition, insurance is required in situations that are unforeseeable. In the case of health insurance, receiving the money that the policy guarantees as soon as possible is vital to the policyholder. Yet this constraint means that the timeline of claims processing must be necessarily as short as possible, hurting the ability of insurance companies to verify claims.

The process for claiming health insurance normally proceeds in several stages. The initial contact between the firm and the policyholder is made via the call center or local office/agent. Often, this step is undertaken by a person close to the holder or by the healthcare provider, because the holder may be incapacitated.

The firm's call center obtains and records basic information from the client, including details regarding the type of policy that the holder owns, the hospital to which the holder has been taken, and the injury/ailment with which the holder has

---

[4]Eynde, Mike Van Den, "Health Insurance Market Overview," State Public Health Leadership Webinar, Deloitte Consulting LLP, August 15, 2013, URL: https://www.cdc.gov/stltpublichealth/program/transformation/docs/health-insurance-overview.pdf (accessed on May 25, 2017).

been afflicted. Armed with this information, the call-center employee forwards the necessary details to the claims processing team.

The claims processing team is responsible for the first line of inquiry into the claim. They check the policyholder's coverage and expenses, verify the hospital and network at which the client has been admitted, and ask for proof such as bills and prescriptions. Upon doing so, the team either classifies the claim as Genuine, Discuss, or Investigate. Genuine cases are processed without any further clarification. Discuss cases are forwarded to a data collection team in order to collect more information and verify details. Investigate cases are forwarded to the claims investigation team and can take a long time to be processed (settled or rejected) further.

At present, the claim processing team at Ideal Insurance examines the following points to create a fraud score:

1. Is the policyholder raising reimbursement claims from a non-network hospital?
2. Are multiple claims raised from a single policy (except group policies) or policyholder?
3. Are there multiple group claims from the same hospital?
4. Is the claim raised close to the policy expiry date?
5. Is the claim raised close to the policy inception date?
6. Is there no pre- or post-claim to the main claim?
7. Is there misrepresentation of material information identified in the report?
8. Was the claim submitted on a weekend?
9. Are there "costlier" investigations?
10. Are there high doctor fees?
11. Was the claim reported one day before discharge?
12. Did the claim intimation happen after 48 h of admission?

Each indicator carries a weight assigned based on prior research and experience of the investigation team. The maximum weighted score is 40. If the weighted score (or fraud score) of a claim is more than 20, then the claim processing team forwards the claims to the investigation team to investigate potential fraud. If the fraud score is between 16 and 20, then the claim processing team seeks additional data from the information collection team and healthcare service provider. The claims with fraud scores of less than or equal to 15 are considered genuine and forwarded to the settlement team for payment to the policyholder or service provider.

Investigating claims requires firms to verify a host of corroborating details. In order to satisfy themselves of the genuine nature of the claim, investigators check the types of medications prescribed, years of patient history, and the specific nature of the ailment. Depending on whether the type of fraud suspected is hard or soft, investigators could choose to examine different levels of data. While soft fraud might be identified by conclusively proving that certain kinds of treatment were not appropriate for the disease diagnosed, hard fraud would need larger and more complex patterns to be uncovered.

Longer, more complicated, claims processes are a double-edged sword for insurance firms. As the number of claims that are investigated in-depth rises, the chance of both inclusion and exclusion errors falls. Yet investigating a large number of claims takes up time, resources, and risks, causing delays to genuine customers.

A claim gets further complicated if the policyholder decides to file a litigation suit due to delay or rejection. Though it may create pressure for quick settlement, providing strong argument for minimizing delay/rejection, a suit by itself does not necessarily mean that the claim is genuine. Litigation is a crucial and important tool for insurance firms. As health insurance is vital to most clients, the decision to classify a claim as fraud can potentially open the door to a host of lawsuits. These lawsuits can be on the behalf of either individual customers or a host of clients. It is usually the company's responsibility to justify its opinion that a claim is fraudulent. Because legal standards for fraud may be different from the company's internal standards, ensuring that the company can win such cases can become complicated. In addition, court costs in themselves can be prohibitive. The company may have to follow different rules as prescribed by the law of the respective land. Large number of cases or long pending cases can also potentially damage the firm's reputation. Avoiding such challenges is the best bet. It is also important not to back off from litigation when the occasion demands to prevent potential fraudsters from taking advantage of the firm.

5 Stakeholders in the Health Insurance Industry

The transformation in the health insurance industry involves and requires influencing numerous stakeholders and market participants. These include[5] the following:

1. Consumers, patients, caregivers, and patient advocacy organizations: These are people experiencing the health problems and who would be beneficiary of various health services and treatments.
2. Clinicians and their professional associations: These are major medical decision-makers; and their skills, experience, and expertise matter the most.
3. Healthcare institutions, such as hospital systems and medical clinics, and their associations: Major healthcare decisions are structured and determined by choices of institutional healthcare providers as they often have a broad view on what is causing a particular health problem.
4. Purchasers and payers, such as employers and public and private insurers: Coverage by insurer and purchaser of healthcare plays an important role in diagnostic and treatment decisions and choices of insured.
5. Healthcare industry, pharmaceutical companies and drug manufacturers, and industry associations: Manufacturers of drugs and treatment devices and their suppliers and distributors influence the quality of healthcare services available in a region.

---

[5] Agency for Healthcare Research and Quality (AHRQ). 2014. Stakeholder Guide. https://www.ahrq.gov/sites/default/files/publications/files/stakeholdr.pdf (accessed on Aug 17, 2018).

6. Healthcare policy makers and regulators: Government policy and regulatory mechanisms including the legal system influence cost and quality of healthcare, health insurance market development, and access to individuals and their families.
7. Healthcare research and research institutions: Local availability of research funds and quality of research institutions play a vital role in the development of the healthcare and medical insurance market in a region.
8. Insurance companies: The insurance companies provide coverage to the policyholder in return of an insurance premium. They underwrite the policy and collect the calculated premium based on the customer's risk profile. They are expected to honor the claims raised by a policyholder in case of unforeseen circumstances. The risks are shared by pooling a large number of customers with diverse risk profiles and with the help of reinsurers.
9. Reinsurance providers: The reinsurers provide coverage to the insurance companies in case of large unforeseen situations such as natural calamities, terrorist attacks, and bankruptcy. They assist in sharing the risk across geographies and a diverse pool of risky customers.
10. Third-party assistance (TPA) service providers: TPAs are the mediators between insurance providers and customers or consumers who assist in providing information to the underwriters and help in smooth processing of claims as and when raised. They assist in the information collection process of insurance providers and in submission of claims by customers.
11. Agents or banks: The insurance service provider either hires the agents/employees to sell various products or collaborates with financial institutions such as banks and third-party providers to cross-sell insurance products to their existing customer base.

The insurance firm acts as a major coordinator between the different players. The insurers help interpret laws and regulations. They are aware of drugs, side effects, treatment schedules, and procedures. They collect data on rates and costs of different services in different places. They also provide information and allied services to the customers. For example, they directly process bills and pay hospitals once the procedure has been approved. In many ways, a strong and competitive insurance industry is necessary to coordinate so many different interests.

6 Fraud in the Insurance Business

Insurance fraud is one of the largest sources of white-collar crime in the world, meaning that significant police effort is also devoted to tracking and eliminating it. However, given limited police resources and a universe of crime that encompasses far more than just the white-collar variety, hard insurance fraud perpetrated by organized criminals tends to be the focus of law enforcement. This leaves unorganized hard fraud and a plethora of soft fraud to remain within the purview of insurance companies.

The health insurance industry is no more immune to fraud than any other insurance subsectors. Experts estimate about 6% of global healthcare spending is lost to fraud annually.[6] In a world in which trillions of dollars are spent on healthcare by governments, nongovernmental organizations, and corporations alike, this amounts to tens of billions lost to criminal enterprises. In the USA alone, fraud is estimated to cause about US$80 billion in losses to the industry annually, with property casualty fraud accounting for US$32 billion.[7] These figures do not include fraud perpetrated on Medicare and Medicaid.

Health insurance fraud is an act of providing misleading or false information to a health insurance company in an attempt to have them pay to a policyholder, another party, or entity providing services (PAIFPA 2017). An individual subscriber can commit health insurance fraud by:

- Allowing someone else to use his or her identity and insurance information to obtain healthcare services
- Using benefits to pay for prescriptions that were not prescribed by his or her doctor

Healthcare providers can commit fraudulent acts (PAIFPA 2017) by:

- Billing for services, procedures, and/or supplies that were never rendered
- Charging for more expensive services than those actually provided
- Performing unnecessary services for the purpose of financial gain
- Misrepresenting non-covered treatments as a medical necessity
- Falsifying a patient's diagnosis to justify tests, surgeries, or other procedures
- Billing each step of a single procedure as if it were a separate procedure
- Charging a patient more than the co-pay agreed to under the insurer's terms
- Paying "kickbacks" for referral of motor vehicle accident victims for treatment
- Patients falsely claiming healthcare costs
- Individuals using false/stolen/borrowed documents to access healthcare

Tackling fraud is critical to the industry, especially with fraud becoming ever more complex. By its nature, insurance fraud is difficult to detect, as its aim is to be indistinguishable from genuine insurance claims. In each of the above cases, identifying the fraud that has been perpetrated can be a laborious process, consuming time and effort. Given that healthcare spending can be sudden, urgent, and unexpected, checking for fraud can be a complicated process. Companies must balance their financial constraints with the reality of healthcare situations.

According to an estimate of the US National Healthcare Anti-Fraud Association (NHCAA),[8] 3% of all healthcare spending is lost to healthcare fraud (LexisNexis

---

[6] "The Health Care Fraud Challenge," Global Health Care Anti-Fraud Network. http://www.ghcan.org/global-anti-fraud-resources/the-health-care-fraud-challenge/ (accessed on Jun 12, 2017).

[7] http://www.insurancefraud.org/statistics.htm (accessed on Jun 12, 2017).

[8] The Challenges of Healthcare Fraud. https://www.nhcaa.org/resources/health-care-anti-fraud-resources/the-challenge-of-health-care-fraud.aspx (accessed on Aug 17, 2018).

2011). Financial fraud including unlawful billing and false claim is the most common type of health insurance fraud and generally tied into aspects of organization and health information management (AHIMA Foundation 2010). The data mining tools and techniques and predictive analytics such as neural network, memory-based reasoning, and link analysis can be used to detect fraud in insurance claim data (Bagde and Chaudhary 2016).

Healthcare fraud leads to higher premium rates, increased expenses to consumers, and reduced coverage. It increases cost to employers for providing healthcare insurance to their employees affecting the cost of doing business. Besides financial losses, fraudulent activities lead to exploitations and exposure of people to unnecessary and unsafe medical procedures, which can have devastating health side effects.

Detecting healthcare insurance fraud is a long drawn-out, complicated process that costs companies time, effort, money, and the goodwill of their customers. Modern technology and statistical software have helped to reduce this cost, but it remains a significant burden on the resources of customer service departments the world over. Healthcare insurance fraud-proofing and management strategies and activities may include "improving data quality, building a data centric culture and applying advanced data analytics."[9] These provide opportunity for significant cost savings by the healthcare industry.

The innovations in insurance products and development in information communication technologies can help to design tailor-made insurance products with improved underwriting and pricing of healthcare insurance and coverage option. Technology and improved information systems can benefit stakeholders and market participants and lead to improved welfare and consumer satisfaction.

In the past, the primary manner in which insurers detected fraud was to employ claims agents who investigated suspicious claims. However, as data analytics software gains prominence and becomes more powerful, firms are becoming more able to identify patterns of abnormal behavior.

Fraud detection, however, must contend with the possibility of misidentifying fraud. Allowing false claims to go through the system hurts the company's profit and increases premiums. Forcing genuine claimants to go through the fraud detection process, however, increases costs and hurts customer satisfaction. As these constraints are diametrically opposed, any attempt to curb one will tend to increase the other.

---

[9]Deloitte (2012), Reducing Claim Fraud – A Cross Industry Issue. http://deloitte.wsj.com/cfo/files/2012/10/reducing_claims_fraud.pdf (accessed on Aug 17, 2018).

7 *Ideal*'s Business

Ideal Insurance Inc. is one of the largest healthcare insurance providers in the USA and other developed nations. It has expansion plans to enter into emerging markets where penetration is much lower. Most of the underwriting work is done in the US or the UK office. It has a large claim processing team located in all the countries of presence. It has back offices in other countries such as India, Singapore, and Thailand. With the increasing competition in the market, the company has had to focus on quick settlement, claims settlement ratio as well as profit margin. The company has been investing significant amount on automating the claim settlement process in order to increase customer satisfaction rate, reduce the length of the settlement cycle, and reduce the loss associated with claim leakage due to potential fraud claims. Table 25.1 shows some of the key performance measures that Sebastian was tracking. Though Ideal offers competitive premiums and maintains a high claim settlement ratio and low repudiation ratio, its net promoter score (NPS)—a metric of customer satisfaction—is significantly lower than the industry average.

The company has an automated system in place that reviews the basic information of all the claims based on prespecified rules set by experts. The rules are used to classify the claims into three categories, namely, Genuine, Discuss, and Investigate. This information is passed to the claims settlement team to act further. The claims classified as "Genuine" are processed on high priority with a settlement turnaround time (TAT) of 7 working days. The claims classified as "Discuss" are forwarded to the data collection team in order to collect more and granular information about the claims. Such claims usually take on an average of up to 30 days. The claims classified as "Investigate" are forwarded to the claim investigation team for thorough enquiry of the stakeholders. These claims usually take between 1 and 2 months or sometimes even more than 3 months for settlement or closure based on the results of the investigations or if the claims are litigated. Some customers file litigation suits in case of rejected claims, and then it is the company's onus to prove that the claim is fraudulent. Anecdotal evidence suggested that Ideal's experienced claim settlement personnel did not completely trust the

Table 25.1 Ideal performance vis-à-vis industry

Parameter	Ideal Insurance[a]	Industry average[a]
Revenue per policy	70	100
Contribution from core operations	20%	35%
Contribution from investments	80%	65%
Claims settlement ratio	92%	87%
Claims repudiation ratio	2.5%	4%
Average settlement period	72 days	30–45 days
Net promoter score (out of 100)	73	86

[a]Scaled to 100 if no metric given

current system. The feeling was that it was somewhat limited due to a "bureaucratic approach." They did their own checks, often uncovering fraud from claims identified as Genuine by the current system. Rachel discussed the same with the more senior personnel to understand the root cause. The argument given was that "data maturity is vital in uncovering fraud pattern and therefore we re-analyze the claims (even if it is identified as Genuine) when more information is populated in the database." Also, their experience suggested that it is difficult to uncover professional soft fraud which is usually identified only by thorough analysis or if there is any lead from an outside stakeholder or by doing network analysis of other fraud claims.

Rachel wanted a feedback on the current processes and hired an independent consultant to examine 100,000 claims from the historical claims data. The consultant with the help of an investigation team did a thorough examination of the provided claims and classified those claims as fraud or non-fraud. Out of 100,000 claims investigated by the consultant, they identified 21,692 as potentially fraudulent claims and 78,308 as genuine claims. Comparing these results with the previous settlement records of 100,000 claims showed that more than 90% of these fraudulent claims were not identified by the existing automated system. Claims of more than 6657 customers were delayed because of the investigation process suggested by the current system. The average claim cost is around US$5000 while the cost of investigation is approximately US$500. The cost of investigation was equally divided between the internal manpower cost and the external cost of obtaining information specific to the claim. Thus, conservatively, Rachel estimated that investigating a genuine claim leads to a loss of US$500 and increases customer dissatisfaction due to the delay in settlement. It also reduces the efficiency of the investigation team. Settling a potential fraudulent claim leads to a loss of US$5000 on average and negatively affects the premium pricing and the effectiveness of the underwriting team.

The management team discussed the report with its claims settlement team and sought advice from them on how to improve the processes. Several excellent suggestions were gathered, such as monitoring each stage of the process instead of the entire process, flagging issues overlooked by the current system, and using past similar-claims data to verify the claim. The claims settlement team also suggested hiring an analytics professional to build a fraud predictive model using advanced data science techniques. They explained that this would not only help in correctly identifying potential fraud claims but also in optimizing the efforts of the claims investigation and settlement teams. They also mentioned that their closest competitor had recently set up an analytics department, which was helping in various aspects of business such as conduct of fraud analytics, predicting claims, review of blacklisted stakeholders, effective underwriting, and developing customized products.

Rachel turned to Raghavan, a recent hire who had graduated with a master's degree in business analytics from a famous school in South Central India. Raghavan had expertise in analytics specifically in the insurance domain. He was charged to hire professionals and supervise the project: to build a predictive model to identify potential fraudulent claims out of reported health insurance claims. This solution, Rachel and Sebastian felt, will help not only in reducing the losses due to fraud but also in improving efficiency and customer satisfaction and reducing the claim settlement cycle.

7.1 Project Details

Raghavan's initial thoughts were to deliver a robust analytical solution that would improve fraudulent claim identification process at *Ideal*'s site without investing much time and effort in the field at the early stage. The potential fraud claims can be investigated more rigorously, while genuine claims can be settled quickly at the same time. He co-opted Caroline Gladys, who had also recently graduated in analytics from one of the premier business schools, who had been working with the business intelligence team and now wanted to switch to the advanced analytics team. Raghavan provided her the opportunity to work on this proof of concept and deliver a solution.

7.2 Data Description

Caroline through her experience within *Ideal Insurance* quickly created a sample dataset at the transaction level for 100,000 health insurance claims. Each observation has up to 33 parameters related to the claim such as policy details, whether there was a third-party administrator, demographic details, and claim details. The complete details are shown in Appendix 1. Tables in Appendix 2 provide the coding of variables such as product type, policy type, and mode of payment.

The data in the tables are collected by the transaction processing system (1) when the policy is issued, (2) when a claim is recorded, and (3) while its progress is tracked. The ERP system did a fairly good job of collecting the necessary data.

Table 25.2 Summary of identified fraud claims (by expert)

Fraud	Number of claims	Proportion (%)
No	78,308	78.3
Yes	21,692	21.7
Total	100,000	

Table 25.3 Summary of current system's recommendation

System's recommendation	Number of claims	Proportion (%)
Genuine (Green)	91,986	92.0
Discuss	6560	6.6
Investigate	1454	1.5
Total	100,000	

Custom software helped put together the information into tables and created reports for further processing. *Ideal* had invested a great deal in automation of transactions in the past and was looking to reap dividends from the reporting system.

Caroline also obtained the classification of claims as Fraud/Not Fraud examined by the expert who had investigated 100,000 claims. The classification is shown in Table 25.2. Additionally, the data in Table 25.3 provide the classification of all 100,000 claims as Genuine, Discuss, and Investigate according to the current automated system.

Caroline put together all the data in a dataset (*idea_insurance.csv*; refer to the website) and also the detailed definitions of the variables available and data description required to decode the categories such as product type, policy type, and claim payment type.

Having collected all this information, Caroline was wondering how to begin the analysis. Was predictive analytics superior to the expert system used by *Ideal*? Would the experts who created the system as well as the senior settlement officers readily accept the changes? She was also worried about the ongoing creation of rules and maintenance of the system. That would cost significant investment in people and technology, not to mention training, obtaining data, etc. She would have to clearly convince the management that this was a worthwhile project to pursue!

Electronic Supplementary Material

All the datasets, code, and other material referred in this section are available in www.allaboutanalytics.net.

- Data 25.1: ideal_insurance.csv

Appendix 1

Sr. no.	Variable name	Description
1	tpa	Third-party administrator ID
2	policy_ref	Policy reference number
3	member_id	Insured member ID
4	sex	Sex of the insured member
5	dob	Date of birth of the insured member
6	policy_start_dt	Date of commencement of policy
7	policy_end_dt	Date of expiry of policy
8	prod_code	Product type
9	policy_type	Policy type
10	sum_insured	Maximum sum insured available to policyholder
11	claim_ref	Claim reference number
12	claim_dt	Date of claim intimation to insurer
13	hospital_id	Unique ID given to hospital
14	hos_zipcode	Zip code of the hospital
15	admit_dt	Date of admission in hospital
16	discharge_dt	Date of discharge from hospital
17	payment_dt	Date of settlement of the claim
18	claim_amt	Amount claimed by the claimant
19	nursing_chg	Nursing charges incurred during treatment
20	surgery_chg	Surgery charges incurred during treatment
21	cons_fee	Doctor consultation charges incurred during treatment
22	test_chg	Investigation charges of medical tests prescribed by the doctor
23	pharmacy_cost	Medicines consumed during treatment
24	other_chg	Any other charges that cannot be categorized in above
25	pre_hosp_exp	Amount claimed for pre-hospitalization treatment
26	post_hosp_exp	Amount claimed for post-hospitalization treatment
27	other_chg_non_hosp	Other non-hospital charges (laundry, paid TV channels, guest F&B, etc.)
28	copayment	Co-payment or excess if applicable
29	settle_amt	Final amount paid to the insured or to the medical service provider on behalf of insured
30	payment_type	The mode of payment (refer description)
31	hosp_type	Whether hospital is networked, Y/N
32	recommendation	Classified as Green/Discuss/Investigate by the claim settlement team
33	fraud	Classified as Fraud/Non-Fraud by an expert consultant

Note: All the amounts are in US$ and dates are in d-mmm-yyyy format. The identity data such as policy number, claim number, and hospital details are masked to maintain the data privacy.

Appendix 2

prod_code	Product type
A	Basic policy
B	Hospital cash plan
C	Outpatient coverage
D	Universal health policy
E	Microinsurance policy
F	Package policy (covering more than one type of health above)
G	Hybrid policy (covering other than health also)
O	Any other product type

policy_type	Policy type details
A	Individual
B	Individual floater
C	Group
D	Group floater
E	Declaration
F	Declaration floater
G	Declaration with group insurance
H	Declaration floater with group insurance
O	Any other cover type

payment_type	Claim payment type description
A	Cashless settlement
B	Reimbursement to claimant
C	Cash benefit
D	Per diem basis
E	Lump-sum basis
F	Cashless to the insured

References

AHIMA Foundation. (2010). *A study of health care fraud and abuse: Implications for professional managing health information*. Retrieved September 15, 2018, from https://www.ahimafoundation.org/downloads/pdfs/Fraud%20and%20Abuse%20-%20final%2011-4-10.pdf.

Bagde, P. R., & Chaudhary, M. S. (2016). Analysis of fraud detection mechanism in health insurance using statistical data mining techniques. *International Journal of Computer Science and Information Technologies, 7*(2), 925–927.

LexisNexis. (2011). *Bending the cost curve: Analytics driven enterprise fraud control*. Retrieved September 15, 2018, from http://lexisnexis.com/risk/downloads/idm/bending-the-cost-curve-analytic-driven-enterprise-fraud-control.pdf.

PA Insurance Fraud Prevention Authority (PAIFPA). (2017). *Health insurance fraud*. Retrieved September 15, 2018, from http://www.helpstopfraud.org/Types-of-Insurance-Fraud/Health.

Singhal, S., Finn, P., Schneider, T., Schaudel, F., Bruce, D., & Dash, P. (2016). *Global private payors: A trillion-euro growth Industry*. New York: McKinsey and Company Retrieved September 15, 2018, from http://healthcare.mckinsey.com/sites/default/files/Global%20private%20payors%20%28updated%29.pdf.

Chapter 26
Case Study: AAA Airline

Deepak Agrawal, Hema Sri Sai Kollipara, and Soumithri Mamidipudi

1 Introduction

Steven Thrush, Chief Revenue Officer of *AAA Airline Corp*, was concerned about his company. The airline industry, buoyed by strong demand and low oil prices, had been on an upswing for the last few years. Rising competition, however, had begun to pressure AAA's operations. Shifting market sentiments and an increasingly complicated market had made travelling to most destinations in the USA dependent for most customers on a number of contrasting factors.

Moreover, the rise of low-cost carriers and online ticket comparison websites had put immense downward pressure on ticket prices, squeezing the margins of companies and forcing them to investigate new avenues of growth in order to maintain their profitability.

Thrush had just returned from a conference focused on the application of data science and analytics in the passenger transport industry. At the conference, researchers and practitioners talked about the rapid advance of big data and its power to understand and predict customer behavior. Thrush grasped that big data

---

Electronic supplementary material The online version of this chapter (https://doi.org/10.1007/978-3-319-68837-4_26) contains supplementary material, which is available to authorized users.

The case study is written by Agrawal, Deepak; Kollipara, Hemasri; and Mamidipudi, Soumithri under the guidance of Professor Sridhar Seshadri. We would like to thank Coldrena et al. (2003) and Garrow (2010) who are the inspiration to develop this case study.

D. Agrawal (✉) · H. S. S. Kollipara · S. Mamidipudi
Indian School of Business, Hyderabad, Telangana, India
e-mail: a.deepak@outlook.com

could help his company move toward new models that took into account a dizzying range of factors in order to make better decisions.

When Linda James, the company's head of route planning, approached him to ask about the feasibility of launching a New York–Boston flight, Thrush immediately thought about employing the customer choice models he had heard about in order to understand the proposition. He asked his data team to use the company's database of its customers to understand the question of how well received a new flight from New York to Boston would be. He knew that to answer such a question, the team would also have to investigate many more issues such as what manner of pricing would be most efficient, what type of aircraft would be most efficient, and how best to reach new customers who might not otherwise fly AAA.

Settling on the correct approach to the problem, Thrush knew, would be the best way to deliver the best service possible to customers while maximizing the profit of his company.

2 AAA Airline Corp

AAA Airline Corp was founded in 2005, amid a sea change in the travel industry. As Internet penetration grew and price comparison websites became increasingly popular, AAA saw an opportunity for a low-cost carrier to capitalize on the increased customer focus on prices.

Like many carriers founded in the wake of the online boom, AAA's philosophy was to compete purely on price. Instead of focusing on specific regional flights and needs, AAA's philosophy was to identify and fill gaps in the market and in doing so carve out a niche for itself. While most of its flights operated in a hub-and-spoke system out of Boston Logan Airport, the company was not averse to operating point-to-point routes that are the hallmark of low-cost carriers worldwide.

AAA's initial method to identify which routes were profitable relied on a mix of market research and intuition. AAA's original management team consisted mostly of industry veterans hailing from Massachusetts, and they were all well acquainted with the needs of local customers. AAA's in-depth expertise in its initial market helped it survive where many of its rivals failed, prompting it to expand its offering and plan for more ambitious growth.

By 2016, the size of AAA's fleet had risen considerably, prompting Thrush's concern regarding its next steps. AAA's history meant that it had access to a large database of its own customers, which it had so far been using to forecast future demand and traffic patterns. Thrush was keen to know, however, what new tools could be used and datasets found in order to analyze the market and help the company stride into the new era of commercial air travel.

3 History and Current Context of the US Airline Industry

The US airline industry had a capacity of more than over 1.1 million available seat miles (accounting for both domestic and international flights) in 2016 and is the largest geography for air travel worldwide. The sector supplies nearly 3500 available seat miles, a measure of carrying capacity, per person in North America, more than double that of the industry in Europe.

The effects of the Airline Deregulation Act of 1978 are still being felt today. Before the Act, American airline companies were strictly constrained by the Civil Aeronautics Board, which was responsible for approving new routes and pricing. The Board could give agreements between carriers anti-trust immunity if it felt it was in the public interest. This resulted in a situation where airlines competed purely on in-flight service and flight timings and frequency.

Legacy carriers—airlines founded before deregulation—are able to offer better service and benefits such as loyalty schemes as a result of the environment in which they operated at their founding. Airlines such as these tend to have larger planes and operate in a hub-and-spoke system that means that their flights are largely based out of a single airport.

After the industry was deregulated, airlines became free to decide what routes to fly and what prices to offer. New low-cost carriers like AAA entered the market and shifted the paradigm by which companies in the industry functioned, forcing full-service airlines to adapt. Since 1978, more than 100 airline carriers have filed for bankruptcy,[1] underscoring the tumultuous nature of the industry.

The proliferation of the Internet was no less disruptive to the airline and travel industries. Customers were more able than ever to compare flights, and their ability to discriminate between a multitude of choices at the tap of a key left companies the world over scrambling to keep up. This meant that companies such as AAA were forced to use ever more complicated models in their attempts to understand and predict customer demand while at the same time keeping track of their costs.

Thrush knew that AAA's spoke-and-hub system helped to keep costs low and enable the airline to fly a large number of passengers. However, he was also aware hub airports were especially hard-hit by the increase in the number of passengers using them, meaning that pressure on his staff and his operations was mounting daily. The industry's domestic load factor, the fraction of available seats that were sold, had risen to 85% in 2016 from 66% in 1995.[2] Domestic ASMs rose 29% in the same period to 794,282. However, the sizes and capacities of hub airports had not risen in line with this explosive growth in passengers due to property, environmental, and financial constraints.

---

[1] http://airlines.org/dataset/u-s-bankruptcies-and-services-cessations/ (accessed on Jul 21, 2018).
[2] http://web.mit.edu/airlinedata/www/2016%2012%20Month%20Documents/ Traffic%20and%20Capacity/Domestic/Domestic%20Load%20Factor%20.htm, accessed on Jul 15, 2017.

The airline industry had so far tackled the problem of being able to supply its customers with the flights they needed by looking to strategic alliances and code sharing deals. Airlines that were part of the same alliance agreed to pool their resources by agreeing to be located in the same terminals in hub airports, operating flights under the banner of more than one carrier, and offering privileges to members of fellow members' loyalty programs. By doing so, companies ensured that they did not have to operate and fly every route their customers demanded.

"We need to consider whether it makes sense to abandon our spoke-and-hub system. Our rivals that use point-to-point routes are eating into demand, and I'm sure passengers are noticing the kind of queues that are building up in the larger airports," James told Thrush.

4 Industry Data Sources

The airline industry uses three main types of data to interpret the environment in which it operates—demand data, such as booking and ticketing; supply data, such as schedules; and operational data, such as delays, cancellations, and check-ins. Thrush found that data scientists used these databases to uncover traveler preferences and understand their behavior.

The demand data in the industry comes from booking and ticketing databases, and detail a plethora of factors that affect customers while booking flights, and take into account exactly what information is available to customers at the time of their purchase. Supply data is usually accessible so that customers are able to identify flights, but the industry's main sources are schedules and guides provided by the Official Airline Guide (OAG). These guides collate information including origin, destination, trip length, and mileage for individual flights globally.[3] Data regarding the operational status of flights is usually available freely, though it is often not granular. AAA, like its competitors, kept detailed records of operational data in order to catch patterns of inefficiency. In addition, The US Department of Transport maintains a databank that consists of 10% of all flown tickets in the country.[4] The databank provides detailed ticketing, itinerary, and travel information and is freely available for research purpose.

---

[3]https://www.oag.com/airline-schedules, accessed on Jul 15, 2017.
[4]https://www.transtats.bts.gov/Tables.asp?DB_ID=125&DB_Name=Airline%20Origin%20and%20Destination%20Survey%20%28DB1B%29&DB_Short_Name=Origin%20and%20Destination%20Survey (accessed on Jun 24, 2018).

5 Meeting with John Heavens, Consultant

Thrush met with John Heavens, a data scientist and airline travel consultant, to inquire further about the possibility of using advanced data models in order to understand and forecast customer behavior.

Heavens told Thrush that the industry's old time-series/probabilistic models had become too outdated. Multinomial logit decision-choice models were the industry's mainstay tools in understanding consumer demand. These models broke itineraries down by assigning utility values to each flight and attempting to determine which factors were most valuable to customers. By observing the factors that affected customer choices for each origin–destination pair, Thrush would be able to predict with confidence where customers were looking to travel next.

However, Heavens also gave Thrush a third option. "Even the decision-choice models are becoming old, and we're moving in new directions now," he said. The consultant pointed out that the industry's MNL models were essentially linear in nature, and were not able to deal with factors that were correlated. In addition, their rigid need for data input meant that they could not predict the demand for new routes and new markets.

Instead, Heavens pointed to groundbreaking artificial intelligence research as the vanguard of an array of new technological tools that could be used to predict future demand. Techniques such as random forests, gradient-boosting machines, and artificial neural networks were able to produce better out-of-sample results without sacrificing in-sample goodness-of-fit. While these techniques lacked the readability and simplicity of MNL models, they were ultimately more efficient.

6 Meeting with Veerabhadram, the Data Scientist

After being presented with the models, Thrush knew he had a difficult decision to make. Moving to new methods of analysis had clear advantages, yet the significant investment in time and effort needed to be justified. Training and hiring employees and conducting ongoing analysis would be a drain on the company's resources.

Thrush looked to Hari Veerabhadram, the newest member of his team, to explain to him exactly which models are best suited to understand customer preferences. Hari knew that he had to explain how the models worked. He started thinking about which variables in the models he would use and what would be the most important. He knew that it would be crucial to explain why particular variables were the most important and which model was better at predicting customer preferences. Management always like visual proof of analysis. Thus, he felt that he would need to explain and compare the models both statistically (mean squared error, percentage variance explained by model, etc.) and through visualization methods (predicted vs. actual fit, training vs. validation results, etc.).

Table 26.1 Itinerary features (data description)

S. no.	Variable name	Description
1	ODPair_ID	Origin–destination pair identifier
2	Itinerary_ID	Itinerary identifier within O–D pair
3	Airline	Identifier for airline (A, B, ..., H, and all others) AAA Airline can be identified as Airline = "A"
4	Aircraft_Type	Type of aircraft in itinerary—propeller, jet, mainline
5	DoW	Day of the week that itinerary flies
6	Service_Level	Level of service of itinerary—non-stop, direct, single connect, double connect
7	Best_Service_Level	Best service level available for itinerary in the O–D Pair Non-stop > Direct > Single connect > double connect
8	Dep_Time	Time of departure at origin (HH:MM)
9	POS_Presence	Point of sale presence of airline at origin and destination airports. This measures the number of flights departing out of airports
10	Code_Share	Whether or not the itinerary is operated as a code-share with another carrier
11	Mileage	Distance travelled by the itinerary (converted into ratio)
12	Fare	Average fare of the itinerary (converted into ratio)
13	Pass_Cnt	Number of passengers who chose itinerary
14	No_of_itinaries	Number of itineraries available for O-D pair
15	Pass_Cnt_ODPair	Number of passengers in O-D Pair (Market Size)
16	Time_Bucket	Hourly time-of-day buckets corresponding to the departure time of the first leg of the itinerary 5 = (4 a.m.–5:59 p.m.); 6 = 6–6:59; ...; 22 = 10 p.m.–11:59 p.m. No departures from 12 midnight to 4 a.m.
17	Airline dummies	Whether the itinerary is with Airline XX
18	Service Level dummies	Whether the itinerary is having specific service level
19	Time bucket dummies	Whether the departure time of the itinerary falls into time bucket XX
20	Aircraft type dummies	Dummy variables for different aircraft type for each the itinerary

Thrush told Veerabhadram to start by looking at the itineraries in which AAA was lagging behind its competitors. Were there any patterns? Which markets did AAA find itself out of the top three? Did AAA need to do anything different? Veerabhadram knew that he would have to recommend changes to AAA's operations and began thinking about how to use the data at his disposal.

Veerabhadram started exploring the available data (refer to Table 26.1 for details) sourced from the company's internal database and third-party data providers. He noticed that the sales data is at the itinerary level for multiple airlines, but not at customer or transactional level. The data consists of 1885 distinct origin–destination (O-D) pairs with all possible itineraries on respective routes (O-D pair). Each row is an itinerary for a given O-D pair and provides additional information such as the identity of the airline, how many passengers chose that itinerary, what was the offered aircraft type, departure day and time, service-level, best service level available on that route, mileage, average fare, etc. For example, say, the O-D pair "5" represents the New York to Los Angeles route, and Airline = "A" represents all the itineraries offered by AAA Airline. Pick Itinerary ID "3." This itinerary offers a Small Propeller service on the route as a single connect option departing at 7 a.m. from New York. Single connect is the best service possible on this route across all the airlines serving that OD pair.

The basic summary statistics (Table 26.2) helped Veerabhadram to understand the variability in the data.[5] He observed that AAA Airline is one of the top performing airlines connecting significant number of cities through single and double connect itineraries. It also means that even a minor change in route can have

Table 26.2 Summary statistics

Basic summary statistics					
Variable name	Distinct Obs	Mean	Median	Minimum	Maximum
ODPair_ID	1885	NA	NA	1	1885
Itinary_ID	82,892	78	63	2	311
POS_Presence	NA	10	7	0	100
Code_Share	NA	NA	0	0	1
Mileage	NA	112	109	100	153
Fare	NA	99	100	10	471
Pass_Cnt	NA	1.7	0	0	534
Market size using Pass_Cnt	NA	212	52	1	3828
Airlines' market share					
Variable name		Total itineraries		Proportion (%)	
Airline A		18,234		22	
Airline B		11,907		14	
Airline C		4463		5	
Airline D		12,593		15	
Airline E		13,849		16	
Airline F		3185		4	
Airline G		12,226		15	
Airline H		5231		6	
Airline others		1564		2	
Flight options offered by all the airlines					
Variable name		Total itineraries		Proportion (%)	
Non-stop		450		1	
Direct		454		1	
Single connect		35,155		42	
Double connect		46,833		57	

---

[5]Refer to the website to download the data (csv and excel version).

a major impact on scheduling and resource allocation. Veerabhadram wondered whether more analysis and better visualization might help represent the current situation in a meaningful way to decision makers.

"Changing the timing of our flights poses serious challenges, Steven. We're better off keeping them the same so that we don't confuse customers and complicate our logistics," James said. Now, Veerabhadram started thinking of how to improve AAA's performance without changing the schedule.

Thrush's second priority was understanding which itineraries have gaps that could be filled by AAA. "Our business was built on flying between places that nobody realized needed more flights. Where do you think we need to go next?" he asked Veerabhadram.

Thrush told him that AAA's fleet currently consists of 120 propellers, 40 jets, and 100 mainline aircraft, all of which were in service. Veerabhadram decided he would pick the five best unserved routes to recommend to Thrush. He also started to consider which routes AAA would have to discontinue in order to start new flights.

At the end of these discussions, Veerabhadram's notebook was full! He decided to start with the following questions:

- Fit a customer choice model to the given data and analyze whether the features, model complexity, and size of data were adequate for answering the business questions.
- Identify variables of importance and make sure they made sound business sense. He thought he would try MNL and other techniques.
- Identify the currently unserved routes and recommend the five best unserved routes where the airline can introduce a new itinerary to increase revenue.
- Identify the O-D pairs where AAA Airline is not among the top three in terms of market share and then:
  - Analyze the offerings of the top three airlines in these O-D pairs.
  - Recommend changes in the itinerary features to improve the market share.
  - Suggest possible routes the airline could drop and explain why.

Electronic Supplementary Material

All the datasets, code, and other material referred in this section are available in www.allaboutanalytics.net.

- Text 26.1: Airline Instruction manual.docx
- Data 26.1: AAA_Airline_dummy.csv
- Data 26.2: AAA_Airline_Template.xlsx

References

Coldrena, G. M., Koppelmana, F. S., Kasturirangana, K., & Mukherjee, A. (2003). Modeling aggregate air-travel itinerary shares—Logit model development at a major US airline. *Journal of Air Transport Management, 9*, 361–369.

Garrow, L. A. (2010). *Discrete choice modelling and air travel demand: Theory and applications.* New York: Routledge.

Chapter 27
Case Study: InfoMedia Solutions

Deepak Agrawal, Soumithri Mamidipudi, and Sriram Padmanabhan

Hui Zhang had just returned from a workshop on sports and media analytics. One of the speakers had described the convergence of media and how it had affected his broadcast business in a very short time span. Hearing others mention the same set of possibilities and with his own experience in the rapidly changing industry, Zhang was convinced that an ever-increasing number of television viewers will, if they haven't done so already, "cut the cord" and move away from traditional viewing platforms. It was this thought that Zhang had at the back of his mind when he read a report predicting that viewership was splintering—more and more specialized channels were sniping away at traditional shows and showtimes. On top of it, the report mentioned that Internet advertising would overtake television and print media in size and spend in the next 5 years. Zhang was concerned that new technologies would threaten the position that his firm had built up in the TV advertising segment.

While millions of dollars were still expected to be spent on advertising on traditional television and cable channels, changes in viewership habits and demographics would surely shift budgets to targeting different audiences through dedicated channels. Moreover, the bundling strategy followed by cable TV companies allowed audience to quickly surf from one show to another! The change in resource allocation of ad spends had not yet happened because of the ongoing debate

---

Electronic supplementary material The online version of this chapter (https://doi.org/10.1007/978-3-319-68837-4_27) contains supplementary material, which is available to authorized users.

D. Agrawal (✉) · S. Mamidipudi
Indian School of Business, Hyderabad, Telangana, India
e-mail: a.deepak@outlook.com

S. Padmanabhan
New York, NY, USA

on tricky issues such as audience measurement, avoiding double counting of viewership, and measuring "reach" (see below for a definition)[1] in cross-platform or cross-channel advertising. These were relevant questions even for traditional media such as newspaper, magazine, radio, and television advertising even without factoring in Internet advertising.

Zhang thought data science might help bring clarity to these questions. Understanding how to use new analytical tools would surely be the key to profitability! As head of advertising strategy for InfoMedia Solutions, Zhang's job was to identify shifts in the market and technology. In order to understand the ways in which his company could better use the tools at its disposal, Zhang approached Luke Gershwin, InfoMedia's head of analytics. He asked Gershwin what data analytics tools InfoMedia might use to better target audience on TV. "Without a clear tech-focused strategy, I'm wary that any attempt to change the way our ad business works might fail," Zhang said.

1 InfoMedia Solutions

InfoMedia's operations spanned the whole gamut of advertising and press-related services. The company offered a full range of services—they were proud of being able to oversee an advertising campaign from conception to its completion. Zhang's role was to advise clients and guide marketing teams, as well as, monitor the advertising campaigns to ensure that they were running smoothly. The ever-increasing complexity and interconnectedness of advertising was not lost on Zhang, and he highlighted to his clients that targeting potential customers would not be a simple proposition in the future.

Zhang's foremost concern was "reach"—the metric every advertiser used to measure their success. Measured as the number of people in a target group who have been exposed to the advertisement at least once, reach was the backbone of every media strategy designed to pull in new customers and impress clients. "The changing context of media," Gershwin said, "had a very simple, obvious impact that was still playing out: the ways that media companies could reach their potential targets had shifted and multiplied."

"Ten years ago, nobody would have imagined that watching 10-s videos would be enough to run a business. Now *Snapchat*'s worth nearly \$30 billion,[2]" Gershwin told Zhang. "You've got to target the right market, and you've got to do it economically."

---

[1] A glossary of advertising business terms can be found at https://www.tubemogul.com/glossary/ or https://www.bionic-ads.com/2016/03/reach-frequency-ratings-grps-impressions-cpp-and-cpm-in-advertising/ (accessed on Aug 22, 2018). Refer to the article for the meaning of terms, such as Reach, CPM, CPV, Cross-Channel, Daypart, GRP, OTT, PTV, and RTB.

[2] http://money.cnn.com/2017/03/02/technology/snapchat-ipo/index.html (accessed on Jun 23, 2018)

The first issue Zhang faced was the problem that all advertisers had to tackle: duplication. Broadcasting an advertisement (ad) ten times would be very efficient if it were watched by ten different people. If only one person saw all ten times, however, the reach of the ad would become much smaller. In order to measure reach effectively, Zhang would have to correctly account for redundant, duplicate impressions that were inevitable in his line of business. Reach-1 measured the number of times a unique viewer saw the ad at least once.

Zhang also knew that most customers would not change their minds regarding a new product the first time they heard about it. In order to create an image in the customer's mind about a product, Zhang would have to reach the same customer multiple times. Thus, Zhang's duplication issue was akin to a Goldilocks problem—too few impressions would not result in a successful sale, but too many would be wasteful and inefficient. Identifying how to deliver the correct number of impressions in order to maximize reach would be at the core of the solution he needed from Gershwin.

Depending on the target audience and the product, Zhang normally tracked Reach-1, -3, or -5, meaning that an impression had been delivered to a potential customer at least three, five, or seven times. Understanding how many impressions would be necessary for each product was crucial to efficiently use resources. Gershwin's next issue would be to identify the inflection point where duplication was no longer useful to expanding reach.

While traditionally, overlap between media was not high enough to have a large impact on duplication, the need to track duplicate impressions on digital, print, and television was growing more and more important. The cross-platform approach, in which the campaign played out in more than one medium in order to gain impressions, was a burgeoning part of the sector that would surely grow. Gershwin felt that the cross-platform duplication could be addressed later.

2 Reach in Advertising

Like most companies in the ad business, InfoMedia thought of reach not as a function of cost but as a function of gross rating points (GRPs) (Goerg 2014). Advertising campaigns might place advertisements in newspapers, broadcast them on the radio, or show them on the television. Rather than simply look at the amount of money spent on the campaign to determine efficiency, Zhang thought of the number of impressions that the campaign would make and how that would translate into reach.

Zhang calculated GRP (gross rating points) by computing in percentage terms the number of impressions made on a target audience relative to the size of the audience. For example, if an advertisement was seen 120 million times by viewers and the size of the target audience is 60 million, then this translates to $(120/60 \times 100) = 200$ GRPs. Time slots that were longer, slots during peak hours, or in popular shows would be able to create more impressions and therefore higher reach. Nothing came free; a time slot that was able to deliver more reach would be more expensive. Zhang thus would have to balance the cost of airing slots against the reach that those slots could give him, by understanding his target audience. Reach-1, -3, -5, and -7 were defined either in percentage terms or in the total number of unique viewers who saw the ad at least one, three, five, and seven times. In the above example, these might be 45, 30, 15, and 7 million viewers or 75%, 50%, 25%, and 11.66%. Obviously, reach will not exceed the size of the population. Reach-1 will be the largest followed by the rest.

Reach-1, -3, -5, and -7 were known to be a concave increasing function of ad spots, which is to say that as the number of spots increased, reach increased, but the rate of that increase for the same increase in spots was diminishing. In an ideal world, Zhang would be able to buy exactly the number of spots at which the rate of decrease of reach times "a dollar value for reach" was balanced by the rate of increase in spots times a "dollar value of a spot." Gershwin thought one could uncover the relationship between ad spots and reach using simulation.

Adding to the complication, a cross-platform approach would need to estimate reach across multiple channels or media. This meant that the function used to calculate reach would have multiple inputs to track.

Guessing the appropriate ad spots target would prove tricky. Gershwin's solution would have to help solve this problem more efficiently if it were to improve InfoMedia's bottom line.

3 Rooster Biscuits

Zhang already had a customer in mind for the first test of the new approach. Rooster Confectionery, a biscuit- and cereal-maker, wanted to launch a new brand of chocolate-coated biscuits. Rooster's chief operating officer, Joni Fernandez, had told Zhang she wanted to focus on a younger target audience (age group 20–35) with the product—exactly the kind of customer who would be shifting to different avenues of media consumption. Zhang felt that if Gershwin's approach could help to optimize Rooster Biscuits' campaign, it would bode well for the use of novel techniques.

However, Fernandez had tight programming and budget constraints. She had already informed Zhang that "Rooster expected that the channels would show at least 20% of its ads during 'prime-time' slots—between 8 pm and 11 pm and at least 30% over weekends." Moreover, while she expected a long-term ad campaign, she wanted Zhang to run a short, 1-week version first to test the market's reaction to the new product. She informed him that the test campaign should target the two biggest cable channels aimed at the 18–34 P (P = person) demographic—The Animation Channel and the Sports 500 Network.

4 Blanc Solutions

Gershwin approached Julien Blanc, founder of data analysis consultancy Blanc Solutions, to understand how to use the new big data techniques. Consultants such as Blanc could quickly evaluate large volumes of data, dig deeper into a problem, and suggest an appropriate solution and data approach. Blanc recommended that InfoMedia use simulation techniques along with historical data to predict reach.

"If your client is planning a new campaign, past campaign performance isn't going to be the best indicator of future success. Using your knowledge of the market to simulate Reach is a better, cheaper solution," Blanc said.

Julien thought it is best to develop the ideas based on data from the two cable channels. Zhang was able to obtain viewership numbers for these channels from a third-party aggregator. In order to run robust simulations, Blanc looked for a past period that resembled the week during which InfoMedia planned to run its campaign. The aggregator was able to provide minute-to-minute data regarding viewership for any given demographic for the two channels.

Zhang was also able to leverage the history that he had with his client in order to provide Blanc with past data on pricing and reach for the company's campaigns. Rooster Biscuits had run an earlier campaign targeting the same demographics, with slightly different parameters. This earlier campaign would provide Blanc with appropriate context and help him make a decision about allocating his slots.

Blanc had found that the Sports 500 Network, which was the more popular of the two channels, tended to charge more for slots compared to the Animation Channel. Moreover, the premiums that Zhang would have to pay for prime-time and weekend slots were higher for the Sports 500 Network.[3] He had run a previous campaign with 150 slots, with 100 in the Sports 500 Network and 50 in the Animation Channel. At least 20% of the slots had to be during prime time and 30% during the weekend. This campaign could be a good test subject for the experiment.

Blanc knew that he had three important tasks to perform.

Task 1: Set up a simulation model to estimate the Reach–GRP curve.

How could Blanc simulate reach? Even if he had viewership data that told him how many people had been watching a channel at any given time, estimating how many of those people had seen an ad before even showing the ad looked like a tricky problem. It was here that new research held promise. For example, researchers at Google had identified a method by which the reach versus GRP relationship could be inferred given only a single data point, since the general shape of the curve was "known" (Goerg 2014). Blanc could expand this method to account for the much larger amount of data he was expecting to obtain from his simulations in order to understand the relationship between reach and number of ads shown. He felt when specializing to a specific campaign a functional form such as a polynomial function could be used, in other words, he could simply fit a polynomial curve through several

---

[3]See Appendix 1 for details.

(*r*,*g*) data points, where *r* is the reach and *g* is the number of spots shown. Based on his experience, Zhang suggested to start with quadratic and cubic fits.

Blanc planned to use the viewership data provided by the third-party aggregator to simulate reach and thus obtain the data points. Zhang had told him that InfoMedia intended to air between 10 and 250 ads in the 7-day period (1 week) of the campaign. The data was collected by a survey of households that viewed the two channels, conducted during a week that was as similar as possible to the target week. In the survey, viewers were asked a number of demographic questions. Blanc would be able to obtain the data regarding what channel they viewed, and for how long, from the cable companies themselves. These two sources, when combined, would give Blanc most of the necessary data.

Broadcasters divided airtime on their channels into 6-min slots, that is, Blanc had 1680 potential slots in the week to air the ad—10 per h * 24 h * 7 days per week. Simulation would involve throwing the ads randomly into these 1680 slots and computing GRP and reach. Thus, Blanc would choose the number of ads to air. Then, Blanc would simulate showing ads for each number of slots and use the viewership data to understand how many people were watching the slot chosen by the simulation. He could also estimate whether a viewer watched at least once, at least thrice, etc. Doing this repeatedly would give a "reach versus spots shown" set of data points. Then, he would fit a curve through these points to obtain the relationship.

Yet there was a key element to be added to his dataset. The viewership data he had was only a *sample* of the total population. It was necessary to add a unique weight to each viewer in his set—a measure of the proportion of viewers in the population that were similar to the selected viewer—to convert the sample numbers to the population number. For example, if a viewer in the 18–34 P had watched at least thrice and if this viewer's weight were 2345, then he would estimate that 2345 viewers had watched at least thrice in the population. Adding these numbers viewer-by-viewer would give an estimate for the GRP and reach. Blanc would thus be able to determine the reach of any combination of slots selected by the simulation by multiplying each viewer who saw the ad by the weight of that viewer. Moreover, by tracking the number of views by the same consumer, multiplying by the weight, and adding up across viewers, he could calculate not just Reach-1 but Reach-3 and Reach-5 as well.

Using multiple simulations,[4] he would be able to obtain a robust set of data that he could use to derive the reach curve. He can then fit a polynomial curve as explained above. The data science team constructed a simulator that produces the reach given the number of ads to be shown and the constraints on when they are to be shown. The help file, interface, and sample outputs are shown in Appendix 3.

Task 2: Develop the estimate.

---

[4]Blanc took help from a simulation expert in his data science team who provided him the code "Infomedia_simulation.R" (refer to Appendix 3) to run the simulation to calculate Reach for each simulation given the constraints.

Blanc decided to partition the data set by time and day of the week and to use that information to improve the prediction. In order to do this, he has multiple approaches—divide each day into 3-h buckets, divide regular and prime time on daily basis, or simply divide regular/prime time over weekday and weekend. The time bucket starting at 2 am would ensure that the effect of the "prime-time" 8 pm–11 pm bucket could be understood separately. Moreover, he also decomposed each day into its own bucket, to better understand the difference in viewership between weekdays and weekends. See Appendix 3 for examples of these data collection methods and how these are reflected in the output produced by the simulator. He reviewed the contract terms and legal notes to understand whether there is any "no show"/"blackout" period and found no such restriction for Rooster's campaign. "No show" or "blackout" period basically restricts the broadcaster from showing commercials when either the customer put the conditions of not broadcasting its ad during a particular time period assuming there will not be target customers or if there are any regulatory restrictions of not showing specific commercials in a particular time zone. Further, in the future, refinements in estimation can be made by using the variance amongst the viewers' demographics such as age and gender and other variables such as the average time spent by segment.

Task 3: Optimal spending.

Once Zhang had the information at hand and built up confidence in the model, the decision he had to make was clear. Blanc's own survey had informed him that the overlap between the viewers of the two channels seemed negligible. Thus, Zhang had to determine an optimal spending pattern for Rooster's campaign. How many ads would he run on each channel? At what times and on which days would he target how many? For demonstration purposes, Blanc thought he could use the previous campaign whose data is shown in Appendix 1 to demonstrate how the new method could work.

He knew he had a difficult task ahead explaining which variables were important in predicting reach. Could he convince his clients about the findings? He thought there were two ways of going about doing this—(a) explain the model very carefully to the client and (b) show how it can be used to increase the reach without increasing the budget.

Electronic Supplementary Material

All the datasets, code, and other material referred in this section are available in www.allaboutanalytics.net.

- Data 27.1: infomedia_ch1.csv
- Data 27.2: infomedia_ch2.csv
- Code 27.1: Infomedia_simulation.R

Exercises: Help Blanc with His Three Tasks

Ex. 27.1 Review the simulator description in Appendix 3 and pseudo-code provided in Appendix 4. Generate data using the simulator for both the channels. Report the model fit as a function of the total number of spots as well as the other explanatory variables. Visualize the reach curve for Reach-1, -3, -5, for both the channels.

Ex. 27.2 Using the campaign information/constraints and the model obtained in Ex. 27.1, demonstrate that a better allocation across channels, weekdays/weekends, and time-of-the-day can yield higher reach. (Maximum allocated budget is $300,000 for 1 week.) Show optimal allocations for each channel separately, as well as together, for Reach-3. Calculate the total spend for each allocation, based on the pricing details given. Use Reach-3 for your final recommendation.

Ex. 27.3 The advertiser realizes that between 2 am and 5 am, there are very few viewers from its target customer who watch TV and therefore adds the blackout window of 3 h every day. Would this change your analysis?

Ex. 27.4 Due to the increasing demand and limited broadcasting slots, the broadcaster is considering offering dynamic pricing. The broadcaster may redefine prime-time concept and significantly change its media marketing strategy. Suggest a new strategy if the broadcaster moves to dynamic pricing.

Ex. 27.5 How can this approach be used when more and more viewers switch to the Internet?

Ex. 27.6 What would happen if the advertiser demands not to broadcast the commercial alongside other similar commercials? What other practical constraints do you see being imposed on a schedule by an advertiser?

Ex. 27.7 What would happen if the broadcaster repeats the ad within the same slot/commercial break (known as a "pod" in the TV ad industry). Discuss how it may impact viewership of target segment and whether you need to change your strategy.

Appendix 1: Rooster Biscuit Campaign 6,412,965—Campaign GRP and Cost

Base rate	Animation channel	Sports 500 network
Weekday	$ 1500 (X)	$ 1200 (Y)

Notes: Base rate corresponds to weekday non-prime-time price for one 30-s spot. Rooster's campaign has an ad of exactly 30 s

Pricing multiples (categorized by spots timing)

Animation channel	Weekday	Weekend
Non-prime time	1X	1.25X
Prime time	1.4X	1.75X

Notes: Price multiples are applied to the base rate to calculate the specific rate for a different time or day of week

Sports 500 network	Weekday	Weekend
Non-prime time	1Y	1.2Y
Prime time	1.25Y	1.5Y

Notes: Price multiples change as per channel

Previous campaign report

Animation channel	Weekday non-prime	Weekday prime	Weekend non-prime	Weekend prime	Total
Spots	58	17	22	3	100
Costs ($)	87,000	35,700	41,250	7875	$ 171,825
Reach-3	929,155	1,527,138	330,370	97,027	2,883,690

Sports 500 network	Weekday non-prime	Weekday prime	Weekend non-prime	Weekend prime	Total
Spots	34	9	6	1	50
Costs ($)	40,800	13,500	8640	1800	$ 64,740
Reach-3	1,210,067	580,777	212,973	53,054	2,056,871

Notes: These were actuals achieved for the given campaign. The actuals are reported 2 weeks after airing

Appendix 2: Data Description

There are two datasets "infomedia_ch1.csv" and "infomedia_ch2.csv" for Animation channel and Sports 500 network, respectively. Each dataset contains ten variables. These variables largely concern when viewers start and stop viewing the channel, the time and date of the broadcast, and demographic data regarding the viewers (Tables 27.1 and 27.2).

Table 27.1 Data description of input dataset

Variable name	Description
day	Day of the broadcast (1-Monday, 2-Tuesday, ..., 7-Sunday)
adtime	Time when the advertisement starts
start	Time when the customer started watching the channel
end	Time when the customer stopped watching the channel
custid	Unique Id of the connection
pid	Unique Id of the family members
age	Age of the viewer
sex	Sex of the viewer
population_wgt	Weightage of the similar type of customer
channel	Channel identifier (1—Animation, 2—Sports 500 Network)

Note: The variables adtime, start, and end are in hour format and carry values between 0 and 2359

Table 27.2 Sample observations from input dataset (top five rows from infomedia_ch1.csv)

Day	Ad_time	Start	End	Cust_id	P_id	Age	Sex	Population_wgt	Channel
2	1342	1308	1355	70,953	1	59	M	3427	1
2	1348	1308	1355	70,953	1	59	M	3427	1
1	2100	2100	2109	79,828	1	23	F	5361	1
1	2106	2100	2109	79,828	1	23	F	5361	1
1	2124	2120	2124	79,828	1	23	F	5361	1

Appendix 3: Simulator Description

Help File

Please refer to the code "Infomedia_simulations.R" to run the simulation for each channel separately. The simulation function will ask for the following information:

`Enter full path of datasource folder:`

< Copy and paste the path name as it is. Please make sure you paste in R console not the editor.>

`Enter dataset name (including .csv):`

< Datasets infomedia_ch1.csv and infomedia_ch2.csv correspond to channels 1 and 2 respectively. Separate simulation is needed for each channel to quickly get results. Enter the source file name (infomedia_ch1 (or 2).csv) including file extension (.csv). Please note that R is case sensitive—spell the file name correctly.>

`Enter the minimum number of slots (typically 5)   :`

< the number of slots to begin the simulation>

`Enter the maximum number of slots (typically 250):`

< the number of slots to end the simulation>

`Enter the incremental number of slots (typically 5):`

< step size, that is, minimum, minimum + stepsize, minimum + 2* stepsize, ... will be simulated>

`Enter the number of simulation to run for each spot (typically 100):`

< the number of replications—too many will slow down the system>

`Minimum percentage slots in prime time [0-100]:`

< must be an integer, typically between 20 and 30>

`Maximum percentage slots in prime time [0-100]:`

< must be an integer, typically between 20 and 30>

`Minimum percentage slots on weekends [0-100]:`

< must be an integer, typically between 20 and 30>

`Maximum percentage slots on weekends [0-100]:`

< Must be an integer, typically between 20 and 30>

Once you enter all the inputs correctly, the simulation function will run the simulation for the requested channel given the constraints and share the two output files—(a) data file that will consist of reach given the number of spots in weekday non-prime, weekday prime, weekend non-prime, and weekend prime time and (b) png file that shows the reach curve against the total number of spots.

Output files (csv and png files) will be saved in the current directory as shown in the code output.

Sample Screenshot (Fig. 27.1)

```
> simulation()
Enter full path of datasource folder: D:\MyData\InfoMedia_Solutions

Current working directory is:
 D:/MyData/InfoMedia_Solutions

Enter dataset name (including .csv): infomedia_ch1.csv

*****************************************************************
Note: Please enter all the values below as positive integer only.
*****************************************************************

Enter the minimum number of slots (typically 5)   : 5
Enter the maximum number of slots (typically 250) : 250
Enter the incremental number of slots (typically 5): 5
Enter the number of simulation to run for each spot (typically 100): 100
Minimum percentage slots in prime time [0-100]: 20
Maximum percentage slots in prime time [0-100]: 30
Minimum percentage slots on weekends [0-100]: 20
Maximum percentage slots on weekends [0-100]: 30

Running 100 simulation for 5 to 250 spots with increment of 5 spots.
Constraints are as follows:
percentage slots during prime time: 20 to 30 percent.
percentage slots on weekend       : 20 to 30 percent.

*****************************************************************
             The simulation is successfully completed.

You can refer to below files (csv and image) in current working directory.

Current working directory:
 { D:/MyData/InfoMedia_Solutions }

Data file name: <simulation_ch1.csv>
Simulation plots file name: <Reach_curve_ch1.png>

Please use this data to fit the curve and for further analysis.
*****************************************************************
>
```

Fig. 27.1 Interface of the simulation function in R

Sample Output (Fig. 27.2)

The simulation function developed by the data science team produces the dataset "simulation_ch1.csv" and "simulation_ch2.csv" for each channel. The sample output is shown in Table 27.3.

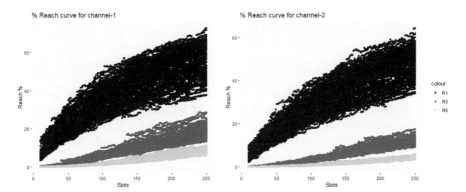

Fig. 27.2 Reach (R-1,-3,-5) vs. spots for channel 1 and 2 (sample output of simulation function)

Table 27.3 Sample output produced by the simulator

	R1	R3	R5	Wkday_NonPrime	WkDay_Prime	WkEnd_NonPrime	WkEnd_Prime
1	596,520	10,660	0	3	1	0	1
2	1,089,675	4438	0	3	1	1	0
3	295,026	0	0	3	0	1	1
4	341,508	0	0	4	0	0	1
5	245,348	0	0	3	0	1	1
..	..	..	..	..	..	..	..

Notes: R1, R3, and R5 are Reach-1, -3, and -5. The rest of the columns correspond to how many slots were shown in weekday non-prime, weekday prime, weekend non-prime, and weekend prime time

Appendix 4 (Pseudo Code)

1. First, we identify the constraints we would place such as prime vs. non-prime time spots, weekday vs. weekend, blackout zone, and number of spots on each channels.
2. Create a data frame with unique spots and days available.
3. Simulation exercise (Task-1): You can either use simulation function provided with case or develop your custom function using the steps below (3a–3c):

    (a) In each run, take random sample based on constraints in step 1, using *sample()* function. *Sample (vector from which we have to choose, no of items to choose)*.

    (b) Merge this data frame with the actual data set depending upon time and day using *merge()* function. Basically, step-3a will generate various samples (to simulate the runs) and will help identify viewers in the next step.

    (c) Now collect the data on customer level how many times a particular customer viewed the ad. By using *count* function in *plyr* library.

4. Fitting the curve (Task-2)

    (a) Now we can calculate the total reach of the ad based on the distinct viewers count who watched the ad repeatedly, that is, at least once (R1), at least thrice (R3), at least five times (R5), where the total reach is the sum of the population weight column which represents the weightage of similar type of customers. Plot the total reach and number of spots (say varying total spots between 5 and 250 in steps of 5 or 10).

    (b) We fit the curve to estimate the average total reach for a given slot size. You will need one curve each for R-1, R-3, R-5, etc.

    $$Reach = f(slots) + \varepsilon$$

5. Error estimation (Task-2)

    (a) Collect the data about the distribution of slots: how the slots have been distributed among the different constraints like day of the week, time bucket, prime time, non-prime, and weekend and weekday.

    (b) We can see if additional data can explain the error in the above fit.

    $$\varepsilon = f\,(daypart, demographics, other\ explanatory\ variables)$$

    (c) Ensemble the above models to better estimate the total reach.

    $$Reach = f\,(slots, daypart, demographics, other\ explanatory\ variables) + \varphi$$

6. Optimization (Task-3): Now, we optimize the total reach using nonlinear optimization techniques (refer to Chap. 11 Optimization):

   (a) Objective function: Maximize appropriate reach
   (b) Constraints
   - Number of weekend spots out of total spots (weekend regular + weekend prime)
   - Number of prime-time spots out of total spots (weekday prime + weekend prime)
   - Available budget for the advertisement (total, channel-wise)

Reference

Goerg, M. (2014). *Estimating reach curves from one data point*. Google Inc. Retrieved June 23, 2018, from https://ai.google/research/pubs/pub43218.

Chapter 28
Introduction to R

Peeyush Taori and Hemanth Kumar Dasararaju

1 Introduction

As data science adoption increases more in the industry, the demand for data scientists has been increasing at an astonishing pace. Data scientists are a rare breed of "unicorns" who are required to be omniscient, and, according to popular culture, a data scientist is someone who knows more statistics than a programmer and more programming than a statistician. One of the most important tools in a data scientist's toolkit is the knowledge of a general-purpose programming language that enables a data scientist to perform tasks of data cleaning, data manipulation, and statistical analysis with ease. Such requirements call for programming languages that are easy enough to learn and yet powerful enough to accomplish complex coding tasks. Two such de facto programming languages for data science used in the industry and academia are Python and R.

In this chapter, we focus on one of the most popular programming languages for data science—R (refer to Chap. 29 for Python). Though we do not aim to cover comprehensively all topics of R, we aim to provide enough material to provide a basic introduction to R so that you could start working with it for your daily programming tasks. A detailed knowledge of R can be gained through an excellent collection of books and online resources. Although prior programming experience is helpful, this chapter does not require any prior knowledge of programming.

P. Taori (✉)
London Business School, London, UK
e-mail: taori.peeyush@gmail.com

H. K. Dasararaju
Indian School of Business, Hyderabad, Telangana, India

1.1 What Is R?

R is a high-level and general-purpose programming language that was first announced in 1992 with a development version being released in 1995. R is essentially an implementation of another programming language called *S* and was developed by *Ross Ihaka* and *Robert Gentleman* (and hence the name R after the initial of both creators of the language). While R is built as a general-purpose programming language, its core usage is in the field of statistics and data science where R enjoys a huge audience and heavy support from scientific communities.

R is an open-source language that allows anyone to contribute to the R environment by creating packages and making them available to other users. R has a fairly large scientific community and is used in a variety of settings such as financial research, algorithms development, options and derivatives pricing, financial modeling, and trading systems. R is written mostly in C, Fortran, and R itself, and one may see that many of the R packages are written in one of these programming languages. This also means that there is good interoperability between R and these programming languages.

1.2 Why R for Data Science?

As stated at the start of this chapter, R is one of the de facto languages when it comes to data science. There are a number of reasons why R is such a popular language among data scientists. Some of those reasons are listed below:

- R is a high-level general-purpose programming language that can be used for varied programming tasks such as web scraping, data gathering, data cleaning and manipulation, and website development and for statistical analysis and machine learning purposes.
- R is a language that is designed mostly for non-programmers and hence is easy to learn and implement.
- R is an open-source programming language. This implies that a large community of developers contributes continually to the R ecosystem.
- R is easily extensible and enjoys active contribution from thousands of developers across the world. This implies that most of the programming tasks can be handled by simply calling functions in one of these packages that developers have contributed. This reduces the need for writing hundreds of lines of codes and makes development easier and faster.
- R is an interpreted language that is platform independent. As compared to some of the other programming languages, you do not have to worry about underlying hardware on which the code is going to run. Platform independence essentially ensures that your code will run in the same manner on any platform/hardware that is supported by R.

1.3 Limits of R

While R is a great programming language meant for general-purpose and scientific computing tasks, R has its own set of limitations. One such limitation is that R has a relatively steeper learning curve than other programming languages such as Python. While this means increased effort for learning the language, once you have a hang of it, the development becomes very fast in R. Another major limitation of R is its inefficiency in handling large datasets. For datasets that are a few hundred MB in size, R can work smoothly, but as soon as datasets size increases, or computation requires creation of intermediate datasets that can take up large memory, then performance of R begins to degrade very fast. While memory management and working with large datasets is indeed a limitation of R, this can be overcome by using commercial offerings of R. Also, for many of data science needs, it might not be necessary to work with large datasets.

2 Chapter Plan

In this section, we describe the R programming language and use the features and packages present in the language for data science-related purposes. Specifically, we will be learning the language constructs, programming in R, how to use these basic constructs to perform data cleaning, processing, and data manipulation tasks, and use packages developed by the scientific community to perform data analysis. In addition to working with structured (numerical) data, we will also be learning about how to work with unstructured (textual) data as R has a lot of features to deal with both the domains in an efficient manner.

We will start with discussion about the basic constructs of the language such as operators, data types, conditional statements, and functions, and later we will discuss specific packages that are relevant for data analysis and research purpose. In each section, we will discuss a topic, code snippets, and exercise related to the sessions.

2.1 Installation

There are multiple ways in which you can work with R. In addition to the basic R environment (that provides R kernel as well as a GUI-based editor to write and execute code statements), most people prefer to work with an Integrated Development Environment (IDE) for R. One such free and popular environment is *RStudio*. In this subsection, we will demonstrate how you can install both R and RStudio.

When you work in a team environment or if your project grows in size, it is often recommended to use an IDE. Working with an IDE greatly simplifies the task of developing, testing, deploying, and managing your project in one place. You can choose to use any IDE that suits your specifics needs.

2.2 R Installation

R can be installed on Windows-, Mac OS X-, and Linux-based machines. In order to install R, go to the following website:
http://cran.r-project.org/

Once at the website, select the R installation specific to your system. Most of the R installations come with a GUI-based installer that makes installation easy. Follow the on-screen instructions to install R on your operating system.

Once you have installed R, an R icon would be created on the Desktop of your computer. Simply double-click the icon to launch R environment.

2.3 R Studio

RStudio is a free and open-source IDE for R programming language. You can install RStudio by going to the following website:
www.rstudio.com

Once at the website, download the specific installation of RStudio for your operating system. RStudio is available for Windows-, Mac OSX-, and Linux-based systems. RStudio requires that you have installed R first so you would first need to install R before installing RStudio. Most of the RStudio installations come with a GUI-based installer that makes installation easy. Follow the on-screen instructions to install RStudio on your operating system.

Once you have installed RStudio, an RStudio icon would be created on the Desktop of your computer. Simply double-click the icon to launch RStudio environment.

There are four major components in RStudio distribution:

1. A text editor at the top left-hand corner. This is where you can write your R code and execute them using the Run button.
2. Integrated R console at the bottom left-hand corner. You can view the output of code execution in this pane and can also write individual R commands here.
3. R environment at the top right-hand corner. This pane allows you to have a quick look at existing datasets and variables in your working R environment.
4. Miscellaneous pane at the bottom right-hand corner. This pane has multiple tabs and provides a range of functionalities. Two of the most important tabs in this pane are Plots and Packages. Plots allow you to view the plots from code execution. In the packages tab, you can view and install R packages by simply typing the package name (Fig. 28.1).

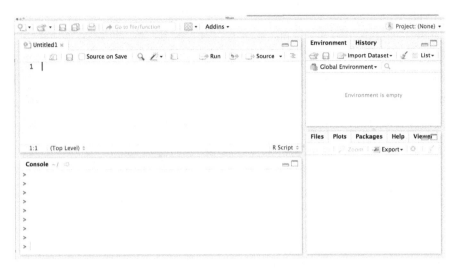

Fig. 28.1 RStudio interface

2.4 R Packages and CRAN

In addition to core R, most of the times you would need packages in R to get your work done. Packages are one of the most important components of the R ecosystem. You would be using packages continuously throughout the course and in your professional lives. Good thing about R packages is that you can find most of them at a single repository: CRAN repository. In RStudio, click on Packages tab and then click on Install. A new window will open where you can start typing the name of the R package that you want to install. If the package exists in the CRAN repository, then you will find the corresponding name. After that, simply click on Install to install the R package and its dependencies as well. This is one of the easiest ways to install and manage packages in your R distribution.

Alternatively, you can install a package from the command prompt as well by using *install.packages* command. For example, if you type the following command, it will install "*e1071*" package in R:

```
> install.packages("e1071")
```

A not so good thing about R packages is that there is not a single place where you will get a list of all packages in R and what they do. In such cases, reading online documentation of R packages is the best way. You can search for specific packages and their documentation on the CRAN website. Thankfully, you will need only a handful of packages to get most of your daily work done.

In order to view contents of the package, type:

```
> library(help=e1071)
```

This will give you a description about the package, as well as all available datasets and functions within that package. For example, the above command will produce the following output:

```
                    Information on package 'e1071'

Description:

Package:            e1071
Version:            1.6-8
Title:              Misc Functions of the Department of
                    Statistics, Probability Theory
                    Group (Formerly: E1071), TU Wien
Imports:            graphics, grDevices, class, stats, methods,
                    utils
Suggests:           cluster, mlbench, nnet, randomForest, rpart,
                    SparseM, xtable, Matrix,
                    MASS
Authors@R:          c(person(given = "David", family = "Meyer",
                    role = c("aut", "cre"),
                    email = "David.Meyer@R-project.org"),
                    person(given = "Evgenia",
                    family = "Dimitriadou", role = c("aut",
                    "cph")), person(given =
                    "Kurt", family = "Hornik", role = "aut"),
                    person(given = "Andreas",
                    family = "Weingessel", role = "aut"), person
                    (given = "Friedrich",
                    family = "Leisch", role = "aut"), person
                    (given = "Chih-Chung", family
                    = "Chang", role = c("ctb","cph"), comment =
                    "libsvm C++-code"),
                    person(given = "Chih-Chen", family = "Lin",
                    role = c("ctb","cph"),
                    comment = "libsvm C++-code"))
Description:        Functions for latent class analysis, short
                    time Fourier transform,
                    fuzzy clustering, support vector machines,
                    shortest path computation,
                    bagged clustering, naive Bayes classifier,
                    ...
License:            GPL-2
LazyLoad:           yes
NeedsCompilation:   yes
Packaged:           2017-02-01 16:13:21 UTC; meyer
Author:             David Meyer [aut, cre], Evgenia Dimitriadou
                    [aut, cph], Kurt Hornik
                    [aut], Andreas Weingessel [aut], Friedrich
                    Leisch [aut], Chih-Chung
                    Chang [ctb, cph] (libsvm C++-code),
                    Chih-Chen Lin [ctb, cph] (libsvm
                    C++-code)
Maintainer:         David Meyer <David.Meyer@R-project.org>
Repository:         CRAN
```

28 Introduction to R

```
Date/Publication:   2017-02-02 12:37:10
Built:              R 3.4.4; x86_64-w64-mingw32; 2018-06-15
                    19:27:40 UTC; windows
```

Index:

allShortestPaths	Find Shortest Paths Between All Nodes in a Directed Graph
bclust	Bagged Clustering
bincombinations	Binary Combinations
bootstrap.lca	Bootstrap Samples of LCA Results
boxplot.bclust	Boxplot of Cluster Profiles
classAgreement	Coefficients Comparing Classification Agreement
cmeans	Fuzzy C-Means Clustering
countpattern	Count Binary Patterns
cshell	Fuzzy C-Shell Clustering
ddiscrete	Discrete Distribution
e1071-deprecated	Deprecated Functions in Package e1071
element	Extract Elements of an Array
fclustIndex	Fuzzy Cluster Indexes (Validity/Performance Measures)
hamming.distance	Hamming Distances of Vectors
hamming.window	Computes the Coefficients of a Hamming Window.
hanning.window	Computes the Coefficients of a Hanning Window.
hsv_palette	Sequential color palette based on HSV colors
ica	Independent Component Analysis
impute	Replace Missing Values
interpolate	Interpolate Values of Array
kurtosis	Kurtosis
lca	Latent Class Analysis (LCA)
matchClasses	Find Similar Classes in Two-way Contingency
matchControls	Find Matched Control Group
moment	Statistical Moment
naiveBayes	Naive Bayes Classifier
permutations	All Permutations of Integers 1:n
plot.stft	Plot Short Time Fourier Transforms
plot.svm	Plot SVM Objects
plot.tune	Plot Tuning Object
predict.svm	Predict Method for Support Vector Machines
probplot	Probability Plot
rbridge	Simulation of Brownian Bridge
read.matrix.csr	Read/Write Sparse Data
rectangle.window	Computes the Coefficients of a Rectangle Window.
rwiener	Simulation of Wiener Process
sigmoid	The Logistic Function and Derivatives
skewness	Skewness

```
stft                    Computes the Short Time Fourier
                        Transform of a Vector
svm                     Support Vector Machines
tune                    Parameter Tuning of Functions Using Grid
                        Search
tune.control            Control Parameters for the Tune Function
tune.wrapper            Convenience Tuning Wrapper Functions
write.svm               Write SVM Object to File

Further information is available in the following vignettes in
directory
'C:/Users/10787/Documents/R/win-library/3.4/e1071/doc':

svmdoc: Support Vector Machines---the Interface to libsvm in
package e1071 (source, pdf)
svminternals: svm() internals (source, pdf)
```

2.5 Finding Help in R

The simplest way to get help in R is to click on the *Help* button on the toolbar. Alternatively, if you know the name of the function you want help with, you just type a question mark "?" at the command line prompt followed by the name of the function. For example, the following commands will give you a description of function solve.

```
> help(solve)
> ?solve
> ?read.table
```

Sometimes you cannot remember the precise name of the function, but you know the subject on which you want help (e.g., data input in this case). Use the *help.search* function (without a question mark) with your query in double quotes like this:

```
> help.search("data input")
```

Other useful functions are "*find*" and "*apropos*." The "*find*" function tells you what package something is in:

```
> find("lowess")
```

On the other hand, "*apropos*" returns a character vector giving the names of all objects in the search list that match your (potentially partial) enquiry:

```
> apropos("lm")
```

3 The R Programming Language

As of date (June 16, 2018), the latest version of R available is version 3.5. However, in this chapter, we demonstrate all the R code examples using version 3.2 as it is one of the most widely used versions. While there are no drastic differences in the two versions, there may be some minor differences that need to be kept in mind while developing the code.

3.1 Programming in R

Before we get started with coding in R, it is always a good idea to set your working directory in R. Working directory in R can be any normal directory on your file system, and it is in this directory that all of the datasets produced will be saved. By default R sets the working directory as the directory where R is installed. You can get the current working directory by typing the following command:

```
> getwd()
```

It will produce the output similar to the one below:

```
[1] "/Users/rdirectory"
```

In order to change working directory, use *setwd()* command with the directory name as the argument:

```
> setwd('/Users/anotherRDirectory')
```

This command will make the new directory as your working directory.

There are two ways to write code in R: script and interactive. The script mode is the one that most of the programmers would be familiar with, that is, all of the R code is written in one text file and the file then executes on a R interpreter. All R code files must have a <dot>R extension. This signals the interpreter that the file contains an R code. In the interactive mode, instead of writing all of the codes together in one file, individual snippets of code are written in a command line shell and executed. The benefit of the interactive mode is that it gives immediate feedback for each statement and makes program development more efficient. A typical practice is to first write snippets of code in the interactive mode to test for functionality and then bundle all pieces of code in a <dot>R file (script mode). RStudio provides access to both modes. The top window in text editor is where you can type code in script mode and run all or some part of it. In order to run a file, just click on the <Run> button in the menu bar, and R will execute the code contained in the file.

The bottom window in the text editor acts as R interactive shell. In interactive mode, what you type is immediately executed. For example, typing $1 + 1$ will respond with 2.

3.2 Syntax Formalities

Let us now get started with understanding the syntax of R. The first thing to note about R is that it is a case-sensitive language. Thus, variable1 and VARIABLE1 are two different constructs in R. While we saw in the other languages such as Python that indentation is one of the biggest changes that users have to grapple with, there is no such requirement of indentations in R. The code simply flows, and you can either terminate the code with a semicolon or simply start writing a new code from a new line, and R will understand that perfectly. We will delve more on these features as we move to further sections.

3.3 Calculations

Since R is designed to be a simple programming language, the easiest way to use R is as a calculator. You can simply type commands and operations in R as you would do with a calculator, and R produces the output. The fundamental idea here is that one should be able to perform most of the processing tasks without worrying about the syntax of a programming language. For example, you can simply type the following commands in R to get the output:

```
> log(50)

[1] 3.912023

> 5+3

[1] 8
```

Multiple expressions can be placed in single line but have to be separated by semicolons.

```
> log(20); 3*35; 5+2

[1] 2.995732
[1] 105
[1] 7

> floor(5.3)

[1] 5

> ceiling(5.3)
[1] 6
```

3.4 Comments

Comments are required in any programming language to improve readability by humans. Comments are those sections of code that are meant for human comprehension and are ignored by R interpreter when executing. In R, you can specify single-line comments with a pound (#) sign.

1. **Single-line comment**

    A single line comment in R begins with a pound (#) sign. Everything after the # sign is ignored by the interpreter until the end of the line.

Code
```
print("This is code line, not a comment line")
#print("This is a comment line")
```
Output
```
This is code line, not a comment line
```

Note that in the above code snippet, the first line is the actual code that is executed, whereas the second line is a comment that is ignored by the interpreter. A strange observation in R is that it does not have support for multiline comments. So if you want to use multiline comments in R, then you have to individually comment each line. Fortunately, IDEs such as RStudio provide work-around for this limitation. For example, in Windows you can use CTRL + SHIFT + C to comment multiple lines of code in RStudio.

3.5 Variables

There are some in-built data types in R for handling different kinds of data: integer, floating point, string, Boolean values, date, and time. Similar to Python, a neat feature of R is that you don't need to mention what kind of data a variable holds; depending on the value assigned, R automatically assigns a data type to the variable.

Think of a variable as a placeholder. It is any name that can hold a value and that value can vary over time (hence the name variable). In other terms, variables are reserved locations in your machine's memory to store different values. Whenever you specify a variable, you are actually allocating space in memory that will hold values or objects in future. These variables continue to exist till the program is running. Depending on the type of data a variable has, the interpreter will assign the required amount of memory for that variable. This implies that memory of a variable can increase or decrease dynamically depending on what type of data the variable has at the moment. You create a variable by specifying a name to the variable and then by assigning a value to the variable by using equal sign (=) operator.

Code

```
variable1 = 100          # Variable that holds integer value
distance  = 1500.0       # Variable that holds floating
                           point value
institute = "ISB"        # Variable that holds a string
```

Output

```
100
1500.0
ISB
```

Code

```
a = 0
b = 2
c = "0"
print(a + b)
print(c)
```

Output

```
2
"0"
```

3.6 Naming Conventions for a Variable

Although a variable can be named almost anything, there are certain naming conventions that should be followed:

- Variable names in R are case-sensitive. This means that Variable and variable are two different variables.
- A variable name cannot begin with a number.
- Remainder of the variable can contain any combination of letters, digits, and underscore characters.
- A variable name cannot contain blank spaces.

The value of the variables can be intialized in two ways:

```
> x <- 5
> y = 5

> print(x)
[1] 5

> print(y)
[1] 5
```

[1] indicates that x and y are vectors and 5 is the first element of the vector.

Notice the use of <- for assignment operator. Assignments in R are conventionally done using <- operator (although you can use = operator as well). For most of the cases, there is no difference between the two; however in some of the specialized cases, you can get different results based on which operator you are using. The official and correct assignment operator that is endorsed is <- operator, and we would encourage the readers to use the same for their coding as well.

3.7 Basic Data Types

In addition to complex data types, R has five atomic (basic) data types. They are Numeric, Character, Integer, Complex, and Logical, respectively. Let us understand them one by one.

Numbers are used to hold numerical values. There are four types of numbers that are supported in R: integer, long integer, floating point (decimals), and complex numbers.

1. **Integer**: An integer type can hold integer values such as 1, 4, 1000, and −52,534. In R, integers have a bit length of 32 bits. This means that an integer data type can hold values in the range of −2,147,483,648 to 2,147,483,647. An integer is stored internally as a string of digits. An integer can only contain digits and cannot have any characters or punctuations such as $.

Code

```
> 120+200
```

```
[1] 320
> 180-42
[1] 138
> 15*8
[1] 120
```

2. **Long Integer**: Simple integers have a limit on the value that they can contain. Sometimes the need arises for holding a value that is outside the range of integer numbers. In such a case, we make use of Long Integer data types. Long Integer data types do not have a limit on the length of data they can contain. A downside of such data types is that they consume more memory and are slow during computations. Use Long Integer data types only when you have the absolute need for it.

Code

```
> 2**32
[1] 4294967296
```

3. **Floating Point Numbers**: Floating point data types are used to contain decimal values such as fractions.
4. **Complex Numbers**: Complex number data types are used to hold complex numbers. In data science, complex numbers are used rarely, and unless you are dealing with abstract math, there would be no need to use complex numbers.

3.8 Vector

Whenever you define a variable in R that can contain one of the above atomic data types, that variable would most likely be a vector. A vector in R is a variable that can contain one of more values of the same type (Numeric, Character, Logical, and so on). A vector in R is analogous to an array in C or Java with the difference that we do not have to create the array explicitly and we also do not have to worry about increasing or decreasing length of array. A primary reason behind having vectors as the basic variable in R is that most of the times, the programmer or analyst would not be working with a single value but a bunch of values in a dataset (think of a column in a spreadsheet). Thus, in order to mimic that behavior, R implements the variable as a vector. A vector can also contain single values (in such a case, it would be a vector of length one). For example, all of the variables below are vectors of length one (since they contain only one element):

```
> a <- 4
> a
[1] 4

> str <- "abc"
> str
[1] "abc"

> boolean <- TRUE
> boolean
[1] TRUE
```

If you want to combine multiple values to create a vector, then you can make use of the c operator in R. c() stands for concatenate operator, and its job is to take individual elements and create a vector by putting them together. For example:

```
> x <- c(1, 0.5, 4)
> x
[1] 1.0 0.5 4.0

> y <- c("a","b","c")
> y
[1] "a" "b" "c"

> z <- vector("numeric",length=50)
> z
 [1] 0 0 0 0 0 0 0 0 0 0 0 0 0 0 0 0 0 0 0 0 0 0 0 0 0 0 0 0 0 0 0 0 0
    0 0 0
[34] 0 0 0 0 0 0 0 0 0 0 0 0 0 0 0 0 0
```

Note that in the last statement, we made use of the vector() function to create a function. Vector is an inbuilt function in R that will create a vector of a specific size (specified by length argument) and type (specified by numeric). If we do not specify default values for the vector, then it will take default values for the specified vector type (e.g., default value for numeric is 0).

You can perform a range of functions on the vector. For example:

#To find the class of a vector, use class function

```
> class(y)
[1] "character"

> #Length of a vector
> length(y)
[1] 3
```

This representation of data in a vector allows you to ask mathematical questions easily. For example:

```
> mean(x)
[1] 1.833333
> max(x)
[1] 4
> quantile(x)
  0%   25%   50%   75%  100%
0.50 0.75 1.00 2.50 4.00
```

Vectors are quite flexible in R and you can create them in a range of ways. One very useful operator in R for vectors is the sequence operator (:). A sequence operator works like an increment operator that will start with an initial value, increment in steps (default is 1), and stop at a terminal value. In doing so, the increment operator will create a vector from initial to terminal value. For example:

```
> x <- 1:50
> x
 [1]  1  2  3  4  5  6  7  8  9 10 11 12 13 14 15 16 17 18 19 20
     21 22
[23] 23 24 25 26 27 28 29 30 31 32 33 34 35 36 37 38 39 40 41 42
     43 44
[45] 45 46 47 48 49 50
> seq(0,8,0.2)
 [1] 0.0 0.2 0.4 0.6 0.8 1.0 1.2 1.4 1.6 1.8 2.0 2.2 2.4 2.6
     2.8 3.0
[17] 3.2 3.4 3.6 3.8 4.0 4.2 4.4 4.6 4.8 5.0 5.2 5.4 5.6 5.8 6.0
     6.2
[33] 6.4 6.6 6.8 7.0 7.2 7.4 7.6 7.8 8.0
```

Note that in the command, we explicitly called the seq() function (it is similar to the sequence operator). The seq() function takes the initial value and terminal value as 0 and 8, respectively, and creates a vector of values by incrementing in the steps of 0.2.

If we want to generate a vector of repetitive values, then we can do so easily by using the rep() function. For example:

```
> rep(4,9)
[1] 4 4 4 4 4 4 4 4 4
> rep(1:7,10)
 [1] 1 2 3 4 5 6 7 1 2 3 4 5 6 7 1 2 3 4 5 6 7 1 2 3 4 5 6 7 1 2
     3 4 5
[34] 6 7 1 2 3 4 5 6 7 1 2 3 4 5 6 7 1 2 3 4 5 6 7 1 2 3 4 5 6
     7 1 2 3
[67] 4 5 6 7
> rep(1:7,each=3)
 [1] 1 1 1 2 2 2 3 3 3 4 4 4 5 5 5 6 6 6 7 7 7
```

In the first case, the rep function repeated the value 4 nine times. In the second command, rep repeated the sequence 1 to 7 ten times. In the third, we created a vector where each value from 1 to 7 was repeated three times.

3.9 Vector Arithmetic and Processing

You can perform the arithmetic operations on vectors in a manner similar to variable operations. Here, the operations are performed on each corresponding element:

```
> x <- c(1, 0.5, 4)
> x
[1] 1.0 0.5 4.0
> y <- c(5,3,2)
> y
[1] 5 3 2
> x+y
[1] 6.0 3.5 6.0
```

What would happen in the following case?

```
> x
[1] 1.0 0.5 4.0
> y <- c(5,3,2,1)
> y
[1] 5 3 2 1
> x+y
[1] 6.0 3.5 6.0 2.0
```

Warning message:

In x + y: longer object length is not a multiple of shorter object length

You would expect that there should be an error since the vectors are not of same length. However, while we received a warning saying that vectors are not of same length, R nevertheless performs the operation in a manner such that when the vector of shorter length finishes, then the whole process starts again from the first element for the short vector. This means that x in our case is the vector with three elements. While first three elements of x are added to three elements of y, but for the fourth element of y, the element from x is the first element (since the process repeats itself for the shorter length vector). This is a peculiar behavior of R that one needs to be careful about. If we are not careful about the length of vectors while performing arithmetic operations, then the results can be erroneous and can go undetected (since R does not produce any errors).

Since a vector can be viewed as an array of individual elements, we can extract individual elements of a vector and can also access sub-vectors from a vector. The syntax for doing so is very similar to what we use in Python, that is, specify the name of the vector followed by the index in square brackets. One point to be careful about is that indexes in R start from 1 (and not from 0 as in Python). For example:

```
> a <- c(1,3,2,4,5,2,4,2,6,4,5,3)
> a
 [1] 1 3 2 4 5 2 4 2 6 4 5 3

> #Extract individual elements of a vector
> a[1]
[1] 1

> #Access multiple values of a vector
> b <- a[c(1,4)]
> b
[1] 1 4

> d <- a[1:4]
> d
[1] 1 3 2 4
```

Let us say that you want to select a subset of a vector based on a condition.

```
> anyvector <- a>3
> a[anyvector]
[1] 4 5 4 6 4 5

> x <- 1:30
```

```
> x[x>5]
 [1]  6  7  8  9 10 11 12 13 14 15 16 17 18 19 20 21 22 23 24 25
      26 27
[23] 28 29 30
```

You can also apply set theory operations (in addition to usual arithmetic operators) on vectors.

```
> setA <- c("a", "b", "c", "d", "e")
> setB <- c("d", "e", "f", "g")

> union(setA, setB)
[1] "a" "b" "c" "d" "e" "f" "g"

> intersect(setA, setB)
[1] "d" "e"

> setdiff(setA,setB)
[1] "a" "b" "c"
```

3.10 Lists

While vectors in R are a convenient way of playing with a number of values at the same time, oftentimes, the need would arise that we need to have values of different types in a vector. For example, we might want to have numeric as well as character values in the same variable. Since we cannot do so with vectors, the data type that comes to our rescue is list. A list in R is nothing but a special type of vector that can contain different types of data. We define list with a list() function in R.

```
> x <- list(1,"c",FALSE)
> x
[[1]]
[1] 1
[[2]]
[1] "c"
[[3]]
[1] FALSE

> x[3]
[[1]]
[1] FALSE

> x[1:2]
[[1]]
[1] 1
[[2]]
[1] "c"
```

In the above case, we defined a list x that contains three elements—numeric 1, character c, and a logical value FALSE.

We can then access individual elements of a list in the similar way we can do so with vectors. In addition to containing basic data types, a list can contain complex data types as well (such as nested lists). For example:

```
> x <- list(col1=1:3, col2 = 4)
> x
$col1
[1] 1 2 3
$col2
[1] 4

> x[1]
$col1
[1] 1 2 3

> x[[1]]
[1] 1 2 3

> x$col[1]
[1] 1 2 3

> x$col1[2]
[1] 2

> x[[1]][2]
[1] 2
```

In the above example, we defined a list x that contains two elements—col1 and col2. Col1 and col2 are lists by themselves—col1 contains numbers 1, 2, and 3; and col2 contains a single element 4. You can access individual elements of a list or elements with the elements by using square brackets and the index of elements.

3.11 Matrices

Lists and vectors are unidimensional objects, that is, a vector can contain a number of values, and we can think of it as a single column in a spreadsheet. But if we need to have multiple columns, then vectors are not a convenient way of work-around. For this R provides two different data structures at our disposal—matrices and data frames. We will first discuss about matrices and then move on to data frames.

A matrix in R is nothing but a multidimensional object where each dimension is an array. There are multiple ways of creating a matrix in R:

```
> m1 <- matrix(nrow=4, ncol=5)

> m1
     [,1] [,2] [,3] [,4] [,5]
[1,]  NA   NA   NA   NA   NA
[2,]  NA   NA   NA   NA   NA
[3,]  NA   NA   NA   NA   NA
```

```
[4,]   NA    NA    NA    NA    NA
> dim(m1)
[1] 4 5

> m1 <- matrix(1:10,nrow=2, ncol=5)
> m1
     [,1] [,2] [,3] [,4] [,5]
[1,]    1    3    5    7    9
[2,]    2    4    6    8   10

> dim(m1)
[1] 2 5

> matrix(data=c(1, 2, 3, 4), byrow=TRUE, nrow=2)
     [,1] [,2]
[1,]    1    2
[2,]    3    4
```

In the first example, we created a 4*5 matrix (where we specified number of rows by **nrow** and number of columns by **ncol** argument, respectively) by calling the matrix function. Since we did not specify any values to be populated for the matrix, it had all **NA** values (default values). If we want to identify the dimensions of a matrix (its rows and columns), then we can make use of dim() function.

In the second example, we created a 2*5 matrix and also specified the values that were to be populated in the matrix (values from 1:10 specified by the sequence). The values would be filled column wise (i.e., the first column will get values followed by the second column and so on). If we wanted to fill values by row then we have to specify the argument byrow=TRUE (as we did in the third example).

Just like you can access individual elements of a vector, we can access rows, columns, and individual elements of a matrix using the similar notation for vectors. For example:

```
> x<- matrix(1:10,2,5)

> x
     [,1] [,2] [,3] [,4] [,5]
[1,]    1    3    5    7    9
[2,]    2    4    6    8   10

> x[1,1]
[1] 1

> x[1,]
[1] 1 3 5 7 9

> x[,2]
[1] 3 4
```

Oftentimes it might happen that we have some vectors at our disposal and we want to create a matrix by combining those vectors. This can be done by making use of **rbind** and **cbind** operators. While rbind will join the columns by row, cbind will join the columns otherwise. For example:

```
> x<- 1:6
> x
[1] 1 2 3 4 5 6

> y <- 12:17
> y
[1] 12 13 14 15 16 17

> cbind(x,y)
     x  y
[1,] 1 12
[2,] 2 13
[3,] 3 14
[4,] 4 15
[5,] 5 16
[6,] 6 17

> rbind(x,y)
  [,1] [,2] [,3] [,4] [,5] [,6]
x    1    2    3    4    5    6
y   12   13   14   15   16   17
```

In addition to the usual arithmetic, matrices come in handy when we have to perform matrix arithmetic. The real use of matrices occurs in those situations where data is numeric in nature and we are dealing with a large set of numbers upon whom we want to perform matrix arithmetic computations. By design matrices have limited scope outside of numbers since they are not designed to be much useful for anything other than numeric data. If we want to exploit the true spreadsheet capability that we experience in Excel, then we need to use data frames in R.

3.12 Data Frame

A primary reason why Excel is very useful for us is that everything is laid out in a neat tabular structure, and this enables us to perform a variety of operations on the tabular data. Additionally, we can also hold string, logical, and other types of data. This capability is not lost for us in R and is instead provided by data frame in R.

Tabular data in R is read into a type of data structure known as data frame. All variables in a data frame are stored as separate columns, and this is different from matrix in the sense that each column can be of a different type. Almost always, when you import data from an external data source, you import it using a data frame. A data frame in R can be created using the function data.frame().

28 Introduction to R

```
> x <- data.frame(col1=1:20, col2 = c(T, F, F, T))
> x
   col1  col2
1     1  TRUE
2     2  FALSE
3     3  FALSE
4     4  TRUE
5     5  TRUE
6     6  FALSE
7     7  FALSE
8     8  TRUE
9     9  TRUE
10   10  FALSE
11   11  FALSE
12   12  TRUE
13   13  TRUE
14   14  FALSE
15   15  FALSE
16   16  TRUE
17   17  TRUE
18   18  FALSE
19   19  FALSE
20   20  TRUE

> nrow(x)
[1] 20
> ncol(x)
[1] 2

> #Check structure of a data frame
> str(x)
'data.frame':    20 obs. of  2 variables:
 $ col1: int  1 2 3 4 5 6 7 8 9 10 ...
 $ col2: logi  TRUE FALSE FALSE TRUE TRUE FALSE ...
```

In the first code snippet, we specified that we are creating a data frame that has two columns (col1 and col2). To find the number of rows and columns in a data frame, we use arguments nrow() and ncol(), respectively. In order to check the structure of a data frame (number of observations, number and types of columns), we make use of the function str().

Similar to matrices, we can select individual columns, rows, and values in a data frame. For example:

```
> x[1]
   col1
1     1
2     2
3     3
4     4
5     5
6     6
```

```
7      7
8      8
9      9
10     10
11     11
12     12
13     13
14     14
15     15
16     16
17     17
18     18
19     19
20     20

> x[1,1]
[1] 1

> x[,2]
 [1]  TRUE FALSE FALSE  TRUE  TRUE FALSE FALSE  TRUE  TRUE FALSE
      FALSE
[12]  TRUE  TRUE FALSE FALSE  TRUE  TRUE FALSE FALSE  TRUE

> x[2:5,1]
[1] 2 3 4 5
```

Additionally, we can also make use of the $ operator to access specific columns of a data frame. The syntax is dataframe$colname.

`x$col1`

3.13 R Operators

Operators in R perform operations on two variables/data values. Depending on what type of data the variable contains, the operations performed by the same operator could differ. Listed below are the different operators in R:

+ **(plus)**: It would add two numbers or variables if they are numbers. If the variables are string, then they would be concatenated. For example:

   4 + 6 would yield 10. "Hey" + "Hi" would yield "HeyHi."

(minus) It would subtract two variables.

(multiply) It would multiply two variables if they are numbers. If the variables are strings/lists, then they would be repeated by a said number of times. For example:

   3 * 6 would yield 18. "ab" * 4 would yield "abababab."

** **(power) It computed x raised to power y.**

4 ** 3 would yield 64 (i.e., 4 * 4 * 4).

/ (divide) It would divide x by y.

%/% (floor division) It would give the floor in a division operation.

7 %/% 2 would yield 3.

% %(modulo) Returns the remainder of the division

8 %% 3 gives 2. −25.5 %% 2.25 gives 1.5.

< (less than) Returns whether x is less than y. All comparison operators return True or False. Note the capitalization of these names.

5 < 3 gives False and 3 < 5 gives True.

> (greater than) Returns whether x is greater than y

5 > 3 returns True. If both operands are numbers, they are first converted to a common type. Otherwise, it always returns False.

<= (less than or equal to) Returns whether x is less than or equal to y

x = 3; y = 6; x <= y returns True.

>= (greater than or equal to) Returns whether x is greater than or equal to y

x = 4; y = 3; x >= 3 returns True.

== (equal to) Compares if the objects are equal

x = 2; y = 2; x == y returns True.
x = 'str'; y = 'stR'; x == y returns False.
x = 'str'; y = 'str'; x == y returns True.

!= (not equal to) Compares if the objects are not equal

x = 2; y = 3; x != y returns True.

! (boolean NOT) If x is True, it returns False. If x is False, it returns True.

x = True; not x returns False.

&& (boolean AND) x and y returns False if x is False, else it returns evaluation of y.

x = False; y = True; x and y returns False since x is False.

|| (boolean OR) If x is True, it returns True, else it returns evaluation of y.

x = True; y = False; x or y returns True.

4 Conditional Statements

After the discussion of variables and data types in R, let us now focus on the second building block of any programming language, that is, conditional statements. Conditional statements are branches in a code that are executed if a condition associated with the conditional statements is true. There can be many different types of conditional statement; however, the most prominent ones are if, while, and for. In the following sections, we discuss these conditional statements.

4.1 The If Statement

We use an if loop whenever there is a need to evaluate a condition once. If the condition is evaluated to be true, then the code block associated with if condition is executed, otherwise the interpreter skips the corresponding code block. The condition along with the associated set of statements is called the if loop or if block. In addition to the if condition, we can also specify an else block that is executed if the if condition is not successful. Please note that the else block is entirely optional.

Code
```
x <- 0
if (x < 0) {
  print("Negative number")
} else if (x > 0) {
  print("Positive number")
} else
  print("Zero")
```

Output
```
[1] "Zero"
```

4.2 The While Loop

Whereas an if loop allows you to evaluate the condition once, the while loop allows you to evaluate a condition multiple number of times depending on a counter or variable that keeps track of the condition being evaluated. Hence, you can execute the associated block of statements multiple times in a while block.

Code
```
a <- 10
while (a>0){
  print(a)
  a<-a-1
}
```

Output

```
[1] 10
[1] 9
[1] 8
[1] 7
[1] 6
[1] 5
[1] 4
[1] 3
[1] 2
[1] 1
```

4.3 For Loop

In many ways, the for loop is similar to a while loop in the sense that it allows you to iterate the loop multiple times depending on the condition being evaluated. However, the for loop is more efficient in the sense that we do not have to keep count of incrementing or decrementing the counter of condition being evaluated. In the while loop, the onus is on the user to increment/decrement the counter, otherwise the loop runs until infinity. However, in a for loop, the loop itself takes care of the increment/decrement.

Code

```
for (j in 1:5){
  print(j)
}
```

Output

```
[1] 1
[1] 2
[1] 3
[1] 4
[1] 5
```

In the code snippet below, we make use of the seq_along() function that acts as a sequence of non-numeric values. The function will iterate through each of the values in the specified vector x, and the print loop will then print the values.

```
x <- c("a","c","d")
for (i in seq_along(x)){
  print(x[i])
}
[1] "a"
[1] "c"
[1] "d"
```

We can alternatively write the same code in the following manner:

```
for (letter in x){
  print(letter)
}
```

```
[1] "a"
[1] "c"
[1] "d"
```

4.4 File Input and Output

Most of the times, in addition to using variables and in-built data structures, we would be working with external files to get data input and to write output to. For this purpose, R provides functions for file opening and closing. There are a range of functions for reading data into R. **read.table** and **read.csv** are the most common for tables, **readLines** for text data, and load for workspaces. Similarly, for writing, use **write.table** and **write.lines**.

read.table is the most versatile and powerful function for reading data from external sources. You can use it to read data from any type of delimited text files such as tab, comma, and so on. The syntax is as follows:

```
> inputdata <- read.table("inputdata.txt",header=TRUE)
```

In the above code snippet, we read in a file inputdata.txt into a data frame input data. By specifying the argument header=TRUE, we are specifying that the input file contains the first line as header.

While you can import data using read.table function as well, there are specific functions for csv and Excel files:

```
> titanicdata <- read.csv("train.csv")
> datafile1 <- read.table("train.csv",header=TRUE,sep=",")
```

Similar to the functions for reading files in R, there are functions for writing back data frames to R. Here are some of the most common examples that you would encounter. This list is not exhaustive, and there are many more functions available for working with different file types.

```
> write.csv(titanicdata,"D:// file1.csv")
```

In the above code snippet, we write the contents of the data frame *titanicdata* to an output file (file1.csv) in the D drive.

5 Function

While R is a great programming language with a number of in-built functions (such as those for printing, file reads, and writes), oftentimes you would need to write your own piece of functionality that is not available elsewhere (e.g., you might want to write a specific piece of logic pertaining to your business). For such cases, rather than writing the same code again at multiple places in code, we make use of functions. Functions are nothing but reusable code pieces that need to be written once and can then be called using their names elsewhere in the code. When we need to create a function, we need to give a name to the function and the associated code block that would be run whenever the function is called. A function name follows the same naming conventions as a variable name. Any function is defined using the keyword function. This tells the interpreter that the following piece of code is a function. After function, we write the arguments that the function would expect within the parenthesis. Following this, we would write the code block that would be executed every time the function is called.

In the code snippet below, we define a function named func1 that takes two arguments a and b. The function body does a sum of the arguments a and b that the function receives.

Code

```
func1 <- function(a,b){
  a+b
}
func1(5,10) #call the function by calling function name and
providing the arguments
```

Output

```
[1] 15
```

Code

```
square.it <- function(x) {
  square <- x * x
  return(square)
}
```

Output

```
square.it(5)
[1] 25
```

In the abovementioned code snippet, we created a function "square.it" using the syntax of the function. In this case, the function expects one argument and does a square of that argument. The **return()** statement will pass the value computed to the calling line in the code (the line or variable that called the **function**). Note that the names given in the function definition are called **parameters**, whereas the values you supply in the function call are called **arguments**.

6 Further Reading

There are plenty of free resources—books and online websites including R documentation itself—available to learn more about R and packages used in R. As mentioned earlier, since R is an open-source platform, many developers keep building new packages and add to the R repository on a frequent basis. The best way to know and learn them is by referring to respective documentation submitted by various package authors.

Chapter 29
Introduction to Python

Peeyush Taori and Hemanth Kumar Dasararaju

1 Introduction

As data science is increasingly being adopted in the industry, the demand for data scientists is also growing at an astonishing pace. Data scientists are a rare breed of "unicorns" who are required to be omniscient and according to popular culture, a data scientist is someone who knows more statistics than a programmer and more programming than a statistician. One of the most important tools in a data scientist's toolkit is the knowledge of a general-purpose programming language that enables a data scientist to perform tasks of data cleaning, data manipulation, and statistical analysis with ease. Such requirements call for programming languages that are easy enough to learn and yet powerful enough to accomplish complex coding tasks. Two such de facto programming languages for data science used in industry and academia are Python and R.

In this chapter, we focus on covering the basics of Python as a programming language. We aim to cover the important aspects of the language that are most critical from the perspective of a budding data scientist. A detailed knowledge of Python can be gained through an excellent collection of books and Internet resources. Although prior programming experience is helpful, this chapter does not require any prior knowledge of programming.

P. Taori (✉)
London Business School, London, UK
e-mail: taori.peeyush@gmail.com

H. K. Dasararaju
Indian School of Business, Hyderabad, Telangana, India

© Springer Nature Switzerland AG 2019
B. Pochiraju, S. Seshadri (eds.), *Essentials of Business Analytics*, International Series in Operations Research & Management Science 264,
https://doi.org/10.1007/978-3-319-68837-4_29

1.1 What Is Python?

Python is a high-level, general-purpose programming language that was first introduced to the world in late 1980s. Although the name of the language comes across as a bit odd at first, the language (or its concepts) does not bear any resemblance to an actual Python and instead was named after its creator Van Guido Rossum's inspiration from a BBC Comedy series named *"Monty Python's Flying Circus."* Initially, Python was received mostly as a general-purpose scripting language, and it was used quite extensively as a language of choice for web programming, and scripting purposes. Over the past decade, it was realized that Python could be a great tool for the scientific computing community, and since then the language has seen an explosive growth in scientific computing and data analytics applications.

Python is an open-source language that allows anyone to contribute to Python environment by creating packages and making them available to other users. Python has a fairly large scientific community and is used in a variety of settings such as financial research, algorithms development, options and derivatives pricing, financial modeling, and trading systems.

1.2 Why Python for Data Science?

As stated at the start, Python is one of the de facto languages when it comes to data science. There are a number of reasons why Python is such a popular language among data scientists. Some of those reasons are listed below:

- Python is a high-level, general-purpose programming language that can be used for varied programming tasks such as web scraping, data gathering, data cleaning and manipulation, website development, and for statistical analysis and machine learning purposes.
- Unlike some of the other high level programming languages, Python is extremely easy to learn and implement, and it does not require a degree in computer science to become an expert in Python programming.
- Python is an *object-oriented programming* language. It means that everything in Python is an object. The primary benefit of using an object-oriented programming language is that it allows us to think of problem solving in a simpler and real-world manner, and when the code becomes too cumbersome then object-oriented languages are the best way to go.
- Python is an open-source programming language. This implies that a large community of developers contribute continually to the Python ecosystem.
- Python has an excellent ecosystem that comprises of thousands of modules and libraries (prepackaged functions) that do not require reinvention of the wheel, and most of the programming tasks can be handled by simply calling functions in one of these packages. This reduces the need for writing hundreds of lines of code, and makes development easier and faster.

- Python is an interpreted language that is platform independent. As compared to some of the other programming languages, you do not have to worry about underlying hardware on which the code is going to run. Platform independence essentially ensures that your code will run in the same manner on any platform/hardware that is supported by Python.

1.3 Limits of Python

While Python is a great programming language meant for general-purpose and scientific computing tasks, there are some limitations associated with Python. For most part, these limitations are not of concern for researchers. Although there are a number of statistical and econometric packages available for Python that make analysis very easy, there might be some specific functionality that might not be available in Python. In such cases, functions can be easily written to implement the functionality and distributed among the community for use. Alternatively, Python can be integrated with other programming languages/platforms, such as R, to make up for any functionality that is available in other platforms.

2 Installation and System Interface

There are multiple ways of installing the Python environment and related packages on your machine. One way is to install Python, and then add the required packages one by one. Another method (recommended one) is to work with an Integrated Development Environment (IDE). Working with an IDE greatly simplifies the task of developing, testing, deploying, and managing your project in one place. There are a number of such IDEs available for Python such as *Anaconda* and *Enthought Canopy*, some of which are paid versions while others are available for free for academic purpose. You can choose to use any IDE that suits your specifics needs. In this particular section, we are going to demonstrate installation and usage of one such IDE: Enthought Canopy. Enthought Canopy is a comprehensive package of Python language and comes pre-loaded with more than 14,000 packages. Canopy makes it very easy to install/manage libraries, and also provides a neat GUI environment for developing applications. In this chapter, we will focus on Python installation using Enthought Canopy. Below are the guidelines on how to install Enthought Canopy distribution.

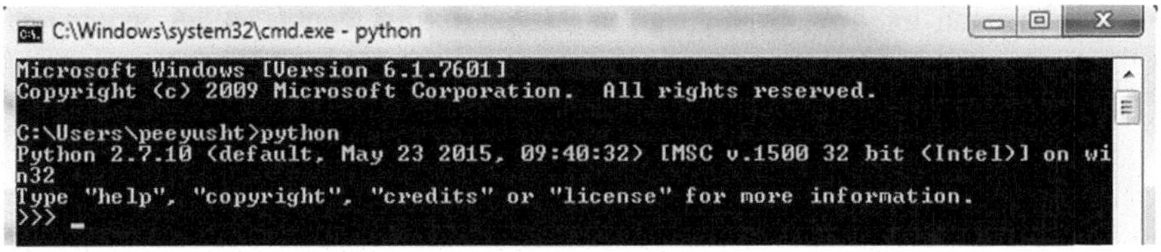

Fig. 29.1 Python working correctly on Windows

2.1 Enthought Canopy Installation

Enthought provides an academic license of Enthought Canopy distribution that is free to use for academic research purpose. You would need to register on the website[1] using your academic email ID, after which you can download and install Canopy.

- Go to https://www.enthought.com/downloads/ (accessed on Jun 19, 2018).
- Download Canopy Full Installer for the OS of your choice (Windows/Mac OS X).
- Run the downloaded file and install it by accepting default settings.
- If you are installing Canopy on OS X, make sure that Xcode is installed on your laptop. You can check if Xcode is installed by launching Terminal app (Application -> Utilities). In the Terminal, type gcc and press enter.

In order to check if Python is installed correctly, open Command Prompt, and type "python". If Python is installed correctly, you should see a message similar to the one shown in Fig. 29.1.

At the time of writing this section (June 19, 2018), the latest version of Python available is version 3.7. Other major version of Python that is used quite extensively is version 2.7. In this chapter, we demonstrate all coding examples using version 2.7 because version 3.7 of Python is not backward compatible. This implies that a number of Python packages that were developed for version 2.7 and earlier might not work very well with Python 3.7. Additionally, 2.7 is still one of the most widely used versions. While there are no drastic differences in the two versions, there are some minor differences that need to be kept in mind while developing the code.

2.2 Canopy Walkthrough

Launch the Canopy icon from your machine. There are three major components in Canopy distribution (Fig. 29.2):

1. A text editor and integrated IPython console.
2. A GUI-based package manager.
3. Canopy documentation.

---

[1]https://www.enthought.com/accounts/register (accessed on Jun 19, 2018)

Fig. 29.2 Canopy interface

We will briefly discuss each of them.

2.2.1 Text Editor

The editor window has three major panes:

1. File Browser Pane: You can manage your Python code files here and arrange them in separate directories.
2. Code Editor Pane: Editor for writing Python code.
3. Python Pane: Contains IPython shell as well as allows you to run code directly from code editor.

2.2.2 Package Manager

The Package Manager allows you to manage existing packages and install additional packages as required. There are two major panes in Package Manager (Fig. 29.3):

1. Navigation Pane: It lists all the available packages, installed packages, and the history of installed packages.

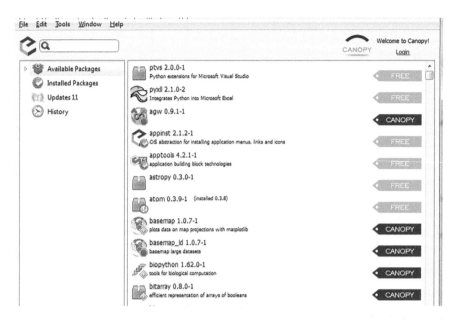

Fig. 29.3 Package Manager

2. Main Pane: This pane gives you more details about each package and allows you to manage the packages at individual level.

2.2.3 Documentation Browser

The documentation browser contains help files for Canopy software and some of the most commonly used Python packages such as Numpy, SciPy, and many more.

3 Hands-On with Python

In this section, we outline the Python programming language, and use the features and packages present in the language for data science related purposes. Specifically, we would be learning the language constructs, programming in Python, how to use these basic constructs to perform data cleaning, processing, and data manipulation tasks, and use packages developed by the scientific community to perform data analysis. In addition to working with structured (numerical) data, we will also be learning about how to work with unstructured (textual) data because Python has a lot of features to deal with both domains in an efficient manner.

We discuss the basic constructs of the language such as operators, data types, conditional statements, and functions, and specific packages that are relevant for data analysis and research purpose. In each section, we discuss a topic, code snippets, and exercise related to the sessions.

3.1 Programming Modes

There are two ways to write the codes in Python: *script* and *interactive*. The scripts mode is the one that most of the programmers would be familiar with, that is, all of the Python code is written in one text file and the file then executes on a Python interpreter. All Python code files must have a "**.py**" extension. This signals the interpreter that the file contains Python code. In the interactive mode, instead of writing all of the code together in one file, individual snippets of code are written in a command line shell and executed. Benefit of the interactive mode is that it gives immediate feedback for each statement, and makes program development more efficient. A typical practice is to first write snippets of code in interactive mode to test for functionality and then bundle all pieces of code in a .py file (Script mode). Enthought Canopy provides access to both modes. The top window in the text editor is where you can type code in script mode and run all or some part of it. In order to run a file, just click on the Run button in the menu bar and Python will execute the code contained in the file.

The bottom window in text editor acts as the Python interactive shell. In interactive mode what you type is immediately executed. For example, typing $1 + 1$ will respond with 2.

3.2 Syntax Formalities

Let us now get started with understanding the syntax of Python. The Python community prides itself in writing the code that is obvious to understand even for a beginner—this specific way is known as "Pythonic" in nature. Although it is true that Python is a very simple and easy language to learn and develop, it has some quirks—the biggest one of which is indentation. Let us first understand the importance of indentation before we start to tackle any other syntax features of the language. Please note that all the codes referred in the following sections are tested on Python 2.7 in the Enthought Canopy console.

3.3 Indentation

Whitespace is important in Python—this is known as *indentation*. Python makes use of whitespace for code structuring and marking logical breaks in the code. This is in contrast with other programming languages such as *R*, *Java*, and *C* that use braces for code blocks. Level of indentation of any code block is used to determine whether the code is part of the main program flow or whether it belongs to a particular branch of the program. Leading whitespaces such as tab and spaces are used for program indentation, and a group of statements that have the same indentation are considered to belonging to the same code block. If a particular code block is indented, then it must belong to a branch of the main program that has to be executed if a certain condition (such as if, for loops; more on them later) associated with the code block is met. Let us understand indentation with the help of simple examples:

Code

```
a = 7
    print ('Value of the variable is {}'.format(a))
        # Error! Look at the space at the beginning of line
print ('This is now correct. Value of variable a is {}'.format(a))
```

Output

```
(Once you comment second line)
This is now correct. Value of variable a is 7
```

In the above piece, each line is a Python statement. In the first statement, we are assigning a value of 7 to the variable "a." In the second statement, notice the space at the beginning. This is considered as indentation by Python interpreter. However, since any indentation block is supposed to have a conditional statement (such as if and for loop), the code here would give an error as the interpreter will consider the second statement as having a separate flow from the first statement. The third statement does not have any indentation (it is in the same block as the first statement) and thus will execute just fine.

It is important to remember that all statements that are expected to execute in the same block should have the same indentation. Presence of indentations improves readability of Python programs tremendously but it also requires a bit getting used to, especially if you are coming from languages such as C and Java where semicolon (;) marks the end of statements. You should also be careful with indentations because if you are not careful with them then they can cause errors in the program to say the least, and if gone undetected they can cause program to behave in an unpredictable manner. Most of the IDEs such as *Canopy* and *Anaconda* have in-built support for indentations that make program development easier.

3.4 Comments

Comments are required in any programming language to improve readability by humans. Comments are those sections of code that are meant for human comprehension, and are ignored by the Python interpreter when executing. In Python, you can write either single-line or multiline comments.

1. **Single-line comment**:
   A single line comment in Python begins with a pound (#) sign. Everything after the # sign is ignored by the interpreter till the end of line.

Code
```
print("This is code line, not a comment line")
#print("This is a comment line")
```

Output
```
This is code line, not a comment line
```

Note that in the above code snippet, the first line is the actual code that is executed, whereas the second line is a comment that is ignored by interpreter.

2. **Multiline comment**:
   Syntax for multiline comments is different from that of single-line comment. Multiline comments start and end with three single quotes ('''). Everything in between is ignored by the interpreter.

Code
```
'''
print("Multi line comment starts from here")
print ("Multi line comment continuing. This will not be printed")
'''
print("Multi line comment ended in above line. This line with
      be printed")
```

Output
```
Multi line comment ended in above line. This line with be printed
```

3.5 Object Structure

With a firm understanding of indentation and comments, let us now look at the building blocks of the Python programming language. A concept central to Python is that of object. Everything in Python, be it a simple variable, function, custom data structure is an object. This means that there is data and certain functions associated with every object. This makes programming very consistent and flexible. However, it does not imply that we have to think of objects every time we are coding in Python.

Behind the scenes everything is an object even if we explicitly use objects or not in our coding. Since we are just getting started, we will first focus on coding without objects and talk about objects later, once we are comfortable with the Pythonic way of programming.

3.6 Variables

There are some in-built data types in Python for handling different kinds of data: integer, floating point, string, Boolean values, date, and time. A neat feature of Python is that you do not need to mention what kind of data a variable holds; depending on the value assigned, Python automatically assigns a data type to the variable.

Think of a variable as a placeholder. It is any name that can hold a value and that value can vary over time (hence the name variable). In other terms, variables are reserved locations in your machine's memory to store different values. Whenever you specify a variable, you are actually allocating space in memory that will hold values or objects in future. These variables continue to exist while the program is running. Depending on the type of data a variable has, the interpreter will assign the required amount of memory for that variable. This implies that memory of a variable can increase or decrease dynamically depending on what type of data the variable has at the moment. You create a variable by specifying a name to the variable, and then by assigning a value to the variable by using equal sign (=) operator.

Code
```
variable1 = 100      # Variable that holds integer value
distance = 1500.0    # Variable that holds floating point value
institute = "ISB"    # Variable that holds a string
print(variable1)

print(distance)
print(institute)
print institute      # print statement has been discontinued
                       from Python3
```

Output
```
100
1500.0
ISB
ISB
```

Code
```
a = 0
b = 2
c = "0"
d = "2"
print(a + b)     # output as integer
print(c + d)     # output as string
print(type(a + b))
print(type(c + d))
```

Output

```
2
02
<type 'int'>
<type 'str'>
```

3.7 Naming Conventions for a Variable

Although a variable can be named almost anything, there are certain naming conventions that should be followed:

- A variable can start with either a letter (uppercase or lowercase) or an underscore (_) character.
- Remainder of the variable can contain any combination of letters, digits, and underscore characters.
- For example, some of the valid names for variables are _variable, variable1. 5Variable, >Smiley are not correct variable names.
- Variable names in Python are case-sensitive. This means that Variable and variable are two different variables.

3.8 Basic Data Types

In addition to complex data types, Python has five atomic (basic) data types. They are Number, String, List, Tuple, and Dictionary, respectively. Let us understand them one by one.

3.8.1 Numbers

Numbers are used to hold numerical values. There are four types of numbers that are supported in Python: integer, long integer, floating point (decimals), and complex numbers.

1. **Integer**: An integer type can hold integer values such as 1, 4, 1000, and -52,534. In Python, integers have a bit length of minimum 32 bits. This means that an integer data type can hold values in the range $-2{,}147{,}483{,}648$ to $2{,}147{,}483{,}647$. An integer is stored internally as a string of digits. An integer can only contain digits and cannot have any characters or punctuations such as $.

Code and output

```
>>> 120+200
320
```

```
>>> 180-42
138
>>> 15*8
120
```

2. **Long Integer**: Simple integers have a limit on the value that they can contain. Sometimes the need arises for holding a value that is outside the range of integer numbers. In such a case, we make use of Long Integer data types. Long Integer data types do not have a limit on the length of data they can contain. A downside of such data types is that they consume more memory and are slow during computations. Use Long Integer data types only when you have the absolute need for it. Python distinguishes Long Integer value from an integer value by character L or l, that is, a Long Integer value has "L" or "l" in the end.

Code and output

```
>>> 2**32
4294967296L
```

3. **Floating Point Numbers**: Floating point data types are used to contain decimal values such as fractions.
4. **Complex Numbers**: Complex number data types are used to hold complex numbers. In data science, complex numbers are used rarely and unless you are dealing with abstract math there would be no need to use complex numbers.

3.8.2 Strings

A key feature of Python that makes it one of the de facto languages for text analytics and data science is its support for strings and string processing. Strings are nothing but an array of characters. Strings are defined as a sequence of characters enclosed by quotation marks (they can be single or double quotes). In addition to numerical data processing, Python has very strong string processing capabilities. Since strings are represented internally as an array of characters, it implies that it is very easy to access a particular character or subset of characters within a string. A sub-string of a string can be accessed by making use of indexes (position of a particular character in an array) and square brackets []. Indexes start with 0 in Python. This means that the first character in a string can be accessed by specifying string name followed by [followed by 0 followed by] (e.g., (stringname[0]). If we want to join two strings, then we can make use of the plus (+) operator. While plus (+) operator adds numbers, it joins strings and hence can work differently depending on what type of data the variables hold. Let us understand string operations with the help of a few examples.

Code

```
newstring = 'Hi. How are you?'
print(newstring)         # It will print entire string
print(newstring [0])     # It will print first character
```

29 Introduction to Python

```
print(newstring [3:7])   # It will print from 4th to 6th character
print(newstring [3:])    # It will print everything from 4th
                           character to end
print(newstring * 3)     # It will print the string three times
print(newstring + "I am very well, ty.") # It will concatenate
                                           two strings
```

Output

```
Hi. How are you?
H
 How
 How are you?
Hi. How are you?Hi. How are you?Hi. How are you?
Hi. How are you?I am very well, ty.
```

Strings in Python are immutable. Unlike other datasets such as lists, you cannot manipulate individual string values. In order to do so, you have to take subsets of strings and form a new string. A string can be converted to a numerical type and vice versa (wherever applicable). Many a times, raw data, although numeric, is coded in string format. This feature provides a clean way to make sure all of the data is in numeric form. Strings are a sequence of characters and can be tokenized. Strings and numbers can also be formatted.

3.8.3 Date and Time

Python has a built-in datetime module for working with dates and times. One can create strings from date objects and vice versa.

Code

```
import datetime
date1 = datetime.datetime(2014, 5, 16, 14, 45, 05)
print(date1.day)
print(date1)
```

Output

```
16
2014-05-16 14:45:05
```

3.8.4 Lists

Lists in Python are one of the most important and fundamental data structures. At the very basic level, List is nothing but an ordered collection of data. People with background in Java and C can think of list as an array that contains a number of elements. The difference here is that a list can contain elements of different data types. A list is defined as a collection of elements within square brackets "[ ]", and each element in a list is separated by commas. Similar to the individual characters in a String, if you want to access individual elements in a list then you can do it by using the same terminology as used with strings, that is, using the indexes and the square brackets.

Code

```
alist = [ 'hi', 123 , 5.45, 'ISB', 85.4 ]
anotherlist = [234, 'ISB']

print(alist)            # It will print entire list
print(alist[0])         # It will print first element
print(alist[2:5])       # It will print 3rd through 5th element in list
print(alist[3:])        # It will print from element 4 till the end
print(anotherlist * 2)  # It will print the list twice
print(alist + anotherlist) # It will concatenate the two lists
```

Output

```
['hi', 123, 5.45, 'ISB', 85.4]
hi
[5.45, 'ISB', 85.4]
['ISB', 85.4]
[234, 'ISB', 234, 'ISB']
['hi', 123, 5.45, 'ISB', 85.4, 234, 'ISB']
```

3.8.5 Tuples

A tuple is an in-built data type that is a close cousin to the list data type. While in a list you can modify individual elements of the list and can also add/modify the number of elements in the list, a tuple is immutable in the sense that once it is defined you cannot change either the individual elements or number of elements in a tuple. Tuples are defined in a similar manner as lists with a single exception—while lists are defined in square brackets "[]", tuples are defined using parenthesis "()". You should use tuples whenever there is a situation where you need to use lists that nobody should be able to modify.

Code

```
tupleone = ('hey', 125, 4.45, 'isb', 84.2)
tupletwo = (456, 'isb')

print(tupleone)         # It will print entire tuple
print(tupleone[0])      # It will print first element of tuple
print(tupleone[1:4])    # It will print 2nd to 4th element of tuple
print(tupleone[3:])     # It will print entire tuple from 4th
                        #   element till last
print(tupletwo * 2)     # It will print tuple twice
print(tupleone + tupletwo) # It will concatenate and print
                           #   the two tuples
```

Output

```
('hey', 125, 4.45, 'isb', 84.2)
'hey'
```

```
(125 , 4.45, 'isb')
('isb', 84.2)
(456, 'isb', 456, 'isb')
('hey', 125, 4.45, 'isb', 84.2, 456, 'isb')
```

If you try to update a tuple, then it would give you an error:

Code
```
tuple = ('hey', 234, 4.45, 'Alex', 81.4)
list = ['hey', 234, 4.45, 'Alex', 81.4]
tuple[2] = 1000   # Invalid (error: tuple object does not
                              support...)
list[2] = 1000    # Valid (it will change 4,45 to 1000)
```

Output
```
['hey', 234, 1000, 'Alex', 81.4]
```

3.8.6 Dictionary

Perhaps one of the most important built-in data structures in Python are dictionaries. Dictionaries can be thought of as arrays of elements where each element is a key–value pair. If you know the key or value then you can quickly look up for corresponding values/key respectively. There is no restriction on what key or values could be, and they can assume any Python data type. Generally, as industry practice we tend to use keys as containing either numbers or characters. Similar to keys, values can assume any data type (be it basic data types or complex ones). When you need to define a dictionary, it is done using curly brackets "{}" and each element is separated by a comma. An important point to note is that dictionaries are unordered in nature, which means that you cannot access an element of a dictionary by using the index, but rather you need to use keys.

Code
```
firstdict = {}
firstdict ['one'] = "This is first value"
firstdict [2]     = "This is second value"
seconddict = {'institution': 'isb','pincode':500111,
              'department': 'CBA'}
print(firstdict ['one'])
print(firstdict [2] )
print(seconddict)
print(seconddict.keys())    # It will print all keys in
                              the dictionary
print(seconddict.values())  # It will print all values in
                              the dictionary
```

Output
```
This is first value
This is second value
```

```
{'institution': 'isb','pincode':500111, 'department': 'CBA'}
['department', 'pincode', 'institution']
['CBA', 500111, 'isb']
```

3.9 Datatype Conversion

Quite often the need might arise where you need to convert a variable of a specific data type to another data type. For example, you might want to convert an int variable to a string, or a string to an int, or an int to a float. In such cases, you use type conversion operators that change the type of a variable. To convert a variable to integer type, use int(variable). To convert a variable to a string type, use str(variable). To convert a variable to a floating point number, use float(variable).

3.10 Python Operators

Operators in Python perform operations on two variables/data values. Depending on what type of data the variable contains, the operations performed by the same operator could differ. Listed below are the different operators in Python:

+ **(plus)**: It would add two numbers or variables if they are numbers. If the variables are string, then they would be concatenated. For example,

   4 + 6 would yield 10. 'Hey' + 'Hi' would yield 'HeyHi'.

− **(minus)**: It would subtract two variables.

* **(multiply)**: It would multiply two variables if they are numbers. If the variables are strings/lists, then they would be repeated by a said number of times. For example,

   3 * 6 would yield 18; '3'*4 would yield '3333'; 'ab*4' would yield 'abababab'.

** **(power)**: It computes x raised to power y. For example,

   4 ** 3 would yield 64 (i.e., 4 * 4 * 4)

/ **(divide)**: It would divide x by y.

// **(floor division)**: It would give the floor in a division operation. For example,

   5 // 2 would yield 2.

% **(modulo)**: Returns the remainder of the division. For example,

   8 % 3 gives 2. −25.5 % 2.25 gives 1.5.

< **(less than)**: Returns whether x is less than y. All comparison operators return True or False. Note the capitalization of these names. For example,

5 < 3 gives False and 3 < 5 gives True.

> (greater than): Returns whether x is greater than y. For example,

5 > 3 returns True. If both operands are numbers, they are first converted to a common type. Otherwise, it always returns False.

<= (less than or equal to): Returns whether x is less than or equal to y. For example,

x = 3; y = 6; x <= y returns True.

>= (greater than or equal to): Returns whether x is greater than or equal to y. For example,

x = 4; y = 3; x >= 3 returns True.

== (equal to): Compares if the objects are equal. For example,

x = 2; y = 2; x == y returns True.
x = 'str'; y = 'stR'; x == y returns False.
x = 'str'; y = 'str'; x == y returns True.

!= (not equal to): Compares if the objects are not equal. For example,

x = 2; y = 3; x! = y returns True.

not (boolean NOT): If x is True, it returns False. If x is False, it returns True. For example,

x = True; not x returns False.

and (boolean AND): x and y returns False if x is False, else it returns evaluation of y. For example,

x = False; y = True; x and y returns False since x is False

or (boolean OR): If x is True, it returns True, else it returns evaluation of y. For example,

x = True; y = False; x or y returns True.

3.11 Conditional Statements and Loops

After the discussion of variables and data types in Python, let us now focus on the second building block of any programming language, that is, conditional statements. Conditional statements are branches in a code that are executed if a condition associated with the conditional statements is true. There can be many different types of conditional statements; however, the most prominent ones are if, while, and for. In the following sections, we discuss these conditional statements.

3.11.1 if Statement

We use an if loop whenever there is a need to evaluate a condition once. If the condition is evaluated to be true, then the code block associated with if condition is executed, otherwise the interpreter skips the corresponding code block. The condition along with the associated set of statements are called the if loop or if block. In addition to the if condition, we can also specify an else block that is executed if the if condition is not successful. Please note that the else block is entirely optional.

Code
```
var1 = 45
if var1 >= 43:
       print("inside if block")
elif var1 <= 40:
       print("inside elif block")
else:
       print("inside else block")
```

Output
```
inside if block
```

3.11.2 while Loop

Whereas an if loop allows you to evaluate the condition once, the while loop allows you to evaluate a condition multiple number of times depending on a counter or variable that keeps track of the condition being evaluated. Hence, you can execute the associated block of statements multiple times in a while block. Similar to an if loop, you can have an optional else loop in the while block, see for loop example below:

Code
```
counter = 0
while (counter < 5):
         print('Current counter: {}'.format(counter))
         counter = counter + 1

print("While loop ends!")
```

Output
```
Current counter: 0
Current counter: 1
Current counter: 2
Current counter: 3
Current counter: 4
While loop ends!
```

3.11.3 for Loop

In many ways for loop is similar to a while loop in the sense that it allows you to iterate the loop multiple times depending on the condition being evaluated. However, for loop is more efficient in the sense that we do not have to keep count of incrementing or decrementing the counter of condition being evaluated. In the while loop, onus is on the user to increment/decrement the counter otherwise the loop runs until infinity. However, in for loop the loop itself takes care of the increment/decrement.

Code

```
for a in range(1,8):
        print (a)
else:
        print('For loop ends')
```

Output

```
1
2
3
4
5
6
7
For loop ends
```

3.11.4 break Statement

Sometimes the situation might arise in which you might want to break out of a loop before the loop finishes completion. In such cases, we make use of the break statement. The break statement will break out of the loop whenever a particular condition is being met.

Code

```
for a in range(1,8):
        if a == 4:
                break
        print(a)

print('Loop Completed')
```

Output

```
1
2
3
Loop completed
```

3.11.5 Continue statement

Whereas the *break* statement completely skips out of the loop, the *continue* statement skips the rest of the code lines in a current loop and goes to the next iteration.

Code

```
while True:
      string = input('Type your input: ')
      if string == 'QUIT':
          break
      if len(string) < 6:
          print 'String is small'
          continue
      print('String input is not sufficient')
```

Output

```
Type your input: 'Hi'
String is small

Type your input: 'abc'
String is small

Type your input: 'verylarge'
String input is not sufficient

Type your input: 'QUIT'
```

3.12 Reading Input from Keyboard

Whenever you need user to enter input from keyboard or you need to read keyboard input, you can make use of two in-built functions—"raw_input" and "input." They allow you to read text line from standard keyboard input.

3.12.1 raw_input Function

It will read one line from keyboard input and give it to the program as a string.

```
string= raw_input('Provide the input: ');
print('Input provided is: {}'.format(string))
```

In the above-mentioned example, the user would get a prompt on the screen with title "Provide the input." The second line will then print whatever input the user has provided.

```
Provide the input: Welcome to Python
Input provided is: Welcome to Python
```

3.12.2 Input Function

"input" is similar to raw_input() with an exception. While "raw_input" assumes that the entered value is a text, "input" would assume that entered text is indeed a Python expression and will proceed to evaluate the Python expression and would provide the output of the expression.

```
string= input('Provide the input: ');
print('Input provided is: {}'.format(string))
```

For example:

```
Provide the input: [a*3 for a in range(1,6,2)]
Input provided is: [1,9,15]
```

3.13 Working with Files

Most of the times, in addition to using variables and in-built data structures, we would be working with external files to get data input and to write output to. For this purpose, Python provides functions for file opening and closing. A key concept in dealing with files is that of a "file" object. Let us understand this in a bit more detail.

3.13.1 open() Function

File object is the handle that actually allows you to create a link to the file you want to read/write to. In order to be able to read/write a file, we first need to use an object of file type. In order to do so, we make use of open() method. When open() executes, it will result in a file object that we would then use to read/write data for external files.

Syntax

```
file_object = open(file_name [, access_mode] [, buffering])
```

Let us understand this function in slightly more detail. The file_name requires us to provide the file name that we want to access. You can specify either an existing file on the filesystem or you can specify a new file name as well. Access_mode tells Python in which mode the file should be opened. There are a number of modes to do so; however, the most common ones are read, write, and append. A more detailed knowledge of each mode type is given in Table 29.1. Finally, buffer mode tells us how to buffer the data. By default, value for buffer is 0. This means that there is no buffering. If it is 1, then it implies that there would be buffering whenever a file is being accessed.

Table 29.1 Access_modes list

Mode	Brief overview
r	Default mode. Opens file in read only mode with pointer at the start of file.
rb	Similar to r. Only difference being that the file being read in binary format.
r+	When this mode is used, then file can be used for both reading and writing.
rb+	Similar to r+ except that reading and writing will happen in the binary format.
w	File can be accessed for writing only. Creates a new file if there is no existing file with the same name.
wb	Same as w except that it is opened in binary format.
w+	Similar to r+.
wb+	Similar to rb+.
a	When this mode is used, data is appended to the file. In w mode, data is being overwritten. Pointer in this mode is at the end of file rather than at the beginning.
ab	Similar to a except that it is in binary mode.
a+	Similar to w+ with append features.
ab+	Similar to wb+ with append features.

3.13.2 close() Function

Once we have opened the file for reading/writing purposes, we would then need to close the connection with the file. It is done using the close() method. close() will flush out any unwritten data to the file and will close the file object that we had opened earlier using open() function. Once the close() method is called, we cannot do any more reads/writes on the file. In order to do so, we would again have to open the file using open() method.

Syntax

```
file_object.close();
```

Code

```
# File Open
file1 = open('transactions.txt', 'wb')
print('File Name: {}'.format(file1.name))

# Close file
file1.close()
```

Output

```
File Name: transactions.txt
```

3.13.3 Reading and Writing Files

While the open() and close() methods allow us to open/close a connection to a file, we need to make use of read() or write() methods to actually read or write data to a file.

3.13.4 write() Function

When we call write(), it will write the data as a string to the file that we opened earlier.

Syntax

```
file_object.write(string);
```

String here is the data that has to be written to the file.

Code

```
# File Open
file1 = open('sample.txt', 'wb')
file1.write('This is my first output.\nIt looks good!!\n');

# Close file
file1.close()
```

When we run the above code, a file sample.txt would be created and the string mentioned in write() function would be written to the file. The string is given by:

```
This is my first output.
It looks good!!
```

3.13.5 read() Function

Just as write() method writes data to a file, read() would read data from an open file.

Syntax

```
file_object.read([counter]);
```

You would notice that we have passed an argument called counter here. When we do this, then it tells the interpreter to read the specified number of bytes. If no such argument is provided, then the reading will read the entire text.

Code

```
# File Open
fileopen = open('sample.txt', 'r+')
string = fileopen.read(10);
print('Output is: {}'.format(string))

# Close file
fileopen.close()
```

Output

```
Output is: This is my
```

3.14 Build Custom Function

While Python is a great programming language with a number of in-built functions (such as those for printing, file reads and writes), oftentimes you would need to write your own piece of functionality that is not available elsewhere (e.g., you might want to write a specific piece of logic pertaining to your business). For such cases, rather than write the same code again at multiple places in code, we make use of functions. Functions are nothing but reusable code pieces that need to be written once, and can then be called using their names elsewhere in the code. When we need to create a function, we need to give a name to the function and the associated code block that would be run whenever the function is called. A function name follows the same naming conventions as a variable name. Any function is defined using keyword def. This tells the interpreter that the following piece of code is a function. After def, we write the function names along with parentheses and any arguments that the function would expect within the parenthesis. Following this, we would write the code block that would be executed every time the function is called.

Code

```
def firstfunc():
    print('Hi Welcome to Python programming!')
        # code block that is executed for the function
# Function ends here #

firstfunc () # Function called first time
firstfunc () # Called again
```

Output

```
Hi Welcome to Python programming!
Hi Welcome to Python programming!
```

In the above-mentioned code snippet, we created a function "*firstfunc*" using the syntax of the function. In this case, the function expects no parameters and that is why we have empty parentheses. Function arguments are the variables that we pass to the function that the function would then use for its processing. Note that the names given in the function definition are called **parameters**, whereas the values you supply in the function call are called **arguments**.

Code

```
def MaxFunc(a1, a2):
    if a1 > a2:
        print('{} is maximum'.format(a1))
    elif a1 == a2:
        print('{} is equal to {}'.format(a1, a2))
    else:
        print('{} is maximum'.format(a2))

MaxFunc (8, 5) # directly give literal values
x = 3
```

```
y = 1
MaxFunc (x, y) # give variables as arguments
MaxFunc (5, 5) # directly give literal values
```

Output

```
8 is maximum
3 is maximum
5 is equal to 5
```

In the above code snippet, we created a function "*MaxFunc.*" MaxFunc requires two parameters (values) a1 and a2. The function would then compare the values and find the maximum of two values. In the first function call, we directly provided the values of 8 and 5 in the function call. Second time, we provided the variables rather than values for the function call.

3.14.1 Default Value of an Argument

If you want to make some parameters of a function optional, use default values in case the user does not want to provide values for them. This is done with the help of default argument values. You can specify default argument values for parameters by appending to the parameter name in the function definition the assignment operator (=) followed by the default value. Note that the default argument value should be a constant. More precisely, the default argument value should be immutable.

Code

```
def say(message, times = 1):
    print(message * times)

say('Hello')
say('World', 5)
```

Output

```
Hello
WorldWorldWorldWorldWorld
```

The function named "say" is used to print a string as many times as specified. If we do not supply a value, then by default, the string is printed just once. We achieve this by specifying a default argument value of 1 to the parameter times. In the first usage of *say*, we supply only the string and it prints the string once. In the second usage of *say*, we supply both the string and an argument 5 stating that we want to say the string message five times.

Only those parameters that are at the end of the parameter list can be given default argument values, that is, you cannot have a parameter with a default argument value preceding a parameter without a default argument value in the function's parameter list. This is because the values are assigned to the parameters by position. For example, def func(a, b=5) is valid, but def func(a=5, b) is not valid.

3.14.2 Return Statement

The "*return*" statement is used to return from a function, that is, break out of the function. You can optionally return a value from the function as well.

Code
```
def maximum(x, y):
      if x > y:
             return x
      elif x == y:
             return 'The numbers are equal'
      else:
             return y
print(maximum(2, 3))
```

Output
```
3
```

3.15 Modules

You can reuse code in your program by defining functions once. If you want to reuse a number of functions in other programs that you write, you can use modules. There are various methods of writing modules, but the simplest way is to create a file with a ".py" extension that contains functions and variables.

Another method is to write the modules in the native language in which the Python interpreter itself was written. For example, you can write modules in the C programming language and when compiled, they can be used from your Python code when using the standard Python interpreter.

A module can be imported by another program to make use of its functionality. This is how we can use the Python standard library as well. The following code demonstrates how to use the standard library modules.

Code
```
import os
print os.getcwd()
```

Output
```
<Your current working directory>
```

3.15.1 Byte-Compiled .pyc Files

Importing a module is a relatively costly affair, so Python does some tricks to make it faster. One way is to create byte-compiled files with the extension ".pyc", which is an intermediate form that Python transforms the program into. This "*.pyc*" file is useful when you import the module the next time from a different program—it will be much faster since a portion of the processing required in importing a module is already done. Also, these byte-compiled files are platform-independent.

Note that these ".pyc" files are usually created in the same directory as the corresponding ".py" files. If Python does not have permission to write to files in that directory, then the ".pyc" files will not be created.

3.15.2 from ... import Statement

If you want to directly import the "*argv*" variable into your program (to avoid typing the sys. everytime for it), then you can use the "from sys import argv" statement. In general, you should avoid using this statement and use the import statement instead since your program will avoid name clashes and will be more readable.

Code

```
from math import sqrt
print('Square root of 16 is {}'.format(sqrt(16)))
```

3.15.3 Build Your Own Modules

Creating your own modules is easy; this is because every Python program is also a module. You just have to make sure that it has a ".py" extension. The following is an example for the same:

Code (save as mymodule.py)

```
def sayhi():
        print('Hi, this is mymodule speaking.')__version__='0.1'
```

The above was a sample module; there is nothing particularly special about it compared to our usual Python program. Note that the module should be placed either in the same directory as the program from which we import it, or in one of the directories listed in sys.path.

Code (Another module—save as mymodule_demo.py)

```
import mymodule
mymodule.sayhi()
print("version {}".format(mymodule.__version__))
```

Output

```
Hi, this is mymodule speaking.
Version 0.1
```

3.16 Packages

In the hierarchy of organizing your programs, variables usually go inside functions. Functions and global variables usually go inside modules. What if you wanted to organize modules? This is where packages come into the picture.

Packages are just folders of modules with a special __init__.py file that indicates to Python that this folder is special because it contains Python modules. Let us say you want to create a package called "world" with subpackages "asia," "africa," etc., and these subpackages in turn contain modules like "india," "madagascar," etc. Packages are just a convenience to hierarchically organize modules. You will see many instances of this in the standard library.

3.16.1 Relevant Packages

There are a number of statistical and econometric packages available on the Internet that can greatly simplify the research work. Following is the list of widely used packages:

1. NumPy: Numerical Python (NumPy) is the foundation package. Other packages and libraries are built on top of NumPy.
2. pandas: Provides data structures and processing capabilities similar to ones found in R and Excel. Also provides time series capabilities.
3. SciPy: Collection of packages to tackle a number of computing problems in data analytics, statistics, and linear algebra.
4. matplotlib: Plotting library. Allows to plot a number of 2D graphs and will serve as primary graphics library.
5. IPython: Interactive Python (IPython) shell that allows quick prototyping of code.
6. Statsmodels: Allows for data analysis, statistical model estimation, statistical tests, and regressions, and function plotting.
7. BeautifulSoup: Python library for trawling the Web. Allows you to pull data from HTML and XML pages.
8. Scikits: A number of packages for running simulations, machine learning, data mining, optimization, and time series models.
9. RPy: This package integrates R with Python and allows users to run R code from Python. This package can be really useful if certain functionality is not available in Python but is available in R.

Chapter 30
Probability and Statistics

Peeyush Taori, Soumithri Mamidipudi, and Deepak Agrawal

1 Introduction

This chapter is aimed at introducing and explaining some basic concepts of statistics and probability in order to aid the reader in understanding some of the more advanced concepts presented in the main text of the book. The main topics that are discussed are set theory, permutations and combinations, discrete and continuous probability distributions, descriptive statistics, and bivariate distributions.

While the main aim of this book is largely beyond the scope of these ideas, they form the basis on which the advanced techniques presented have been developed. A solid grasp of these fundamentals, therefore, is crucial to understanding the insights that can be provided by more complex techniques.

However, in explaining these ideas, the chapter briefly sketches out the core principles on which they are based. For a more comprehensive discussion, see *Complete Business Statistics* by Aczel and Sounderpandian (McGraw-Hill, 2009).

P. Taori (✉)
London Business School, London, UK
e-mail: taori.peeyush@gmail.com

S. Mamidipudi · D. Agrawal
Indian School of Business, Hyderabad, Telangana, India

2 Foundations of Probability

2.1 Axioms and Set Theory

In order to understand the mathematical study of probability, it is important to first define some axioms of the field and introduce some basic set theory.

A set is a collection of objects. For example, the set of all single-digit whole numbers is {1,2,3,...,9}; the set of single-digit even numbers would be {2,4,6,8}, while the set of single-digit prime numbers would be {2,3,5,7}. A subset is a set of elements that is wholly included in some other set. So the set of all odd numbered single-digit primes {3,5,7} is a subset of the set of single-digit primes.

We can also use some basic notation to denote operations performed on two or more sets. Let us define the set of single-digit even numbers as A: {2,4,6,8}, and the set of single-digit prime numbers as B: {2,3,5,7}. A union of the two sets would include all elements of both sets, denoted by the symbol ∪. So A∪B would be {2,3,4,5,6,7,8}. An intersection of the two sets would include only the objects, or elements, present in both sets, denoted by the symbol ∩. Thus, A∩B would be {2}. A complement to a subset (usually denoted with the symbol ′) is a subset that contains all elements not present in the original subset. So A′, the complement to A, would be {1,3,5,7,9}. (It is important to point out that the complementation operation requires the definition of the full set or universal set; in this case, we assumed the set of single-digit whole numbers is the universal set.) It is possible to use these operations to include more sets—for example, we could denote the intersection of four sets called W, X, Y, and Z by writing W∩X∩Y∩Z.

In the study of probability, we can use set theory to define the possible outcomes of an experiment. We call the set of all possible outcomes of some experiment the "sample space" of that experiment. The sample space of rolling a die, for example, would be {1,2,3,4,5,6}. An event is the set of outcomes (a subset of the sample space) for which the desired outcome occurs. Thus, the event "roll an even number" would be described by the subset {2,4,6}. The event "roll an odd number" would be described by the subset {1,3,5}. The intersection of these two sets does not contain any elements. We call such sets "disjoint." The union of these two sets describes the sample space. We call such sets a "partition" of the sample space (they are said to be mutually exclusive and collectively exhaustive).

If we have a subset A that contains our outcomes, we denote the probability of that event as $P(A)$. To denote the probability of event A occurring given that event B has occurred, we write $P(A|B)$. If A and B are disjoint, $P(A|B) = 0$. If A and B are independent events, which means that the likelihood of one occurring does not affect the likelihood of the other, then $P(A|B) = P(A)$ and $P(B|A) = P(B)$. From this we can see that two events can only be both disjoint and independent if one of the events has a probability of 0. What is probability of a set exactly? In the simple world of frequencies, it is relative count of the event defined by the set. For example, how often will we see the number 1 while rolling a dice? If dice were fair one would say 1/6—on average once in every six tosses.

The aim of studying probability is to understand how likely an event is to occur. Given some number of observations of an experiment (such as drawing a card from a pack), probability can tell us how likely some outcome of that experiment is (such as drawing a king, or a diamond). Set theory enables us to study these questions by supplying us with a mathematical vocabulary with which we can ask these questions.

There are three main axioms of probability:

1. The probability of any event occurring must be between zero and one.

$$0 <= P(A) <= 1$$

2. Every experiment must result in an event. The probability of nothing (denoted as null or $\emptyset$) happening is zero. The probability of sample space (denoted here by S—but is not a standard notation for something) happening is one.

$$P(\emptyset) = 0$$
$$P(S) = 1$$

3. If two or more events are mutually exclusive (the subsets that describe their outcomes are disjoint), then the probability of one of them happening is simply the sum of the individual probabilities.

$$P(A \cup B \cup C) = P(A) + P(B) + P(C).$$

2.2 Bayes' Theorem

Bayes' theorem is one of the most powerful tools in probability. The theorem allows us to relate *conditional* probabilities, or the likelihood of an event occurring given that some other event has occurred, to each other.

Say that P(A| B) is the probability of an event A given that event B has occurred. Then, the probability of A and B occurring together is the probability of B occurring times the probability of A occurring given B has occurred (this is like a chain rule).

$$P(A \cap B) = P(B).P(A|B)$$

This is also true in reverse.

$$P(B \cap A) = P(A).P(B|A)$$

But A∩B and B∩A are the same!

Therefore, Bayes' theorem of conditional probability provides the foundation for one of the most important machine learning algorithms—the naïve machine learning algorithm. The algorithm postulates the likelihood of an event occurring (the prior), absorbs and analyzes new data (the likelihood), and then updates its analysis to reflect its new understanding (the posterior).

We can use Bayes' theorem to analyze a dataset in order to understand the likelihood of certain events given other events—for example, the likelihood of owning a car given a person's age and yearly salary. As more data is introduced into the dataset, we can better compute the likelihood of certain characteristics occurring in conjunction with the event, and thus better predict whether a person with a random set of characteristics may own a car. For example, say 5% of the population is known to own a car—call this A. This can be inferred from your sample data. In your sample, 12% are between 30 and 40 years of age—call this B. In the subset of persons that own a car, 25% are between age 30 and 40—this is (B| A). Thus, P(A) = 0.05. P(B) = 0.12. P(B| A) = 0.25. Thus, P (A|B) = $\frac{(0.25 \times 0.12)}{0.05}$ = 0.60. In other words, 60% of those that are between 30 and 40 years of age own a car.

2.3 Random Variables and Density Functions

Until now we have discussed probability in terms of sample spaces, in which the likelihood of any single outcome is the same. We will now consider experiments in which the likelihood of some outcomes is different than others. These experiments are called random variables. A random variable assigns a numerical value to each possible outcome of an experiment. These variables can be of two types: discrete or continuous.

Discrete random variables are experiments in which there are a finite number of outcomes. We might ask, for example, how many songs are on an album. Continuous random variables, on the other hand, are experiments that might result in all possible values in some range. For example, we might model the mileage driven of a car as a continuous random variable.

Normally, we denote an experiment with a capital letter, such as X, and the possibility of an outcome with a small letter, such as x. Therefore, in order to find out the likelihood of X taking the value x (also known as x occurring), we would write P(X = x). From the axioms of probability, we know that the sum of all P(x) must be 1. From this property, we can construct a probability mass function (PMF) P for X that describes the likelihood of each event x occurring.

Consider Table 30.1, which describes the results from rolling a fair die.

The PMF for each outcome P(X = x) (x = 1,2,...,6) is equal to 1/6.

Now consider Table 30.2, which describes a die that has been altered.

In this case, the PMF tells us that the likelihood for some outcomes is greater than the likelihood for other outcomes. The sum of all the PMFs is still equal to one, but we can see that the die is no longer equally likely to produce each outcome.

Table 30.1 Probability from rolling a fair die

Outcome (x)	Probability (p)	PMF: P(X = x) = p
1	1/6	P(X = 1) = 1/6
2	1/6	P(X = 2) = 1/6
3	1/6	P(X = 3) = 1/6
4	1/6	P(X = 4) = 1/6
5	1/6	P(X = 5) = 1/6
6	1/6	P(X = 6) = 1/6

Table 30.2 Probability from rolling an altered die

Outcome (x)	Probability (p)	PMF: P(X = x) = p
1	1/12	P(X = 1) = 1/12
2	3/12	P(X = 2) = 1/6
3	1/6	P(X = 3) = 1/6
4	1/6	P(X = 4) = 1/6
5	3/12	P(X = 5) = 1/4
6	1/12	P(X = 6) = 1/6

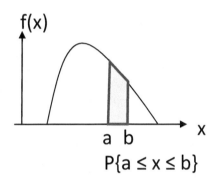

Fig. 30.1 Computing probability of x between a and b

A second useful function is the cumulative distribution function (CDF), which is defined as $P(X \leq x)$. When x is at its greatest, the CDF is equal to one. For the fair die, $P(X \leq 5) = 5/6$. For the unfair die, $P(X \leq 5) = 11/12$.

Continuous random variables are experiments in which the result can be any value in some range. For example, we might say that the mileage of a car may be between 0 and 10,000 miles. In this case, the PMF is not ideal as there are a large number of possible outcomes, each with a small chance of occurring. Instead, we can use a probability density function (PDF)—a function that tells us the area of the CDF in the range we are looking for. So if we wanted to know the likelihood of the mileage of a car being between 6000 and 8000 miles, we can find it by subtracting the likelihood of the mileage being below 6000 miles (point a) from the likelihood of the mileage being below 8000 miles (point b) (Fig. 30.1).

More generally, $P(a \leq X \leq b) = P(X \leq b) - P(X \leq a)$. In this figure, the function f(x) measures the height at every point of the curve. Therefore, f is called the probability density function or the density function.

Like any PMF, a PDF should also satisfy two conditions:

(a) $f(x) \geq 0$ for every x.
(b) $\int_{-\infty}^{+\infty} f(x)dx = 1$. (In general, this integral need extend only over the range over which f is defined. This range is also called the support of the probability distribution.)

2.4 Mean and Variance

However, describing the probability density function of a random variable is often cumbersome. As random variables can take any number of possible values, visualizing a function can be difficult. In order to make such a process simpler, we use two main tools of summarization: the mean and the variance. The mean is a measure of central tendency—the expected or average value of the distribution. The variance is a measure of dispersion—how clustered together the outcomes are. These two measures can give us an idea of the distribution and its relation to the experiment.

The mean of a random variable is also called its expected value—the probability weighted average that we "expect" will occur. It is calculated as the sum of the products of each outcome x and the likelihood of that outcome $P(X = x)$, and is denoted by μ. (In general, the value of a function, say $G(x)$, computed using the PDF f, is written as $E[G] = \int_{-\infty}^{+\infty} G(x)f(x)dx$. This is called the expected value of G under f. Thus, E[X] is the expected value of X, which is also referred to as mean.)

In mathematical terms,

$$\mu = E(X) = \sum (x.P(X = x))$$

Here, the symbol $\sum$ stands for summation over all values of x. For a continuous distribution,

$$E[X] = \int_{-\infty}^{+\infty} xf(x)dx$$

In the case of a fair die, the expected value is:

$$\mu = 1*1/6 + 2*1/6 + 3*1/6 + 4*1/6 + 5*1/6 + 6*1/6 = 3.5$$

This tells us that the "expected" value of an experiment may not actually be equal to a value that the experiment can take. We cannot actually ever roll 3.5 on a die, but we can expect that on average, the value that any die will take is 3.5.

In the case of a continuous random variable, the mean of the PDF cannot be computed using discrete arithmetic. However, we can use calculus to derive the same result.

By using the integral function $\int$ to replace the additive function $\sum$, we can find:

$$\mu = \int x.f(x) \, dx$$

where f(x) is the PDF.

Here, the limits of the integral are assumed to be the range over which f is defined—and omitted in the sequel below.

The second important summary is the variance. The variance of an experiment is a measure of how far away on average any outcome is from the mean. Knowing the variance of a function allows us to understand how spread out the outcomes are relative to the mean. In order to find this, we can measure the distance between each outcome and the mean: $(x - \mu)$, and add them. By definition, however, some values are below the mean, while other values are above the mean. Simply summing the distances of these outcomes from the mean will lead us to cancel out some outcomes. In order to circumvent this, we add the squares of the distances: $(x - \mu)^2$.

The variance of a discrete random variable, therefore, can be defined as:

$$Var(x) = E(x - \mu)^2$$

It can also be calculated as:

$$Var(X) = E\left(X^2\right) - E(X)^2$$

For continuous distributions,

$$Var(X) = \int [x - E(X)]^2 . f(x) dx = \int [x - \mu]^2 . f(x) dx$$

As variance is measured in terms of the square of the random variable, it is not measured in the same units as the distribution itself. In order to measure dispersion in the same units as the distribution, we can use the standard deviation (denoted as σ), which is the square root of the variance.

Example: Read the sample cars data, preloaded in R datasets. To print first five lines type the following:

```
> head(cars)
  speed dist
1     4    2
2     4   10
3     7    4
4     7   22
5     8   16
```

Fig. 30.2 Scatter plot—distribution of cars dataset

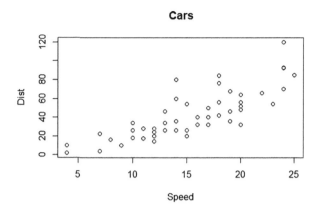

The R command to obtain the summary of descriptive statistics given a dataset is given by (Fig. 30.2):

```
> summary(cars)
     speed            dist
 Min.   : 4.0    Min.   :  2.00
 1st Qu.:12.0    1st Qu.: 26.00
 Median :15.0    Median : 36.00
 Mean   :15.4    Mean   : 42.98
 3rd Qu.:19.0    3rd Qu.: 56.00
 Max.   :25.0    Max.   :120.00

> plot(cars)
```

2.5 Bernoulli Distribution

The Bernoulli distribution is a type of discrete random variable which models an experiment in which one of two outcomes can occur. For example, a question with a yes/no answer or a flip of a coin can be modeled using Bernoulli distributions.

As there are only two outcomes (say x_1 and x_2), knowing the likelihood of one outcome means that we know the likelihood of the other outcome, that is, if $P(X = x1) = p$, then $P(X = x2) = 1 - p$. This is denoted by X ~ Bernoulli (p). The symbol ~ stands for "distributed as."

A fair coin will have $P(X = \text{heads}) = 0.5$. An unfair coin may have $P(X = \text{heads}) = 0.45$, which would mean $P(X = \text{tails}) = 0.55$.

But what if we have repeated trials? Say, many Bernoulli trials? See further.

2.6 Permutations and Combinations

In this case, we can use combinatorics to identify how many ways there are of picking combinations. Combinatorics deals with the combinations of objects that belong to a finite set. A permutation is specific ordering of a set of events. For example, the coin flipping heads on the first, third, and fourth flip out of five flips is a permutation: HTHHT. Given "n" objects or events, there are n! (n factorial) permutations of those events. In this case, given five events: H, H, H, T, and T, there are 5! ways to order them. 5! = 5*4*3*2*1 = 120. (There may be some confusion here. Notice that some of these permutations are the same. The 120 number comes up because we are treating the three heads as different heads and two tails as different tails. In one other way of saying this, the five events are each different—we would have been better if we had labeled the events 1,2,3,4,5.)

However, sometimes we may want to choose a smaller number of events. Given five events, we may want a set of three outcomes. In this case, the number of permutations is given by 5!/(5 − 3)! = 5*4*3 = 60. That is, if we have "n" events, and we would like to choose "k" of those events, the number of permutations is n!/(n − k)! If we had five cards numbered 1–5, the number of ways that we could choose three cards from them would be 60. (In another way of seeing this, we can choose the first event in five ways, the second in four ways, and the third in three ways, and thus 5 * 4 * 3 = 60.)

A combination is the number of ways in which a set of outcomes can be drawn, irrespective of the order in which the outcomes are drawn. If the number of permutations of k events out of a set of n events is n!/(n − k)!, the number of combinations of those events is the number of permutations, divided by the number of ways in which those permutations occur: n!/((n − k)!k!). (Having drawn k items, they themselves can be permuted k! times. Having drawn three items, we can permute the three 3! times. The number of combinations of drawing three items out of five equals 5!/((5 − 3)!3!) = 60/6 = 10.)

Using the theory of combinations, we can understand the binomial distribution.

2.7 Binomial Distribution

When we have repeated trials of the Bernoulli experiment, we obtain the binomial distribution. Say we are flipping the unfair coin ten times, and we would like to know the probability of the first four flips being heads.

$$P(HHHHTTTTTT) = (0.45)^4 * (0.55)^6 = 0.0011351.$$

Consider, however, the probability of four out of the ten flips being heads. There are many orders (arrangements or sequences) in which the four flips could occur, which means that the likelihood of $P(X = 4)$ is much greater than 0.0011351. In

Table 30.3 Binomial distribution with n = 10 and p = 0.25

X	P
1	0.187712
2	0.281568
3	0.250282
4	0.145998
5	0.058399
6	0.016222
7	0.003090
8	0.000386
9	0.000029
10	0.000001

this case, it is given by: $10!/[(10 - 4)! * 4!] * (0.45)^4 * (0.55)^6 = 0.238$ (= 210*0. 0.0011351, where 210 represents the number of combinations of drawing four out of ten items.)

In general, a binomial distribution has two outcomes: 1 or 0, with probability p and (1 − p) respectively—we write this as $X \sim B(n,p)$. If there are n independent trials, the PMF describes the likelihood of an event occurring x times as:

$$P(X = x) = n!/[(n - x)!x!] * p^x * (1 - p)^{n-x}$$

For $X \sim B(n,p)$, the mean E(X) is n*p, and Var(x) is n*p*(1 − p). (One can verify that these equal n times the mean and variance of the Bernoulli distribution.) A sample probability distribution for n = 10 and p = 0.25 is shown in Table 30.3. In Excel, the command is BINOMDIST(x,n,p,cumulative). In R, the command is DBINOM(x, n, p).

2.8 Poisson Distribution

The Poisson distribution is an extension of the binomial distribution for situations in which the number of trials is very large, the likelihood of an event occurring is very small, and the mean (n*p) of the distribution is finite.

In this case, we can use the Poisson distribution, which has the PMF

$$P(X = x) = \frac{\left[e^{-n*p} . (n * p)^x\right]}{x!}$$

We use λ to denote nxp, the mean, We write this as $X \sim \text{Poisson}(\lambda)$.

Here, for the Poisson distribution, the mean and the variance are both λ. A sample Poisson distribution with λ = 2.5 (compare with the binomial distribution) is shown in Table 30.4. The Excel command is POISSON(number of successes, mean, cumulative (0/1)). In R, the command is DPOIS(x, λ).

Table 30.4 Poisson distribution with mean = 2.5

X	P
1	0.205212
2	0.256516
3	0.213763
4	0.133602
5	0.066801
6	0.027834
7	0.009941
8	0.003106
9	0.000863
10	0.000216

2.9 Normal Distribution

The normal distribution is one of the most important continuous distributions, and can be used to model a number of real-life phenomena. It is visually represented by a bell curve.

Just as the binomial distribution is defined by two parameters (n and p), the normal distribution can also be defined in terms of two parameters: μ (mean) and sigma (standard deviation). Given the mean and standard deviation (or variance) of the distribution, we can find the shape of the curve. We can denote this by writing $X \sim N(\mu, \text{sigma})$.

The curve of normal distribution has the following properties:

1. The mean, median, and mode are equal.
2. The curve is symmetric about the mean.
3. The total area beneath the curve is equal to one.
4. The curve never touches the x-axis.

The mean of the normal distribution represents the location of centrality, about which the curve is symmetric. The standard deviation specifies the width of the curve.

The shape of the normal distribution has the property that we can know the likelihood of any given value falling within one, two, or three standard deviations from the mean. Given the parameters of the distribution, we can confidently say that 68.2% of data points fall within one standard deviation from the mean, 95% within two standard deviations of the mean, and more than 99% fall within three standard deviations of the mean (refer Fig. 30.3). A sample is shown below with mean = 10 and standard deviation = 1. In Excel, the command to get the distribution is NORMDIST(x, μ, sigma, cumulative (0/1)). In R, the command is PNORM(x, μ, sigma) (Fig. 30.4).

However, computing the normal distribution can become difficult. We can use the properties of the normal distribution to simplify this process. In order to do this, we can define the "standard" normal distribution, denoted Z, as a distribution that

Fig. 30.3 Shape of the normal distribution

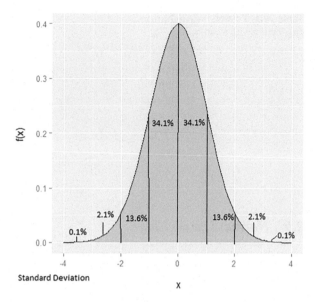

Fig. 30.4 Normal distribution with mean = 10 and standard deviation = 1

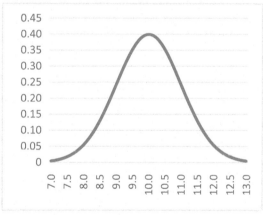

has mean 0 and standard deviation 1. For any variable X described by a normal distribution, $z = (X - \mu)/\text{sigma}$. The z-score of a point on the normal distribution denotes how many standard deviations away it is from the mean. Moreover, the area beneath any points on a normal distribution is equal to the area beneath their corresponding z-scores. This means that we only need to compute areas for z-scores in order to find the areas beneath any other normal curve.

The second important use of the properties of the normal distribution is that it is symmetric. This means that:

1. $P(Z > z) = 1 - p(Z < z)$
2. $P(Z < -z) = P(Z > z)$
3. $P(z1 < Z < z2) = P(Z < z2) - P(Z < z1)$

Standard normal distribution tables provide cumulative values for P(Z < z) until z = 0.5. Using symmetry, we can derive any area beneath the curve from these tables.

The normal distribution is of utmost importance due to the property that the mean of a random sample is approximately normally distributed with mean equal to the mean of the population and standard deviation equal to the standard deviation of the population divided by the square root of the sample size. This is called the central limit theorem. This theorem plays a big role in the theory of sampling.

3 Statistical Analysis

Merriam-Webster defines statistics as a "branch of mathematics dealing with the collection, analysis, interpretation, and presentation of masses of numerical data."[1] Statistics, therefore, is concerned with recording occurrences that take place, and attempting to understand why those occurrences do so in that manner. A statistic, however, is also a "quantity (such as the mean of a sample) that is computed from a sample." Thus, we may have several statistics, or facts, about a set of data that we have gathered, which we have found through the use of statistics.

Let us define some useful terms.

- A "dataset" is simply a recording of all the pieces of information we have collected. If we were examining cars, our dataset might include each car's color, age, model, place of manufacture, and so on.
- A "population" is the sum total of all pieces of data in the field we are examining. For example, if we wanted to investigate the mileage of every car in the world, our population would consist of each and every car that has ever been made.
- A "sample" is a subset of the population, which we have actually recorded. Often statistics must rely on samples as it is infeasible to record the entire population—finding the mileage of every car ever made sounds like an impossible task.

The difference between a sample and a population is key to statistical analysis. If we use a dataset that consists of the entire population of cars in the world (imagining for a moment that we have been able to collect it) we can know for sure that we have accounted for every possible recording that is available. However, if we are using a sample that we have drawn from the population, we cannot know for sure that there are other findings that we have missed that may drastically change the nature of our dataset. Refer Chap. 2 for more details.

This is important because collection is only one part of statistics. After collecting data, we must analyze it in order to find insights about the dataset we have obtained,

---

[1] https://www.merriam-webster.com/dictionary/statistics (accessed on Jun 22, 2018).

and thus about the world that we have recorded in our dataset. These tools of analysis, despite being very simple, can be incredibly profound and inform the most advanced computational tools.

The use of data analysis that helps to describe, show, or summarize data in a way that helps us identify patterns in the dataset is known as *descriptive statistics*. The tools we use to make predictions or inferences about a population are called *inferential statistics*. There are two main types of statistical analysis. The first is *univariate analysis*, which describes a dataset that only records one variable. It is mainly used to describe various characteristics of the dataset. The second is *multivariate analysis*, which examines more than one variable at the same time in order to determine the empirical relationship between them. *Bivariate analysis* is a special case of multivariate analysis in which two variables are examined.

In order to analyze a dataset, we must first summarize the data, and then use the data to make inferences.

The first type of statistic that we can derive from a variable in a numerical dataset is measures of "central tendency," or the tendency of data to cluster around some value. The *arithmetic mean*, or the average, is the sum of all the values that the variable takes in the set, divided by the number of values in the dataset. This mean corresponds to the expected value we find in many probability distributions.

The *median* is the value in the dataset above which 50% of the data falls. It partitions the dataset into two equal halves. Similarly, we can divide the data into four equal quarters, called *quartiles*, or 100 equal partitions, called *percentiles*.

If there is a value in the dataset that occurs more times (more often) than any other, it is called the *mode*.

The second type of statistic is measures of dispersion. *Dispersion* is a measure of how clustered together data in the dataset are about the mean. We have already encountered the first measure of dispersion—*variance*. The variance is also known as the *second central moment* of the dataset—it is measured by the formula:

$$\frac{\sum (Data\ value - Mean)^2}{n}$$

where n is the size of the sample.

In order to find higher measures of dispersion, we measure the expected values of higher powers of the deviations of the dataset from the mean. In general,

$$r - th\ central\ moment = \mu_r = \frac{\sum (Data\ value - Mean)^r}{n} = E\left[(X - \mu)^r\right].$$

Mainly, the third and fourth central moments are useful to understand the shape of the distribution. The third central moment of a variable is useful to evaluate a measure called the *skewness* of the dataset. The skewness is a measure of symmetry, and usually the mode can indicate whether the dataset is skewed in a certain direction. The coefficient of skewness is calculated as:

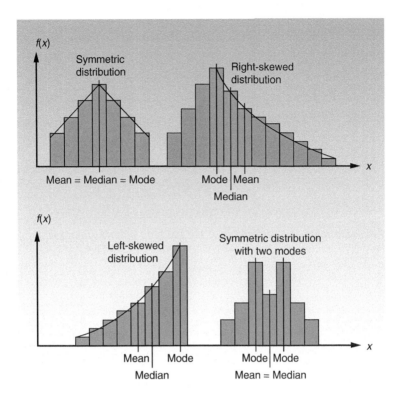

Fig. 30.5 Different types of distributions

$$\beta_1 = \frac{\mu_3^2}{\mu_2^3}$$

As skewness proceeds from negative to positive, it moves from being left skewed to right skewed. At zero it is a symmetric distribution (Fig. 30.5).

The fourth central moment is used to measure *kurtosis*, which is a measure of the "tailedness" of the distribution. We can think of kurtosis as a measure of how likely extreme values are in the dataset. While variance is a measure of the distance of each data point from the mean, kurtosis helps us understand how long and fat the tails of the distribution are. The coefficient of kurtosis is measured as (Fig. 30.6):

$$\beta_2 = \frac{\mu_4}{\mu_2^2}$$

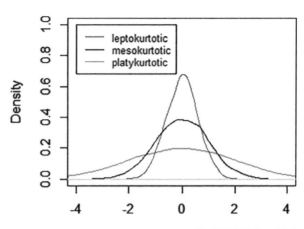

Fig. 30.6 Normal distributions with different kurtosis. The red line represents a frequency curve of a long tailed distribution. The blue line represents a frequency curve of a short tailed distribution. The black line is the standard bell curve

Table 30.5 Frequency distribution table

Color	Frequency
Red	10
Green	14
Black	12
White	19
Blue	11
Orange	2
Purple	1

4 Visualizing Data

Visualizing data can be extremely important as good visualization can clarify patterns in the set, while poor visualization can obscure characteristics of the data. A basic method of visualizing data is the frequency table. A frequency table merely lists each value in the dataset and counts the frequency with which those values have occurred. For example, consider Table 30.5, which lists the color of cars driving down a road:

Graphs and charts can also be an effective tool to portray characteristics of data. In a pie chart, a circle is divided into various "pies" that denote the ratio of the dataset's composition. In a bar graph, the size of the variable for several categories is portrayed as a vertical bar. In a scatter plot, datapoints that consist of two values are plotted on a two-dimensional graph that can portray a relationship between the two variables.

Using bar graphs to represent datasets can become visually confusing. In order to avoid this, we can use box plots, which simplify the datasets by dividing them into partitions of equal size. Then, by drawing a box plot, we can understand intuitively whether the dataset is skewed and how the data is concentrated.

In order to draw a box plot, we:

1. Divide the data into quartiles
2. On an axis representing the variable, draw a box of length equal to Q_3-Q_1

Fig. 30.7 Box plot of distance variable in cars dataset

Fig. 30.8 Q-Q plot of distance variable in cars dataset

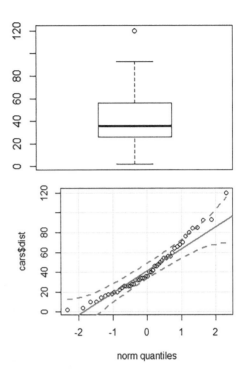

3. From each side of the box, extend a line to the maximum and minimum values
4. Indicate the median in the box with a solid line

In R, the box plot is created using the function boxplot. The syntax is boxplot(variable name). For example, let us draw a box plot for the distance variable in the cars dataset (Fig. 30.7):

```
> boxplot(cars$dist)
```

We can use box plots to understand whether a distribution is normal. In order to do this, we plot two sets of quintiles in the same graph. If they are both from the same distribution, they should lie on the same line. (This approach can be applied to any two distributions. The x-axis plots the points at which say 5, 10, 15, ... 100% of observations lie. The y-axis does the same for the comparison distribution. If the 5 and 5%, the 10 and 10%, etc. points are the same then we get a straight line.) In R, the following commands will plot a Q-Q plot and also a confidence interval for distance variable in the cars dataset (Fig. 30.8).

```
> qqPlot(cars$dist)
```

In general, in R, the command qqplot(x,y) will produce the quantile–quantile plot for x and y variables. For example, the command qqplot(cars$dist, cars$speed) produces the plot shown in Fig. 30.9.

Fig. 30.9 Q-Q plot of distance versus speed variable in cars dataset

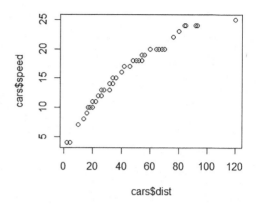

5 Bivariate Analysis

Bivariate analysis is among the most basic types of analysis. By using the tools developed to consider any single variable, we can find correlations between two variables. Scatter plots, frequency tables, and box plots are frequently used in bivariate analysis.

The first step toward understanding bivariate analysis is extending the idea of variance. Variance is a measure of the dispersion of a variable. *Covariance* is a measure of the combined deviation of two variables. It measures how much one variable changes when another variable changes. It is calculated as:

$$\text{Cov}(X, Y) = E[(X - E(X))(Y - E(Y))] = E(XY) - E(X)E(Y)$$

However, using covariance to compare the relationships between two variables is difficult, as the units of covariance are dependent on the original variables. Moreover, since covariance depends on scale, comparing different covariances must take scale into account. In order to do this, we can standardize the measure. This measure is called *correlation*. The correlation coefficient (also written as Corr) of two variables X and Y is denoted as ρ_{xy}.

$$\rho_{xy} = \frac{Cov(X, Y)}{\sigma_x . \sigma_y}$$

The coefficient of correlation always lies between -1 and $+1$. As it moves from negative to positive, the variables change from moving perfectly against one another to perfectly with one another. At 0, the variables do not move with each other (to be perfectly honest we need to say in an average sense). Independent variables are uncorrelated (but uncorrelated variables are not independent with some exceptions such as when both variables are normally distributed).

Some properties of covariance and correlation:

1. Corr(X,X) = 1 (X is perfectly correlated with itself)
2. Cov(X,X) = Var(X) (The dispersion of X compared to itself is the variance)
3. Var(X + Y) = Var(X) + Var(Y) + 2Cov(X,Y)
4. Var(X − Y) = Var(X) + Var(Y) − 2Cov(X,Y)

It is important to note that correlation is a powerful but simple tool, which may not catch nuanced observations. For example, if X and Y are related in a quadratic manner, their correlation will be 0 even though we know that there is a relationship between the two variables. Moreover, the existence of rogue datapoints, or outliers, can change the value of correlation dramatically. Most importantly, it is critical to remember that *correlation does not imply causation*. Simply because two variables are correlated in some way does not give us enough evidence to infer a relationship between them. More details are given in the Chap. 7.

In R, the functions cov(x,y) and cor(x,y) produce the covariance and correlation, respectively. If there are more than two variables, giving the name of the dataset produces the covariance and correlation matrices. For example, these commands on the cars dataset produce the following output:

```
> cov(cars$dist, cars$speed)
[1] 109.9469
> cor(cars$dist, cars$speed)
[1] 0.8068949
```

For the variables cars$dist and cars$speed, covariance = 109.95 and correlation = 0.8068.

Index

A

A/B testing, 617, 728, 739, 755, 757, 758
Academic Medical Centers, 765–771, 785, 787–788
ACF, *see* Auto-covariance function (ACF)
Activation Forward Propagation, 538
Ad Exchange, 726, 728
Ad servers, 725
Adaptive-response-rate single exponential smoothing (ARRSES), 392, 409, 411, 415
Adipose tissue problem, 180–181, 189
Adjusted R^2, 200–204, 220–222, 239
Advanced inventory models, 839
Affordable Care Act (ACA), 13, 770
Aggregate planning, 337
Aila, 617
Airbnb, 652
Akaike information criterion (AIC), 221–223, 263–265, 268, 397, 428, 434, 453, 455
Amazon, 78, 105, 106, 509, 556, 565, 602, 652, 728, 744, 803, 804, 824, 826
Amazon Go, 605
Amazon Prime, 550, 652
Amazon S3 (Simple Storage Service), 78, 105
Amazon Web Services (AWS), 105, 106
AmazonDB, 80
AMC, *see* Asset management company (AMC)
Anaconda, 919, 924
Analysis of variance (ANOVA), 4, 137, 166–169, 171, 427, 728, 730–733, 735–738
ANN, *see* Artificial neural network (ANN)
Anscombe's quartet, 114, 115, 243
Apache OpenNLP, 298
Apple, 102, 108, 510, 824, 830
Apple Watch, 766
Application programming interface (API), 22, 34, 36, 37, 80, 96, 97, 99, 106, 285
Arbitrage-free drift, 670
Arbitrage pricing theory (APT), 12, 662
ARCH, *see* Auto-regressive conditional heteroscedasticity (ARCH) model
ARIMA, *see* Auto-regressive integrated moving average (ARIMA)
Arithmetic mean, 27, 958
ARMA, *see* Auto-regressive moving average (ARMA)
ARRSES, *see* Adaptive-response-rate single exponential smoothing (ARRSES)
Artificial intelligence (AI), 71, 509, 527, 533, 848, 867
Artificial neural network (ANN), 10, 11, 383, 537, 540, 570–576, 579, 607, 867
ASSESSOR model, 626
Asset management, 659, 660, 685, 849
Asset management company (AMC), 660, 685, 771–774, 782–784, 787, 788
Asset price estimation, 12, 663–672
Assortment optimization, 603, 613, 614
Assortment planning, 600, 654, 825
Atlas of health and climate, 769
Autocorrelation coefficient, 389
Auto-covariance function (ACF), 389
Auto-encoders, 540, 575
Automated essay grading, 584

Automatic Region Server failover, 80
Auto-regressive conditional heteroscedasticity (ARCH) model, 396, 397, 667
Auto-regressive integrated moving average (ARIMA), 8, 389, 395–399, 409, 410, 414, 417, 605
Auto-regressive models, 389, 625, 667
Auto-regressive moving average (ARMA), 395–397, 667, 700–702
Axioms of probability, 947, 948
Axis Bank, 744

B
Babylon, 767
Back propagation, 573, 592
Backward feature selection, 519
Bagging, 10, 546, 547, 607
Bag-of-words, 6, 286, 297, 532, 563
Bank for International Settlements, 686
Banner advertisement, 13, 723
Basel Committee on Banking Supervision (BCBS), 672
Baseline hazard model, 625
Basis, 24, 34, 104, 218, 291, 292, 295, 354, 359, 360, 402, 404, 421, 440, 510, 534, 540, 625, 644, 650, 662, 676, 690, 714, 716, 721, 739, 751, 752, 757, 758, 767, 806, 818, 829, 830, 834, 835, 879, 915, 945
Bass model, 626
Batch-wise gradient descent, 573, 583
Bayesian classifier, 10, 488, 493, 529–549
Bayesian estimation, 669
Bayesian information criterion (BIC), 221, 222, 264, 265, 268, 397
Bayes' theorem, 947–948
BCBS, *see* Basel Committee on Banking Supervision (BCBS)
BeautifulSoup, 944
Bernard Madoff, 684
Bernoulli distribution, 491, 952, 954
Best subset regression, 180, 220, 222–223
Best-worst conjoint, 643
Bfonics, 617
Bias-variance-trade off, 520
BIC, *see* Bayesian information criterion (BIC)
Big data, 1, 3, 32, 42, 71–108, 289, 382, 602, 663, 741, 794, 863
Big data techniques, 766–770, 794, 877
Binary hierarchical classifiers, 547–549
Bivariate analysis, 958, 962–963
Bivariate distributions, 945

Black and Scholes, 12
Black-Box, 403
Blended recommendations, 564–565
Blending problem, 337
Blippar, 617
Bloomberg, 685, 687
Bluetooth beacons, 11, 604
Blue Yonder, 618
Boosting, 10, 546–547
Bootstrap, 401, 523, 607, 634
Bootstrap aggregation, 546, 607
Bounding, 366, 367
Box plots, 185, 186, 194, 196, 197, 202, 238, 239, 960–962
Box–Cox transformation, 180, 232–233
Branch-and-bound method, 7, 365–367
Branching, 366–367, 369
Brand specific conjoint, 643
Brickstream, 617
Brownian motion, 660

C
Cancer, 570, 766
Capital asset pricing model (CAPM), 12, 661, 662
Cartesian coordinate system, 344–345
Cassandra, 80
Categorical features, 526, 527
Categorical response variable, 248
Category management, 825
Causation, 963
Celect, 618
Censoring, 8, 439–444, 457, 602
Center for Disease Control (CDC), 284
CentOS Linux, 87
Central limit theorem, 4, 137, 140–141, 148, 316, 957
2nd Central moment, 958
Certainty assumption, 343
CF, *see* Collaborative filtering (CF)
CGE, *see* Computable general equilibrium (CGE)
Chain ratio method, 626
Chi-Square (χ^2) distribution, 4, 149, 150, 174, 254, 255
Choice analytics, 625
Choice based conjoint, 643, 644
CISCO, 827
Citibank, 660
CIV, *see* Customer influence value (CIV)
Class label, 461, 510, 511, 514, 515, 524, 528, 535, 545, 571

Index

Classification, 5, 101, 264, 265, 269, 273, 292, 293, 373–374, 402, 432, 461, 478, 479, 484, 511–513, 515, 517, 518, 520, 522–549, 554, 570, 571, 574, 576, 577, 579, 588, 859
  plot, 259–260, 265
  table, 259–260, 265
Classifier, 10, 510, 511, 522–523, 567, 571, 579, 592
Clickstream log, 73
Cloud computing, 103–108
Cloudera, 3, 86–90
Cluster analysis, 625
CLV, see Customer lifetime value (CLV)
CNNs, see Convolutional neural networks (CNNs)
Coefficient of determination, 191–192, 194
COGS, see Cost of goods sold (COGS)
Cold start problem, 557, 559, 562
Cole bombing suspects, 722
Cole Haan, 617
Collaborative filtering (CF), 9, 101, 555, 557, 559, 561, 564
Collinearity, 5, 180, 215–224, 228, 239, 241, 243
Combinations, 28, 60, 118, 167, 238, 295, 298, 325, 337, 355, 380, 388, 395, 397, 418, 424, 432, 437, 466, 477, 481, 488, 497, 514, 518, 526, 535, 536, 549, 554, 562, 565, 604, 629, 643, 645, 675, 735, 736, 739, 749, 750, 774, 807, 809, 810, 832–834, 878, 900, 927, 945, 953, 954
Commodity procurement, 14, 829
Common identifier, 31
Comparison of two populations, 161–162
Complement, 33, 234, 630, 720, 744, 761, 946
Complementary slackness conditions of optimality, 361
Complete enumeration, 23, 366
Completely randomized design (CRD), 13, 729–732, 735
Component capital, 679–682
Computable general equilibrium (CGE), 829
Computer simulation, 6, 307, 333
Concordant pairs, 5, 261–262
Conditional mean model, 642, 700–701
Conditional probabilities, 444, 532, 674, 947
Conditional variance, 396, 397, 430, 701
Conditional variance models, 701–702
Confidence intervals, 4, 137, 141–151, 161, 169, 170, 192–195, 203, 204, 233, 256, 271, 315, 316, 329, 388, 451, 634, 961
Conjoint analysis, 12, 624–626, 643–650, 655
Connected cows, 828

Constant return to scale (CRS), 637, 638
Consumer's ratings, 643
Continuous random variables, 948, 949, 951
Contour plots, 344, 348
Control variables, 340
Convex hull, 362, 364, 380
Convexity constraint, 638
Convolution layer (CONV), 10, 579–582
Convolutional neural networks (CNNs), 10, 540, 570, 574, 576–583, 588, 589, 591, 594
Co-occurrence graphs (COG), 289–290, 296, 298, 501
Cook's distance, 212, 213, 430
Copula functions, 12, 672, 673, 676
Correlation, 208–211, 219, 273, 300, 389, 504, 509, 514, 564, 579, 607, 661, 668, 672–677, 679, 683, 696, 715, 716, 739, 826, 962, 963
Cosine similarity, 476, 482, 499, 545, 555, 556, 562
Cost efficiency, 823, 824, 830, 831, 842, 844
Cost of goods sold (COGS), 290, 292, 298–301, 670, 826
Count data regression model, 421–437
Covariance, 187, 188, 212, 242, 492–494, 530, 532–534, 661, 690, 696, 962, 963
Covariates, 9, 183, 188, 427, 428, 441, 446–448, 450, 453, 455–457, 606, 625
Cox and Snell R^2, 257–258, 264, 268
Cox proportional hazard model, 440, 447–449, 456
CRAN repository, 801
CRD, see Completely randomized design (CRD)
CreditMetrics™, 670
Crew scheduling, 337
Cross domain recommendations, 565–566
Cross-sectional data, 24, 238, 383, 625, 667
Croston's approach (CR), 8, 400, 401
CRS, see Constant return to scale (CRS)
CRV, see Customer referral value (CRV)
Cumulative distribution function (CDF), 442, 949
Cumulative hazard, 442, 457
Curvilinear relationships, 12, 630–635
Custom/algorithmic attribution, 747–750
Customer analytics, 12, 75–76, 626, 650–654
Customer identity, 32, 33
Customer influence value (CIV), 12, 653–654
Customer lifetime value (CLV), 12, 440, 624, 650–652, 654
Customer referral value (CRV), 12, 626, 652–654

Customer retention, 600, 651
Customer satisfaction, 14, 623, 637, 823, 828, 830–832, 842–844, 847, 848, 855, 856, 858
Customer service, 284, 306, 308, 824, 825, 830, 844, 855
Cyclical, 388–390

D
Damped exponential smoothing, 394, 413
Data collection, 2, 11, 19–38, 101, 138, 213, 285, 384, 440, 600, 603–605, 607, 616, 617, 644–645, 654, 767, 772, 774, 776–778, 851, 856, 879
Data dimensions, 119, 120, 122, 134
Data driven marketing, 741
Data envelopment analysis (DEA), 12, 624, 625, 636–641, 655
Data mining, 75, 498, 855
Data relationships and graphs, 119
Data warehousing, 600
Database administrator (DBA), 48
Database management system (DBMS), 42–43
Data control language (DCL), 47, 48
Data definition language (DDL), 47–48
Dataframes, 101
Data manipulation language (DML), 47–48
DataNodes, 78, 81–84
DBA, *see* Database administrator (DBA)
DBMS, *see* Database management system (DBMS)
DBN, *see* Deep belief networks (DBN)
DCL, *see* Data control language (DCL)
DCO, *see* Dynamic creative optimization (DCO)
DDL, *see* Data definition language (DDL)
DEA, *see* Data envelopment analysis (DEA)
Decision boundary, 527, 533, 534, 540, 542, 547, 549, 567
Decision making under uncertainty, 7, 305, 317–327
Decision-making units (DMUs), 637, 638
Decision trees (DTs), 10, 11, 182, 273, 519, 524, 526, 527, 535, 545–547
Decision variables, 305, 317–319, 321, 324–327, 338–344, 356–358, 360–369, 375–378, 605, 608, 812, 840
Deep belief networks (DBN), 575
Deep content based recommendations, 562–563
Deep learning, 6, 10, 106, 374, 516, 534, 540, 569–594, 605, 607, 661
Deep neural networks (DNNs), 570, 607

Deep Q-network (DQN), 575
Degenerate, 353–355, 377
Degrees of freedom, 145–147, 149, 150, 160, 164–169, 176, 191, 193–195, 199, 200, 206, 211, 252, 255–257, 263, 264, 268, 270, 426, 427, 477, 733
Deletion diagnostics, 180, 206, 211–215
Delphi method, 626
Demand forecasting, 11, 14, 15, 400, 416, 607, 654, 704, 798, 811, 813, 814, 819, 824–826
Demand side platform (DSP), 726
Demand-side software (DSP), 725
Denoising auto-encoder (DA), 575
Dependent variable, 183, 257, 263, 266, 269, 272, 273, 576, 583, 625–627, 630, 633, 635, 737
Depth-first-search process, 368
Depth-threshold, 525
Derivative security, 660
Descriptive classifiers, 488, 522, 523, 534
Descriptive statistics, 179, 185–186, 769, 945, 952, 958
Designed experiment, 22
Deviance R^2, 25
DFFITS, 212
DHL, 602
Dickey–Fuller test, 305, 666, 693–695
Dictaphones, 768
Digital attribution, 654, 722, 744–750
Digital Diaries, 768
Dimensionality, 29, 122, 286, 344, 345, 467, 482, 531, 559, 572, 606, 616, 661, 663, 683, 685, 690, 691, 696
Dimensionality reduction, 101
Discordant pairs, 261, 262
Discovery, 524, 549, 550, 570
Discrete event simulation (DES), 13, 770, 772, 786
Discrete random variables, 139, 948, 951, 952
Discriminant analysis, 533, 625, 747
Discriminative classifiers, 522, 523, 534, 535
Disjoint, 946, 947
Dispersion, 8, 185, 238, 428, 707, 708, 950, 951, 958, 962
Distributed computing, 77, 78
Distribution centers (DCs), 827, 828, 830, 831, 833, 834, 839, 840, 843
Divide and conquer, 365, 390, 545
Divisibility assumption, 343
DML, *see* Data manipulation language (DML)
DMUs, *see* Decision-making units (DMUs)
DNA, 768
DNNs, *see* Deep neural networks (DNNs)

Document, 6, 283, 286–292, 294–301, 476, 482, 484, 510, 511, 517–519, 532, 545, 670, 852
Document classifier, 510
Document-term matrix (DTM), 288, 292, 296, 300
Dor, 318, 617
DQN, *see* Deep Q-network (DQN)
Dropbox referral program, 652
Dual values, 355, 358
Dummy variables, 5, 180, 224–233, 239, 247, 248, 263, 437, 647, 730, 731, 734
Dynamic creative optimization (DCO), 728

E
EBay, 728, 804
E-commerce, 73, 331, 507, 549, 553, 554, 563, 599, 601, 616, 743, 803, 804
Economies of scale, 343, 757, 833, 834
EDA, *see* Exploratory data analysis (EDA)
Efficiency model, 830
Efficient market hypothesis (EMH), 662, 665
EGARCH, 685, 701–703, 705, 707–709, 713, 716
Eigenvalues, 470, 471, 590, 688–690
Eigenvectors, 470, 688–690
Elliptical distribution, 662
EMH, *see* Efficient market hypothesis (EMH)
Engagement ads, 724
Engagement bias, 564
Engagement matrix, 551, 556–561
Ensemble, 10, 527, 545–547, 549, 607, 886
Enthought Canopy, 919, 920, 923
Entity integrity, 43
Entropy, 525, 587, 589
Epidemiology, 766
ERP systems, 825, 858
Error back propagation, 539
Estimates, 5, 72, 74, 137–151, 163, 164, 167, 169, 170, 179, 180, 183, 190–195, 199–208, 211–215, 218–221, 223, 225, 228, 229, 232, 233, 239, 241, 242, 247, 249–256, 258, 261–268, 273–276, 278, 284, 292, 296, 309, 310, 313, 315–320, 328, 331, 332, 369–371, 373, 381, 389, 390, 394, 396–398, 408, 409, 425, 427, 428, 431, 432, 434, 436, 440, 442–445, 448, 449, 451, 453, 455, 488, 489, 524, 529, 531, 532, 556, 571, 584, 599, 601, 605, 606, 608, 613, 614, 616, 628, 629, 633, 634, 638, 640–642, 645, 647–649, 652, 662, 663, 665, 668–670, 683–686, 690, 692, 702, 708, 709, 715, 728, 732, 735, 736, 739, 747, 748, 776, 797, 801, 805, 806, 811, 816, 817, 819, 826, 829, 833, 834, 840, 841, 850, 854, 857, 876–878, 886
Estimation, 7, 9, 12, 33, 139, 180, 186, 189, 193, 195, 199, 211, 223, 233, 250–251, 263, 264, 275, 276, 278, 279, 328, 338, 369, 371, 383, 388, 397, 398, 441, 443, 449, 459, 462, 488–491, 494, 504, 508, 512, 577, 602, 603, 610, 611, 616, 617, 628, 641, 642, 645, 652, 663–665, 667–669, 683, 685, 696, 702, 805, 806, 829, 879, 886, 944
Estimation of parameters, 189, 199, 223, 250–251, 275
Estimators, 139–141, 148, 149, 190, 191, 199, 223, 396, 444, 490, 606, 616
Euclid analytics, 617
Euclidian distances, 476, 528
Eugene Fama, 662
Excessive zeros, 430
Exogeneity, 188
Expectation-maximization (EM) algorithm, 480, 613
Expected value, 140, 147, 148, 151, 186, 307, 328, 424, 425, 429, 660, 950, 958
Experiment, 13, 22, 87, 422, 434, 608, 729–733, 736, 737, 739, 752–754, 773, 775, 778–781, 783, 787–789, 877, 946–948, 950–953
Experimental design, 644, 647, 649, 723, 728, 729, 735, 739, 750, 785
Explanatory variables, 182, 247–249, 251–253, 255, 257, 262–266, 268–270, 421, 422, 427, 435, 447–449, 451, 642, 886
Exploratory data analysis (EDA), 427
Exponential smoothing, 391–395, 401, 410, 413, 605
Exponomial choice (EC) model, 610
Exposure variable, 430
ExtenSim version 9.0, 786
External data, 23, 908
Extreme point, 351, 352

F
Face detection, 570
Facebook, 13, 26, 71, 72, 75, 76, 284, 285, 478, 499, 512, 545, 550, 553, 565, 570, 601, 654, 719, 721, 724, 726, 728, 739, 748–750, 762
Factor analysis, 295, 625, 663, 690–691
Factorial designs, 644, 736, 739
Fama–French three-factor model, 667

Fashion retailer, 6, 7, 306, 309, 317, 318, 321, 323, 336, 826
F distribution, 4, 165, 167, 176, 195, 212, 335, 685, 698, 806, 959
Feature centric data scientists, 516
Feature distributions, 514
Feature engineering, 10, 516–519, 532, 537, 574, 577
Feature normalization, 517
Feature selection, 519, 606
Feature space, 486, 488, 522, 524–526, 531, 533, 537, 545, 562
Feature transformation, 517
Federal Reserve Bank (Fed), 679, 686
FedEx, 825, 843
Filter methods, 519
Financial instrument, 660, 661
First interaction/first click attribution model, 746
First normal form (1NF), 44
Fisher discriminant analysis, 468, 515
FitBit, 766
Fit-for-purpose visualization, 124
Flash, 723, 826
Flipkart, 556, 744
Flume, 101
Footmarks, 617
Ford, 132, 828
Forecast/prediction intervals, 7, 193, 195, 204, 385, 406, 408
Forecasting, 7, 8, 11, 14, 15, 75, 180, 338, 381–418, 512, 518, 521, 601, 605–607, 625, 626, 647, 654, 796, 805–806, 814, 824–826
Forecasting intermittent demand, 399–400
Foreign key, 43, 64, 66–67
Forward feature selection, 519
Forward method, 263
Fourier transformations, 517
Fraudulent claims detection, 76
Full profile conjoint, 643
Fully connected layers (FC), 579, 582, 592
Function, 5, 7–9, 61, 102, 149, 185, 191, 210, 234, 249–263, 266, 269, 270, 275, 276, 279, 300, 308–310, 312, 317, 319, 320, 323, 325, 336, 337, 339, 341, 342, 344, 346–359, 370–378, 382, 389, 390, 396, 397, 408, 411–414, 426, 431, 434, 440, 442, 443, 446–448, 451, 453, 455, 456, 463, 464, 470, 472, 476, 480, 482, 485–490, 492–494, 524, 528–537, 540–547, 552, 563, 571, 574, 576, 581, 582, 586, 587, 591, 592, 601, 607, 609, 610, 638, 640–642, 662, 667, 669, 672–724, 761, 767, 771, 796, 797, 805, 829, 833, 834, 839, 875–877, 880, 883–887, 896, 902, 903, 905, 907–909, 913–915, 925, 936–942, 944, 948–951, 961
Functional magnetic resonance imaging, 34
Fundamental theorem of asset pricing, 660
Fundamental theorem of linear programming, 352

G

Gaana, 550
Gasoline consumption, 181–182, 201, 206–208, 210
Gaussian copulas, 672–676, 715
GE, 307
Gender discrimination, 182, 224, 229
Generalized autoregressive conditional heteroscedasticity (GARCH) model, 397, 667, 668, 684, 685, 696–699, 701–706, 708–713, 715
Generalized linear models (GLM), 425, 536, 544
Generating random numbers, 333–334
Generative adversarial networks, 540
Generic conjoint, 643
Geo-fencing, 655, 722
Geometric Brownian motion (GBM), 660
Ginni index, 524
GJR, 685, 700–704, 706–709, 712, 716
GLM, *see* Generalized linear models (GLM)
Goldman Sachs, 660
Google, 6, 37, 38, 77, 105, 106, 284, 301, 383, 498, 499, 509, 520, 570, 601, 720, 722–725, 728, 741, 742, 744, 750, 755, 760, 762, 769, 877
Google Adwords, 720, 741, 762
Google Analytics, 720
Google Big Query, 106
Google Brain, 769
Google Cloud Platform (GCP), 106
Google Compute Engine, 106
Google Display Networks, 724
Google DoubleClick, 725
Google Maps, 37, 38, 520, 723
Google Prediction API, 106
Google Traffic, 722
Google Trends, 769
Gradient descent, 573, 581, 583, 594, 607
Graph processing, 98, 101
Graphics processing units (GPUs), 482, 516, 575

Index

GraphX, 101
Gumbel distribution, 610

H
H1N1, 6, 284
Hadoop, 3, 72, 77–85, 87, 89, 90, 92–99, 101, 105–108
Hadoop architecture, 80–82
Hadoop distributed file system (HDFS), 3, 78, 80–85, 94, 101
Hadoop ecosystem, 78–s81
Hadoop streaming, 89, 90, 92, 93
Harry Markowitz, 661
Harvard, 684, 685
HARVEY, 307
Hat matrix, 205
Hazard ratio, 447, 449, 451
HBase, 80, 94
HBO, 724
HDFS, *see* Hadoop distributed file system (HDFS)
Healthcare operations, 766, 768
Heat maps, 125, 126, 617
Hermitian matrix, 677
Heterogeneous, 292, 483, 495, 498
Heteroscedasticity, 5, 180, 215, 233–237, 239, 249, 667–669
Hive, 80, 94, 101
Holdout data, 387
Holt exponential smoothing (HES), 392–393, 410
Holt's method, 394, 413
Holt–Winters' trend and seasonality method, 393–394
Homogenous, 291, 666, 817
Hortonworks, 87
Hosmer and Lemeshow test, 252, 257, 263, 264, 271
Human visual perception, 115
Hurdle models, 422, 434–435
Hypothesis testing, 4, 137, 151–156, 169–171

I
IAAS, *see* Infrastructure as a service (IAAS)
IBM, 43, 47, 78, 87, 570
IBM Watson, 570
ICICI bank, 636, 637
IDC Manufacturing Insights, 827
IDE, *see* Integrated development environment (IDE)
Ideal sample size, 144, 148, 149
Image classification, 570, 577, 579, 588

ImageNet challenge, 570, 578
Imputation techniques, 519
Incremental capital, 679–681
Indentation, 897, 923–925
Independent events, 946
Independent samples, 163–166, 169, 170
Independent variables, 6, 183, 251, 252, 254, 255, 266, 272, 275, 347, 373, 424, 434, 625–628, 630, 655, 962
Infeasible region implies, 346
Inferential statistics, 958
Influential observation, 211–213, 232, 239
Information extraction, 510
Infrastructure as a service (IAAS), 104, 105
In-memory computing, 96, 97
Innovation analytics, 626
Input features, 467, 511, 512, 517, 518, 536, 537, 547, 569
Instagram, 654
Instantaneous hazard, 442
Insurance, 8, 15, 71, 76, 213, 284, 338, 421–423, 430–434, 437, 440, 441, 659, 766, 769, 770, 843, 847–861
Integer optimization, 614
Integer programming, 7, 343, 362, 365, 374
Integrated development environment (IDE), 891, 892, 919
Integrated distribution, 824, 831
Integrates R with Python, 944
Interaction effect, 11, 227, 626–632, 655, 736–739, 755, 759
Interactive mode, 897, 923
Interactive Python (IPython), 944
Internet of things (IoT), 73, 105, 460, 507, 604
Interpolation, 240, 273, 557, 584, 688
Interpreted language, 890, 919
Intersection, 32, 125, 345, 346, 348, 350, 353, 503, 504, 946
Interval, 2, 4, 7, 8, 24, 27, 83, 98, 131, 137, 139, 141–151, 157, 161, 169, 170, 192, 193, 195, 203, 204, 233, 256, 271, 313, 315, 316, 329, 373, 385, 388, 399, 400, 406, 408, 414, 429, 441, 442, 444, 451, 574, 619, 634, 661, 663, 674, 961
Invariant features, 517
Inventory management, 824, 826, 827
Inventory optimization, 14, 337, 609, 826–827, 829
Inventory turnover, 826, 830
IoT, *see* Internet of things (IOT)
IPad, 776
IRIS dataset, 467, 514
Irregular/error components, 388
Item-item collaborative filtering, 555, 556

J

JavaScript, 723
Java virtual machine (JVM), 99
JioCinema, 556
Joint probability density function, 371, 531
JPMorgan, 660, 681, 682

K

Kafka, 101
Kaplan–Meier estimates, 444–445
Kate Spade, 617
Kernel functions, 489, 490, 528, 544, 545
Key performance indicators (KPIs), 600
Klarna, 601
Kline, 605
K-nearest neighbor classifier (k-NN), 10, 274, 527–529, 532, 535, 540, 545
Kohl's, 603
Kurtosis, 698–700, 714, 959, 960

L

Lagrange multiplier, 541–543
Language processing, 6, 283, 286, 294, 297–299, 569, 570
Language translation, 570
Laplacian smoothing, 524, 532
Large-sample confidence intervals, 147–148
Last AdWords click attribution model, 746
Last interaction/last click attribution model, 13, 745, 746
Last non-direct Click attribution model, 745
Least absolute shrinkage and selection operator (LASSO) regression, 11, 180, 220, 223, 224, 338, 606
Leverage values, 205–206
LexisNexis, 769, 854
Likelihood ratio test, 5, 252, 255, 448
Linear attribution model, 13, 746
Linear discriminant analysis (LDA), 10, 533–535
Linear inequalities, 345
Linear optimization, 7, 338, 339, 614, 828
Linear programming (LP), 7, 339–347, 349, 351–369, 374, 375, 377, 378, 380, 610, 637, 638, 809, 812–814, 817
  formulation, 340–342, 353
Linear regression, 4, 5, 7, 114, 179, 180, 184–186, 189, 193, 198, 201, 204, 205, 208, 211, 215, 230, 231, 234–244, 247, 248, 250, 258, 264, 273, 275, 369, 371, 373, 421–424, 429, 437, 535, 606, 626, 645
Lines of business (LoB), 681, 716
LinkedIn, 652
Linux, 87, 892
List-wise deletion, 272
Location-based data capture, 32, 34, 603
Location bias, 565
Log odds, 251, 535
Logistic regression, 7, 249–252, 254, 256–258, 260–264, 266, 269–271, 274, 441, 517, 518, 535–537, 540, 544, 545, 547, 594, 607, 747
Logistics planning, 831, 844
Log-logistic distribution, 454
Long short-term memory (LSTM), 574, 590
Lord & Taylor, 617
Loyalty cards, 600
LP formulation, *see* Linear programming (LP)
LSTM, *see* Long short-term memory (LSTM)

M

Machine learning (ML), 9–11, 15, 71, 95, 98, 101, 106, 374, 459–505, 507–570, 600, 601, 605–607, 618, 769, 848, 944, 947
Machine translation, 10, 583, 584, 588–590
Macy's, 602, 605
MANOVA, 625
Map Advertisements, 723
MapR, 87
MapReduce, 3, 77, 78, 80, 81, 84–98, 108, 501–504
Market share forecasting, 626, 647–648
Markov chain, 613
Master–Slave architecture, 78, 81
Matplotlib, 944
Matrix factorization approaches, 557–559, 561
Mattel, 297
Maximum likelihood estimate (MLE), 5, 12, 247, 250, 266, 275–279, 371–372, 397, 398, 425, 427, 432, 445, 448, 449, 616, 634, 668, 669, 685
MBC, *see* Multimodal Bayesian classifier (MBC)
McFadden R2, 257–258, 268
Measure of central tendency, 185, 238, 950, 958
Measure of dispersion, 185, 238, 708, 950, 951, 958, 962
Measures of purity, 524, 525
Measure the efficiency, 636, 637, 639
Median, 27, 32, 108, 185, 193, 194, 201, 208, 209, 214–216, 223, 225, 227, 229, 230, 235, 237, 273, 405, 425, 427, 433, 435, 454, 456, 488, 808, 952, 955, 958, 961

Index 973

Mediation analysis, 12, 625, 633, 635, 656
Medicinal value in a leaf, 182–183
Memory-based recommendation engine, 552, 555, 557
Mesos, 99
Meta-rules of data visualization, 4, 116–133
Method of least squares, 4, 180, 189, 191, 199, 223, 232, 233, 236, 275
Microsoft, 43, 46, 47, 106, 297, 344, 375, 405, 570, 589
Microsoft Azure, 106
Microsoft Cognitive Toolkit, 106
Millennials demographic, 602
MIN price, 62
Missing data, 5, 31, 247, 272–274, 372, 384, 526, 687
Missing features, 519, 526
Mitigation strategy, 828
Mixed integer optimization, 614
Mixture-of-Gaussians (MoG), 483, 489, 492–494, 530, 531, 535, 545
ML, see Machine learning (ML)
MLE, see Maximum likelihood estimate (MLE)
MLPs, see Multi-layered perceptrons (MLPs)
MNIST data, 10, 471, 515, 575–576, 582–583
MNL, see Multinomial logit (MNL)
Mobile advertising, 725
Mobile auto-correct, 584
Mode, 27, 30, 99, 219, 273, 549, 550, 611, 719, 723, 858, 897, 923, 937, 938, 955, 958
Model centric data scientists, 516–517
Model validation, 237–239
Moderation, 625, 628, 635
MoG, see Mixture-of-Gaussians (MoG)
Monte Carlo simulation, 6, 12, 306, 669–672
Mordor Intelligence, 766
Morgan Stanley, 660
Movie recommendation, 563
Moving average (MA) methods, 8, 388, 389, 391, 396, 397
MR, see Multiple regression (MR)
Multi-layered perceptrons (MLPs), 10, 570–576, 579, 581–585, 592–594
Multimodal Bayesian classifier (MBC), 530–531
Multinomial logistic regression (MNL), 5, 15, 266–269
Multinomial logit (MNL), 610–613, 806, 817, 867, 870
Multiple regression (MR), 225, 252, 396, 626, 627, 634
Multiple variance ratio test, 666
Multivariate analysis, 625, 958

Multivariate Cauchy distribution, 662
Multivariate exponential, 662
Multivariate normal, 662
Multivariate statistical analysis, 625
Multivariate student t-distribution, 662
MySQL, 1, 3, 42, 43, 47–51, 53, 54, 68, 80, 102, 106

N

Nagelkerke R2, 257–258, 268
Naïve Bayes classifier, 10, 531–532
Naïve method (NF), 8, 390–391
Named entity recognition (NER), 286, 298, 300
NameNode, 81–84
Namespace, 83
Natural language processing (NLP), 6, 283, 286, 287, 294, 297–300, 569
Natural language toolkit (NLTK), 298
Near field communication (NFC), 604, 751
Negative binomial distribution, 8, 431
Negative binomial regression model, 431
Neocognitron, 579
NER, see Named entity recognition (NER)
Nested logit model, 11, 613
Netflix, 19–20, 509, 512, 550, 556, 826
Net present value (NPV), 12, 19, 661
Network
  analytics, 722
  and influence diagrams, 722
  planning, 337
Neural networks (NN), 10, 11, 338, 374, 383, 537–540, 545, 582, 583, 586, 591, 607, 855
New product
  design, 626
  development, 11, 75, 402, 624, 643
Newspaper Problem, 181
News recommendation, 563
Nextel, 830
Next word prediction, 10, 584, 585, 587–588, 594
NFC, see Near field communication (NFC)
NLP, see Natural language processing (NLP)
9/11, 722
Nobel prize, 660, 662
Nominal, 2, 24, 25, 27, 132, 247, 266, 670, 730
Non-linear analysis, 369–374
Non-linear optimization, 369–374
Non-negative matrix factorization, 558
Nonparametric classifier, 527
Nonparametric envelopment frontier, 637
Nonparametric resampling procedures, 634

Nonparametric tools, 11, 625
Non-random walk hypothesis, 666
Non-sampling errors, 23
Normal distribution, 141, 149, 156, 157, 185, 187, 193, 231, 249, 275, 276, 278, 310, 311, 313, 319, 371, 430, 466, 491, 529, 669, 670, 673, 685, 696–698, 702, 708, 816, 839, 955–957, 960
Normality, 194, 196, 202, 208, 215, 231–233, 238, 239, 466, 634, 685, 698, 699
Normalization, 43–46, 517, 528
NoSql database, 80
NPV, *see* Net present value (NPV)
Null deviance, 425, 427, 428, 433, 435
Null hypothesis, 151–154, 156–161, 163–167, 170, 171, 193, 195, 206, 253, 255–257, 426, 434, 694
Numerical Python (NumPy), 944

O

Oak Labs, 617
Objective function, 7, 9, 337, 339, 341, 342, 344, 346–352, 354, 355, 357, 375, 376, 378, 463–465, 470, 472, 480, 486, 487, 490, 493, 535, 536, 540–543, 640, 887
Objective function coefficients, 341, 357, 378
Object-oriented programming, 918
Object recognition, 510
Observational data, 636
Observational equations, 187
Occam's razor, 520, 574
Odds ratio, 249, 251, 535
Office365, 105
Offset regressor, 430
Offset variable, 422, 430, 432
OLS, *see* Ordinary least square (OLS)
Omni-channel retail, 11, 616
One-tailed hypothesis test, 159–161
Oozie, 80
Open source, 1, 3, 77, 78, 80, 286, 298, 606, 890, 892, 915, 918
   language, 890, 918
Operational efficiency, 640, 830
Operations research (OR), 339, 773, 774, 778, 785
Optical illusion, 126
Optimality, 7, 338, 344, 356, 359–362, 364, 365, 527
Optimization, 7, 9, 14, 21, 34, 49, 97, 325–327, 337–380, 386, 391, 393, 397, 398, 414, 463–465, 470–475, 479, 480, 489, 490, 493, 497, 521, 535, 541, 542, 549–551, 568, 587, 603, 608–610, 613, 614, 672, 747, 748, 750, 793–798, 806, 808–811, 815, 817, 824, 826–829, 887, 944
Ordinal, 2, 24, 25, 27, 180, 230, 266, 269, 512
Ordinal logistic regression models, 266
Ordinary least square (OLS), 4, 8, 236, 249, 369–371, 373, 634
ORION, 307
Orthogonal GARCH, 683, 684, 696
Orthogonality, 468, 469, 739–740
Outliers, 5, 8, 205–206, 213, 243, 384, 388, 403, 404, 462, 488, 490, 493, 517, 545, 963
Output feature ratios, 518
Over-defined, 354
Overdispersion, 8, 430–433, 436
Ozon.ru, 602

P

PACF, *see* Partial autocorrelation function (PACF)
Paired-observation comparisons, 162–163
Pairwise classifier, 547–549
Pair-wise Deletion, 272
Palisade DecisionTools Suite, 310, 311
Pandas, 944
Parallel computing, 3, 77, 81, 84
Parameters, 4, 5, 30, 33, 138–139, 141, 142, 145, 151, 156, 164, 179, 183, 184, 187–191, 199–201, 204, 212, 220, 223, 232, 241, 247, 249–251, 264, 266, 275, 276, 278, 369–371, 374, 377, 392, 397, 398, 400, 410–414, 422, 423, 425–428, 430, 431, 433–435, 442, 443, 447–449, 453, 455, 463–465, 473, 476, 480–482, 486, 489, 490, 492, 493, 517, 524, 530, 531, 535, 536, 542, 544, 545, 558, 572, 574–576, 579–583, 585–587, 592, 606, 607, 610, 613, 615, 629, 642, 650, 667–669, 685, 686, 701, 702, 708, 715, 726, 728, 742, 773, 778, 779, 787, 797, 801, 805, 806, 833–835, 858, 877, 896, 915, 940, 941, 955
Parametric density function, 489, 490
Parametric methods, 456, 634
Parametric model(ing), 443, 447, 449, 453–455, 457, 535
Partial autocorrelation function (PACF), 8, 389, 395–397
Partial profile conjoint, 644
Partial regression coefficients, 202, 203, 241
Parzen Window classifier (PWC), 10, 492, 528–529, 531, 540
PaaS, *see* Platform as a Service (PaaS)

Index 975

PayPal, 652
PayTM, 565
PCA, *see* Principal components analysis (PCA)
PDF, *see* Probability density function (PDF)
Pearson's correlation, 555–557
Percentiles, 193, 212, 517, 958
Perceptron, 10, 534–536, 540, 570–572, 575, 591, 592
    algorithm, 534–536, 570, 571, 592
    based classifiers, 535
Perceptual Maps, 625
PERCH, 617
Performance analysis, 659
Performance index (PI), 828
Permutations, 749, 750, 895, 945, 953
Personalization, 33, 549, 550, 566, 600, 618, 654, 741
Personalized education, 549, 566
Pfizer, 768
Physical simulation, 6, 307
Pig, 80
Pig Latin, 80
PIMCO, 660
Platform as a Service (PaaS), 105, 106
Platform independent, 890, 919, 943
PMF, *see* Probability mass function (PMF)
Point-forecast, 7, 384
Point-of-Sale (POS), 11, 30, 75, 138, 476, 477, 495, 511, 603, 617, 744, 824, 827
Poisson distribution, 8, 423, 424, 430, 431, 491, 611, 614, 954–955
Poisson histogram, 423
Poisson regression model, 8, 421–425, 431, 434, 435, 437
Poisson variance, 423
Polytomous models, 266
Ponzi, 684
Pooling layers (POOL), 10, 579–583, 594
Popularity bias, 565
Population, 4, 13, 23, 31, 132, 137–151, 156, 160–167, 169, 183, 195, 203, 213, 237, 240, 276, 278, 316, 320, 423, 466, 730, 769–771, 805, 829, 830, 850, 876, 878, 886, 948, 957, 958
Population mean, 138–144, 146, 147, 149, 151, 156, 163, 164, 166, 167, 276, 316
Population proportion, 138, 147, 148, 161
Portfolio management, 659, 663, 664, 683
Portfolio selection, 337
POS, *see* Point-of-Sale (POS)
Position based attribution model, 746–747
Posterior probability, 492, 532, 549
Power Curve, 155, 156

P-quants, 12, 659–663, 685, 709, 714
Predictive analytics, 75, 338, 381, 382, 407, 601, 824, 855, 859
Preference bias, 564
Price and revenue optimization, 337, 796
Primal perceptive processes, 113
Primary data, 21, 31, 284, 683
Primary key, 33, 43, 53, 64–65
Prime numbers, 946
Principal components analysis (PCA), 9, 462, 466, 468–472, 475, 480, 515, 516, 535, 663, 683, 688, 690–692, 696, 709
Probability, 2, 5, 138–140, 142, 143, 147, 152–157, 159, 161, 163, 185, 186, 203, 242, 248–252, 254, 255, 257–262, 265, 269, 272, 275–277, 295, 296, 299, 309, 310, 313, 331, 334, 336, 371–373, 399, 431, 442, 444, 445, 448, 449, 462, 464, 482, 488, 490, 492, 498–500, 512, 524, 525, 529, 531–533, 549, 552, 583–585, 610, 612, 619, 652, 662, 665, 667–669, 671–676, 694, 697, 701, 709, 715, 736, 801, 802, 808, 809, 816, 817, 819, 828, 839, 848, 945–963
Probability density function (PDF), 276, 371, 372, 431, 442, 531–533, 839, 949–951
Probability distributions, 138, 140, 242, 299, 334, 336, 373, 482, 585, 662, 665, 672, 676, 701, 945, 954, 958
Probability mass function (PMF), 948–950, 954
Probit, 625
Process map of clinic visit, 13
Production function, 640, 641
Production planning, 337, 379
Product life cycle analysis, 626, 804
Profile based recommendation engine, 559–567
Programmatic advertising, 13, 725–728
Programmatic display, 725–728
Proportional hazard model, 9, 440, 447–449, 456, 625
Proportionality assumption, 343
Pruning, 366, 367, 496, 497
Pure regions, 522, 524, 533, 540
Purity-threshold, 525
p-value, 157–159, 161, 163, 165, 166, 168, 193–195, 203, 206–209, 214–216, 221, 223, 225, 227, 229, 230, 235, 237, 255–257, 263, 268, 270, 426, 431–434, 451, 733, 738
PWC, *see* Parzen Window classifier (PWC)
P-world, 659–663, 665, 714

Python, 1–3, 47, 49, 68, 84, 87–90, 92, 93, 96, 99, 101, 108, 298, 344, 889, 891, 897, 899, 904, 917–944

Q
QDA, *see* Quadratic discriminant analysis (QDA)
Q-quants, 12, 659–661, 663
QR codes, *see* Quick response (QR) codes
Quadratic discriminant analysis (QDA), 10, 533–535
Quantitative finance, 12, 659, 663
Quantitative supply chain analysis, 844
Quartiles, 185, 196, 231, 958, 960
Query length bias, 518
Quick response (QR) codes, 604, 750
Q-world, 659–661, 663, 669, 672

R
R^2, 191, 192, 194, 200–204, 212, 215, 220–222, 237, 238, 240, 257–258, 263, 268
Radio frequency identification (RFID), 11, 31, 34, 604
Rakuten, 825
Random forest, 10, 527, 546, 607, 867
Random sample, 137, 138, 181, 308–310, 312, 315, 317, 319, 327, 336, 401, 443, 886, 957
Random variables, 140–142, 147, 149, 162, 308, 309, 312, 313, 315, 317, 318, 321, 323, 324, 327, 334, 343, 370, 372, 672–674, 772, 787, 801, 948–952
Random walk hypothesis, 665
Rank based conjoint, 643
Rating based conjoint, 643
Ratio, 2, 24, 27, 122, 131, 164, 191, 194, 212, 249, 251–253, 255, 267, 330, 426, 433, 440, 448, 449, 451, 518, 520, 535, 536, 573, 626, 637, 638, 640, 666, 682, 750, 802, 848, 850, 856, 960
Ratio features, 518
RBI, *see* Reserve Bank of India (RBI)
RDBMS, *see* Relational data base management systems (RDBMS)
RDD, *see* Resilient Data Distribution (RDD)
Read–Eval–Print Loop (REPL), 97
Real time bidding (RTB), 13, 725–727, 874
Real-time decision making, 827
Real-time location systems (RTLS), 777
Real time translation, 570

Receiver operating characteristics (ROC) curve, 5, 261
Recency, frequency, monetary value (RFM) analysis, 626, 650
Recommendation for Cross-Sell, 564
Recommendation for lifetime value, 564
Recommendation for loyalty, 564
Recommendation for Preventing Churn, 564
Recommendation for Upsell, 564
Recommendation paradigm, 461, 512–513
Recommendation score, 521, 552, 556, 557, 562–565, 567
Recommendation systems, 33, 550, 553, 555, 570, 859
Rectified linear units layer (RELU), 10, 579–581
Recurrent neural networks (RNNs), 10, 540, 570, 571, 574, 583–590, 593, 594
RedHat Linux, 87
Redshift, 105
Referential integrity, 43, 64, 66
Regressand, 183
Regression, 5, 183–184, 222
  analysis, 114, 179–245, 247–281, 381, 384, 441, 606, 730, 731, 734, 735, 738
  models, 5, 8, 11, 13, 179, 180, 184, 185, 187, 195, 252–257, 373, 423–425, 441, 446–447, 512, 518, 527, 625, 730, 734
  paradigm, 512, 513
Regressors, 4, 5, 179–180, 182–185, 187–206, 208, 210–212, 216–223, 228, 232–234, 238–241, 426, 430, 436, 449, 454, 456
Regularization, 558, 606, 610
Reinforcement learning, 10, 460, 508, 575
Relational data base management systems (RDBMS), 3, 41–68, 76
Reliability, 31, 81, 441, 830
RELU, *see* Rectified linear units layer (RELU)
Reserve Bank of India (RBI), 679, 686, 714
Residual(s), 4, 179, 190, 191, 193–194, 196, 199–202, 204–206, 209, 212–216, 220, 221, 223, 225, 227, 229–232, 234–239, 241, 243, 258, 370, 397, 425–427, 429, 430, 432, 433, 435, 449, 470, 558, 665, 667, 682
  deviance, 425–428, 432, 433, 435
  plots, 179, 197, 205–209, 230, 239, 241, 243
Resilient Data Distribution (RDD), 99–102
Response variable, 5, 183–185, 188, 189, 191–195, 197, 200, 202, 205, 208, 211, 219, 220, 231–233, 237–241, 244, 247–249, 251, 252, 369, 371–373, 424, 435, 439, 606, 739

Index

Restricted Boltzmann machines (RBM), 575
Retail analytics, 11, 12, 337, 599–619, 654, 826
RetailNext, 617
Retrieval paradigm, 461, 513
RFID, *see* Radio frequency identification (RFID)
RFM analysis, *see* Recency, frequency, monetary value (RFM) analysis
Ridge regression, 180, 220, 223–224, 606
Right defaults, 519
@Risk, 310–313, 316, 319, 325, 334, 335
Risk aggregation, 672
Risk-averse, 12, 659, 661, 677
Risk management, 12, 659, 660, 663, 664, 672–683, 686
RiskMetrics™, 681
Risk-neutral world, 659, 661
Rite Aid, 605
RNNs, *see* Recurrent neural networks (RNNs)
ROC curve, *see* Receiver operating characteristics (ROC) curve
Ronald Fisher, 728, 736
RPy, 944
RStudio, 291, 292, 296, 891–893, 897, 899
RTB, *see* Real time bidding (RTB)
Rudyard Kipling, 660
Rue La La, 826
Rule-based classifiers, 523–524

S
Saavn, 550
Sales analytics, 626
Salesforce, 27, 28, 626
Samarthan, 826
Sample, 138–140
  distributions, 140–142, 147, 149, 156–158, 163
  error, 23, 309, 316
  estimates, 138, 170, 433
  mean, 4, 39, 138, 140–143, 145, 146, 148, 152, 156, 167, 310, 316
  proportion, 138, 140, 147, 148, 161
  size determination, 144, 148–149
  space, 946, 948
  statistic, 139, 140, 151, 156, 157, 160, 161
  survey, 23
Samsung, 744, 824
SAP HANA, 827
Satellite imagery, 570
Saturated model, 426, 432, 434, 437
SBA, *see* Syntetos and Boylan approximation (SBA)

Scale effects, 638
Scatter Plots, 4, 31, 188, 189, 197, 198, 205, 210, 211, 216–218, 466–467, 477, 514, 515, 952, 960, 962
Scikits, 944
SciPy, 922, 944
Scoring throughput, 520, 526
Scree plot, 292
Scripts mode, 897, 923
Search engine optimization (SEO), 721
Search methods, 338, 365, 393
Seasonal, 8, 37, 331, 388–391, 393, 395, 517
Seasonal and cyclical adjustment, 389
SEBI, *see* Securities and Exchange Board of India (SEBI)
Secondary data, 2, 21–23, 31
Second normal form (2NF), 44–45
Securities and Exchange Board of India (SEBI), 679, 686
Securities and Exchange Commission (SEC), 679, 686
Seed, 312, 314–316, 319, 322, 333, 336, 493, 566, 783
Segmentation, 11, 30, 283, 291–295, 577, 624, 625, 655, 793, 794
Selfie walls, 602
Semantic feature, 518
Semi-structured data, 73, 74
Sensitivity, 7, 260, 261, 355–360, 376, 377, 379, 480–481, 783, 787, 795, 813
Sensitivity analysis, 7, 355–360, 787, 813
Sentiment analysis, 6, 13, 283, 291–295, 298, 570, 601, 721
Sequential processing, 95, 98
Serendipity, 549, 550
Service level optimization, 337
SES, *see* Simple exponential smoothing (SES)
Set theory, 496, 905, 945–947
SFA, *see* Stochastic frontier analysis (SFA)
Shadow prices, 7, 355–361, 364, 375–378, 813, 814, 816
Shakespeare, 286
Shapiro Wilk test, 698
Shared-nothing architecture, 84
Shelfbucks, 617
Shiny, 290–293, 296, 299
Shipping cost rate, 833, 834
Signal-to-noise-ratio, 520
Simple exponential smoothing (SES), 391–392, 395, 399, 400, 409–411, 415
Simple linear regression model, 273, 369
Simplex algorithm, 339–340, 351, 354, 355, 361

Simulation, 6, 7, 12–15, 305–337, 614, 615, 619, 669–672, 678, 686, 687, 765–770, 772–774, 780–785, 876–878, 883–886, 944
Single regressor, 4, 188–197, 199, 200, 202–205, 210, 234
Singular value decomposition (SVD), 557
Size-threshold, 525
Skewness, 234, 238, 426–427, 958, 959
Sklar's theorem, 672
Sku IQ, 617
Slack variables, 353, 357, 542–544
Sliding motion, 349
Smoothing, 8, 390–395, 400, 401, 410, 413, 414, 524, 532, 557, 584, 605
Social bias, 565
Social graph, 32, 33, 565
Software as a Service (SaaS), 105
Solution space, 345
Song recommendation, 563
Spark, 3, 71, 72, 94–102, 105, 108
Spark ecosystem, 99–100
SPARK MLib, 101
SPARK SQL, 101, 102
SPARK Streaming, 101
Specificity, 260–261
Speech recognition, 517, 569, 570, 584
Speech tagger, 510
Spell checkers, 584
Spotify, 601
Sprint, 830
SQL, *see* Structured query language (SQL)
Sqoop, 80
Standalone capital, 680
Standard conjoint, 643
Standard error, 4, 114, 140, 141, 144, 146, 148, 152, 160, 161, 163, 166, 180, 191–194, 199, 201, 209, 211, 212, 214–217, 219, 223, 225, 227, 229, 230, 235, 237, 256, 268, 270, 426, 428, 429, 431, 445, 451, 634, 708, 782
Standard normal distribution, 143, 145–147, 163, 171, 231, 670, 674, 955, 957
Stationarity, 389, 395, 397, 692–695
Stationary, 389, 396, 397, 665
Statistical sampling, 309, 316, 327
Statsmodels, 944
Stem-word, 287
Stephen Ross, 662
Stepwise method, 221, 263
Stepwise regression, 180, 220, 222, 223
Stitch Fix, 618
Stochastic frontier analysis (SFA), 625, 626, 640–642

Stochastic gradient descent, 573
Stress test(ing), 159, 170, 676
Structural equation modeling (SEM), 634
Structured data, 73, 76
Structured query language (SQL), 43, 47–68, 80, 99, 101
Subjective-bias, 523, 553
Subset selection problem, 519
Supervised learning, 9–10, 374, 460–461, 464, 468, 478, 504–505, 507–513, 521, 527, 570, 571, 574, 575, 607, 608, 618
Supply chain disruption, 14, 827–828
Supply chain management, 14, 823, 824, 829
Supply side platform (SSP), 726
Support vector machines (SVM), 10, 518, 540–542, 544, 545, 747, 896
Survival analysis, 8, 9, 439–457, 625, 748
Survival curve, 439, 443–446, 449, 451, 454, 456, 457
Survival distribution, 439
SVD, *see* Singular value decomposition (SVD)
SVM, *see* Support vector machines (SVM)
Swachha Bharat (clean India) program, 826
Swasthya Slate, 767
Syntetos and Boylan approximation (SBA), 8, 400, 410, 414, 418
System thinking, 844

T
Tableau, 101
Target, 11–13, 20, 27, 33, 292, 316, 331, 375, 405, 512, 535, 557, 589, 604, 617, 624, 625, 636, 643, 655, 720–722, 724, 725, 727, 728, 741, 747, 757, 794, 818, 826, 839, 873–880
Tata Motors, 652
TCL, *see* Traction control language (TCL)
t distribution, 145–147, 160, 163, 173, 193, 206, 662, 702, 703, 707–709
Teaching content recommendations, 563
Tech bubble burst, 827
Telemedicine, 13, 766, 767
Tensor product representation (TPR), 574
Teradata, 87
Term-document matrix (TDM), 6, 286, 288, 291, 294, 296, 298, 299
Term Frequency—Inverse Document Frequency (TFIDF), 288, 289, 300, 476, 482, 545
Testing of hypotheses, 192–193
Tests of significance, 200
Text analytics, 5–6, 283–301, 928
Text corpus, 287, 291, 295, 584, 595

Theta model, 395, 410, 413
Third industrial revolution, 71
Third normal form (3NF), 45–46
Thomson Reuters, 687
Threshold persistence (TP), 664, 676, 685–687, 703, 707, 709
Timberland, 617
Time Decay Attribution Model, 746–747
Time-series, 7, 8, 24, 132, 383–384, 386–388, 403, 404, 605, 625, 667, 692, 694, 867
Times Inc, 724
Time-to-event, 8, 439–441, 443
Time to recovery (TTR), 828
Tobin's separation theorem, 662
Tobit, 625
Tokenization, 286, 287, 298, 300
Topic mining, 6, 295–297, 299
Traction control language (TCL), 47, 48
Trade-off analysis, 626, 647–648
Trend, 13, 76, 113, 132, 208, 210, 220, 273, 388–395, 397, 403, 404, 408, 410, 412, 565, 601, 602, 604, 607, 630, 666, 667, 692, 694, 720, 721, 724, 741, 768, 794, 795, 815, 824, 826, 847
Trilateration, 722
Trimmed means, 388
TripAdvisor, 721
Twitter, 13, 36, 37, 71, 550, 553, 565, 601, 719, 721, 724, 739, 748, 756, 762
Two-fund theorem, 662
Two-level factorial designs, 736, 739
Type I error, 4, 152, 153, 156, 157, 159, 171
Type II error, 4, 153, 156

U
Uber, 20, 21, 35, 75, 479, 652, 794, 803
Unbalanced, 24
Unbounded feasible region, 345–347
Uncertainty modeling, 308
Unconstrained optimization, 338
Unidimensional objects, 906
Unimodal Bayesian classifier (UBC), 530
Union, 495, 946
Unit root, 389, 395, 694
Univariate analysis, 625, 958
Universal set, 946
University of California, Berkeley's AMP Lab, 95
Unpunctuality, 772, 780, 781, 787, 788
Unstructured data, 23, 31, 73, 74, 284, 604, 605, 607, 752, 891, 922
Unsupervised learning, 9, 10, 292, 295, 459–505, 508, 510, 530, 532, 575, 688

UPS, 307, 825
Urban Outfitters, 617
User–user collaborative filtering, 555

V
Validation, 5, 31–32, 180, 237–239, 244, 270, 521, 574, 867
Valuation, 12, 19–21, 440, 624, 659, 663, 796, 802, 804–805, 815
Value at Risk (VaR), 12, 664, 676–680, 685, 686, 696, 702, 703, 707–709, 714–716
VaR, *see* Value at Risk (VaR)
Variable return to scale (VRS), 637, 638
Variance, 4, 5, 12, 27, 32, 147–151, 161, 164–167, 170, 180, 186, 188, 189, 192, 193, 195, 198, 200, 201, 204, 205, 211–213, 218, 233, 239, 252, 273, 310, 337, 371, 389, 396, 397, 400, 423, 427, 428, 430, 431, 433, 434, 462, 468–471, 474, 475, 493, 515, 517, 518, 546, 606, 607, 611, 642, 653, 662, 665, 666, 668, 694, 697, 699–702, 709, 776, 777, 784, 799, 817, 867, 879, 950–952, 954, 955, 958, 959, 962
Variance decomposition proportions (VP), 218, 219, 241
Variance inflation factors(VIF), 218, 219, 239
Variation between samples, 168
Variation within samples, 168
Variety, 30, 31, 73–75, 117, 291, 330, 387, 459, 460, 465, 467, 477, 484, 494, 498, 504, 507, 508, 534, 536, 540, 545, 557, 567, 577, 582, 588, 600, 611, 660, 766, 769, 786, 809, 853, 890, 908, 918
VASTA, 14, 830–844
Vector auto regression (VAR) models, 625, 667
Velocity, 31, 72–74
Velocity feature, 518
Veracity, 73, 74
Verizon, 830
Video
  analytics, 617
  captioning, 570
  search, 570
  surveillance, 570
VIF, *see* Variance inflation factors(VIF)
Virtual machine (VM), 3, 86–90, 99
Virtual mirror, 306, 308
Virtual Physiological Human Institute, 768
Vistaprint, 601
Visual cortex system, 579
VM, *see* Virtual machine (VM)

Volume, 31, 73, 74, 76–79, 181, 202, 203, 307, 309, 318, 323, 384, 399, 507, 531, 580, 581, 593, 601, 663, 682, 720, 782, 784, 793, 824, 832–837, 839, 840, 849, 877
VRS, *see* Variable return to scale (VRS)

W
Wage balance problem, 182
Wald's test, 5, 252, 255–256, 270
Wallpaper advertisements, 724
WalMart, 300, 604, 794, 823, 843
WarbyParker.com, 616
Wavelet transformations, 517
Wearable technologies, 13, 766, 767
Web scraping, 35, 300, 890, 918
Website development, 890, 918
Weibull distribution, 455, 457
Whole numbers, 946
Wikipedia, 300, 383, 570
Winsorized means, 388
Wordcloud, 6, 288, 289, 292, 294, 296, 297, 299–301
WordNet, 576, 577
Word2vec, 584, 587
Workflow based recommendations, 566–567
Workforce optimization, 337
Work units, 772

World Health Organization, 769
World War II, 307
Wrapper methods, 519
Write-once Read-many (WORM), 81

X
XGBoost, 527, 547

Y
Yale, 684, 685
Yet Another Resource Negotiator (YARN), 3, 78, 80, 81, 93–95, 98, 99
Yield to maturity, 665, 666
Youden's Index, 5, 259–261, 271
YouTube, 71, 74, 291, 478, 495, 509, 512, 513, 550, 553, 564, 744

Z
Zero-coupon bond, 665
Zero-inflated models, 422
Zero inflated Poisson (ZIP) and zero inflated negative binomial (ZINB) Models, 434–435
Zomato, 509